Sedimentary Responses to Forced Regressions

Special Publication reviewing procedures

The Society makes every effort to ensure that the scientific and production quality of its books matches that of its journals. Since 1997, all book proposals have been refereed by specialist reviewers as well as by the Society's Publication Committee. If the referees identify weaknesses in the proposal, these must be addressed before the proposal is accepted.

Once the book is accepted, the Society has a team of series editors (listed above) who ensure that the volume editors follow strict guidelines on refereeing and quality control. We insist that individual papers can only be accepted after satisfactory review by two independent referees. The questions on the review forms are similar to those for *Journal of the Geological Society*. The referees' forms and comments must be available to the Society's series editors on request.

Although many of the books result from meetings, the editors are expected to commission papers that were not presented at the meeting to ensure that the book provides a balanced coverage of the subject. Being accepted for presentation at the meeting does not guarantee inclusion in the book.

Geological Society Special Publications are included in the ISI Science Citation Index, but they do not have an impact factor, the latter being applicable only to journals.

More information about submitting a proposal and producing a Special Publication can be found on the Society's web site: www.geolsoc.org.uk.

It is recommended that reference to all or part of this book should be made in one of the following ways.

HUNT, D. & GAWTHORPE, R. L. (eds) 2000. *Sedimentary Responses to Forced Regression*. Geological Society, London, Special Publications, **172**.

AINSWORTH, R. B., BOSSCHER, H. & NEWALL, M. J. 2000. Forward modelling of forced regressions: evidence for the genesis of attached and detached lowstand systems. *In*: HUNT, D. & GAWTHORPE, R. L. (eds) *Sedimentary Responses to Forced Regression*. Geological Society, London, Special Publications, **172**, 1–383.

GEOLOGICAL SOCIETY SPECIAL PUBLICATION NO. 172

Sedimentary Responses to Forced Regressions

EDITED BY

D. HUNT and R. L. GAWTHORPE
The University of Manchester, UK

2000
Published by
The Geological Society
London

THE GEOLOGICAL SOCIETY

The Geological Society of London was founded in 1807 and is the oldest geological society in the world. It received its Royal Charter in 1825 for the purpose of 'investigating the mineral structure of the Earth' and is now Britain's national society for geology.

Both a learned society and a professional body, the Geological Society is recognized by the Department of Trade and Industry (DTI) as the chartering authority for geoscience, able to award Chartered Geologist status upon appropriately qualified Fellows. The Society has a membership of 8600, of whom about 1500 live outside the UK.

Fellowship of the Society is open to persons holding a recognized honours degree in geology or a cognate subject and who have at least two years' relevant postgraduate experience, or not less than six years' relevant experience in geology or a cognate subject. A Fellow with a minimum of five years' relevant postgraduate experience in the practice of geology may apply for chartered status. Successful applicants are entitled to use the designatory postnominal CGeol (Chartered Geologist). Fellows of the Society may use the letters FGS. Other grades of membership are available to members not yet qualifying for Fellowship.

The Society has its own Publishing House based in Bath, UK. It produces the Society's international journals, books and maps, and is the European distributor for publications of the American Association of Petroleum Geologists (AAPG), the Society for Sedimentary Geology (SEPM) and the Geological Society of America (GSA). Members of the Society can buy books at considerable discounts. The Publishing House has an online bookshop (*http://bookshop.geolsoc.org.uk*).

Further information on Society membership may be obtained from the Membership Services Manager, The Geological Society, Burlington House, Piccadilly, London W1V 0JU (Email: *enquiries@geolsoc.org.uk;* tel: +44 (0)171 434 9944).

The Society's Web Site can be found at *http://www.geolsoc.org.uk/*.The Society is a Registered Charity, number 210161.

Published by The Geological Society from:
The Geological Society Publishing House
Unit 7, Brassmill Enterprise Centre
Brassmill Lane
Bath BA1 3JN,
UK
(*Orders*: Tel. +44(0)1225 445046
Fax +44(0)1225 442836)
Online bookshop: *http://bookshop.geolsoc.org.uk*

First published 2000

British Library Cataloguing in Publication Data
A catalogue record for this book is available from the British Library.

ISBN 1-86239-063-0
ISSN 0305-8719

Typeset by Type Study, Scarborough, UK

Printed in Great Britain by Alden Press, Oxford, UK

Distributors

USA
AAPG Bookstore
PO Box 979
Tulsa
OK 74101-0979
USA
(*Orders*: Tel. +1 918 584-2555
Fax +1 918 560-2652)
Email bookstore@aapg.org

Australia
Australian Mineral Foundation
63 Conyngham Street
Glenside
South Australia 5065
Australia
(*Orders*: Tel. +61 88 379-0444
Fax +61 88 379-4634)
Email bookshop@amf.com.au

India
Affiliated East-West Press PVT Ltd
G-1/16 Ansari Road, Daryaganj,
New Delhi 110 002
India
(*Orders*: Tel. +91 11 327-9113
Fax +91 11 326-0538)

Japan
Kanda Book Trading Co.
Cityhouse Tama 204
Tsurumaki 1-3-10
Tama-shi
Tokyo 206-0034
Japan
(*Orders*: Tel. +81 (0)423 57-7650
Fax +81 (0)423 57-7651)

Contents

Preface

An increasing number of studies in recent years have demonstrated that significant progradation of shallow marine systems occurs under conditions of base-level fall. These new data are forcing many sedimentary geologists to critically re-evaluate many aspects of sequence stratigraphy relating to erosion and deposition during base-level (lake - or relative sea-level) fall, and the intrinsic link made between stratal geometries and base-level change. For the first time, this volume brings together a collection of articles that focus solely on forced regressions, providing a more complete picture of the development, formation, variability and preservation of the surfaces and deposits generated during base-level fall.

There were three main stimuli for bringing this volume to fruition. The first was interest expressed in the stratigraphic surfaces and stratal units developed during base-level fall, and the processes responsible for their formation and preservation. The second was the controversy concerning the position of the sequence boundary with respect to forced regressive deposits, and suggestions that sediments deposited during base-level fall should be incorporated within a fourth systems tract. The third objective was to provide a discussion forum dedicated to new ideas and data that could address the conceptual and practical problems related to the recognition and differentiation of the stratal surfaces and units generated during forced regression from those formed during base-level rise. Thus, the volume was conceived to try and resolve controversial issues, but more importantly aimed to emphasize the significant progress being made in understanding sedimentary responses to forced regression, and the important implications of these findings have for the understanding and interpretation of the rock record.

The volume comprises three natural groups of papers. The first group contains two papers that give an overview of the main concepts, models and practical issues related to deposition during base-level fall and provide important background for those readers unfamiliar with the subject. The second uses mainly sedimentological and geometrical criteria to identify forced regressive deposits and infer base-level changes. This group of papers contains an article from northern Greenland, two studies of Triassic and Jurassic strata from northern Europe, and a collection of three articles from the Late Cretaceous Western Interior Seaway of North America. The latter complement the first overview paper that also presents and utilizes data from the Western Interior Seaway. The third group begins with an exploration of forced regressive deposits in active tectonic settings. The main thrust of papers in this section focuses on the Late Pliocene–Recent where biostratigraphic and radiometric dating allows direct comparison of the stratigraphic units and the bounding surfaces formed against a well-constrained high-frequency, high-amplitude glacio-eustatic signal once the subsidence/uplift history of an area is known. It is in these settings that sequence stratigraphic concepts and models related to base-level fall can be most rigorously tested.

In an attempt to provide coherence between the wide range of geological settings and age of the strata discussed in this volume, authors were requested to address at least one of seven important issues related to forced regressions: (i) criteria for the recognition of forced regressive deposits and for their differentiation from strata formed during base-level rise, (ii) the expressions of the bounding surfaces to forced regressive strata and their variability, (iii) changes to facies and facies stacking patterns during forced regression, (iv) controls on the preservation potential of the surfaces and strata formed during base-level fall, (v) along strike and down-dip variability in forced regressive deposits as a function of differences in relative sea-level change, physiography and sediment supply, (vi) the placement of the 'main' or 'master' sequence boundary with respect to forced regressive deposits and (vii) implications for existing sequence stratigraphic models and concepts.

Collectively, the articles in this volume clearly show that sediments deposited during base-level fall can play a significant role in the outbuilding of continental margins and in the progradation of depositional systems in general. They provide an important discussion of the practical issues related to the recognition of key stratal surfaces and sediments formed during forced regression both outcrop and subsurface datasets. Significantly, many of the papers challenge the notion that there is a simple relationship between stratal geometry and base-level change, and provide important insights as to why the importance of sediments formed during forced regression has often been overlooked in the past. The reasons for this oversight appear to be due to practical problems related to the recognition of strata deposited during forced regression, apparently resulting from the formation/preservation of non-diagnostic stratal geometries, combined with the effects of postdepositional tilting, deformation and incorrect choice of datum. The results of the studies published here will be of interest to all geologists attempting to understand the relationship between changes in base-level and stratigraphy, and to all who use sequence stratigraphy as a method of stratigraphic correlation and interpretation at outcrop and in the subsurface.

As with any volume, the generous donation of time and financial help from many sources is essential. In this regard, we would like to thank Elf Aquitaine, Esso Exploration & Production UK Ltd, BP Exploration Operating Company Ltd and Norsk Hydro for their generous financial support, and to the Geological Society and the British Sedimentological Research Group for their invaluable logistical and financial contribution. At the University of Manchester we thank Marina Raven for secretarial help, past PhD students including Fiona Burns, Matt Docherty, Pierre Eliet, Matt Hall, Lesley McMurray, Andrew Quallington and Andrew Thurlow for their superb help in organising the original conference, and Dave Owens for his special projection skills. Finally, it would have been impossible to compile this volume without the invaluable contribution of time and effort by the referees, to whom we wish to extend sincere thanks on behalf of ourselves and the authors herein. They are: Bruce Ainsworth, Hubert Arnaud, William Fitchen, Bob Carter, Francesco Chiocci, Richard Collier, Trevor Elliott, Evan Franseen, Bruce Fouke, Mark Harris, Bruce Hart, William Helland-Hansen, Francisco Hernández-Molina, John Howell, Peter Johannessen, Steve Johnson, Tim Naish, David Piper, Philip Playford, Guy Plint, Andy Pulham, Ian Sharp, Don Swift, Kevin Taylor, Maurice Tucker, Tjeerd van Andel, Dave Waltham and several others who wished to remain anonymous. Cath Hunt, Ian Sharp and Mike Young are thanked for carefully reading through various edited versions of some of the papers included here. At the Geological Society we would thank Angharad Hills and Andrew Morton for their editorial assistance and advice.

We dedicate this volume to the memory of our friend and colleague, Marina Raven, who assisted with the organization of the conference and who sadly passed away during the preparation of this volume.

D. Hunt
R. L. Gawthorpe

The falling stage systems tract: recognition and importance in sequence stratigraphic analysis

A. GUY PLINT[1] & DAG NUMMEDAL[2,3]

[1]*Department of Earth Sciences, University of Western Ontario, London, Ontario, N6A 5B7, Canada*

[2]*Department of Geology and Geophysics, Louisiana State University, Baton Rouge, Louisiana, 70803, USA*

[3]*Present address: Unocal Corporation, 14141 Southwest Freeway, Sugar Land, Texas, 77478, USA*

Abstract: Until recently, sequence stratigraphic models have attributed systems tracts to periods of relative sea-level rise, highstand and lowstand. Recognition of a discrete phase of deposition during relative sea-level fall has been limited to a few studies, both in clastic and carbonate systems. Our work in siliciclastic ramp settings suggests that deposition during relative sea-level fall produces a distinctive *falling stage systems tract* (FSST), and that this is the logical counterpart to the transgressive systems tract. The FSST lies above and basinward of the highstand systems tract, and is overlain by the lowstand systems tract. The FSST is characterized by stratal offlap, although this is likely to be difficult or impossible to recognize because of subsequent subaerial or transgressive ravinement erosion. The most practical diagnostic criteria of the FSST is the presence of erosive-based shoreface sandbodies in nearshore areas. The erosion results from wave scouring during relative sea-level fall, and the stratigraphically lowest surface defines the base of the FSST. Further offshore, shoaling-upward successions may be abruptly capped by gutter casts filled with HCS sandstone, reflecting increased wave scour on the shelf during both FSST and LST time. The top of the FSST is defined by a subaerial surface of erosion which corresponds to the sequence boundary. This surface becomes a correlative submarine conformity seaward of the shoreline, where it forms the base of the lowstand systems tract. Differentiation of the FSST and LST may be difficult, but the LST is expected to contain gradationally-based shoreface successions because it was deposited when relative sea level was rising. Internally, the FSST may be an undifferentiated body of sediment or it may be punctuated by internal regressive surfaces of marine erosion and ravinement surfaces which record higher-frequency sea-level falls and rises superimposed on a lower-frequency sea-level fall. The corresponding higher-order sequences are the building blocks of lower-order sequences. The addition of a falling stage systems tract results in a significant reduction in the proportion of strata within a sequence that are assigned to the classical highstand and lowstand systems tracts.

Many outcrop and subsurface cross-sections use an overlying ravinement, or maximum flooding surface as datum. Those surfaces might be flat, but they are not horizontal. Both dip seaward at slopes that generally are steeper than the fluvial system responsible for creating the sequence boundary. When a section is restored to such a datum, the falling stage systems tract will appear to record stratigraphic climb, whereas in fact it does not.

The issue of systems tracts

Historical perspective

Before presenting our reasons for suggesting the formalization of a falling stage systems tract, it is appropriate to review briefly the evolution of systems tract systematics. Early sequence stratigraphic models (e.g. Mitchum *et al.* 1977) focused primarily on the recognition of the boundaries of seismic sequences, and the lap-out patterns between sequence boundaries. The subdivision of sequences into component systems tracts was first presented by Vail (1987) and elaborated by Posamentier & Vail (1988) who partitioned depositional sequences into four systems tracts; transgressive, highstand, lowstand and shelf margin. Lowstand and shelf margin systems tracts can be considered as variants on a theme, both representing deposition at relatively low sea level. Although this tripartite scheme assigned sediments deposited during rapid relative sea-level rise to the transgressive systems tract, there was no corresponding recognition of deposition during relative sea-level fall; highstand was immediately followed by lowstand. This may be due to the fact that it is

From: HUNT, D. & GAWTHORPE, R. L. (eds) *Sedimentary Responses to Forced Regressions*. Geological Society, London, Special Publications, **172**, 1–17. 1-86239-063-0/00/$15.00

commonly difficult to recognize offlapping strata in seismic sections (because of subsequent regressive or transgressive erosional modification, or because it is below seismic resolution (e.g. Vail *et al.* 1977, fig. 8). The limited recognition of sediments deposited during relative sea-level fall led to the early representation of relative sea-level fall as 'instantaneous', based on seismic lap-out geometry (i.e. the 'saw-tooth' sea-level curve of Vail *et al.* 1977, fig. 13).

Further elaboration of the sequence stratigraphic model by Posamentier & Vail (1988) included some discussion of the possibility of shelf deposition during relative sea-level fall, resulting in shelf-perched lowstand deposits that were assigned to the lowstand fan systems tract. However, neither specific examples nor detailed criteria for recognition of shelf-perched lowstand deposits, either in outcrop or subsurface, were provided.

On the basis of subsurface and outcrop work in the Cretaceous Western Interior of Canada, Plint (1988, 1991) and Plint & Norris (1991), emphasized the occurrence and significance of erosive-based shoreface deposits, and interpreted them to record deposition on a ramp-type shelf during relative sea-level fall (a process termed 'forced regression' by Plint 1991). However, these papers adhered to the existing tripartite systems tract terminology by assigning forced regressive deposits to the late highstand or early lowstand systems tracts.

Van Wagoner *et al.* (1990, pp. 35–36) also discussed the development of shelf-perched shoreface deposits, which they assigned to the early lowstand systems tract. The authors argued that the sequence boundary should be placed at the subaerial erosion surface *on top of* the perched shoreface deposits, yet at the same time, they placed the same sequence boundary *beneath* deep-water sediments deposited while relative sea level was falling. In other words, coeval shelf and deep-water deposits were separated by a sequence boundary, or they implied that deposition of deep-water strata in fact post-dated deposition of the shelf-perched shoreface.

In a discussion of deposition during relative sea-level fall (forced regression), Posamentier *et al.* (1992) argued that all strata deposited *after the onset* of relative sea-level fall should be assigned to the lowstand systems tract. The onset of relative sea-level fall was considered to correspond to the formation of a wave-cut regressive surface of marine erosion. This surface is likely to be well-developed only on the inner shelf; further basinward it grades into a correlative conformity. This surface and correlative conformity was assigned sequence boundary status by Posamentier *et al.* (1992) despite the inherent practical difficulties of recognizing such a surface beyond the inner shelf. Thus the sequence boundary extended basinward *beneath* the package of shelf-perched falling sea-level deposits, and also beneath any deep-water strata deposited while relative sea-level was falling. Unlike Van Wagoner *et al.* (1990), this scheme preserved the chronostratigraphic equivalency of shelf and deep water deposits.

Hunt & Tucker (1992) highlighted the inconsistency inherent in the terminology of Van Wagoner *et al.* (1990). In order to avoid separating falling sea-level shelf deposits from coeval deep-water strata, they proposed that a new 'forced regressive wedge' systems tract (FRWST) be defined, later renamed the 'forced regressive' systems tract (Hunt & Tucker 1995). This systems tract would include both shelf and deep water strata deposited between the onset of relative sea-level fall, and relative sea-level lowstand. The lower boundary of the FRWST was termed the basal surface of forced regression (equivalent to the sequence boundary of Posamentier *et al.* (1992), and the upper surface was the sequence boundary. Both surfaces formed progressively throughout relative sea-level fall, the former through submarine erosion at wavebase, the latter as a result of subaerial processes.

The new systems tract scheme was criticized by Kolla *et al.* (1995) who argued that the existing tripartite scheme was sufficiently flexible to accommodate local variations without recourse to a new systems tract. Hunt & Tucker (1995) defended their new systems tract, and provided arguments that we feel justify the need to recognize a fourth systems tract. This need is underscored by other recent studies (e.g. Ainsworth & Pattison 1994), who document classic examples of falling stage systems tracts, but, constrained by the existing terminology, are obliged to use the term 'attached lowstand' systems tract to describe deposits that are neither classical highstand, nor obviously detached 'lowstand' deposits!

At about the same time that Hunt & Tucker (1992) were striving to apply sequence and systems tract schemes to carbonate platform and basin deposits, the authors of this paper were, independently, working on similar problems in siliciclastic ramp deposits in the Cretaceous Western Interior of North America. In particular, it was apparent from our own work, and that of others, that extensive, thin, erosive-based shoreface sandstone bodies were widely developed on the shelf, and that these must represent a significant period of relative sea-level fall when marine accommodation was limited (e.g. Plint 1988, 1991, 1996; Plint & Norris 1991, Walker & Plint 1992; Nummedal *et al.* 1992, 1993; Ainsworth 1994; Hart & Long 1996; Tirsgaard 1996). In some instances, sharp-based sandstones simply pinch out downdip into offshore siltstones. In other cases a discrete, sometimes isolated sandbody lay basinward of the sharp-based shoreface sandstones and evidently represented deposition following maximum sea-level lowstand (e.g. as summarized in fig. 25 of Walker & Plint 1992; fig. 3 of Hunt & Tucker 1992).

Our present contribution builds upon the idea, independently formulated by Hunt & Tucker (1992) and Nummedal *et al.* (1992) of the need for a new systems tract that corresponds to the time between the onset of relative sea-level fall, and sea-level lowstand. Although the terminology varies (forced regressive wedge systems tract; Hunt & Tucker 1992; forced regressive systems tract, Hunt & Tucker 1995; Helland-Hansen & Gjelberg 1994; falling sea-level systems tract, Nummedal *et al.* (1992), or falling stage systems tract (this paper), these studies all address a similar problem, and arrive at a similar conclusion.

Existing definitions

We do not take lightly the issue of formalizing a falling stage systems tract; the sequence stratigraphic nomenclature is already sufficiently complex. The formal proposal of this new systems tract is justified, however, because case studies (e.g. Hunt & Tucker 1992) show that 'highstand' and 'lowstand' systems tracts overlap in time. These inconsistencies stem from unfortunate attributes of existing definitions. In current usage (Posamentier *et al.* 1988; Van Wagoner 1995*a*) each systems tract is 'defined objectively by stratal geometries at bounding surfaces, position within the sequence, and internal parasequence stacking patterns. Each systems tract is interpreted to be associated with specific segments of the eustatic curve, although not defined on that basis'. In classical sequence stratigraphy, systems tract definitions are all based on stratal relations and the tie to causative sea-level change always will remain tenuous, not least because stacking pattern is strongly influenced by sediment supply (e.g. Schlager 1993). Yet, in nearly all sequence stratigraphic studies, reconstructing past sea-level changes is a major objective. Moreover, a clear understanding of what part of the sea-level curve was responsible for deposition of a given systems tract is crucial to the prediction of the distribution and character of related systems tracts. The contrasts between Van Wagoner's (1995*b*) and our (Nummedal *et al.* 1995) interpretations of the Cretaceous Castlegate and Desert Sandstones of Utah, illustrate the point.

The highstand systems tract (HST) is characterized by 'parasequences [that] onlap onto the sequence boundary in a landward direction and downlap onto the top of the transgressive or lowstand systems tract in the basinward direction' (Van Wagoner *et al.* 1988). This definition makes it clear that the highstand systems tract consists mostly of deposits that formed during relative sea-level rise. Constrained by the assumptions used in the Exxon model, there is no onlap at the landward margin of the strata deposited during relative sea-level fall (Christie-Blick 1991; Christie-Blick & Driscoll 1995).

The lowstand systems tract (LST), 'if deposited in a basin with a ramp margin, consists of a relatively thin lowstand wedge that may consist of two parts. The first part is characterized by stream incision and sediment bypass of the coastal plain, interpreted to occur during a relative fall in sea-level during which the shoreline steps rapidly basinward until the relative fall ceases. The second part is characterized by a slow rise in relative sea-level, the infilling of incised valleys, and continued shoreline progradation' (Van Wagoner *et al.* 1988). The first component of the LST was named by Vail *et al.* (1991) 'the lower lowstand prograding complex'.

The term 'lower lowstand' is inappropriate because deposition of this package begins as soon as relative sea-level fall commences, i.e. immediately following the relative sea-level highstand. The 'lower lowstand' component of the LST, therefore, is deposited below the subaerial surface of erosion that is created as a consequence of the sea-level fall. With few exceptions, this erosion surface is widely accepted as 'the' sequence boundary. Within the constraints of the existing terminology, the other alternative is to consider the 'lower lowstand' package to be part of the 'highstand' systems tract, as has been done in several recent publications. This is even less appropriate, because strata in the 'lower lowstand' complex do not onlap onto the sequence boundary, nor do they record relative sea-level highstand: indeed, they are deposited during relative fall, the end of which occurs at lowstand.

Definition

The falling stage systems tract (FSST) is defined in terms of the stratal geometries at bounding surfaces, position within the sequence, internal

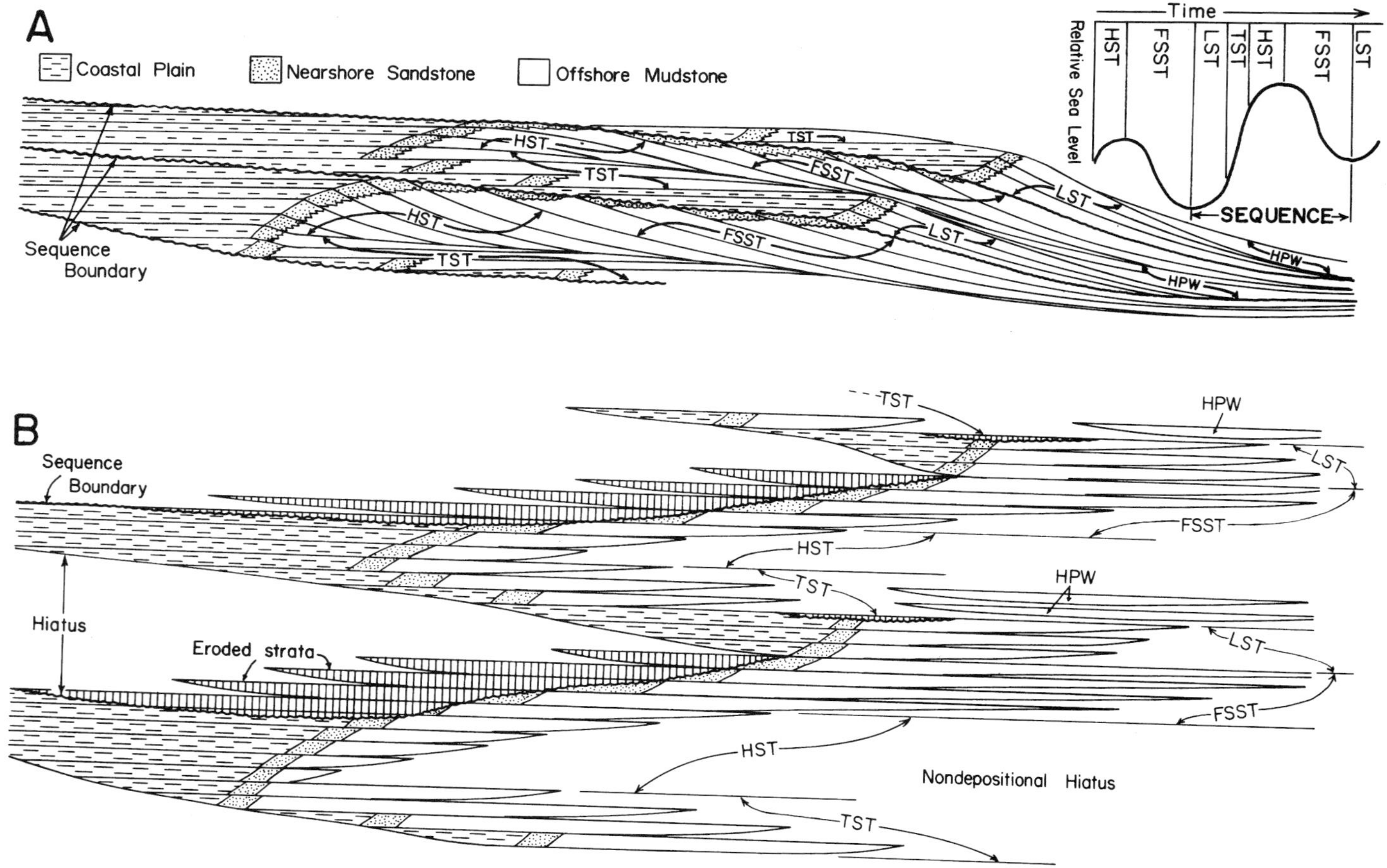

Fig. 1. (**a**) The stacking pattern of all four systems tracts in a ramp setting sequence. Corresponding relative sea-level curve on the right. The falling stage systems tract (FSST) is bounded below by the lowest (oldest) surface of stratal offlap and at the top by the first surface onto which strata onlap. In practice, recognition of offlap is likely to be difficult or impossible, and sedimentological criteria, such as erosive-based shoreface successions, provide the best evidence of relative sea-level fall and development of the FSST. HPW on the right refers to the healing phase wedge (Posamentier & Allen 1993), which is a basinward expression of the early transgressive systems tract. (**b**) Chronostratigraphic chart projected directly from (a) above. Note the diachronous development of the sequence boundary during deposition of the FSST, the updip cannibalization of high-order sequences in the lower-order FSST, and localized ravinement of the LST.

stacking patterns, and character of bounding surfaces, as follows (Fig. 1).

(1) The FSST is characterized by offlap. Offlap was originally defined by Grabau (1913; Christie-Blick 1991) as a stratal termination pattern where successively younger strata extend less far landward. The AGI Glossary (Gary *et al.* 1972) provides an additional definition: 'The progressive offshore degression of the updip termination of the sedimentary units within a conformable sequence of rocks . . . in which each successively younger unit leaves exposed a portion of the older unit on which it lies'. These definitions distinguish offlap from onlap, where the opposite is true. The other three systems tracts, highstand, lowstand, and transgressive, are all characterized by onlap.

(2) The FSST lies above the highstand and below the lowstand systems tracts. The lower boundary is the first offlapping stratum. This is interpreted to correspond in time with commencement of relative sea-level fall. In practice, the beginning of offlap will be difficult or impossible to recognize in outcrop or well logs, and in practice, we suggest that the beginning of relative fall is commonly expressed as the stratigraphically lowest shoreface succession that has a regressive surface of marine erosion at its base. Good examples of such regressive surfaces of marine erosion are described by Plint (1988, 1991, 1996); Plint & Norris (1991), Dam & Surlyk (1992), Posamentier *et al.* 1992, Hadley & Elliott (1993) Hart & Plint 1993, Ainsworth (1994), and papers in this volume. The upper boundary is the subaerial erosion surface (the sequence boundary) and its correlative downdip, subaqueous conformity. Seismically, the upper boundary of the FSST is characterized by renewed onlap of overlying strata onto the sequence boundary. In outcrop, well logs, and core, the upper boundary of the FSST (the subaerial erosion surface), is easy to pick landward of the lowstand shoreline. Farther downdip, the upper boundary of the FSST is defined at the top of the youngest shoreface succession that has a regressive surface of marine erosion at its base.

(3) If it can be resolved, the stacking pattern in the FSST is one of forestepping higher-order (i.e. higher-frequency) sequences.

(4) We infer that the FSST is produced during a phase of relative sea-level fall (Fig. 1), a hypothesis that is testable through observation of the geometric relations developed below.

This definition of the FSST follows that suggested by Nummedal *et al.* (1992) and Nummedal & Molenaar (1995), and incorporates the sedimentological attributes considered diagnostic of relative sea-level fall (e.g. Plint 1988, 1991, 1996; Plint & Norris 1991; Hart & Plint 1993). The definition is also consistent with the ideas of Hunt & Tucker (1992, 1993, 1995). The process that drives the formation of the FSST is that of forced regression (Plint 1991; Plint & Norris 1991; Posamentier *et al.* 1992; Posamentier & Morris this volume).

Although the definition of the FSST is probably no longer particularly controversial, considerable debate still surrounds the issue of where to place the sequence boundary (e.g. Posamentier & Morris this volume). The upper surface of the FSST, which is a surface of subaerial erosion formed throughout the period of relative sea-level fall, is objectively the most easily mappable surface, and we choose this as the sequence boundary. The surface of the delta front and correlative shelf at the time of relative sea-level lowstand is the correlative conformity. From a practical point of view, this marine surface will be difficult or impossible to identify (e.g. see Fitzsimmons & Johnson; Hernández-Molina *et al.*, this volume). The shoreline will continue to prograde above the sequence boundary, while building a lowstand systems tract (Figs 1, 2). Because relative sea-level is no longer falling at this time, the resultant progradational shoreface successions should have 'normal' gradational bases. Fluvial channels will continue to feed these lowstand systems tract deltas. Theoretically, these channels will be younger than the sequence boundary, but in practice they are very difficult to separate from those that form on the sequence boundary.

If deposition takes place in a basin with a well-developed shelf-slope physiography, the FSST will form as forced regression drives the delta shoreline from its highstand position to the shelf break (e.g. Sydow & Roberts 1994 and Chiocci; Hernández-Molina *et al.*; Kolla *et al.*; Trincardi & Correggiari this volume). Once that position has been attained, the basin floor fan of the lowstand systems tract starts to receive more, and probably coarser-grained sediment. Thus, the subaerial erosion surface on top of the shelf-phase deltas will be the surface of bypass across which sediment to feed the lowstand fan is carried. Therefore, the upper boundary of the FSST correlates with a surface that lies at (or a little way above?) the base of the basin floor fan (cf. Hunt & Tucker 1993, 1995).

Formation of the falling stage systems tract

Figure 2 is a geologically realistic rendition of a simple forward model of what is expected to happen during highstand, fall, lowstand, and rise of a single relative sea-level cycle. The relative sea-level cycle is represented by the curve on the right side of the diagram. The curve is divided into 14 equal time increments and 14 corresponding sediment surface profiles are represented in the cross-section to the left. The model was constructed by progressively shifting subaerial and submarine equilibrium surfaces vertically in response to relative sea-level change and laterally in response to sedimentation and accommodation development.

As rivers incise in response to relative sea-level fall, they will cause truncation of earlier highstand strata, first in the incising valleys and later as the entire land surface becomes graded to lower sea-level. Christie-Blick (1991, 1995) and Christie-Blick & Driscoll (1995) have stressed that the offlap geometry that characterizes the upper surface of the FSST is 'fundamentally due to bypass during progradation, implying that sequence boundaries develop gradually over a finite interval of geologic time'. The seaward expansion of the zone of bypass ceases once relative sea-level rise at the shoreline begins. The sea bed at that point in time constitutes the correlative conformity (time 7, Fig. 2).

Concurrent with the seaward expansion of the zone of subaerial sediment bypass, a zone of bypass is also expanding basinward across the shelf because the pre-existing (highstand) offshore profile now lies above the marine equilibrium profile because relative sea-level is falling (Plint 1988, 1991, 1996; Dominguez & Wanless 1991; Posamentier *et al.* 1992; Hart & Long 1996). The resulting erosion surface is a regressive surface of marine erosion (RSME; Fig. 3a–d). This surface will be more prominent if the rate of relative sea-level fall is high and the shelf slope is very gentle. Formation of the RSME will tend to be suppressed if rates of subsidence are high. We emphasize that the RSME, like the subaerial erosion surface above, is a product of relative sea-level fall. However it is neither a logical nor practical surface at which to place the sequence boundary. This conclusion is contradictory to that forwarded by Posamentier *et al.* (1992) and Posamentier & Morris (this volume).

Seaward of the toe of the shoreface, the RSME will gradually change into a surface of bypass and eventually into an area of uninterrupted deposition (Plint 1991). Our observations in outcrop suggest that the zone of bypass is commonly characterized by abundant, mutually-erosive, shore-normal gutter casts, sometimes associated with 'starved' hummocks (Fig. 3d, e, f). The gutter casts record storm-related scour of a semi-consolidated muddy shelf floor by waves and possibly also by storm-driven downwelling flows (Snedden *et al.* 1988; Plint 1991, 1996; Plint & Norris 1991; Myrow & Southard 1996). Gutters are filled with hummocky laminated sandstone and are separated by a veneer of fair-weather deposits. In a landward direction, the degree of amalgamation of the scouring events progressively increases, ultimately merging into a single regressive surface of erosion (Fig. 3a–d).

Following lowstand (Fig. 2, time 7), relative sea level at the shoreline starts to rise, resulting in the onset of alluvial aggradation. The shoreline continues to prograde, but at a diminishing rate as more and more sediment is partitioned into the aggrading valley fill to landward, leaving progressively less material available to nourish the delta. In consequence, the LST develops an increasingly aggradational geometry (Fig. 2, time 7–9). Between time 9 and 10 in our model (Fig. 2), the rate at which new accommodation is provided exceeds the ability of the system to fill the space with sediment, and transgression begins. Landward translation of the shelf equilibrium profile as the shoreline moves landward results in ravinement erosion of the upper part of the LST, significantly modifying its final geometry. Behind the transgressive barrier lies a lagoon and coastal plain depositional system, the surface of which is essentially horizontal. During relative sea-level rise, these flat-lying backbarrier deposits will progressively onlap onto the seaward-sloping surface of the underlying alluvial plain deposits (Fig. 2). Concomitantly, a wedge of shelf muds will onlap the lowstand clinoform, and this constitutes the 'healing phase' wedge of the transgressive systems tract (Fig. 1; Posamentier & Allen 1993). The transition to highstand systems tract occurs at time 12 in the model (Fig. 2). The marine maximum flooding surface will have a terrestrial correlative conformity that lies at the top of the youngest estuarine deposits, containing the bay-head and flood-tidal deltas (Zaitlin *et al.* 1994). The facies-stacking patterns generated by the remainder of the relative sea-level cycle are generally consistent with other recent studies of transgressive and highstand stratigraphy (Dalrymple *et al.* 1992; Posamentier & Allen 1993; Allen & Posamentier 1994; Zaitlin *et al.* 1994), and will not be further elaborated here.

Higher-order sequences

A second simple model illustrates the development and resultant geometry of a forestepping and downstepping set of higher-order sequences within a lower-order FSST (Fig. 4a). The development of discernible internal structure

Fig. 3 Field examples of forced regressive deposits, mainly from allomember F of the Dunvegan Formation. See Fig. 7 for the stratigraphic context of these photographs. (**a**) Seven metre thick swaley and trough cross-stratified shoreface sandstone rests on a sharp, gutter-casted surface cut in laminated dark mudstones and very fine sandstones typical of an offshore setting. Section exposed in the Smoky River railway cut, illustrated in Fig. 7b. (**b**) View looking upward at the basal surface of the shoreface sandstone shown in (a) showing large bathtub-shaped gutter casts, oriented normal to the local shoreline (see Fig. 7a). Larger gutter cast is 0.8 cm wide. (**c**) Sharp, gutter-casted regressive surface of marine erosion beneath shoreface sandstone that is probably attributable to Dunvegan allomember G exposed near Deadhorse Meadows, 50 km east of Beaverdam Creek (see Fig. 7a). (**d**) Detail of nested gutter casts cutting HCS and wave rippled storm beds immediately below the main RSME illustrated in (c); scale bar is 0.2 m. (**e**) Side view of a large, isolated, shore-perpendicular gutter cast filled with HCS very fine sandstone at the top of a sandier-upward succession of rippled and HCS storm beds at the top of allomember F at Flood Creek (Fig. 7b); scale bar is 0.2 m.(**f**) Detail of groove casts cut into cohesive mudstone forming the wall of the gutter cast illustrated in (e).

within a higher-order FSST will tend to be favoured by relatively high rates of subsidence, sediment supply, and/or high rates of eustatic change, and suppressed by low rates of both subsidence and sediment supply. For simplicity, this model was constructed by shifting only the equilibrium shoreface profile in response to the relative sea-level curve on the right; it did not attempt to incorporate the linked fluvial and marine systems, as shown in Fig. 2.

The punctuated relative sea-level fall will produce a FSST consisting of an offlapping succession of shoreface sand-body wedges, each bounded below by a regressive surface of marine erosion, and above by a ravinement surface. The regressive surfaces of marine erosion, however, may not extend along the entire shoreface base; downdip, and along strike they may change to correlative conformities depending on subsidence rate and slope. This is a practical reason for not defining the sequence boundary at the RSME.

The deposits shown in Fig. 4a constitute a set of shoaling-upward facies successions that classically would be termed parasequences. However, these apparent parasequences contain an internal regressive surface of marine erosion, and were generated in response to higher-order sea-level cycles of relative rise and fall. The resulting stratal packages must therefore be *sequences* (e.g. Fitzsimmons & Johnson this volume). The stratal succession between two regressive surfaces of marine erosion goes from a regressive shoreface at the base, across a ravinement surface/sequence boundary, into transgressive inner shelf mudstones, across a *high-order* maximum flooding surface, and finally up into a new regressive shoreface system that may have another RSME at its base. The building blocks of the low-order FSST, therefore, are a series of higher-order *sequences* because the presence of regressive surfaces of marine erosion at the base of many shoreface successions violates the original definition of a parasequence (cf. Plint 1991; Plint & Norris 1991; Martinsen 1993; Fitzsimmons & Johnson this volume). A fore- and downstepping set of these higher-order sequences constitute the falling stage systems tract of a lower-order composite sequence (e.g. Figs 1, 5, 6). The sequence set will be bounded above by a subaerial surface comprising an amalgamation of higher-order sequence boundaries. The component sequences are individually characterized by regressive surfaces of marine erosion, ravinement, and maximum flooding surfaces.

The *high-order* sequences may look like parasequences only if one's 'window to the world' is the basinward margin of the higher-order sequence where the RSME failed to develop, or where the correlative conformity to

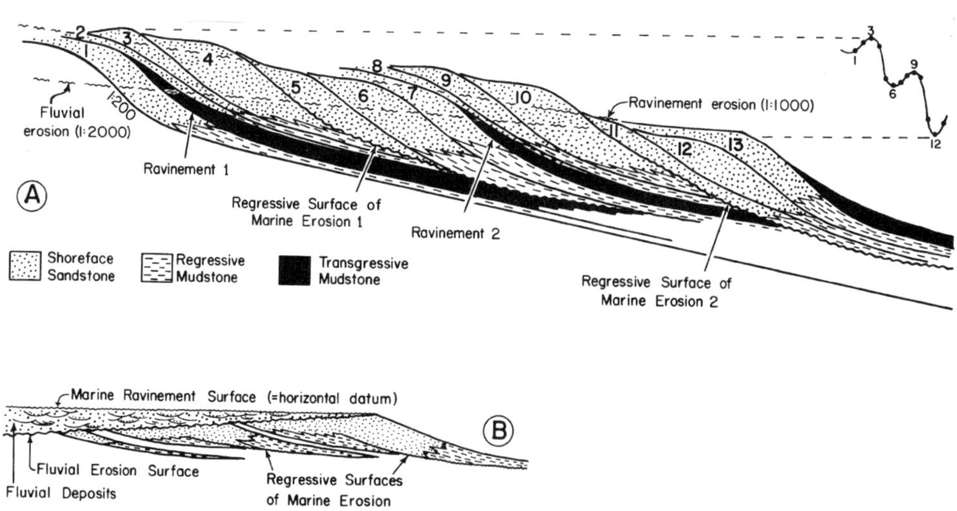

Fig. 4. (**a**) Schematic evolution of a falling stage systems tract in response to a longer-term sea-level fall, punctuated by smaller-scale rises and falls. Units 1–13 represent arbitrary temporal subdivisions. The stratigraphically-significant surfaces that result from the relative sea-level curve on the right will consist of an alternation of ravinement surfaces and regressive surfaces of marine erosion. The horizontal datum in this model diagram is sea-level. (**b**) Simplified version of (a) with the overlying ravinement surface as datum. Note the appearance of stratigraphic climb when this datum is set horizontal.

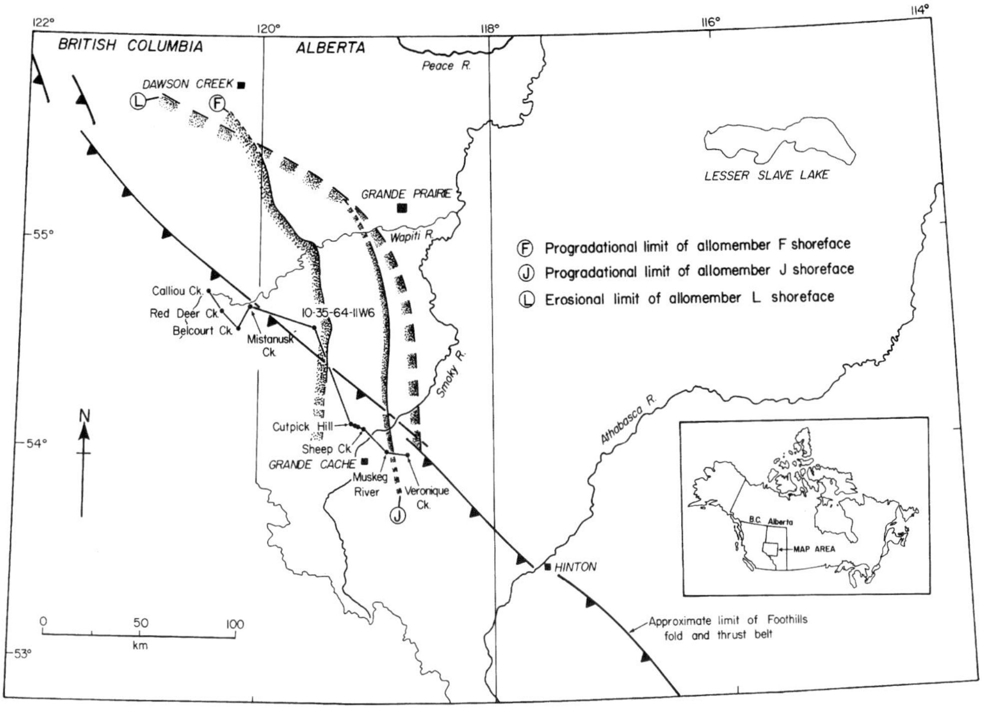

Fig. 6. Location of stratigraphic sections shown in Fig. 5, and distribution of shoreface sandstones in sequences F, I+J and K+L. Inset map shows location of study area. Based on Plint & Norris (1991).

the subaerial sequence boundary merges with the subsequent flooding surface (as shown in Fig. 2). The results of this model experiment are consistent with observations by Wright-Dunbar (1992), Bhattacharya (1993) and Arnott (1995) who stress that parasequences generally do in fact contain transgressive deposits, albeit thin, in their upper parts.

Although our models emphasize sedimentation in response to relative sea-level *oscillations*, one must not loose sight of the likelihood that some shallow marine shoaling-upward successions may be entirely of autogenic origin, such as delta lobe switching. The differentiation of such autogenic cycles (of more *localized* distribution) from those controlled by relative sea-level cycles (of more regional distribution) is likely to be difficult in offshore areas where evidence of relative sea-level fall is lacking. Distinction may only be possible if data are available from a complete distal-proximal transect, or where the geometry of the sediment body (e.g. a downstepping, offlapping pattern) can be mapped out in detail (e.g. Bhattacharya & Walker 1991; Plint 1996).

The choice of datum

The succession of high-order sequences in Fig. 4a descends relative to an original horizontal datum. The model was constructed using a horizontal sea-level datum. In constructing stratigraphic cross-sections from real rocks, however, we normally no longer know what was horizontal at the time of deposition. As a datum, we tend to use 'practical' surfaces that are reasonably flat. Outcrop sections are commonly hung on a ravinement surface, because it separates cliff-forming sandstone below from recessive mudstone above. Subsurface cross-sections may use a ravinement or maximum flooding surface because these are distinctive in logs. Both of these surfaces may be flat, commonly they are also nearly parallel, but they are certainly not horizontal (e.g. see high-resolution seismic profiles of late Quaternary margins of Chiocci; Hernández-Molina *et al.*; Kolla *et al.*; Trincardi & Correggiari this volume). Ravinement and maximum flooding surfaces dip seaward, typically at gradients that are steeper than the slope of the alluvial plain that prograded during building of the FSST.

In Fig. 4b, the FSST geometry of part 4a has been simplified and rotated to make the ravinement a 'horizontal' datum. With respect to this datum the FSST will appear to climb basinward. Several misinterpretations of relative sea-level history have been made due to failure to recognize this geometric reality.

Examples from the geological record

Below, we briefly illustrate two examples of high-order sequences that exhibit characteristics typical of the FSST.

Marshybank Formation

Figure 5 illustrates an oblique dip section through the mid–late Coniacian Marshybank Formation exposed in outcrop in the Alberta and British Columbia foothills. Correlation (Fig. 5) and palaeogeography (Fig. 6) are constrained by over 1500 well logs from the adjacent subsurface, as documented by Plint (1990) and Plint & Norris (1991).

The original allostratigraphic nomenclature of Plint (1990) and Plint & Norris (1991) is shown in Fig. 5. In this scheme, the Marshybank Formation was divided into 12 discontinuity-bounded units, A–L, mappable in subsurface and outcrop. In Fig. 5, a more genetic interpretation of the allostratigraphic units is emphasized by attaching an interpretation to each key bounding surface (sequence boundary, flooding surface, regressive surface of marine erosion). Thus, allomembers A, B, C, F and H are each interpreted to record deposition during a relative sea-level rise and fall and constitute high-order sequences. Allomembers D+E, I+J, and K+L are grouped into three more high-order sequences. Even higher-order sequences are present in allomembers B, J and L, (bounding surfaces are shown in Fig. 5) but these are of limited mappability and were not distinguished as separate units.

Sequence A (Fig. 5) can be traced throughout the basin in well logs and outcrop. The top surface forms both a higher-frequency flooding surface/sequence boundary, and a lower-frequency downlap surface. Sequences B, C, and D+E downlap onto sequence A in a basinward direction. Sequences B and C comprise sandier-upward successions, bounded above by flooding surfaces with, in sequence B, a veneer of chert pebbles that probably records a cryptic, wholly reworked lowstand deposit. In sequence D+E, allomember D contains abundant gutter casts in a bioturbated sandy siltstone matrix, and is interpreted as an inner shelf expression of the FSST (e.g. Red Deer Creek, Mistanusk Creek, Fig. 5); the TST and HST are not developed, or are unrecognizable. The overlying swaley cross-stratified shoreface sandstone of allomember E, represents a FSST and possible LST. Spectacular gutter casts are developed on the RSME at the base of the sandstone (Fig. 3, and cf. Fig. 3d). This RSME can be traced for tens of kilometres in well logs. Basinward, sequence D+E grades laterally into a simple, sandier-upward succession, capped locally (e.g. Muskeg River, Fig. 5) by gutter casts.

Relative sea-level rise terminated deposition of sequence D+E, and transgressive ravinement cut an erosional surface at least as far landward as Calliou Creek (Figs 5, 6), where a thin veneer of transgressive/highstand mudstone is preserved. Between Red Deer Creek and well 10–35–64–11W6, shoreface sandstone of sequence F rests erosively on the eroded remnant of sequence D+E: the transgressive and regressive surfaces of marine erosion are coplanar and are marked by a veneer of chert pebbles. Like D+E, sequence F grades seaward into a sandier-upward succession of HCS and rippled sandstone beds in a bioturbated siltstone matrix, capped by an interval of isolated gutter casts (e.g. Cutpick Hill, Cutpick Creek; Fig. 5). This shoaling-upward succession embodies transgressive, highstand and falling stage/lowstand systems tracts; it is the downdip expression of a high-order sequence.

Marine deposits of sequence H onlap southwestward onto sequence F and onto nonmarine deposits of allomember G. Allomember G is interpreted to represent aggradation of the coastal plain behind a transgressive barrier associated with transgression at the base of sequence H. The local presence of a chert pebble lag on the top of sequence H is similar to sequence B and suggests that sequence H records a high-frequency sea-level cycle terminated by lowstand erosion and winnowing. Sequences I+J, and K+L provide two more examples of high-frequency sequences in which the delta front sandstone represents the FSST, deposited above an extensive RSME. Although allomembers J and L were originally defined as units suitable for mapping in well logs (Plint 1990), the detail afforded by outcrop shows that allomember J in the Cutpick Creek-Sheep Creek area (Fig. 5) comprises two very thin sequences locally separated by a veneer of mud, and likewise allomember L may contain at least two thin sequences. These thin units suggest that by late Marshybank time, long-term relative sea-level fall was able to negate most of the accommodation generated by the high-order sea-level cycles.

Dunvegan Formation

Our second example is based on one unit of the Cenomanian Dunvegan Formation of the Alberta foreland basin. This formation has been divided into ten allomembers, labeled J–A in ascending order, on the basis of regional flooding surfaces (Bhattacharya & Walker 1991; Plint 1996), and examples of FSST deposits from northeastern British Columbia were illustrated by Plint (1996). The allomember-bounding surfaces are, in general, composite, including a component of subaerial erosion, followed by transgressive reworking, and can be considered as equivalent to type 1 sequence boundaries. Figure 7 illustrates two cross-sections through Dunvegan allomember (sequence) F, based on outcrop sections exposed in the Foothills in the vicinity of Grande Cache, Alberta. The relative position of palinspastically restored outcrop sections is shown as an inset map in Fig. 7a, which also shows the progradational limit of allomember F shoreface sandstones, based on both outcrop and subsurface mapping.

The base of allomember F is picked at a distinctive decimetre-thick spherulitic concretionary siderite bed resting sharply on dark laminated marine mudstone of allomember G. The top is a regional flooding surface, locally marked by an intraclast lag or an eroded, lithified surface. The bulk of allomember F consists of dark, laminated mudstones, with minor bioturbated sandy siltstone units, which are broadly representative of transgressive and highstand deposition. The upper part of allomember F comprises one or more, metre-scale, erosive-based shoreface sandstone bodies (e.g. Fig. 3a, c). The base of each sandstone typically displays spectacular gutter casts that are consistently oriented normal to the regional shoreline, which has been independently mapped using subsurface data (Fig. 7a, inset map). The sharp-based shoreface sandstones grade laterally seaward, over only 1–2 km, into a set of HCS sandstone beds typified by large gutter casts (Fig. 3e) which also have a shore-normal orientation (Fig. 7a, b). The gutter-casted sandstones commonly appear rather abruptly at the top of a succession of centimetre-scale very fine sandstone and siltstone storm beds and appear to record a relatively abrupt increase in storm energy at the seabed.

Although there are local variations in the stratigraphic succession of allomember F, and it is impossible to verify exact correlations, due to lack of exposure, there is a fairly consistent pattern in which landward areas are characterized by one or more, closely spaced, sharp-based shoreface sandstones interbedded with laminated shelf mudstones. More seaward areas are typified by a succession of centimetre-scale, muddy storm beds, rather abruptly overlain by decimetre-scale gutter-casted HCS sandstone beds. These observations suggest that repeated minor relative sea-level falls produced sharp-based, forced regressive sandstone sheets, attributable to the FSST. Relative sea-level fall was also recorded on the shelf, perhaps up to 10–15 km from the shoreline, by emplacement of gutter-casted HCS storm sandstone beds. The single RSME beneath the shoreface records sediment bypassing on the lower shoreface/inner shelf. Further seaward, where a little more accommodation was available, the RSME divides into an array of locally guttered surfaces that record alternating storm-related erosion and bypass, and limited fair-weather aggradation. In this offshore setting, it is impossible to place the lower boundary of the FSST at a single surface, but, for practical purposes, it can be approximated at the base of the first gutter-casted sandstone capping much thinner-bedded shelf facies. Similarly, it is impossible to differentiate a discrete lowstand systems tract in these examples, and the forced regressive sandstone tongues are best considered as composite FSST/LST deposits.

Conclusions

The falling stage systems tract is characterized by offlap and a basinward shift in facies. It

Fig. 7 (overleaf). Two stratigraphic cross-sections through outcrop sections of allomember F of the Dunvegan Formation exposed in various thrust slices in the vicinity of Grande Cache, Alberta Foothills. The inset map in (**a**) shows the sections palinspastically restored. The base of allomember F is placed at a very distinctive spherulitic sideritic horizon recognizable in most sections. The top of the allomember is a widely-mappable flooding surface, which is unusual in this formation in being locally overlain by a lag of sideritic intraclasts and inoceramid shell debris. The sections are interpreted in terms of a series of higher-frequency sequences, producing an offlapping set of sharp-based shoreface sandstones. The discrete basal RSME typical of the nearshore area (e.g. Fig. 3a,b), merges seaward into an array of mutually-erosive gutter-casted HCS storm beds that record intermittent local erosion of the more distal shelf during relative sea-level fall. Note that guttor casts beneath the shoreface sandstone, and in offshore strata, have a shore-normal orientation, suggesting that oscillatory wave action was the dominant process in their formation. This is supported by the lack of clear polarity in small-scale erosional structures on gutter walls (Fig. 3f).

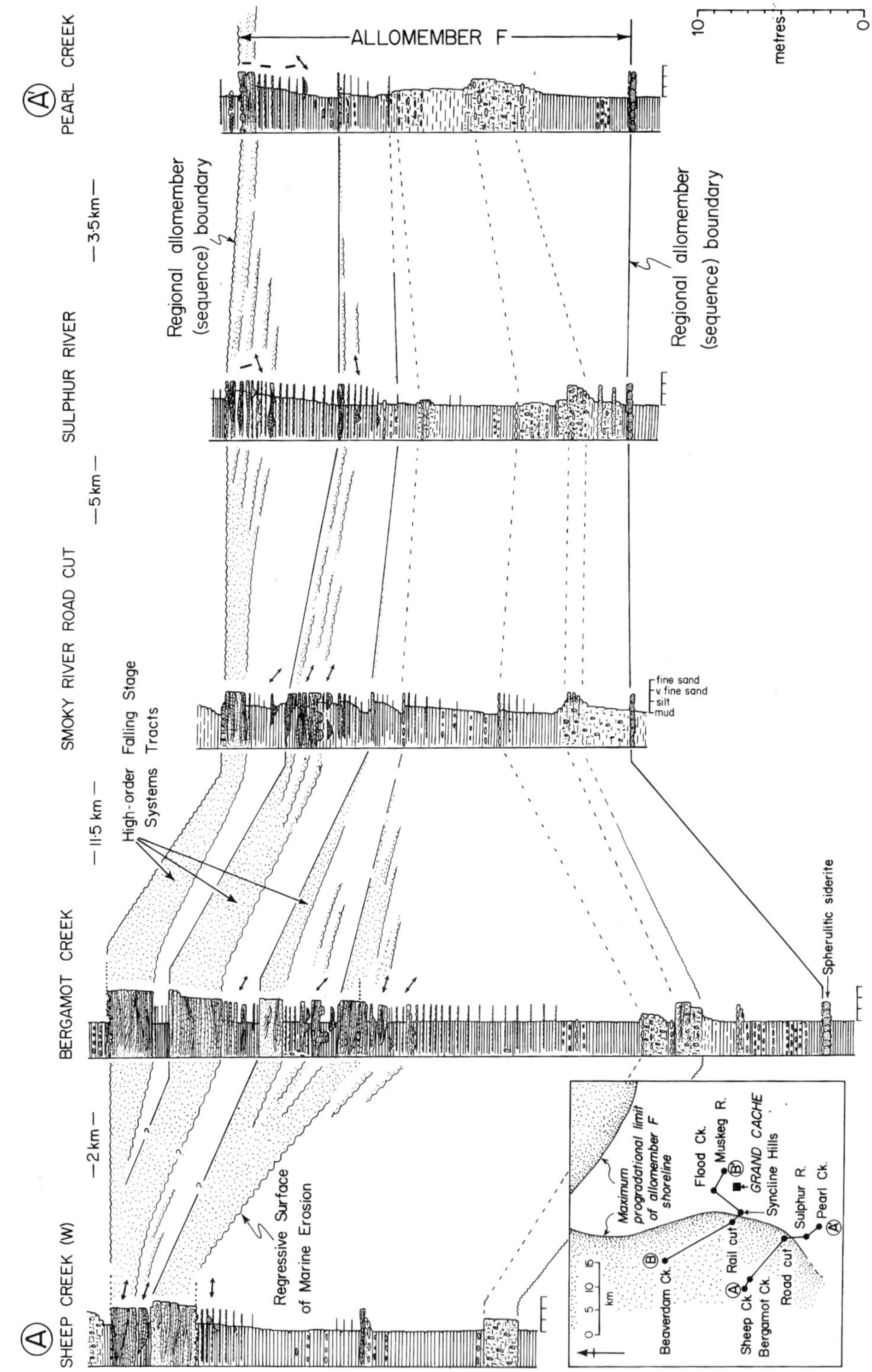

A SHEEP CREEK (W)
—2 km—
BERGAMOT CREEK
— 11·5 km —
SMOKY RIVER ROAD CUT
—5 km—
SULPHUR RIVER
—3·5 km—
A' PEARL CREEK
ALLOMEMBER F
Regional allomember (sequence) boundary
Regional allomember (sequence) boundary
High-order Falling Stage Systems Tracts
Regressive Surface of Marine Erosion
Spherulitic siderite
fine sand
v. fine sand
silt
mud
10
metres
0
Maximum progradational limit of allomember F shoreline
Flood Ck.
Muskeg R.
B'
GRAND CACHE
Syncline Hills
Rail cut
Beaverdam Ck.
B
Sulphur R.
Pearl Ck.
A'
Road cut
A
Sheep Ck
Bergamot Ck.
0 5 10 15
km

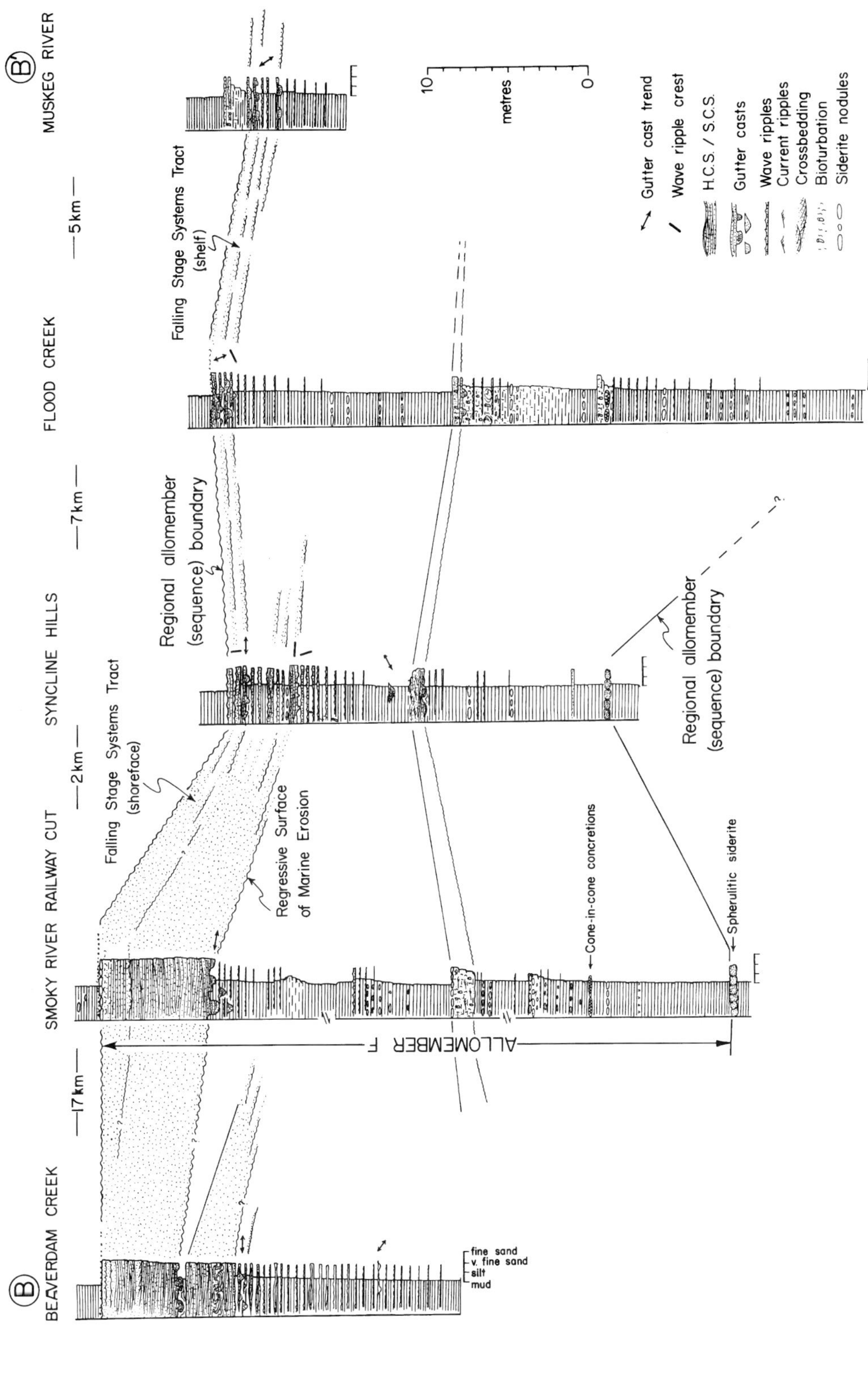
B
BEAVERDAM CREEK
17 km
SMOKY RIVER RAILWAY CUT
2 km
SYNCLINE HILLS
7 km
FLOOD CREEK
5 km
MUSKEG RIVER
B'
Falling Stage Systems Tract (shoreface)
Regressive Surface of Marine Erosion
Falling Stage Systems Tract (shelf)
Regional allomember (sequence) boundary
Regional allomember (sequence) boundary
ALLOMEMBER F
Cone-in-cone concretions
Spherulitic siderite
fine sand
v. fine sand
silt
mud
10
metres
0
Gutter cast trend
Wave ripple crest
H.C.S. / S.C.S.
Gutter casts
Wave ripples
Current ripples
Crossbedding
Bioturbation
Siderite nodules

replaces the upper part of the highstand systems tract (as currently defined), and lies below the lowstand systems tract. It may contain a forestepping set of higher-order sequences, and is inferred to be produced during relative sea-level fall. Compared to the standard systems tract scheme (Posamentier *et al.* 1988), our introduction of a falling stage systems tract results in a significant change in the way systems tracts are defined. The HST now terminates at relative sea-level highstand. The FSST is deposited during the period of relative sea-level fall. The LST begins at relative sea-level lowstand and ends when transgression initiates the TST.

In nearshore areas, the lower boundary of the FSST is the stratigraphically lowest regressive surface of marine erosion. This discrete surface, commonly with gutter casts, may merge seaward into a zone of amalgamated gutter casts that collectively record relative sea-level fall and limited erosion of the shelf by storm waves. Even further seaward, the base of the FSST may be indicated by an abrupt coarsening of the sediment but without development of an erosional surface.

The upper boundary of the FSST is the sequence boundary. This surface consists of an updip regional subaerial unconformity produced as a surface of sediment bypass during relative sea-level fall, and a correlative conformity that is the sea floor at the time of relative sea-level lowstand. Transgressive erosion of FSST deposits can result in an apparently 'detached' lowstand deposit that appears to represent an abrupt basinward shift in facies (e.g. Plint 1988, 1996; Walker & Plint 1992; Ainsworth *et al.* this volume).

At high frequency, the sequence boundary is characterized by minor onlap (Fig. 2), but in a lower-frequency FSST, any onlapping deposits are likely to be eroded during the long-term relative sea-level fall (Figs. 1b & 5).

The FSST may consist of a forestepping set of higher-frequency sequences. These are the building blocks of the lower-frequency sequence. The shoreface sand bodies observed within higher-frequency sequences of the FSST display regressive surfaces of marine erosion at their bases and coeval gutter cast horizons in more distal shelf settings. The top of the FSST is defined by an erosion surface (originally subaerial but usually modified by ravinement), and its correlative offshore flooding surface.

Outcrop and subsurface cross sections commonly use an overlying ravinement or maximum flooding surface as datum. When a section is restored to such a datum, the falling stage systems tract will appear to record stratigraphic climb, whereas in fact it does not.

Lively discussions over several years with H. Posamentier, N. Christie-Blick, D. James and J. C. Van Wagoner have provided stimulus for the ideas presented here. We also express our gratitude to D. Hunt and R. Gawthorpe for organizing the special meeting on 'Sedimentary response to forced regressions' at the Geological Society of London in September 1995. DN acknowledges partial support from NSF grant EAR–9205811; AGP was funded by NSERC grant A1917, with additional support from Canadian Hunter Exploration, ESSO Canada, Home Oil, Texaco Canada and Unocal Canada Ltd. AGP is grateful to M. McMechan for help with palinspastic restoration of sections in the Alberta Foothills, and to A. Noon for photography. We thank W. Fitchen, A. Pulham and D. Hunt for their constructive reviews of this manuscript.

References

AINSWORTH, R. B. 1994. Marginal marine sedimentology and high-resolution sequence analysis; Bearpaw-Horseshoe Canyon transition, Drumheller, Alberta. *Bulletin of Canadian Petroleum Geology*, **42**, 26–54.

—— & PATTISON, S. A. J. 1994. Where have all the lowstands gone? Evidence for attached lowstand systems tracts in the Western Interior of North America. *Gelogy*, **22**, 415–418.

——, BOSSCHER, H. & NEWALL, M. J. 2000. Forward modelling of forced regressions: evidence for the genesis of attached and detached lowstand systems. *This volume*.

ALLEN, G. P. & POSAMENTIER, H. W. 1994. Transgressive facies and sequence architecture in mixed tide- and wave-dominated incised valleys: example from the Gironde estuary, France. *In:* DALRYMPLE, R. W., ZAITLIN, B. A. & BOYD, R. (eds) *Incised valley systems: origin and sedimentary sequences*: Society of Economic Paleontologists and Mineralogists, Special Publications, **51**, 225–240.

ARNOTT, R. W. C. 1995. The parasequence definition – are transgressive deposits inadequately addressed? *Journal of Sedimentary Research* **B65**, 1–6.

BHATTACHARYA, J. P. 1993. The expression and interpretation of marine flooding surfaces and erosional surfaces in core; examples from the Upper Cretaceous Dunvegan Formation, Alberta foreland basin, Canada. *In:* POSAMENTIER, H. W., SUMMERHAYES, C. P., HAQ, B. U. & ALLEN, G. P. (eds) *Sequence Stratigraphy and Facies Associations*. International Association of Sedimentologists, Special Publications, **18**, 125–160.

—— & WALKER, R. G. 1991. Allostratigraphic subdivision of the Upper Cretaceous, Dunvegan, Shaftesbury and Kaskapau formations in the subsurface of northwestern Alberta. *Bulletin of Canadian Petroleum Geology*, **39**, 145–164.

CHIOCCI, F. L. 2000. Depositional response to Quaternary fourth-order sea-level falls on the Latium margin (Tyrrhenian Sea, Italy). *This volume*.

CHRISTIE-BLICK, N. 1991. Onlap, offlap, and the origin of unconformity-bounded depositional sequences. *Marine Geology*, **97**, 35–56.

—— 1995. Forced logic: sequence boundary development in ramp and shelf settings. *In:* HUNT, D, GAWTHORPE, R. & DOCHERTY, M. (convenors) *Sedimentary Responses to Forced Regressions.* Geological Society of London, Abstract Volume, 45–48.

—— & DRISCOLL, N. W. 1995. Sequence stratigraphy. *Annual Review of Earth and Planetary Science*, **23**, 451–478.

DALRYMPLE, R. W., ZAITLIN, B. A. & BOYD, R. 1992. Estuarine facies models: conceptual basis and stratigraphic implications: *Journal of Sedimentary Petrology*, **62**, 1130–1146.

DAM, G. & SURLYK, F. 1992. Forced regressions in a large wave- and storm-dominated anoxic lake, Rhaetian-Sinemurian Kap Stewart Formation, East Greenland. *Geology*, **20**, 749–752.

DOMINGUEZ, J. M. L. & WANLESS, H. R. 1991. Facies architecture of a falling sea-level strandplain, Doce River coast, Brazil. *In:* SWIFT, D. J. P., OERTEL, G. F., TILLMAN, R. W. & THORNE, J. A. (eds) *Shelf sand and sandstone bodies – geometry, facies and sequence stratigraphy.* International Association of Sedimentologists, Special Publications, **14**, 259–281.

FITZSIMMONS, R. & JOHNSON, S. 2000. Forced regressions: recognition, architecture and genesis in the Campanian of the Bighorn Basin, Wyoming. *This volume.*

GARY, M., MCAFEE, R. & WOLF, C. L. 1972. *Glossary of Geology*. American Geological Institute. Washington, DC.

GRABAU, A. W. 1913. *Principles of Stratigraphy*. Seiler and Co., New York.

HADLEY, D. F. & ELLIOTT, T. 1993. The sequence-stratigraphic significance of erosive-based shoreface sequences in the Cretaceous Mesaverde Group of northwest Colorado. *In:* POSAMENTIER, H. W., SUMMERHAYES, C. P., HAQ, B. U. & ALLEN, G. P. (eds) *Sequence Stratigraphy and Facies Associations*. International Association of Sedimentologists, Special Publications, **18**, 521–535.

HART, B. S. & LONG, B. F. 1996. Forced regressions and lowstand deltas: Holocene Canadian examples. *Journal of Sedimentary Research*, **A66**, 820–829.

—— & PLINT, A. G. 1993. Origin of an erosion surface in shoreface sandstones of the Kakwa Member (Upper Cretaceous Cardium Formation, Canada): importance for reconstruction of stratal geometry and depositional . *In:* POSAMENTIER, H. W., SUMMERHAYES, C. P., HAQ, B. U. & ALLEN, G. P. (eds) *Sequence Stratigraphy and Facies Associations*. International Association of Sedimentologists, Special Publications, **18**, 451–467.

HELLAND-HANSEN, W. & GJELBERG, J. G. 1994. Conceptual basis and variability in sequence stratigraphy: a different perspective. *Sedimentary Geology*, **92**, 31–52.

HERNÁNDEZ-MOLINA F. J., SOMOZA, I. & LOBO, F. 2000. Seismic stratigraphy of the Gulf of Cádiz continental shelf: a model for late Quaternary very high-resolution sequence stratigraphy and response to sea-level fall. *This volume.*

HUNT, D. & TUCKER, M. E. 1992. Stranded parasequences and the forced regressive wedge systems tract: deposition during base-level fall. *Sedimentary Geology*, **81**, 1–9.

—— & —— 1993. Sequence stratigraphy of carbonate shelves with an example from the mid-Cretaceous (Urgonian) of southeast France. *In:* POSAMENTIER, H. W., SUMMERHAYES, C. P., HAQ, B. U. & ALLEN, G. P. (eds) *Sequence Stratigraphy and Facies Associations*. International Association of Sedimentologists, Special Publications, **18**, 307–341.

—— & —— 1995. Reply to Discussion. *Sedimentary Geology*, **95**, 147–160.

KOLLA, V., BIONDI, P., LONG, B. & FILLION, R. 2000. Sequence stratigraphy and architecture of the late Pleistocene Lagniappe delta complex, northeast Gulf of Mexico. *This volume.*

——, POSAMENTIER, H. W. & H. EICHENSEER, H. 1995. Discussion: Stranded parasequences and the forced regressive wedge systems tract: deposition during base level fall. *Sedimentary Geology*, **95**, 139–145.

MARTINSEN, O. J. 1993. Namurian (late Carboniferous) depositional systems of the Craven-Askrigg area, northern England: implications for sequence-stratigraphic models. *In:* POSAMENTIER, H. W., SUMMERHAYES, C. P., HAQ, B. U. & ALLEN, G. P. (eds) *Sequence Stratigraphy and Facies Associations*. International Association of Sedimentologists, Special Publications, **18**, 247–281.

MIALL, A. D. 1991. Stratigraphic sequences and their chronostratigraphic correlation. *Journal of Sedimentary Petrology,* **61**, 497–505.

MITCHUM, R. M., VAIL, P. R. & THOMPSON, S. 1977. Seismic stratigraphy and global changes of sea level, Part 2: The depositional sequence as a basic unit for stratigraphic analysis. *In*: PAYTON, C. E. (ed.) *Seismic stratigraphy – application to hydrocarbon exploration.* American Association of Petroleum Geologists, Memoirs, **26**, 53–62.

MYROW, P. M. & SOUTHARD, J. B. 1996. Tempestite deposition. *Journal of Sedimentary Research*, **66**, 875–887.

NUMMEDAL, D., RILEY, G. W., COLE, R. D. & TREVENA, A. S. 1992. The falling sea level systems tract in ramp settings (Abstract). *In: Mesozoic of the Western Interior.* Society of Economic Paleontologists and Mineralogists, Theme Meeting, Fort Collins, Colorado, August 17–19, 1992, p. 50.

NUMMEDAL, D., RILEY, G. W. & TEMPLET, P. L. 1993. High-resolution sequence architecture: a chronostratigraphic model based on equilibrium profile studies. *In:* POSAMENTIER, H. W., SUMMERHAYES, C. P., HAQ, B. U. & ALLEN, G. P. (eds) *Sequence Stratigraphy and Facies Associations.* International Association of Sedimentologists, Special Publications, **18**, 55–68.

NUMMEDAL, D., GUPTA, S., PLINT, A. G. & COLE, R. D. 1995. The falling stage systems tract: definition, character and expression in several examples from the Cretaceous from the U. S. Western Interior. *In.* HUNT, D., GAWTHORPE, R. & DOCHERTY, M. (convenors) *Sedimentary*

Responses to Forced Regressions. Geological Society of London, Abstract Volume, 45–48.
NUMMEDAL, D. & MOLENAAR, C. M. 1995. Sequence stratigraphy of the Gallup Sandstone, *In:* VAN WAGONER, J. C. & BERTRAM, G. T. (eds) *Sequence Stratigraphy of Foreland Basin Deposits* American Association of Petroleum Geologists Memoirs, **64**, 277–310.
PLINT, A. G. 1988. Sharp-based shoreface sequences and 'offshore bars' in the Cardium Formation of Alberta: their relationship to relative changes in sea level. *In:* WILGUS, C. K., HASTINGS, B. S., KENDALL, C. G. ST.C., POSAMENTIER, H. W., ROSS, C. A. & VAN WAGONER, J. C. (eds) *Sea-level changes: An integrated approach* Society of Economic Paleontologists and Mineralogists Special Publications, **42**, 357–370.
—— 1990. An allostratigraphic correlation of the Muskiki and Marshybank formations (Coniacian-Santonian) in the Foothills and subsurface of the Alberta Basin. *Bulletin of Canadian Petroleum Geology*, **38**, 288–306.
—— 1991. High frequency relative sea level oscillations in Upper Cretaceous shelf clastics of the Alberta foreland basin: possible evidence of a glacio-eustatic control? *In:* MACDONALD, D. I. M. (ed.) *Sedimentation, tectonics and eustasy* International Association of Sedimentologists Special Publications, **12**, 409–428.
—— 1996. Marine and nonmarine systems tracts in fourth-order sequences in the Early-Middle Cenomanian, Dunvegan Alloformation, northeastern British Columbia, Canada. *In:* HOWELL, J. A. & AITKEN, J. F. (eds) *High Resolution Sequence Stratigraphy: Innovations and applications.* Geological Society Special Publications, **104**, 159–191.
PLINT, A. G. & NORRIS, B. 1991. Anatomy of a ramp margin sequence: facies successions, paleogeography and sediment dispersal patterns in the Muskiki and Marshybank formations, Alberta foreland basin. *Bulletin of Canadian Petroleum Geology*, **39**, 18–42.
POSAMENTIER, H. W. & ALLEN, G. P. 1993. Variability of the sequence stratigraphic model: effects of local basin factors. *Sedimentary Geology*, **91**, 91–109.
—— & MORRIS, W. R. 2000. Aspects of the stratal architecture of forced regressive deposits. *This volume.*
—— & VAIL, P. R. 1988. Eustatic controls on clastic deposition II – sequence and systems tract models. *In:* WILGUS, C. K., HASTINGS, B. S., KENDALL, C. G. ST. C., POSAMENTIER, H. W., ROSS, C. A. & VAN WAGONER, J. C. (eds) *Sea-Level Changes: An integrated approach* Society of Economic Paleontologists and Mineralogists Special Publications, **42**, 125–154.
——, ALLEN, G. P., JAMES, D. P. & TESSON, M. 1992. Forced regressions in a sequence stratigraphic framework: concepts, examples and exploration significance. *American Association of Petroleum Geologists Bulletin*, **76**, 1687–1709.
——, JERVEY, M. T. & VAIL, P. R. 1988. Eustatic controls on clastic deposition I – conceptual framework: *In:* WILGUS, C. K., HASTINGS, B. S., KENDALL, C. G. ST.C., POSAMENTIER, H. W., ROSS, C. A. & VAN WAGONER, J. C. (eds) *Sea-Level Changes: An integrated approach* Society of Economic Paleontologists and Mineralogists Special Publications, **42**, 109–124.
SCHLAGER, W. 1993. Accommodation and supply – a dual control on stratigraphic sequences. In: CLOETINGH, S., SASSI, W., HORVATH, F. & PUIGDEFABREGAS, C. (eds) *Basin Analysis and Dynamics of Sedimentary Basin Evolution. Sedimentary Geology*, **86**, 111–136.
SNEDDON, J. W., NUMMEDAL, D. & AMOS, A. F. 1988. Storm- and fair-weather combined flow on the central Texas continental shelf. *Journal of Sedimentary Petrology*, **58**, 580–595.
SYDOW, J. & ROBERTS, H. H. 1994. Stratigraphic framework of a Late Pleistocene shelf-edge delta, northeast Gulf of Mexico. *American Association of Petroleum Geologists, Bulletin*, **78**, 1276–1312.
TIRSGAARD, H. 1996. Cyclic sedimentation of carbonate and siliciclastic deposits on a late Precambrian ramp: The Elisabeth Bjerg Formation (Eleonore Bay Supergroup), East Greenland. *Journal of Sedimentary Research*, **66**, 699–712.
TRINCARDI, F. & CORREGGIARI, A. 2000. Quaternary forced-regression deposits in the Adriatic Basin and the record of composite sea-level cycles. *This volume.*
VAIL, P. R. 1987. Seismic stratigraphy interpretation using sequence stratigraphy. *In*: BALLY, A. W. (ed.) *Atlas of Seismic Stratigraphy, Vol. 1.* American Association of Petroleum Geologists, Studies in Geology, **27**, 1–10.
VAIL, P. R., AUDEMARD, F., BOWMAN, S. A., EISNER, P. N. & PEREZ-CRUZ, G. 1991. The stratigraphic signature of tectonics, eustasy, and sedimentation. *In:* EINSELE, G., RICKEN, W. & SEILACHER, A. (eds) *Cycles and events in stratigraphy*. Springer-Verlag, 617–659.
——, MITCHUM, R. M. & THOMPSON, S. 1977. Seismic stratigraphy and global changes of sea level, Part 3: Relative changes of sea level from coastal onlap. *In*: PAYTON, C. E. (ed.) *Seismic stratigraphy – application to hydrocarbon exploration.* American Association of Petroleum Geologists, Memoirs, **26**, 63–81.
VAN WAGONER, J. C. 1995*a*. Overview of sequence stratigraphy of foreland basin deposits. *In:* VAN WAGONER, J. C. & BERTRAM, G. T. (eds) *Sequence Stratigraphy of Foreland Basin Deposits* American Association of Petroleum Geologists Memoirs, **64**, ix-xxi.
—— 1995*b*. Sequence stratigraphy and marine to nonmarine facies architecture of foreland basin strata, Book Cliffs, Utah, U. S. A. *In:* VAN WAGONER, J. C. & BERTRAM, G. T. (eds) *Sequence Stratigraphy of Foreland Basin Deposits.* American Association of Petroleum Geologists Memoirs, **64**, 137–223.
——, MITCHUM, R. M., CAMPION, K. M. & RAHMANIAN, V. D. 1990. *Siliciclastic sequence stratigraphy in well logs, cores and outcrop.* American Association of Petroleum Geologists, Methods in Exploration Series, **7**.

——, POSAMENTIER, H. W., MITCHUM, R. M. JR., VAIL, P. R., SARG, J. F., LOUTIT, T. S. & HARDENBOL, J. 1988. An overview of the fundamentals of sequence stratigraphy and key definitions. *In:* WILGUS, C. K., HASTINGS, B. S., KENDALL, C. G. ST. C., POSAMENTIER, H. W., ROSS, C. A. & VAN WAGONER, J. C. (eds) *Sea-level changes: An integrated approach*. Society of Economic Paleontologists and Mineralogists Special Publications, **42**, 39–46.

WALKER, R. G. & PLINT, A. G. 1992. Wave- and storm-dominated shallow marine systems. *In:* WALKER, R. G. & JAMES, N. P. (eds) *Facies Models – response to sea level change*. Geological Association of Canada, St. John's, 219–238.

WRIGHT-DUNBAR, R. 1992. Shoreline cyclicity and the transgressive record: a model based on Point Lookout Sandstone exposures at San Luis, New Mexico. *In:* LUCAS, S. G., KUES, B. S. (eds) *San Juan Basin IV*. New Mexico Geological Society Guidebook, 12–16.

ZAITLIN, B. A., DALRYMPLE, R. W. & BOYD, R. 1994. The stratigraphic organization of incised valley systems associated with relative sea-level change. *In:* DALRYMPLE, R. W., ZAITLIN, B. A. & BOYD, R. (eds) *Incised valley systems: origin and sedimentary sequences*. Society of Economic Paleontologists and Mineralogists, Special Publications, **51**, 45–60.

Aspects of the stratal architecture of forced regressive deposits

HENRY W. POSAMENTIER & WILLIAM R. MORRIS

ARCO ARII, PO Box 260 888, Plano, Texas 75026-0888, USA

Abstract: Forced regression refers to the process of seaward migration of a shoreline in direct response to relative sea-level fall. Recognition criteria for forced regressive deposits include: (1) presence of a significant zone of separation between successive shoreface deposits, (2) the presence of sharp-based shoreface/delta front deposits, (3) the presence of progressively shallower clinoforms going from proximal to distal, (4) the occurrence of long-distance regression, (5) the absence of fluvial and/or coastal plain/delta plain capping the proximal portion of regressive deposits, (6) the presence of a seaward-dipping upper bounding surface at the top of the regressive succession, (7) the presence of increased average sediment grain size in regressive deposits going from proximal to distal and (8) the presence of 'foreshortened' stratigraphic successions.

The principal factors driving the stratal architecture of forced regressive deposits include: (1) the gradient of the sea floor progressively exposed by falling relative sea-level, (2) the ratio of the sediment flux to the rate of relative sea-level fall, (3) the 'smoothness' of relative sea-level fall, (4) the variability of sediment flux and (5) the changes of sedimentary process that occur as sea-level falls and progressively more of the shelf is subaerially exposed.

Forced regressive deposits are grouped into attached v. detached, and smooth-topped v. stepped-topped. Attached deposits are defined as successive downstepped stratigraphic units whose shoreface/delta front deposits are generally in contact with each other. In contrast, detached deposits are defined as successive downstepped stratigraphic units whose shoreface/delta front deposits are generally not in contact with each other. Rather, in this instance a zone of sedimentary bypass exists. Stepped-top forced regressive deposits are characterized by a succession of horizontally topped though downstepping stratigraphic units. In contrast, smooth-topped forced regressive deposits are characterized by a seaward-dipping, albeit smooth, upper bounding surface. The bounding surfaces of forced regressive deposits commonly are expressed as a ravinement surface at the top and an unconformity to correlative conformity at the base.

Transgression and regression refer to landward and seaward shifts of the shoreline, respectively. Regression occurs either during a time of relative sea-level rise when sediment flux is sufficient to exceed the rate at which accommodation is created, or during a time of relative sea-level fall, when accommodation is lost. In the former instance, the shoreline migrates seaward as a result of the progressive infill of the available accommodation. However, in the latter instance, regression invariably occurs regardless of how much or how little sediment is delivered to the shoreline. For this reason, this type of regression, forced by relative sea-level fall and independent of sediment flux variations, has been referred to as forced regression (Posamentier *et al.* 1992). Forced regression is characterized by different facies and stratigraphic relationships from other, or normal regressions, and it is these differences that justify separation of regression type in this way (Fig. 1). In contrast, transgression commonly occurs when the rate of relative sea-level rise is sufficiently high so as to create space (i.e. accommodation) for sediment to fill at a higher rate than sediment can fill that space. In response, the shoreline migrates in a landward direction by the process of shoreface retreat/ravinement or in-place drowning. In some instances transgression can occur under conditions of a relative sea-level stillstand. This will happen where the coastline is influenced by high-energy waves or currents, so that erosion of the beach and adjacent dunes and coastal plain results in a landward migration of the shoreline.

The distinction between 'normal' and 'forced' regression is significant because there are fundamentally different processes active during formation of each type of associated deposit. Alluvial/coastal plain aggradation as well as shoreface/delta-front progradation commonly accompany normal regression. In marked contrast, fluvial downcutting and sedimentary bypass typically occur during forced regression. Thus, progradation as well as aggradation in most instances accompanies normal regression, whereas progradation without aggradation commonly characterizes forced regression. This forced regressive process results in cannibalization of the substrate and hence, possible changes

From: Hunt, D. & Gawthorpe, R. L. (eds) *Sedimentary Responses to Forced Regressions*. Geological Society, London, Special Publications, **172**, 19–46. 1-86239-063-0/00/$15.00

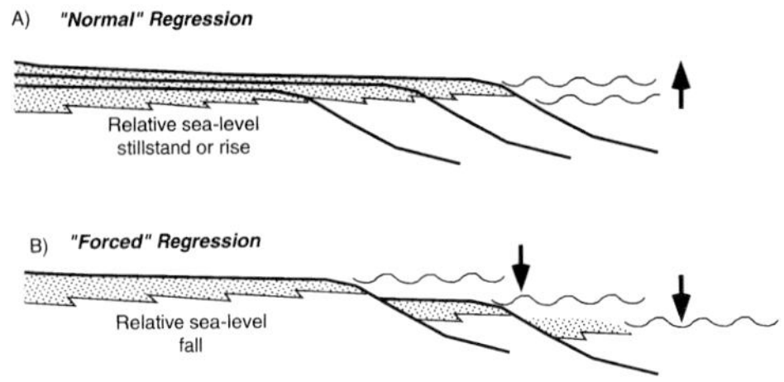

Fig. 1. Schematic depiction of (**a**) normal regression and (**b**) forced regression. Normal regression is associated with a situation wherein an excess sediment flux exists relative to the rate at which marine accommodation is created in the nearshore marine environment. Forced regression is associated with relative sea-level fall and decreasing accommodation.

in depositional systems, sediment grain-size distribution (i.e. increase of sand to mud ratio), as well as an increase of sediment flux (Morris *et al.* 1995; Posamentier *et al.* 1995). Moreover, because of the virtual elimination of significant active floodplains during and immediately after the period of downcutting, channel gradients, channel patterns (e.g. braided, meandering etc.), and sediment type, likely will be affected as well.

The concept of forced regression also is useful to help explain long-distance regression of shorelines across a shelf. During normal regression, shorelines migrate into progressively deeper water as they build across shelfal areas. As such, progressively more sediment is required to fill both the ever-increasing accommodation of these deeper-water settings in addition to the ever-expanding coastal/alluvial plain. Eventually, given constant sediment flux, the sediment supply would be insufficient to keep up with this increasing accommodation necessary to be filled in order to allow progradation to continue, causing regression to halt. Also mitigating against long-distance normal regression is (1) the fact that with continued progradation, the strike length of a shoreline increases, so that with each additional increment of progradation, ever more sediment would be required, and (2) the increased tendency of up-dip regional-scale avulsion due to the creation of shorter, steeper routes to the basin margin (Elliott pers. comm. 1996). Another possible factor of lesser importance is the progressive grain size decrease with the increasingly distant provenance (all other factors, such as climate, remaining constant). Each of the aforementioned factors would be of far less concern if sea-level fall occurs; sea-level fall would have the effect of decreasing the water depth, and hence accommodation, as regression extends across a shelf. And, although regression would still result in the shoreline being ever farther from the provenance, substrate cannibalization would re-introduce a relatively coarser grain-size fraction even to distal settings. In this way relative sea-level fall serves to facilitate long-distance regression.

Key surfaces associated with forced regressive deposits are shown in Fig. 2. The lower bounding surface has varied expression ranging from an erosive surface formed by wave action associated with lowering wavebase during relative

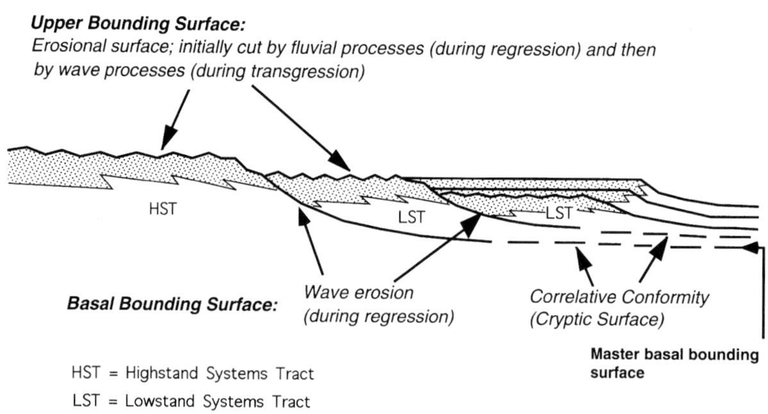

Fig. 2. Schematic depiction of key bounding surfaces associated with forced regressive deposits. The upper bounding surface commonly is well developed and typically is expressed as a ravinement surface. The lower bounding surface can be expressed as a sharp-based shoreface/delta front deposit that commonly grades into a correlative conformity. Note, as shown, there exist multiple lower bounding surfaces that form with each successive downstepping of relative sea-level. We place the sequence boundary at the base of the first downstepped unit and refer to that as the master bounding surface.

sea-level fall, to a correlative conformity expressed as a cryptic surface where lowered wavebase does not touch the sea floor. The upper bounding surface can be expressed as a ravinement surface or a subaerial exposure surface. The upper bounding surface will be expressed as a subaerial exposure surface if the sediments deposited during the period of slow relative sea-level rise subsequent to the period of relative sea-level fall (i.e. late lowstand systems tract deposits) onlap the top of the forced regressive deposits (i.e. the early lowstand systems tract). In those areas where the late lowstand systems tract deposits do not onlap the forced regressive wedge, the upper bounding surface will be expressed as a ravinement surface.

In general, forced regressive deposits will be preserved only where they are deposited in sufficient thickness so as to be able to survive the erosive processes acting upon them both during periods of sea-level fall as well as subsequent sea-level rise. In areas of low sediment flux, it is possible that minimal forced regressive deposits might be preserved (e.g. Tropeano & Sabato this volume). Another situation that would result in minimal forced regressive deposition is an environment characterized by a sea-floor gradient that is too steep to provide a stable substrate for prograding deposits. In such a setting, commonly observed at the shelf edge, active mass movement processes will effectively preclude the preservation of forced regressive deposits.

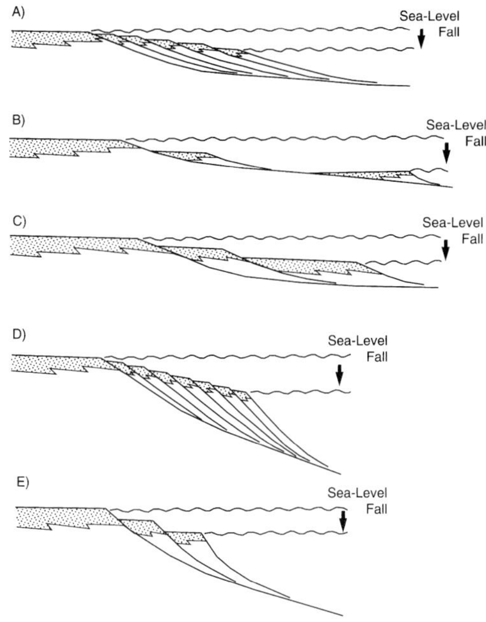

Fig. 3. Stratal architecture of forced regressive deposits. These include (**a**) smooth-topped and attached, forming on a gently inclined shelf, (**b**) stepped-top and detached, forming on a gently inclined shelf, (**c**) stepped-top and attached, forming on a gently inclined shelf, (**d**) smooth-topped and attached, forming on a steeply inclined shelf and (**e**) stepped-topped and attached, forming on a steeply inclined shelf. In this illustration, we show the smooth-topped style of forced regression as consisting of numerous small steps. In the real world situation, progradation during forced regression probably proceeds as a succession of numerous small steps consistent with the notion that progradation likely is associated with small scale catastrophic events. Thus, on a very small scale the steps may represent alternations of normal (i.e. during the catastrophic periods of sedimentation) and forced (i.e. between catastrophic events) regression. However, these small steps may be unrecognizable as discrete steps, instead producing a smooth topped forced regressive wedge, referred to as an accretionary forced regressive wedge by Helland-Hansen & Martinsen (1996).

Major types of forced regressive deposits

Attached v. detached

Two end-member types of forced regressive deposits are observed: attached and detached regressive deposits, as illustrated in Fig. 3 (a & b) (Ainsworth & Pattison 1994; Ainsworth *et al.* this volume). Both types are commonly observed, in many instances even within the same regressive stratigraphic complex (e.g. McMurray & Gawthorpe this volume). Attached forced regressive deposits are, as the term implies, attached to immediately preceding highstand regressive deposits. These deposits are attached in the sense that there is contact between the shoreface sediments of the two successive types of regressive units. In contrast, where the shoreface sediments of these two types of regressive wedges are not in contact, rather there exists a significant zone of separation between the two, the forced regressive wedge is said to be detached (Fig. 4).

Figure 5 illustrates an example of a detached forced regressive deposit. These shelf-edge palaeo-Rhône deposits clearly pinch out landward and are detached from the modern Rhône deltaic deposits. One can infer that a zone of sedimentary bypass, characterized by fluvial channels that were possibly incised (e.g. note possible incision near the northeast end of the seismic profile, just above the scale bar in Fig. 5) characterizes the space between these two features. This aspect of detachment is in marked

Fig. 4. Shoreline of the Great Salt Lake showing a highstand (i.e. Late Pleistocene) shoreline separated from the current (lowstand) shoreline. If these two shoreface deposits were in the subsurface, they would be isolated from each other from a fluid flow perspective. Note, however, that the two shoreface deposits (i.e. highstand and lowstand) would be characterized by similar facies and log response, and without the awareness that a zone of separation exists between the sand-prone deposits of the two shoreface, such deposits would readily be incorrectly correlated.

contrast to the attached forced regressive deposits shown on the left side of the seismic profile shown in Fig. 6 (see discussion below). Note, however, that forced regressive deposits can be detached in the sense that they are separated from the preceding highstand deposits, but attached with respect to each other. For example, although the gross relationship shown in Fig. 3b illustrates detached forced regressive deposits, the detached forced regressive wedge shown on the right of Fig. 3b may in its own right consist of a succession of attached forced regressive deposits.

Corner *et al.* (1990) illustrated an example of a succession of attached forced regressive deposits for a small fjord delta in northern Norway. Figure 7 summarizes the mapped distribution of the successive forced regressive wedges (a) as well as the downstepping aspect of these deposits (b). In this instance reworking by waves and tidal currents have modified and partially cannibalized each abandoned (i.e. raised deltaic terrace) forced regressive deposit so that any random profile may not encounter every terrace. Bardaji *et al.* (1990) illustrated another example of attached forced regressive deposits from their outcrop studies of the Cope Basin, southeast Spain. They documented a distinct seaward downstepping succession of coastal plain, shoreface, and fan delta deposits associated with fluctuations of sea level controlled in part by glacioeustasy and in part by tectonic uplift. Each successive downstepped stratigraphic unit is clearly attached to the unit that preceded it.

In the examples of detached forced regression discussed above, the attribute of detachment was in each instance in a proximal to distal sense. Figure 8 illustrates detachment of forced regressive deposits in three dimensions, i.e. along strike as well as dip (Hill *et al.* 1997). Shown here are a Late Pleistocene to Holocene succession of downstepping shoreface deposits. These deposits formed as isostatic rebound following continental glaciation caused a significant relative sea-level fall. Figure 8 clearly shows deposition of a highstand delta, interpreted to be of Late Pleistocene to early Holocene age (greater

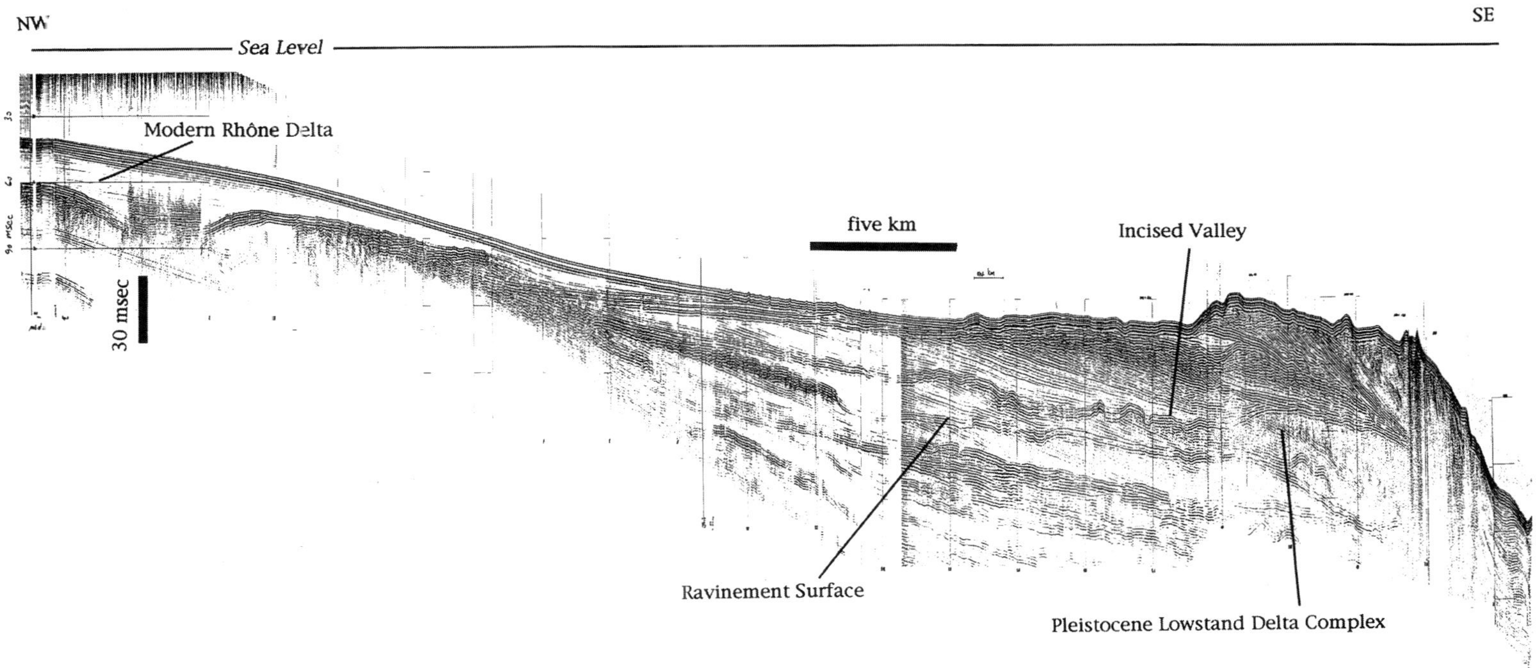

Fig. 5. Seismic reflection profile oriented parallel to dip, of a forced regressive complex offshore the Rhône delta. This wedge constitutes a forced regressive wedge clearly detached from the underlying highstand Rhône delta. Note that this entire regressive complex pinches out in the landward direction; landward of this location, extending to the provenance area, lies a zone of sedimentary bypass that existed during the time of formation of this shelf-edge complex (seismic section courtesy of M. Tesson).

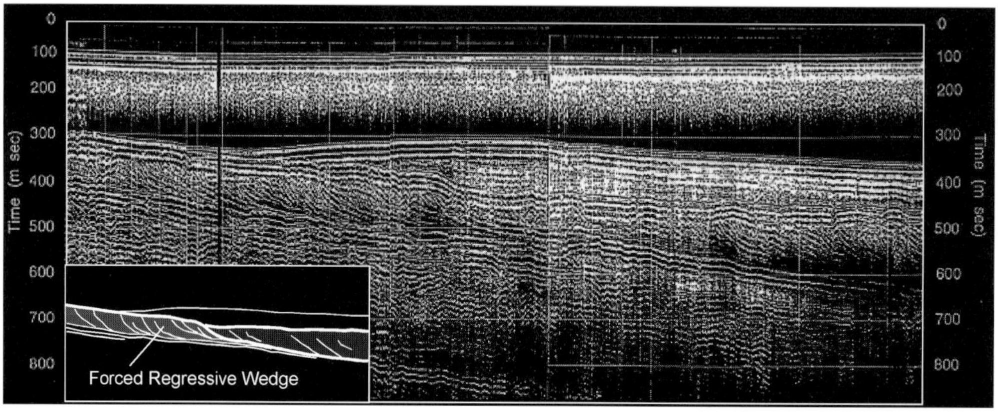

Fig. 6. Shallow-penetration seismic reflection profile offshore Lagniappe Delta (Mississippi Delta complex), Gulf of Mexico (from Kolla *et al.* this volume). On the left side of the profile there exists an apparently smooth-topped forced regressive wedge, characterized by concave-up clinoforms and whose top bounding surface dips seaward. This is in contrast to the progradational unit on the right side of the profile whose top bounding surface appears horizontal. This latter unit appears downstepped relative to the stratigraphic unit on the left side of the profile and is inferred to have been deposited during a relative sea-level stillstand.

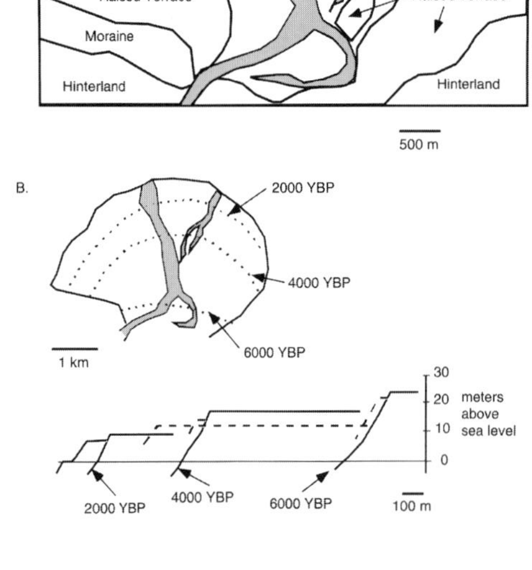

Fig. 7. Forced regression associated with progradation of a modern delta: Alta delta, Norway. This area had been glaciated during the late Pleistocene and since that time has been characterized by isostatic uplift resulting in a prolonged period of relative sea-level fall throughout the Holocene. As a result, progradation of the Alta delta during this time has been characterized by forced regression. (**a**) The Alta delta is shown in plan view characterized by a number of terraces that have been modified by tide and wave processes and only partially preserved. These terraces represent earlier-formed and subsequently uplifted delta plain deposits. (**b**) A profile through these terraces shows the successive downstepping of the delta through time (after Corner *et al.* 1990).

than 10.7 ka BP; Hill *et al.* 1997) followed by deposition of isolated downstepping and seaward-stepping shoreface deposits. Because of the isolated aspect of the forced regressive deposits, along any given transect the forced regressive deposits may or may not be observed. Strike variability of forced regressive wedges also is documented in the northern Peloponnese peninsula, Greece where McMurray &

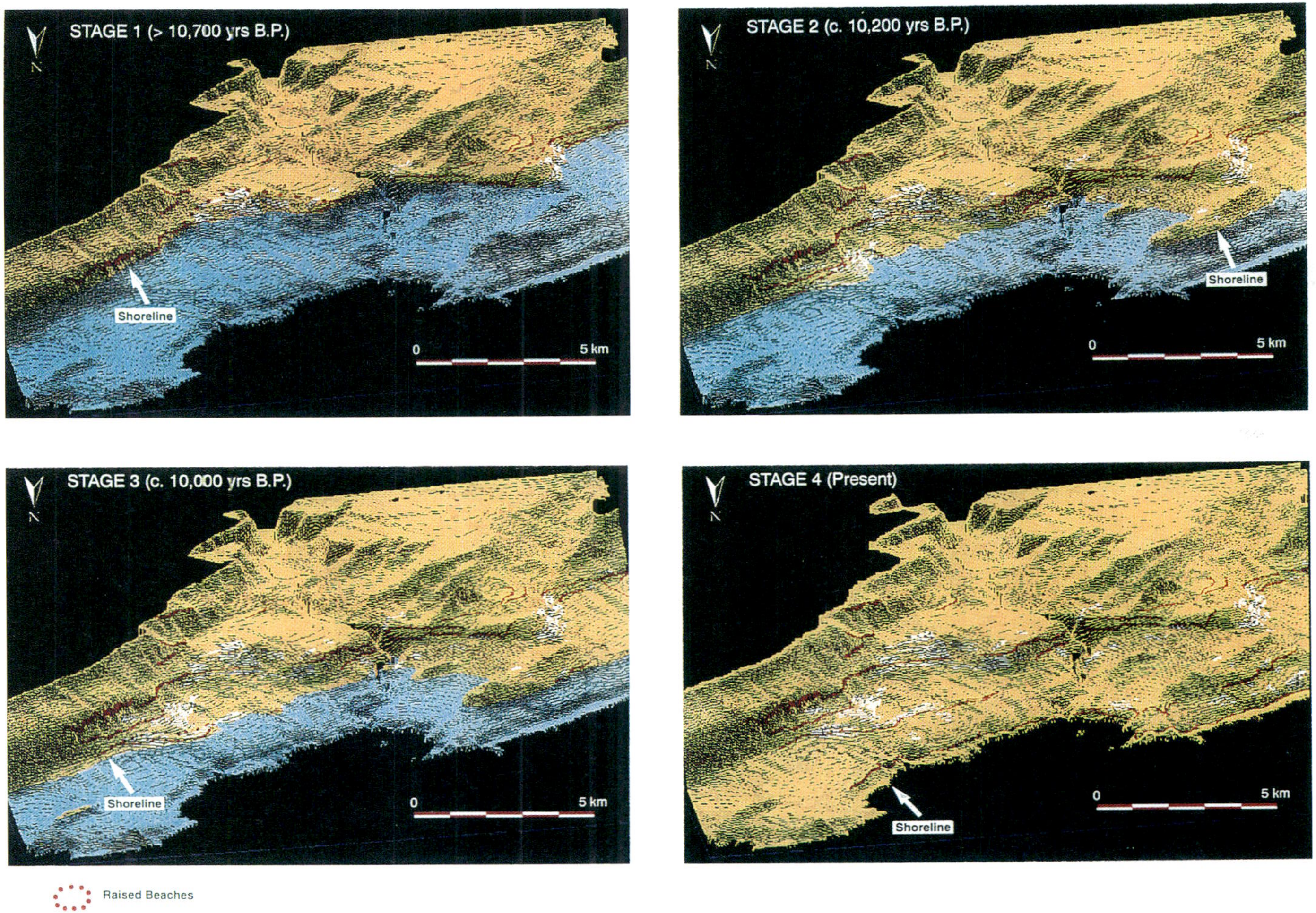

Fig. 8. Forced regressive deposits associated with the Metis River, New Brunswick (from Hill 1997). This area is characterized by isostatic rebound occurring in response to deglaciation during the late Pleistocene. Consequently, relative sea-level fall has characterized this area since that time. Four stages are illustrated, representing the shoreline position at *c.* 10.7 ka BP (top left), 10.2 ka BP (top right), 10 ka BP (bottom left) and present (bottom right). Each successive seaward shift of the shoreline is associated with a downstep. Note that the Metis River forms a significant incised valley only where it passes through the coastal prism (Posamentier & Allen 1994) consisting of the delta formed *c.* 10.7 ka BP. Detached forced regressive deposits as a result of partial preservation are associated with each downstepped shoreline. Note that beaches formed at each stage are not in contact with each other because of partial preservation and are detached along strike as well as dip as shown in red.

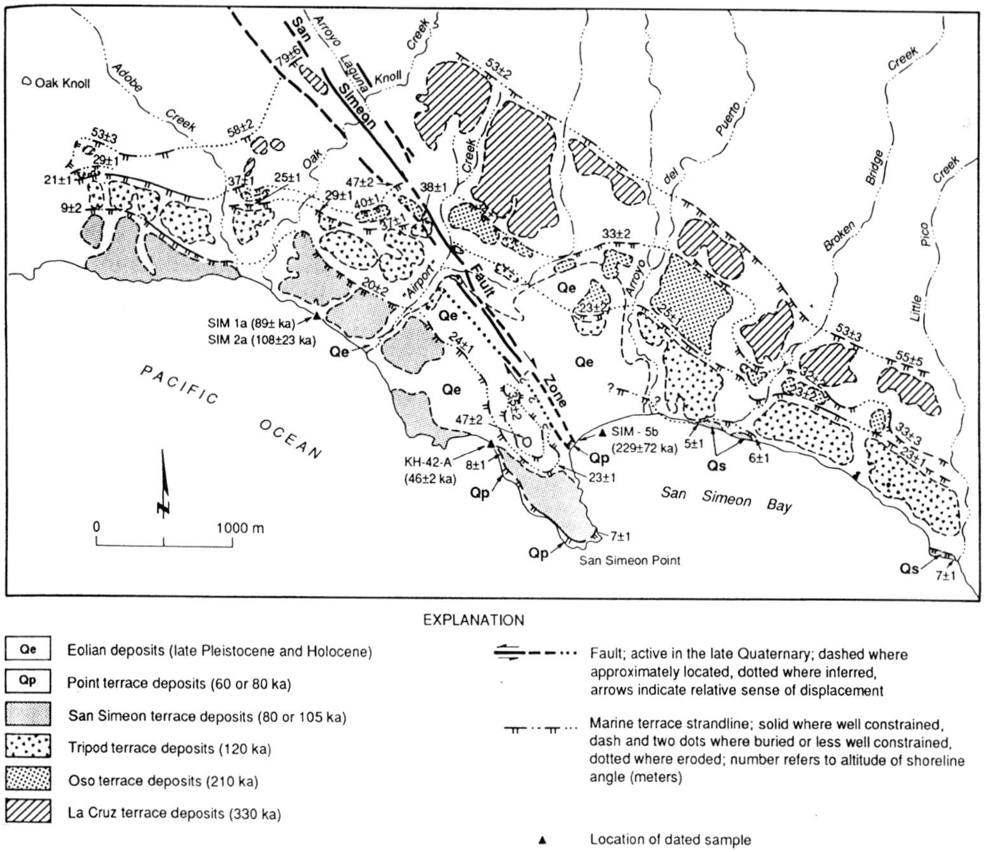

Fig. 9. Geological map of marine terraces along the southern onshore reach of the San Simeon fault zone (Hanson *et al.* 1992). Five beach terraces, each with overlying shoreface deposits corresponding to deposition during successive lowering of relative sea-level, have been mapped. These sediments are not in contact with each other and therefore constitute detached forced regressive deposits. In this instance, forced regression is driven by tectonic uplift associated with San Simeon fault activity (Hanson *et al.* 1992) and results in relative sea-level fall.

Gawthorpe (this volume) documented contemporaneous development of attached and detached deltaic and shoreface sequences deposited in response to long-term tectonic uplift.

Figure 9 illustrates another example of detached forced regressive deposits. In this instance relative sea-level fall is driven by tectonically-associated uplift (Hanson *et al.* 1992). In response to the episodic uplift, five beach terraces have formed, ranging in age from 330 ka to 60 ka. With each relative sea-level fall event, a beach terrace initially forms, followed by deposition of shoreface deposits on the terrace. Insofar as the shoreface deposits seem to be confined to the terraces, the individual terrace deposits do not seem to be in contact with each other and hence would constitute detached forced regressive deposits.

Controls. The principal factors that determine whether attached or detached forced regression deposits form include; (1) the rate of relative sea-level fall, (2) the rate of sediment supply, (3) the energy of the nearshore/fluvial environments and (4) the gradient of the sea floor. In general the formation of attached forced regression deposits is favoured by a low rate of relative sea-level fall, a high rate of sediment supply, a high-energy nearshore system and a presence of relatively steep shelfal gradients. In contrast, when the opposite conditions exist, the formation of a detached forced regression deposits is favoured.

When the rate of relative sea-level fall is low, given a constant sediment flux, the rate at which the coastline is forced to migrate seaward is commensurably low. This relatively slow regression allows for the formation of successive shoreface deposits, including the previous highstand shoreline, to be in contact with each other. When the rate of relative sea-level fall is high, the coastline migrates relatively rapidly seaward. If the rate of sediment supply is relatively high, then the rate of shoreface sedimentation may be sufficient to keep up with the rapid rate of regression so that successive shoreface deposits will be in contact with each other (i.e. attached forced regressive deposits). If the rate of sediment supply is not sufficiently high, then the rate of shoreface sedimentation may not be sufficient to keep up so that the rapid rate of regression will produce detached forced regressive deposits.

The energy of the nearshore/fluvial system controls the depth and seaward extent of the nearshore environment at any given time. High wave-energy shorefaces tend to be characterized by higher relief (i.e. from top of beach to base of shoreface) and therefore, given that the

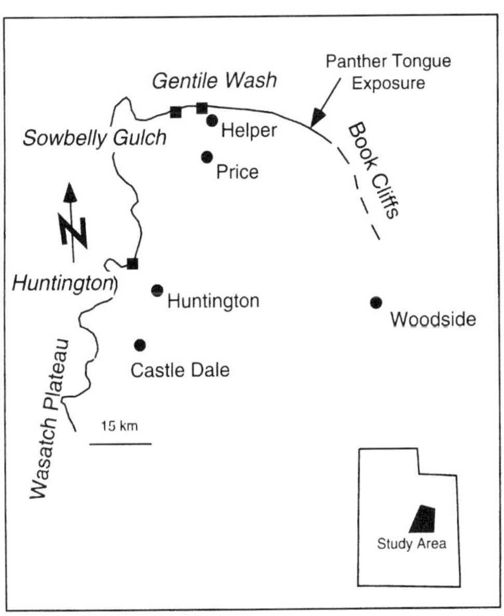

Fig. 10. (**a**) Outcrop exposure of the Panther Tongue Member, Star Point Formation, at Sowbelly Gulch, near Helper, Utah (**b**). Total thickness of the section shown is 15 m. A ravinement surface (i.e. transgressive surface of erosion) caps the outcrop (note, arrows on the photo). The rocks immediately underlying the ravinement surface comprise marine distributary mouth bar facies with no fluvial, coastal plain, or delta plain deposits preserved there. Over the entire dip-oriented outcrop exposure of this member, *c.* 52 km, no fluvial, coastal plain, or delta plain deposits are anywhere observed beneath the ravinement surface. This outcrop is located nearly at the most proximal location (stratigraphic dip is from north to south) of the Panther Tongue outcrop.

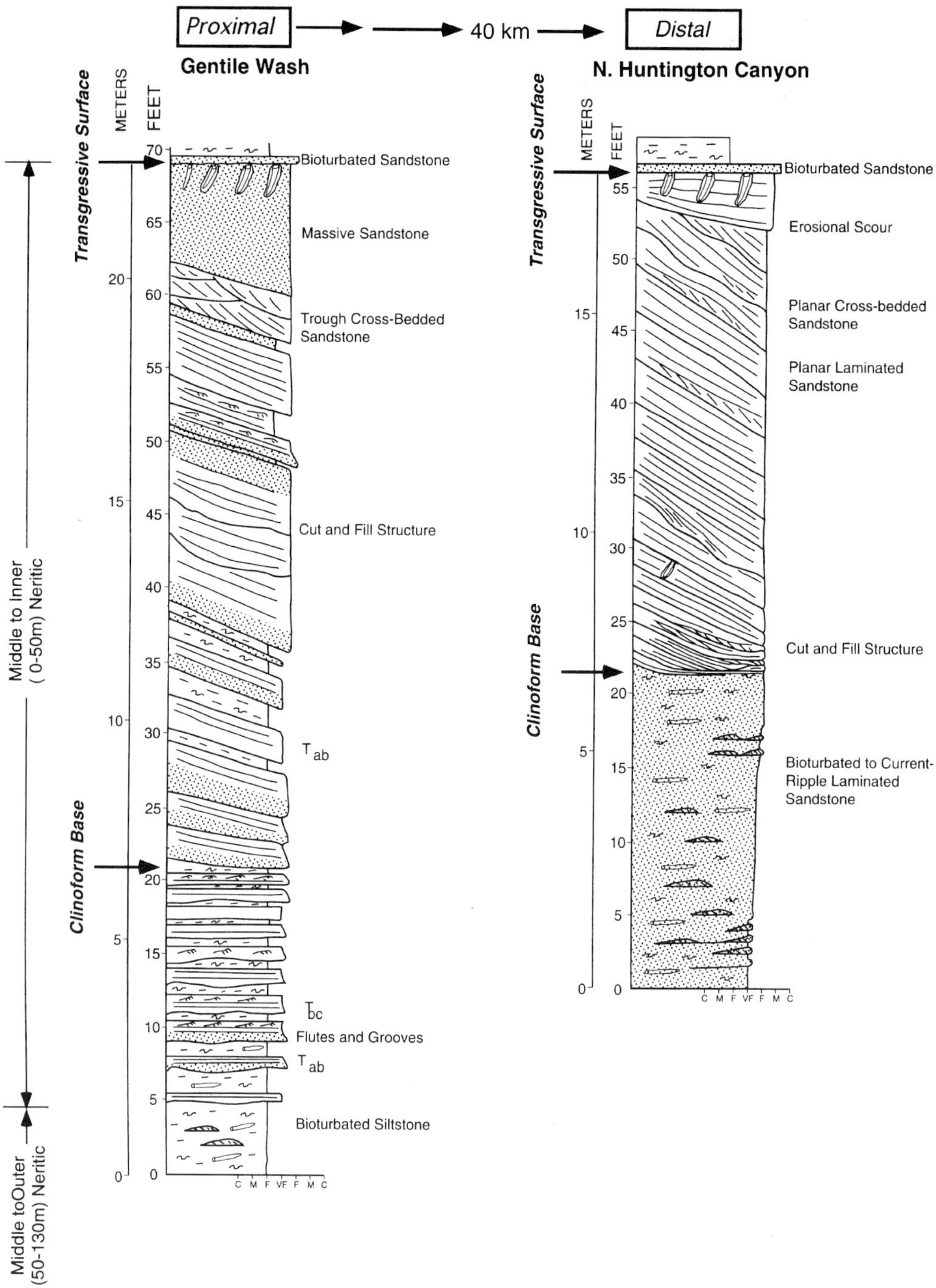

Fig. 11. Two measured sections from the Campanian Panther Tongue Member of the Star Point Formation, Wasatch Plateau, Utah. These measured sections are *c.* 35 km apart along dip and at both locations, the Panther Tongue is characterized by clinoform geometry. The clinoforms at the Gentile Wash section near Helper, Utah, are *c.* 15 m (47 ft) high, where as the clinoforms at the North Huntington Canyon, Utah, section, *c.* 35 km downdip, are *c.* 10 m (34 ft) high. Palaeobathymetric information is based on interpretation of foraminifera assemblages (P. Thompson pers. comm. 1997) (see Fig. 10 for location).

nearshore marine sandstone facies extend to greater depth, the likelihood of developing attached forced regressive deposits increases. In deltaic settings, different fluvial systems with different sediment load characteristics will have a similar effect insofar as resulting in low- or high-relief delta fronts. In mixed or suspended load fluvial systems, the prodelta aggrades rapidly, creating a broad, relatively shallow platform across which the delta front subsequently progrades. In bedload-dominated fluvial systems, the aggradation of the prodelta is greatly diminished and the delta front is potentially characterized by higher relief as the system builds out into deeper water. For example, the delta front of the bedload-dominated Panther Tongue Sandstone (Morris *et al.* 1995; Posamentier *et al.* 1995) discussed below, is characterized by slopes of from 7–8° to as much as 27° with relief in excess of 15 m (50 ft) in places, Figs 10 and 11). Thus, systems with higher wave energy in the nearshore environment or systems that are characterized by bedload-dominated rivers will be associated with increases in the depth/thickness of the nearshore system, thereby favouring the development of attached forced regressive deposits.

It is important to note that significant lobe switching, or lateral shifting of point sources, cannot be invoked to account for detachment of successive deltaic lobes typical of the process of forced regression. This is because with successive sea-level falls, distributary channels become incised and therefore fixed in their location, thus not permitting lateral shifting to occur.

The gradient of the sea floor, in combination with other factors, can influence whether attached or detached forced regressive deposits form. Attached forced regressive deposits are more likely where relative sea-level falls expose a relatively steep sea floor. With the same amount of relative sea-level fall, forced regressive deposits will be in closer proximity to each other in a relatively steep sea floor setting (Fig. 12b) than in a gentle sea floor setting (Fig. 12a). All else being equal, therefore, the likelihood of forced regressive deposits being attached is enhanced in relatively steep sea floor settings.

From an oil and gas exploration perspective, each of these two types of forced regressive deposits has exploration and field development significance. Attached forced regressive deposits will be in direct fluid communication with each other as well as with the preceding highstand deposits. Baffles and possible barriers to flow will likely exist, and will be parallel to bedding planes that are oriented parallel to the shoreface/delta front profile at any given time. However, if the attached deposits are associated with forced regression occurring in discrete steps rather than gradationally (see discussion below), then there is a stronger possibility that more readily definable reservoir compartments associated with these steps will form (Fig. 3c and e). Detached forced regressive shoreface deposits will be isolated from each other, separated by a zone of offshore/pro-delta muds (Figs 3b and 4). Such detached forced regressive deposits will form discrete reservoir compartments potentially separated by pronounced barriers to flow.

Forced regressive deposits: stepped v. smooth-topped

The upper bounding surface of attached forced regression deposits can be characterized as ranging from stepped to smooth-topped (Fig. 3). The degree to which such discrete steps will be recognizable (i.e. what will be their preservation potential) subsequent to later transgression back across the top of these early lowstand deposits will depend on (1) how far apart each of these equal steps was, and how much of a downstep characterized each successive step, (2) how much of the upper part of each downstepped unit was removed by subaerial erosional processes during the lowest sea-level stand and (3) how much of the upper part of each downstepped unit was removed by wave-associated erosional processes during the subsequent transgression.

Subaerial erosional process can be effective in removing what may have started out as a downstepped upper bounding surface of the forced regressive deposits. During a protracted period of sea-level lowstand, following a period of forced regression, fluvial cannibalization of the substrate can be effective laterally as well as vertically. In fact, given a long enough period of time, laterally-eroding incised valleys can coalesce resulting in elimination of interfluves and complete removal of any downstepping geometry. Transgressive erosion also is capable of eroding significant amounts of sediment off the top of the subjacent forced regressive deposits. For example, along the Canterbury Bight on the southeast coast of South Island, New Zealand, Leckie (1994) has documented that high wave-energy conditions have resulted in transgressive erosion of up to 40 m. It should be noted that the erosive effect of the high wave-energy is enhanced by their undercutting and slumping of sea cliff faces. Whereas this is probably an end-member situation in terms of the amount of

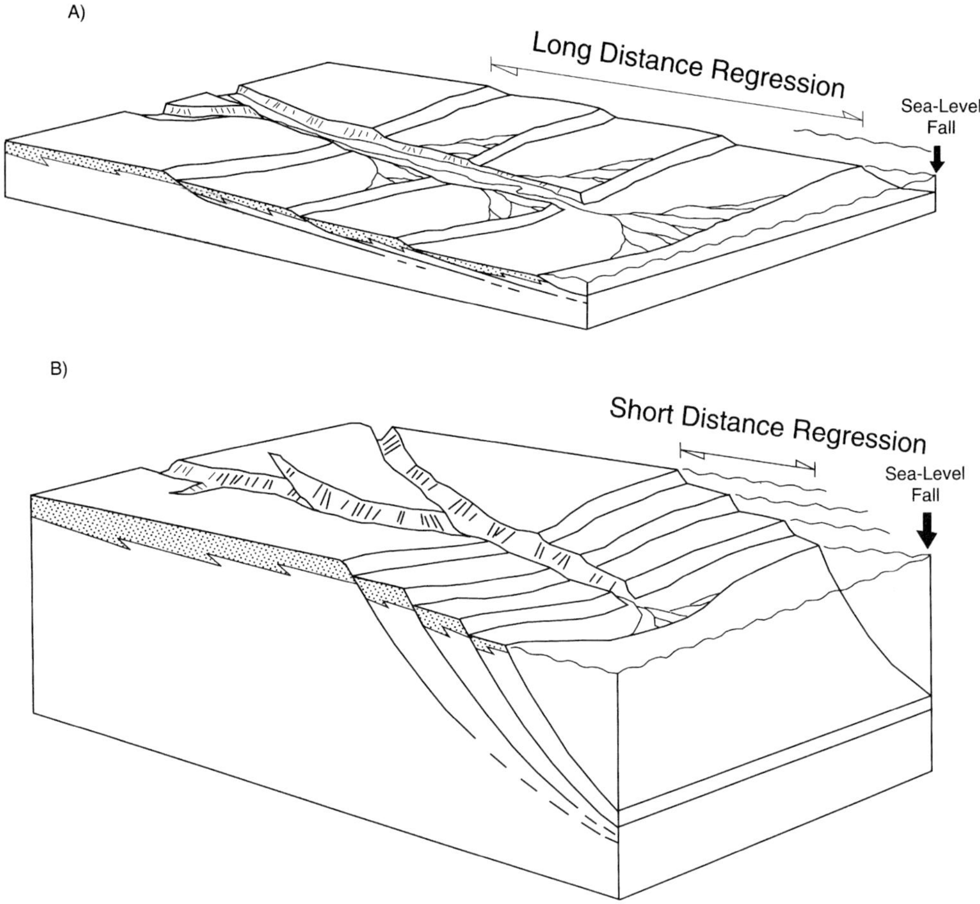

Fig. 12. (**a**) Forced regression stratal architecture associated with gentle shelf/slope gradients. Note the relatively long-distance regression. (**b**) Forced regression stratal architecture associated with steep shelf/slope gradients. Note the relatively short-distance regression.

erosion, it is nonetheless indicative of the potential efficacy of transgressive erosion.

The degree to which the initial upper boundary of a forced regressive succession will be characterized by a stepped morphology is a function of primarily two factors: (1) the variability or 'smoothness' of the relative sea-level fall and (2) the variability of the sediment flux (Helland-Hansen & Martinsen 1996). The development of smooth-topped forced regressive deposits will be favoured by uniform rates of relative sea-level fall in concert with uniform sediment flux (Fig. 13). Under these conditions, the rate of forced regression likely will proceed at a relatively uniform rate. The result will be the formation of an upper bounding surface that will be characterized by a succession of equal steps (Fig. 3a and d). If these steps are small ones, then a continuum of downstepping will have occurred and the forced regressive deposits will appear smooth-topped. This upper-bounding surface will not be horizontal, but rather will be characterized by a seaward dip. Lowstand and transgressive erosion will further enhance the smoothed aspect of this surface.

The development of stepped-topped forced regressive deposits will be favoured by irregular rates of relative sea-level fall in concert with variable sediment flux (Fig. 14). In this situation forced regression will proceed episodically. During times of sea-level stillstand (or slow rise) punctuating an overall sea-level fall (i.e. normal regression alternating with forced regression), a series of discrete horizontal steps will form. Alternatively, if sediment supply is erratic, then even with a uniform relative sea-level fall,

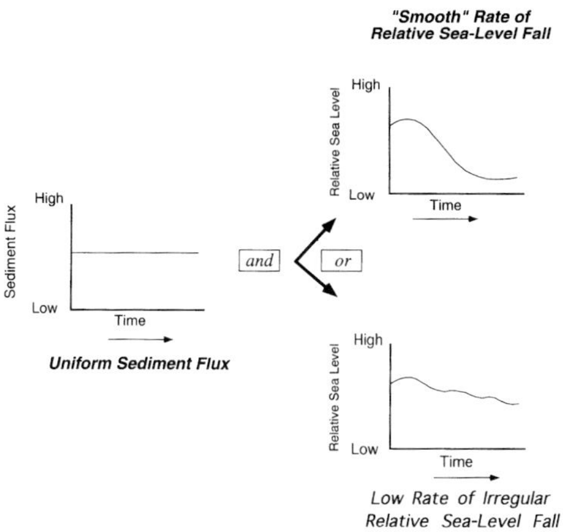

Fig. 13. Conditions favourable for development of smooth-topped forced regressive deposits. These include a uniform sediment flux coupled with either a smooth rate of relative sea-level fall or a low rate of irregularly falling relative sea level.

discrete steps can develop. In all likelihood, most forced regressive deposits are step-topped when they initially form insofar as sea-level change as well as sediment supply rarely are characterized by uniform rates. Such a step-topped surface will be preserved in the rock record only if subsequent erosion, either during sea-level fall or during later sea-level rise, is minimal (Fig. 15).

Recognition criteria for forced regressive deposits

Distinguishing forced regressive deposits from normal regressive deposits can be difficult, especially where available data sets comprise only small windows on the world. Most commonly, of the criteria listed below (some of which are shown schematically in Fig. 16), only a subset of this list will be observed at any given locality. Some of these criteria comprise directly observable evidence, whereas other criteria represent de facto evidence, or evidence by omission. Each of these criteria, however, should be an indication of the possibility of the presence of forced regressive deposits, and should lead to the search for other converging lines of evidence. Table 1 summarizes the usefulness of various types of geological data in recognizing each of the following criteria.

Separation of shoreface deposits

The presence of a zone of separation between shoreface deposits located on basin margins and shoreface deposits located farther seaward (Fig. 16a). This relationship is indicative of a zone of sedimentary bypass that has produced basin-isolated sandstone deposits that are typical of detached forced regressive deposits (e.g. Plint 1988; Ainsworth & Pattison 1994). An example of such a zone of sedimentary bypass between highstand and lowstand deposits is shown in Fig. 5. Note the pinchout of the lowstand deposits in the landward direction, detached from the modern Rhône delta.

An outcrop-based example of sedimentary bypass and inferred detachment is shown in Fig. 17. The outcrop photo shows coastal plain deposits of the Fruitland Formation, New Mexico, directly overlying the offshore marine deposits of the Lewis Shale. The surface between the two is inferred to have been initially modified by lowering wavebase during relative sea-level fall. During this time, decreasing

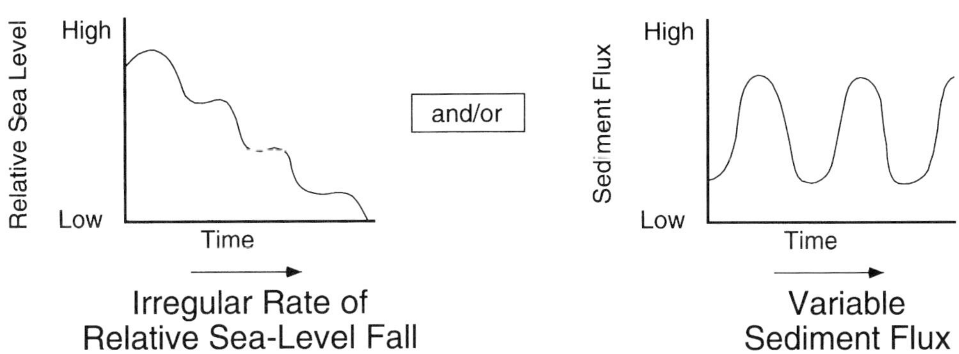

Fig. 14. Conditions favourable for development of stepped-topped forced regressive deposits. These include a highly irregular rate of relative sea-level fall and/or a highly variable sediment flux.

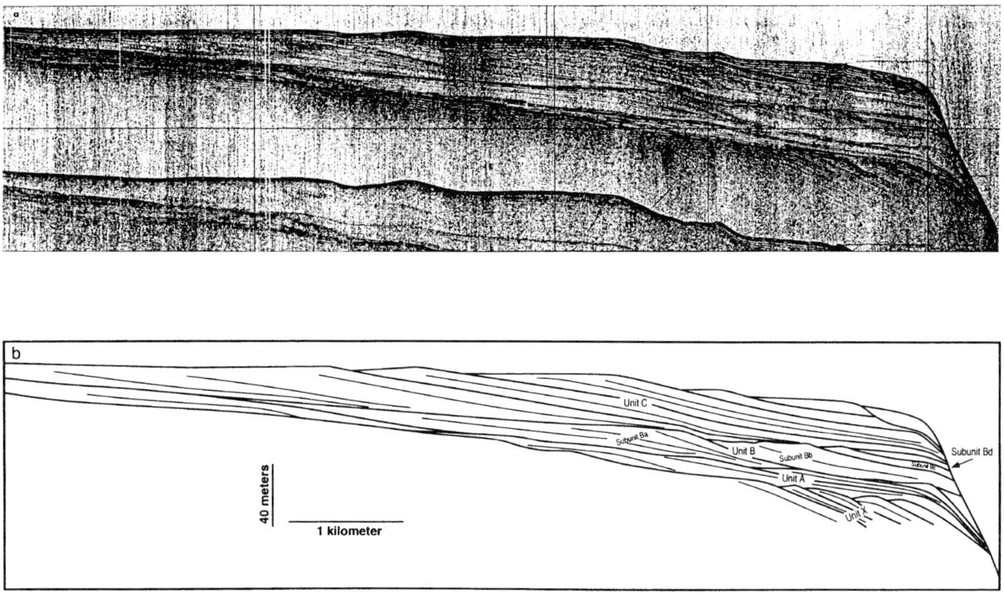

Fig. 15. Seismic line on the shelf seaward of the Rhône delta, showing three successive stepped-top forced regressive events (from Posamentier *et al.* 1992). Each successive unit (labelled Units A, B, and C) constitutes a forced regressive lowstand wedge in its own right, and each is characterized by a stepped top.

accommodation likely resulted in non-deposition of nearshore deposits on this surface at this location. Ultimately, this surface was subaerially exposed and sedimentary bypass of this area was inferred to have been associated with shoreface deposition seaward of this location. The geological map shown in Fig. 17b shows the location of the outcrop photo and illustrates that landward of this location (to the S SW) shoreface deposits (of the Pictured Cliffs Sandstone/Lewis Shale) underlie the same surface shown in the photo (Shomaker *et al.* 1971). These shoreface deposits are interpreted as highstand deposits that get progressively less sandy and shale out landward of where the lowstand shoreface deposits of the Fruitland Formation pinchout (i.e. in the landward direction). Thus the two successive shoreface deposits may be said to be detached.

Another example of sedimentary bypass inferred to have been associated with forced regression is the C Member of the Kuparuk

Table 1. *Utility of different data sets for identifying forced regression recognition criteria*

	Outcrop	Core	Well log	Seismic
(1) Presence of a significant zone of separation between successive shoreface deposits	Good	Fair	Fair	Fair
(2) Sharp-based shoreface/delta front deposits	Good	Good	Good	Poor
(3) Progressively shallower clinoforms going from proximal to distal	Good	Poor	Poor	Good
(4) Occurrence of long-distance regression	Fair	Poor	Poor	Good
(5) Absence of fluvial and/or coastal plain/delta plain capping the proximal portion of regressive deposits	Good	Good	Fair	Poor
(6) Presence of a seaward-dipping upper bounding surface	Fair	Poor	Fair	Good
(7) Increased average sediment grain size in regressive deposits going from proximal to distal	Good	Good	Fair	Poor
(8) Presence of 'foreshortened' stratigraphic successions	Good	Good	Poor	Poor

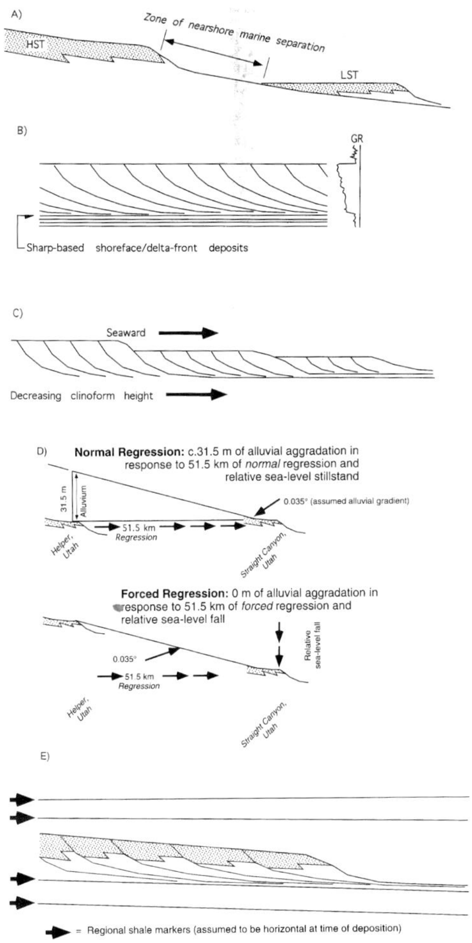

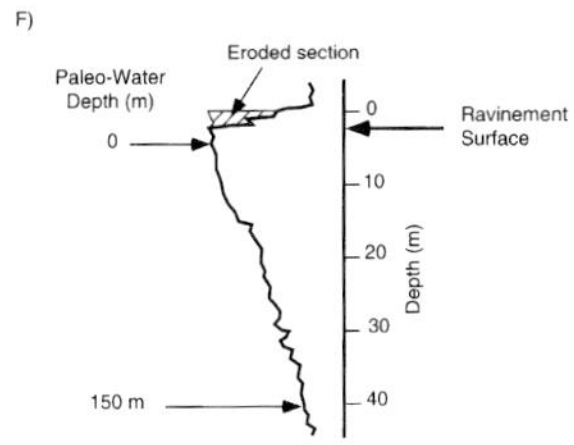

Fig. 16. Schematic depiction of physical stratigraphic criteria for the recognition of forced regression. (**a**) The presence of a zone of sedimentary bypass between a wedge of basinally-isolated nearshore marine sediments and immediately preceding highstand nearshore marine sediments. (**b**) The presence of sharp-based shoreface/delta front deposits. (**c**) The presence of progressively lower-relief clinoforms going from proximal to distal. (**d**) The absence of fluvial and/or coastal plain/delta plain facies capping the proximal portion of regressive deposits. (**e**) The presence of a seaward-dipping upper bounding surface atop a mid to outer shelf progradational unit, where the dip exceeds that which would be reasonably expected of a non-marine environment. (**f**) The presence of a foreshortened stratigraphic section such that the palaeobathymetric change from base to top of the regressive succession is significantly greater than the decompacted thickness of that regressive succession.

Formation, North Slope, Alaska (Fig. 18). These deposits are interpreted to overlie a ravinement surface associated with a regional transgressive event. This surface also is a major unconformity with several million years of section absent due to erosion associated with falling relative sea level and subsequent sea level lowstand. We infer that this erosion was largely subaerial due to its regional extent, as well as the fact that hundreds of metres of section have been removed at this surface. Thus, these conclusions, coupled with the interpretation of transgression across this surface shown in Fig. 18, lead us to infer that a shoreline and associated shoreface deposits must have existed seaward of this location, being deposited as a response to major relative sea-level fall (probably tectonically driven), which therefore existed as a detached or isolated forced regressive deposit.

Long-distance regression

The occurrence of long-distance regression across a shelf. With increased distance of regression during periods of relative sea-level stillstands or slow rise, progressively more sediment is required to fill the ever expanding space across an ever-deepening shelf. Eventually, the rate of regression will undoubtedly slow and ultimately give way to transgression (P. McCabe & K. Shanley pers. comm. 1994). The optimal way to ensure long distance regression in the face of ever-deepening water going from proximal to distal, is to lower relative sea-level, thus suppressing the space (i.e. accommodation) that sediments need to fill in order to continue the regression. However, merely observing a long distance regression is alone insufficient evidence insofar as it is circumstantial, nonetheless, this

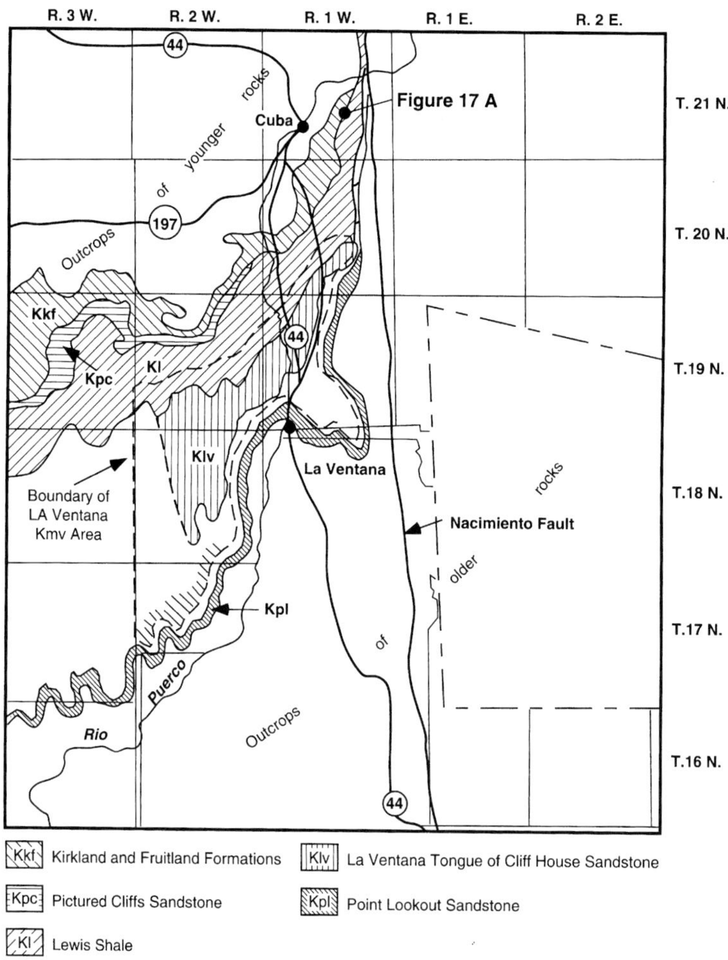
R. 3 W.
R. 2 W.
R. 1 W.
R. 1 E.
R. 2 E.
T. 21 N.
T. 20 N.
T.19 N.
T.18 N.
T.17 N.
T.16 N.
44
197
Cuba
Figure 17 A
Outcrops of younger rocks
Kkf
Kpc
Kl
Klv
La Ventana
Boundary of LA Ventana Kmv Area
Nacimiento Fault
Outcrops of older rocks
Kpl
Rio Puerco
Kkf Kirkland and Fruitland Formations
Kpc Pictured Cliffs Sandstone
Kl Lewis Shale
Klv La Ventana Tongue of Cliff House Sandstone
Kpl Point Lookout Sandstone

observation should at least raise the awareness of forced regression as a possible working hypothesis to be tested.

Sharp-based shorefaces/deltas

The presence of sharp-based shoreface/delta front deposits (Figs 16b and 19) is indicative of a missing transitional facies (e.g. Plint 1988). The absence of this transitional facies would be associated with erosion occurring in response to lowering of wavebase as relative sea-level falls (Plint 1988). It is not clear, however, how widespread this sharp-based attribute is relative to a forced regressive wedge. In a core and well-log based study of a forced regressive deposits, it has been noted that the extent to which forced regressive deposits are characterized by a sharp base is limited to only the most proximal 2–4 km of the detached forced regressive wedge (Posamentier & Chamberlain 1993). However, the extent of this sharp-based may be significantly greater for more extensive forced regressive wedges (the forced regressive wedge documented by Posamentier & Chamberlain was only *c.* 20 km wide). Other papers in this volume document in detail and discuss the length scale of sharp-based shorefaces interpreted to form in response to relative sea-level fall (e.g. Ainsworth *et al.*; Fitzsimmons & Johnson; Gawthorpe *et al.*; Mellere & Steel; McMurray & Gawthorpe; Plint & Nummedal; Trincardi & Correggiari).

Clinoform relief

The presence of progressively lower-relief clinoforms going from proximal to distal (Fig. 11 and 16c). Typically, the shelf is characterized by a seaward-sloping profile. Consequently, with stable or slowly rising relative sea level, prograding nearshore deposits (i.e. associated with normal regression) progressively build into ever deeper water, which, all else being equal, results in progressively higher-relief clinoforms. Thus, if progressively lower-relief clinoforms are observed in a seaward direction, it implies that the progradational depositional system is building into progressively shallower water (see Fig. 11).

Absence of non-marine aggradation

The absence of fluvial and/or coastal plain/delta plain facies capping the proximal portion of regressive deposits (Figs 10 and 16d). During periods of sea-level stillstand or slow rise, normal regression commonly is associated with a progressive aggradation of fluvial and/or coastal plain/delta plain facies in proximal areas. This occurs in response to subaerial accommodation that develops in association with normal regression. The surface atop the regressive deposits must develop a gradient so as to maintain the flow of distributary and fluvial systems. Delta plain environments may prevail initially, followed eventually by fluvial environments. The absence of these facies, especially in the most proximal areas, suggests either that extensive erosion of these facies has occurred during transgression or that forced regression has taken place. As discussed above, forced regression will produce an upper bounding surface with a seaward-dipping gradient, thus negating the need to aggrade a fluvial and/or coastal plain/delta plain so as to maintain a fluvial grade.

Seaward-dipping surfaces

The presence of a seaward-dipping upper bounding surface atop a mid to outer shelf progradational unit, where the dip exceeds that which would be reasonably expected of a non-marine environment (Fig. 12). This seaward dipping surface may be either smooth-topped or stepped (see discussion above). Figure 6 illustrates an excellent example of such a forced regression deposit. Note the seaward-dipping top characterizing the clinoform geometry on the left side of the seismic profile, in sharp contrast to the immediately adjacent horizontal top characterizing the right side of the profile. The horizontal aspect of the right hand clinoform package indicates the present orientation of what at the time of deposition was a horizontal surface, clearly suggesting that the seaward

Fig. 17. Outcrop photo (**a**) of coastal plain deposits of the Upper Cretaceous Fruitland Formation overlying offshore marine deposits of the Lewis Shale, near Cuba, New Mexico. This surface is interpreted to have been within a zone of sedimentary bypass during the time that shoreface deposits were forming seaward of this location. The shoreface deposits coeval to the upper part of the Lewis Shale (i.e. the Pictured Cliffs Sandstones) constitute the highstand systems tract and as shown on the geological map (**b**), shale out in the seaward direction (i.e. NNW). The coastal plain deposits of the Fruitland Formation constitute the late lowstand systems tract onlapping the sedimentary bypass surface shown in the photo (**a**). Forced regressive shoreface deposits of the Fruitland Formation outcrop to the north northwest of the area shown. With the exception of the surface at the base of the Fruitland Formation, all contacts are characterized by interfingering.

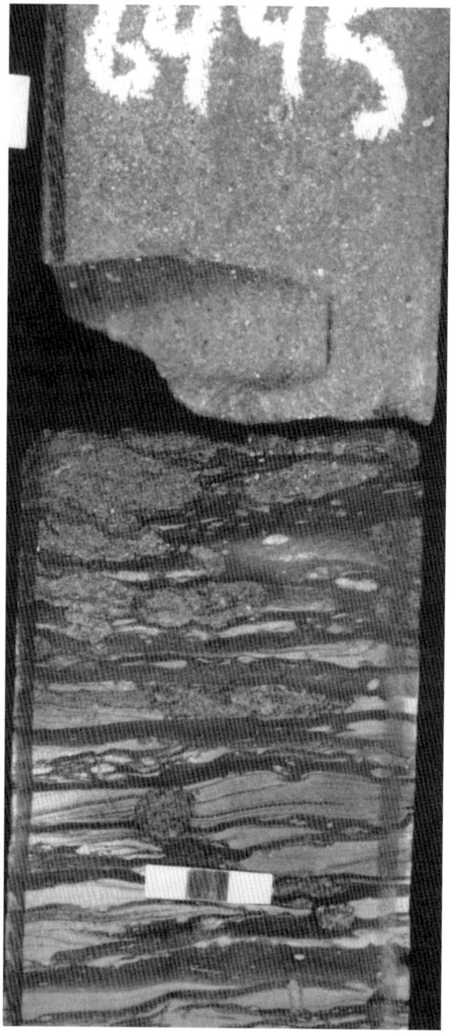

Fig. 18. Photo of core illustrating the contact between the Kuparuk B and Kuparuk C Members of the Kuparuk River Formation. This surface constitutes a regional unconformity across which there exists a hiatus of several million years. The contact pictured is inferred to have formed as a subaerial erosion surface during times of relative sea-level lowstand, and was subsequently modified by transgressive erosion. The Kuparuk B Member is expressed in this area as a lower shoreface to offshore silty sandstone, whereas the overlying Kuparuk C Member is expressed here as a transgressive lag deposit associated with the passage of a shoreface environment.

dipping left hand clinoform package represents a surface formed by successive small downsteps of offlap wedges rather than a profile that has been tilted.

Grain size

The presence of an increased average sediment grain size in regressive deposits going from proximal to distal. This increased grain size is caused by the cannibalization and winnowing of earlier-deposited highstand and potentially increased fluvial gradients. In some instances, this phenomenon may not result in an increase of grain size seaward, but only in a diminishment of the tendency for grain size decrease seaward.

Foreshortened stratigraphy

The presence of foreshortened stratigraphic sections. Stratigraphic sections where the decompacted thickness of a shoaling-upward section is significantly less than the palaeowater depth difference from base to top (where the top is at or near sea-level) (Fig. 16f). For example, the palaeo-water depth near the base of a shoaling upward section for the Panther Tongue Sandstone, Utah, (Fig. 7) is 75–100 m (estimated on the basis of sedimentary structures and biostratigraphic information; P. Thompson pers. comm. 1997), and the total decompacted thickness of the section between the base and the top of this section is only 25 m. The forced regressive process must be invoked to account for such a foreshortened section.

It is possible that many if not most cyclothems may have formed in association with forced regression (P. Heckel pers. comm. 1996). Figure 20 illustrates a generic cyclothem with its associated palaeobathymetry. Note that the water depth goes from relatively deep water to non-marine over a section of 5 m. Thus, because deep water implies depths significantly greater than 5 m, this section appears foreshortened and forced regression must have accompanied deposition of these successions. Moreover, long distance regression across a broad seaway, such as that which characterized these cyclothems, would have been facilitated by forced regression (see discussion above).

Another example of a foreshortened section is shown in Fig. 21. The well information shown here is from a well bore that penetrates a Late Pleistocene shelf-edge delta offshore Louisiana, Gulf of Mexico (Kolla *et al.* this volume). The well is drilled through a clinoform package (as observed on seismic data; Kolla *et al.* this volume) and is characterized by a coarsening-upward and shallowing-upward lithology, consistent with the presence of progradational architecture. The palaeo-water depth at approximately 65 m (*c.* 215 ft) below the ravinement surface and 45 m (*c.* 148 ft) below the depth at which zero palaeo-water depth is interpreted,

Fig. 19. Outcrop exposure of the Panther Tongue Member of the Star Point Formation near Helper, Utah (see Fig. 11 for location). Note the sharp, well-defined top as well as the relatively well-expressed base of this sandstone unit. The top is expressed as a ravinement surface and interpreted as a transgressive surface separating lowstand deposits below, from transgressive deposits above (solid arrow). The base is expressed as a rapid transition from massively-bedded, intensely burrowed silty sandstone below, to a tabular-bedded, less intensively burrowed sandstone above, and is interpreted as a sequence boundary associated with the process of forced regression (hollow arrows). It is not clear, however, whether this basal bounding surface constitutes the master basal boundary or whether the master basal boundary exists here as a correlative conformity somewhat below the surface shown here (compare with Fig. 2).

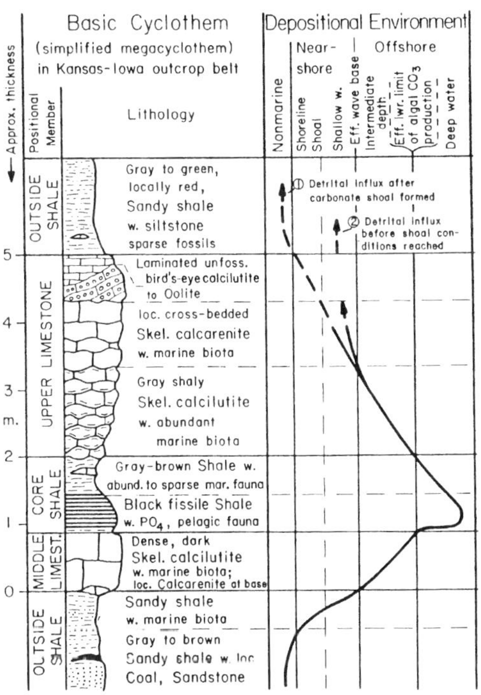

Fig. 20. Basic vertical sequence of an individual Kansas Cyclothem (after Heckel 1977). Note that the interpreted depositional environment for the so called core shale is deep water, and that the depositional environment for the so called outside shale, four meters above, is non-marine. Thus, it would appear that to go from deep water to non-marine over four meters would suggest the presence of a foreshortened section indicative of a relative sea-level fall during progradation and hence a forced regressive event.

is estimated at 135 m (c. 450 ft). Even taking compaction into account, this section seems significantly foreshortened, suggesting progradation in the presence of forced regression.

Position of the sequence boundary

Surfaces bounding forced regressive deposits

The upper bounding surface of forced regressive deposits is affected by erosive processes both during the period of relative sea-level fall, as well as during the period of subsequent relative sea-level rise and transgression. During the period of relative sea-level fall, the top of early-deposited forced regressive deposits are acted upon by fluvial (Corner *et al.* 1990; Hart & Long 1996) and other subaerial erosive processes, as well as wave and tidal processes in some instances (Corner *et al.* 1990). The amount of sediment removed can be highly variable. The degree of fluvial valley entrenchment and valley widening will be a function of (1) the gradient of the sea-floor exposed during sea-level fall (i.e. the higher the sea-floor gradient, the greater the likelihood of significant valley incision, Posamentier *et al.* 1992), (2) the discharge of the fluvial system, (3) the degree of induration of the substrate, (4) the type of vegetative cover, (5) the magnitude of lowstand fluvial discharge and (6) the amount of environmental wave and tidal energy acting at the shoreline near the mouths of fluvial systems.

During the subsequent period of relative sea-level rise-induced transgression, the tops of

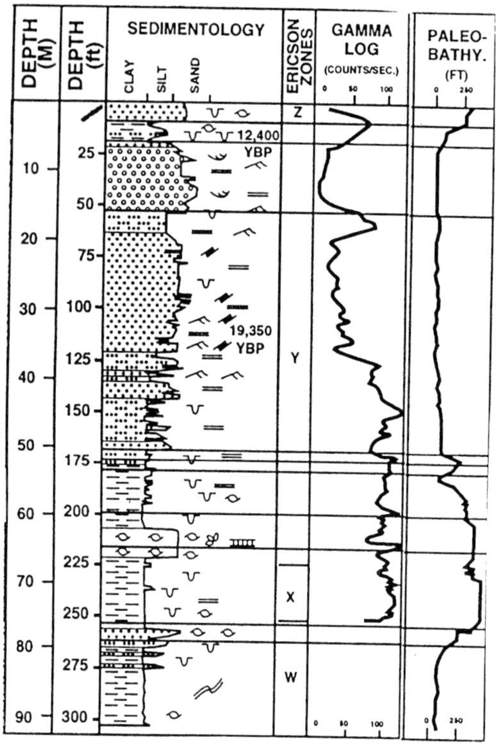

Fig. 21. Core description, gamma-ray log, and palaeobathymetric interpretation from borehole MP 303, Gulf of Mexico (from Kolla *et al.* this volume). This borehole penetrates a progradational stratigraphic unit offshore Louisiana, Gulf of Mexico. At the base of the progradational succession, the palaeo water depth is interpreted at *c.* 135 m (450 ft) and at the top the palaeo water depth is zero. The progradational succession is 65 m (200 m) thick thus suggesting a foreshortened section so that palaeobathymetry goes from 135 m to 0 m over a section only 65 m thick. This foreshortening is indicative of progradation under the influence of falling relative sea-level and therefore is evidence for forced regression (See Kolla *et al.* this volume for details).

these same forced regressive deposits are again acted upon by erosive forces, this time by wave and tidal processes associated with transgressing shorelines. The amount of sediment removed during transgression is again a function of several factors: (1) the degree of induration of the substrate, (2) the energy of wave and tidal processes acting along the coastline, (3) the rate of transgression, (4) the vegetative cover of the coastal/delta plain and (5) the grain size of the deposits that comprise the substrate. The amount of erosion has been estimated to be approximately equivalent to the height of the shoreface (Colquhoun 1969). Colquhoun (1969) estimates that the amount of erosion is approximately 10 m for the South Carolina coastline of the US, whereas Anderson (pers. comm. 1996) estimates erosion to be in the order of 9–11 m for coastlines along the Texas Gulf Coast.

In contrast to the well-defined tops of forced regressive deposits, the base can commonly be characterized as an extensive correlative conformity where forced regressive deposits are preserved. This surface may have little objective expression aside from existing as a bedding surface that can be correlated with a coeval unconformity surface. This coeval unconformity surface can be observed in some instances to lie below forced regressive deposits, being expressed as a sharp-based near-shore sandstone directly overlying offshore marine mudstones (Plint 1988; Posamentier *et al.* 1988, 1992; and papers in this volume by Ainsworth *et al.*; Fitzsimmons & Johnson; Gawthorpe *et al.*; Mellere & Steel; McMurray & Gawthorpe; Plint & Nummedal; Trincardi & Correggiari). In other instances, the coeval unconformity extends only as far as the subaerially-exposed surface atop immediately preceding highstand systems tract deposits.

Discussion

Clearly, forced regressive deposits are distinctly different from those of the immediately preceding highstand systems tract. Where highstand deposits commonly are characterized by both progradation and aggradation, forced regressive deposits are characterized dominantly by progradation. Moreover, the downstepping of the forced regressive deposits' tops commonly causes incision of fluvial systems atop the earlier-deposited highstand systems tract or at least non-deposition in that part of the system (Corner *et al.* 1990; Hart & Long 1996). The net effect is sedimentary bypass of the area inboard of the forced regressive wedge shoreline. Consequently, the sediment flux as well as the sediment calibre delivered to the near-shore environment can be significantly modified during this period of relative sea-level fall.

As a consequence of the distinctive nature of these deposits, some have suggested that these stratigraphic units should be considered a separate systems tract. Posamentier & Allen (1993) refer to these deposits as the early lowstand systems tract; Plint & Nummedal (this volume) refer to these deposits as the falling stage systems tract; Hunt & Tucker (1992, 1993) refer to these deposits as the forced regressive wedge systems tract, later shortened to the forced regressive systems tract (Hunt & Tucker 1995) also used by Helland-Hansen & Martinsen (1996). Still other workers choose to include these deposits with the underlying aggradational/progradational deposits of the highstand systems tract and call these deposits the late highstand systems tract (Van Wagoner 1995). A review of these different systematics is given by Hunt & Gawthorpe (this volume).

In the presence of forced regressive deposits, the choice of which surface constitutes the master sequence boundary has been the subject of some debate (Vail *et al.* 1977; Posamentier & Vail 1988; Galloway 1989; Hunt & Tucker 1992, 1995; Kolla *et al.* 1995; Van Wagoner 1995). Figure 22 illustrates the two possible surfaces. These are (1) the contact between the normal and the first forced regressive wedge (e.g. Posamentier *et al.* 1992) and (2) the top of the forced regressive wedge (e.g. Hunt & Tucker 1992, 1993, 1995; Helland-Hansen & Gjelberg 1994; Plint & Nummedal this volume). The principal arguments favouring placement of the sequence boundary at the top of the wedge are that this surface is the most easily recognizable, its expression as an unconformity (as opposed to a correlative conformity) is widespread, and as a result it constitutes the most readily mappable surface in this succession (e.g. Hunt & Tucker 1992, 1993, 1995; Van Wagoner 1995; Helland-Hansen & Gjelberg 1994; Plint & Nummedal this volume). This surface commonly is expressed as a sharply defined erosional interface formed by a combination of the process of fluvial, tidal, or wave erosional processes. Thus, from the point of view of ease of recognition (at least locally, on the shelf), the upper bounding surface would be the surface of choice.

Nonetheless, we argue that whereas the upper bounding surface may be the easiest to identify, it constitutes a diachronous surface and comprises an amalgamation of higher frequency sequence boundaries that form during the overall fall of relative sea-level. We will also argue that whatever surface is selected as the master sequence boundary must have relevance in a broad range of coeval physiographic settings where sedimentation rates may be higher or lower, and in physiographic settings ranging from shelf to basin. In other words, the surface selected as the sequence boundary should have universal significance and not just local or provincial significance.

We favour placing the master sequence boundary at the base of the forced regressive wedge. This represents the surface that exists at the time of initiation of sea-level fall (surface A, Fig. 22). Subsequent to this time, downstepping sea-level results in sediment bypass of the previously deposited highstand sediments. Depending upon the gradient of the surface that is exposed by this earliest sea-level fall, incised valleys with associated abandoned flood plains (i.e. interfluves) may begin to form at this time. On the seaward side of the last highstand shoreline, there will be a relatively abrupt seaward shift of facies assemblages. It is important to note that this surface which we refer to here as the master sequence boundary is expressed in part as an unconformity and in part as a correlative conformity. This varied expression of the sequence boundary is consistent with the earliest definitions of the sequence boundary concept (Mitchum 1977; Posamentier & Vail 1988; Van Wagoner *et al.* 1988) wherein it was recognized that sequence boundaries can be expressed as subaerial erosional surfaces, the base of incised valleys, correlative conformities, etc. In addition, it is important to note that this bounding surface has chronostratigraphic significance insofar as it represents the palaeogeography at a moment in time. As sea-level fall continues, successive forced regressive wedges form (3–6, Fig. 22). In some instances they can form what Ainsworth & Pattison (1994) refer to as attached lowstand deposits, and it is this scenario that we will assume in Fig. 2.

With each successive sea-level downstepping, higher-order sequence boundaries form (surfaces B, C, and D, Fig. 22) (see Posamentier *et al.* 1992, figs 12 and 13). When illustrated on a Wheeler diagram (Fig. 22b), these surfaces are depicted as time lines. On the depth section (Fig. 22a), these surfaces merge in a landward direction so that the surface at the top of unit 2 represents a composite surface comprising sequence boundaries A+B+C+D. The surface at the top of unit 3 represents a composite surface comprising sequence boundaries B+C+D, and so on. The surface at the base of incised valley fluvial/estuarine deposits at the top of the forced regressive wedge, is coeval with the deposits of the seaward-most wedge (i.e. unit 6) and corresponds to time line D (unfortunately, the Wheeler Diagram shows only when deposition and non-deposition occur, but not when erosion occurs). The physical surface at the top of the forced regressive wedge as shown in Fig. 22a clearly represents an amalgam of surfaces that began to form earlier proximally and later distally. It is, of course an excellent lithostratigraphic boundary, but as Fig. 22b illustrates it is clearly diachronous. It is, in fact, similar diachroneity that leads us to reject the transgressive surface of erosion (i.e. ravinement surface) as a candidate for a sequence boundary.

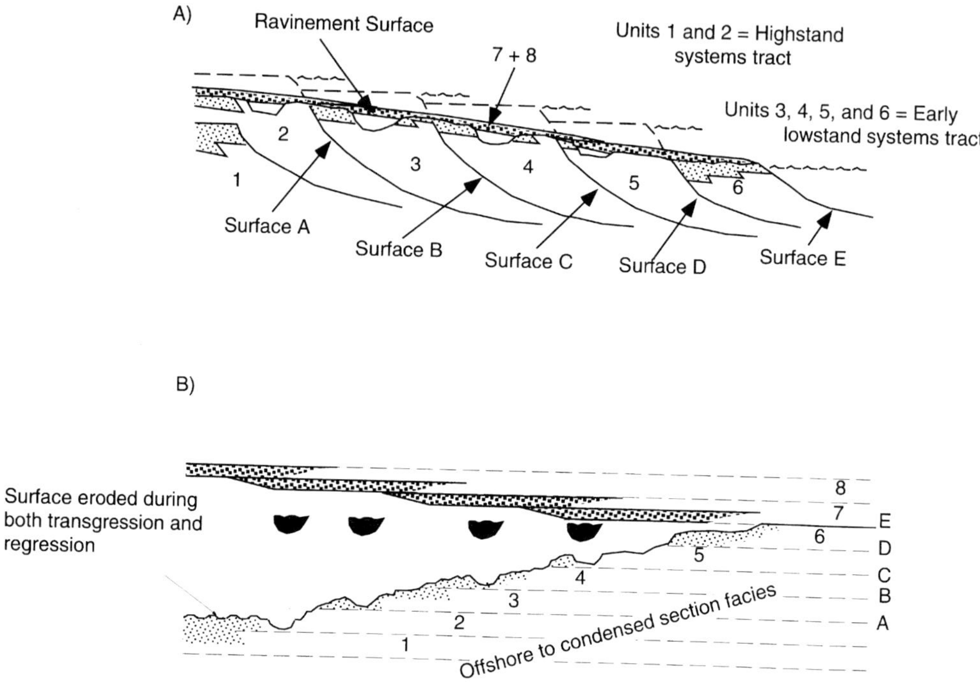

Fig. 22. Illustration in time and depth of sediments deposited during falling and subsequent rising relative sea-level. In this illustration the turnaround from regression to transgression occurs after a short-lived stillstand during time 6. Note the time transgressive nature of the top of the forced regressive wedge complex (i.e. units 3, 4, and 5). The base of the backstepping transgressive deposits (i.e. units 7 and 8) is also time transgressive as shown in (B). We interpret surface A to represent the master sequence boundary (following Posamentier *et al.* 1992), rather differently to the systematics of Hunt & Tucker (1992, 1993, 1995), Helland-Hansen & Martinsen (1996) and Plint & Nummedal (this volume) who place their major surface at the top of sediments deposited during relative sea-level fall.

As shown in Fig. 22b, the age of the sequence boundary, if it is placed at the top of the forced regressive wedge would be equivalent to the age of the youngest underlying sediments. These youngest sediments would be observed at interfluve locations, where the erosional vacuity would be minimal. As Fig. 22b clearly shows, this surface is time transgressive and would be a poor choice for serving as the basis of a palaeogeographic map. Nonetheless, from a lithostratigraphic perspective, this surface is a readily identifiable surface. This leads us to pose the question: should ease of recognition lead us to select this time transgressive (i.e. lithostratigraphic) surface as the sequence boundary, when the essence of sequence stratigraphy and the heart of sequence stratigraphic analyses has involved the recognition of time synchronous surfaces?

There is no question that surface A (Fig. 22) is a more difficult surface to identify than the hybrid, time diachronous surface at the top of the forced regressive wedge (Posamentier *et al.* 1992; Hunt & Tucker 1995). The reason for this is that surface A is characterized as an unconformity over part of the area and a correlative conformity over the remaining area, and while this is true also of the surface at the top of the forced regressive wedge (i.e. it is expressed as a correlative conformity seaward of unit 6; Fig. 22), for surface A the correlative conformity covers a proportionally greater area. As Fig. 22 shows, the criterion for determining the age of surface A is the age of the oldest preserved interfluve strata, usually observed in the proximal regions of the cross-section.

Figures 23 and 24 show specific situations that illustrate problems with selecting the top of the forced regressive wedge as the sequence boundary. Figure 23 schematically depicts the situation described by McMurray & Gawthorpe (this volume) for the northern Peloponnese peninsula, Greece. Along this coastline of the Gulf of Corinth, there exist areas characterized by high sediment flux and high shelf gradient on the one hand, and low sediment flux and low shelf gradient on the other. The sequence architecture in these two areas is markedly different highlighting the need to consider the effects of strike variability on sequence architecture (Gawthorpe *et al.* 1994; Martinsen & Helland-Hansen 1995). In the area characterized by high sediment flux and high shelf gradient, attached forced regressive deposits in the form of fan-deltas occur (Fig. 23a). In contrast, in the area characterized by low sediment flux and low shelf gradient, detached forced regressive deposits in the form of shallow marine shorefaces are found (Fig. 23b). Both areas have been influenced by the same relative sea-level change.

Using the Hunt & Tucker (1992, 1993, 1995), Plint & Nummedal (this volume) or Helland-Hansen & Martinsen (1996) approach of placing the sequence boundary at the top of the forced regressive wedge causes problems as one moves down the coast from the site characterized by Fig. 23a to the site characterized by Fig. 23b. At site A, the sequence boundary would be the surface capping units 2, 3, and 4. At site B, the position of the sequence boundary would certainly be above unit 2, but then it could be either below unit 3 or above it depending upon whether the data set is limited to the window marked as 'X' or as 'Y'. If the data set were restricted to the 'X' window, then the sequence boundary would be below unit 3, insofar as this wedge would be inferred to be the seaward-most, or lowstand, wedge. If the data set were restricted to the 'Y' window, then the sequence boundary would be above unit 3, insofar as this wedge would be part of the forced regressive wedge, deposited en route to the seaward-most, or lowstand wedge, unit 4. If the window of data were to include the entire profile, then the position of the sequence boundary would change yet again; the sequence boundary would be placed above units 2, 3, and 4, and below unit 5. Thus, depending on the extent of the data set (i.e. the window to the world for the geologist), the timing of the sequence boundary within a given profile can change dramatically if the Hunt & Tucker (1992, 1993, 1995), Plint & Nummedal (this volume) or Helland-Hansen & Martinsen (1996) criteria are employed. Should the position of the sequence boundary depend on the extent of data coverage? We believe that this would be inadvisable. Likewise, the placement of the sequence boundary on coeval profiles at different locations along a coastline can potentially be radically different as a function such local factors as sediment flux and shelf gradient. In this instance, one should ask whether local factors should play a major role in determining the position of the sequence boundary.

Using the approach advocated by Posamentier & Vail (1988) and Posamentier *et al.* (1992) the sequence boundary would be placed at the top of unit 2 and the base of unit 3 on both profiles (Fig. 23a and b). This choice of sequence boundary placement is independent of any local factors of physiography or sediment flux. In other words, this choice of position is not provincial in nature and affords greater confidence in the accuracy of palaeogeographic maps based on this surface. It would seem a marked shift back to the realm of lithostratigraphy and away from all that sequence stratigraphy represents in the way of chronostratigraphy not to utilize this surface.

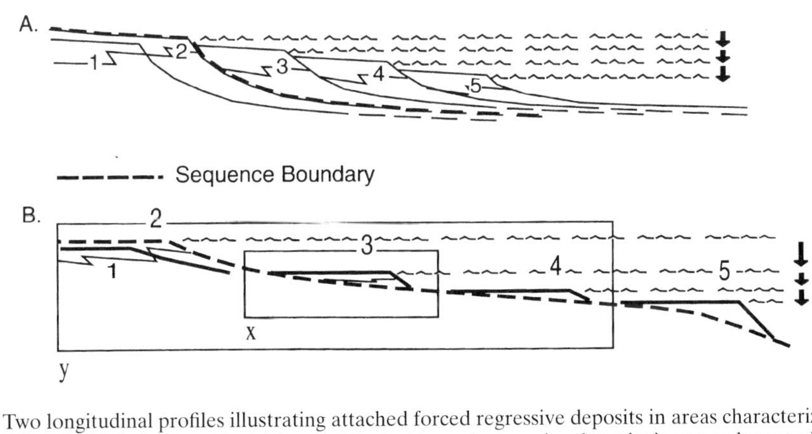

Fig. 23. Two longitudinal profiles illustrating attached forced regressive deposits in areas characterized by high sediment flux and high shelf gradient (**a**) and detached forced regressive deposits in areas characterized low sediment flux and low shelf gradient (**b**). Both areas are assumed to have been influenced by the same relative sea-level change.

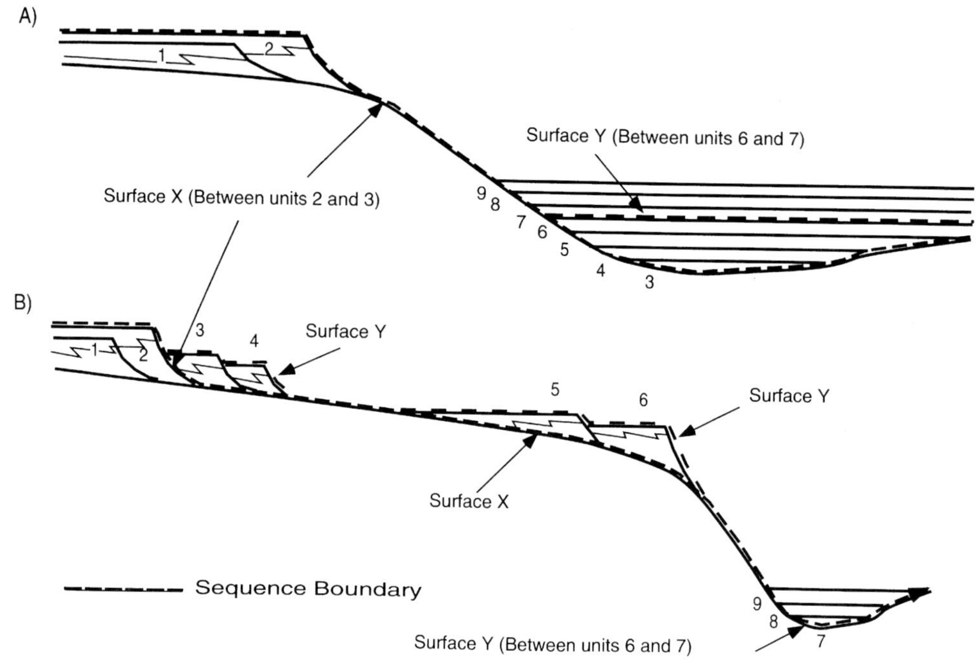

Fig. 24. Two longitudinal profiles illustrating forced regressive stratal architecture in two different physiographic settings. (**a**) A situation where highstand deposition has reached the shelf margin just prior to the initiation of relative sea-level fall. (**b**) A situation elsewhere along the coast where highstand deposition reaches only a mid-shelf position just prior to the initiation of relative sea-level fall. The resulting positions of the sequence boundary are Surface X following the approach of Posamentier *et al.* (1992) and Surface Y following the approach of Hunt & Tucker (1992, 1993, 1995), Helland-Hansen & Martinsen (1996) and Plint & Nummedal (this volume).

Figure 24 illustrates another situation where the choice of where to place the sequence boundary will be greatly influenced by local physiography. These two profiles are patterned after the modern day physiography of the Louisiana (Fig. 24a) and Texas Gulf Coast (Fig. 24b). For the purposes of this discussion we will assume that relative sea-level fall begins at the end of time 2 at both locations. At location A, highstand progradation has nearly reached the shelf edge when sea-level fall begins, whereas at the same time at location B, highstand progradation is restricted to the inner shelf. At location A, with the depocentre at the shelf edge, canyon cutting and incised valley formation begin almost immediately in response to the initiation of relative sea-level fall. Progradation and shoreline regression are minimal because of the relatively steep sea-floor at this location. This steep gradient results in instability and mass wasting of sediment delivered to this area. At the same time, at location B with a significantly gentler sea-floor gradient, forced regression is initiated. Deep-water sedimentation begins at location A at time 3, but not until time 7 at location B. Thus the issue is clear; if the Hunt & Tucker (1992, 1993, 1995), Plint & Nummedal (this volume) or Helland-Hansen & Martinsen (1996) approach is employed, the sequence boundary at location A is observed along surface X and at location B surface along surface Y. Surface Y, which would be interpreted as the sequence boundary at location B would be observed within the lowstand deposits at location A. Clearly, correlation of the sequence boundary from location A to location B would be problematic and palaeogeographic maps based on this diachronous surface would be meaningless. In contrast, using the Posamentier *et al.* (1992) approach, the sequence boundary would at both locations be observed along surface X. Palaeogeographic maps would correctly show that in part of the region (i.e. at location B) the expression of the early lowstand systems tract (i.e. forced regressive systems tract of Hunt & Tucker 1992, 1993, 1995; Helland-Hansen & Martinsen 1996 or falling stage systems tract of Plint & Nummedal this volume) would be forced regressive deposition on the shelf whereas in another part of the region (i.e. at location A) the expression of early lowstand systems tract would be deep-water deposition on the slope and in the basin.

Conclusions

The process of forced regression is a relatively common process, occurring on the shelf as an invariable consequence of relative sea-level fall. Every lowstand of relative sea-level is preceded by a period characterized by forced regression. This can be simply illustrated by examining successive strand lines ringing a lake or reservoir where water level has fallen (Figs 4 and 25). Note that along both of these lakeshores shown in Figs 4 and 25, successive shorelines have formed progressively farther offshore in response to lowering of water-level. This constitutes a regression, and, in fact, regression without significant accompanying progradation (note that there has been minor shoreface progradation during what must have been a lake-level stillstand during deposition of the highest-formed shoreface deposit). In every instance of relative sea-level (or water-level) fall, forced regression must occur. Whether or not there will be any preservation of sediments deposited during forced regression depends on a number of factors. These factors include: (1) how much erosion of these deposits occurs firstly during the period of sea-level fall, when these deposits may be eroded by fluvial and other subaerial processes, as well as by wave and tidal action, and secondly the period of subsequent transgression, when these deposits may be eroded by wave action or other submarine processes, (2) the sediment flux in this area so that if sediment flux is low, such as is the situation for the lake shown in Fig. 25, then it is likely there will be little or no preservation of forced regressive deposits and (3) the gradient of the sea floor; if the sea floor is too steep to provide a stable substrate for prograding forced regressive deposits, such as can be the situation at the shelf edge, then active mass movement processes will preclude the preservation of forced regressive deposits.

A variety of criteria have been identified that can be used to determine the presence of forced regressive deposits. These include: (1) the presence of a zone of separation between shoreface deposits located on basin margins and shoreface

Fig. 25. Forced regressive shorelines associated with lowering of the level of the Millsite Reservoir near Ferron, Utah. Note that there is no evidence of significant progradation that accompanied this regression. The vertical relief from the highest to the lowest shorelines shown is approximately 18 m.

deposits located farther seaward, (2) the presence of sharp-based shoreface/delta front deposits, (3) the presence of progressively shallower clinoforms going from proximal to distal, (4) the occurrence of long-distance regression across a shelf, (5) the absence of fluvial and/or coastal plain/delta plain facies capping the proximal portion of regressive deposits, (6) the presence of a seaward-dipping upper bounding surface atop a mid to outer shelf progradational unit, where the dip exceeds that which would be reasonably expected of a non-marine environment, (7) the presence of an increased average sediment grain size in regressive deposits going from proximal to distal and (8) the presence of foreshortened stratigraphic sections. Clearly, the more of these criteria that can be verified the greater the confidence level in a forced regressive interpretation.

A number of factors control the stratal architecture within deposits associated with the process of forced regression. These factors include: (1) the gradient of the sea floor progressively exposed by falling relative sea level, (2) the ratio of the sediment flux to the rate of relative sea-level fall, (3) the smoothness of relative sea-level fall, (4) the variability of sediment flux and (5) the changes of sedimentary process that occur as sea level falls and progressively more of the shelf is subaerially exposed. Thus, the stratal architecture of forced regressive deposits can be highly varied depending upon local conditions. Figure 3 summarizes the principal types of forced regressive deposits grouped according to whether the sea floor is characterized by a gentle or steep gradient. Those deposits that form where the sea floor is gentle include smooth-topped attached, and stepped top attached or detached (Fig. 3a, b, and c); those deposits that form where the sea floor is steep include attached smooth- and stepped-top (Fig. 3d and e).

We are indebted to numerous colleagues who have shared with us their thoughts on the forced regressive deposits. These include G. Allen, D. James, D. Leckie, J. Bhattacharya, V. Kolla, P. McCabe, K. Shanley, and D. Nummedal, among others. We also acknowledge the insightful reviews of W. Helland-Hansen, T. Elliott and D. Hunt. Their comments (especially those of Helland-Hansen) resulted in significant improvements of the text. Thanks also go to ARCO Exploration and Production Technology for permission to publish this paper.

References

AINSWORTH, R. B. & PATTISON, S. A. J. 1994. Where have all the lowstands gone? Evidence for attached lowstand systems tracts in the western interior of North America. *Geology*, **22**, 415–418.

AINSWORTH, R. B., BOSSCHER, H. & NEWALL, M. J. 2000. Forward modelling of forced regressions. Evidence for the genesis of attached and detached lowstand systems. *This volume*.

BARDAJI, T., DABRIO, C. J., GOY, J. L., SOMOZA, L. & ZAZO, C. 1990. Pleistocene fan deltas in southeastern Iberian peninsula: sedimentary controls and sea-level changes. *In*: COLELLA, A. & PRIOR, D. B. (eds) *Coarse-Grained Deltas*. International Association of Sedimentologists, Special Publications, **10**, 129–151.

COLQUHOUN, D. J. 1969. Coastal plain terraces in the Carolinas and Georgia, U.S.A. *Quaternary Geology and Climate*, National Academy of Sciences, Washington, DC, **1701**, 150–162.

CORNER, G. D., NORDAHL, E., MUNCH-ELLINGSON, K. & ROBERTSEN, K. R. 1990. Morphology and sedimentology of an emergent fjord-head Gilbert-type delta: Alta delta, Norway. *In*: COLELLA, A. & PRIOR, D. B. (eds) *Coarse-Grained Deltas*. International Association of Sedimentologists, Special Publications, **10**, 155–168.

FITZSIMMONS, R. & JOHNSON, S. 2000. Forced regressions, architecture and genesis in the Campanian of the Bighorn Basin, Wyoming. *This volume*.

GALLOWAY, W. E. 1989. Genetic stratigraphic sequences in basin analysis I: architecture and genesis of flooding surface bounded depositional units. *American Association of Petroleum Geologists Bulletin*, **73**, 125–143.

GAWTHORPE, R. L. G., HALL, M., SHARP, I. & DREYER, T. 2000. Tectonically enhanced forced regressions: examples from growth folds in extensional and compressional settings, the Miocene of the Suez rift and the Eocene of the Pyrenees. *This volume*.

——, FRASER, A. J. & COLLIER, R. E. LI. 1994. Sequence stratigraphy in active extensional basins: implications for the interpretation of ancient basin fills. *Marine and Petroleum Geology*, **11**, 642–658.

HANSON, K. L., LETTIS, W. R., WESLING, J. R., KELSON, K. I. & MEZGER, L. 1992. Quaternary marine terraces, south central coastal California: implication for crustal deformation and coastal evolution. *In:* FLETCHER, C. H. III & WEHMILLER, J. F. (eds) *Quaternary coasts of the United States: marine and lacustrine Systems*. Society of Economic Paleontologists and Mineralogists, Special Publications, **48**, 323–332.

HART, B. S. & LONG, B. F. 1996. Forced regressions and lowstand deltas: Holocene Canadian examples. *Journal of Sedimentary Research*, **66**, 820–829.

HECKEL, P. H. 1977. Origin of phosphatic black shale facies in Pennsylvanian cyclothems of Mid-continent North America. *American Association of Petroleum Geologists Bulletin*, **61**, 1045–1068.

HELLAND-HANSEN, W. & GJELBERG, J. B. 1994. Conceptual basis and variability in sequence stratigraphy: a different perspective. *Sedimentary Geology*, **92**, 31–52.

—— & MARTINSEN, O. 1996. Shoreline trajectrories

and sequences: description of variable deposition -dip scenarios. *Journal of Sedimentary Research*, **4**, 670–685.

HILL, P. R., ROBERGE, M. & BAECHTOLD, F. 1997. Holocene analogs for forced regression sand and gravel bodies (abstract). *In*: *Official Program, American Association of Petroleum Geologists*, Dallas, April 6–9, 1997, A51.

HUNT, D. & GAWTHORPE, R. L. G. 2000. Sedimentary responses to forced regressions: an Introduction. *This volume*.

—— & TUCKER, M. E. 1992. Stranded parasequences and the forced regressive wedge systems tract: deposition during base-level fall. *Sedimentary Geology*, **81**, 1–9.

—— & —— 1993. Sequence stratigraphy of carbonate shelves with an example from the mid-Cretaceous (Urgonian) of southeast. *In:* POSAMENTIER, H. W., SUMMERHAYES, C. P., HAQ, B. U. & ALLEN, G. P. (eds) *Sequence Stratigraphy and Facies Associations*. International Association of Sedimentologists, Special Publications, **18**, 307–341.

—— & —— 1995. Stranded para sequences and forced regressive wedge systems tract: deposition during base-level fall- reply. *Sedimentary Geology*, **95**, 147–160.

KOLLA, V., BIONDI, P., LONG, B. & FILLON, R. 2000. Sequence stratigraphy and architecture of the late Pleistocene Lagniappe delta complex, northeast Gulf of Mexico. *This volume*.

——, POSAMENTIER, H. W. & EICHENSEER, H. 1995. Stranded parasequences and the forced regressive wedge systems tract: deposition during base-level fall – discussion. *Sedimentary Geology*, **95**, 139–145.

LECKIE, D. A. 1994. Canterbury Plains, New Zealand – implications for sequence stratigraphic models. *American Association of Petroleum Geologists Bulletin*, **78**, 1240–1256.

MCMURRAY, L. S. & GAWTHORPE, R. L. G. 2000. Along-strike variability of forced regressive deposits: late Quaternary, northern Pelopnnesos, Greece. *This volume*.

MARTINSEN, O. & HELLAND-HANSEN, W. 1995. Strike variability of clastic depositional systems; does it matter for sequence-stratigraphic analysis. *Geology*, **23**, 439–442.

MELLERE, D. & STEEL, R. 2000. Style contrast between forced regressive and lowstand/transgressive wedges in the Campanian of south-central Wyoming *This volume*.

MITCHUM, R. M. 1977. Seismic stratigraphy and global changes of sea level, part 1: glossary of terms used in seismic stratigraphy. *In:* PAYTON, C. E. (ed.) *Seismic stratigraphy – applications to hydrocarbon exploration*. American Association of Petroleum Geologists, Memoirs, **26**, 117–143.

MORRIS, W. R., POSAMENTIER, H. W., LOOMIS, K. B., BHATTACHARYA, J. P., KUPECZ, J. A. WU, C. , LOPEZ-BLANCO, M., THOMPSON, P. R., SPEAR, D. B., LANDIS, C. R. & KENDALL, B. A. 1995, Cretaceous Panther Tongue sandstone outcrop case study II: evolution of delta type within a forced regression (abstract). *In*: *Official Program, American Association of Petroleum Geologists*, Houston, USA, March 5–8, 68A.

PLINT, A. G. 1988. Sharp-based shoreface sequences and 'offshore bars' in the Cardium Formation of Alberta; their relationship to relative changes in sea level. *In*: WILGUS, C. K., HASTINGS, B. S., KENDALL, C. G. ST. C., POSAMENTIER, H. W., ROSS, C. A. & VAN WAGONER, J. C. (eds) *Sea Level Changes – An Integrated Approach*. Society of Economic Paleontologists and Mineralogists, Special Publications, **42**, 357–370.

—— & NUMMEDAL, D. 2000. The falling stage systems tract: recognition and importance in sequence stratigraphic analysis. *This volume*.

POSAMENTIER, H. W. & ALLEN, G. P. 1993. Variability of the sequence stratigraphic model: effects of local basin factors. *Sedimentary Geology*, **86**, 91–109.

—— & —— 1994. *Siliciclastic Sequence Stratigraphy – Concepts And Applications.* American Association of Petroleum Geologists, Short Course Notes.

—— & CHAMBERLAIN, C. J. 1993. Sequence stratigraphic analysis of Viking Formation lowstand beach deposits at Joarcam Field, Alberta, Canada. *In*: POSAMENTIER, H. W., SUMMERHAYES, C. P., HAQ, B. U. & ALLEN, G. P. (eds) *Sequence Stratigraphy and Facies Associations*. International Association of Sedimentologists, Special Publications, **18**, 469–485.

—— & VAIL, P. R. 1988. Eustatic controls on clastic deposition II – sequence and systems tract models. *In*: WILGUS, C. K., HASTINGS, B. S., KENDALL, C. G. ST. C., POSAMENTIER, H. W., ROSS, C. A. & VAN WAGONER, J. C. (eds) *Sea level change – an integrated approach*. Society of Economic Paleontologists and Mineralogists, Special Publications, **42**, 125–154.

——, ALLEN, G. P., JAMES, D. P. & TESSON, M. 1992. Forced regressions in a sequence stratigraphic framework: concepts, examples, and exploration significance. *American Association of Petroleum Geologists Bulletin*, **76**, 1687–1709.

——, MORRIS, W. R., BHATTACHARYA, J. P., KUPECZ, J. A., LOOMIS, K. B., LOPEZ-BLANCO, M., WU, C., KENDALL, B. A., LANDIS, C. R., SPEAR, D. B. & THOMPSON, P. R. 1995. Cretaceous Panther Tongue sandstone outcrop case study I: regional sequence stratigraphic analysis (abstract). *In*: *Official Program, American Association of Petroleum Geologists*, Houston, USA, March 5–8, 1995, 77A.

SHOMAKER, J. W., BEAUMONT, E. C. & KOTTLOWSKI, F. E. 1971. *Strippable Low Sulfur Coal Resources of the San Juan Basin in New Mexico and Colorado*. New Mexico Bureau of Mines and Mineral Resources Memoir, **25**.

TRINCARDI, F. & CORREGGIARI, A. 2000. Quaternary forced-regression deposits in the Adriatic Basin and the record of composite sea-level cycles. *This volume*.

TROPEANO, M. & SABATO, L. 2000/Response of late Pliocene–Early Pleistocene mixed carbonate–clastic temperate-water systems to forced regressions: the Calcarenite di Gravina Formation, Puglia, SE Italy. *This volume*.

VAIL, P. R., MITCHUM, R. M. & THOMPSON, S. 1977. Seismic stratigraphy and global changes of sea level, Part 3: Relative changes of sea level from coastal onlap. *In*: PAYTON, C. E. (ed.) *Seismic stratigraphy – application to hydrocarbon exploration.* American Association of Petroleum Geologists, Memoirs, **26**, 63–81.

VAN WAGONER, J. C. 1995. Sequence stratigraphy and marine to nonmarine facies architecture of foreland basin strata, Book Cliffs, Utah, U.S.A., *In:* VAN WAGONER, J. C. & BERTRAM, G. T. (eds) *Sequence Stratigraphy of Foreland Basin Deposits.* American Association of Petroleum Geologists Memoirs, **64**, 137–223.

——, POSAMENTIER, H. W., MITCHUM, R. M., VAIL, P. R., SARG, J. F., LOUTIT, T. S. & HARDENBOL, J. 1988. An overview of the fundamentals of sequence stratigraphy and key definitions. *In:* WILGUS, C. K., HASTINGS, B. S., KENDALL, C. G. ST. C., POSAMENTIER, H. W., ROSS, C. A. & VAN WAGONER, J. C. (eds) *Sea level change – an integrated approach.* Society of Economic Paleontologists and Mineralogists, Special Publications, **42**, 39–45.

Carbonate megabreccias in a sequence stratigraphic context; evidence from the Cambrian of North Greenland

JON R. INESON[1] & FINN SURLYK[2]

[1]*Geological Survey of Denmark and Greenland (GEUS), Thoravej 8, 2400 Copenhagen NV, Denmark (e-mail: ji@geus.dk)*

[2]*Geological Institute, University of Copenhagen, Øster Voldgade 10, 1350 Copenhagen K, Denmark*

Abstract: In carbonate sequence stratigraphy, carbonate megabreccias have acquired particular significance, being deemed characteristic of the lowstand systems tract (LST) or the forced regressive systems tract (FRST). Large-scale mass-wastage can, however, result from factors other than sea-level change and it is rare that the sequence stratigraphic significance of megabreccias can be rigorously tested. In the Cambrian of North Greenland, erection of a robust sequence stratigraphic framework is facilitated by extensive fjord-wall exposures of the platform to deep shelf transect and by a well-developed carbonate–siliciclastic reciprocal sedimentation pattern within off-platform strata. On the basis of this independent framework, megabreccias are represented locally *within* the LST and the highstand systems tract (HST), but occur systematically above the HST. These HST-capping megabreccias are composite sheets tens of metres thick that extend up to 50 km distally and flank the platform for up to 400 km along strike. They comprise debris derived from the highstand platform margin and slope and are directly overlain by mixed carbonate–siliciclastic sediments of the succeeding LST. The HST-capping megabreccias are assigned to the FRST; they record extensive failure of the platform margin and upper slope during relative fall of sea-level and prior to the onset of lowstand sedimentation. Although the LST megabreccias are compositionally distinctive, the sole example of an intra-HST megabreccia differs from those of the FRST only in terms of areal extent. In the absence of an independent framework, therefore, the sequence stratigraphic affinities of megabreccias may be ambiguous.

One of the most hotly debated topics within sequence stratigraphy in recent years has been the significance and affinities of sediments deposited during falling sea-level – the so-called forced regressive deposits (Posamentier *et al.* 1992; Hunt & Tucker 1992, 1993, 1995; Kolla *et al.* 1995; Mellere & Steel 1995; Plint & Nummedal this volume; Posamentier & Morris this volume). Although some workers maintain that such deposits can be adequately classified within the three-fold subdivision of the classic Exxon sequence (see Kolla *et al.* 1995; Posamentier & Morris this volume), this view has not found universal acceptance. Hunt & Tucker (1992), in a largely theoretical discussion, highlighted certain logical inconsistencies in the Exxon scheme. They recognized a fourth systems tract – the forced regressive wedge systems tract (FRWST), later shortened to the forced regressive systems tract (FRST; Hunt & Tucker 1995). Similar ideas were proposed independently by Nummedal (1992), Helland-Hansen & Gjelberg (1994), Pomar & Ward (1994) and Plint & Nummedal (this volume). Subsequently, a number of field-based studies particularly from the Cretaceous of the Western Interior USA (e.g. Nummedal & Molenaar 1995; Mellere & Steel 1995, this volume; Plint 1996; Plint & Nummedal this volume; Fitzsimmons & Johnson this volume), have demonstrated that deposition during falling sea-level can create sediment packages that are geometrically and sedimentologically dissimilar from the deposits of the preceding highstand and those of the subsequent slow relative sea-level rise (the lowstand prograding wedge systems tract of Hunt & Tucker 1992).

Much of the theoretical and field-based discussion of this problem has been centred around siliciclastic deposits (particularly coastal deposits in ramp settings). However, some of the more dramatic and illustrative examples of sedimentation during relative fall in sea-level have come from carbonate successions (e.g. Dabrio *et al.* 1981; Franseen & Mankiewicz 1991; Pomar & Ward 1994; Mutti *et al.* 1996; Pomar *et al.* 1996). Hunt & Tucker (1992, 1993) suggested that rimmed-shelf carbonate settings may respond to relative sea-level fall by failure of the margin and upper slope and the deposition of megabreccia sheets; they referred such deposits to the FRST. Carbonate megabreccias had

From: HUNT, D. & GAWTHORPE, R. L. (eds) *Sedimentary Responses to Forced Regressions*. Geological Society, London, Special Publications, **172**, 47–68. 1-86239-063-0/00/$15.00

previously been considered to characterize the lowstand systems tract (LST) of the three-fold Exxon sequence as applied to carbonate systems (e.g. Sarg 1988).

In addressing this debate, it is important first to investigate the relationship between major sediment wastage events and sea-level change since sediment failure and mass transport may be the result of a range of factors acting independently of, or in concert with, sea-level change. These include depositional oversteepening (Yose & Heller 1989; Harris 1994) and seismicity associated with tectonic activity (Hine *et al.* 1992). As stressed by a number of workers (e.g. Hine *et al.* 1992; Grammer *et al.* 1993; Handford & Loucks 1993; Hunt & Tucker 1993), the occurrence of coarse-grained redeposited carbonates within deeper-water successions is not *diagnostic* of falling sea-level or lowstand. Spence & Tucker (1997) recently reviewed the mechanisms that may promote instability and the generation of megabreccias. These workers also stressed the non-diagnostic nature of megabreccias in sequence stratigraphy but concluded that conditions are particularly favourable for megabreccia genesis during falling or lowstand of sea-level.

In considering the potential sequence stratigraphic significance of megabreccia sheets, therefore, certain basic questions arise.

(1) Can it be demonstrated that certain major mass flow events coincided with times of fall or lowstand of relative sea-level, based on an independent sequence stratigraphic framework?

(2) If so, to which systems tract are these megabreccias best assigned – the forced regressive systems tract of the four-part sequence stratigraphic scheme (e.g. Hunt & Tucker 1995) or the lowstand systems tract of the three-part Exxon scheme (e.g. Sarg 1988)?

(3) Can these megabreccias be differentiated intrinsically from their counterparts that occur in other systems tracts?

The Cambrian in North Greenland is exposed in vertical fjord walls that provide continuous dip-oriented sections up to a kilometre high and many tens of kilometres long. Such sections document the transition from platform interior through margin and foreslope to slope apron and deep shelf. A well-developed carbonate–siliciclastic reciprocal sedimentation pattern in the off-platform succession provides a sequence stratigraphic framework that is independent of the stratigraphic position of carbonate megabreccia sheets. This, then, allows us to address the questions posed above and to contribute to the debate concerning the sequence stratigraphic status of forced regressive deposits. In the context of forced regression we consider all sediments deposited during periods of relative sea-level fall regardless of depositional environment, to be forced regressive deposits, following the broad definition of Hunt & Tucker (1995) rather than the restrictive view of Posamentier *et al.* (1992) and Posamentier & Morris (this volume).

Geological setting

During the early Palaeozoic, the Franklinian Basin covered much of present-day North Greenland and extended westwards into the Canadian Arctic Islands (Fig. 1). From the Early Cambrian to the Early Silurian, the basin consisted of two discrete depositional settings: a broad shelf to the south bordering the craton passing northward into a deep-water trough (see Surlyk & Hurst 1984; Higgins *et al.* 1991; Surlyk 1991 for reviews of basin evolution). From the late Early Cambrian to the earliest Ordovician, the shelf displayed a stepped or terraced profile (Fig. 2). For most of this period, the shallow-water carbonate platform in the south was of 'rimmed-shelf' type and was fringed seawards by a high-energy belt of carbonate sands, periodically associated with microbial mounds. The shallow-water margin passed northward via steeply dipping foreslopes (15–30°) into the outer shelf which extended some 50–80 km farther north to the shelf-slope break at the southern margin of the deep-water trough (Fig. 2). This Cambrian shelf profile showing two discrete breaks of slope, at the platform margin and at the continental shelf margin, is comparable in overall morphology to the Miocene of the central west Florida continental shelf (see Mullins *et al.* 1988, their fig. 17).

The lithostratigraphy of the Cambrian shelf strata is summarized in Fig. 3 (Ineson & Peel 1997); platform interior strata are referred to the Ryder Gletscher Group (uppermost Lower Cambrian–Middle Ordovician) whereas the

Fig. 2. Cambrian highstand palaeogeography in North Greenland, viewed from the north, showing the terraced shelf profile and the prograding platform and carbonate slope apron. Relief at the platform edge was 100–150 m; the elevation of the Cambrian shelf edge above the basin floor is poorly constrained but was probably 500–1000 m. BF, Buen Formation; BFG, Brønlund Fjord Group; TIG, Tavsens Iskappe Group; RGG, Ryder Gletscher Group. Approximate horizontal scale: E–W, 400 km; N–S, 150 km. From Ineson *et al.* (1994).

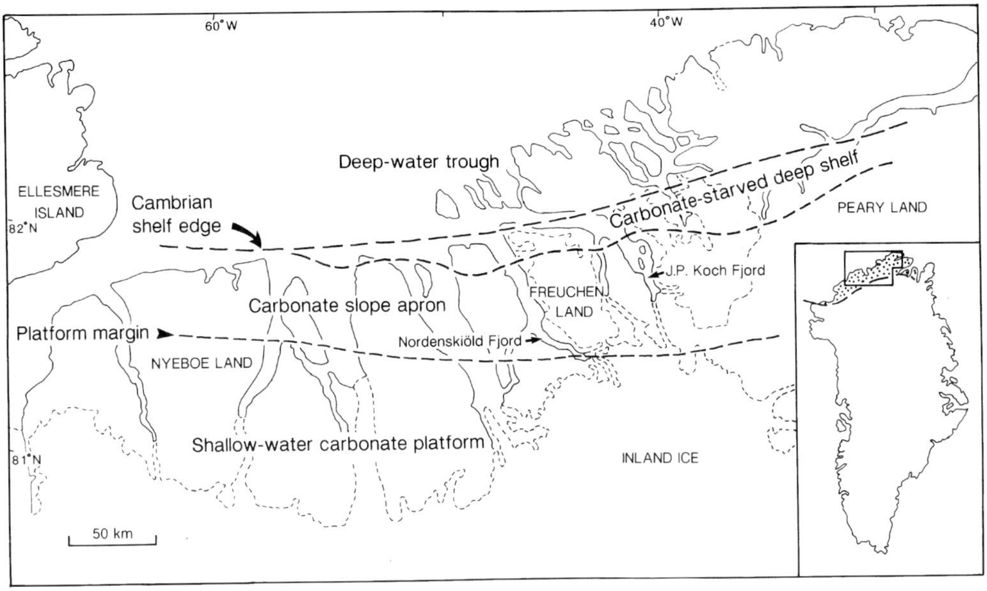

Fig. 1. Map of North Greenland (see inset; present-day distribution of the Franklinian Basin stippled) showing the major palaeogeographic elements of the Franklinian Basin during the late Early Cambrian. The position of the shelf edge was fixed during the Cambrian but the margin of the shallow-water platform prograded northwards; the palaeogeography shown corresponds to the late Early Cambrian highstand (Sequence 2, HST of Fig. 9). Note the excellent fjord control and strike extent of the platform margin – approximately 400 km.

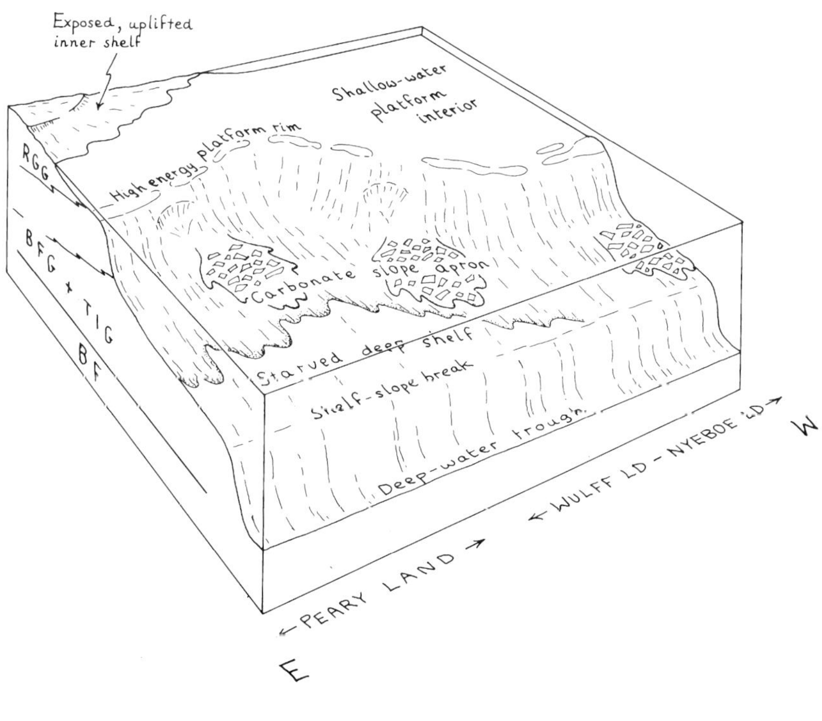

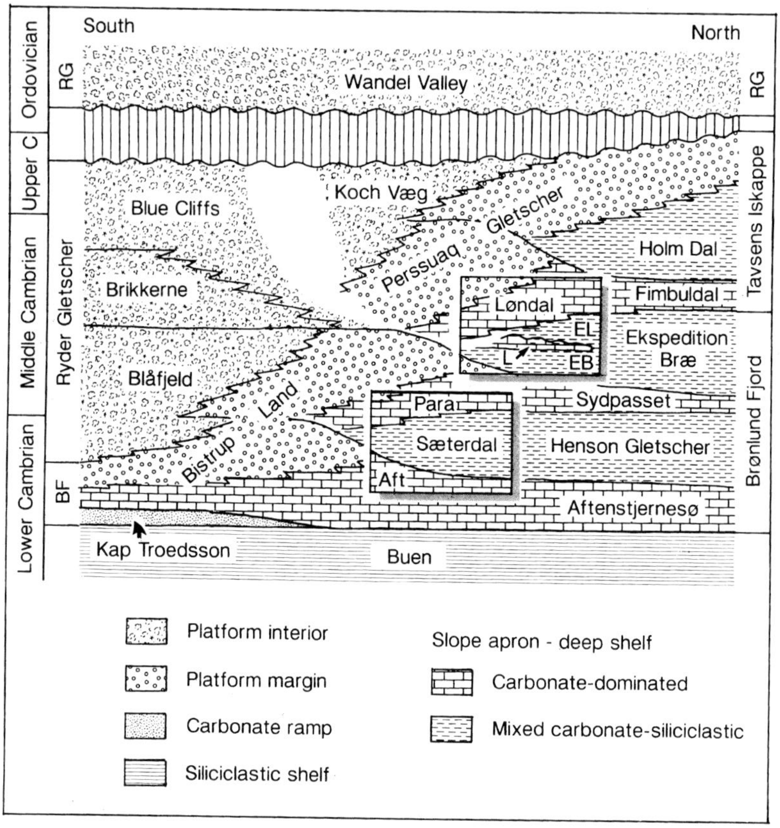

Fig. 3. Lithostratigraphy of the Brønlund Fjord (BF), Tavsens Iskappe and Ryder Gletscher (RG) groups in the southern parts of central and western North Greenland (from Ineson & Peel 1997). The upper and lower insets show the lithostratigraphic subdivision in central-south Peary Land and southeast Peary Land, respectively. The Blue Cliffs and Koch Væg formations are geographically isolated from each other and the detailed stratigraphic relationships are unknown, hence the missing area of tone in the diagram. No vertical or lateral scale implied. Aft, Aftenstjernesø Formation; Para, Paralleldal Formation; EB, Ekspedition Bræ Formation; L, Lønelv Formation; EL, Erlandsen Land Formation.

platform margin and slope apron strata are assigned to the Brønlund Fjord and Tavsens Iskappe groups (uppermost Lower Cambrian–Lower Ordovician). Platform interior and margin facies are largely dolomites; limestones form less than 10% of the platform succession. The proportion of dolomite decreases northward from over 80% at the toe of the platform fore-slope to less than 10% on the outermost shelf.

Off-platform cyclicity

Off-platform deposits of the Brønlund Fjord and Tavsens Iskappe groups show a well-developed

Fig. 4. Cliffs overlooking western Henson Gletscher (head of J. P. Koch Fjord, see Fig. 1), exposing off-platform strata of late Early to Middle Cambrian age. This section illustrates the large-scale cyclicity defined by cliff-forming carbonates (green) capped by carbonate megabreccia (red) alternating with recessive-weathering mixed carbonate–siliciclastic deposits (yellow). Note the light-coloured sandstone bands in the mixed carbonate–siliciclastic Henson Gletscher Formation (H) and the southward thinning and basal onlap displayed by the argillaceous limestones of the Ekspedition Brae Formation (E). The carbonate tongue picked out in orange consists of cross-bedded skeletal grainstones; it thins rapidly to the north (basinward), ultimately wedging out within the upper levels of the mixed carbonate–siliciclastic Henson Gletscher Formation. A, Aftenstjernesø Formation; S, Sydpasset Formation; F, Fimbuldal Formation; Fig. 3 illustrates the regional stratigraphic context of these formations.

cyclicity defined by an alternation of carbonate-dominated intervals and mixed carbonate–siliciclastic intervals (Fig. 4). Each cycle, comprising a lower mixed carbonate–siliciclastic half-cycle and an upper carbonate half-cycle, is typically 150–200 m thick close to the coeval platform and thins northward. This cyclic pattern of off-platform sedimentation persisted from the late Early Cambrian to the early Late Cambrian. A major influx of siliciclastic detritus, related to

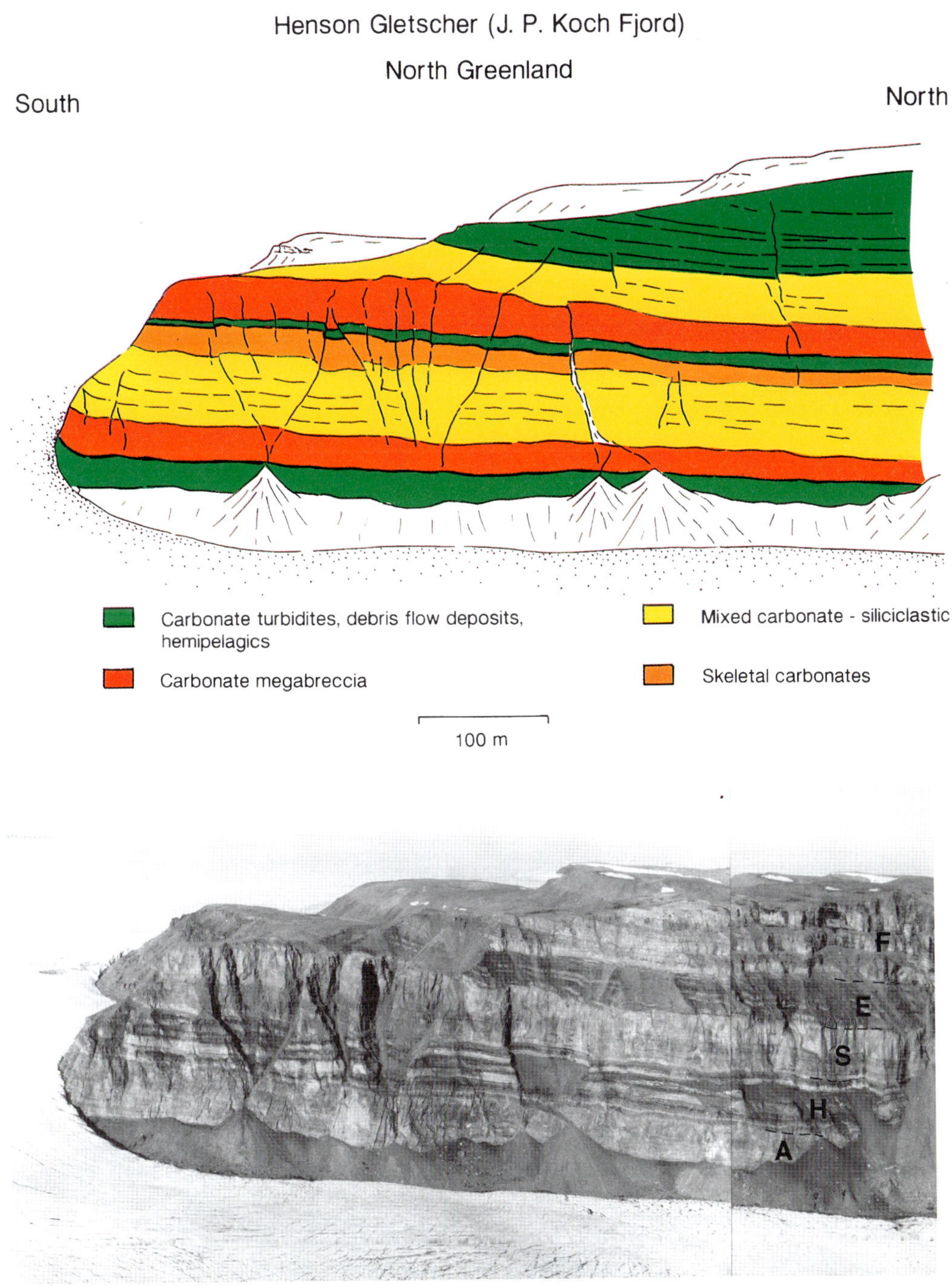

tectonic uplift in eastern North Greenland, overwhelmed the carbonate system in eastern areas in the Late Cambrian (Hurst & Surlyk 1983; Surlyk & Ineson 1987; Bryant & Smith 1990).

The cyclic nature of the off-platform strata provides a robust and readily correlatable stratigraphic framework that is reflected in both the lithostratigraphy (Fig. 3; Ineson & Peel 1997) and the sequence stratigraphy (see below; Ineson & Surlyk 1995). The typical facies development, lateral relationships and sequence stratigraphic interpretation of a single cycle are described below, illustrated in particular by the Henson Gletscher and Sydpasset formations (uppermost Lower Cambrian–medial Middle Cambrian; Fig. 5). Detailed documentation of the sequence stratigraphy of the Cambrian of North Greenland is beyond the scope of this paper and will be presented elsewhere; a preliminary description was given by Ineson & Surlyk (1995).

Mixed carbonate–siliciclastic half-cycles

Facies, processes and environment. The carbonate and siliciclastic sediments forming the lower part of the off-platform cycles abruptly overlie carbonate megabreccia sheets that cap the underlying cycle (Figs 4 and 5). Typically fine-grained and dark-coloured, these sediments are dominated by calcareous mudstones, marlstones and thin-bedded lime mudstones (Fig. 6a). Commonly parallel-laminated and bituminous, this facies contains significant organic carbon at certain levels (total organic carbon values typically 1–2% in the Henson Gletscher Formation; Christiansen *et al.* 1987). Bioturbation is rare; about 10 km from the coeval platform margin, discrete beds display *Chondrites* traces (Fig. 5). Farther basinward (north), this facies is finely-laminated and non-bioturbated. Black chert is common in certain formations (e.g. Henson Gletscher Formation) and becomes increasingly important towards the north (basinward), concomitant with a decrease in the proportion of fine-grained carbonate. Silicious sponge spicules, partially to wholly replaced by calcite, are common at certain levels, as are agnostoid and other trilobites. Phosphoritic hardground surfaces occur locally. Interbedded with these fine-grained deposits are rare thin graded skeletal grainstones/packstones and clast-supported limestone breccia sheets. The latter are typically less than a metre thick, may be impersistent laterally (over a few tens of metres) and are composed of lime mudstone or skeletal wackestone clasts.

Passing south towards the coeval platform, within 5–10 km of the platform edge, marlstones and argillaceous lime mudstones become subordinate to skeletal packstones and grainstones interbedded with bioturbated wackestones. Slumped strata and draped slump scars are associated with these facies. The shelly grainstones commonly contain glauconite, locally show low-angle (hummocky?) and trough cross-stratification and, in some sections, cap shallowing-upward units, 1–3 m thick.

Although the siliciclastic component of these half-cycles is largely of mud-grade, fine- to very fine-grained sandstones are commonly present as isolated beds and form a prominent sand-dominated packet within the Henson Gletscher Formation (Figs 4 and 5). This sandstone unit is up to 80 m thick in southern, more proximal outcrops in central North Greenland and is persistent along depositional strike (east–west) for over 150 km. It thins basinward and essentially pinches out some 30 km north of the coeval platform margin, although rare beds have been recorded up to 50 km from the margin. Sheet sandstone beds are dominant, from 0.1 to several metres thick. Although typically parallel-sided and structureless in distal sections, these sand sheets are commonly laterally impersistent in proximal settings (within about 10 km of the platform edge) and may display parallel- and hummocky cross-stratification (Figs 5 and 6b). In this proximal belt, interbedded silty sandstones and siltstones are commonly bioturbated or show ripple cross-lamination, locally of inferred wave origin (Christiansen *et al.* 1987).

Although not a feature of the remaining mixed carbonate–siliciclastic half-cycles, a prominent northward-prograding carbonate body is developed in the upper part of the half-cycle represented by the Henson Gletscher Formation (Fig. 4). At Nordenskiöld Fjord (Fig. 7), this unit is up to 150 m thick and shows northward-dipping clinoforms (see also Fig. 12). Cross-bedded skeletal and intraclastic grainstones and packstones are the dominant facies; redeposited carbonates, including laterally impersistent megabreccias (see below), commonly occur at the toe of the clinoforms. Thinning rapidly basinward (see Fig. 4) to only a few metres in sections 10–15 km farther north (Fig. 5), this carbonate wedge pinches out within the dark fine-grained lime mudstones and marlstones of the upper Henson Gletscher Formation. Laterally, along depositional strike, the carbonate wedge is recognized over a distance of more than 150 km.

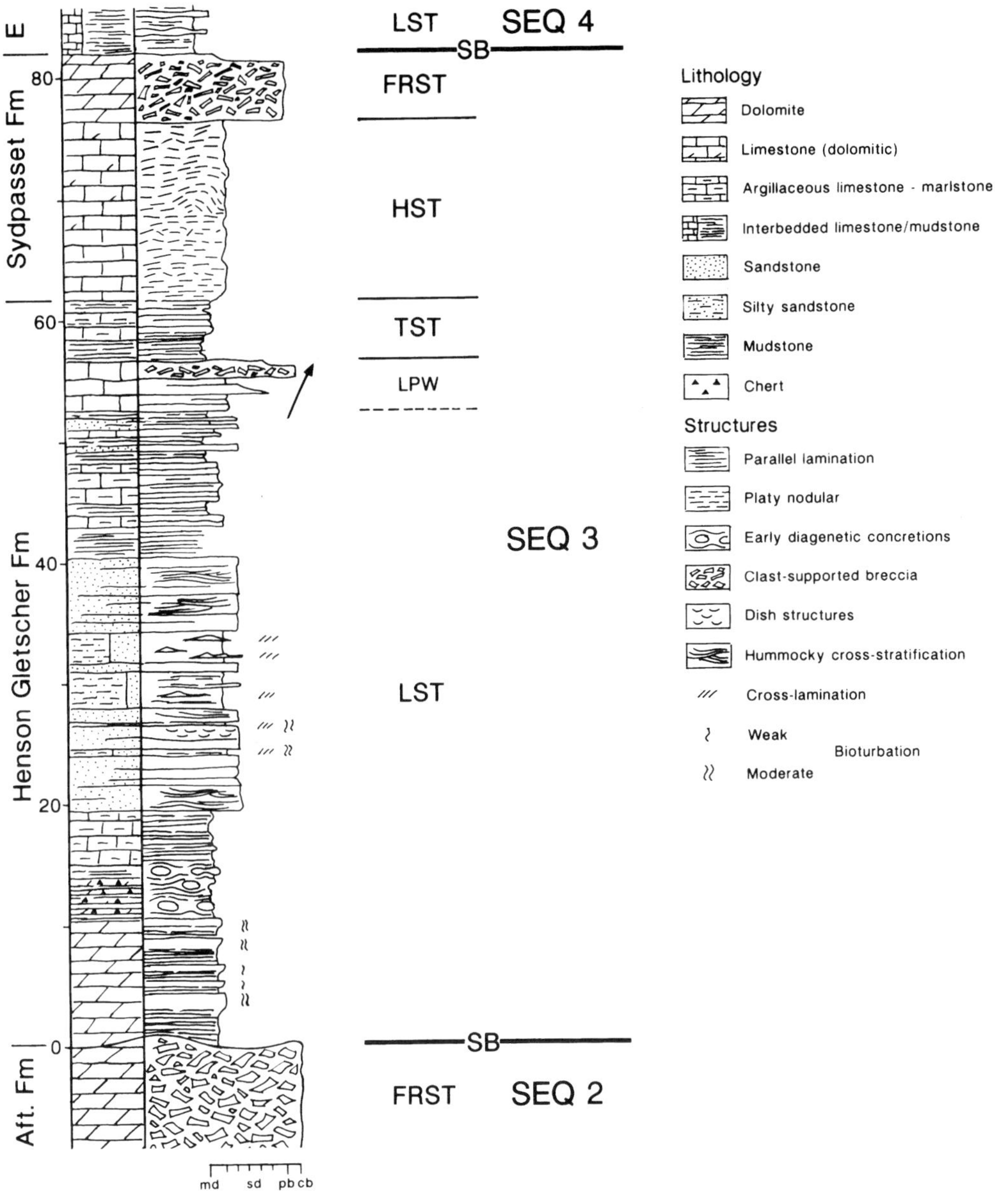

Fig. 5. Sedimentological log (thickness in metres) through a single off-platform cycle comprising a lower mixed carbonate–siliciclastic half-cycle (Henson Gletscher Formation) and an upper carbonate-dominated half-cycle (Sydpasset Formation). The thin coarsening-upward carbonate unit (arrow) in the upper levels of the Henson Gletscher Formation represents the distal toe of a progradational carbonate tongue that is prominent about 10 km farther south at Henson Gletscher (Fig. 4) and to the southwest at Nordenskiöld Fjord (see Figs 7 and 12). Note the debris-flow breccia beds capping the Aftenstjernsø (Aft.; see Fig. 10a) and Sydpasset formations. Location: 2 km southwest of the head of J. P. Koch Fjord. E, Ekspedition Bræ Fm; LST, lowstand systems tract; LPW, lowstand prograding wedge; TST, transgressive systems tract; HST, highstand systems tract; FRST, forced regressive systems tract; SB, sequence boundary.

Fig. 6. (**a**) Parallel-laminated, organic-rich marlstones; Henson Gletscher Formation, north Nyeboe Land. Lens cap for scale, 49 mm across. (**b**) Fine-grained sandstones; note the sharp bed boundaries and the pronounced pinch-and-swell exhibited by single beds (e.g. immediately above the figure). Henson Gletscher Formation, Nordenskiöld Fjord.

The mixed carbonate–siliciclastic half-cycles record deposition primarily of fine-grained sediment from suspension, accumulating below wavebase in a low-energy and typically poorly-oxygenated environment. Low rates of sedimentation are indicated by the high organic content, the concentration of chert and the presence of glauconite and phosphorite-impregnated hardgrounds. Graded limestones and limestone breccia beds record deposition from turbidity currents and debris flows respectively. Excluding the *c.* 5–10 km wide proximal belt, therefore, the off-platform portion of the shelf was sediment-starved and accumulated a thin succession of spicular marls and argillaceous lime muds with occasional incursions of coarser siliciclastic detritus, carried basinward by density flows. In the proximal belt, sedimentation was influenced, at least periodically, by shallow-marine processes, as testified by the cross-bedded glauconitic grainstones and the presence of wave-ripples and hummocky cross-stratification in sandstones. Indeed, the prograding carbonate wedge in the upper Henson Gletscher Formation represents a tongue of shoal-water carbonates that temporarily invaded the outer shelf setting.

Lateral relationships. The mixed carbonate–siliciclastic half-cycles occur sandwiched between carbonate units (Fig. 4). Traced south towards the coeval platform, they ultimately thin and wedge out between clinoform-bedded foreslope strata of the subjacent and overlying carbonate half-cycles. In Fig. 4, southward thinning of the Ekspedition Bræ Formation is evident between thickening tongues of the carbonate half-cycles and the lower beds of the formation onlap southward onto the underlying carbonate wedge. This formation shows similar relationships in the vicinity of Nordenskiöld Fjord. The Henson Gletscher Formation half-cycle also wedges out abruptly at the platform edge but the critical onlap relationships are not seen due to recent erosion. However, the sand-rich siliciclastics of this formation are not represented within the shallow-water platform interior deposits, suggesting that this and subsequent mixed carbonate–siliciclastic half-cycles represent basin-restricted wedges. Basinward, these half-cycles thin and become carbonate-poor, typically comprising a few tens of metres of black cherty mudstones in outermost shelf sections, 50–80 km from the platform.

Fig. 7. Lower–Middle Cambrian strata at Nordenskiöld Fjord, central North Greenland; the outlined area on the sketch is shown on the accompanying photograph. This section illustrates the proximal portion of the off-platform succession, *c.* 5 km north of the coeval platform edge. Note: (a) the prominent megabreccia sheet capping the Aftensternesø Formation (A (photograph); sequence 2) and containing large pale blocks of platform margin carbonate, (b) two syndepositional normal faults (A, B (sketch)) that were active in the latest Early to early Middle Cambrian (during deposition of sequence 3, LST) and (c) the well-developed progradational carbonate wedge shown in orange (sequence 3, LPW), displaying northward-dipping clinoforms and thinning rapidly to the north (basinward). The foreslope carbonates capping the section (sequence 3, HST) grade northward into an extensive fringe of slope carbonates represented by the Sydpasset Formation (Fig. 5). The northeastern portion of this transect is shown in Fig. 12. H, Henson Gletscher Formation; B, Bistrup Land Formation. Note that the relationship between the clinoformed foreslope strata and the platform topsets is not observed at this location. Modified from Ineson & Surlyk (1995).

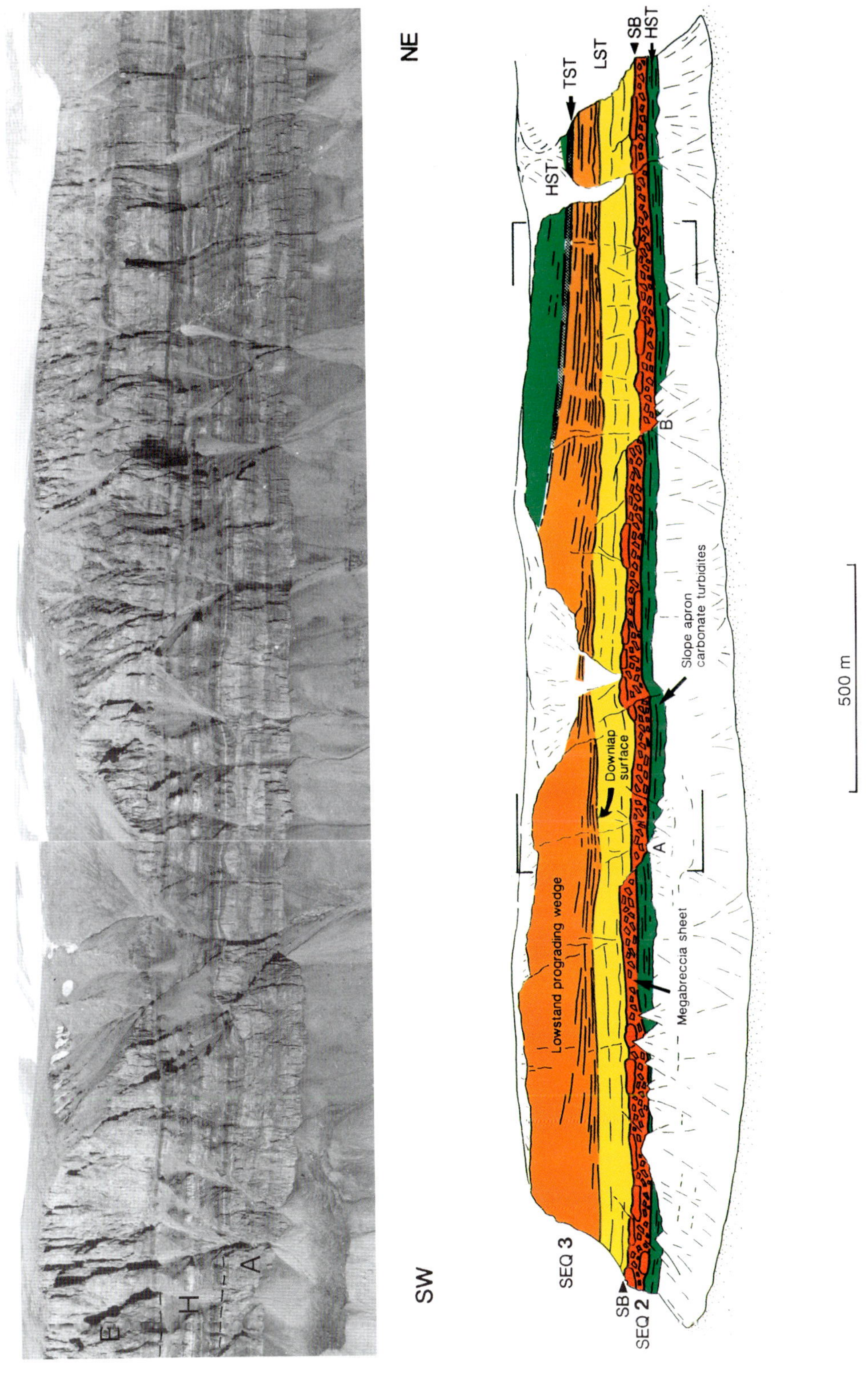
NE
SW
E
H
A
HST
TST
LST
SB
HST
B
A
Slope apron carbonate turbidites
Downlap surface
Lowstand prograding wedge
Megabreccia sheet
SEQ 3
SB
SEQ 2
500 m

Sequence stratigraphy. On the basis of the evidence of low sedimentation rates, the influx of siliciclastics and the lap-out relationship with the shallow-water platform, the mixed carbonate–siliciclastic half-cycles are interpreted broadly to represent lowstand conditions. Indeed, the evidence of periodic impingement of wavebase on the proximal portion of the off-platform region confirms the relative low sea-level stand; such evidence is not observed in the carbonate half-cycles in equivalent palaeogeographic positions (see below).

The proportion of the siliciclastic component (both mud and sand) of these half-cycles increases to the south towards the coeval platform, although ultimately wedging out at the platform margin. This, together with unpublished palaeocurrent data (Ineson 1985), indicates derivation from the south. The scarcity of siliciclastics within the shallow-water platform succession and recent sequence stratigraphic analysis of this succession (unpublished field data) suggests bypass of the platform during times of platform emergence.

The prograding carbonate body in the upper Henson Gletscher Formation wedges out basinward within the mixed carbonate–siliciclastic half-cycle and is interpreted to represent a lowstand prograding wedge (LPW). Where the LPW is recognizable, the mixed carbonate–siliciclastic half-cycle is subdivided into the lowstand systems tract (LST) and the transgressive systems tract (TST; Figs 5 and 7). In such sections, the TST is typically dominated by dark grey to black organic-rich (TOC up to 4%) argillaceous lime mudstones (or dolomites) that are finely laminated and commonly rich in agnostoid trilobites. Where the LPW and thus the TST are not recognized, the off-platform mixed carbonate–siliciclastic half-cycles are assigned broadly to the lowstand systems tract while recognizing that the upper levels may correlate with strata of transgressive character on the coeval platform.

Carbonate half-cycles

Facies, processes and environment. The boundary with the underlying mixed carbonate–siliciclastic half-cycle forms a readily mappable horizon but is gradational in detail (Figs 4 and 5). The carbonate half-cycles are composed of two elements: a lower succession of thin-medium bedded carbonates capped by a thick and laterally persistent carbonate megabreccia bed (Figs 4, 5 and 7). The latter component is described in a subsequent section.

Three main facies make up the carbonate half-cycles. Most characteristic are lime mudstones showing very thin platy nodular bedding (5–20 mm thick), superimposed on a parallel-laminated fabric. This facies shows abundant evidence of early differential cementation and downslope creep (Fig. 8a). Pull-aparts, boudins, interstratal breccia lenses, creep folds and discrete slides are common features (see Ineson & Surlyk 1995, figs 9.5 and 9.6) and result in large-scale hummocky or chaotic stratal patterns that are comparable to those observed on modern carbonate slopes (e.g. Mullins & Neumann 1979). Peloidal, intraclastic and ooidal grainstones occur interbedded with the platy nodular facies and dominate certain formations (e.g. Aftenstjernesø Formation, see Fig. 10a). Typically 50–100 mm thick, such beds are parallel-sided, commonly normally graded and may show the Bouma sequence of sedimentary structures (Fig. 8b). Carbonate breccia beds up to several metres thick are also well-represented in the carbonate half-cycles. They are typically

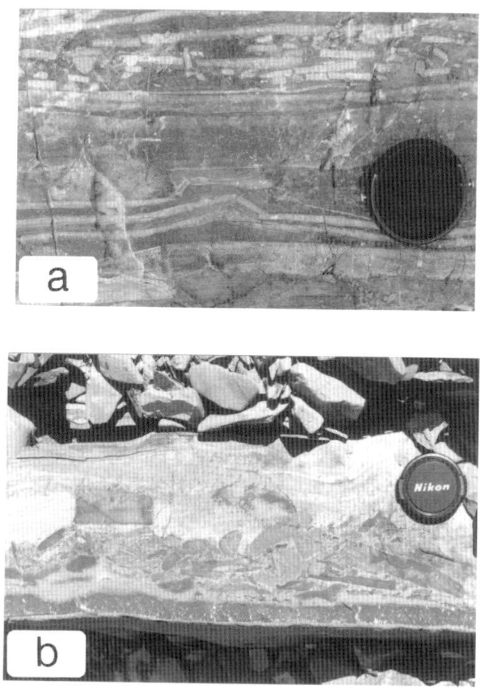

Fig. 8. (**a**) Platy nodular lime mudstones showing buckling (lower centre) and interstratal brecciation (top) – evidence of downslope creep within differentially cemented sediment; Sydpasset Formation, J. P. Koch Fjord. (**b**) Coarse-grained carbonate turbidite showing a pebbly base grading abruptly into cross-laminated grainstone; Kap Stanton Formation (Tavsens Iskappe Group), north Nyeboe Land.

clast-supported and composed of platy, slope-derived lime mudstone clasts up to several hundreds of millimetres across in a lime mudstone matrix. Megabreccia beds are rare, being restricted to a single occurrence (see below). Passing south towards the coeval platform, the platy nodular facies is replaced by wavy or irregular thin-bedded bioturbated nodular wackestone and mudstone; this facies is locally slumped or dissected by slump scars and interdigitates with the toes of clinoform-bedded platform foreslope strata.

The off-platform carbonate half-cycles represent extensive carbonate slope aprons deposited in a low energy, often poorly-oxygenated outer shelf setting (Ineson & Surlyk 1995). Even in proximal settings, within a few kilometres of the coeval platform, these half-cycles show no evidence of shallow-marine processes, in contrast to the mixed carbonate–siliciclastic half-cycles. Rather, they record deposition from suspension, from turbidity currents and a range of mass-flow processes.

Lateral relationships. The carbonate half-cycles thin basinward (i.e. northward), tapering from 50–100 m at the toe of the foreslope (Fig. 4) and wedging out 50–70 km farther north, near the Cambrian shelf edge (Fig. 2). North of the pinch-out of the carbonate half-cycles, the outermost shelf section is a thin, condensed succession of cherty black mudstones in which the cyclicity described here is no longer recognizable (Higgins *et al.* 1991). Along depositional strike, the carbonate half-cycles and their intervening mixed carbonate–siliciclastic half-cycles can be recognized for up to 450 km (Ineson & Surlyk 1995). However, in contrast to the carbonate–siliciclastic half-cycles, the carbonate half-cycles pass southward directly into the toes of the platform foreslope. For example, the carbonate half-cycle represented by the Sydpasset Formation (Fig. 5) can be traced directly into the prograding foreslope strata exposed at Nordenskiöld Fjord (Fig. 7). The toplap relationships are not observed in this case but other off-platform carbonate half-cycles correlate with platform margin carbonates that typically show sigmoidal clinoforms with preserved topsets.

Sequence stratigraphy. The carbonate half-cycles record times of significant export of carbonate sediment, largely lime mud, from the carbonate platform to the deeper-water outer shelf. The resultant carbonate slope aprons flanked the length of the carbonate platform and extended up to 70 km out onto the outer shelf. As noted by Ineson & Surlyk (1995), the wedge-out of the slope apron deposits represents the effective basinward limit of dispersal of platform-derived carbonate by means of debris flows, turbidity currents and in suspension.

The development of such extensive carbonate aprons, in association with evidence of coeval progradation and aggradation of the platform and the scarcity of siliciclastic sediment, indicates deposition during relative highstands of sea-level when the shallow-water platform was submerged and productive (see Schlager 1991). The carbonate half-cycles are thus referred to the highstand systems tract (HST).

Sequence stratigraphic framework

The off-platform cyclicity described above is an illustrative example of the principle of reciprocal sedimentation in a mixed carbonate–siliciclastic system (Meissner 1972). On the basis of the facies analysis, together with the large-scale relationships between the platform and off-platform strata, the cyclicity can be shown to follow the accepted model of reciprocal sedimentation proposed by Meissner (1972) and adapted to sequence stratigraphy by Sarg (1988, see also Schlager 1991; Handford & Loucks 1993; Brown & Loucks 1993; Southgate *et al.* 1993; Sonnenfeld & Cross 1993). The mixed carbonate–siliciclastic half-cycles thus record lowstands (and transgressive periods in many cases) whereas the carbonate half-cycles record highstands of sea-level. With the exception of sequence 2, in which siliciclastics are absent, the sequence stratigraphic framework presented schematically in Fig. 9 is based on the reciprocal sedimentation pattern, stacking patterns, facies analysis and stratal relationships. This is a robust framework that, within the off-platform section, is constrained by detailed trilobite biostratigraphy (Robison 1984, 1988, 1994; Blaker 1986, 1991; Babcock 1990, 1994; Blaker & Peel 1997). Correlation from the off-platform succession to the platform interior is difficult in detail due to limited biostratigraphic data within the platform succession and to problems of physical correlation of key surfaces through massive, poorly stratified platform margin facies. The sequence stratigraphy of the platform interior is presently under study and further discussion is premature; in general, sequence boundaries are defined on the basis of truncation, karstification or abrupt, widespread changes in the style of platform sedimentation. Biostratigraphic confirmation of inferred hiatal surfaces is not possible.

The six depositional sequences show an offlapping stacking pattern, sequences stepping progressively basinward with time (Fig. 9).

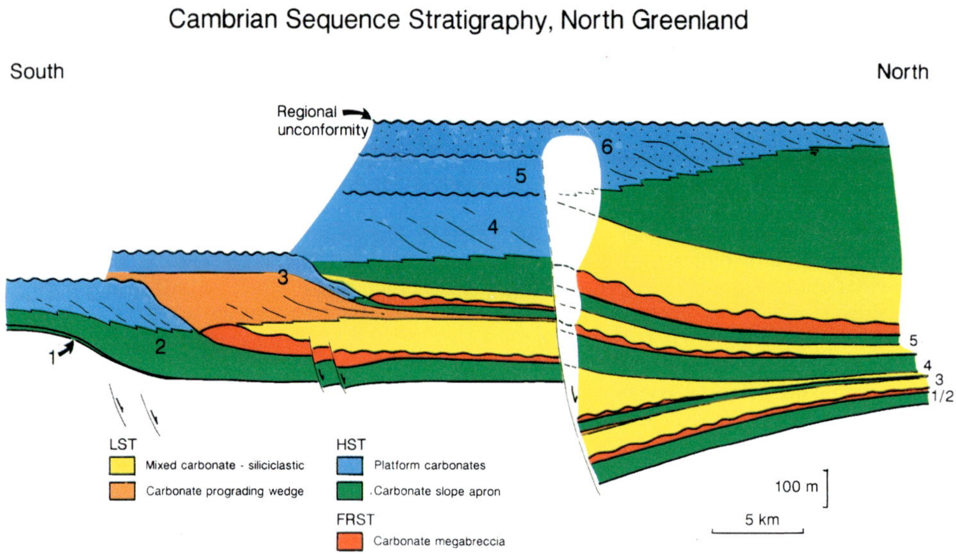

Fig. 9. Schematic sequence stratigraphic framework; see text for discussion. Sequence boundaries within platform carbonates (wavy lines) are indicated by truncation, karstification or abrupt, widespread changes in the style of platform sedimentation. Note that due to the recognition of an additional sequence (sequence 1), sequences 1 and 2 of Ineson & Surlyk (1995) are re-numbered sequences 2 and 3, respectively.

Sequence geometry was, however, influenced by observed and inferred syndepositional down-to-basin normal faults that became progressively active from south to north with time. The southernmost fault (or fault zone) is inferred to have been active in late Early Cambrian times and influenced the geometry of the platform margin and proximal slope apron in sequence 2. The minor faults 5–10 km farther north were active in the latest Early–early Mid-Cambrian (see Fig. 7) whereas the northernmost structure, which held up progradation of sequences 4 and 5, was probably active from the medial Mid-Cambrian to the early Late Cambrian.

Megabreccias: characteristics and depositional processes

Following Cook *et al.* (1972), megabreccia sheets are understood here as laterally persistent mass-flow deposits that contain conspicuous angular clasts over 1 m across. Such deposits are a striking feature of the Cambrian off-platform deposits in North Greenland (Figs 7 & 10a). They range in thickness from 5 m to 50 m and are sheet-like in form, although commonly showing highly irregular hummocky upper surfaces where large rafted blocks protrude up to 20 m above the top of the deposit (Fig. 7). Bed bases are typically flat and non-erosional. Internally, the beds comprise clast-supported breccia,

Fig. 10. Carbonate megabreccias: (**a**) Pale megabreccia bed (base and top arrowed, 20 m thick) capping HST carbonate turbidites of sequence 2 (Aftenstjernesø Formation, see Fig. 5). Note that the irregular top of this megabreccia bed is abruptly overlain by dark argillaceous carbonates and sandstones of the succeeding sequence (Henson Gletscher Formation; sequence 3) and that the megabreccia bed is underlain by deformed but essentially *in situ* HST carbonates. J. P. Koch Fjord. (**b**) Upper levels of a 10 m thick megabreccia sheet showing normal coarse-tail grading in the uppermost few metres and a light-coloured carbonate turbidite cap (0.9 m thick). Aftenstjernesø Formation, sequence 1, J. P. Koch Fjord. (**c**) Clast-supported breccia; note the irregular, wavy and nodular outlines of individual clasts (e.g. above scale with 10 mm divisions). Such fabrics resulted from sliding and disaggregation of differentially cemented nodular slope carbonates. Aftenstjernesø Formation, sequence 1, southeast Peary Land. (**d**) Intra-HST megabreccia sheet dominated by slope-derived breccia (dark) but also including blocks of light-coloured ooid grainstone derived from the platform margin. A few kilometres away, such clasts attain house-size in this megabreccia bed. Fimbuldal Formation, sequence 4, J. P. Koch Fjord.

dominated by angular platy clasts of coarse pebble to cobble size with an interstitial mud-grade carbonate matrix. In proximal sections (within *c.* 10 km of the coeval platform margin), large rafts (up to 100 × 30 m in cross-section) of bedded slope carbonate are prominent, together with equidimensional or rectangular blocks of pale platform margin carbonate up to 100 m across (Fig. 7). The former show all stages of disaggregation of differentially cemented, thinly stratified fine-grained carbonate and clearly represent the source of both the platy clasts and the fine-grained breccia matrix (Fig. 10c). Close to the contemporaneous platform, the megabreccias are chaotic and non-graded with a lack of recognizable clast organization. Traced basinward, the megabreccia sheets typically become more parallel-sided, contain fewer 'megaclasts' and may show weak normal, coarse-tail grading in their upper levels (Fig. 10b). Tabular clasts are preferentially oriented parallel to bedding and discrete grainstone turbidite beds, up to a metre thick, commonly cap the deposit (Fig. 10b).

The megabreccia beds record catastrophic failure, sliding and mass flow of platform margin and/or slope strata. The ubiquitous fine-grained matrix and clast-supported framework suggest that the flow process was intermediate between cohesive debris flow and density modified grain flow (terminology of Lowe 1979), clast interactions being an important supporting mechanism (Ineson 1980, 1985). The coarse-tail graded upper portion probably resulted from a loss of competence due to shear above the rigid plug (Surlyk 1978; Naylor 1980; Nemec & Steel 1984) and/or matrix dilution and the onset of weak turbulence due to the incorporation of seawater. The grainstone caps represent deposition from turbulent flows developed at the mass flow-seawater interface (see Hampton 1972; Krause & Oldershaw 1979).

Megabreccias and sequence stratigraphy

Given the sequence stratigraphic framework outlined above, based on a well-developed reciprocal sedimentation pattern, stacking patterns, facies analysis and stratal relationships, the megabreccias can be described further in terms of their sequence stratigraphic position: lowstand megabreccias, intra-highstand megabreccias and highstand-capping megabreccias (Fig. 11).

LST megabreccias

Mass-flow deposits within the LST are typically thin (1–2 m) and composed solely of tabular, pebble-cobble sized, lime mud-rich clasts derived from the off-platform setting. The beds are commonly laterally impersistent on the scale of tens of metres and may grade laterally into deformed/slumped but essentially *in situ* strata. Megabreccia sheets are rare. In sequence 3, megabreccia beds are prominent at the toes of clinoforms within the lowstand prograding wedge (Figs 7 and 12); they are up to 30 m thick and include rafts up to 20 × 50 m in cross-section. They typically wedge out within a few kilometres of the clinoform toes and are composed almost exclusively of slabs of bedded skeletal, intraclastic packstone and grainstone, facies that are characteristic of the shallow-water portion of the LPW.

A more persistent megabreccia sheet occurs enveloped within argillaceous LST carbonates of sequence 4 in southern Peary Land where it is recognized lithostratigraphically as the Lønelv Formation (Fig. 3; Ineson & Peel 1997). This bed is 15–30 m thick and is composed almost exclusively of angular blocks of cream cross-bedded ooid grainstone (typical highstand margin facies). Clasts range in size up to 30 m across and the sheet can be mapped over an area of 5 × 13 km; the minimum volume of this deposit is estimated to be 1.3 km^3.

Intra-HST megabreccias

As noted earlier, mass-flow breccias are an important component of the highstand systems tract, contributing to the extensive slope aprons shed over the deep shelf during high sea-level stand (Ineson & Surlyk 1995). Typically, these beds are 1–5 m thick, sheet-like in form and composed of platy, slope-derived, lime mudstone clasts of pebble to cobble size. A single intra-HST megabreccia sheet has been recognized, occurring within the HST of sequence 4. It is 5–15 m thick and is dominated by penecontemporaneous highstand slope debris, commonly as rafted slabs several tens of metres across; blocks of platform margin grainstone up to 30 × 75 m in cross-section are prominent in this bed (Fig. 10d). Lack of continuous exposure at this level, especially basinward, precludes an accurate estimate of the lateral extent and volume of this deposit; the sheet is observed over an area of about 5 × 5 km, giving a minimum volume of 0.25 km^3. Regional correlation suggests that the along-strike extent of this sheet does not exceed 20 km.

HST-capping megabreccias

Spectacular megabreccias occur systematically atop the carbonate-dominated slope apron deposits of the HST (Figs 7, 10a and 12), where they are directly overlain by the mixed carbonate–siliciclastic facies of the succeeding LST. Such megabreccia sheets may comprise a single bed or several amalgamated beds, totalling up to 50 m in thickness. They contain equidimensional or rectangular, pale-coloured blocks of platform margin grainstone up to 100 m across and tabular slabs of slope carbonate of similar magnitude. The former commonly protrude high above the megabreccia top and are draped by lowstand deposits (Figs 7 and 12), often ostensibly resembling carbonate buildups (see Cook *et al.* 1972). The breccias themselves are clast-supported and are composed of cobble to coarse pebble-sized, tabular clasts; angular and irregular, nodular outlines are typical (Fig. 10c). The matrix is of mud-grade carbonate, although largely dolomitized; siliciclastic sand is not observed. All stages of disaggregation of slope carbonates are preserved in the debris sheets, from coherent bedded slabs to chaotic clast-supported breccia, indicating that both breccia clasts and matrix were derived by mass wastage of differentially lithified highstand slope carbonates.

In all these respects, these megabreccia sheets are identical to the example described above from within the carbonate-dominated HST of sequence 4. They differ significantly only in their lateral persistence. The megabreccia sheets capping sequences 2 and 3 are well-exposed and can be confidently traced across much of western and central North Greenland. The first of these, of late Early Cambrian age, stretches at least 50 km north of the coeval platform edge and extends some 400 km parallel to the platform (Fig. 13). In proximal sections, within 5 km of the coeval margin, the sheet is 30–40 m thick, thinning to 10–15 m some 20–25 km distant from the platform margin and to about 5 m at a distance of 50 km. Adopting conservative values ($400 \times 50 \times 0.01$ km), the megabreccia sheet has a depositional volume of 200 km^3. The megabreccia sheet capping sequence 3 (medial Middle Cambrian) is up to 25 m thick proximally, extends at least 100 km along depositional strike and wedges out some 20 km north of the coeval platform margin. Assuming an average thickness of 5 m, the megabreccia sheet has a minimum volume of 10 km^3. The extent of megabreccia sheets capping subsequent highstand systems tracts is uncertain due to impersistent exposure at these higher stratigraphic levels.

The two examples provided above, however, demonstrate the regional extent of the megabreccia sheets that occur sandwiched between HST slope apron carbonates and the mixed carbonate–siliciclastic sediments of the overlying LST. In the case of the Lower Cambrian example, the volume of debris involved is two orders of magnitude greater than the estimated volumes of individual megabreccia sheets occurring within the highstand and lowstand systems tracts. The HST-capping megabreccia sheets are commonly composite and vary laterally along depositional strike, both in terms of thickness and composition (relative proportion of slope- versus platform derived clasts). It is not envisaged, therefore, that these deposits represent single flow events but rather record a series of genetically related events along the length of the platform margin and slope, depositing a laterally composite sheet constructed of overlapping debris lobes. Individual flows were probably comparable in magnitude to the individual intra-HST and LST megabreccia sheets described above. It is notable that the areal dimensions of one such flow unit are comparable to those of individual megabreccia beds mapped seismically off modern platforms on the Nicaraguan Rise (Hine *et al.* 1992).

Deposition of such extensive composite megabreccia sheets requires a regional mechanism affecting the length of the carbonate platform margin (at least 400 km demonstrable in the Lower Cambrian). The sequence stratigraphy (see Figs 5, 7, 9 and 12) demonstrates that each of these events coincided with a regional relative fall in sea level. It is logical, therefore, to suggest that the widespread failure of the platform margin and upper slope was related, directly or indirectly, to relative sea-level fall. Failure and mass flow of the highstand platform margin and slope carbonates may have been favoured both by the steep foreslopes (up to 30°) generated during highstand progradation and by the inherent instability of cemented platform margin carbonates overlying differentially cemented, well-stratified slope facies. Sliding may also have been promoted by excess pore fluid pressures in the uncemented layers of the differentially-cemented slope sediments, both due to compaction and to pore overpressure following relative sea-level fall and consequent decrease in the ambient hydrostatic load pressure (Hilbrecht 1989; Spence & Tucker 1997).

It must be emphasized that the influence of relative sea-level fall was primarily to render the highstand carbonate edifice prone to failure, as outlined above. It is not certain, however, that the changing sea-level stand was directly responsible for the extensive failure events along the margin and slope. Although cyclical storm-wave loading may have triggered mass failure, other agents are equally likely. Periodic tectonic instability of the shelf is suggested by the presence of syndepositional faults and the potential role of mild earthquakes in triggering failure of the platform margin and slope should not be overlooked. The question of the triggering mechanism is, however, overshadowed in importance in the context of this paper by the clear correlation between times of falling relative sea level and major mass failure events. The relationship may be indirect but is significant nonetheless.

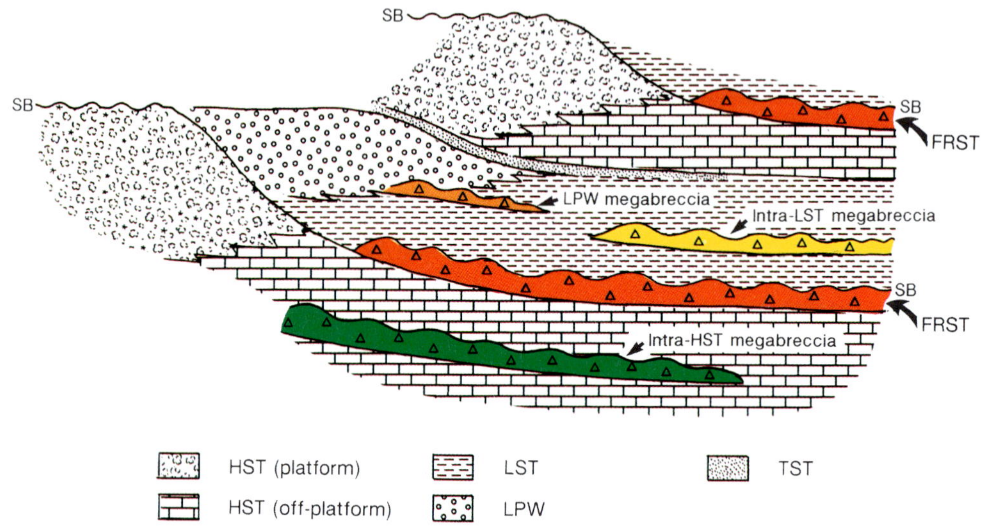

Fig. 11. Conceptual diagram showing the location of megabreccia sheets within the sequence stratigraphic framework, based on the Cambrian of North Greenland.

Discussion

Carbonate megabreccias are becoming increasingly utilized in the definition of systems tracts and depositional sequences (e.g. Pujalte *et al.* 1993; Garcia-Mondéjar & Fernández-Mendiola 1993; Strohmenger & Strasser 1993). Thus, consideration of the distribution and character of such deposits within independently defined sequences is clearly timely. In the Cambrian of North Greenland, the well-developed reciprocal sedimentation pattern and spectacular exposure of platform margin to deep shelf strata permit erection of a robust sequence stratigraphic framework that allows us to address the questions posed earlier.

(1) Can certain megabreccias be related to sea-level fall or lowstand and (2) if so, which systems tract do they characterize?

In this case study, megabreccia sheets occur systematically at the boundary between the HST and the succeeding LST, indicating a relationship, direct or indirect (see above), between relative fall in sea level and extensive sediment failure of the platform margin and upper slope. These mass-flow deposits are composed solely of platform margin and slope carbonate derived from the underlying HST. Mixed carbonate–siliciclastic facies of the succeeding LST immediately overlie the megabreccia sheets and drape the hummocky upper surface (Figs 7, 10a, 12), yet are not present beneath, nor as clasts within, the megabreccias. The affinities of the megabreccia sheets are thus with the underlying sequence. Indeed, in proximal sections where the upper HST slope carbonates are extensively disrupted by slope creep and slumping, the boundary between essentially *in situ* strata and the overlying megabreccia sheet can be difficult to locate (see Fig. 10a). The top of the megabreccia sheet, however, is a readily identifiable and correlatable surface that defines a marked shift in sedimentation style from the actively prograding carbonate system of the highstand to the carbonate-starved, siliciclastic-influenced lowstand system. Furthermore, on a practical level, the upper surface of the megabreccia sheet marks a significant lithological boundary (from carbonate to mixed carbonate–siliciclastic) that

Fig. 12. View of the northeastern end of the cliff-section illustrated in Fig. 7, Nordenskiöld Fjord. The thick megabreccia sheet (red) capping the lower sequence (sequence 2, see Fig. 7), contains both pale-coloured platform margin blocks (P) and extensive rafts of bedded slope carbonate. Note the hummocky, discontinuous megabreccia beds at the downlapping foreslope toes of the LPW. Modified from Ineson & Surlyk (1995).

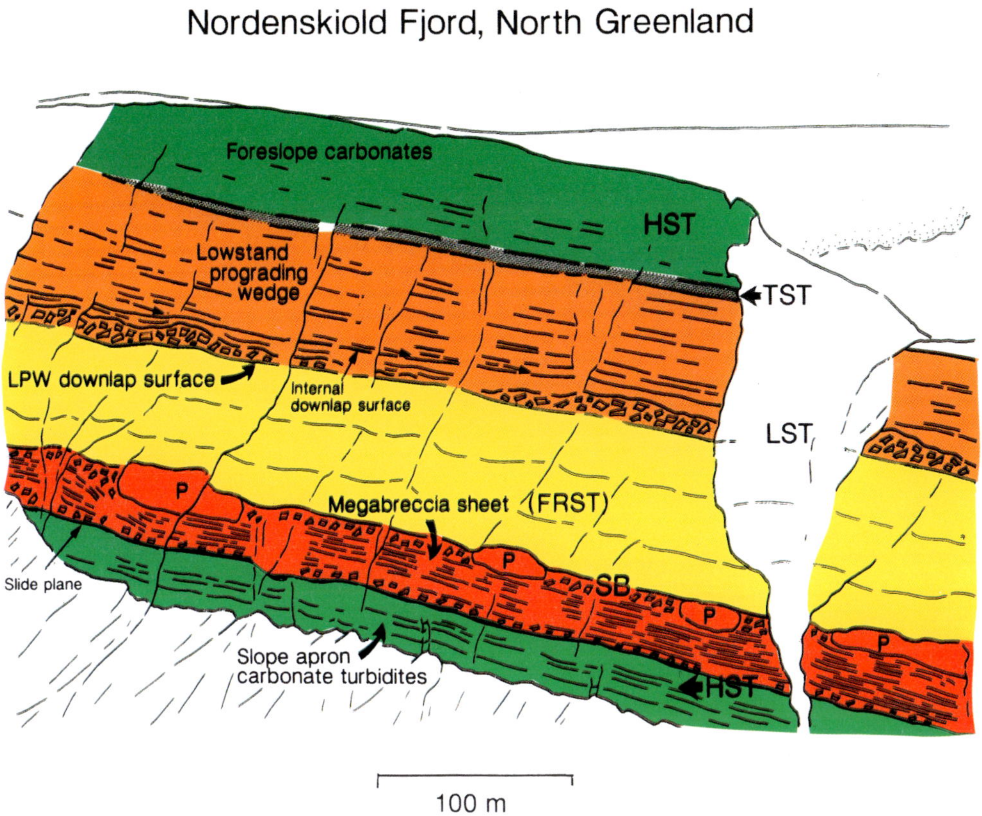
Nordenskiold Fjord, North Greenland
Foreslope carbonates
HST
Lowstand prograding wedge
TST
LPW downlap surface
Internal downlap surface
LST
P
Megabreccia sheet (FRST)
Slide plane
SB
Slope apron carbonate turbidites
HST
100 m

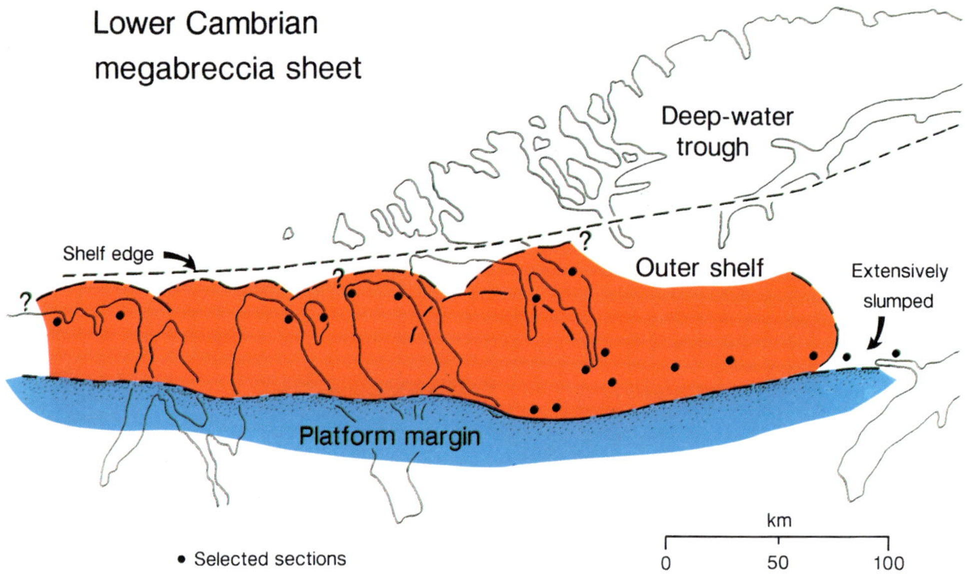

Fig. 13. Distribution of the megabreccia (FRST) capping sequence 2 (see Figs 4, 7, 10a, 12), based on measured sections linked by observations along extensive fjord and glacier cliff sections.

is likely to be recorded on reflection seismic data and thus will represent an important seismic stratigraphic surface.

This surface is thus considered to define the sequence boundary in the off-platform setting; it probably records the maximum fall in relative sea-level (Hunt & Tucker 1992, 1993). The megabreccia sheets that occur sandwiched between HST slope carbonates and mixed carbonate–siliciclastic LST deposits are interpreted to have been shed during falling relative sea-level and are referred to the FRST of Hunt & Tucker (1995) and Helland-Hansen & Gjelgerg (1994), also termed the falling stage systems tract (FSST) by Nummedal (1992) and Plint & Nummedal (this volume), or the offlapping systems tract (OST) by Pomar & Ward (1994).

In the Cambrian of North Greenland, then, there is a clear relationship (whether direct or indirect) between sea-level fall and widespread sediment failure yet caution should be exercised in applying these results to other ancient carbonate platforms. The North Greenland succession represents one end-member amongst a range of platform types, i.e. a progradational rimmed shelf with steep foreslopes grading basinward into differentially cemented, well-stratified slope facies. Widespread failure of the platform margin and upper slope may represent the typical response of such a platform but may not be applicable to other platform types (see discussion by Hunt & Tucker 1993).

(3) On what basis can megabreccias shed during sea-level fall be identified?

In this study, the most prominent and widespread megabreccia sheets occur systematically between the HST slope apron carbonates and the overlying mixed carbonate–siliciclastic deposits of the LST; they can thus be related, directly or indirectly, to the relative fall of sea-level. However, megabreccia beds have also been recognized, albeit rarely, within both the LST and the HST. This illustrates the point stressed by Hine *et al.* (1992) and Grammer *et al.* (1993) that, in the absence of additional criteria to identify systems tracts in off-platform settings, the presence of megabreccia sheets *per se* is of little value.

In the Cambrian of North Greenland, the few megabreccias recognized within the lowstand systems tract are compositionally distinctive. The laterally impersistent megabreccias at the toe of the foreslopes of the lowstand prograding wedge (LPW) are composed solely of skeletal and intraclastic grainstone clasts derived from the coeval shallow-water portion of the LPW. Angular blocks of cross-bedded grainstone also form the bulk of the clasts in the intra-LST megabreccia in sequence 4; the platy nodular slope-derived clasts that typify the HST and FRST megabreccias are not conspicuous. On the basis of these few examples, then, it appears that the lowstand megabreccias in this succession can be differentiated compositionally from those of the highstand and forced regressive systems tracts.

In contrast, the megabreccias of the FRST are indistinguishable in nearly all respects from the example observed within the HST of sequence 4; they differ significantly only in terms of scale. In order to document the lateral extent of these megabreccia sheets, however, one is reliant on the broad lithostratigraphic and sequence stratigraphic framework. In the absence of such an independent framework, areal extent (i.e. regional significance) is unlikely to be a conclusive criterion.

A number of other criteria have been proposed in the literature to aid in the differentiation of lowstand and highstand megabreccias (see Yose & Hardie 1990). Sarg (1988) suggested that highstand mass-flow deposits can be traced back up foreslope clinoforms and thus can be distinguished from onlapping lowstand deposits. As noted by Schlager & Camber (1986), however, such geometric criteria are highly dependent on the style of platform margin development (i.e. depositional versus bypass/erosional; McIlreath & James 1978) and are thus equivocal. Furthermore, Brown & Loucks (1993) and Melim & Scholle (1995) have demonstrated the role of sediment fabric (grain-size variation) in dictating foreslope processes and thus geometric relationships with basinal strata.

An additional line of evidence is the matrix composition of the slide mass or megabreccia sheets. As demonstrated by Haak & Schlager (1989; see also Reijmer *et al.* 1991), the composition of sediment dispersed into deeper-water can be related to the sea-level stand: ooids and peloids dominate during highstands whereas lowstand off-platform carbonates are typically rich in skeletal detritus. Similarly, siliciclastic sediment, if available, typically bypasses the platform during sea-level lowstands to be shed into deeper-water. Thus, the matrix composition may aid broad differentiation between highstand and lowstand megabreccia sheets. This criterion was applied by George *et al.* (1995) to demonstrate that megabreccia sheets of assumed lowstand origin were in fact shed under highstand conditions. As noted above, however, the extensive megabreccia sheets assigned to the FRST in the Cambrian of North Greenland consist solely of highstand margin and slope debris and thus cannot be differentiated from intra-highstand megabreccia sheets on this basis.

Diagenetic evidence of exposure in megabreccia clasts (e.g. karstic dissolution features, meteoric cements) is suggestive of lowstand derivation but is also equivocal since a number of platform sequences may be sampled during extensive failure of the margin, a problem noted by Brown & Loucks (1993). In the North Greenland succession, extensive dolomitization of proximal off-platform strata, particularly debris beds, precludes recognition of primary cement fabrics. Karstic dissolution features have not been noted in platform-derived clasts.

Conclusions

The excellent fjord-wall exposures and well-developed reciprocal carbonate–siliciclastic sedimentation pattern exhibited by the Cambrian off-platform succession in North Greenland permit erection of a robust sequence stratigraphic framework. Megabreccias occur only rarely within lowstand and highstand systems tracts but are present systematically atop the carbonate-dominated highstand systems tract (HST). Composed solely of highstand debris derived from both the platform margin and the upper slope, these megabreccia sheets are assigned to the forced regressive systems tract (FRST). They record extensive failure of the platform margin and upper slope during relative fall of sea-level and prior to the onset of lowstand deposition.

It is unclear to what extent such behaviour can be extrapolated to carbonate platforms in general. It is likely that certain features of this Cambrian platform, i.e. steep progradational foreslopes and ubiquitous diffential cementation of slope fines, rendered it prone to failure during a fall in sea-level. Given such a propensity to failure, the potential role of minor intrabasinal tectonics in triggering catastrophic mass-flow events should not be underestimated.

In this Cambrian succession, extensive dolomitization of proximal off-platform carbonates precludes the use of detailed fabric evidence to develop further criteria to distinguish megabreccias deposited during falling sea-level or low sea-level stand from those shed during highstand. Although the lowstand megabreccia sheets display distinctive clast compositions, the FRST and HST megabreccias are intrinsically similar, differing only in terms of areal extent.

In the absence of an independent sequence stratigraphy, therefore, the value of megabreccia sheets in the identification of systems tracts is limited. Criteria that have been suggested to aid differentiation between megabreccias shed during highstands and lowstands include large-scale geometric relationships, clast and matrix composition and clast diagenetic history. Individually, these criteria are ambiguous but in association may contribute to a sequence stratigraphic interpretation.

The authors thank J. Lautrup for photographic work and M. E. Tucker, P. E. Playford and D. Hunt for their constructive reviews. The paper was completed under a project entitled 'Resources of the sedimentary basins of North and East Greenland', supported financially by the Danish Natural Science Research Council. The paper is published with the permission of the Geological Survey of Denmark and Greenland.

References

BABCOCK, L. E. 1990. *Biogeography, phylogenetics, and systematics of some Middle Cambrian trilobites from open-shelf to basinal lithofacies of North Greenland and Nevada.* PhD Thesis, University of Kansas.

—— 1994. Systematics and phylogenetics of polymeroid trilobites from the Henson Gletscher and Kap Stanton formations (Middle Cambrian), North Greenland. *Grønlands Geologiske Undersøgelse Bulletin*, **168**, 79–127.

BLAKER, M. R. 1986. Notes on the trilobite faunas of the Henson Gletscher Formation (Lower and Middle Cambrian) of central North Greenland. *Grønlands Geologiske Undersøgelse Rapport*, **132**, 65–73.

—— 1991. *Early Cambrian trilobites from North Greenland.* PhD Thesis, University of Keele.

—— & PEEL, J. S. 1997. *Lower Cambrian trilobites from North Greenland.* Meddelelser om Grønland Geoscience, **35**.

BROWN, A. A. & LOUCKS, R. G. 1993. Influence of sediment type and depositional processes on stratal patterns in the Permian basin-margin Lamar Limestone, McKittrick Canyon, Texas. *In*: LOUCKS, R. G. & SARG, J. F. (eds) *Carbonate Sequence Stratigraphy: Recent Developments.* American Association of Petroleum Geologists, Special Publications, **57**, 133–156.

BRYANT, I. D. & SMITH, M. P. 1990. A composite tectonic-eustatic origin for shelf sandstones at the Cambrian-Ordovician boundary in North Greenland. *Journal of the Geological Society, London*, **147**, 795–801.

CHRISTIANSEN, F. G., NØHR-HANSEN, H. & NYKJÆR, O. 1987. The Cambrian Henson Gletscher Formation: a mature to postmature hydrocarbon source rock sequence from North Greenland. *Grønlands Geologiske Undersøgelse Rapport*, **133**, 141–157.

COOK, H. E, MCDANIELS, P. M., MOUNTJOY, E. W. & PRAY, L. C. 1972. Allochthonous carbonate debris flows at Devonian 'bank' margins, Alberta, Canada. *Bulletin of Canadian Petroleum Geology*, **20**, 439–497.

DABRIO, C. J., ESTEBAN, M. & MARTIN, J. M. 1981. The coral reef of Níjar, Messinian (Uppermost Miocene), Almeria Province, S. E. Spain. *Journal of Sedimentary Petrology*, **51**, 521–539.

FITZSIMMONS, R. & JOHNSON, S. 2000. Forced regressions: recognition, architecture and genesis in the Campanian of the Bighorn Basin, Wyoming. *This volume.*

FRANSEEN, E. K. & MANKIEWICZ, C. 1991. Depositional sequences and correlation of middle(?) to late Miocene carbonate complexes, Las Negras and Níjar areas, southeastern Spain. *Sedimentology*, **38**, 871–898.

GARCIA-MONDÉJAR, J. & FERNANDEZ-MENDIOLA, P. A. 1993. Sequence stratigraphy and systems tracts of a mixed carbonate and siliciclastic platform-basin setting: the Albian of Lunada and Soba, northern Spain. *American Association of Petroleum Geologists Bulletin*, **77**, 245–275.

GEORGE, A. D., PLAYFORD, P. E. & POWELL, C. MCA. 1995. Platform-margin collapse during Famennian reef evolution, Canning Basin, Western Australia. *Geology*, **23**, 691–694.

GRAMMER, G. M., GINSBURG, R. N. & HARRIS, P. M. 1993. Timing of deposition, diagenesis and failure of steep carbonate slopes in response to a high-amplitude/high-frequency fluctuation in sea level, Tongue of the Ocean, Bahamas. *In*: LOUCKS, R. G. & SARG, J. F. (eds) *Carbonate Sequence Stratigraphy: Recent Developments.* American Association of Petroleum Geologists, Special Publications, **57**, 107–131.

HAAK, A. B. & SCHLAGER, W. 1989. Compositional variations in calciturbidites due to sea-level fluctuations, late Quaternary, Bahamas. *Geologische Rundschau*, **78**, 477–486.

HAMPTON, M. A. 1972. The role of subaqueous debris flow in generating turbidity currents. *Journal of Sedimentary Petrology*, **42**, 775–793.

HANDFORD, C. R. & LOUCKS, R. G. 1993. Carbonate depositional sequences and systems tracts – responses of carbonate platforms to relative sea-level changes. *In*: LOUCKS, R. G. & SARG, J. F. (eds) *Carbonate Sequence Stratigraphy: Recent Developments.* American Association of Petroleum Geologists, Special Publications, **57**, 3–41.

HARRIS, M. T. 1994. The foreslope and toe-of-slope facies of the Middle Triassic Latemar buildup (Dolomites, northern Italy). *Journal of Sedimentary Research*, **B64**, 132–145.

HELLAND-HANSEN, W. & GJELBERG, J. G. 1994. Conceptual basis and variability in sequence stratigraphy: A different perspective. *Sedimentary Geology*, **92**, 31–52.

HIGGINS, A. K., INESON, J. R., PEEL, J. S., SURLYK, F. & SØNDERHOLM, M. 1991. Lower Palaeozoic Franklinian Basin of North Greenland. *Grønlands Geologiske Undersøgelse Bulletin*, **160**, 71–139.

HILBRECHT, H. 1989. Redeposition of Late Cretaceous pelagic sediments controlled by sea-level fluctuations. *Geology*, **17**, 1072–1075.

HINE, A. C., LOCKER, S. D., TEDESCO, L. P., MULLINS, H. T., HALLOCK, P., BELKNAP, D. F., GONZALES, J. L., NEUMANN, A. C. & SNYDER, S. W. 1992. Megabreccia shedding from modern, low-relief carbonate platforms, Nicaraguan Rise. *Bulletin of the Geological Society of America*, **104**, 928–943.

HUNT, D. & TUCKER, M. E. 1992. Stranded parasequences and the forced regressive wedge systems tract: deposition during base-level fall. *Sedimentary Geology*, **81**, 1–9.

—— & —— 1993. Sequence stratigraphy of carbonate shelves with an example from the mid-Cretaceous (Urgonian) of southeast France. *In*: POSAMENTIER, H. W., SUMMERHAYES, C. P., HAQ, B. U. & ALLEN, G. P. (eds) *Sequence stratigraphy and Facies Associations.* International Association of Sedimentologists, Special Publications, **18**, 307–341.

—— & —— 1995. Stranded parasequences and the forced regressive wedge systems tract: deposition during base-level fall – reply. *Sedimentary Geology*, **95**, 147–160.

HURST, J. M. & SURLYK, F. 1983. Initiation, evolution and destruction of an early Paleozoic carbonate shelf, eastern North Greenland. *Journal of Geology*, **91**, 671–691.

INESON, J. R. 1980. Carbonate debris flows in the Cambrian of south-west Peary Land, eastern North Greenland. *Grønlands Geologiske Undersøgelse Rapport*, **99**, 43–49.

—— 1985. *The stratigraphy and sedimentology of the Brønlund Fjord and Tavsens Iskappe Groups (Cambrian) of Peary Land, eastern North Greenland.* PhD Thesis, University of Keele.

—— & PEEL, J. S. 1997. Cambrian shelf stratigraphy of North Greenland. *Geology of Greenland Survey Bulletin*, **173**, 1–120.

—— & SURLYK, F. 1995. Carbonate slope aprons in the Cambrian of North Greenland: geometry, stratal patterns and facies. *In*: PICKERING, K. T., HISCOTT, R. N., KENYON, N. H., RICCI LUCCHI, F. & SMITH, R. D. A. (eds) *Atlas of Deep Water Environments*. Chapman & Hall, London, 56–62.

——, ——, HIGGINS, A. K. & PEEL, J. S. 1994. Slope apron and deep shelf sediments of the Brønlund Fjord and Tavsens Iskappe Groups (Lower Cambrian–Lower Ordovician) of North Greenland. *Grønlands Geologiske Undersøgelse Bulletin*, **169**, 7–25.

KOLLA, K., POSAMENTIER, H. W. & EICHENSEER, H. 1995. Stranded parasequences and the forced regressive wedge systems tract: deposition during base-level fall – discussion. *Sedimentary Geology*, **95**, 139–145.

KRAUSE, F. F. & OLDERSHAW, A. E. 1979. Submarine carbonate breccia beds – a depositional model for two-layer, sediment gravity flows from the Sekwi Formation (Lower Cambrian), Mackenzie Mountains, Northwest Territories, Canada. *Canadian Journal of Earth Sciences*, **16**, 189–199.

LOWE, D. R. 1979. Sediment gravity flows: their classification and some problems of application to natural flows and deposits. *In*: DOYLE, L. J. & PILKEY, O. H. (eds) *Geology of Continental Slopes*. Society of Economic Paleontologists and Mineralogists, Special Publications, **27**, 75–82.

MCILREATH, I. A. & JAMES, N. P. 1978. Facies models – 13. Carbonate slopes. *Geoscience Canada*, **5**, 189–199.

MEISSNER, F. F. 1972. Cyclic sedimentation in Middle Permian strata of the Permian basin, west Texas and New Mexico. *In*: ELAM, J. C. & CHUBER, S. (eds) *Cyclic sedimentation in the Permian Basin.* Texas Geological Society, Midland Texas, 203–232.

MELIM, L. A. & SCHOLLE, P. A. 1995. The forereef facies of the Permian Capitan Formation: the role of sediment supply versus sea-level changes. *Journal of Sedimentary Research*, **B65**, 107–118.

MELLERE, D. & STEEL, R. 1995. Variability of lowstand wedges and their distinction from forced-regressive wedges in the Mesaverde Group, southeast Wyoming. *Geology*, **23**, 803–806.

—— & —— 2000. Style contrast between forced regressive and lowstand/transgressive wedges in the Campanian of south-central Wyoming. *This volume*.

MULLINS, H. T. & NEUMANN, A. C. 1979. Deep carbonate bank margin structure and sedimentation in the northern Bahamas. *In*: DOYLE, L. J. & PILKEY, O. H. (eds) *Geology of Continental Slopes.* Society of Economic Paleontologists and Mineralogists, Special Publications, **27**, 165–192.

——, GARDULSKI, A. F., HINCHEY, E. J. & HINE, A. C. 1988. The modern carbonate ramp slope of central west Florida: *Journal of Sedimentary Petrology*, **58**, 273–290.

MUTTI, M., BERNOULLI, D., EBERLI, G. P. & VECSEI, A. 1996. Depositional geometries and facies associations in an Upper Cretaceous prograding carbonate platform margin (Orfento Supersequence, Maiella, Italy). *Journal of Sedimentary Research*, **66**, 749–765.

NAYLOR, M. A. 1980. The origin of inverse grading in muddy debris flow deposits – a review. *Journal of Sedimentary Petrology*, **50**, 1111–1116.

NEMEC, W. & STEEL, R. J. 1984. Alluvial and coastal conglomerates: their significant features and some comments on gravelly mass-flow deposits. *In*: KOSTER, E. H. & STEEL, R. J. (eds) *Sedimentology of Gravels and Conglomerates.* Canadian Society of Petroleum Geologists, Memoirs, **10**, 1–31.

NUMMEDAL, D. 1992. The falling sea level systems tract in ramp settings. Abstract. *In*: *Mesozoic of the Western Interior.* Society of Economic Paleontologists and Mineralogists Theme Meeting, Fort Collins Colorado, Abstract volume, 50.

—— & MOLENAAR, C. M. 1995. Sequence stratigraphy of ramp-setting strand plain successions: the Gallup Sandstone, New Mexico. *In*: VAN WAGONER, J. C. & BERTRAM, G. T. (eds) *Sequence Stratigraphy of Foreland Basin Deposits – Outcrop and Subsurface Examples from the Cretaceous of North America.* American Association of Petroleum Geologists, Memoirs, **64**, 277–310.

PLINT, A. G. 1996. Marine and nonmarine systems tracts in fourth-order sequences in the Early-Middle Cenomanian, Dunvegan Alloformation, northeastern British Columbia, Canada. *In*: HOWELL, J. A. & AITKEN, J. F. (eds) *High resolution sequence stratigraphy: Innovations and applications*. Geological Society, London, Special Publications, **104**, 159–191.

—— & NUMMEDAL, D. 2000. The falling stage systems tract: recognition and importance in sequence stratigraphic analysis. *This volume*.

POMAR, L. & WARD, W. C. 1994. Response of a late Miocene Mediterranean reef platform to high-frequency eustacy. *Geology*, **22**, 131–134.

POMAR, L., WARD, L. C. & GREEN, D. G. 1996. Upper Miocene reef complex of the Llucmajor area, Mallorca, Spain. *In*: FRANSEEN, E. K., ESTEBAN, M., WARD, W. C. & ROUCHY, J.-C. (eds) *Models For Carbonate Stratigraphy From Miocene Reef Complexes Of Mediterranean Regions*. Society For Sedimentary Geology Concepts In Sedimentology And Paleontology, **5**, 191–226.

POSAMENTIER, H. W. & MORRIS, W. S. 2000. Aspects of the stratal architecture of forced regressive deposits. *This volume*.

——, ALLEN, G. P. JAMES, D. P. & TESSON, M. 1992. Forced regressions in a sequence stratigraphic framework: concepts, examples and exploration significance. *American Association of Petroleum Geologists Bulletin*, **76**, 1687–1709.

PUJALTE, V., ROBLES, S., ROBADOR, A., BACETA, J. I. & ORUE-ETXEBARRIA, X. 1993. Shelf-to-basin Palaeocene palaeogeography and depositional sequences, western Pyrenees, north Spain. *In*: POSAMENTIER, H. W., SUMMERHAYES, C. P., HAQ, B. U. & ALLEN, G. P. (eds) *Sequence stratigraphy and Facies Associations*. International Association of Sedimentologists, Special Publications, **18**, 369–395.

ROBISON, R. A. 1984. *Cambrian Agnostida of North America and Greenland; part 1, Ptychagnostidae*. Paleontological Contributions of the University of Kansas, Paper **109**.

—— 1988. Trilobites of the Holm Dal Formation (late Middle Cambrian), central North Greenland. *Meddelelser om Grønland Geoscience*, **20**, 23–103.

—— 1994. Agnostoid trilobites from the Henson Gletscher and Kap Stanton formations (Middle Cambrian), North Greenland. *Grønlands Geologiske Undersøgelse Bulletin*, **169**, 25–77.

REIJMER, J. J. G., TEN KATE, W. G. H. Z., SPRENGER, A. & SCHLAGER, W. 1991. Calciturbidite composition related to the exposure and flooding of a carbonate platform (Triassic, Eastern Alps). *Sedimentology*, **38**, 1049–1074.

SARG, J. F. 1988. Carbonate sequence stratigraphy. *In*: WILGUS, C. K., HASTINGS, B. S., KENDALL, C. G. ST. C., POSAMENTIER, H. W., ROSS, C. A. & VAN WAGONER, J. C. (eds) *Sea-level changes – an integrated approach*. The Society of Economic Paleontologists and Mineralogists, Special Publications, **42**, 155–181.

SCHLAGER, W. 1991. Depositional bias and environmental change – important factors in sequence stratigraphy. *Sedimentary Geology*, **70**, 109–130.

—— & CAMBER, O. 1986. Submarine slope angles, drowning unconformities and self-erosion of limestone escarpments. *Geology*, **14**, 762–765.

SONNENFELD, M. D. & CROSS, T. A. 1993. Volumetric partitioning and facies differentiation within the Upper Permian San Andres Formation of Last Chance Canyon, Guadalupe Mountains, New Mexico. *In*: LOUCKS, R. G. & SARG, J. F. (eds) *Carbonate Sequence Stratigraphy: Recent Developments*. American Association of Petroleum Geologists, Special Publications, **57**, 435–474.

SOUTHGATE, P. N., KENNARD, J. M., JACKSON, M. J., O'BRIEN, P. E. & SEXTON, M. J. 1993. Reciprocal lowstand clastic and highstand carbonate sedimentation, subsurface Devonian reef complex, Canning basin, Western Australia. *In*: LOUCKS, R. G. & SARG, J. F. (eds) *Carbonate Sequence Stratigraphy: Recent Developments*. American Association of Petroleum Geologists, Special Publications, **57**, 157–179.

SPENCE, G. H. & TUCKER, M. E. 1997. Genesis of limestone megabreccias and their significance in carbonate sequence stratigraphic models: a review. *Sedimentary Geology*, **112**, 163–193.

STROHMENGER, C. & STRASSER, A. 1993. Eustatic controls on the depositional evolution of Upper Tithonian and Berriasian deep-water carbonates (Vocontian Trough, SE France). *Centres de Recherces Exploration-Production. Elf-Aquitaine, Bulletin*, **17**, 183–203.

SURLYK, F. 1978. Submarine fan sedimentation along fault scarps on tilted fault blocks (Jurassic – Cretaceous boundary, East Greenland). *Grønlands Geologiske Undersøgelse Bulletin*, **128**, 1–108.

—— 1991. Tectonostratigraphy of North Greenland. *Grønlands Geologiske Undersøgelse Bulletin*, **160**, 25–47.

—— & HURST, J. M. 1984. The evolution of the early Paleozoic deep-water basin of North Greenland. *Geological Society of America Bulletin*, **95**, 131–154.

—— & INESON, J. R. 1987. Aspects of Franklinian shelf, slope and trough evolution and stratigraphy in North Greenland. *Grønlands Geologiske Undersøgelse Rapport*, **133**, 41–58.

YOSE, L. A. & HARDIE, L. A. 1990. The significance of carbonate megabreccias in sequence stratigraphy: examples from the Triassic of the Dolomites, northern Italy (abstract). *American Association of Petroleum Geologists Bulletin*, **74**, 795.

YOSE, L. A. & HELLER, P. L. 1989. Sea-level control of mixed-carbonate-siliciclastic, gravity-flow deposition: lower part of the Keeler Canyon Formation (Pennsylvanian), southeastern California. *Geological Society of America Bulletin*, **101**, 427–439.

Shingled, sharp-based shoreface sandstones: depositional response to stepwise forced regression in a shallow basin, Upper Triassic Gassum Formation, Denmark

LARS HAMBERG[1] & LARS HENRIK NIELSEN[2]

[1]*Dansk Olie- & Naturgas A/S, Agern Allé 24-26, DK-2970 Hørsholm, Denmark (e-mail: ham@dopas.dk)*

[2]*Geological Survey of Denmark and Greenland (GEUS), Thoravej 8, DK-2400 Copenhagen NV, Denmark*

Abstract: Sharp-based marine shoreface sandstones interpreted as forced regressive deposits are a characteristic feature of the Gassum Formation in the intracratonic Danish Basin. Detailed process-based sedimentological and a high-resolution, sequence-stratigraphic interpretation of cores from closely-spaced wells has led to improved understanding of the erosional and depositional processes active during the formation of the sharp-based sandstones. Each sandstone shows an internal stacking of forced regressive shoreface units separated by thin muddy offshore facies. This stacked pattern records low-amplitude but widespread changes in relative sea-level during the overall progradation due to low depositional gradients. Laterally, the stacked forced regressive shoreface deposits show a seaward-dipping, shingled geometry indicating seaward displacement of the shoreline through stepwise, forced regressions during overall fourth-order relative sea-level fall. Thereby each sharp-based shoreface sandstone records deposition resulting from interaction of from two scales of superimposed relative sea-level fluctuations: a lower fourth-order fall responsible for the overall seaward shoreface displacement, and a higher fifth-order oscillation that resulted in repeated forced regression within the lower-order sequences. Although these stepwise, forced regressive deposits dynamically resemble 'stranded'parasequences, they differ from the conceptualized picture of 'stranded' parasequences as simple downstepping of forced regressive deposits, because of their gently dipping shingled geometry and distinctive deposition component resulting from intervening, high-order drowning.

For both the fifth-order forced regressive units and the lower-order forced regressive sharp-based sandstones it is possible to differentiate between: (1) deposits formed during falling sea level as part of the forced regressive systems tract and (2) the last, forced regressive to progradational part formed at sea-level lowstand representing the lowstand systems tract. Accordingly, the sequence boundary, whether of high- or low-order, is placed below the last, forced regressive deposits and associated lowstand progradational deposits, but above the deposits formed during falling relative sea-level. Thus the sequence boundary is placed at the surface of subaerial exposure passing seaward into a marine regressive surface of erosion reflecting maximum regression. The basal, regressive surface of erosion below the fourth-order forced regressive systems tract is demonstrated to consist of coalesced fifth-order forced regressive surfaces. Therefore, the fourth-order regressive surface is a composite surface reflecting a series of forced regressions and intervening drowning and as such is diachronous. The basinwide dominance of sharp-based, forced regressive shoreface deposits in Upper Triassic of the Danish Basin is interpreted to reflect the interaction between low gradient and shallow palaeobathymetry, sediment supply and low-amplitude relative sea-level changes.

The simplest forced regressive deposits likely occur in response to a single continuous fall in relative sea level, where no deposition takes place during the sea-level fall, and deposition only results from shoreline progradation at the lowest point of sea-level, i.e. during lowstand (Plint 1988; Posamentier *et al.* 1992). This is the case of non-accretionary forced regression (Helland-Hansen & Gjelberg 1994; Helland-Hansen & Martinsen 1996; Posamentier & Morris this volume). Smooth and continuous relative sea-level fall is probably the exception rather than a rule. If the overall fall in relative sea-level is comprised of higher-frequency sea-level oscillations, then alternating deceleration and acceleration of the overall sea-level fall is expected to result in a downward stepping series of discrete, forced regressive deposits (Helland-Hansen & Gjelberg 1994; Hunt & Tucker 1995; Kolla *et al.* 1995; Posamentier & Morris this volume). Deposition takes

From: HUNT, D. & GAWTHORPE, R. L. (eds) *Sedimentary Responses to Forced Regressions.* Geological Society, London, Special Publications, **172**, 69–89. 1-86239-063-0/00/$15.00

place during deceleration of the overall sea-level fall, and when the overall sea level continues to fall the newly formed deposits are exposed. The next forced regressive deposit is thus formed a step lower than the preceding. Such discrete forced regressive deposits are also referred to as 'stranded' parasequences (Van Wagoner *et al.* 1990). The time of deposition of such a series of forced regressive deposits is during falling sea-level, which led Hunt & Tucker (1992, 1995), Helland-Hansen & Gjelberg (1994), Helland-Hansen & Martinsen (1996) and Plint (1996) and Plint & Nummedal (this volume) respectively to include such deposits in a forced regressive or falling stage systems tract beneath a common sequence boundary.

In this study of Triassic well-data from Denmark, we describe forced regressive shoreface sandstones that are interpreted to have formed during fourth-order sea-level falls modulated by fifth-order fluctuations in a shallow, intracratonic basin. The resulting sharp-based shoreface successions are composed of two or more, stacked, forced regressive shoreface deposits. Laterally, the stacked forced regressive deposits show a seaward-dipping, shingled geometry indicating seaward displacement of the shoreline through stepwise, forced regressions during an overall fall in sea-level. The shingled fifth-order forced regressions dynamically resemble 'stranded' parasequences (Van Wagoner *et al.* 1990). But in contrast to 'stranded' parasequences, these forced regressive units (i) are separated by thin offshore deposits recording fifth-order relative rises and drowning of the shoreface and (ii) show a more complex, vertical and lateral stratigraphy than the simple downstepping commonly envisaged for forced regressive deposits.

The aim of this study is to provide a detailed high-resolution sequence stratigraphic interpretation of two composite, sharp-based sandstones based on data from closely spaced wells and regional data. Based on this interpretation we discuss depositional timing, position and significance of sequence boundaries within the fifth and forth-order deposits, and discuss practical application of the sequence stratigraphic systematics forwarded by Hunt & Tucker (1992, 1995), Helland-Hansen & Gjelberg (1994), Helland-Hansen & Martinsen (1996), Plint (1996) and Plint & Nummedal (this volume) against those of Posamentier *et al.* (1992) and Posamentier & Morris (this volume).

Geological setting

The Danish Basin is a WNW–ESE-trending intracratonic basin located in the eastern part of the North Sea rift system (Fig. 1). It was formed by Late Palaeozoic rifting followed by Mesozoic thermal subsidence (Vejbæk 1989). The basin is bounded to the NW by the Skagerrak–Kattegat Platform, and to the south by a WNW–ESE-trending high of basement blocks, the Ringkøbing–Fyn High (Fig. 1). In Rhaetian times, the Danish Basin formed a narrow and semi-enclosed basin covering *c.* 60 000 km^2. During lowstands the Ringkøbing–Fyn High formed a topographic barrier so that the basin was only connected to open seas through a narrow western passage toward the North Sea rift basins. Tectonic and thermal subsidence was focused along the northern margin of the half-graben bounded by the Skagerrak–Kattegat Platform. This northern side of the basin experienced higher rates of subsidence and accommodation development compared to southern margin (Fig. 2). However, high sediment fluxes from the northern margin balanced subsidence so that the basin was shallow and almost flat-based, with it deepest part located near its centre. A depositional shelf break was never developed within the basin. Locally, halokinetic movements influence deposition, mainly by controlling drainage pattern during extreme (third-order) lowstands.

Stratigraphy

Sharp-based shoreface sandstones discussed in this paper occur in the Upper Rhaetian part of the Norian–Lower Sinemurian Gassum Formation

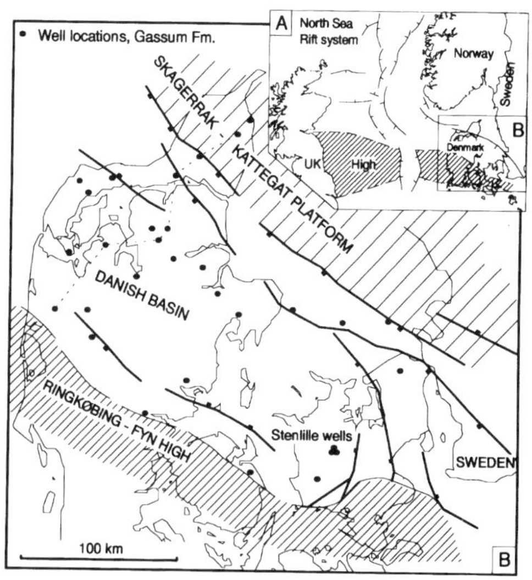

Fig. 1. Location map of the Danish Basin within the North Sea Rift System showing the dominant WNW–ESE structural grain of the Danish Basin and location of wells utilized in this study. Dashed line marks cross-section in Fig. 2 in NE Denmark.

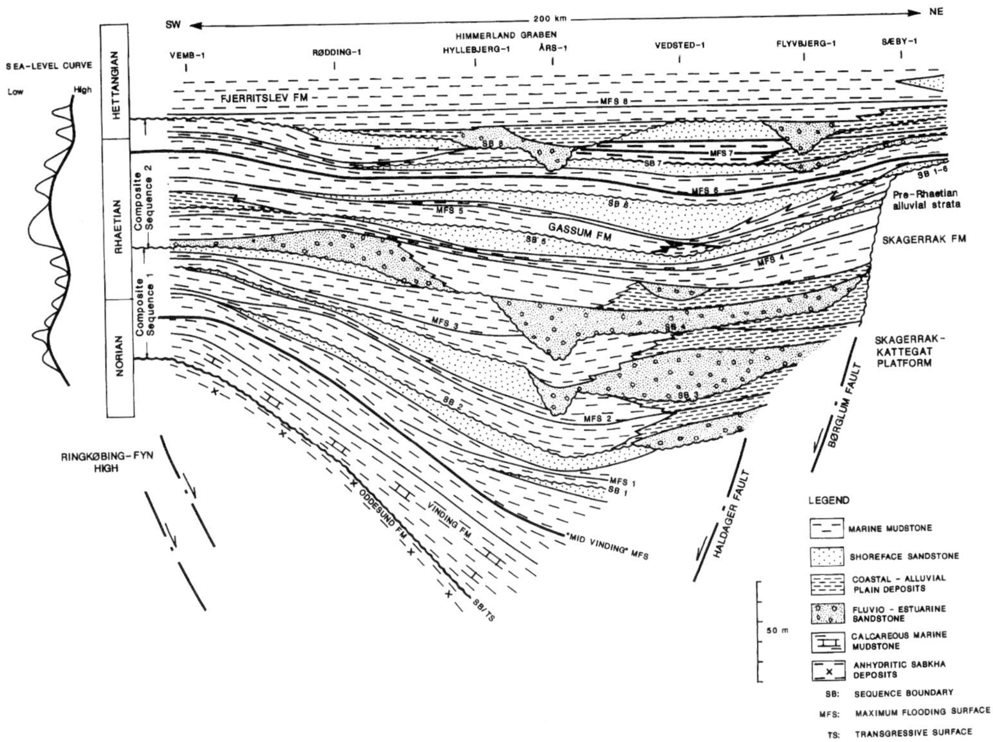

Fig. 2. Schematic, northeast-southwest cross-section across the western part of the Danish Basin illustrating the sequence stratigraphic subdivisions of the Gassum Formation comprising 8 fourth-order sequences (labelled 1–8 from oldest-youngest) that form part of two third-order composite sequences (see Fig. 1 for location; adopted from Nielsen *et al. in* Hamberg 1994). Note that almost all of the shorefaces have basinwide extent and are sharp-based, due to the shallow palaeobathymetry and gradients of the basin. Locally, however gradational bases to shoreface sandstones are however observed (e.g. sandstone 6 adjacent to the NE margin and sandstone 8 in the southwest of the basin). The sea-level curves show the interpreted third-order changes (heavy line) and the superimposed fourth-order fluctuations (thin line). The cross-section is hung on a Hettangian maximum flooding surface (MFS 8). Notice that sedimentation was capable of constantly levelling the basin despite the asymmetrical, half-graben subsidence.

(Fig. 2). The 100–250 m thick succession has been penetrated by many wells (Fig. 1), and is extensively cored and logged. It consists of cyclically interbedded sharp-based shoreface sandstones and offshore marine mudstones locally interrupted by fluvio-estuarine and lagoonal deposits (Figs 2 and 3; Hamberg 1994; Nielsen 1995). The Gassum Formation represents part of the general long-term second-order transgression of the Danish Basin, starting from the continental to shallow marine deposits of the underlying Upper Triassic Oddesund and Vinding Formations (Fig. 2, lower), and ending in the overlying fully marine claystones of the Lower Jurassic Fjerritslev Formation (Bertelsen 1978; Fig. 2, upper).

A detailed process-based sedimentological and sequence stratigraphic study of the Gassum Formation by Nielsen *et al.* (1994) has demonstrated an overall conformable stacking pattern of eight fourth-order sequences (bounded by sequence boundaries 1–8; Fig. 2) that comprise two composite third-order sequences. In detail, the regressive part of the fourth-order sequences primarily consist of basinwide shoreface sandstones composed of shingled fifth-order sequences and parasequences (Hamberg *et al.* 1994). The basin-fill thus reflects four orders of superimposed, relative sea-level changes. Nielsen *et al.* (1994) and Nielsen (1995) compared the third-order changes to the published eustatic sea-level charts, the details of which are beyond the scope of this paper. However, it is emphasized that the hierarchy fifth- to second-order used herein is primarily a number system utilized for easing communication, and as such is not intended to covey a specific duration of individual sequences.

Sharp-based sandstones

The sharp-based shoreface sandstones of the Gassum Formation are 4–30 m thick and correlate over distances of 100–200 km from the basin margins into the basin centre (Figs 2 and 3). They consist of clean to slightly carbonaceous, fine- to medium-grained and upward-coarsening sandstone successions. The sharp-based sandstones may be subdivided into two types: (i) thin, 4–15 m thick bodies and (ii) thick 20–30 m thick bodies (Fig. 3) that are found to differ in terms of their depositional dynamics and timing of deposition. Basically, the thin types are identical to the basal 12–15 m of the thick examples. The main difference exist in the upper-half of the thick sandstones interpreted to record aggradational shoreface deposition during early rise in relative sea-level after the lowest stance has been reached (Hamberg 1994). As such the upper-half of the thick sandstones are considered to correspond to the upper part of the originally defined *lowstand wedge* (Van Wagoner *et al.* 1988), the late lowstand of Posamentier *et al.* (1992) and Posamentier & Morris (this volume) or *lowstand prograding wedge systems tract* (of Hunt & Tucker 1992). The latter being equivalent to the modified versions of the lowstand systems tract systematics as discussed by Helland-Hansen & Gjelberg (1994), Helland-Hansen & Martinsen (1996), Plint (1996) and Plint & Nummedal (this volume).

In this paper we focus on the thin sharp-based sandstones referred to as sandstone 5 and sandstone 6 that occur below fourth-order sequence boundaries SB 5 and SB 6 (Figs 2, 3 and 4). These two regressive sandstones were chosen because they lie close to the maximum flooding surface 6 (MFS 6) of the third-order composite sequence 2 (Figs 2 and 4) that ensures (1) choice of a reliable datum, (2) a reliable basinwide correlation of the sandstones and (3) a good resolution of the subtle modulating fifth-order sea-level rises, as these were superimposed on the rising limb of the third-order sea-level curve and therefore enhanced.

The lateral distribution of regressive shoreface sandstones in the Danish Basin decreases from sandstone 5 to 6 (Fig. 5a, b). Sandstone 6 is generally thinner and muddier than sandstone 5, only shows evidence of exposure along the northern basin margin and in general consists of distal shoreface facies. This backstepping from sandstone 5 to 6 is part of the transgressive system of the third-order composite sequence 2, ending in basinwide drowning represented by MFS 6 (Figs 2 and 4). The thin sandstones 5 and 6 typically show a two-fold subdivision into a lower and upper sandstone unit separated by a muddy offshore heterolithic facies marked by a gamma-ray spike (Fig. 4). Two- or three-fold subdivisions are also seen in the other sharp-based sandstones including the overlying sandstone 7 and 8 (Fig. 4). For sandstones 5 and 6, the two-fold character is most pronounced along the southeastern basin margin and in the central part of the basin (Fig. 4). Toward the northern basin margin sandstone intervals are thicker due to higher subsidence rates and maintain a two-fold division, or alternatively either display a

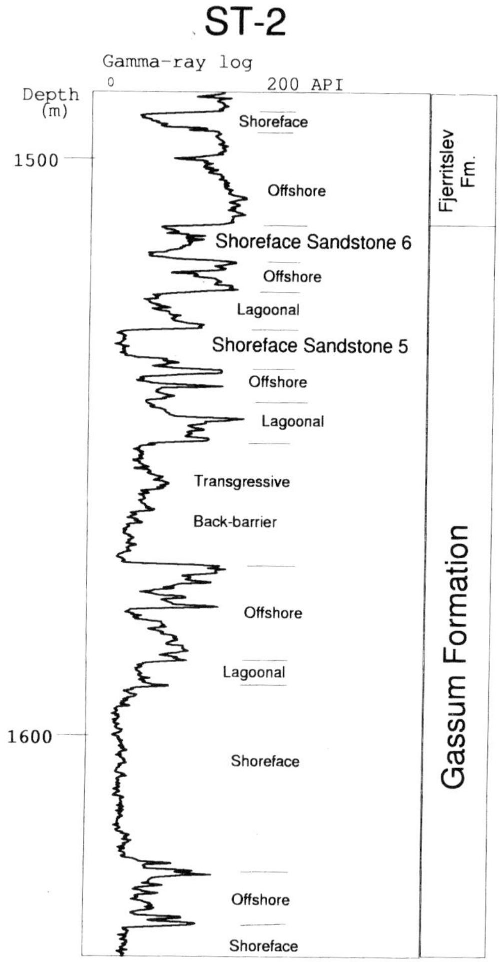

Fig. 3. Close-up of a gamma-ray log through the shallow marine Gassum Formation in the Stenlille-2 well (see Fig. 1 for location). An example of a thick, sharp-based shoreface sandstone is seen near the base (e.g. 1600 m) whereas the two thin sharp-based shoreface sandstones 5 and 6 examined in detail in this study are located in the upper part of the well-log (e.g. 1540 m and 1525 m).

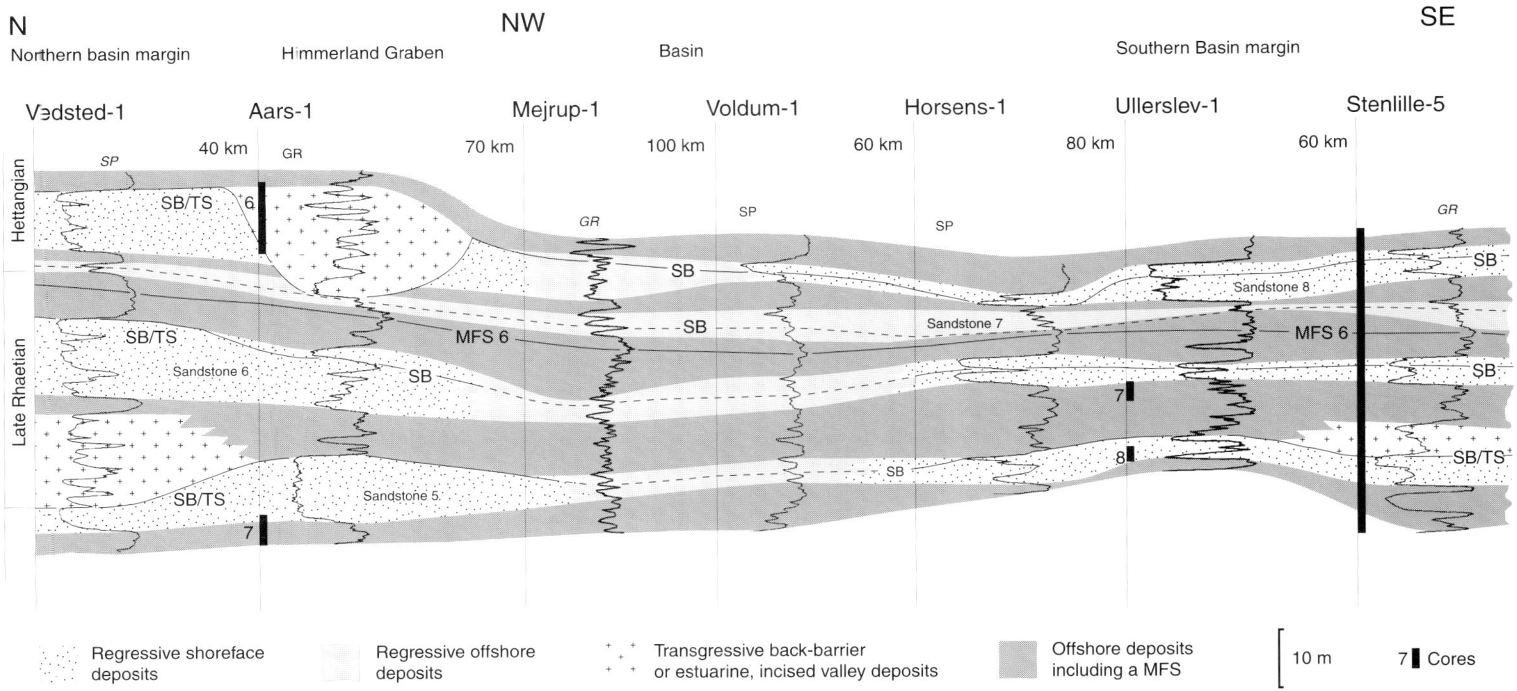

Fig. 4. Example of gamma-ray log and SP correlation of sharp-based sandstone 5, 6, 7 and 8 in the Danish Basin as located on Figure 5. Each sharp-based sandstone is interpreted to represent a fourth-order forced regression. Notice the typical, two-fold subdivision of the sandstones, also shown by the muddy, basinal deposits, only locally replaced by a tree-fold or blocky motif. These subdivisions reflect stepwise and shingling progradation through repeated forced regression related to higher-order fluctuations in sea level superimposed on an overall sea-level fall. As most of these sandstones record deposition during overall falling sea level, the sequence boundaries SB (fourth-order) are placed at subaerial surfaces of exposure (or later ravinement surfaces and transgressive surfaces TS), but below the last, high-order forced regressive deposits representing the lowstand systems tracts (Modified from Nielsen *et al. in* Hamberg 1994).

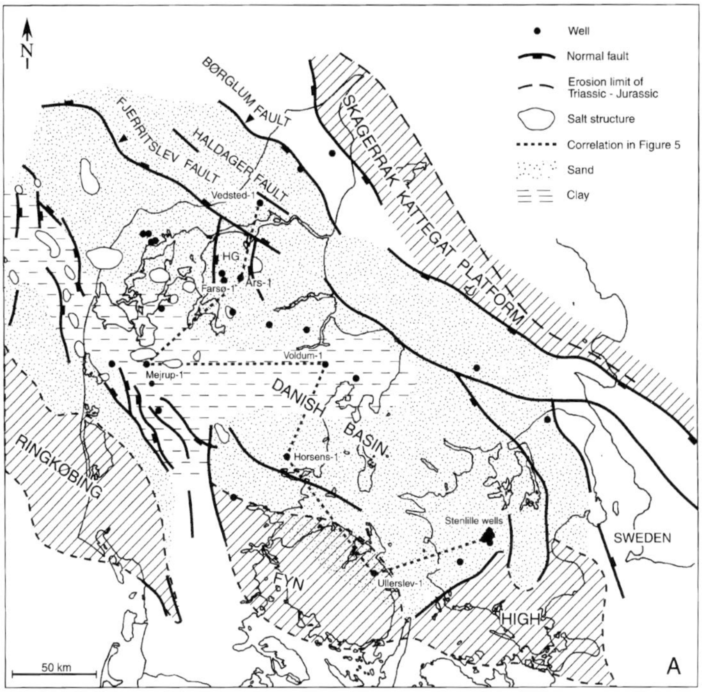
N
Well
Normal fault
Erosion limit of Triassic - Jurassic
Salt structure
Correlation in Figure 5
Sand
Clay
BØRGLUM FAULT
FJERRITSLEV FAULT
HALDAGER FAULT
SKAGERRAK KATTEGAT PLATFORM
Vedsted-1
HG
Farsø-1
Års-1
Voldum-1
Mejrup-1
DANISH BASIN
Horsens-1
RINGKØBING
Stenlille wells
SWEDEN
Ullerslev-1
FYN
HIGH
50 km
A

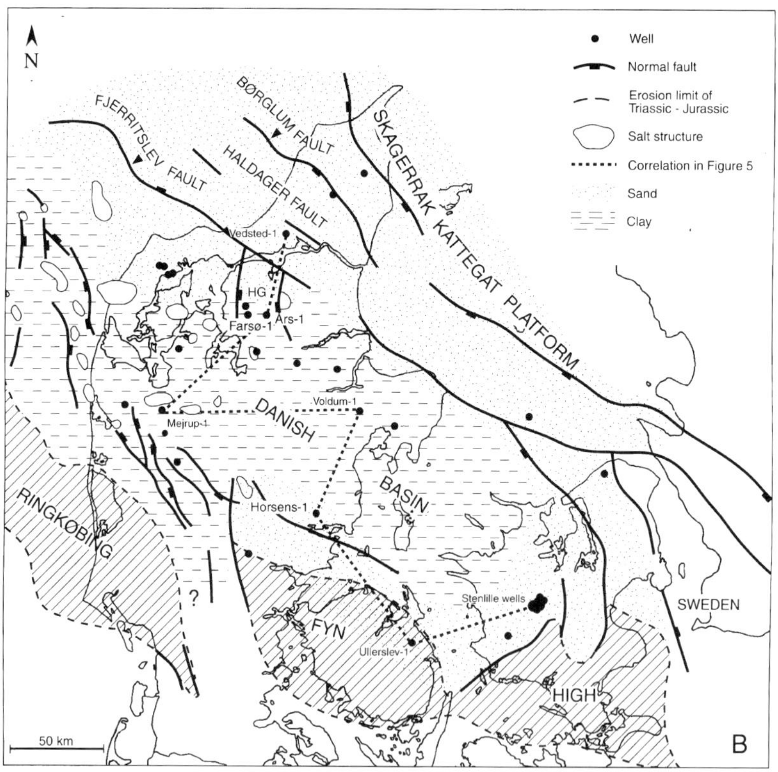
N
Well
Normal fault
Erosion limit of Triassic - Jurassic
Salt structure
Correlation in Figure 5
Sand
Clay
BØRGLUM FAULT
FJERRITSLEV FAULT
HALDAGER FAULT
SKAGERRAK KATTEGAT PLATFORM
Vedsted-1
HG
Farsø-1
Års-1
Voldum-1
Mejrup-1
DANISH BASIN
Horsens-1
RINGKØBING
?
Stenlille wells
SWEDEN
Ullerslev-1
FYN
HIGH
50 km
B

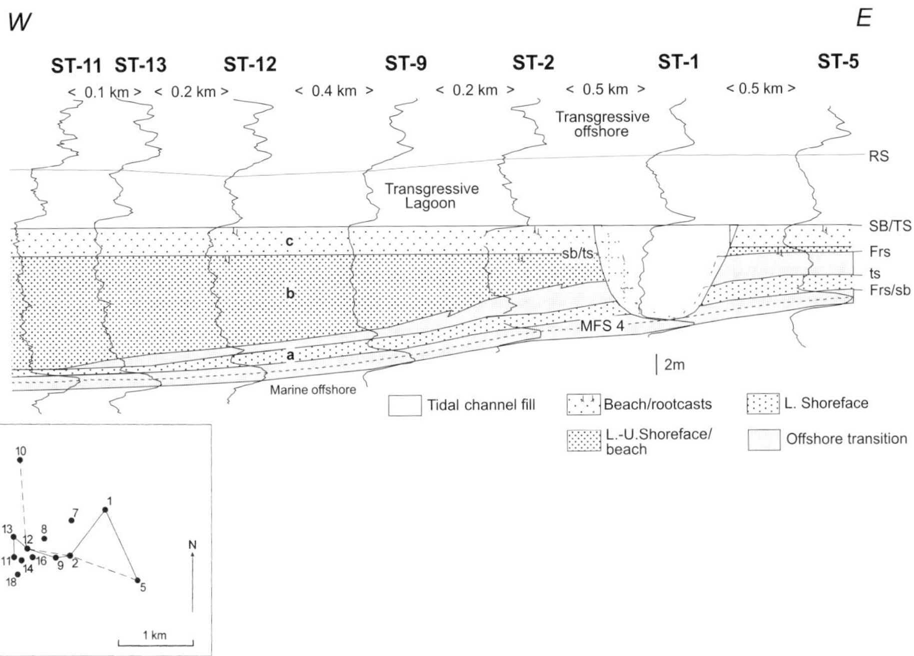

Fig. 6. Gamma-ray log cross section parallel to depositional dip of the thin sharp-based shoreface sandstone 5 across Stenlille Gas Storage site (full line on insert map). The shoreface succession erosionally overlies a thin offshore mudstone unit containing maximum flooding surface 4 (**MFS**). Units **a**, **b** and **c** each represents a forced regression. The base of b and **c** are locally erosive and place shoreface sandstones directly on top of offshore mudstones (e.g. Fig. 7, logs for ST 2, -10, -12). **Frs** – forced regressive surface; **ts/TS** – transgressive surface of high and lower-order; **sb/SB** – sequence boundary of high- and lower-order. In ST-5, the shoreface sandstones are cut-out by a lowstand to transgressive tidal channel fill. The forced regressive shoreface deposit is overlain by a transgressive succession of first lagoonal then offshore marine deposits above a ravinement surface (**RS**).

three-fold subdivision or appear as blocky units (Fig. 4). As will be demonstrated below, a two-fold sandstone may laterally change into a blocky or three-fold sandstone as a consequence of its shingled depositional nature. A possible interpretation of this recurrent two-fold or three-fold subdivision is presented below, based on closely spaced well data from the Stenlille gas storage site located near the southeastern basin margin (Figs 1, 6 and 7).

Sandstone 5

At Stenlille, sandstone 5 is an upward-coarsening sandstone succession forming a wedge-shaped body which thickens basinward (west; Fig. 6). It rests erosively on an 1 m thick offshore marine mudstone that contains maximum flooding surface 4 and can be correlated over the entire basin (Figs 6 and 8, MFS 4). Internally, sandstone 5 can be subdivided by an offshore heterolithic interval into a lower sandstone unit **a**, separated from two upper sandstone units **b** and **c** (Figs 6 and 7). Units **b** and unit **c** are separated by an erosional surface truncating root casts. Unit **c** also show root-casts in the top. Besides this subdivision, a tidal channel fill cuts-out the entire shoreface succession of sandstone 5 in well ST-1 (Fig. 6). Sandstone 5 is erosively overlain by transgressive lagoonal deposits and

Fig. 5. Palaeogeographic reconstructions of the Upper Rhaetian lowstand situations mapped and maximum extent of shoreface sandstones for A) sandstone 5, and B) sandstone 6 in the Danish intracratonic basin. Between shorefaces 5 & 6 the shoreline steps landward so reducing the basinward extent of sandstone deposition. Sandstone 6 is generally thinner and muddier than its precursor and only shows evidence of subaerial exposure on the northern margin of the basin. Backstepping, fining and thinning of sandstone 6 is interpreted to reflect the composite nature of relative sea-level and its development within a long-term 3 rd-order TST (e.g. see Fig. 2). The location of the profile illustrated in Figure 4 is also shown.

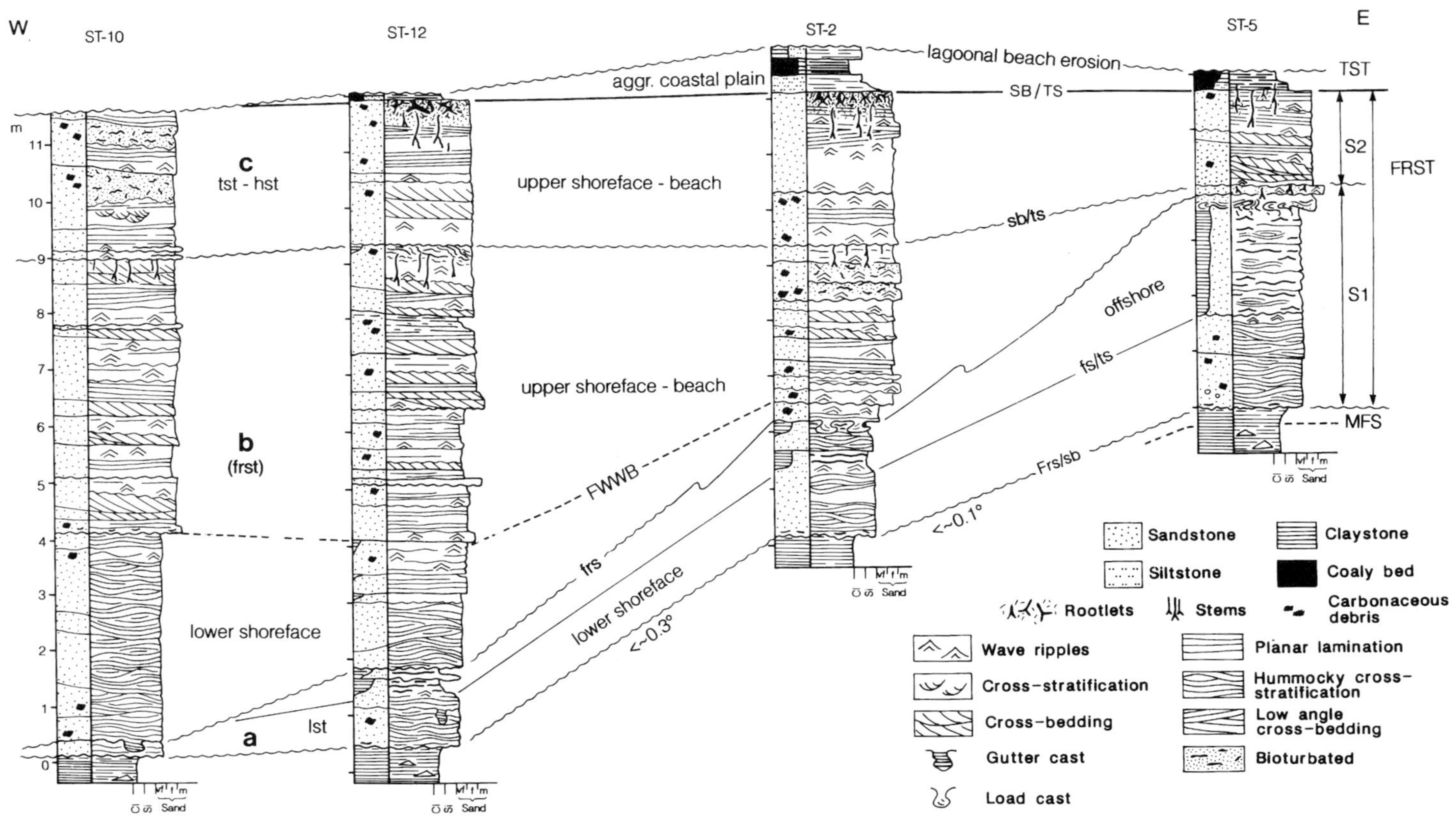

Fig. 7. Core-log cross section parallel to depositional dip of sandstone 5 (dashed line on insert map in Fig. 6). Notice well developed two-fold nature and very thin shoreface successions in the proximal wells ST-2 and -5. Seaward, from well ST-12 to ST-10, marine erosion of the interbedded offshore interval beneath unit **b** resulted in amalgamation of the shoreface succession. When reconstructed as here, the basal erosional surface dips less than 0.1–0.3 degrees. Annotation of boundaries as in Figure 6. **lst** – lowstand systems tract; **tst** – transgressive systems tract; **hst** – highstand systems tract; **frst/FRST** – forced regressive systems tract of high and lower-order respectively, **S1-S2** – high-order (fifth-order) sequence 1 and 2. **FWWB** – fair weather wavebase. The overlying regional transgressive systems tract (TST) was initiated by aggradation on the lowstand coastal plain seen as coaly and lacustrine mudstone deposits. **SB** – major 4th-order sequence boundary marking most basinward and interrupted lowest point of relative sea-level; **sb** – higher-order sequence boundary.

by transgressive offshore deposits overlying a ravinement surface (RS in Fig. 6).

Sedimentology

*Lower sandstone unit **a**.* The lower unit **a** consists of sharp-based hummocky cross-stratified fine-grained sandstones that are 1.5 m thick in well ST-5 and thin basinward (west; Figs 7 and 8). The basal erosional surface dips 0.1–0.3° basinward and follows and locally cuts into the gradient of the underlying marine mudstone of MFS 4 (Figs 6–8). The erosion surface is overlain by abundant plant and wood fragments, sideritic pebbles and rounded rip-up clasts from the underlying marine mudstones (Fig. 7). Water escape structures and slumping are occasionally seen, especially in the distal settings, indicating that deposition of sand occurred upon only slightly consolidated mudstones. Gutter casts are commonly observed at the base of the hummocky beds (Figs 7, 8).

Interpretation. The hummocky cross-stratified sandstones of unit **a** are interpreted as storm deposits accumulated below fair-weather wave-base in the lower shoreface zone. The underlying mudstones of MFS 4 represent distal offshore deposits formed below storm-wave-base, and the contact to the overlying sandstones of unit **a** indicates an abrupt shallowing and a fall in relative sea-level. The basal erosional surface records wave scouring before and/or during deposition and unit **a** is interpreted as a forced regressive unit overlying offshore mud (cf. Plint 1988, 1996; Posamentier *et al.* 1992; and papers by Ainswoth *et al.*, Fitzsimmons & Johnson, Mellere & Steel, Plint & Nummedal and Posamentier & Morris this volume).

Muddy heterolithic facies. The lower sandstone unit **a** is abruptly or gradually overlain by burrowed, heterolithic mudstones and sandstones grading to siltstones (Figs 6–8). The muddy heterolithic facies wedge out basinward from 2 m in well ST-5 to nothing between ST-12 and 10 where it has been eroded (Figs 6 and 7). The sandstone and siltstone layers show horizontal lamination to small scale hummocky cross-stratification draped by black mudstone (Fig. 8). Horizontal trace fossils dominate but *Teichichnus* traces are locally abundant. A palynological sample of the mudstone layers shows abundant marine dinoflagellate cysts (*Dapcodinium priscum*).

Interpretation. The muddy heterolithic facies is interpreted as storm-dominated offshore sediments deposited close to storm-wavebase within the offshore-shoreface transition. It represents relatively deeper water deposition than the underlying shoreface sandstones of unit **a**, indicating a deepening and drowning of the former shoreface.

*Upper sandstone unit **b/c**.* The upper sandstone is an upward-coarsening succession comprised of two units, **b** and **c** (Figs 6 and 7). The upper sandstone erosionally overlies the offshore heterolithic facies in the proximal wells ST-5, -2 and -12 although basinward (e.g. in wells ST-10 and -11) it directly truncates the lower sandstone unit **a** (Figs 6 and 7). In well ST-5, overturned and slumped bedding is seen in the heterolithic beds right below the erosional surface (Figs 7 and 8b).

Unit **b** is very thin (0.25 m) in well ST-5 but basinward (west) it shows a dramatic change in thickness increasing to a 9 m thick upward-coarsening succession in well ST-10 (Figs 6 and 7). The thick succession includes basal fine-grained and hummocky cross-stratified sandstones, erosionally truncated by cross-bedded, wave-rippled and low-angle cross-stratified fine-medium-grained sandstones (Fig. 7). A characteristic feature of these sandstones are the repetitive alternation of fine- and medium-grained beds. The coarser beds are erosive, wave rippled and sometimes burrowed. Locally, large 4 cm wide irregular burrows are seen, probably related to burrowing activity of larger crustaceans. The fine-grained beds drape the underlying coarser beds and show wave-ripple to horizontal stratification with scattered, vertical burrows and a high content of carbonaceous debris. Stem impressions and some root casts are seen in top of unit **b** and these are truncated by an erosional surface at the base of the overlying unit **c** (Figs 7 and 8c).

Unit **c** forms a tabular ca. 2.5 m thick package that consists of fine- to medium-grained sandstones, with local coarse-grained beds (e.g. Fig. 7). The sedimentary facies are comparable to unit **b,** but low-angle cross bedding and burrowing are more pronounced. The top of unit C shows a dense network of tenuous and thicker root casts as well as some hollow moulds of stems (Figs 7 and 8a, c). The top of unit **c** is cut by an erosional surface truncating the layers with root casts (Figs 7 and 8). This erosional surface is draped by silty to coaly mudstones, less than 1 m thick (Figs 7 and 8). The palynoflora of the coaly mudstones show abundant firn spores and *Botryococcus*, but no marine dinoflagellate cysts. Upward, the coaly mudstones are truncated by an upward-coarsening succession of transgressive lagoonal beach and fill deposits (Figs 6 and 7; Hamberg 1994). Laterally, in

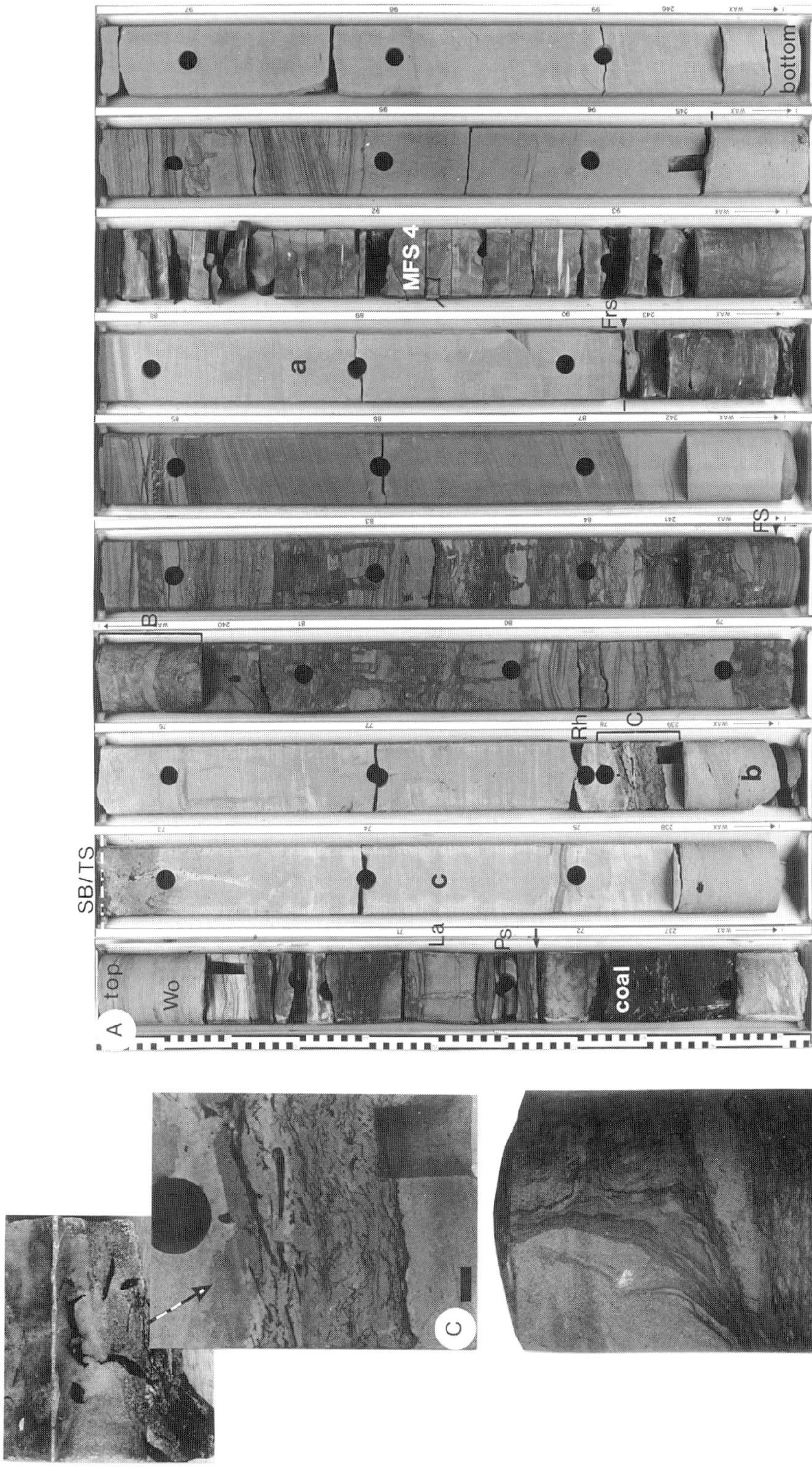
A
top
Wo
La
Ps
coal
SB/TS
c
Rh
C
b
B
FS
a
Frs
MFS 4
bottom
B
C

well ST-1 sandstone 5 is replaced by an upward-fining and muddier sandstone succession characterized by a bell-shaped gamma-ray log motif (Fig. 6). In cores, it consists of gently-inclined sandy heterolithic beds with regular mud-draping on cross-beds burrowed by amphipods as well as larger crustaceans. This succession is interpreted as representing a tidal channel fill (Hamberg 1994).

Interpretation. In unit **b**, the basal hummocky cross-stratified sandstones represent lower shoreface deposition (Fig. 7). The overlying coarser, cross-bedded and wave rippled to low angle cross-stratified sandstones with root casts were deposited in the high-energy zone above fair-weather wavebase and are interpreted as upper shoreface to beach deposits. Repetitive alternation of fine-grained and coarser sandstone beds reflects alternating storm- and fair-weather deposition in the upper shoreface zone (cf. Clifton 1981). The erosional contact between the hummocky cross-stratified sandstones and the overlying, coarse-grained sandstones may represent the fair-weather wavebase (Fig. 7, FWWB). The upward-coarsening succession of unit **b** is interpreted as representing shoreline progradation, but because unit **b** is characterized by a basal, erosion surface, it differs from normal, gradationally based, progradational shoreline successions (cf. Plint 1988, 1996; Posamentier *et al.* 1992, and papers by Ainsworth *et al.*, Fitzsimmons & Johnson, Mellere & Steel, Plint & Nummedal and Posamentier & Morris this volume). At least three conditions along this basal erosive contact are indicative of an abrupt shallowing and a fall in relative sea-level: (1) the close, vertical juxtaposition of the offshore muddy heterolithic facies and overlying sandstones with root casts in well ST-5; (2) the rapid vertical transition from the offshore muddy heterolithic facies into coarse-grained, upper shoreface sandstones in well ST-2 and (3) the erosional contact showing progressively deeper truncation of the underlying offshore heterolithic facies basinward and finally into unit **a**, implying a successive lowering of average storm-wavebase during deposition. Indeed, the thick offshore heterolithic facies in well ST-5 suggests little basal erosion here, so that the deeper and deeper seaward truncation must have taken place after exposure of the more proximal areas around well ST-5. In conclusion, unit **b** is interpreted to be the result of a forced regression with active erosion and deposition during falling relative sea-level.

Slumping and other soft sediment deformations beneath the erosive surface suggest abrupt deposition due to forced regression and rapid transition from mud to sand deposition (e.g. Fitzsimmons & Johnson this volume). Stem impressions in the top of unit **b** indicate exposure and vegetation on a coastal plain formed when sea-level fall caused exposure of the Stenlille area. The stem impressions are interpreted as representing reclamation by possible halophytic vegetation (*Equisetites*?), comparable to a modern marine reed swamp. If this interpretation is correct, vegetation in top of unit **b** may have acted as a sediment trap responsible for the baffling and accumulation of sediment during stationary or slightly rising sea-level. Therefore, the uppermost part of unit **b** with stem impressions is thought to possibly record a later rise in relative sea-level and aggradation over the strand plain.

In unit **c**, the dominance of low-angle swash-type cross-bedding, wave ripples and root casts records deposition in the upper-shoreface and beach zones. In comparison to the exposed top of the underlying unit **b**, deposition of unit **c** involves a relative sea-level rise to create accommodation space over the former coastal plain. The initial phase of this relative rise in sea-level is probably recorded by the reed swamp like vegetation interpreted in top of unit **b**. The erosional surface separating unit **b** and **c** is a ravinement type surface separating shoreface deposits from a vegetated coastal plain. Unit **c**

Fig. 8. Core photo of sandstone 5 in well ST-5 (same as logged in Fig. 7). (A) The distinct, two-fold subdivision of the shoreface unit **a** and **b/c** by offshore heterolithic facies is apparent. Unit **a** shows a basal, forced regressive surface **Frs** (4th core box from right) and on top a flooding surface **FS** (base of 6th core box from right), below offshore heterolithic and bioturbated facies reflecting drowning of the shoreface. Forced regression in connection to unit **b** is indicated by 1) overturned underlying offshore strata (**B**, located at the top of the 4th core box from left), 2) truncation of the underlying strata, and 3) close juxtaposition to root casts and stem marks in the sandstone right above, at **Rh**, close-up in (**C**, located near base of 3rd core box from left). Erosional contact to overlying unit **c** is seen below arrow in close-up (**C**). The coarse, upper shoreface sandstone of unit **c** is overlain by coaly mudstones above a transgressive surface, **TS**, followed by a transgressive succession seen in the leftmost core box, initiated by a lagoonal paralic shoreline erosion, **Ps**, followed by lagoonal drowning, **La**, and washover deposition, **Wo**. The transgressive surface is also the overall sequence boundary **SB** (fourth-order), here marking the lowest position of sea level for the forced regressive progradation of sandstone 5 (e.g. compare with Fig 6 and 7). For scale and facies description see the core log in Figure 7.

being only 2.5 m thick and with dense root casts in the top reflects a rapid upward shallowing and later subaerial exposure, indicating that the relative sea-level rise was followed by a fall. It is likely that most of unit **c** was deposited during the relative rise in sea level, when accommodation space was increasing. During a succeeding fall the strand plain was rapidly exposed and the channel in ST-1 was cut by a river. The root casts in top of unit **c** are much more dense compared to those in top of unit **b**, and root casts dominate rather than moulds of stems indicating longer time of exposure and a more established vegetation.

Instead of a ravinement surface, the erosional base of unit **c** may be interpreted as the proximal part of a forced regressive surface of erosion eroding most of the deposits formed during a preceding sea-level rise. However, the close occurrence of the root casts in top of unit **b** and unit **c**, as well as the tabular geometry of unit **c** suggest very shallow water at all times, and any lowering of sea level would rapidly expose the area and leave little time and space for deposition.

The overlying coaly mudstones and siltstones are interpreted to have accumulated in shallow lakes, a typical evolution for the onset of base-level rise over a coastal plain area (e.g. Surlyk *et al.* 1995). The lake fill and overlying transgressive lagoonal and offshore deposits mark a progressive transgression and drowning of the coastal plain. The tidal channel fill in well ST-1 is directly truncated by the transgressive lagoonal deposits (Fig. 6). This channel was most likely filled during early sea-level rise by transformation of a small river into a tidal drainage channel behind a lagoon. Later the channel was abandoned and transgressed by the retreating lagoon.

Deposition through repeated forced regression

Correlation of sandstone 5 as shown in Figs 6 and 7 indicates a seaward dipping, shingled geometry of sandstone unit **a** and **b**/**c** and the separating heterolithic offshore deposits. Units **a** and **b** as well as the upper part of unit **c**, are interpreted to represent individual forced regressive sandstones formed during successive sea-level falls that were separated by sea-level rises. As such, deposition of unit **a** through the muddy heterolithic facies to unit **b** and later unit **c** describes three sea-level cycles of a high-frequency oscillation (Fig. 9). The shingled geometry record seaward displacement of the shoreline through stepwise forced regressions, a situation only to be accomplished if the overall (fourth-order) sea-level trend shows a fall. When superimposed on an overall lower-frequency fall in sea-level, the high-frequency falls represented by unit **a**, **b** and **c** will be enhanced and the intervening rises diminished (cf. Mitchum & Van Wagoner 1991), with the result of a seaward and downward stepping of the shorelines following the trend of the overall sea-level fall. In conclusion, sandstone 5 records deposition from two scales of relative sea-level fluctuations; a high-frequency oscillation responsible for repeated forced regressions that was superimposed on a low-frequency fall that led to the seaward displacement and overall deposition of sandstone 5. This situation is schematically illustrated in Fig. 9, steps 1 to 6.

Steps 1–2; deposition of shoreface unit a. Before start of the overall fall in sea-level and shingled progradation, the shoreline prograded slowly under conditions of steady sediment supply, a typical highstand scenario (Fig. 9, step 1). Subsequent high-frequency fall in relative sea level and lowering of wavebase to force the shoreline seaward led to deposition of a thin forced regressive shoreface sand on a wave-scoured surface (Fig. 9, step 2). Unit **a** of sandstone 5 in the Stenlille wells is interpreted to correspond to the distal lower shoreface of such a forced regressive deposit (e.g. compare Figs 6–8 with Fig. 9, step 2). The landward transition from the beach deposits equivalent to unit **a** into the preceding highstand deposits is unknown. The interpretation of step 2 illustrated in Figure 9 is based on the interpretation of the overlying unit **b** of sandstone 5 recording forced regressive, upper shoreface to beach deposits. Unit **b** indicates that (1) during falling sea level, erosion and deposition was associated with a downstepping of the shoreline and (2) preservation of unit **b** also reflects a high sediment supply during regression with the result of an accretionary-type forced regression where bedding planes of the prograding shoreface prism downlap the basal wave scoured surface (e.g. Plint 1988, 1996; Dominguez & Wanless 1991; Nummedal *et al.* 1993; Helland-Hansen & Gjelberg 1994, papers by Ainsworth *et al.*, Fitzsimmons & Johnson, Mellere & Steel, Plint & Nummedal and Posamentier & Morris this volume). The distal forced regressive deposits of unit **a** are therefore not believed to represent isolated lowstand deposits but are attached to a set of downstepping shoreline deposits formed during falling sea-level (as defined by Ainsworth & Pattison 1994), probably merge landward with the highstand deposits (Fig. 9, step 2).

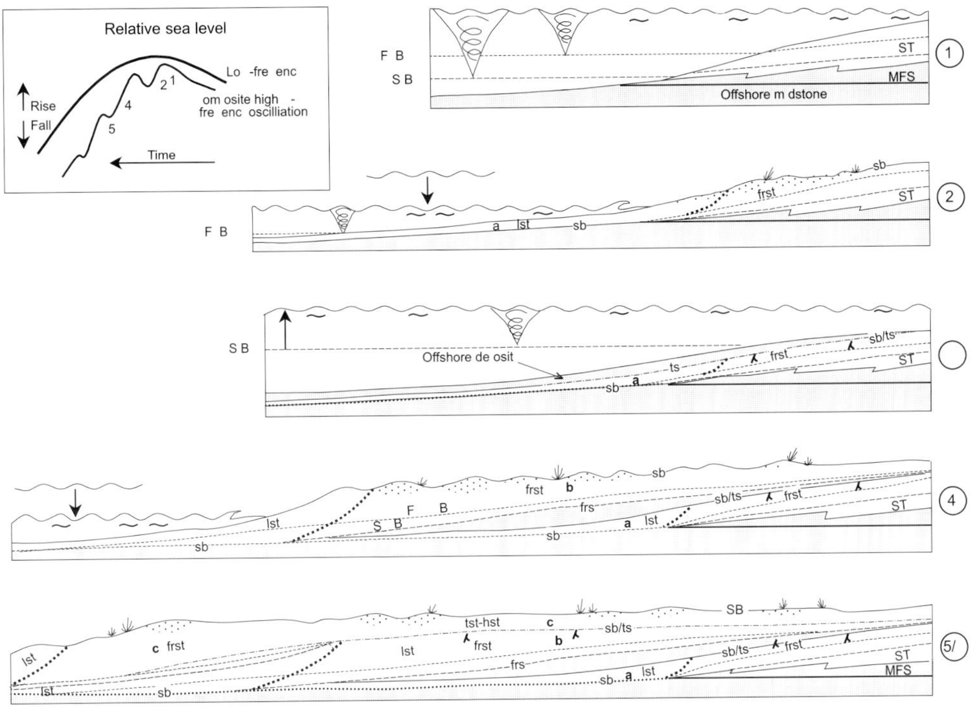

Fig. 9. Depositional model of shoreline progradation through stepwise, forced regression and intervening drowning controlled by high-frequency sea-level oscillations superimposed on an overall, lower-frequency fall. Position of step 1–6 are marked on the composite interpreted high-frequency sea-level curve. Annotation of boundaries and systems tracts as in Figs 6–8. **FWWB** – fair-weather wavebase, **SWB** – storm wavebase. Notice, that both SWB and FWWB are believed to be lowered successively during the sea-level falls. Dotted line at the end of the subaerial shoreface prisms represents the last bedding plane separating shoreface progradation during falling sea level from shoreface accumulations formed at sea-level lowstand.

As unit **a** represents deposition during the last part of this high-order forced regression and later, lowest sea-level (Fig. 9), unit **a** and the corresponding upper shoreface to beach deposits represent a high-order (fifth-order) lowstand shoreline deposit and lowstand systems tract (Figs 7 and 9, step 2). This is basically comparable to the situation discussed by Posamentier *et al.* (1992) and a high-order sequence boundary (sb) is placed at the basal surface of forced regressive erosion (Figs 7 and 9, step 2). Landward, the equivalent sequence boundary is the surface of exposure developed over the highstand deposits and the deposits formed during falling sea-level marking maximum regression as discussed by Hunt & Tucker (1992, 1993, 1995). The subaerial part of the sequence boundary is likely to merge with the subaqueous part along the last bedding plane of the shoreface progradation, basically corresponding to a surface of maximum regression (Helland-Hansen & Martinsen 1996) (stippled in Fig. 9, step 2).

Step 3; flooding and ravinement. Step 3 in Fig. 9 illustrates the ensuing sea-level rise and drowning of the shoreline resulting in reworking and draping of the deposits of the former shoreline by offshore muddy heterolithic deposits. These muddy heterolithic facies represent transgressive and perhaps highstand deposition and overlie a combined flooding and transgressive surface (Figs 7 and 9, step 3). Landward, the subaerial surface of exposure is eroded during transgression and replaced by a ravinement surface. The ravinement surface is a composite surface representing the transgressive surface and the sequence boundary (Helland-Hansen & Gjelberg 1994; Hunt & Tucker 1995; Plint 1996; Plint & Nummedal this volume).

Step 4; deposition of shoreface unit b. As sea-level ceases to rise, a new high-frequency sea-level fall begins resulting in wave-scouring of the former offshore area and deposition of a new forced regressive shoreface unit, comparable to unit **b** (Fig. 9, step 4). Unit **b** is a seaward thickening wedge reflecting both the natural dip of the shoreface profile and the deeper and deeper basal erosion that occurred during lowering of the wavebase. The result is a basal forced regressive surface dipping gently seaward (compare Figs 6, 7 and 9, step 4). The shoreface to beach deposits of unit **b** were primarily deposited during falling sea-level and forced regression. As the basal erosional surface of unit **b** truncates unit **a** basinward, this surface is believed to merge with the basal erosional surface of unit **a** further seaward, and therefore the high-order sequence boundary below unit **a** will only be of local extension and similarly so will the lowstand deposits of unit **a** (compare Figs 7 and 9, step 4). Significantly, in this respect the basal erosion surface is a composite one and diachronous across its length (see also Fitzsimmons & Johnson this volume, figs 11 and 12).

For unit **b**, the lowstand shoreline deposited at lowest sea-level must exist-basinward (Fig. 9, step 4). As such, unit **b** is regarded as representing a forced regressive systems tract overlying a basal, forced regressive surface (following Hunt & Tucker 1992, 1995) otherwise known as a regressive surface (e.g. Plint 1996; Plint & Nummedal this volume). The top of unit **b** is a surface of subaerial exposure and a high-order (fifth-order) sequence boundary (Fig. 7), correlating basinward with the offshore marine expression of the sequence boundary below the predicted lowstand deposits (Fig. 9, step 4). Later, this subaerial sequence boundary was eroded during transgression and deposition of unit **c**, forming a combined transgressive surface and sequence boundary (sb/ts; Fig. 7). A fifth-order sequence, S1 (Fig. 7, right), is interpreted to exist between the two high-order sequence boundaries and so includes unit **a** and unit **b** (Fig. 7).

Steps 5–6; deposition of shoreface unit c. The subsequent deposition of unit **c** involved renewed drowning of the shoreface and deposition of offshore facies prior to renewed fall of relative sea level (Fig. 9, step 5/6). At this stage of increased rate of low-frequency sea-level fall, the intervening high-frequency drowning at the base of unit **c** was subdued (see sea-level curves in Fig. 9). It only resulted in a minor base-level rise accompanied by a landward retreat of the shoreface and deposition of most of unit **c**. The next fall in sea level, reflecting the superimposed high- and low-frequency falls (Fig. 9, step 6 at the composite sea-level curve), rapidly exposed the former beach and shoreface, and a deep channel was cut now preserved in well ST-1 (Fig. 6). Basinward and downward where space were available, a new forced regressive shoreface sand was deposited (Fig. 9, far left in step 5/6). The top of unit **c** is a surface of subaerial exposure (Figs 7, 8) reflecting maximum regression and is here interpreted as a the upper (3rd) high-order (fifth) sequence boundary of sandstone 5. Consequently, an upper fifth-order sequence S2 composed of unit **c** can be distinguished (Fig. 7).

Visualizing continued shingled progradation from step 6 and onwards, the last lowstand shoreline of sandstone 5 (i.e. the final lowstand unit of the overall, composite sea-level fluctuation) is predicted to have developed some 100 km basinward according to the mapped extension of sandstone 5 (Fig. 5a). But at Stenlille, this final lowstand and 'point of lowest sea level' is represented by a single surface, the subaerial exposure surface at the top of unit **c**. This is interpreted to represent a surface of maximum regional lowstand and the overall composite fourth-order sequence boundary to sandstone 5 (SB; Figs 6 and 7). The internal shingled stratigraphy of sandstone 5 suggest that this sandstone was formed during a fourth-order sea-level fall, and is here separated as a fourth-order forced regressive systems tract (FRST) following the systematics of Hunt & Tucker (1992, 1995), equivalent to the falling stage systems tract following the systematics of Plint & Nummedal (this volume).

Sandstone 6

Sandstone 6 at Stenlille is a tabular body (Fig. 4) that erosionally overlies a 1–2 m thick marine mudstone containing a fourth-order maximum flooding surface (MFS 5), (Fig. 10). It can be divided into a lower and upper sandstone units **a** and **b**, separated by an offshore heterolithic facies (Figs 10 and 11). At its top sandstone 6 is truncated by a marine ravinement surface and offshore muddy deposits of the Fjerritslev Formation comprising MFS 6 (Figs 2 and 12).

Sedimentology

*Lower sandstone unit **a**.* The erosional surface at the base of unit **a** is planar with local relief less than 0.1 m. In two wells the surface is overlain by a basal lag of intra-formational, mudstone/siltstone chips (Fig. 11). In the proximal wells ST-2 and ST-5, the surface truncates thin and silty heterolithic facies. Down-dip this facies thins out, and in well ST-4 sandstone unit **a** rests directly on offshore mudstones close to MFS 5 (Figs 10 and 11). The heterolithic facies is

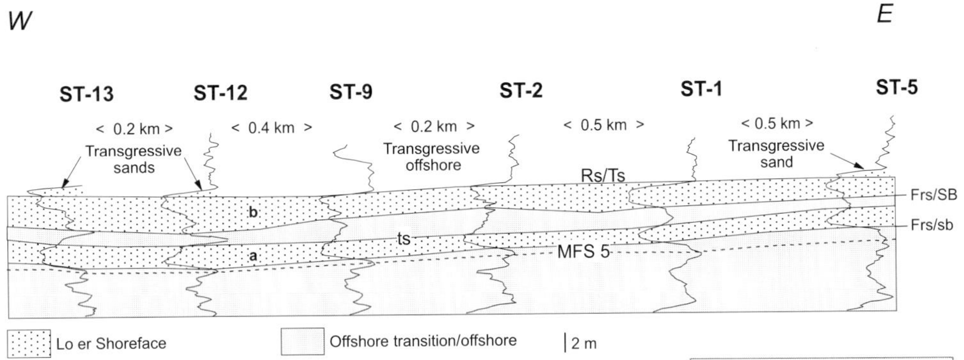

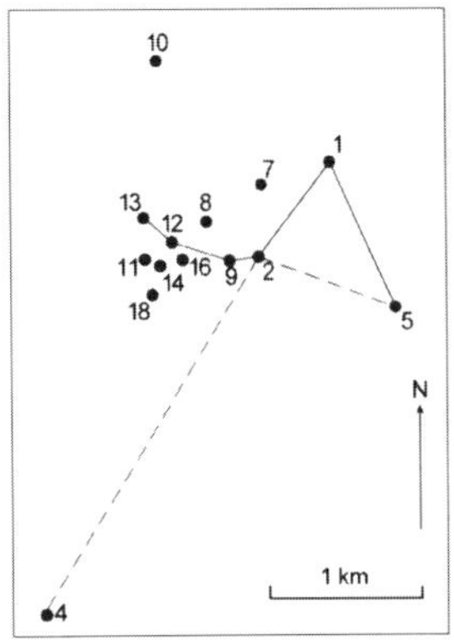

Fig. 10. Gamma-ray log cross section sub-parallel to depositional dip of the thin, sharp-based shoreface sandstone 6 from the Stenlille Gas Storage site (solid line on insert map). The shoreface succession erosionally overlies a thin offshore mudstone unit containing a maximum flooding surface (**MFS**). Notice the well developed two-fold subdivision and overall tabular geometry of the shoreface sandstone. Annotation of boundaries and systems tracts as in Figs 6 and 7. the dashed line on the insert map shows the profile shown in Figure 11.

characterized by small-scale soft-sediment deformation and water escape structures. The sandstones of unit **a** overlying the erosion surface are composed of very fine-grained, hummocky cross-stratified beds (Fig. 11). Unit **a** thins down-dip and seaward from 2 m in well ST-5 to 1 m in ST-4 (Figs 10 and 11).

Interpretation. The hummocky cross-stratified sandstones of unit **a** were formed in the lower shoreface zone above storm-wavebase. The heterolithic facies below the erosion surface represents deposition close to storm-wavebase on the soft, offshore mud of MFS 5 and represents an initial gradual shallowing possibly representing the highstand systems tract. The erosive contact to the overlying lower shoreface sandstones of unit **a** however marks an abrupt shallowing and suggests a forced regression and erosion by high-energy processes near the base of the shoreface zone during a relative sea-level fall (Fig. 11).

Muddy heterolithic facies. This facies abruptly overlies the hummocky cross-stratified sandstones of unit **a** (Figs 10 and 11). Across the Stenlille area its thickness varies from 1 to 2 m, but in general the muddy heterolithic facies thickens seaward toward well ST-4 (Fig. 11). It consists of thinly interbedded siltstone to very fine-grained sandstone and mudstone layers grading basinward into a siltstone-mudstone heterolithic facies (Figs 11 and 12). Sandstone and siltstone layers are graded and parallel laminated to hummocky cross-stratified. This facies is slightly burrowed by *Teichichnus*, *Chondrites* and *Rhizocorallium* traces as well as equilibrichnion traces and crypto-bioturbation (by amphipods?). Syneresis cracks and water escape features occur sporadically. Palynological samples of mudstone layers have yielded the marine dinoflagellate cysts *Dapcodinium priscum* and *Rhaetogonyaulax rhaetica*. In the more proximal wells ST-2 and ST-5, the upper part of this facies shows a gentle upward-coarsening and thickening of siltstone and sandstone layers toward the base of sandstone unit **b** (Figs 11 and 12).

Interpretation. The muddy heterolithic facies was deposited in the offshore transition zone near storm wavebase. Relative to the underlying lower shoreface sandstones of unit **a**, these facies

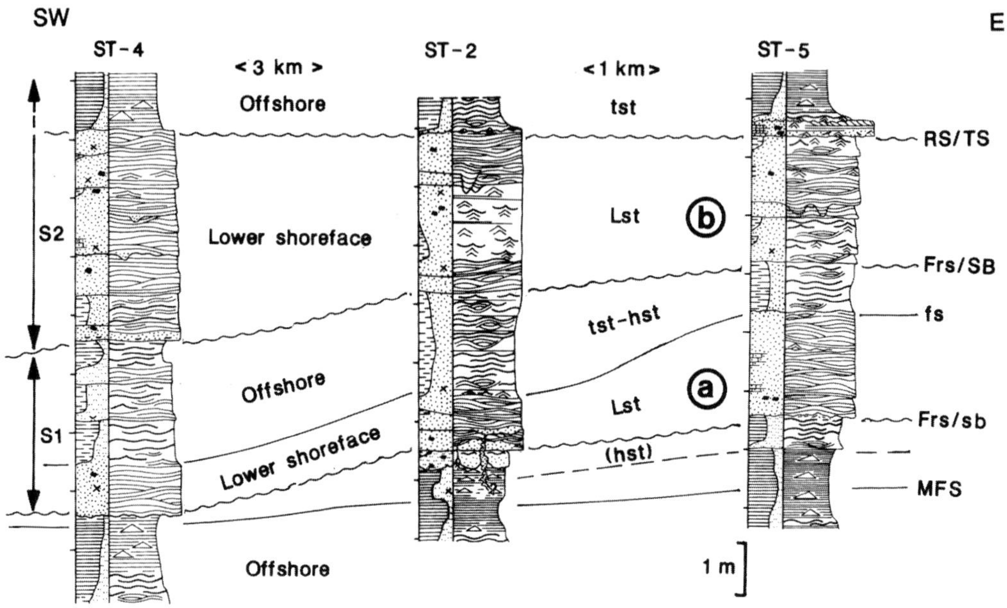

Fig. 11. Core-log cross section of sandstone 6 (dashed line on insert map in Fig. 10). This cross-section is parallel to depositional dip controlled by the major faults east of the Stenlille area (see map in Fig. 5B). Unit **a** thins basinward (SW), whereas unit **b** thickens in this direction. Annotation of boundaries and systems tracts as in Figs 6 and 7. **S1-S2** – high-order (fifth-order) sequence 1 and 2.

record a deepening and a relative sea-level rise. The upward-coarsening toward the base of unit **b** indicates renewed progradation of the shoreface toe.

*Upper sandstone unit **b**.* The base of unit **b** is a thin, fine-grained and hummocky cross-stratified sandstone bed with a sharp contact to the underlying muddy heterolithic facies (Figs 11 and 12). The sandstone bed is overlain by a sandy heterolithic facies dominated by equilibrichnion trace fossils. The sandy heterolithic facies is coarser than the underlying heterolithic facies and characterized by thicker, hummocky cross-stratified sandstone beds and silty mudstone layers. Upward, the sandy heterolithic facies rapidly passes into interbedded, hummocky cross-stratified sandstone beds with gutter casts and thin mudstone layers or heterolithic layers, and finally into amalgamated hummocky cross-stratified sandstones (Figs 11 and 12). The top of unit **b** is truncated by a thin erosive and wave-rippled sandstone layer, overlain by muddy heterolithic facies grading into mudstones. These mudstones contain abundant marine dinoflagellate cysts (*D. priscum* and *R. rhaetica*) and can be correlated over the entire basin marking MFS 6 (Fig. 4). In well ST-5 the wave-rippled sandstone layers is 0.2 m thick and distinctly coarser, consisting of a fine to medium-grained sandstone with abundant carbonaceous debris (Figs 11 and 12). Similarly coarse beds are also seen in the wells ST-12 and 13 (Fig. 10), and seems to occur in isolated lenses across the Stenlille field.

Interpretation. The interbedded hummocky cross-stratified sandstones and sandy heterolithic facies in unit **b** are interpreted as deposited near the base of the shoreface zone above storm-wavebase. The sharp and relatively rapid transition between unit **b** and the deeper water, muddier heterolithic facies below marks a somewhat abrupt shallowing. This interpretation assumes a normal shoreface–shelf transition at the time of deposition, where normal progradation caused by sediment supply would produce a more gradationally based, upward-coarsening succession. The shallowing at the base of unit **b** is interpreted as the result of a lowering of relative sea level. The sharp contact at the base is interpreted as the seaward and subtle expression of a forced regressive surface of erosion below mean storm wavebase (Fig. 11) comparable with the distal expression of such surfaces described by Plint (1996). The erosional truncation in top of unit **b** and overlying mudstones represent a shift toward deeper water and a drowning of the shoreface. The erosive surface is a ravinement surface (Figs 10–12, Rs). The coarser sand layers which occur locally are interpreted as transgressive sandstones formed as wave-reworked lags of material eroded from coarser shoreface and beach deposits during transgression and transported offshore by storm-related processes (e.g. Swift *et al.* 1991).

Timing of deposition and systems tracts

Sandstone 6 is a sharp-based sandstone succession in the Stenlille area as well as in all the wells along the basin margins (Fig. 4) and represents deposition from an overall forced regression. The depositional significance of unit **a** versus unit **b** is less obvious. They both consist of amalgamated, hummocky cross-stratified beds and record lower shoreface deposition above a lower bounding forced regressive erosion surface. As in sandstone 5, unit **a** of sandstone 6 is interpreted as the distal toe of a forced regressive shoreface deposits, and is regarded as part of the high-order (fifth-order) lowstand systems tract (Fig. 11, compare Fig. 9, step 2). The basal, forced regressive surface is interpreted as the distal expression of a high-order (fifth-order) sequence boundary.

After the drowning represented by the overlying offshore heterolithic facies, shoreline progradation resumed but was overtaken by a new forced regression causing deposition of unit **b**. The offshore heterolithic facies belongs to the high-order (fifth-order) transgressive systems tract and early(?) progradation of the high-order highstand (Fig. 11). Unit **b** shows no evidence of exposure and no third, forced regressive deposit is interpreted to be present basinward. Unit **b** seems to record only one episode of progradation during a fall in sea level, and is interpreted as deposited during the last forced regression and sea-level lowstand (Fig.11). As unit **b** is thicker than unit **a** and thickens westward and basinward. Whereas unit **a** thins in this direction, the shoreface of unit **b** extended more basinward than unit **a** and probably ended some 5–15 km west of Stenlille (Fig. 5b). In the context of this interpretation, units **a** and **b** form a gently seaward-dipping shingled geometry, and unit **b** reflects maximum regression within sandstone 6. The basal erosion surface of unit **b** is interpreted as both a high-order (fifth-order) sequence boundary (sb) and the overall sequence boundary of fourth-order (SB; Figs 4, 10 and 11). The high-order sequence, S1 occurs between the two high-order sequence boundaries and consists of unit **a** and the overlying muddy heterolithic facies (Fig. 11). The high-order sequence boundary beneath unit **b** forms the base of the overlying S2 high-order sequence (Fig. 11). The ravinement surface on top of unit **b** represents the transgressive surface (TS) related to a later lower-order overall sea-level rise.

In conclusion, the two-fold sandstone 6 at Stenlille consists of two forced regressive deposits representing two high-order lowstands; the upper regressive unit also represents the overall fourth-order lowstand systems tract of the sandstone 6. The different development of sandstone 6 compared to sandstone 5 is believed to reflect that the forced regressive deposits of sandstone 6 were formed during the late part of a composite third-order relative sea-level rise (see Fig. 2). The high-order falls were therefore diminished relative to the sea-level falls during formation of sandstone 5.

Sequence stratigraphic implications

Controls on the depositional response to forced regression

The interpretations of sandstone 5 and 6 suggest that these thin and widespread sharp-based shoreface successions are not simple forced regressive deposits related to a single continuous relative sea-level fall. Instead, they consist of shingled forced regressive shoreface deposits and record stepwise progradation of a shoreline under a general sea-level fall modulated by higher-frequency oscillations of relative sea level. It was in response to stepwise progradation related to repeated forced regressions that sandstones 5 and 6 come to resemble successions of 'stranded' parasequences (cf. Van Wagoner *et al.* 1990) and multiple or stepwise forced regressive deposits (e.g. Posamentier *et al.* 1992; Hunt & Tucker 1992, 1993, 1995; Kolla *et al.* 1995). However, the stepwise progradation of 'stranded' parasequences and multiple forced regressive deposits are envisaged as simple downstepping caused by punctuations and deceleration and acceleration of the overall fall (e.g. Hunt & Tucker 1995, fig. 1b). In contrast to this conceptual picture, the forced regressive shoreface units **a**, **b** and **c** of sandstone 5, and units **a** and **b** of sandstone 6 are separated by muddy, offshore deposits recording intervening drowning events of the superimposed high-frequency sea-level changes similar to the situation described by Hernández-Molina *et al.* (this volume). Because aggradation occurred during the small-scale sea-level rises in proximal areas, the fifth-order sequence boundaries climb vertically basinward, despite being formed during longer-term fourth-order relative sea-level fall (e.g. Fig. 9).

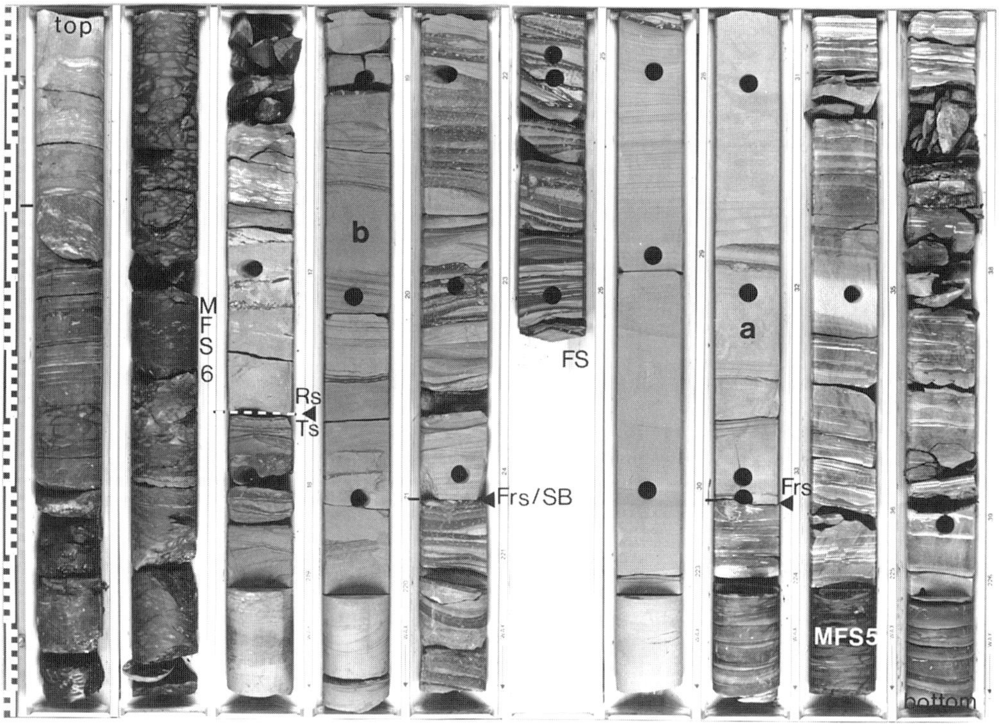

Fig. 12. Core photo of sandstone 6 in well ST-5 (same as logged in Fig. 11). Hummocky cross-stratified sandstone of unit **a** shows a basal, forced regressive surface **Frs** (3rd core box from right), and on top a flooding surface **FS** (5th core box from right) overlain by offshore muddy heterolithic facies. The basal forced regressive surface of unit **b** is marked by a shift to hummocky cross-stratified sandstone and sandy heterolithic facies and the forced regressive surface represents the overall sequence boundary SB of sandstone 6 (fourth-order) below the last, high-order forced regression recorded by unit **b**. On top the coarser sandstone overlying a ravinement surface **Rs** (3rd core box from left) represents a transgressive lag accumulation. The ravinement surface is also the overall (fourth-order) transgressive surface, **TS**. For vertical scale and facies description refer to log of ST-5 in Figure 11.

Sandstones 5 and 6 also show a different, more stretched and gently inclined, shingled depositional geometry (Fig. 9, step 5/6) than the simple down stepping depositional architecture of 'stranded' parasequences. The differences in depositional record and geometry between 'stranded' parasequence and the forced regressive deposits of sandstone 5 and 6 probably reflects differences in depositional gradients and palaeobathymetry of the basins discussed. The models and concepts of stranded parasequences and multiple forced regressive deposits basically originate from foreland, ramp-style margins. In the intracratonic and rifted Danish Basin, the large lateral extent of the thin, sharp-based shoreface sandstones 5 and 6 as well as their gently inclined shingled geometry demonstrated in the Stenlille area, indicate a very gently sloping to an almost flat basin-floor and a shallow palaeobathymetry. In this shallow basin, even a minor fall in sea-level would result in a rapid progradation of the shoreline far into the basin. Similarly, the intervening drowning events, although of small amplitude, will also be rapid and widespread.

During sea-level falls, the shallow palaeobathymetry and large sediment supply resulted in a very broad and gently inclined shoreface trajectory in the Triassic of the Danish Basin. Progradation of such a gently inclined shoreface will continuously face shallow water during sea-level fall and tend to overlie a wave-scoured surface far into the basin (cf. Plint 1988, 1996; Helland-Hansen & Gjelberg 1994). Sandstone 5 and 6 show long-distance progradation over 100 km into the basin (Figs 4 and 5). Without the fifth-order sea-level fluctuations superimposed on fourth-order falls, long-distance shoreline regression in such a shallow basin will probably be characterized by erosion and only limited sedimentation. In addition, the Rhaetian Danish Basin was receiving sand supplied from three sides of the basin, and the record of stepwise forced regressions shown by unit **a**, **b** and **c** probably demands a system rich in sand-prone sediment. Comparable forced regressive shingled deposits are interpreted to have formed on the flat, epicontinental shelf developed over Southern Sweden in Cambrian times (Hamberg 1994).

Position of the Sequence Boundary

Position of the sequence boundary over or below stepwise, forced regressive deposits has been the subject of much debate (e.g. Van Wagoner *et al.* 1990; Posamentier *et al.* 1992; Hunt & Tucker 1992, 1995; Kolla *et al.* 1995; Plint 1996 and papers in this volume). As discussed by Hunt & Tucker (1995), the position of the sequence boundary will partly reflect what is a practical, correlatable surface, but also rely on whether deposition takes place from a single, continuous sea-level fall or a sea-level fall punctuated by higher-frequency changes. In our interpretation we have been able to differentiate between forced regressive deposits formed during falling sea level, the *forced regressive systems tract*, and deposits related to the last, forced regressive to progradational part formed at sea-level lowstand. Such division often proves difficult (Fitzsimmons & Johnson and Plint & Nummedal this volume). This differentiation and subdivision is possible for both (i) the fifth-order, forced regressions and (ii) the fourth-order, forced regressive progradation of sandstone 5 and 6. The sequence boundary is placed at the surface of subaerial exposure over the deposits formed during falling sea-level, e.g. unit **b** of sandstone 5, here reflecting maximum regression corresponding to lowstand in sea-level as discussed by Hunt & Tucker (1992, 1993, 1995), Helland-Hansen & Gjelberg (1993), Helland-Hansen & Martinsen (1996), Plint (1996), and Plint & Nummedal (this volume). In a similar way the subaerial surface overlying sandstone 5 is interpreted as the sequence boundary recording maximum lowstand of the fourth-order sea-level fall. Although this surface is often eroded and represented by a transgressive erosion surface (cf. Walker & Plint 1992), it is a reliable and correlatable surface within the Gassum Formation (Nielsen *et al.* 1994). The basal, regressive surface of erosion below the forced regressive systems tract is a composite surface consisting of coalesced fifth-order forced regressive surfaces (sequence boundaries). Thereby the fourth-order regressive surface is broken up and less useful for basinwide correlations in genetic interpretations.

As discussed for sandstone 6 the fourth-order lowstand deposits of sandstone 5 and 6 are represented by the last, fifth-order forced regression (i.e. the last shingle) and the sequence boundary is placed at the basal forced regressive surface (e.g. as below unit **b** of sandstone 6 in Fig. 11), similar to the situation described by Hunt & Tucker (1995, fig. 1b) and Plint & Nummedal (this volume). Within the fifth-order forced regressive units, separation is more subtle. Each forced regressive unit reflects simultaneous deposition and basal erosion during the sea-level fall. The last part of a regressive unit represents sea-level lowstand and overlies a sequence boundary, e.g. as below unit **a** in sandstone 5 and 6 (Figs 7 and 11). Examples of a basinwide correlation of sandstone 5, 6, 7 and 8 where the fourth-order sequence boundaries are placed to separate shingled progradational units formed during falling relative sea-level from the last, forced regressive units and lowstand are shown in Fig. 4.

Conclusions

The detailed process-based sedimentology and sequence stratigraphy of two, sharp-based shoreface sandstones from the Upper Triassic Gassum Formation has been described and interpreted. The recurrent subdivision of these sharp-based sandstones into two or three, forced regressive deposits separated by offshore facies precludes the inference that deposition took placed during one forced regression. Based on a high-resolution sequence-stratigraphic interpretation the following conclusions can be drawn.

(1) Each sharp-based shoreface sandstone records deposition from two scales of relative sea-level fluctuations, a high-order (fifth-order) oscillation superimposed on a lower-order fall (fourth-order).

(2) High-order oscillations are recorded by an internal stratigraphy of alternating forced regressive shoreface sandstones and transgressive offshore deposits.

(3) The low-order falls resulted in widespread progradation and a seaward-dipping, shingled depositional geometry of the high-order deposits.

(4) Both the high-order forced regressive deposits and the overall, lower-order, forced regressive sharp-based deposits are dominated by forced regressive sandstones deposited during falling sea level and therefore included in *forced regressive systems tracts.*

(5) The sequence boundary is the surface of subaerial exposure, which may be substituted by a ravinement surface during subsequent sea-level rise, and is located at the top of both the high-order and lower-order forced regressive deposits representing maximum lowstand in sea level.

(6) For each fourth-order, sharp-based sandstone, the lowstand systems tract is the last, high-order forced regressive unit and the associated lowstand progradation. The sequence boundary is placed at the basal surface of forced regression distally, continuing in a landward direction into the subaerial surface of exposure.

(7) Within the high-order forced regressive units, the lowstand systems tract is subtle consisting of the last part formed at sea-level lowstand, and as such may be hard to differentiate in other systems. The high-order sequence boundary is placed at the forced regressive surface, continuing in the landward direction along the bedding plane below the associated shoreface to beach deposits, and finally merging with the subaerial sequence boundary over the forced regressive systems tract.

(8) The stepwise, high-order forced regressive deposits dynamically resemble 'stranded' parasequences, but differ from the conceptualized picture of 'stranded' parasequences as simple downstepping of forced regressive deposits by showing: (i) a gently seaward-dipping shingled geometry, (ii) distinctive sediments deposited during the intervening high-order drowning of the shorefaces that punctuated the sea-level falls and (iii) high-order sequence boundaries that can climb vertically because of deposition during these high-order drowning deposits.

(9) The thin and widespread sharp-based shoreface sandstones with internal, shingled forced regressive units interpreted to be the result of deposition in a very gently dipping, shallow intracratonic basin. Records of high-order, stepwise forced regressive units like unit **a**, **b** and **c** probably also demand a sediment rich system.

(10) The Triassic sharp-based shoreface successions demonstrate the importance of deposition during falling relative sea level in shallow and low-angle, intracratonic basins.

This paper draws on the authors PhD projects supervised by F. Surlyk and G. K. Pedersen at the University of Copenhagen. The supervisors and the Basin Research Group at University of Copenhagen are thanked for their encouragement and many fruitful discussions. P. N. Johannessen (Geological Survey of Denmark and Greenland) is thanked for critical reviews of early versions. Dansk Olie- & Naturgas A/S, University of Copenhagen, the Danish Research Academy and Geological Survey of Denmark and Greenland are thanked for financial and technical support.

References

AINSWORTH, R. B. 1994. Marginal marine sedimentology and high-resolution sequence analysis; Bearpaw-Horseshoe Canyon transition, Drumheller, Alberta. *Bulletin of Canadian Petroleum Geology*, **42**, 26–54.

——, BOSSCHER, H. & NEWALL, M. J. 2000. Forward modelling of forced regressions. Evidence for the genesis of attached and detached lowstand systems. *This volume*.

BERTELSEN, F. 1978. *The Upper Triassic-Lower Jurassic Vinding and Gassum Formations of the Norwegian-Danish Basin*. Danmarks Geologiske Undersøgelser Series B, **3**.

CLIFTON, H. E. 1981. Progradational sequences in Miocene shoreline deposits, Southeastern Caliente Range, California. *Journal of Sedimentary Petrology*, **51**, 165–184.

DOMINGUEZ, J. M. L. & WANLESS, H. R. 1991. Facies architecture of a falling sea-level strandplain, Doce River coast, Brazil. *In*: SWIFT, D. J. P., OERTEL, G. F., TILLMAN, R. W. & THORNE, J. A. (eds) *Shelf Sand and Sandstone Bodies*. International Association of Sedimentologists Special Publications, **14**, 259–281.

FITZSIMMONS, R. & JOHNSON, S. 2000. Forced regressions: recognition, architecture and genesis in the Campanian of the Bighorn Basin, Wyoming. *This volume*.

HAMBERG, L. 1994. *Anatomy of clastic coastal sequences of the Rhaetian Gassum Formation, Stenlille, Denmark*. PhD Thesis, University of Copenhagen, Denmark.

——, NIELSEN, L. H. & KOPPELHUS, E. B. 1994. Dynamics and timing of shoreface deposition in the intracratonic Danish basins: An example of Norian-Hettangian deposition controlled by high-frequency sea-level fluctuations (abstract). *In*: JOHNSON, S. (ed.) *High Resolution Sequence Stratigraphy: Innovations and Applications*, Liverpool, March 1994, 335–337.

HELLAND-HANSEN, W. & GJELBERG, J. C. 1994. Conceptual basis and variability in sequence stratigraphy: a different perspective. *Sedimentary Geology*, **92**, 31–52.

—— & MARTINSEN, O. J. 1996. Shoreline trajectories and sequences: Description of variable depositional-dip scenarios. *Journal of Sedimenary Research*, **66**, 670–688.

HERNÁNDEZ-MOLINA, F. J., SOMOZA, I. & LOBO, F. 2000. Seismic stratigraphy of the Gulf of Cádiz continental shelf: a model for late Quaternary very high-resolution sequence stratigraphy and response to sea-level fall. *This volume*.

—— & TUCKER, M. E. 1992. Stranded parasequences and the forced regressive wedge systems tract: deposition during base-level fall: *Sedimentary Geology*, **81**, 1–9.

—— & —— 1993. Sequence stratigraphy of carbonate shelves with an example from the mid-Cretaceous (Urgonian) of southeast France. *In*: POSAMENTIER, H. W., SUMMERHAYES, C. P., HAQ, B. U. & ALLEN, G. P. (eds) *Sequence Stratigraphy and Facies Associations*. International Association of Sedimentologists, Special Publications, **18**, 307–341.

—— & —— 1995. Stranded parasequences and the forced regressive wedge systems tract: Deposition during base-level fall – reply. *Sedimentary Geology*, **95**, 147–160.

KOLLA, V., POSAMENTIER, H. W. & EICHENSEER, H. 1995. Stranded parasequences and the forced regressive wedge systems tract: Deposition during base-level fall – discussion. *Sedimentary Geology*, **95**, 139–145.

MELLERE, D. & STEEL, R. 2000. Style contrast between forced regressive and lowstand/transgressive wedges in the Campanian of south-central Wyoming. *This volume*.

MITCHUM, R. M. & VAN WAGONER, J. C. 1991. High-frequency sequences and their stacking patterns: Sequence-stratigraphic evidence of high-frequency eustatic cycles. *Sedimentary Geology*, **70**, 131–160.

NIELSEN, L. H. 1995. *Genetic Stratigraphy of the Upper Triassic-Middle Jurassic deposits of the Danish Basin and Fennoscandian Border Zone*. PhD Thesis, University of Copenhagen, Denmark.

——, HAMBERG, L. & KOPPELHUS, E. B. 1994. Sequence development of a shallow marine – non-marine, intra-cratonic basin-fill; the Norian-Hettangian of the Danish Basin. *In*: HAMBERG, L. *Anatomy of clastic coastal sequences of the Rhaetian Gassum Formation, Stenlille, Denmark (Part 2)*. PhD Thesis, University of Copenhagen, Denmark.

NUMMEDAL, D., RILEY, G. W. & TEMPLET, P. L. 1993. High-resolution sequence architecture: a chronostratigraphic model based on equilibrium profile studies. *In*: POSAMENTIER, H. W., SUMMERHAYES, C. P., HAQ, B. U. & ALLEN, G. P. (eds) *Sequence Stratigraphy and Facies Associations*. International Association of Sedimentologists Special Publications, **18**, 55–68.

PLINT, A. G. 1988. Sharp-based shoreface sequences and 'offshore bars' in the Cardium Formation of Alberta: Their relationship to relative changes in sea level. *In*: WILGUS, C. K., HASTINGS, B. S., KENDALL, C. G. ST. C., POSAMENTIER, H. W., ROSS, C. A. & VAN WAGONER, J. C. (eds) *Sea-Level Changes – An Integrated Approach*. Society of Economic Paleontologists and Mineralogists, Special Publications, **42**, 357–370.

—— 1996. Marine and nonmarine systems tracts in fourth-order sequences in the Early-Middle Cenomanian, Dunvegan Alloformation, northeastern British Columbia, Canada. *In*: HOWELL, J. A. & AITKEN, J. F. (eds) *High Resolution Sequence Stratigraphy: Innovations and Applications*. Geological Society, London, Special Publications, **104**, 159–191.

—— & NUMMEDAL, D. 2000. The falling stage systems tract: recognition and importance in sequence stratigraphic analysis. *This volume*.

POSAMENTIER, H. W. & MORRIS, W. S. 2000. Aspects of the stratal architecture of forced regressive deposits. *This volume*.

——, ALLEN, G. P., JAMES, D. P. & TESSON, M. 1992. Forced regressions in a sequence stratigraphic framework: Concepts, examples, and exploration significance. *American Association of Petroleum Geologists Bulletin*, **76**, 1687–1709.

SURLYK, F., ARNDORFF, L., HAMANN, N.-E., HAMBERG, L., JOHANNESSEN, P. N., KOPPELHUS, E. B., NIELSEN, L. H., NOE-NYGAARD, N., PEDERSEN, G. K. & PETERSEN, H. I. 1995. High-resolution sequence stratigraphy of a Hettangian-Sinemurian paralic succession, Bornholm, Denmark. *Sedimentology*, **42**, 323–354.

SWIFT, D.J. P., PHILLIPS, S. & THORNE, J. A. 1991. Sedimentation on continental margins, IV: lithofacies and depositional systems. *In*: SWIFT, D. J. P., OERTEL, G. F., TILLMAN, R. W. & THORNE, J. A. (eds) *Shelf Sand and Sandstone Bodies*. International Association of Sedimentologists Special Publications, **14**, 89–152.

VAN WAGONER, J. C. V., MITCHUM, R. M., CAMPION, K. M. & RAHMANIAN, V. D. 1990. *Siliciclastic sequence stratigraphy in well logs, cores and outcrops: Concepts for high-resolution correlation of time and facies*. AAPG Methods in Exploration Series, **7**.

——, POSAMENTIER, H. W., MITCHUM, R. M., VAIL, P. R., SARG, J. F., LOUTIT, T. S. & HARDENBOL, J. 1988. An Overview of the Fundamentals of Sequence Stratigraphy and Key Definitions. *In*: WILGUS, C. K., HASTINGS, B. S., KENDALL, C. G. ST. C., POSAMENTIER, H. W., ROSS, C. A. & VAN WAGONER, J. C. (eds) *Sea-Level Changes – An Integrated Approach*. Society of Economic Paleontologists and Mineralogists, Special Publications, **42**, 39–46.

VEJBÆK, O. V. 1989. Effects of asthenospheric heat flow in basin modelling exemplified with the Danish Basin. *Earth and Planetary Science Letters*, **95**, 97–114.

WALKER, R. G. & PLINT, A. G. 1992. Wave- and storm-dominated shallow-marine systems. *In*: WALKER, R. G. & JAMES, N. P. (eds) *Facies Models: Response to Sea Level Changes*. Geological Association of Canada, 219–238.

The significance of the Etive Formation in the development of the Brent system: distinction of normal and forced regressions

TINA R. OLSEN[1] & RON J. STEEL[2]

Geological Institute, University of Bergen, Allégaten 41, N-5007 Bergen, Norway

[1]Present address: BP Amoco Norge AS, PO Box 197, N-4065 Stavanger, Norway

[2]Present address: Deptartment of Geology/Geophysics, University of Wyoming, Laramie, WY 82071, USA

Abstract: Recent sequence stratigraphic debate on the Brent system have focused on the interpreted nature of the progradational trajectory (horizontal, slightly upwards or downwards) of the shoreline (Rannoch/Etive Formations) through time, as this gives a direct measure of how late Aalenian–Bajocian relative sea level changed during regression. Early interpretations emphasized the unified shallowing-upward nature of the Rannoch–Etive–Ness depositional system, and implicitly accepted a uniform shoreline progradation, i.e. a shoreline trajectory that was horizontal or slightly rising, implying a stable or slightly rising relative sea level. No irregularities of the trajectory were noted, and unusual shifts in facies, grain size etc. were normally related to autocyclic processes. More recent work has suggested that in some instances there *is* evidence for more irregular shoreline progradation at certain times, and for fall(s) in relative sea level and forced regression. This evidence comes from incised valleys and deep erosion/subaerial exposure surfaces from the landward (Etive–Ness boundary) and basinward (Rannoch–Etive) reaches of the Brent system respectively. However, it is currently unclear if any of these downshift surfaces recognized in the strandplain/coastal plain and shoreface environments are in time-equivalent strata.

Current debate is mostly handicapped by a lack of agreement on the origin and depositional facies of the Etive Formation. There is significant debate about the relative amounts of fluvial, tidal and wave influence detected in the strata of this formation, with some authors arguing for a dominance of fluvial distributaries and mouth-bar deposits, whereas others propose either tidal-channel and inlet deposits or wave-dominated shoreface and strandplain settings. The stratigraphy is impacted by this disagreement. The character and sharp base of the Etive Formation can be argued to be consistent with normal shoreline processes, where wave or tidal conditions can produce significant erosion in the shoreface, without the necessity of any forced regression. Other interpretations, particularly where the Etive Formation is seen in terms of fluvial facies and processes, require a significant basinward shift of the shoreline to explain the Rannoch–Etive superposition, and a fall of sea level to cause the erosive boundary between the two formations.

However, there is now ample evidence, including new evidence presented here, that both of the end-member scenarios for the progradation of the Brent system are incorrect. The notion that the overall progradation was entirely a product of normal regression, during stable and/or slightly rising relative sea level, is negated by local evidence of incised valleys, of subaerial exposure and plant growth in lower shoreface strata in the Rannoch Formation, and of repeated erosion surfaces with coarse-grained lags at the base of the Etive Formation. On the other hand, the idea of continuous sea level fall or of a single, late-stage fall, such that there was regional valley incision of the Etive into the Rannoch Formation and that the former is entirely younger than the latter, is negated by local evidence of gradual upward facies change between the formations, of stratigraphic interfingering between the formations, and of time lines passing through the Etive into the Rannoch Formation. It is perhaps not surprising that the system's overall regressive trajectory varied in time from being forced to being normally regressive, and that further detailed local studies are required before regional generalisations can be made.

The Middle Jurassic Brent Group forms one of the most extensive and prolific hydrocarbon reservoir horizons in the British and Norwegian oil and gas fields located between 60ºN and 62ºN in the Viking Graben, northern North Sea (Fig. 1). Numerous studies on the Brent Group over the last 20 years (see Richards 1992) have revealed controversy regarding virtually every aspect of the group, with particular disparity of views on the sedimentological and sequence

From: HUNT, D. & GAWTHORPE, R. L. (eds) *Sedimentary Responses to Forced Regressions*. Geological Society, London, Special Publications, **172**, 91–112. 1-86239-063-0/00/$15.00

stratigraphic interpretations of the Etive Formation, and its relationship to the underlying Rannoch Formation.

Presently, many of the Brent province oil and gas fields have reached a mature stage where increased recovery is of prime importance. One possible way to improve late-stage field production is to critically re-examine already gathered data and literature, emphasising key aspects of interest. In this paper we review literature published on the Etive Formation in terms of (i) depositional environment and, partly dependent on this, (ii) the sequence stratigraphic relationships of the Etive to the underlying Rannoch Formation, as well as facies relationships internally within the Etive. Before addressing the Etive Formation specifically, we firstly highlight a few obvious trends and tendencies in the published literature on the Brent Group in addition to those already emphasized by Richards (1992).

(1) Pioneer papers focused on sedimentary facies analysis of a few scattered wells and presented facies models placed in the context of relatively simple lithostratigraphical subdivisions (Budding & Inglin 1981; Parry *et al.* 1981; Simpson & Whitley 1981). In contrast, papers published since 1990 are based on more complete integration of sedimentological, palynological, petrographical, structural and seismic data. In general, such studies have aimed towards producing a regional or subregional sequence stratigraphic model for the Brent system (Helland-Hansen *et al.* 1992; Mitchener *et al.* 1992; Johannessen *et al.* 1995; Olsen & Steel 1995; Fjellanger *et al.* 1996). Papers in this latter category have aimed towards spatial and temporal exploration scale prediction of sandstone distribution in both drilled and especially undrilled areas.

(2) The need for improved recovery of hydrocarbons has been reflected in the need for more complete sedimentary descriptions integrated with field production and petrophysical data. An example of changing emphasis on detail comes from the Murchison Field, UK sector. The early work of Simpson & Whitley (1981) emphasized the simple nature and homogeneity of the Etive Formation. In contrast, more recent work by Daws & Prosser (1992) led to the recognition of four orders of permeability heterogeneity that, on the basis of detailed examination of core and wireline logs, can be related to facies transitions, sedimentological bounding surfaces, laminations and tectonic structures.

(3) Traditionally, the Rannoch and Etive Formations have been viewed in terms of a single sand-prone upward-coarsening unit, reflecting relatively uniform coastline behaviour and progradation (e.g. Graue *et al.* 1987, Fält *et al.* 1989). Deviations or breaks in this simple trend, such as marked vertical grain-size jumps, erosion surfaces or sharp-based fining-upward units were usually interpreted in terms of autocyclic processes such as the shifting of distributary channels (e.g. Brown *et al.* 1987; Brown & Richards 1989) or tidal channels (e.g. Scott 1992). In the last few years, some of these anomalies are being recognized as more widespread features that have been related to relative sea-level changes during the overall progradation of the system.

In terms of sequence stratigraphy, the origin of erosion surfaces, abrupt grain-size changes and unusual vertical facies changes, have been attributed to: (1) major basinward shifts of the system and the forced progradation of alluvial plain (Van Wagoner *et al.* 1993; Reynolds 1995) or braidplain systems (Elliott 1989) across the area which previously had been the shoreline, (2) local development of incised valleys (Jennette & Riley 1996), (3) increased accommodation space to sediment supply ratio, during intervals of more rapid rise of relative sea level or when the system prograded into deeper water to the north (Olsen & Steel 1995) or (4) minor forced regression alternating with transgressive episodes (Olsen & Steel 1995).

These few points illustrate clearly that both the description of depositional features and environments, and the sequence stratigraphic interpretations of the Etive Formation should be improved through re-examination and unification of ideas on the behaviour of the progradational part of the Brent system in time and space. In the review below, we pay particular attention to the growing evidence for relative sea-level variations during progradation of the Brent system. It is apparent that intervals of both normal (sea level stability or rise) and forced regression (sea-level fall) of the shoreline can be documented within the Brent system at various times.

Geological setting

The 'Brent Province' of hydrocarbon discoveries is geographically coincident with the northern part of the Viking Graben and its flanking terraces to the west and east, the East Shetland Basin and the Horda Platform, respectively. Thickness of the Brent Group reflects structural relief of the North Sea graben system varying from less than 100 m in the western part of the East Shetland Basin and on the Horda Platform to more than 600 m in the centre of the Viking trough.

The Viking Graben is an extensional basin

floored by thin (12–15 km) pre-Mesozoic basement (Yielding *et al.* 1992). The pre-Mesozoic fabric of the northern Viking Graben is characterized by two main lineament trends: NE–SW trends of predominantly Caledonian origin and N–S trends of Permian/Triassic origin (Eynon 1981; Threlfall 1981). The major crustal thinning of Permian and early Triassic age (Roberts *et al.* 1993) caused tilting of basement fault blocks (Badley *et al.* 1984; Steel & Ryseth 1990; Yielding *et al.* 1992). By mid-Triassic times, a post-rift thermal subsidence basin had been established (Steel 1993).

The factors leading to the establishment of the Brent system are still debated, although its initiation is generally believed to have been related to Toarcian–?early Aalenian domal uplift of the North Sea rift dome (Underhill & Partington 1993). Development of the dome caused uplift and a low-order relative fall of sea-level (Ziegler 1982; Yielding *et al.* 1992), ultimately leading to widespread subaerial exposure, erosion (Underhill & Partington 1993), and hence an increase in the sediment supply leading to deposition of a major clastic wedge in the North Sea basin (Steel 1993). Nevertheless, the general structural control on the deposition of the Brent Group was thermal subsidence following early Triassic crustal stretching, although in many areas there is increasing evidence for extensional block-rotation in the late Bajocian and Bathonian (Johannessen *et al.* 1995; Fjellanger *et al.* 1996). In this context, it is important to note that faulting during progradation of the Brent system (e.g. during the main growth of the Rannoch and Etive Formations) was subtle, with only few facies changes observed across the main faults (Graue *et al.* 1987; Fjellanger *et al.* 1996). This seemingly indicates that during progradation, the rate and amount of sediment supply to the shoreline was capable of adjusting to the accommodation space created by the faulting and thermal subsidence. In contrast, the upper Ness and the Tarbert Formations reflect syndepositional movements by a marked thickening across some of the major lineaments (Graue *et al.* 1987; Johannessen *et al.* 1995). Thus, the Rannoch, Etive and lower Ness Formations can be regarded as a post-rift succession, whereas the upper Ness and Tarbert Formations, would represent early syn-rift deposition (Johannessen *et al.* 1995; Nøttvedt *et al.* 1995; Ravnås *et al.* 1997).

Following the deposition of the Brent Group, extension peaked during the late Jurassic deposition of the Heather and Draupne Formations of the Humber Group (Yielding *et al.* 1992). During this time the footwalls to major normal faults were uplifted and eroded, and the main structural traps (tilted fault blocks) for the Brent Province oil and gas fields were created. Restricted conditions in the grabens formed at this time led to anoxic conditions that resulted in deposition of Humber Group source rocks (Fjaeran & Spencer 1991).

Stratigraphy and development of the Brent Group

The Brent Group is Aalenian to early Bathonian in age, although late Bajocian strata are commonly missing because of sub-regional unconformity(ies) along the flanks of the basin at this level (Johannessen *et al.* 1995). Deegan & Scull (1977) divided and formalized the Brent Group into five lithostratigraphic formations; the Broom, Rannoch, Etive, Ness and Tarbert Formations. Graue *et al.* (1987) suggested the incorporation of a new unit, the Oseberg Formation, to include the basal sandstones in the marginal areas of the Norwegian Sector.

Prior to the outbuilding of the Brent system, a shallow sea, dominated by deposition of fine-grained sediments (Dunlin Group), extended across large areas of the present northern North Sea (Marjanac 1995; Marjanac & Steel 1997). In this sea, a series of transverse fan-deltas built out from the basin margins (Oseberg and Broom Formations) and subdued a late Toarcian-early Aalenian, fault-controlled topography (Helland-Hansen *et al.* 1992). Fan deltas of the Oseberg Formation were drowned during the latest Aalenian in a transgression believed to be of regional significance (Graue *et al.* 1987; Helland-Hansen *et al.* 1992). The transgression produced an extensive marine basin that opened northwards. The Rannoch–Etive–lower Ness system, then located at a position close to the present 60° N, prograded towards the north across the foundations created by the Oseberg and Broom Formations.

Seen on a very gross scale, the Rannoch–Etive Formations form a variably thick coarsening-upwards sandstone succession, that has been interpreted in terms of a storm-wave-dominated, delta-front or barrier-island shoreface succession (Budding & Inglin 1981; Brown *et al.* 1987; Graue *et al.* 1987; Fält *et al.* 1989; Helland-Hansen *et al.* 1989, 1992). The Ness Formation forms a variably thick heterolithic interval of mudstones, siltstones and sandstones interpreted as coastal plain deposits that are partly the terrestrial time-equivalents of the Rannoch–Etive shallow marine facies (Ryseth 1989; Helland-Hansen *et al.* 1992). The system switched to an overall transgressive mode during the late Bajocian that resulted in the deposition of the Tarbert Formation. Because of important transgressive ravinement, marine sandstones of the Tarbert Formation almost invariably display a sharp and erosional base over back-barrier and lagoonal deposits of the Ness Formation. The contact between the Tarbert and the overlying Heather Formation is also relatively abrupt but more often represents an erosional topography related to footwall uplift during the earliest phases of late Cimmerian extension and block rotation (Johannessen *et al.* 1995). The Heather Formation was deposited in an offshore environment in front of the Tarbert Formation shoreline systems. This unit represents a major flooding event in the Bathonian separating the Brent and the Krossfjord megasequences (Steel 1993).

The Brent shoreline: key issues

Two fundamental issues are critical to interpretation of the regressive components of the Brent system, and both are related to the Etive Formation. These issues are: (i) the correct identification of depositional structures and hence environments of deposition in the Etive Formation, since existing disagreements impact dramatically on dynamic stratigraphic interpretations; (ii) the nature and sequence stratigraphic significance of the relationship(s) between the Etive and the underlying Rannoch Formation. Various interpretations of the Etive Formation as either an integral part of a normal regressive shoreline, a widespread valley-fill cut down into the Rannoch Formation, or a variably forced regressive/normal regressive shoreline respectively depend on whether the boundary between these units is gradational, unconformable or variable along its length.

Etive depositional environments

The stratigraphic position of the Etive Formation helps to constrain its overall depositional and environmental setting. It is positioned above the very fine- to fine-grained, micaceous Rannoch Formation, generally accepted to have been deposited in lower shoreface to offshore transitional environments (Budding & Inglin 1981; Graue *et al.* 1987; Scott 1992). In contrast, the overlying Ness Formation is coal-bearing and of coastal plain origin (Livera 1989; Ryseth 1989). Given this depositional context, there are several possible alternatives for the depositional environments of the Etive Formation, depending on the absence or presence of discontinuities that may indicate significant basinward shifts of the depocentre. In the simplest solution, assuming an absence of discontinuities, strata between the Rannoch and Ness Formations would represent upper shoreface-strandplain (along barrier-island reaches), or uppermost delta front (along wave-dominated delta reaches) environments. However, if discontinuities exist, for example at the base and top of the Etive, then different kinds of channel deposits might dominate. However, literature review shows that the more detailed subenvironmental interpretations vary greatly depending on the areas studied; some of these are well-documented and backed up by core-descriptions, others are not. Some of the more detailed sedimentological analyses are given by Daws & Prosser (1992), Scott (1992), Johannessen *et al.* (1995), Olsen & Steel (1995) and Reynolds (1995).

In essence, the main debate concerns the relative amounts of fluvial, tidal and wave influence detected in the strata. Whereas some authors argue for fluvial distributaries and mouth bar deposits (Brown & Richards 1989; Johannessen *et al.* 1995), others propose tidal-channel and inlet deposits (Daws & Prosser 1992; Scott 1992), or mainly wave-dominated shoreface, foreshore, barrier and strandplain settings (Cannon *et al.* 1992; Mitchener *et al.* 1992; Jennette & Riley 1996) (Fig. 1). The Appendix summarizes these various depositional models, which are considered in more detail below.

Fluvial-dominated environments

Stacked fluvial (braided) distributaries. The entire Etive Formation has been interpreted to have originated as a series of stacked fluvial (braided) distributary channel deposits in the

Fig. 1. The Etive Formation (Brent Group) has been interpreted quite differently by various researchers in terms of both dominant sedimentary processes and response to relative sea-level change, as summarized here. The various interpretations are here organised into six groups, and have been plotted accordingly to the geographical location of the studied Brent oil and gas fields. A complete list of author(s), area/field/wells, sedimentary facies description, sedimentary and sequence stratigraphic interpretations are given in the Appendix. (**a**) Shoreface-barrier bar complex interpretation; (**b**) shoreface-barrier bar complex dominated by longshore drift; (**c**) shoreface-barrier bar complex with identified tidal inlets; (**d**) proximal delta-front with distributary channels with mouth bars; (**e**) incised valley infill with braided fluvial distributaries; (**f**) areas where forced regression has been discussed.

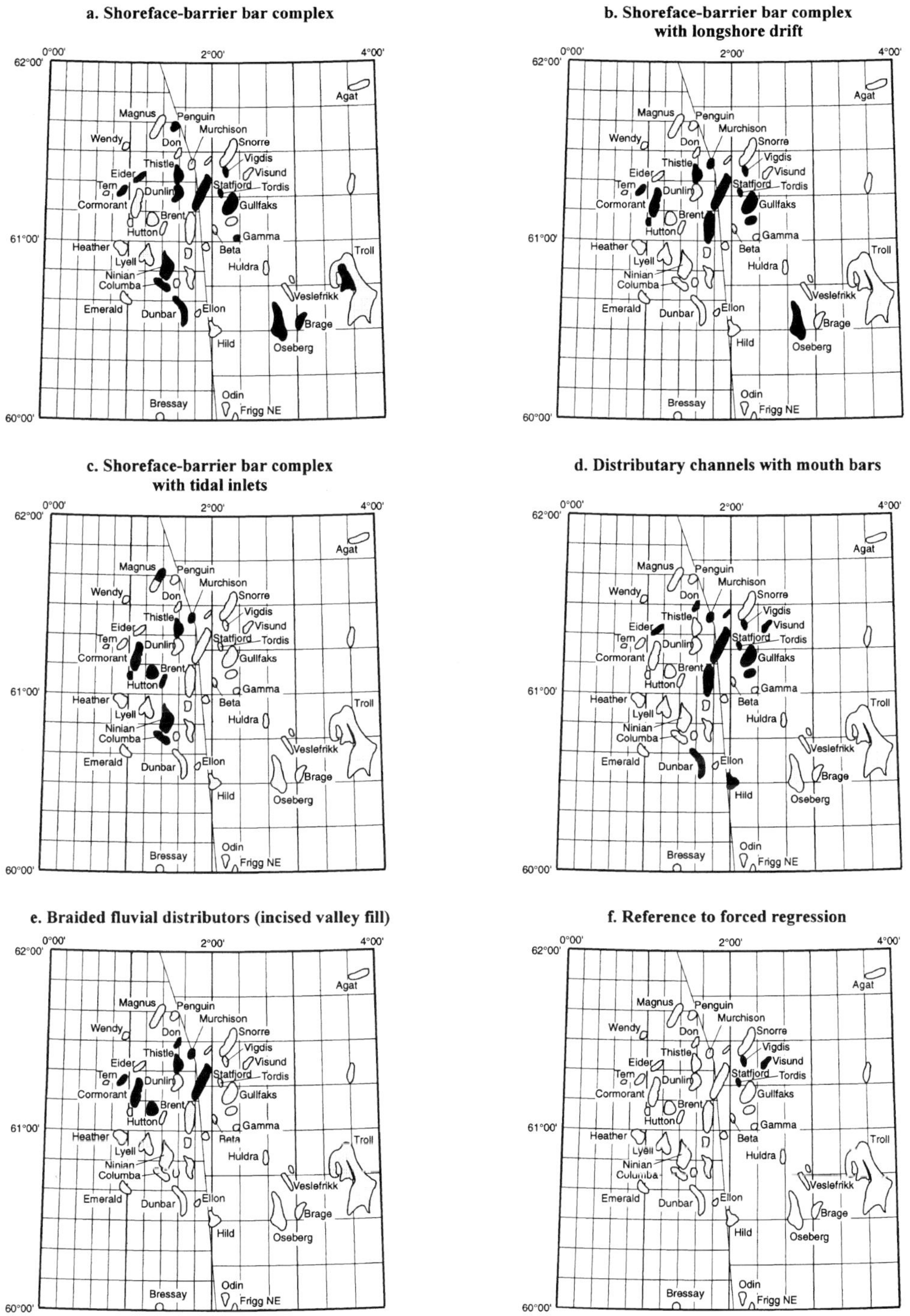
a. Shoreface-barrier bar complex
b. Shoreface-barrier bar complex with longshore drift
c. Shoreface-barrier bar complex with tidal inlets
d. Distributary channels with mouth bars
e. Braided fluvial distributors (incised valley fill)
f. Reference to forced regression
0°00'
2°00'
4°00'
62°00'
61°00'
60°00'
Agat
Magnus
Penguin
Murchison
Wendy
Don
Snorre
Vigdis
Thistle
Visund
Eider
Tern
Dunlin
Statfjord
Tordis
Cormorant
Gullfaks
Brent
Hutton
Gamma
Heather
Beta
Troll
Lyell
Huldra
Ninian
Columba
Veslefrikk
Emerald
Dunbar
Ellon
Brage
Hild
Oseberg
Bressay
Odin
Frigg NE

Don, Murchison and Statfjord Fields. This interpretation is made on the basis of stacked fining-upwards units, some with erosive bases, of fine to rarely coarse-grained sandstones with various types of cross bedding and ripple laminations (Parry *et al.* 1981; Brown & Richards 1989; Van Wagoner *et al.* 1993). A similar origin has been suggested for the Thistle Field on the basis of poorly sorted, stacked fining-upward units with moderate or low-angle cross-stratification/plane parallel lamination, and the absence of marine trace fossils (Reynolds 1995). Importantly, Reynolds (1995) noted the presence of several marine trace fossils attesting a high-energy marine environment in the upper part of the Etive Formation and attributed this to an increasing marine influence in the fluvial channels. Elliott (1989) noted that channel sandstones are extremely common in the lower part of the Etive Formation, and interpreted these deposits in terms of a braid-plain system, noting that as being atypical of a wave-dominated system.

Fluvial-dominated channel/mouth bar complex. Parts of the Etive Formation within the Murchison Field and the Tampen Spur area have been interpreted in terms of fluvial distributary and mouth bar deposits (Simpson & Whitley 1981; Graue *et al.* 1987; Johannessen *et al.* 1995; Olsen & Steel 1995; Fjellanger *et al.* 1996). The deposits comprise medium- to coarse-grained, sometimes pebbly, poorly sorted, current rippled, planar and trough cross-stratified sandstones arranged in single or composite coarsening-upward units. Sharp-based, fining-upward units also occur. The poor sorting, lack of typical wave-generated structures and coarsening-upward pattern is taken to reflect out-building of mouth bars in an upper shoreface setting, whereas the fining-upward trends may reflect fluvial-channel fill in erosional contact with the mouth bar sands (e.g. Johannessen *et al.* 1995).

Wave-dominated environments

Upper shoreface/foreshore (barrier bar or upper delta front) and strandplain setting. The most common interpretation of the Etive Formation is one of deposition within (i) upper shoreface-strandplain (along barrier reaches) or (ii) uppermost delta front areas, albeit along wave-dominated reaches (Budding & Inglin 1981; Graue *et al.* 1987; Livera & Caline 1990; Daws & Prosser 1992; Scott 1992; Johannessen *et al.* 1995; Olsen & Steel 1995; Jennette & Riley 1996; Fjellanger *et al.* 1996). The upper shoreface/foreshore environment is represented by fine to medium-grained, moderately to well-sorted clean sandstones with small-scale trough or low-angle cross-bedded strata, plane parallel lamination and minor current ripple lamination, arranged in slightly fining- or coarsening-upward grain-size trends. The observed facies are interpreted as the product of storm-waves and fairweather currents within the surf- and swash zones of the upper shoreface/foreshore environments. The foreshore zone passes gradationally into the subaerial backshore zone within the uppermost part of the Etive Formation (Scott 1992; Jennette & Riley 1996). Here, the sandstones have fabrics that vary between vaguely stratified to mottled to homogeneous. Each stratification type characteristically shows evidence of minor soft sediment deformation and dewatering structures. Burrowing and root traces are common. The succession described above is interpreted in terms of a prograding barrier beach complex (Olsen & Steel 1995). Livera & Caline (1990) noted that the Etive Formation was not always a barrier system, but at times distributaries supplied sediment onto the shoreface, forming cuspate wave-dominated deltas.

Tide-influenced environments

Barrier shoreline with tidal inlets, and other tidal channels. Tidal-channel deposits have been recognized locally, and are best described from the Murchison and Cormorant Fields. Here tidal channels are represented by stacked upward-fining units of fine- to medium-grained sandstones dominated by trough cross-stratification and clay/silty draped ripples towards the top, as well as rare bi-directional sets of cross-bedding (e.g. Cannon *et al.* 1992; Daws & Prosser 1992; Olaussen *et al.* 1992; Scott 1992). Carbonaceous rip-up clasts are often present at the base of the channels, perhaps indicating proximity to a subaerial coastal plain (Cannon *et al.* 1992; Daws & Prosser 1992; Scott 1992). However, although Cannon *et al.* (1992) and Scott (1992) acknowledged that unequivocal evidence of tidal processes are absent, they noted that the presence of the bi-directional cross-bedding and clay/siltstone-draped ripples, and their position with respect to the barrier suggest that the channels represent tidal inlets. Scott (1992) took the paucity of tidal features and the scarce development of inlet facies in the southern Cormorant area to suggest that at least locally the Etive Formation developed on a microtidal coast.

Uncertainty in facies and surface interpretations

Although many of the above interpretations are well argued and generally plausible, a great deal of uncertainty exists (Appendix). For example one of the most common facies of the Etive Formation, stacked fining-upward units with low-angle trough cross-stratification commonly associated with plane parallel laminated sandstones, has apparently been variably interpreted as (i) the product of storm-waves and fair-weather currents on the upper shoreface (Olsen & Steel 1995), (ii) evidence of fluvially influenced channels (Reynolds 1995) or (iii) as tidal channel-fill sandstones (Daws & Prosser 1992). Consequently, the sharp-based nature of some of the fining-upward units within the Etive Formation (especially at its base) has variously been taken as proof of the importance of erosion by autocyclic processes related to longshore troughs, bars and rip channels on a normally prograding upper shoreface (Jennette & Riley 1996) as evidence of the importance of tidal channel incision (Scott 1992) or of distributary channel erosion (Livera & Caline 1990; Daws & Prosser 1992). Alternatively others interpreted these sharp surfaces as evidence of a major basinward shift of the shoreface (Van Wagoner *et al.* 1993), or of more minor forced regression on the shoreface (e.g. Olsen & Steel 1995).

Similarly, the associated lag deposits have been interpreted to represent storm lags (Brown & Richards 1989), a series of laterally migrating channel deposits or as evidence of transgressive ravinement (Cannon *et al.* 1992). The fining-upward motifs which are common in the Etive Formation are similarly interpreted as resultant from tidal-inlet infill (Daws & Prosser 1992; Scott 1992) from fluvial channel-infill (Elliott 1989; Reynolds 1995), or from processes dominating in the upper shoreface/foreshore environment (Olsen & Steel 1995; Jennette & Riley 1996). This spectrum of interpretations may be construed to reflect Brent shoreline variability, but probably rather demonstrates the inadequate and equivocal nature of many data sets.

However in general, most authors agree that the lower progradational part of the Brent system was deposited within a high-energy, wave- and storm-dominated environment. In such an environment, distributary channels would represent a minor preserved component of the prograding package as compared to a fluvial-dominated shoreline system (Bhattacharya & Walker 1992). Although some workers have argued that either parts of, or the indeed entire Etive Formation, originated as braided fluvial deposits, emplaced after a major fall in relative sea level and drainage reorganisation (e.g. Elliott 1989; Van Wagoner *et al.* 1993; Reynolds 1995) these interpretations have to contend with: (1) the results of a petrographical study by Morton (1992) who on the basis of the garnet assemblages of the Rannoch, Etive and Ness Formations in the Tern, North Cormorant, Cormorant, Thistle, Murchison, Dunlin, Brent, Statfjord, Gullfaks and Oseberg fields convincingly argued that the Etive sands were derived longshore from the same source as the Oseberg and Rannoch sandstones, and (2) palynological studies of the Rannoch, Etive and Ness Formation which have shown that time lines pass obliquely from the Ness Formation into the Etive and then through the Rannoch Formation (Helland-Hansen *et al.* 1992; Whitaker *et al.* 1992; Johannessen *et al.* 1995). Critically, these lines of evidence imply a genetic relationship between the Rannoch, Etive and Ness formations as further discussed below.

Sequence stratigraphy of the Brent system: distinction of progradation in response to normal and forced regression

Debate concerning sequence stratigraphy of the Brent system is focused on the *nature of the trajectory* followed by the prograding shoreline. As the Brent progradational phase lasted from latest Aalenian through late Bajocian times (*c.* 4 Ma), it is unlikely that a single regressive trajectory would have characterised the system during this entire interval. However, from the available subsurface data, where large scale seismic geometries are not resolved, and most information comes from wells in widespread hydrocarbon fields, reconstruction of the shoreline trajectory is far from easy, and can only be indirectly reconstructed. The type and density of the data available has implications for understanding if relative sea-level was rising, stable or falling during the regression of the Brent system. Only where there is evidence of falling relative sea level can the shoreline trajectory be described as 'forced regressive' because stable or rising relative sea level can produce a range of 'normal regressive' trajectories (Helland-Hansen & Gjelberg 1994).

Development of an understanding of the nature and stratigraphic relationship between the Etive and Rannoch Formations may be distilled into the following questions.

(1) Is the Rannoch–Etive relationship always transitional, implying that the system's progradational trajectory was a result of continuous normal regression?
(2) In places where there is a clear erosion surface associated with an abrupt grain-size increase at the base of the Etive Formation, can a hypothesis of normal regression still be sustained? How do the alternative facies interpretations for Etive Formation further impact this hypothesis?
(3) Where are there additional features (major basinal shift of facies at base of the Etive Formation, presence of incised valleys within Etive, or evidence of subaerial exposure at top Rannoch) that *do* indicate relative sea level fall and forced regression of the system, how widespread do the erosive effects need to be?

There are clearly two end-member situations for the relationship between the Rannoch and Etive Formations. In the first, there would everywhere be a gradual upwards-shoaling into shallower water facies, the product of normal shoreline regression under conditions of sea-level stillstand or rise. The other extreme involves a major discontinuity and an implied major basinward shift at the boundary between the two formations as a result of forced regression, implying a relative sea-level fall. These two extreme cases are well illustrated by Johannessen *et al.* (1995, fig. 24). Each of these scenarios involves quite different progradational trajectories for the regressing shoreline, and imply quite different relative sea level changes during the latest Aalenian–early Bajocian in the Brent basin. Although these are two extreme models, they are sometimes portrayed as the only two options for the whole 4 Ma of Brent deposition, a most unlikely situation.

Continuous gradational relationship

Facts. Most shallow marine deposits are characterized by a prograding clinoform geometry and the clinoform model provides a norm that predicts that surfaces should dip gently seaward as facies become increasingly fine-grained (e.g. Bhattacharya & Walker 1992). The clinoform model has been applied to the Brent system by workers who consider there to be a close genetic relationship between the Rannoch, Etive and Ness Formations (e.g. Helland-Hansen *et al.* 1992; Johannessen *et al.* 1995; Olsen & Steel 1995; Fjellanger *et al.* 1996). The two main descriptive features indicative of gradual upward changes across the Rannoch–Etive boundary are (i) a large-scale vertical repetition of the Rannoch and Etive Formations in some wells and (ii) a fairly gradual upward change in grain-size.

The vertical repetition of Rannoch and Etive lithosomes is well recorded in the Statfjord, Gullfaks and Visund Fields (Johannessen *et al.* 1995, figs 8–10; Olsen & Steel 1995, fig. 3). Such repetition implies stratigraphic interfingering of the two formations, where they are genetically linked in clinothem units as part of the large-scale progradation of the Brent shoreline (see also Olsen & Steel 1995, fig. 3). Gradual vertical change of grain size and facies related to upward-coarsening (Fig. 2), is a classic indicator of a genetically related succession, and such a relationship is apparent in some Rannoch–Etive profiles in northern North Sea wells (Johnson & Stewart 1985; Johannessen *et al.* 1995, fig. 21). In addition to this grain-size change, there are descriptions of upward-decreasing mica content (Olsen & Steel 1995), heavy mineral data indicating derivation of sediment supply from a constant provenance (Morton 1992), and biostratigrahic timelines passing down from the Ness Formation through the Etive Formation and into the Rannoch Formation in some areas (Helland-Hansen *et al.* 1992, figs 2 & 7). All of these data and features are supportive of a gradual upward transition and a genetically related facies succession.

Implications. The above features of the combined Rannoch–Etive Formations emphasize the uniform, shallowing-upward nature of the Rannoch–Etive–Ness depositional trend in some areas, and, in line with the older Brent Group literature, lead to an interpretation of the succession as part of a normal and uniformly prograding shoreline system (Fig. 5a; Graue *et al.* 1987; Brown & Richards 1987; Fält *et al.* 1989; Helland-Hansen *et al.* 1992; Eschard *et al.* 1993). Implicit in this interpretation is the view that the formations under discussion comprise part of a low-order, highstand systems tract, where relative sea level was stable or rose very slightly during progradation. It is important to take into account the fact that most of these were mainly early regional studies, where the well spacing was such that there was little possibility of identifying coastline segments with a downward shift in the Rannoch/Etive clinoform trajectory.

It should also be added, that a number of studies do show some variation in the mode of normal regression within the Brent system. This is particularly apparent where shoreface sandbodies begin to split and pinch-out in the northernmost areas where a highly aggradational

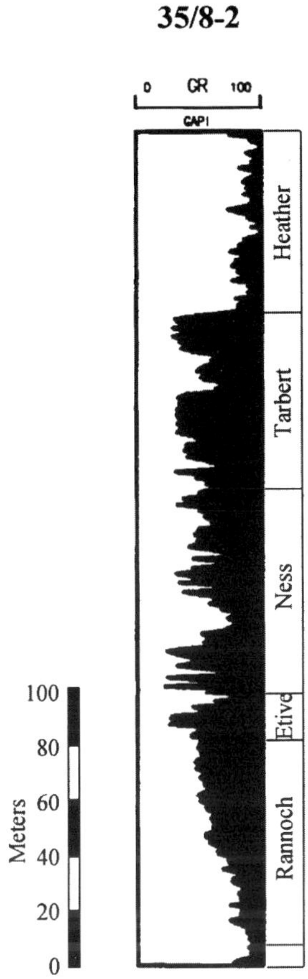

Fig. 2. An example of gradational vertical change upwards from the Rannoch into the Etive Formation, as seen in the gamma ray log expression of well 35/8–2.

stacking pattern of the Rannoch–Etive lithosomes has been observed (e.g. Cannon *et al.* 1992; Mitchener *et al.* 1992; Johannessen *et al.* 1995). The implication is that in these areas of climbing regressive shoreface trajectories, there was no longer a clear excess of sediment supply in relation to the accommodation created. Retrogradational phases that intervene between the regressive sandbodies tended to be marked by thin, coarse-grained lithosomes, with sharp basal ravinement surfaces, as well as time-equivalent 'transgressive' wedges of coastal plain sediments (Ness Formation). However, in the studies of these sandstone tongues only changes (increases) in the rate of rise of relative sea level are implied, and shoreface progradation in response to forced regression has not been recognized todate.

Sharp-based Etive Formation: without major discontinuity

A marked erosion surface alone (e.g. Fig. 3), is insufficient evidence of a major discontinuity at the base of the Etive Formation. The sedimentological interpretation of the Etive Formation also has a significant impact on that of sequence stratigraphy. As already discussed, the facies of the Etive Formation have been variously interpreted in terms of shoreface, tidal or fluvial processes. These quite different interpretations, coupled with a marked erosive boundary between the Rannoch and Etive Formations, can lead to rather different sequence stratigraphic interpretations. Van Wagoner *et al.* (1993) and Reynolds (1995) favoured a fluvial interpretation for the Etive Formation, which coupled with marked erosion into the underlying lower/middle shoreface deposits of the Rannoch Formation, implies a major discontinuity and basinward facies shift. Of course, such an implication is much less significant if the Etive Formation is interpreted in terms of upper shoreface processes.

An erosively based Etive Formation, dominated by shoreface deposits can be interpreted in terms of 'normal' shoreline regression. Studies on normally prograding high-energy barred coastline systems by Davidson-Arnott & Greenwood (1976), Howard & Reineck (1979), Hunter *et al.* (1979) and Wright *et al.* (1979) very clearly demonstrated that sharp erosive contacts and grain-size shifts can be a natural part of the shoreface profile. Based on studies of the Oregon coast, Hunter *et al.* (1979, fig 12) presented a vertical model of facies produced by progradation of an oblique bar/rip channel system. Their facies succession comprises fine-grained planar to hummocky cross-stratified sandstone cut by a subhorizontal erosional surface that is overlain by coarse-grained sandstone with trough and planar cross-beds. The locally sharp contact is interpreted to have resulted from the migration of longshore troughs, bars and rip channels in the upper shoreface zone driven by variations in wave energy (shore-normal oscillatory motion, longshore and rip currents) caused by major storms as well as seasonal changes (Hunter *et al.* 1979; Wright *et al.* 1979; McCubbin 1982).

In a similar way, the same basal Etive erosion surface has been taken as evidence for the importance of tidal channel incision (e.g. Daws & Prosser 1992; Scott 1992) or of distributary channel erosion (e.g. Brown & Richards 1989; Livera & Caline 1990; Mitchener *et al.* 1992) into the shoreline. Both of these interpretations are

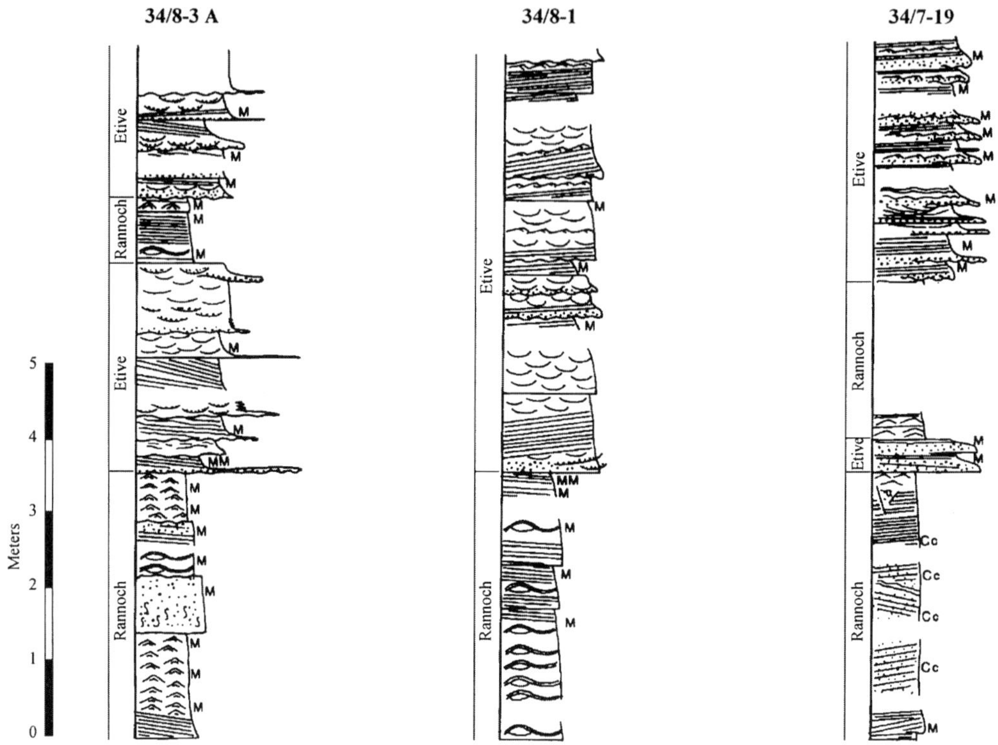

Fig. 3. Three examples of a grain-size jump across an abrupt and erosive contact between the Rannoch and the Etive Formations in the area of the Vigdis/Visund fields, respectively at approximately 2°20′ E, 61°22′ N and 2°25′ E, 61°20′ N, as located on Fig. 1 (modified from Olsen & Steel 1995).

consistent with normal regression of the Brent shoreline, because distributary channels and tidal channels can be expected to cut down, at times, into their own shoreline deposits.

Major discontinuity between the Rannoch and Etive Formations

Facts. Increasingly, recent studies have suggested that there *is* evidence for forced regression and fall of relative sea level during the main regression of the Brent system, even though these occurrences may be of relatively local spatial and temporal extent. This evidence has taken the form of: (a) the presence of an abrupt grain-size change and marked erosion at base of the Etive Formation, (b) the presence of erosively based incised valleys within the Ness and Etive Formations in updip areas and (c) the presence of a significant basinward shift of facies across the Rannoch–Etive boundary.

Erosion and abrupt coarsening of grain-size at the base of the Etive Formation (Fig. 3) is widespread and has been reported from many areas (Olsen & Steel 1995, figs 5–7; Reynolds 1995, figs 3 & 4; Johannessen *et al.*, figs 8–10). This observation has been used, together with an abrupt upward change from lower shoreface (Rannoch) to fluvial facies (Etive), as evidence of valley incision in response to a relative sea-level fall. It has been suggested that this type of vertical change occurs across much of the Statfjord (Van Wagoner *et al.* 1993) and Thistle Fields (Reynolds 1995). Smaller scale valley incisions, that imply at least two episodes of sea level fall across the Tern–Eider–Pelican–Cormorant Field areas of the East Shetland Basin have been recognized by Jennette and Riley (1996), who also provided evidence for downward facies shifts the in Rannoch/Etive shoreface/shoreline, through the recognition of estuarine units within the clinoformed shoreface profiles.

New evidence. The interpretation of the facies at the base of the Etive Formation is critical to the strength of the argument that there is a major discontinuity at the base of the Etive Formation.

Fig. 4. Cores from well 35/11–6, showing not only the erosive base of Etive Formation (just below 481) and the grain-size jump (very fine-grained below to granule sandstones above) across this boundary, but also the development of plant roots near top Rannoch and 1m below top Rannoch (below 485). Each of the 3 core lengths shown are some 60 cm long. Note that markings on the core are in feet.

Proponents of widespread valley incision into the Rannoch Formation argue strongly for a fluvial facies interpretation (Van Wagoner *et al.* 1993; Reynolds 1995) as fluvial facies superimposed on lower shoreface facies denotes a far greater basinward shift than upper shoreface on lower shoreface. New evidence indicating the existence of a relative fall of sea level at the Rannoch–Etive boundary has been documented from the Lomre Terrace (Norwegian Sector) in the northernmost North Sea. Most of the wells in block 35/11 show coarse-grained, cross-stratified granule sandstones that abruptly and erosively cut into well-laminated and massive, very fine-grained sandstones of the uppermost Rannoch Formation. And, for example, well 35/11–6 shows the development of plant roots just below the base of the Etive Formation (Fig. 4).

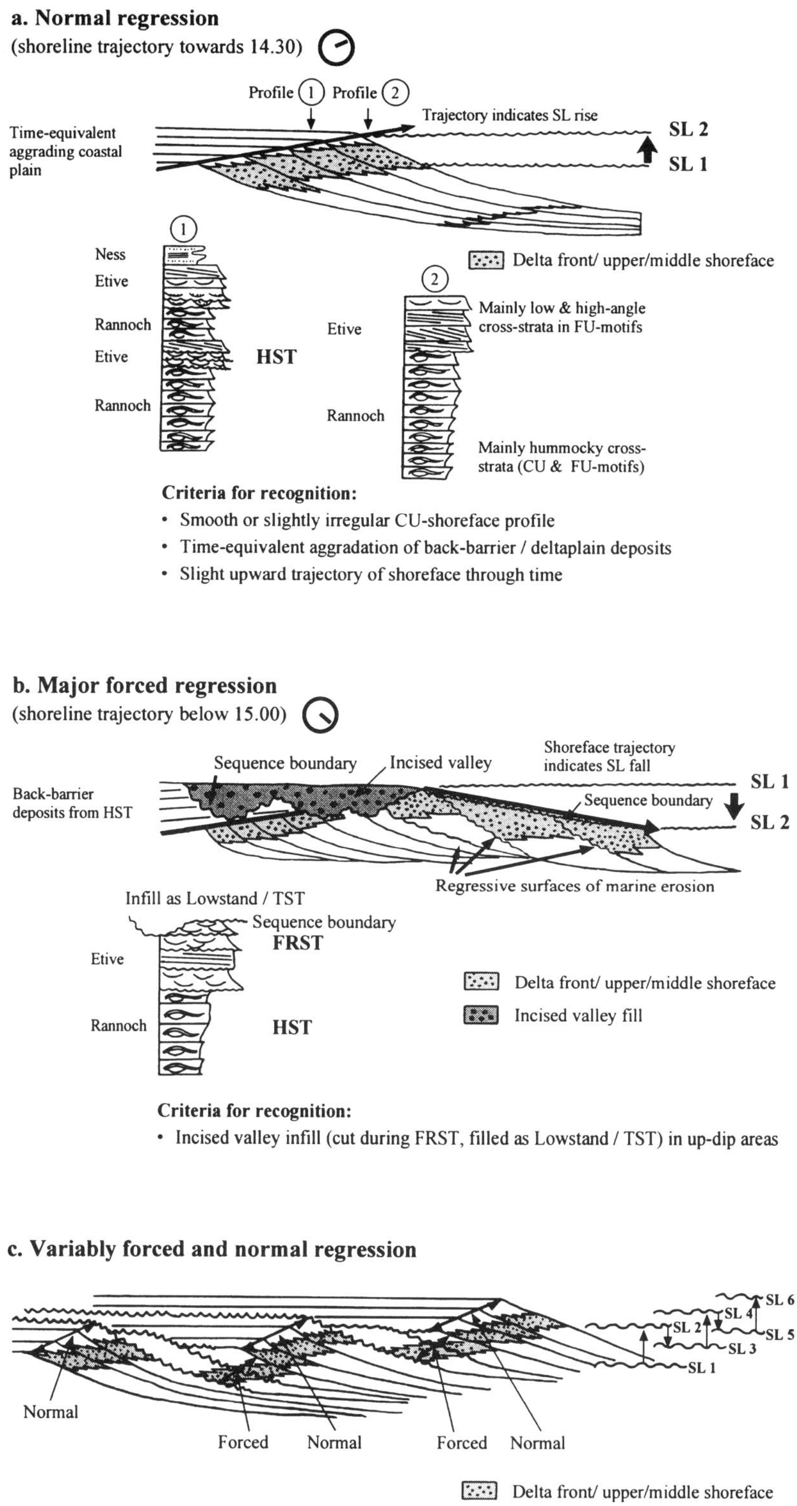
a. Normal regression
(shoreline trajectory towards 14.30)
Profile 1
Profile 2
Trajectory indicates SL rise
Time-equivalent aggrading coastal plain
SL 2
SL 1
Ness
Etive
Rannoch
Etive
Rannoch
HST
Delta front/ upper/middle shoreface
Etive
Rannoch
Mainly low & high-angle cross-strata in FU-motifs
Mainly hummocky cross-strata (CU & FU-motifs)
Criteria for recognition:
• Smooth or slightly irregular CU-shoreface profile
• Time-equivalent aggradation of back-barrier / deltaplain deposits
• Slight upward trajectory of shoreface through time
b. Major forced regression
(shoreline trajectory below 15.00)
Sequence boundary
Incised valley
Shoreface trajectory indicates SL fall
Back-barrier deposits from HST
SL 1
Sequence boundary
SL 2
Regressive surfaces of marine erosion
Infill as Lowstand / TST
Sequence boundary
FRST
Etive
Rannoch
HST
Delta front/ upper/middle shoreface
Incised valley fill
Criteria for recognition:
• Incised valley infill (cut during FRST, filled as Lowstand / TST) in up-dip areas
c. Variably forced and normal regression
SL 6
SL 4
SL 2
SL 5
SL 3
SL 1
Normal
Forced
Normal
Forced
Normal
Delta front/ upper/middle shoreface

Implications. The new evidence from the Lomre Terrace, in particular that showing shoreface deposits of the Rannoch Formation to have been subaerially exposed, provides clear proof that the Rannoch–Etive boundary is, at least locally (on the scale of individual hydrocarbon fields and greater), an unconformable surface associated with a fall of relative sea level. This implies that reaches of the Brent shoreline *were* subject to forced regression. However, such evidence of relative sea level fall is localized, and so does not necessarily imply that the Rannoch–Etive boundary is a major regional unconformity (Fig. 5b). Indeed local evidence elsewhere, as discussed above, negates this.

Helland-Hansen *et al.* (1992) and Johannessen *et al.* (1995) have provided ample documentation of mild extension and slight block rotation in Bajocian times, and at least locally the base-level fall and subaerial exposure of the Rannoch Formation may well have been generated by local or sub-regional uplift. However, there is now no doubt that the trajectory of the Brent shoreline was not everywhere the product of normal regression. High rates of accommodation to sediment supply at times caused the shoreline trajectory to climb upwards, and shoreface units to stack sub-vertically; whereas relative sea-level fall at times forced the trajectory downwards as well as outwards (Fig. 5c). It remains to be seen if the episodes of forced regression have resulted in significant accumulations of sand farther basinward that the known extent of the Brent shoreline, as such forced regressive and lowstand sandstones would form a new exploration target.

Conclusions

Any determination of possible forced regression of the Brent system during its late Aalenian–late Bajocian progradation, requires the following evaluation at any locality.

(1) Is there a gradual coarsening upwards of the vertical profile, with transition from offshore up through shoreface to shoreline and coastal plain facies? This, as a first working hypothesis, would suggest 'normal' regression at this location. The implication here would be a progradational trajectory which was horizontal or rising upwards and basinwards, driven by a stable or rising relative sea level.

(2) Does the upward-coarsening of the shoreface profile show significant irregularity, such as from a combination of abrupt grain-size jump, abrupt facies shift, or a marked erosion surface? There may then be some possibility of a relative fall of sea level and forced regression during progradation of the system. Where the abrupt vertical change (Rannoch–Etive) is caused by an erosively based upper shoreface unit, or a tidally influenced channelled unit, the hypothesis of forced regression is tenuous, on the basis of this evidence alone. Where the abrupt upward change is to fluvial deposits, the notion of forced regression is much more likely, but additional proof is still desirable.

(3) Where the latter scenario in (2) above can be combined with updip evidence of incised valley(s), evidence of subaerial exposure in the shoreface deposits below the level of abrupt bas-inward shift, or evidence (where wells are tightly spaced) of shoreface units stepping progressively downwards as well as basinward (see Mellere & Steel, this volume), only then a relative fall of sea level and forced regression of Brent shoreline is demonstrated.

Determination that there has been forced regression of the Brent system at times, and normal regression at other times, *does* tend to negate both of the end-member scenarios (perhaps the most commonly expressed viewpoints). These are that progradation of the Brent system was either (a) continuously normal or was (b) subject to a major late-stage or continuous fall of sea level, such that the Etive Formation lies everywhere incised into the Rannoch Formation. The 4 Ma interval of progradation, in itself, makes both of these scenarios unlikely. The clear local evidence for both types of regression, along different reaches of the progradational trajectory, describes a variably stable, rising and falling relative sea level during the interval in question (Fig. 5c).

We are grateful to very many colleagues in the Norwegian Oil industry for countless discussions, but more recently to J. Crabaugh, E. Fjellanger, R. Knarud and T. Olsen as well as to J. Gjelberg and an anonymous reviewer. We wish to express appreciation to Total Norge AS and Saga Petroleum ASA for their continued support of this project and their encouragement to publish this paper.

Fig. 5. Schematic cross-sections illustrating shoreline architecture and key sequence stratigraphic surfaces associated with shorelines undergoing (**a**) normal regression; (**b**) major forced regression and (**c**) variably normal and forced regression. The bold arrow to the right in each figure indicates the direction of change in relative sea level (SL). The sedimentary log(s) show the appearance of the uppermost and lowermost few metres of the Rannoch and Etive Formations respectively, as well as the character of the boundary between them.

References

BADLEY, M. E., EGEBERG, T. & NIPEN, O. 1984. Development of rift basins illustrated by the structural evolution of the Oseberg feature, block 30/6, offshore Norway. *Journal of the Geological Society, London*, **145**, 455–472.

BHATTACHARYA. J. P. & WALKER, R. G. 1992. Deltas. *In*: WALKER, R. J. & JAMES, N. P. (eds) *Facies Models – Response to sea level change*. Geological Association of Canada, 157–177.

BROWN, S. & RICHARDS, P. C. 1989. Facies and development of the Middle Jurassic Brent delta near the northern limit of its progradation, UK North Sea. *In*: WHATELEY, M. K. G, & PICKERING, K. T. (eds) *Deltas: Sites and Traps for Fossil Fuels*. Geological Society, London, Special Publication, **41**, 253–267.

——, —— & THOMPSON, A. R. 1987. Patterns in the deposition of the Brent Group (Middle Jurassic) UK North Sea. *In*: BROOKS, J. & GLENNIE, K. W. (eds) *Petroleum Geology of North West Europe*. Graham & Trotman, London, 890–913.

BUDDING, M. C. & INGLIN, H. F. 1981. A reservoir geological model for the Brent Sands in the Southern Cormorant. *In*: ILLING, L. V. & HOBSON, G. D. (eds) *Petroleum Geology of the Continental Shelf of the Northwest Europe*. Heyden, London, 326–334.

BUZA, J. W. & UNNEBERG, A. 1987. Geological and reservoir engineering aspects of the Statfjord Field. *In*: KLEPPE, J. ET AL. (eds) *North Sea Oil and Gas Reservoirs*. The Norwegian Institute of Technology, Graham and Trotman, 23–38.

CANNON, S. J. C., GILES, M. R., WHITAKER, M. F., PLEASE, P. M. & MARTIN, S. V. 1992. A regional reassessment of the Brent Group, UK sector, North Sea. *In*: MORTON, A. C., HASZELDINE, R. S., GILES, M. R. & BROWN, S. (eds) *Geology of the Brent Group*. Geological Society, London, Special Publications, **61**, 81–107.

CLIFTON, H. E. & FIRTH, I. R. 1993. Sequence stratigraphy and reservoir characteristics of the Brent Group of the Huldra Field. (Abstract). *Norwegian Petroleum Society*.

DAVIDSON-ARNOTT, R. G. D. & GREENWOOD B. 1976. Facies relationships on a barred coast, Kouchibouguac Bay, New Brunswick, Canada. *In*: DAVIS, R. A. & ETHINGTON, R. L. (eds) *Beach and nearshore sedimentation*. Society of Economic Paleontologists and Mineralogists Special Publications, **24**, 149–168.

DAVIES, E. J. & WATTS, T. R. 1977. The Murchison oilfield. *In*: *Mesozoic North Sea Symposium*. Norwegian Petroleum Society, Oslo, MNNSS/17.

DAWS, J. A. & PROSSER, D. J. 1992. Scales of permeability heterogeneity within the Brent Group. *Journal of Petroleum Geology*, **15**, 397–418.

DEEGAN, C. E. & SCULL, B. J. 1977. *A proposed standard lithostratigraphic nomenclature for the Central and Northern North Sea*. Norwegian Petroleum Directorate Bull. Report of the Institute of Geological Sciences **77/25**.

ELLIOTT, T. 1989. Deltaic systems and their contribution to an understanding of basin-fill successions. *In*: WHATELEY, M. K. G. & PICKERING, K. T. (eds) *Deltas, Sites and Traps for Fossil Fuels*. Geological Society, London, Special Publications, **41**, 3–10.

ERICHSEN, T, HELLE, M., HENDEN, J. & ROGNEBAKKE, A. 1987. Gullfaks. *In*: SPENCER, A. M. ET AL. (eds) *Geology of the Norwegian Oil and Gas Fields*. Graham & Trotman, 273–286.

ESCHARD, R., TVEITEN, B., DESAUBILIAUX, G., LECOMTE, J. C. & VAN BUCHEM, F. S. P. 1993. High resolution sequence stratigraphy and reservoir prediction of the Brent Group (Tampen Spur area) using an outcrop analogue (Mesaverde Group, Colorado). *In*: ESCHARD, R. & DOLIGEZ, B. (eds). Editions Technip, Paris, 35–52.

EYNON, G. 1981. Basin development and sedimentation in the Middle Jurassic of the northern North Sea. *In*: ILLING, L. V. & HOBSON, G. D. (eds) *Petroleum Geology of the Continental Shelf of North-west Europe*. Heyden, London, 196–209.

FJAERAN, T. & SPENCER, A. M. 1991. Proven hydrocarbon plays, offshore Norway. *In*: SPENCER, A. M. (ed.) *Generation, accumulation, and production of Europe's hydrocarbons*. Special Publications of the European Association of Petroleum Geoscientists, **1**. Oxford University Press, Oxford, 25–48.

FJELLANGER, E., OLSEN, T. R. & RUBINO, J. L. (1996). Sequence stratigraphy and regional palaeogeography of the middle Jurassic Brent delta system, Northern North Sea. *Norsk Geologisk Tidsskrift*, **76**, 2, 75–106.

FÄLT, L. M., HELLAND, R., WIIK JACOBSEN, V. & RENSHAW, D. 1989. Correlation of transgressive-regressive depositional sequences in the Middle Jurassic Brent/Vestland Group megacycle, Viking Graben, Norwegian North Sea. *In*: COLLINSON, J. D. (ed.) *Correlation in Hydrocarbon Exploration*. Norwegian Petroleum Society. Graham & Trotman, 191–200.

GRAUE, E., HELLAND-HANSEN, W, STEEL, R., NAKAYAMA, K. & KENDALL, C. G. 1987. Advance and retreat of the Brent Delta System, Norwegian North Sea. *In*: BROOKS, J. & GLENNIE, K. (eds) *Petroleum Geology of North West Europe*. Graham & Trotman, London, 315–325.

HALLETT, D. 1981. Refinement of the Geological Model of the Thistle Field. *In*: ILLING, L. V. & HOBSON, G. D. (eds) *Petroleum Geology of the Continental Shelf of North-west Europe*. Heyden, London, 196–209.

HAZEU, G. J. A. 1981. 34/10 Delta structure. Geological evaluation and appraisal. *In*: *Norwegian symposium on Exploration* (NSE81). Norsk Petroleumsforening, Bergen, **NSE/13**

HELLAND-HANSEN, W. & GJELBERG, J. G. 1994. Conceptual basis and variability in sequence stratigraphy: a different perspective. *Sedimentary Geology*, **92**, 31–52.

——, ASHTON, M., LØMO, L. & STEEL, R. 1992. Advance and retreat of the Brent delta: recent contributions to the depositional model. *In*: MORTON, A. C., HASZELDINE, R. S., GILES, M. R. &

BROWN, S. (eds) *Geology of the Brent Group.* Geological Society, London, Special Publications, **61**, 109–127.

——, STEEL, R., NAKAYAMA, K. & KENDALL, C. G. ST. C. 1989. Review and computer modelling of the Brent Group stratigraphy. *In*: WHATELEY, M. K. G. & PICKERING, K. T. (eds) *Deltas, Sites and Traps for Fossil Fuels.* Geological Society, London, Special Publications, **41**, 237–252.

HOWARD, J. D. & REINECK, H.-E. 1979. Sedimentary structures of 'high-energy' beach-to-offshore sequence; Ventura-Port Hueneme area, California (abs). *American Association of Petroleum Geologists, Bulletin*, **63**, 468–469.

HUNTER, R. E., CLIFTON, H. E. & LAWRENCE PHILLIPS, R. 1979. Depositional processes, sedimentary structures, and predicted vertical sequences in barred nearshore systems, southern Oregon coast. *Journal Sedimentary Petrology*, **49**, 3, 0711–0726.

JENNETTE, D. C. & RILEY, C. O. 1996. Influence on relative sea level on facies and reservoir geometry of the Middle Jurassic lower Brent Group, UK North Viking Graben. *In*: HOWELL, J. A. & AITKEN, J. F. (eds) *High Resolution Sequence Stratigraphy: Innovations and Applications.* Geological Society, London, Special Publications, **104**, 87–113.

JOHANNESSEN, E. P., MJØS, R., RENSHAW, D., DALLAND, A. & JACOBSEN, T. 1995. Northern limit of the 'Brent Delta' at the Tampen Spur – a sequence stratigraphic approach for sandstone prediction. *In*: STEEL, R. J., FELT, V .L., JOHANNESSEN, E. P. & MATHIEU, C. (eds) *Sequence Stratigraphy on the Northwest European Margin.* Norwegian Petroleum Society Special Publications, **5**, 213–256.

JOHNSON, H. D. & STEWART, D. J. 1985. Role of clastic sedimentology in the exploration and production of oil and gas in the North Sea. *In*: BRENCHLEY, P. J. & WILLIAMS, B. P. J. (eds) *Sedimentology: Recent Developments and Applied Aspects.* Geological Society, London, Special Publications, **18**, 249–310.

LIVERA, S. E. 1989. Facies associations and sand-body geometries in the Ness Formation of the Brent Group, Brent Field. *In*: WHATELEY, M. K. G. & PICKERING, K. T. (eds) *Deltas: Sites and Traps for Fossil Fuels* Geological Society, London, Special Publications, **41**, 269–286.

—— & CALINE, B. 1990. The sedimentology of the Brent Group in the Cormorant block IV oilfield. *Journal of Petroleum Geology*, **13**, 367–396.

—— & GDULA, J. E. 1990. Brent Oil Field. *In*: BEAUMONT, E. A. & FOSTER, N. H. (eds) *Structural Traps II: Traps associated with tectonic faults.* AAPG Treatise on Petroleum Geology, Atlas of Oil and Gas Fields, **A–017**, 21–63.

MARJANAC, T. 1995. Architecture and sequence stratigraphic perspectives of the Dunlin Group formations and proposals for new type- and reference-wells. *In*: STEEL, *ET AL.* (eds) *Sequence stratigraphy on the northwest European margin.* NPF Special Publications, **5**, Elsevier, Amsterdam, 143–167.

—— & STEEL, R. J. 1997. Dunlin Group sequence stratigraphy in the northern North Sea: a model for Cook Sandstone deposition. American Associations of Petroleum Geologists Bulletin, **81**, 276–292.

MCCUBBIN, D. G. 1982. Barrier Island and Strand-Plain Facies. *In*: SCHOLLE, P.A. & SPEARING, D. (eds) *Sandstone depositional environments.* American Associations of Petroleum Geologists, Memoirs, **31**, 247–279.

MEARNS, E. W. 1989. Neodymium isotope stratigraphy of Gullfaks oilfield. *In*: COLLINSON, J. D. (ed.) *Correlation in Hydrocarbon Exploration.* Norwegian Petroleum Society. Graham & Trotman, 201–215.

—— 1992. Samarium-Neodymium isotopic constraints on the provenance of the Brent Group. *In*: MORTON, A. C., HASZELDINE, R. S., GILES, M. R. & BROWN, S. (eds) *Geology of the Brent Group.* Geological Society, London, Special Publications, **61**, 213–225.

MELLERE, D. & STEEL, R. 2000. Style contrast between forced regressive and lowstand/transgressive wedges in the Campanian of south-central Wyoming. *This volume.*

MITCHENER, B. C., LAWRENCE, D. A., PARTINGTON, M. A., BOWMAN, M. B. J. & GLUYAS, J. 1992. Brent Group: sequence stratigraphy and regional implications. *In*: MORTON, A. C., HASZELDINE, R. S., GILES, M. R. & BROWN, S. (eds) *Geology of the Brent Group.* Geological Society, London, Special Publications, **61**, 45–80.

MORTON, A. C. 1992. Provenance of Brent Group sandstones: heavy mineral constraints. *In*: MORTON, A. C., HASZELDINE, R. S., GILES, M. R. & BROWN, S. (eds) *Geology of the Brent Group.* Geological Society, London, Special Publications, **61**, 227–244.

—— & HUMPHREYS, B. 1983. The petrology of the Middle Jurassic sandstones from the Murchison Field, North Sea. *Journal of Petroleum Geology*, **5**, 245–260.

——, STIBERG, J. P., HURST, A. & QUALE, H. 1989. Use of heavy minerals in lithostratigraphic correlation, with examples from the Brent sandstones of the Northern North Sea. *In*: COLLINSON, J. D. (ed.) *Correlation in Hydrocarbon Exploration*, Norwegian Petroleum Society. Graham & Trotman, 217–230.

NAGY, J., DYPVIK, H. & BJAERKE, T. 1984. Sedimentological and paleontological analysis of Jurassic North Sea deposits from deltaic environments. *Journal of Petroleum Geology*, **7**, 2, 169–188.

NIPEN, O. 1987. Oseberg. *In*: SPENCER, A. M. *ET AL.* (eds) *Geology of the Norwegian Oil and Gas Fields.* Graham & Trotman, 379–387.

NØTTVEDT, A., GABRIELSEN, R. H. & STEEL, R. J. 1995. Tectonostratigraphy and sedimentary architecture of rift basins, with reference to the northern North Sea. *Marine and Petroleum Geology*, **12**, 881–901.

OLAUSSEN, S., BECK, L., FÄLT, L.-M., JACOBSEN, K. G., MALM, O. A. & SOUTH, D. 1992. Gullfaks Field-Norway. East Shetland Basin, Northern North Sea. *In*: FOSTER, N. H. & BEAUMONT, E. A. (eds)

Structural Traps VI. AAPG Treatise of Petroleum Geology, Atlas of Oil and Gas Fields, **A-24** 55–83.

OLSEN, T. R. & STEEL, R. J. 1995. Shoreface pinch-out style on the front of the Brent delta in the easterly Tampen Spur area. *In*: STEEL, R. J., FELT, V. L., JOHANNESSEN, E. P. & MATHIEU, C. (eds) *Sequence Stratigraphy on the Northwest European Margin*. Norwegian Petroleum Society, London, Special Publications, **5**, 273–289.

PARRY, C. C., WHITLEY, P. K. J. & SIMPSON, R. D. H. 1981. Integration of Palynological and Sedimentological Methods in Facies Analysis of the Brent Formation. *In*: ILLING, L. V. & HOBSON, G. D. (eds) *Petroleum Geology of the Continental Shelf of North-west Europe*. Heyden, London, 205–215.

PEVERARO, R. C. A. & RUSSEL, K. J. 1984. Interpretation of wireline log and core data from a mid-Jurassic sand/shale sequence. *Clay Minerals*, **19**, 483–505.

RAVNÅS, R., BONDEVIK, K., HELLAND-HANSEN, W., LØMO, L., RYSETH, A. & STEEL, R. J. 1997. Sedimentation history as an indication of rift initiation and development: the Late Bajocian-Bathonian evolution of the Oseberg-Brage area, northern North Sea. *Norsk Geologisk Tidsskrift*, **77**, 202–222.

REYNOLDS, A. D. 1995. Sedimentology and sequence stratigraphy of the Thistle field. *In*: STEEL, R. J., FELT, V. L., JOHANNESSEN, E. P. & MATHIEU, C. (eds) *Sequence stratigraphy on the Northwest European Margin*. Norwegian Petroleum Society Special Publications, **5**, 257–271.

RICHARDS, P. C. 1990. The early to mid-Jurassic evolution of the northern North Sea. *In*: HARDMAN, R. F. P. & BROOKS, J. (eds) *Tectonic Events Responsible for Britain's Oil and Gas Reserves*. Geological Society, London, Special Publications, **55**, 191–205.

—— 1992. An introduction to the Brent Group: A literature review. *In*: MORTON, A. C., HASZELDINE, R. S., GILES, M. R. & BROWN, S. (eds) *Geology of the Brent Group*, Geological Society, London, Special Publications, **61**, 15–26.

ROBERTS, A. M., YIELDING, G., KUSZNIR, N. J., WALKER, I. & DORN-LOPEZ, D. 1993. Mesozoic extension in the North Sea: constraints from flexural backstripping, forward modelling and fault populations. *In*: PARKER, J. R. (eds) *Petroleum Geology of Northwest Europe: proceedings on the 4th conference*. The Geological Society, London, 1123–1136.

RYSETH, A. 1989. Correlation of depositional patterns in the Ness Formation, Oseberg area. *In*: COLLINSON, J. D. (ed.) *Correlation in Hydrocarbon Exploration*. Norwegian Petroleum Society, Graham and Trotman, 313–326.

SCOTT, E. S. 1992. The palaeoenvironments and dynamics of the Rannoch–Etive nearshore and coastal successions, Brent Group, northern North Sea. *In*: MORTON, A. C., HASZELDINE, R. S., GILES, M. R. & BROWN, S. (eds) *Geology of the Brent Group*. Geological Society, London, Special Publications, **61**, 129–148.

SIMPSON, R. D. H. & WHITLEY, P. K. J. 1981. Geological input to reservoir simulation of the Brent Formation. *In*: ILLING, L. V. & HOBSON, G. D. (eds) *Petroleum Geology of the Continental Shelf of North-west Europe*. Heyden, London, 310–314.

STEEL, R. J. 1993. Triassic–Jurassic megasequence stratigraphy in the Northern North Sea: rift to post-rift evolution. *In*: PARKER, J. R. (ed.). *Petroleum Geology of Northwest Europe: Proceedings of the 4th Conference*. Geological Society, London, 299–315.

—— & RYSETH, A. 1990. The Triassic-Early Jurassic succession in the northern North Sea: megasequence stratigraphy and intra-Triassic tectonics. *In*: HARDMAN, R. F. P. & BROOKS, J. (eds) *Tectonic Events Responsible for Britain's Oil and Gas Reserves*. Geological Society, London, Special Publications, **55**, 139–168.

THRELFALL, W. F. 1981. Structural framework for the central and northern North Sea. *In*: ILLING, L. V. & HOBSON, G. D. (eds) *Petroleum Geology of the Continental Shelf of North-West Europe*. Heyden, London, 98–103.

UNDERHILL, J. R. & PARTINGTON, M. A. 1993. Jurassic thermal doming and deflation in the North Sea: implications of the sequence stratigraphic evidence. *In*: PARKER, J. R. (ed.) *Petroleum Geology of Northwest Europe: Proceedings of the 4th Conference*. Geological Society, London, 337–345.

VAN WAGONER, J. C., JENNETTE, D. C., TSANG, P., HAMAR, G. P. & KAAS, I. 1993. Applications of High Resolution Sequence Stratigraphy and Facies Architecture in mapping potential additional Hydrocarbon Reserves in the Brent Group, Statfjord Field (Abstract). *In*: *Sequence Stratigraphy: Advances and applicaions for exploration and production in North West Europe*. Norwegian Petroleum Society, Stavanger Forum, Norway, 1–3 February 1993.

VOLLSET J. & DORÉ, A. G. (eds) 1984. A revised Triassic and Jurassic lithostratigraphic nomenclature for the Norwegian North Sea. *Norwegian Petroleum Directorate Bulletin*, **3**.

WHITAKER M. F., GILES M. R. & CANNON S. J. C. 1992. Palynological review of the Brent Group, UK Sector, North Sea. *In*: MORTON, A. C., HASZELDINE, R. S., GILES, M. R. & BROWN, S. (eds) *Geology of the Brent Group*. Geological Society, London, Special Publications, **61**, 169–202.

WILLIAMS, G. 1992. Palynology as a palaeoenvironmental indicator in the Brent Group, northern North Sea. *In*: MORTON, A. C., HASZELDINE, R. S., GILES, M. R. & BROWN, S. (eds) *Geology of the Brent Group*. Geological Society, London, Special Publications, **61**, 203–212.

WRIGHT, L. D., CHAPPELL, J., THOM, B. G., BRADSHAW, M. P. & COWELL, P. 1979. Morphodynamics of reflective and dissipative beach and inshore systems: southeastern Australia. *Marine Geology*, **32**, 105–140.

YIELDING, G., BADLEY, M. E. & ROBERTS, A. M. 1992. The structural evolution of the Brent Province. *In*: MORTON, A. C., HASZELDINE, R. S., GILES, M. R. & BROWN, S. (eds) *Geology of the Brent Group*. Geological Society, London, Special Publications, **61**, 27–44.

ZIEGLER, P. A. 1982. *Geological Atlas of Western and Central Europe*. Shell, The Hague.

Appendix: Interpretations of the Etive Formation: list of authors and their sedimentological and sequence stratigraphical (if any) interpretations of the Etive Formation

Davies & Watts (1977)

Area/field/wells. Murchison Field

Facies description. Top: fine–medium-grained sandstone, vague cross-bedding. Base: coarse–very coarse sandstone, some erosion surfaces, alternating fine–coarse beds.

Interpretation. Distributary *mouth bar* sequence; *tidal channel.*

Budding & Inglin (1981)

Area/field/wells. Southern Cormorant Field 211/21–1A, – 8, 211/26–1, –5, –6

Facies description. Fine–coarse-grained sandstone, alternating decimetre-scale cross-bedding and parallel-laminated sandstones at the base, partly stratified mottled or rippled sandstones above, mud, coal clasts and occasional clay drapes.

Interpretation. Deposition within the upper shoreface, foreshore and *barrier* top (aeolian) environments.

Hallett (1981)

Area/field/wells. Thistle Field

Facies description. None; sandstone coarser than below.

Interpretation. Distributary *mouth bar deposits* cut by distributary channels; east-west tidal channel (cut by rip currents) with flanking barrier bars.

Hazeu (1981)

Area/field/wells. Statfjord Field

Facies description. Medium-grained, clean sandstone with thin interbeds of coarse-grained sandstone.

Interpretation. Beach deposits

Parry et al. (1981)

Area/field/wells. Murchison Field (211/19–4) & Statfjord Field (211/24–4)

Facies description. Coarse-grained sandstone, well sorted, clean scoured bases, large-scale trough cross-bedding common, some foresets, massive initially.

Interpretation. Transverse & longitudinal bar sand deposits representing *braided distributary channel* deposits

Simpson & Whitley (1981)

Area/field/wells. Murchison Field 211/19–2, –3, –4

Facies description. Coarse-grained sandstone, relatively massive.

Interpretation. Fluvially dominated *distributary mouth system*

Morton & Humphreys (1983)

Area/field/wells. Murchison Field 211/19–3, –4

Facies description. Coarse–fine-grained sandstone, non-micaceous with moderate to good sorting.

Interpretation. Barrier-bar complex. Deposition of the Etive Formation is ended by a minor transgressive event.

Nagy et al. (1984)

Area/field/wells. East of Statfjord Field 33/9–3

Facies description. Fining-upwards units of cross-bedded sandstone.

Interpretation. Distributary channel

Peveraro & Russell (1984)

Area/field/wells. Northern North Sea

Facies description. Fine–medium-grained sub-arkosic sandstone, cross-bedded with minor thin micaceous interbeds in lower part, becoming finer grained in upper part. The two parts are separated by heavy minerals (zircon) concentration.

Interpretation. Barrier bar

Vollset & Doré (1984)

Area/field/wells. Northern North Sea

Facies description. Fine–coarse grained, occasionally pebbly, massive grey-brown to clear sandstone, cross-bedding; mica-poor.

Interpretation. Upper shoreface, *barrier-bar, mouth bar* and distributary channel

Johnson & Stewart (1985)

Area/field/wells. North Sea

Facies description. Coarse-grained, well-sorted and non-micaceous sandstone. Rannoch–Etive contact is sharp.

Interpretation. Upper shoreface/foreshore distributary channel and beach-ridge environments; distributary and barrier *inlet channel* cut into finer grained shoreface sandstone

Brown et al. (1987)

Area/field/wells. East Shetland Basin 210/24–2, 210/25–4, 211/11–1, 211/17–2, 211/18–2, –7, –21, 211/19–5, 211/21–7, 211/22–1, 211/23–2, 211/28–1A, –5, 3/3–3, 3/8A–5A

Facies description. Medium-grained, rather massive sandstone, locally with thin basal coarser lag deposits. Sedimentary structures: uneven lamination at the base, cross-bedding or indistinct lamination. Varying grain-size trends. Rannoch–Etive contact is sharp erosive or gradational.

Interpretation. Composite polygentic character; interpreted as a *barrier-bar complex* (genetically linked to the Rannoch and Ness Formations)

Buza & Unneberg (1987)

Area/field/wells. Statfjord Field 33/9–1, 33/12–1, 211/24–1

Facies description. Fine–coarse-grained sandstone with some disseminated mica, thin coal beds near the top of Etive.

Interpretation. Beach barrier complex

Erichsen et al. (1987)

Area/field/wells. Gullfaks Field block 34/10

Facies description. Top: medium–coarse-grained, well-sorted sandstone with minor mica and clay matrix. Base: fining-upwards sandstone.

Interpretation. Beach deposits; *channel-fill* deposits

Graue et al. (1987)

Area/field/wells. Tampen Spur area

Facies description. Medium–coarse-grained sandstone with low-angle laminations and trough cross-stratification, overall coarsening-upwards (0.5–3 m) capped by thin shale/coal.

Interpretation. Upper shoreface/foreshore environments (*barrier bar*); figure indicate mainly barrier and *mouth bar deposits* (and distributary channel)

Nipen (1987)

Area/field/wells. Oseberg Field 30/6–1, –2, –3, –4, –6, –9, –10, 30/9–1

Facies description. Coarse-grained, poorly sorted pebbly sandstone, massive. Three coarsening-upwards sequences; each cycle: wave influenced micaceous, fine-grained sandstone pass up into coarser-grained less micaceous sandstone.

*Interpretation. Delta-front/*lower delta-plain deposits

Hurst & Morton (1988)

Area/field/wells. Oseberg Field 30/6–1 to 11, 30/9–1, –2

Facies description. Medium–coarse-grained sandstone.

Interpretation. Deposited in *shoreline setting*, lack any evidence for fluvial input. In well 30/6–9 the Etive Fm is cut through by a fluvial channel belonging to the Ness Formation.

Brown & Richards (1989)

Area/field/wells. Don and Murchison Fields 211/13–7, 211/18A–21, –22, 211/19–2, –3, –4, –6

Facies description. Top: medium–very fine-grained sandstone with abundant wispy mud intercalations and pervasive bioturbation (*Scoyenia & Planolites).* Structureless/indistinct lamination. Mud/bioturbation increases upwards. Or, fine to rarely coarse-grained sandstone. Characterized by stacked fining-upwards units, indistinct cross-laminated sandstone with rare mica particles and *in situ* coal. Base: fine to rarely coarse-grained relatively well-sorted sandstone characterized by stacked fining-upwards with indistinct even parallel laminations or structureless; traces of medium-scale cross-bedding and ripple cross-laminations. No argillaceous intercalations. Rannoch–Etive contact is sharp.

Interpretation. Top: Sand-dominated delta-plain deposits *distributary channel sands* that were gradually drowned by marine incursions (overextended fluvial system drowned when sediment supply was insufficient to keep pace with rising sea level). Or: Stacked *distributary channel* fill.

Base: *Stacked distributary channel* deposits.

Elliott (1989)

Area/field/wells. Northern North Sea

Interpretation. Channel sandstone extremely common in lower part, forms an extensive multistorey, multilateral channel-belt sandstone body (i.e. *braid-plain* deposit).

Helland-Hansen et al. (1989)

Area/field/wells. Northern North Sea

Facies description. Medium–coarse-grained sandstone, low-angle laminations, trough cross-stratification or structureless, overall coarsening-upwards, but smaller-scale fining-upwards trends common, with pebbles and erosional bases. Rannoch–Etive contact is sharp or gradational.

Interpretation. Barrier-bar and *mouth-bar* setting.

Mearns (1989)

Area/field/wells. Gullfaks Field 34/10–1, –7, –8, –13

Facies description. Medium–coarse-grained sandstone, laminated and trough cross-stratified. Upper Etive: provenance ages ranging from 1800 to 1170 Ma. Lower Etive: provenance ages ranging from 1550 to 1650 Ma.

Interpretation. Mouth bar, distributary channel or upper shoreface deposits *Upper Etive*: distributary channel from the south, *Lower Etive*: detritus from several rivers transported by longshore drift.

Morton et al. (1989)

Area/field/wells. Statfjord, Gullfaks and Oseberg Fields

Interpretation. Shoreface sandstones derived by *longshore drift* from the east.

Livera & Caline (1990)

Area/field/wells. Cormorant Field, block IV, 211/21–9S, – CN11, – CN27

Facies description. Marked grain-size shift, cross-bedded, planar horizontal bedding and massive towards the top, some roots. Coarser grained (than Rannoch), poorly sorted, organic debris lag, cross-bedded. Rannoch–Etive contact is sharp.

Interpretation. Upper shoreface, *barrier* top and barrier attached deposits. Non-channelized in the north, sediment supply by longshore drift; major *distributary channels* south of block IV.

Livera & Gdula (1990)

Area/field/wells. Brent Field 211/29–1, –2, –6

Facies description. Medium–coarse-grained trough cross-stratified sandstone with organic debris, capped by mica-free, well-sorted sandstone. Overall coarsening upwards. Rannoch–Etive contact is sharp.

Interpretation. Upper shoreface and channel sands capped by coastal dune sediments (i.e. *barrier).*

Richards (1990)

Area/field/wells. East Shetland Basin

Facies description. Fine–medium-grained sandstone, fining-upwards over sharp base. Coarsening upwards over gradational base. Rannoch–Etive contact is sharp or gradational.

Interpretation. Composite *barrier bar* and shoreface system.

Cannon et al. (1992)

Area/field/wells. East Shetland Basin 211/7–1, 211/12–1, 211/17–2, 211/18–5, –9, –11, 211/12–1, –2, 211/27–6, 211/28–1, 3/3–2, –3, –5A, 3/7–1, –2

Facies description. (a) Upper boundary marked by *in situ* coal, which is included in the Ness Fm. (b) Low-angle, cross-lamination, grain fall and flow structures, mottled sandstone, weak bioturbation, wavy lamination and roots. (c) Massive sandstone, homogenized, vague parallel laminations, local trough cross-laminations, bi-directional cross-bedding. (d) Base: coarse-grained, possibly erosive lag overlain by decimetre-scale planar and trough cross-laminated sandstone.

Interpretation. (a) Evidence of plant colonization. (b) Aeolian reworking of barrier top and different barrier top environments. *Barrier system.* (c) *Tidal influenced channels.* (d) Lag probably represents migrating channels or a ravinement surface channelized beach plain deposits.

Daws & Prosser (1992)

Area/field/wells. Murchison Field (block 211/19) MS3, M04, M05, M10, M14, M14Y, M18Z, M19, M23, M27

Facies description. Top: (a) Fine–medium-grained well-sorted clean sand, planar, low-angle cross-stratification, undulose, discontinuous mica laminations (lumpy bedding), fluid escape structures, convoluted laminations, heavy mineral concentrations. (b) Interbedded with trough cross-stratified fine–medium-grained moderate/well-sorted sandstone, mica, clay or organic drapes, erosional set base. Middle: (c) Fine-grained, poorly sorted micaceous sandstone with clay and carbonaceous matter, deformed and wavy bedding. (d) It overlies a fine- to medium-grained sandstone with trough cross-stratification and clay draped ripples. Carbonaceous rip-up clasts. (e) Base: Coarse-grained sandstone, heterogeneous with trough cross-stratification and mica- and clay-draped laminations. Scoured, pebbly bases and well-developed fining-upwards trends. Micaceous and argillaceous top.

Interpretation. (a) Deposits within the swash-zone of the *foreshore* (shallow water, high current velocity) = beach/foreshore sandstone, disrupted by bioturbation. (b) Minor channels which locally rework and migrate across the beach/foreshore. (c) Inactive channel fill. (d) Tidal channel fill sandstone. (e) Stacked tidal channel deposits or stacked channel (fluvial) fill sandstone.

Helland-Hansen et al. (1992)

Area/field/wells. Troll, Brage, Oseberg, Huldra, Gullfaks Gamma Fields 30/2–2, 30/6–8, 30/9–1, –2, –3, 31/2–2, –8, 31/4–3, –6, –9, 34/10–23, 35/8–1, –2, 35/11–1

Facies description. Coarser-grained (than very fine–fine), mica-poor sandstone of variable character, ranging from massive to low-angle laminated, to rough and planar cross stratified; burrowing is rare.

Interpretation. Polygenetic origin within upper shoreface/foreshore *barrier bar* or upper delta front realm. A few wells probably contain mouth bar and distributary channel facies (increasing thickness of Etive towards the north).

Howe (1992)

Area/field/wells. Cormorant Field 211/21, 211/26

Facies description. Fine–medium-grained sandstone, cross-bedded and horizontal stratified at the base, partly stratified mottled or rippled sandstone above. Mud, coal clasts and clay drapes occasionally present, fining-upwards units. Rannoch–Etive contact is gradational. Etive–Ness contact in sharp.

Interpretation. Upper shoreface and foreshore areas (beach, dune, channel/*tidal inlet*) barrier attached wash-overs. Laterally extensive distributaries.

Mearns (1992)

Area/field/wells. Gullfaks Field

Facies description. Upper Etive: provenance ages of *c.* 1700–1800 Ma, ending with 1300 Ma. Lower Etive: provenance ages of *c.* 1550–1650 Ma.

Interpretation. Upper Etive: derived from a proximal southwesterly source. *Lower Etive*: sandstone derived from a source form east. Sandstone transported by *longshore* currents.

Mitchener et al. (1992)

Area/field/wells. UK and Norwegian sector, northern North Sea. 450 wells

Facies description. (a) Erosionally based sequence of medium–coarse grained sandstone. Single/stacked fining-upwards sequence with small-scale planar cross-sets, irregular mud drapes and vertical burrows. Erosionally cutting into the top. (b) Top: finer-grained (than medium), laminated sediments, capped by rootlet bed and thin coal. Medium–fine grained, well-sorted sandstone, coarsening – or fining-upwards trends, massive or poorly laminated. (c) Base: coarser small-scale cross-bedded sandstone. Rannoch–Etive contact is sharp and erosive.

Interpretation. (a) *Estuarine/tidal inlet* association. (b) Backshore to aeolian environment. (c) Beach-*barrier*/upper shoreface environment (stack shoreface sequence in Don/Thistle area).

Morton (1992)

Area/field/wells. Tern 210/20–1, 210/25–2, N Cormorant 211/21–3, Cormorant 211/26–1, Thistle 211/18-A33, Murchison 211/19–4, Dunlin 211/23–2, Brent 211/29–2, Statfjord, Gullfaks; Oseberg 30/6–7, –9, –10A

Facies description. No sedimentary descriptions. The Etive Formation is dominated by C*de* type garnet assemblages similar to those of the Oseberg Fm.

Interpretation. Shoreface and barrier system. Most of the shoreface sequence were *sourced longshore,* with material carried westwards from the Norwegian source.

Olaussen et al. (1992)

Area/field/wells. Gullfaks Field 34/10-A–5H, –9H, –10, –11, –19

Facies description. Some interfingering between Rannoch and Etive. Top: medium–fine-grained sandstone, cross-bedded, generally fining upwards; composite sequence of fining – and coarsening-upwards trends do occur. Base: medium–coarse-grained, occasionally very coarse-grained, pebbly sandstone.

Interpretation. A blending of *distributary channels, mouth bar deposits, barrier islands* and shoreline deposited in a proximal delta-front setting.

Scott (1992)

Area/field/wells. Southern Cormorant Field

Facies description. (a) Top: vaguely defined sedimentary structures. (b) Fine-grained, clean, well sorted sandstone, with low-angle planar lamination.

(c) Coarsening-upwards sequence: very fine–fine-grained, ripple-laminated sandstone with mud drapes, to planar laminated sandstone. (d) The sequence above is gently coarsening upwards; the sequence is periodically cut by fining-upwards sequence. (e) Sharp erosionally based fine–medium-grained sandstone with angular mud clasts and carbonaceous debris, high angle tabular cross-bedding, succeeded by climbing ripple laminations. (f) Parallel and ripple laminated sandstone. Base: trough cross-bedded sandstone.

Interpretation. (a) Backshore zone. (b) Swash – backwash laminations (foreshore zone). (c) Nearshore bar. (d) Sequence is interpreted as prograding *barrier beach.* (e) Small *channels* cutting the barrier, probably *tidal* but no unequivocal evidence. (f) Longshore bar and trough system rip channel.

Williams (1992)

Area/field/wells. Thistle (211/18A-A31) and Ninian Field (3/3–5A)

Facies description. Highly impoverished assemblage of black wood with rare pollen.

Interpretation. High energy, possibly *barrier sand* environment.

Clifton & Firth (1993)

Area/field/wells. Huldra Field 30/2–1, –2, 30/3–1

Facies description. None; 20 m thick fining-upwards sequence.

Interpretation. Open coast shoreface *large distributary channel.*

Eschard et al. (1993)

Area/field/wells. Tampen Spur area

Facies description. Coarse-grained sandstones. Rannoch–Etive contact is sharp or gradational with some interfingering between them.

Interpretation. Foreshore, distributary channels or *tidal complexes.*

Van Wagoner et al. (1993)

Area/field/wells. Statfjord Field

Interpretation. Braided stream deposits.

Sequence stratigraphic interpretation. The Etive Formation is interpreted as an incised valley fill, bounded by two sequence boundaries.

Johannessen et al. (1995)

Area/field/wells. Tampen Spur area 33/9–14, 33/12-B37, – B41, 34/7–13, –19, 34/8–1, –5, 34/10–1, –3, – A–9H, –14, –16, –23, –34

Facies description. Three facies associations. (1) Fine–coarse-grained, well-sorted, trough cross-stratified–massive and low-angle cross-stratified sandstone, with pebbles on basal scour surface. Rootlets. (2) Medium–coarse-grained sometimes pebbly, poorly sorted sandstone, current ripples, planar and trough cross-stratified sandstone with single/complex coarsening-upwards units. (3) Medium–coarse-grained sandstone, often pebbly above sharp erosive base. Fining-upwards to fine-grained sandstone and siltstone, planar and trough cross-bedded to massive in coarse-grained part and ripple-lamination in fine-grained part.

Interpretation. Upper shoreface and foreshore environment (troughs in surf zone with scour and deposition by longshore currents). Rannoch–Etive contact is gradational *mouth bars and distributary channels* in an upper deltafront setting. Distributary channels on the delta plain.

Sequence stratigraphic interpretation. Uses a T-R sequence stratigraphic model where the sequence boundary is = transgressive surface. A sequence boundary is located within the Etive Formation in the Gullfaks and Statjford fields, dipping down into the Rannoch Formation in the Visund field. Base Etive may be interpreted as a regressive surface of erosion in the Visund area (i.e. forced regression).

Olsen & Steel (1995)

Area/field/wells. Visund Field 34/8–1, –3A, –5, –6. block 34/7

Facies description. (a) Coarse–medium-grained, poorly sorted sandstone, deformed and massive bedded fining-upwards units (facies E2). (b) Medium-grained sandstone, small-scale trough and low-angle cross-bedding and planar parallel-laminated, current ripple lamination, roots and disseminated carbonaceous matter (facies E3). (c) Base: sharp, erosionally based, pronounced, stacked fining-upwards units of very coarse–fine grained sandstone, moderately to well-sorted, massive to trough cross-stratification or low-angle cross-bedding to climbing ripple-lamination. Overlain by 1–3 cm of micaceous/mud lamination (facies E1).

Interpretation. (a) Deposits in *subaqueous mouth bars.* (b) Upper shoreface/foreshore (surf zone succeeded by foreshore processes (swash zone, wave wash-up and backwash). (c) Base represents probably down- and outwards shift of deposition on the delta front, linked to multiple erosion episodes.

Sequence stratigraphic interpretation. The basal (multiple) erosional surfaces and pronounced fining upwards-units are interpreted in terms basinward/outward shift of deposition on the delta front (*forced regression*) related to several small-scale drops in relative sea level.

Reynolds (1995)

Area/field/wells. Thistle Field 211/18-A5, – A7, – A10, – A33, 211/19–1

Facies description. (a) Top: lower fine–lower medium grained laminated bioturbated sandstone with *Macaronichus, Diplocraterion, Skolithus linearis, Ophiomorpha, Teichichnus, Palaeophycos, Planolites, Rhizocarallium, Helminthopsis.* (b) lower medium–upper fine grained sandstone to granule size, poorly sorted sandstone with low to moderate-angle cross-stratification, rare planar cross-bedding, low angle scours. (c) Upper fine to lower medium-grained, poorly sorted sandstone with massive to crude planar lamination.

Interpretation. (a) High energy marine setting and marginal *mouth bar*, estuary fill and transgressive sandsheet-deposits. Deposition on landwards side of a *tidal inlet* or in a *tidal-influenced distributary system* migrating dunes that range in scale and form. (b) *Fluvial* stage plane bed flow. (c) *Channels.*

Sequence stratigraphic interpretation. The stacked fluvial channels reflects a downward shift of facies belts due to relative sea level fall. The Etive Formation is interpreted as a valley fill above a sequence boundary (Exxonian, type I). Significant sandy lowstand deposits basinwards of the Thistle Field.

Jennette & Riley (1996)

Area/field/wells. Hudson, Osprey, Pelican, Cormorant, Eider & Tern Fields 210/25–3ST2, – TA02, –2, TA03, –32, –5, – TA08, – TA11, TA19, TA28, 211/16–2, –6, – EA08, – EA09, – EA18, EA19s1, – EA22, 211/23–7, 211/24–12, 211/26–1, – CAUP1, – CAUP2, – CAUP07, – CA07, – CA31, – CA35s1, – CN20

Facies description. (a) Pedogenized mudstone and coal. The Etive Fm (described below) is truncated by an erosional surface with a coarse-grained lag in the Cormorant and Tern Fields. (b) Top: well-sorted, small-scale trough cross-bedding, gently wedging cross-lamination and planar parallel lamination, minor soft sediment deformation and dewatering structures. Burrows common, vague root traces. (c) Middle: upper fine to lower medium-grained laminated, better-sorted sandstone. Wavy to sub-parallel lamination and shallow scour and fill. Low amplitude current ripples. (d) Base: Medium-grained, moderately well-sorted sandstone with low angle, trough cross-beds and planar parallel lamination. Rannoch–Etive contact is gradational to sharp.

Interpretation. (a) Subaerial coastal plain facies stacked channel fill succession. (b) *Top: strand plain* environment. *Middle*: Foreshore envelope (deposited during variations in wave swash. Deposition by bar and runnel systems and ephemeral creeks). (d) *Base:* Upper shoreface envelope (deposits within longshore, troughs, bars, and rip channels) (Rannoch and Etive genetically related).

Sequence stratigraphic interpretation. The stacked channel fill deposits incising into the upper shoreface – strand plain succession of the Etive Formation is interpreted as an incised valley fill (lowstand systems tract) based by a sequence boundary and formed when fluvial and coastal plain facies infilled the space created after a sea level drop (Cormorant & Tern Fields). Outside the limits of the incised valleys on the interfluves, both sequence boundary and associated flooding surface merge to form a single hiatal surface.

Fjellanger et al. (1996)

Area/field/wells. Northern North Sea between 59° and 61.50'° N

Facies description. (a) Coarse–medium-grained poorly sorted sandstone, deformed and massive bedding, fining-upwards units (facies E2). (b) Medium-grained sandstone small-scale trough and low-angle cross-bedding and planar parallel lamination, current ripple lamination, roots and disseminated carbonaceous matter (facies E3). (c) Base: sharp, erosionally based, pronounced stacked fining-upwards units of very coarse–fine grained sandstones, moderately well sorted, massive to trough cross-stratification or low-angle cross bedding to climbing ripple lamination. Overlain by 1–3 cm of micaceous/mud laminae (facies E1).

Interpretation. (a) Deposits in subaqueous *mouth bars*. Upper shoreface/foreshore (surf zone succeeded by foreshore processes (swash zone, wave wash-up and backwash). (c) Base represents probably down-and outwards shift of deposits on the delta front, linked to multiple erosional episodes. Contact Rannoch/Etive: sharp in north, gradational in south.

Sequence stratigraphic interpretation: Base Etive is interpreted as a type II sequence boundary (Exonian) of the shelf margin systems tract in the northern part of the Brent delta system. In the southern part of the Brent delta the Etive Formation was deposited within the high stand systems tract (gradational base to the Rannoch Formation).

Forced regressions: recognition, architecture and genesis in the Campanian of the Bighorn Basin, Wyoming

ROY FITZSIMMONS[1,2] & STEVE JOHNSON[1,3]

[1]*Department of Earth Sciences, University of Liverpool, LE11 3QU, UK*

[2]*Present address: Conoco Inc., 600 North Dairy Ashford, Houston, TX 77252-2197, USA*

[3]*Present address: Statoil Research Centre, Postutak, 7005 Trondheim, Norway*

Abstract: The Cretaceous Mesaverde Group of the Bighorn Basin, northwestern Wyoming, is comprised of two major clastic wedges that record the progradation and retrogradation of deltaic depositional systems within the Cretaceous Western Interior Seaway. Within the Campanian Virgelle and Judith River Formations 16 sand-rich clastic tongues, deposited in mixed wave/storm-dominated shallow marine shoreface depositional environments, have been studied and traced into equivalent updip non-marine and downdip offshore facies. Each tongue is typically a massive shoreface sandbody that pinches-out and correlates basinward (east) with fine-grained offshore heterolithic progradational parasequences. Regional correlation reveals the sandstone tongues to be sharp-based. Their lower bounding surfaces are characterized by: (1) a marked basinwards shift in facies, (2) an abrupt increase in sand: shale ratio, (3) missing/eroded facies below, (4) a change in parasequence stacking patterns, (5) local development of *Glossifungites* firmground ichnofabrics, (6) deposition of precursor gutter cast facies, (7) widespread soft-sediment deformation and growth faulting, (8) changes in palaeocurrent orientation, (9) regional truncation of older parasequences and systems tracts, (10) regional depositional-dip correlation of 20–40 km. The basal sharp-based surfaces of the shoreface sandstones are interpreted to be regressive surfaces of marine erosion (RSME) formed during falls in relative sea-level, with the shallow marine successions deposited during forced regression of the shoreline.

Internal hetrogeneities and erosion surfaces within the massive shoreface sandstones are interpreted to record stepwise progradation during relative sea-level falls. These relatively steep seaward-dipping erosion surfaces reflect the overall trend of shoreface deposition and amalgamation during periods of decreasing accommodation space. Downdip, the internal erosion surfaces amalgamate with the basal RSME and provide evidence that this basal surface is composite in nature and as such diachronous in its development.

Within the studied interval, four examples of forced regressive deposits are confidently correlated updip to correlative incised valley fills. In each case, the basal erosion surface to the incised valleys truncates strandplain deposits, and ties laterally to a subaerial exposure surface (interfluve) developed across the top of the strandplain. These surfaces, formed in response to a fall in relative sea-level, are interpreted as sequence boundaries. Traced basinward their expression is commonly lost as the upper strandplain and capping interfluve are eroded by transgressive ravinement at the base of tidal inlets. However, the interfluve is thought to correlate downdip to the final seaward dipping erosion surface that separates the massive amalgamated shoreface sandstones of the falling stage system tract, deposited during overall relative sea-level fall, from the more heterolithic parasequences of the lowstand systems tract, deposited under conditions of stillstand to relative sea-level rise. Within sediments deposited during relative sea-level fall the critical transition from the subaerial to submarine expression of the sequence boundary is recognized as the main factor in the ongoing controversy regarding the identification of a finite chronostratigraphic sequence-bounding surface. Drawbacks exist in making a simple choice between the subaerial exposure surface or the RSME as the sequence boundary because they are normally diachronous and at the same time form contemporaneously.

In updip areas two separate important stratigraphic surfaces may be distinguished; the RSME and an overlying subaerial exposure surface. In these areas the subaerial exposure surface must be regarded as the main sequence bounding surface. Basinward exists a critical transition zone where (i) storm-related erosion surfaces and shoreface amalgamation during deposition of the strandplain inhibit correlation and (ii) transgressive ravinement may erode part of the subaerial expression of the sequence boundary. In this area choice of surfaces proves difficult. In constrast, basinward of the last sharp-based shoreface, the RSME would be the principal (and most obvious) stratal surface, and must be interpreted as a sequence boundary. By considering the evolution of the RSME to occur at the same time as the fluvial erosion/subaerial exposure surfaces, the massive sharp-based shorefaces and their distal equivalents can be observed to be the response of an linked and dynamic system to relative sea-level fall.

From: HUNT, D. & GAWTHORPE, R. L. (eds) *Sedimentary Responses to Forced Regressions*. Geological Society, London, Special Publications, **172**, 113–139. 1-86239-063-0/00/$15.00

As sequence stratigraphic interpretations of sedimentary systems have become more widespread, research has focused on the variations in depositional style that are apparent within individual systems tracts. For example, the processes that control the formation of incised valleys and their fills have been rigorously documented (Dalrymple *et al.* 1994). However, the contemporaneous shoreline deposits to which the incised valleys supply sediment have only received similar documentation and attention more recently (Plint 1988, 1996; Posamentier *et al.* 1992; Hunt & Tucker 1992, 1995; Ainsworth & Pattison 1994; Mellere & Steel 1995*a*, *b* and this volume; and papers by Ainsworth *et al.*; Hamberg & Nielsen; Plint & Nummedal; Posamentier & Morris this volume). This paper aims to provide a process-based understanding of shoreface sandbody development during falling relative sea level, focusing on their linkage to equivalent up- and down-dip deposits, and origin of their bounding surfaces. Such an approach can ultimately lead to better (i) correlation and linkage of contemporaneous deposits and surfaces, (ii) palaeogeographic reconstructions, (iii) up- and downdip facies prediction and ultimately to greater exploitation as hydrocarbon reservoirs.

Here, we describe a series of shallow marine, storm- and wave-dominated sandbodies exposed in the Bighorn Basin of northwestern Wyoming, USA (Fig. 1). From these sandbodies a series of generic features are recognized that characterize the basal surfaces of the forced regressive sandbodies and reveal the processes responsible for their development. In particular, we focus on the lower and upper bounding surfaces of the marine sandbodies, their internal hetrogeneities and relationships to coeval updip incised valley systems. These relationships enable comparison to be made with the existing sequence stratigraphic models, concepts and systematics specific to forced regressive deposits (e.g. Hunt & Tucker 1992, 1995; Posamentier *et al.* 1992; Plint 1996; Plint & Nummedal this volume; Posamentier & Morris this volume), and provide important insights as to the nature and timing of depositional and erosional processes during relative sea-level fall.

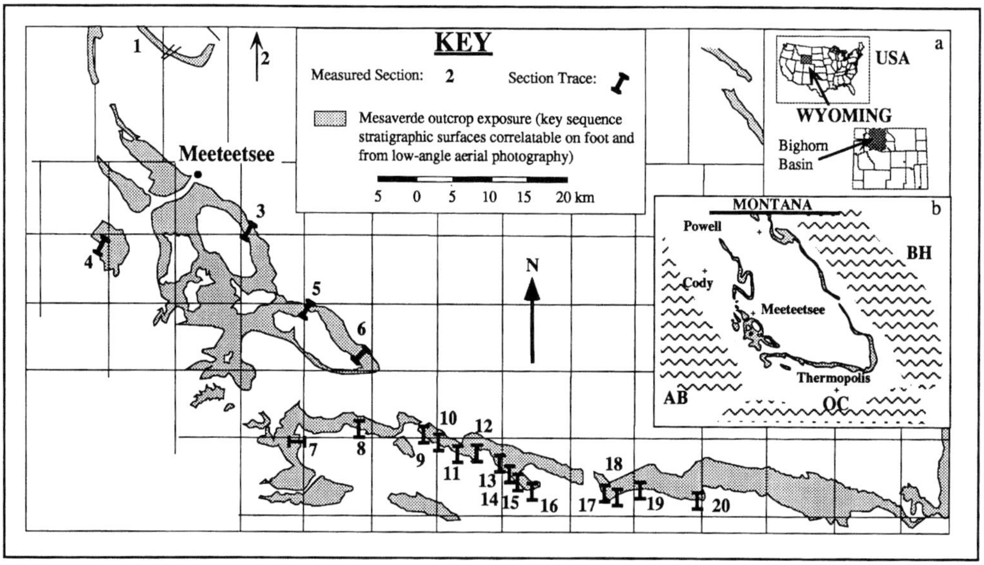

Fig. 1. Location map of Bighorn Basin (**a**) and the main outcrop belt of the Mesaverde Group (**b**). The main map focuses on the outcrop belt in the southern region of the basin near to Meeteetsee, where the main depositional dip profile is exposed, as summarized in Fig. 2. A listing of the abbreviations to individual measured sections (as also used in Figs 2, 4, 12) follows: AB, Abrasoka Mountains; OC, Owl Creek Mountains; BH, Bighorn Mountains. Measured sections: 1, Oregon Basin (OB); 2, Elk Basin (EB, north of Powell); 3, Little Buffalo Basin (LBB); 4, Sunshine Reservoir (SR); 5, North Grass Creek Basin (NGC); 6, South Grass Creek Basin (SGC); 7, Wagonhound Draw (WH); 8, Hamilton Dome (HD); 9, Cottonwood Creek (CWC); 10, Little Sand Draw (LSD); 11, Gloin Reservoir (GR); 12, Sand Draw(SD); 13, Mountain (MTN); 14, Syncline Draw (SYD); 15, Ronoco Mine (RM); 16, Gebo; 17, Cowboy Mine (CM1); 18, Cowboy Mine #2 (CM2); 19, Double Draw (DD); 20, Zimmerman Butte (ZB).

Geological setting and sedimentary facies associations

A detailed study of the late Cretaceous Mesaverde Group (Campanian to early Maastrichtian), Bighorn Basin of northwest Wyoming (Fig. 1) has been undertaken with a view to undertand better the relationship between surfaces and deposits formed in contemporaneous marine and non-marine strata in response to changes in relative sea level (Fitzsimmons 1994*a*; Johnson 1995). Deposition of the Mesaverde Group occurred in a retroarc-foreland basin within the Western Interior Seaway to the east of the Sevier orogenic belt (Severn 1961; Gill & Cobban 1966*a*, *b*, 1973; Asquith 1970, 1974; McGookey *et al.* 1972; Weimer 1984). The modern day Bighorn Basin is a Tertiary Laramide structure, with the surrounding uplifts of the adjacent Abrasoka, Owl Creek, Bighorn and Crazy Horse ranges, producing a simple syncline within which the Mesozoic foreland basin stratigraphy is exposed. The combination of shallow dips, typically between 10° and 20°, within the main Mesaverde outcrop belt, and the numerous valleys which dissect it, provide near-continuous exposure from the down-dip pinch-out of the shallow marine sandstone tongues, through the various shoreface environments, and into their up-dip correlative coastal plain deposits.

The Mesaverde Group is classically divided into four formations, the Eagle, Claggett, Judith River and Teapot Sandstone, as indicated on Fig. 2. The Eagle Formation is further subdivided into the Fishtooth, Telegraph Creek, Virgelle and Gebo Members (Figs 2 and 3). Our recent work has led to the development of a detailed sequence stratigraphic framework for the marine and fluvial strata of the Mesaverde Group (Fig. 2; Fitzsimmons 1994; Johnson 1995). This framework incorporates observations from both depositional dip- and strike-oriented profiles and measured sections. Figure 2 shows a dip-oriented correlation panel constructed along the southern flank of the outcrop belt, and provides a synthesis of our observations concerning the overall architecture of the Mesaverde wedge.

Deposition of the Mesaverde Group spanned a range of approximately 83–76 Ma (Fig. 1; Gill & Cobban 1966*a*, *b*, 1973; Hicks 1993; Obradovitch 1993). The group is comprised of four third-order sequences (*sensu* Van Wagoner *et al.* 1990), bounded by four major basinward dislocations of facies representing the Fishtooth, Virgelle, Judith River and Teapot low-order sequence boundaries (LOSB), respectively denoted FTLOSB, VLOSB, JRLOSB and TLOSB (e.g. Fig. 2). No biostratigraphic or chronostratigraphic data was available that would have enabled higher-resolution dating of the individual clastic tongues (i.e. Fig. 2; V1 to V8 etc.). Indivdually, each clastic tongue is tentatively interpreted to represent a high- or fourth-order (*sensu* Van Wagoner *et al.* 1990) cycle on the basis of equal division of the time-span represented by the low-order composite sequences.

Facies and facies relationships

Clastic tongues of the Mesaverde Group dominantly composed of shallow-marine lithofacies are present in the Virgelle and Gebo Members of the Eagle Formation. The Virgelle and Gebo Members are respectively comprised of eight and three tongues, whereas the Claggett Formation is divided into four tongues and the Judith River Formation is represented by single tongue (Fig. 2). During deposition of these sandstone tongues, the palaeoshoreline had a general northeast–southwest orientation, with the majority of the fluvial systems feeding sediments to the shoreface systems from the northwest. During transgression, the shorelines generally retreated westward, towards the region of the present day Abrasoka Mountains (Severn 1961; Asquith 1970, 1974; Gill & Cobban 1973).

Seven major facies associations are recognized within these marine and non-marine strata, and these have been assigned to specific depositional environments based on their sedimentary and biogenic structures and lateral facies relationships mapped in the field, as summarized in Table 1. The associations are:

(1) multistorey, fluvio/tidal channels – CHT;
(2) tidal inlet – T1;
(3) upper shoreface/foreshore – SF/FS;
(4) middle shoreface – SF2;
(5) lower shoreface – SF1;
(6) offshore transition zone – S2;
(7) offshore marine – S1.

Detailed descriptions of the storm and wave dominated (S1, S2, SF1, SF2, & SF/FS), and fluvio/tidal (CHT & T1) facies of the Western Interior Seaway have been comprehensively described by many previous workers (e.g. Balsley 1980; Elliott 1986; Van Wagoner *et al.* 1990; Devine 1991; Walker & Plint 1992; Brenchley *et al.* 1993). Rather than detail individual facies, a resume of which is presented in Table 1, a brief description of the main architectural elements of the shallow marine successions follow. The relationships of the shallow-water

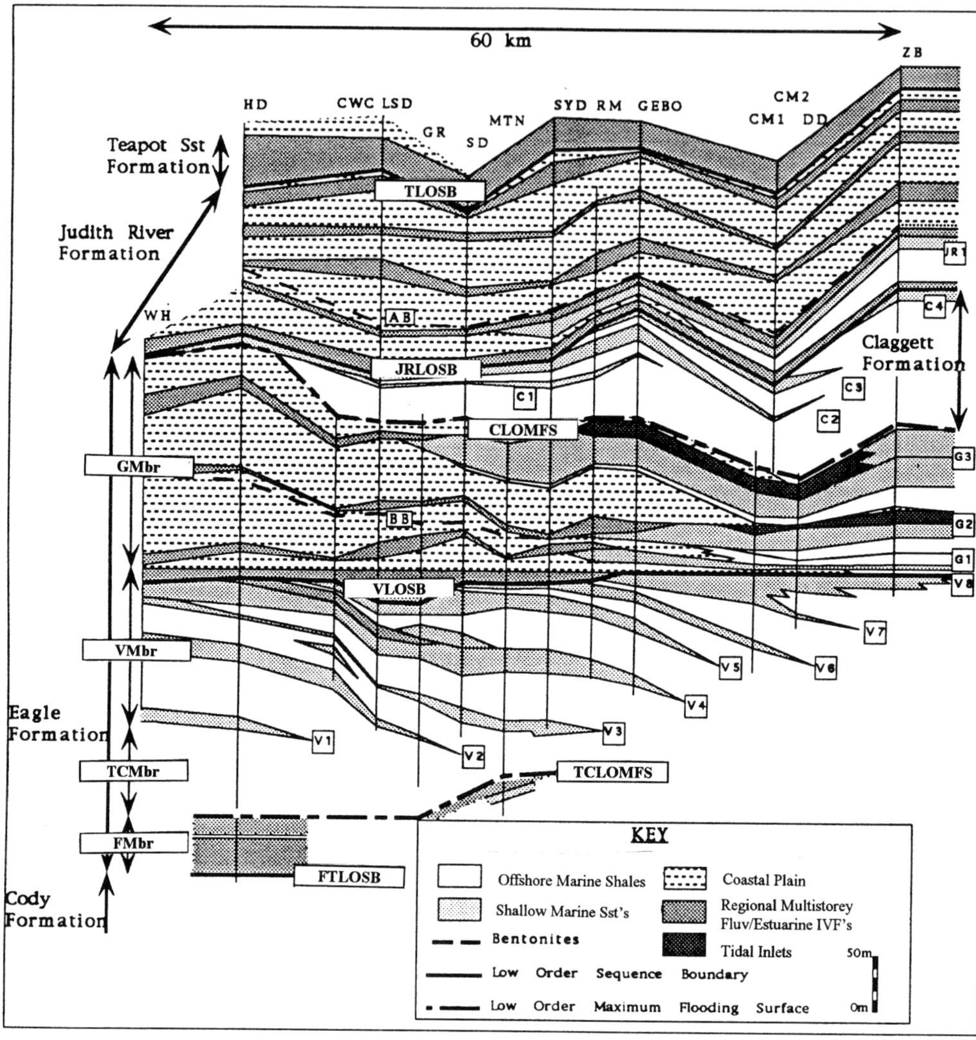

Fig. 2. Generalized architecture and facies relationships within the Campanian strata of the southern Bighorn Basin. The 57. 5 km correlation panel is traced southeasterly from Wagonhound Draw (WH, Fig. 1) to Zimmerman Buttes (ZB) and roughly parallels the depositional dip profile of the Mesaverde deltaic wedges. The principal third- or low-order sequence boundaries (LOSB) and maximum flooding surfaces (MFS) are identified on the basis of the regional studies of Fitzsimmons (1994*a*) and Johnson (1995). The down-dip pinch out of the 16 shallow marine 'tongues' can clearly be seen. Of these, eight are developed in the Virgelle Member (V1–V8), three in the Gebo Member (G1–G3), four in the Claggett Formation (C1–C4) and one in the Judith River Formation (JR1). The datum for this correlation is the low order initial flooding surface which overlies the Virgelle incised valley/interfluve low order sequence boundary. This was selected as it enabled the best graphical portrayal of the easterly prograding clastic wedge. FTLOSO, Fish Tooth low-order sequence boundary; TCLOMFS, Telegraph Creek low-order sequence boundary; VLSOB, Virgelle low-order sequence boundary; CLOMFS, Claggett low-order maximum flooding surface; JRLOSB, Judith River low-order sequence boundary; TLSOB, Teapot low-order sequence boundary.

strata with multistorey fluvio/tidal channels (CHT) and tidal inlets (T1) is described in a later section.

Within strata of the Mesaverde Group in the Bighorn Basin, a similar cyclical deposition pattern is observed within each of the coarse-grained shallow-marine clastic tongues (e.g. V1–8, G1–3, C1–4, JR1–4, T1, Fig. 2) and the

Table 1. *Sedimentology of the principal facies associations developed within the clastic 'tongues' of the Mesaverde Group of the Bighorn Basin, as summarized from Fitzsimmons (1994) and Johnson (1995)*

Lithofacies association	Description	Palaeocurrents	Ichnofossils (% bioturbated)
CHT: multistorey, fluvio/tidal channels (valley fill)	Multistorey, channelized, fine- to medium-grained sandstones (2–30 m thick). Individual channels 3–15 m thick. Trough (complex and sigmoidal), and planar-tabular cross-sets. Paired mud drapes, reactivation surfaces and inclined heterolithic strata. Parallel lamination, current and rare wave ripples	Bi-directional, dominantly at 90° (N–S) to shoreline progradation direction	Rootlets, *Ophiomorpha*, *Teredolites*, *Macronichnus* **(<20%)**
T1: tidal inlet (deposition on inner shelf)	Sharp/erosive based, fine sandstones (5–12 m thick). Mud-draped, trough cross-sets, reactivation surfaces and herring bone cross-strata	Bi-directional, at high angle (dominantly SW) to palaeo-shoreline progradational trend	*Skolithos*, *Ophiomorpha* **(<30%)**
SF/FS: upper shoreface-foreshore (deposition on inner shelf)	Fine- to medium-grained, sandstones. Trough and rare tabular cross-sets. Low angle. Low-angle, planar laminations. Very rare swaley cross-stratification	Shore sub-parallel cross-sets (NE–SW). Low angle, planar laminae directed seaward	Rare rootlets, *Ophiomorpha* and *Skolithos* **(<30%)**
SF2: middle shoreface (deposition on inner shelf)	Amalgamated, lower to upper fine-grained sandstones, 5 to 15 m thick; dominated by swaley cross-stratification, rare hummocks, planar laminations and wave ripples; internal, erosive amalgamation surface highlighted by intraformational rip up clasts	Rare wave ripples have shore parallel crestlines (NE–SW)	*Ophiomorpha* (*nodosa*, sp.), *Thalassinoides*, *Diplocraterion* and *Skolithos* **(<40%)**
SF1: lower shoreface (deposition on inner shelf)	Interbedded, very fine- to fine-grained sandstones and shales; sandstones (0.1 to 2 m thick) sharp based dominated by hummocky cross-stratification; rare planar lamination; tops of sandstones commonly reworked by ripples; erosive, sharp bases commonly result in amalgamation of individual beds, truncating shale units (1–11 cm thick)	Mixed wave/current ripples and grooves show transport to SE	*Skolithos*, *Ophiomorpha* (*nodosa*, sp.), *Terebelina*, *Arenicolites*, *Diplcraterion*, *Rosellia*, *Chondrites* and *Thalasinoides* in sands; *Planolites*, *Palaeophycus* and *Terebelina* in shales. **(<60%)**
S2: offshore transition Zone (deposition on mid to inner shelf)	Thin, 1–10 cm thick, fine-grained sandstones, interbedded with 1–20 cm thick, silts, and shales; mixed wave/current ripples and planar laminae	Current ripples, groves and primary current lineation offshore to SE; wave ripple crests shore parallel	Intense bioturbation, *Skolithos*, *Chondrites* and *Arenicolites* in sands; *Terebelina*, *Planolites*, *Palaeophycus* and *Chondrites* in shales **(<80%)**
S1: offshore marine (deposition on outer to middle shelf)	Homogeneous, shales and silts, with poorly preserved planar/ripple lamination; occasional limestones and spherical calcareous concretions	Poorly preserved, offshore (SE) transport direction from ripples	Intense bioturbation, *Terebelina*, *Planolites*, *Palaeophycus* and *Chondrites* **(>80%)**

intervening fine-grained deposits. Figure 5 serves as an example of the cyclic sedimentation observed within individual tongues, and was derived from measured sections within the Virgelle Member of the Eagle Formation. The upper surface to each progradational tongue is bounded by a 5–10 km landward shift of facies tracts, and is interpreted as an extensive marine initial flooding surface (IFS) (i.e. Fig. 5, surface SEQ6 IFS). In proximal positions, the tongues are dominantly composed of massive sandstones of middle to upper shoreface/foreshore facies associations (SF2 & SF/FS; e.g. Fig. 5, SEQ6 at the SD locality). When traced down-dip, the shoreface succession becomes heterolithic, and is comprised of two to three distinctive pro-gradationally stacked shoaling-upwards cycles (S2, SF1 & SF2; e.g. Fig. 5, SEQ6 at SYN, RA and GEBO locations). These units commonly have a rather 'top-heavy' appearance, as they have a higher net to gross ratio and contain facies associations deposited in shallow(er) environments than those developed below the tongues (e.g. Fig. 5, at the MTN-GEBO sections,

AGE	STAGE	GROUP	FORMATION	FACIES ASSOCIATION	AGE(ma)	AMMONITE ZONE
CRETACEOUS	CAMPANIAN	MESAVERDE	TEAPOT	ESTUARINE	76 Ma	
			JUDITH RIVER	DELTAIC		Bacculites Scottii
					79.5 Ma	Bacculites obtusus
			CLAGGETT	OFFSHORE/ OPEN SHELF		Bacculites Sp (weak ribs)
					80.54 (+/- 0.55) Ma	Bacculites Sp (smooth)
			EAGLE	DELTAIC		
		CODY	CODY	OFFSHORE/ OPEN SHELF	83. 1 (+/- 0.5) Ma	Scaphites hippocrepis 2 Scaphites hippocrepis1

Fig. 3. Stratigraphy of the Mesaverde Group within the Bighorn Basin, illustrating the general depositional environments, and specific radiometric dates and principal ammonite zones. Data based on Gill & Cobban (1996*a*, *b*, 1973; Hicks 1993; Obradovitch 1993).

below surface SEQ6 HOSB (high-order sequence boundary)). The flooding surfaces that cap the individual cycles are a magnitude smaller than those bounding the upper surface to individual tounges, but are still areally extensive enough to merit interpretation as parasequence flooding surfaces (*sensu* Van Wagoner *et al.* 1990).

Above the coarse-grained progradational shoaling-upwards cycles more distal fine-grained lithofacies are observed (S1 & S2). They are bounded by flooding surfaces and have a retogradational stacking pattern so that parasequences become mudstone-prone (e.g. see Fig. 5, cycles below surface SEQ6 HOSB, between the MTN-GEBO sections). The upper bounding surface to these retrogradational cycles is coincident with the maximum landward dislocation of facies (10–30 km), which apparently terminated the deposition of sand grade material across the entire study area. In Fig. 5 the maximum floding surfaces (MFS) to sequences 5 and 6 (bold, dashed lines) are truncated up dip below the high-order sequence boundaries to sequences 6 and 7, respectively. Overlying the MFS, the succession typically becomes dominantly aggradational and is composed of dark grey/black, offshore marine shales (S1), with rare distal sandstones (S2). The cycle ends with the abrupt return to the strongly progradational trend of the next coarse-grained and normally sharp-based clastic tongue, as clealy seen in Fig. 5 at the base of high-order sequences 6 and 7 (HOSB6 and HOSB7).

To summarize, a complete cycle through progradational, to retrogradational, to aggradational, back to progradational could be interpreted as a single genetic sequence bounded by maximum flooding surfaces (*sensu* Galloway 1989*a*, *b*). However, this interpretation does not address the fact that many of the coarse-grained tongues are sharp- and erosionally-based and have a major discontinuity and dislocation of facies at their base. As such, the stratigraphy of the Mesaverde Group in the Big Horn basin is not easily explained or accommodated by means of simple regressive/transgressive model as advocated by Galloway (1989*a*, ***b***).

Identification of surfaces related to high-order falls in relative sea-level and forced regression

The shallow marine clastic tongues of the Mesaverde Group differ subtly from the traditional model of normal shoreline progradation, as shown in Fig. 6a. In the idealized waltherian-type model of progradation, progressive basinward regression of the shoreline is recorded in the sedimentary succession by gradual upward passage from open marine

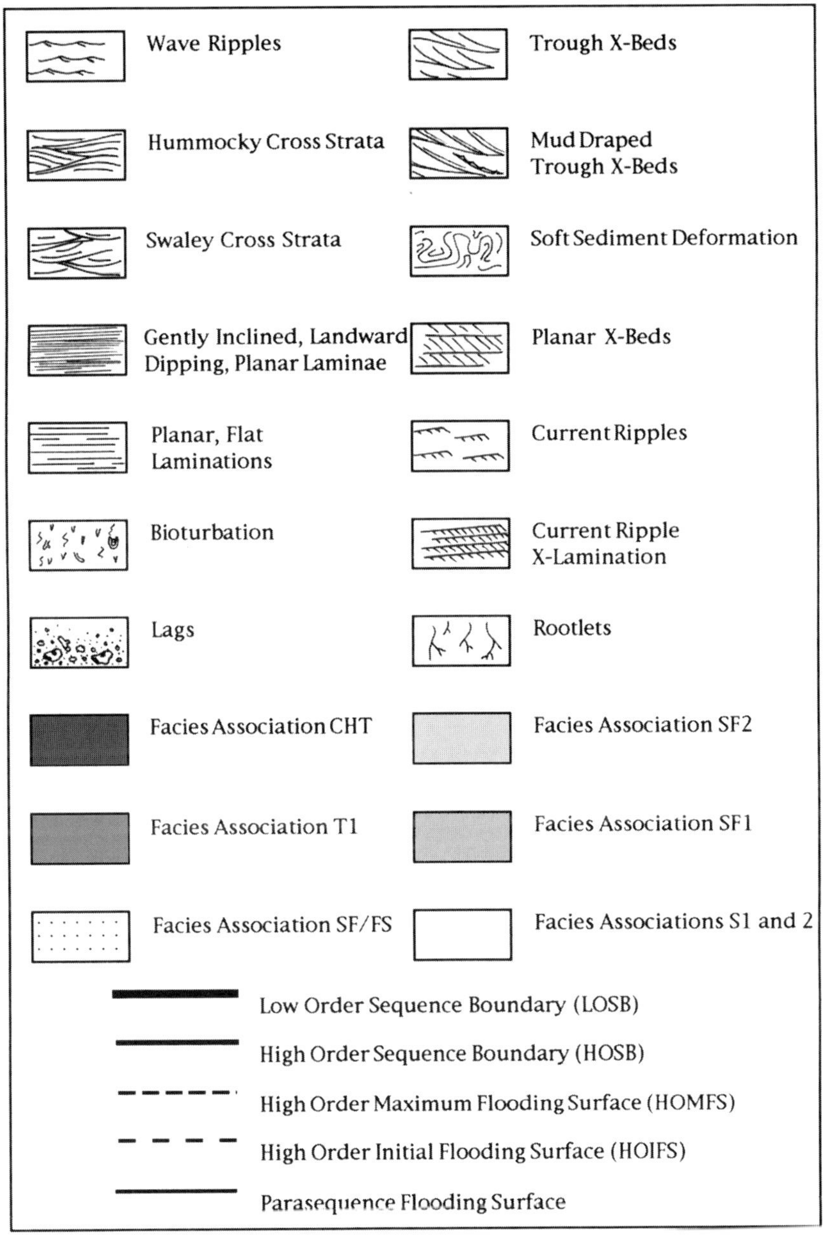

Fig. 4. Key to measured sections in Figures 5, 6 and 12.

facies, through offshore transition zone, lower-, middle- to upper-shoreface and foreshore environments into coastal plain sediments (Fig. 6a). However, within the Mesaverde Group, disruptions to this trend are commonplace. Typically, a distinctive and sharp surface of considerable lateral extent is observed across which facies belts of the expected shoreface succession are *omitted*, foreshortening the expected shoreface profile (e.g. Fig. 6 b–d). These are distinctive and important surfaces across which major changes in sandbody architecture and stacking patterns occur *across* the study area. The surfaces are of undoubted marine origin

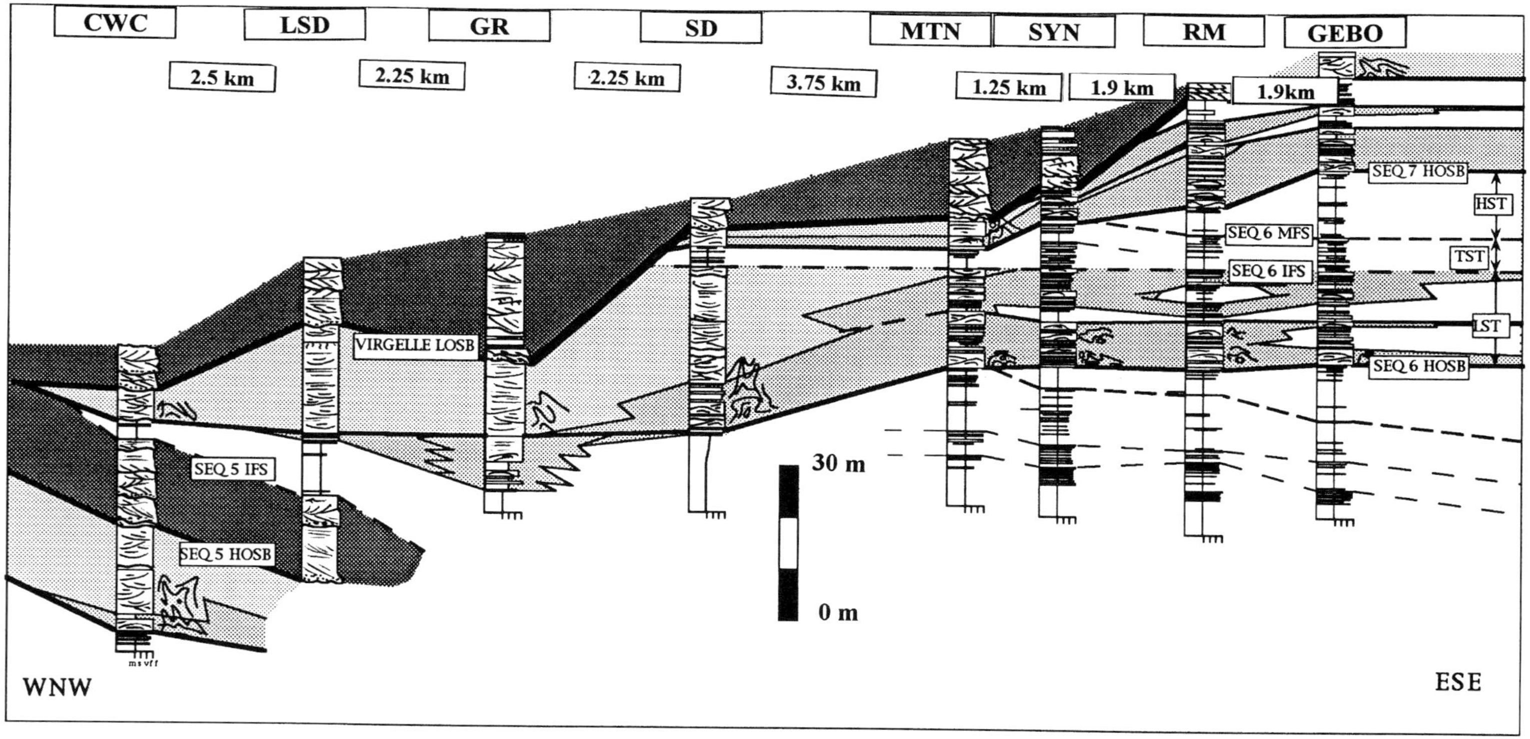

Fig. 5. Dip-oriented cross section extending approximately 17 km down depositional dip from Cottonwood Creek (CWC) to Gebo, detailing the facies and stratal relationships within a single clastic tongue developed within the Virgelle Member (Tongue V5, Fig. 2). Note that in up-dip proximal positions the tongue is a relatively massive and homogeneous unit, but when traced down-dip it splits into two upward-shoaling parasequences. However, across its length of more than 15.8 km, the tongue retains its sharp base. For key to facies see Fig. 4, and Fig. 1 for location of sections.

Table 2. *Summary of the features associated with the expression of regressive surfaces of marine erosion, within the shallow marine successions of the Mesaverde Group of the Bighorn Basin*

Characteristics of regressive erosion surfaces
1. Marked basinward shift in facies
2. An abrupt increase in sand:shale ratio across the surface
3. Missing/eroded facies below the surface
4. Change in parasequence stacking patterns across the surface
5. Development of '*glossifungites*' ichnofabric
6. Development of precursor gutter cast facies
7. Development of widespread soft sediment deformation and growth faulting
8. Change in local palaeocurrents across the surface
9. Regional truncation of older parasequences and systems tracts below the surface
10. Surfaces can be traced regionally

and are interpreted as regressive surfaces of marine erosion (RSME) produced during periods of relative sea-level fall (e.g. see also Plint 1988, 1996; Walker & Plint 1992; Plint & Nummedal this volume).

The main characteristics of the regressive surfaces of marine erosion are summarized in Table 2, and discussed in the following section. Their regional extent and association with major reorganisation of stacking patterns in proximal and distal settings and other attributes indicate that their development *is not* due to autocyclic processess as has been suggested elsewhere (e.g. Van Wagoner 1995*b*). Instead, as is argued below, these features are associated with relative base-level falls and the destructtion of accommodation space. As such, their characteristic features may form useful criteria for the recognition of forced regressive shoreface profiles elsewhere.

Dislocation of facies tracts

An abrupt basinward dislocation of facies associations is encountered at the base of each shallow marine clastic tongue, and is the clearest indication of relative sea-level fall. Facies associations, normally many metres thick in a shoreface profile resulting from normal regression, are condensed or absent. In many cases the basinward facies shift is expressed as a sharp-based shoreface sand body with thick- to massively-bedded, lower to middle shoreface facies resting directly upon open marine shales (e.g. Figs 6, 7a, b). Three such surfaces are displayed in Fig. 4, at the base of high-order sequences 5, 6 and 7.

In detail, the base of the sharp-based shoreface may be locally planar, or highly erosive (Fig. 7a, b), indicating either an interval of non-deposition or erosion prior to deposition of the overlying sandstones, that can account for the missing facies below the sharp-based shoreface. Detailed mapping and walking-out of the shoreface sandbodies shows that the abrupt shift of facies is of regional extent, and is observed even the most distal portions of each tongue (e.g. Fig. 5, RM and GEBO sections). In such locations, an abnormally thick sandstone bed, in relation to those exposed above and below, initiates deposition of the overlying progradational cycles. These beds commonly have an increased abundance of well developed, small-scale erosive features upon their bases (large grooves, flutes and scours). This is interpreted to be due to deposition after significant scouring of the underlying succession (see also Plint 1996; Mellere & Steel this volume; Plint & Nummedal this volume).

Changes in sand to shale ratios

In association with the basinward shift in facies, it is typical that the sand: shale ratio of the overall succession typically shows dramatic increases across the sharp basal surfaces to the clastic tongues. This may be spectacular where for example, 20 m thick massive- and thick-bedded upper shoreface sands rest upon finer-grained deposits in proximal settings (e.g. Fig. 5 LSD-SD locations; Figs 7a, 8a). In distal settings, although more subtle, the increase in the coarser-grained sand fraction is still pronounced, producing characteristic top-heavy upward-shoaling cycles, as shown in Fig. 4 (MTN-GEBO sections).

Stacking patterns

Within each cycle, taken from base of tongue to base of tongue (a choice of cycle boundary discussed further later), three distinctive stacking patterns are recognized within the upward-shoaling parasequences (Fig. 5). The cycles

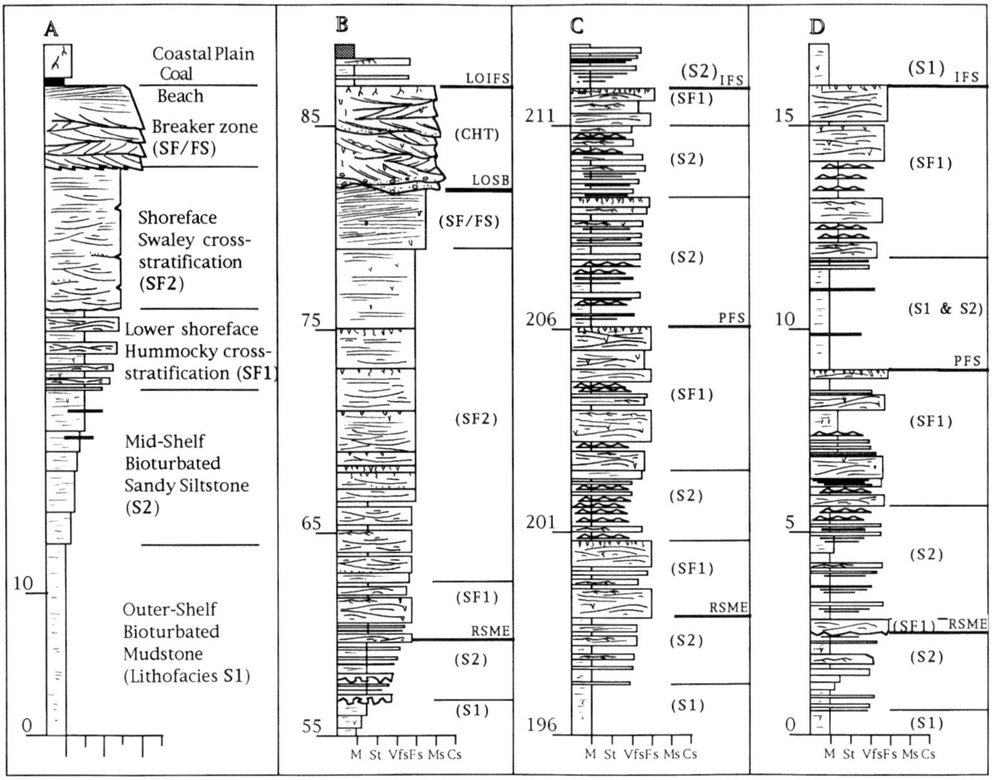

Fig. 6. Comparison the model expression of a gradational shoreface profile (**a**) expected from normal regression under conditions of stillstand or slight relative rise, as modified from Walker & Plint (1990) with three profiles from the Virgelle Member of the Eagle Formation; (**b**) Tongue V8, Elk Basin; (**c**) Tongue V3, Hamilton Dome; (**d**) Tongue V4, Ronoco Mine. The three profiles from the Virgelle illustrate increasingly distal positions along the depositional profile, yet all have abrupt bases, and do not reflect the smooth model profile. For key to facies see Fig. 4. Scale in metres.

overlying the sharp-base of each tongue are sand-prone and strongly progradational (Fig. 8a). An extensive marine flooding surface, interpreted as the initial flooding surface (IFS, e.g. Fig. 5) separates these progradational cycles from the overlying retrogradationally stacked ones (Fig. 8b). The upper bounding surface of the backstepping cycles is the maximum flooding surface (MFS; e.g. Fig. 5, SEQ6 MFS), marking the maximum landward facies dislocation as shown in Fig. 8b. The final unit, comprising fine-grained marine shales and rare distal sandstones, has an aggradational to slightly progradational stacking pattern.

Distinctive trace fossil assemblages

Frequently, distinctive trace fossil assemblages are observed along the base of the sharp-based storm successions within the Mesaverde Group.

Fig. 7. The planar to erosive nature of high order sequence boundaries at the bases of shoreface successions is well illustrated at Elk Basin (located north of the town of Powell (section number 2; Fig. 1). Here, Tongue V8 is always expressed as a massive shoreface, however, locally the basal expression changes from being (**a**) planar and sharp to (**b**) highly erosive with 2 to 3 m of relief. (**c**) The shorefaces erosively overlie precursor gutter cast facies. (**d**) Lags of reworked burrows (A) are often found along the sharp-based shoreface profiles. These are interpreted to be the remains of distal shoreface deposits, cannibalised during the continued cycle of relative sea-level fall. The partial lithification of the burrows during early diagenesis (perhaps due to high faecal concentrations), results in their preservation as clasts, whereas the bed matrix is reincorporated into the massive shoreface.

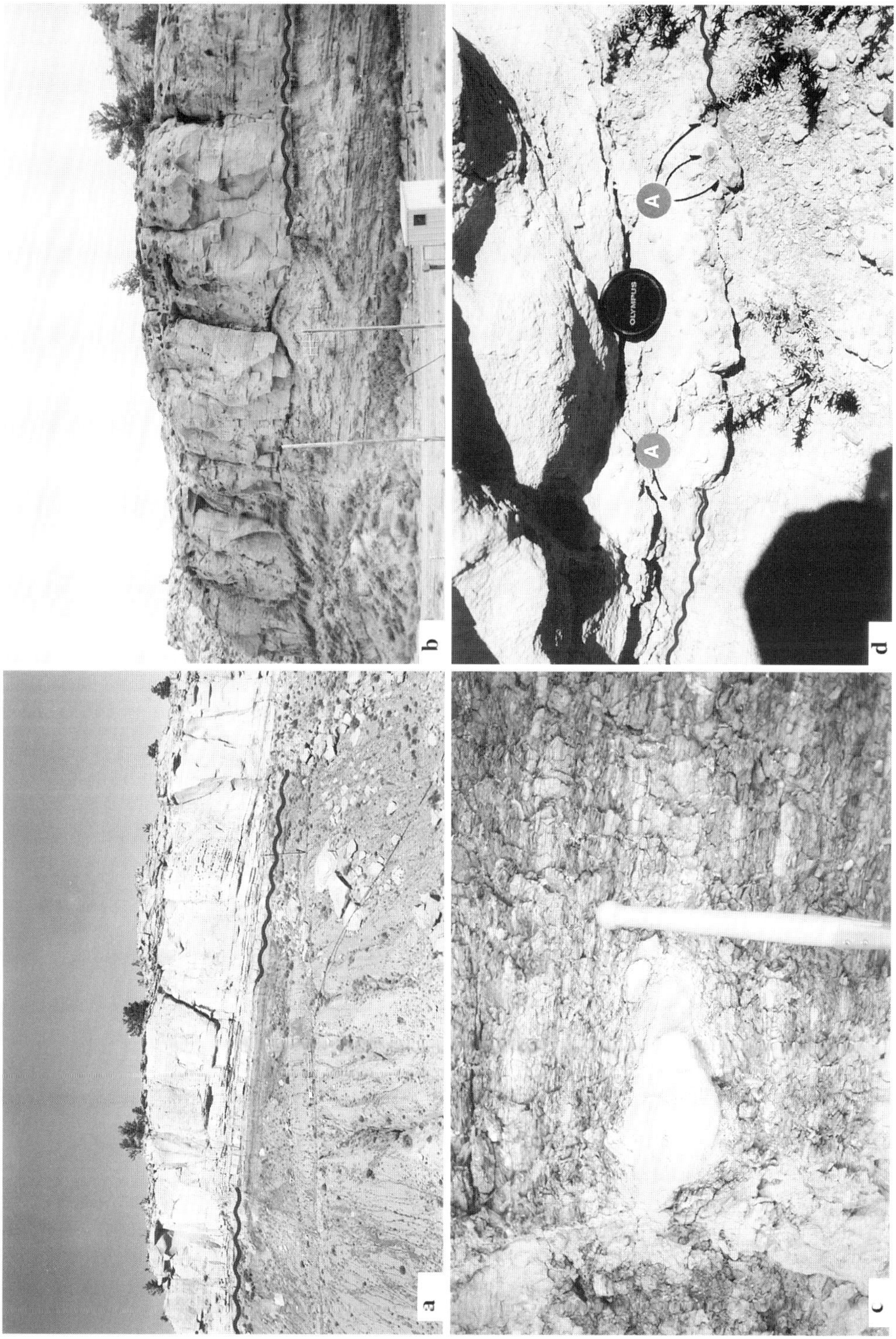
a
b
c
d
A
A
OLYMPUS

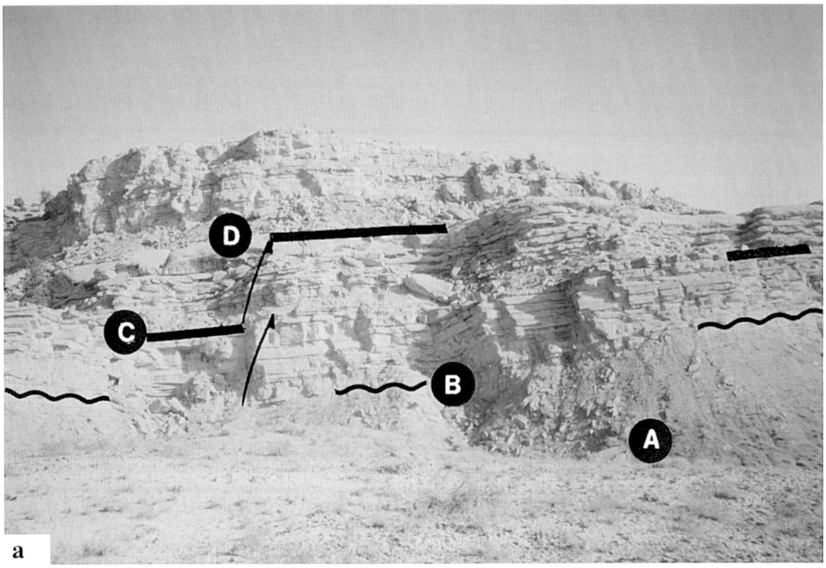

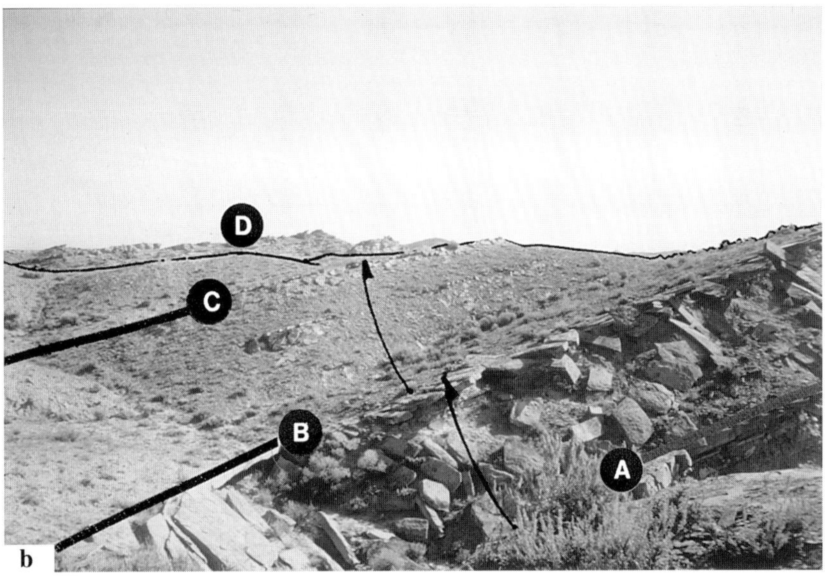

Fig. 8. Comparison between the progradational and retrogradational components of a clastic tongue. (**a**) Fine-grained shales (A) are abruptly overlain by a succession of coarser-grained proximal facies. The contact is characterised by a distinctive sandstone bed (B), which is in turn overlain by two (C & D) progradationally stacked parasequences (Tongue V4, Sand Draw). (**b**) Two retrogradationally stacked parasequences (B & C) are developed above the initial flooding surface (A) which caps the progradational component of Tongue V3 (Sand Draw). The top of the second parasequence (C) is coincident with the maximum flooding surface, which terminated coarse-grained clastic deposition within this cycle. Above this surface, fine-grained, aggradational offshore mud- and siltstones (D) are developed. Note that the coarser grained deposits forming the distant horizon are those of Tongue V4, as detailed in (**a**).

Characteristically, they consist of a low-diversity assemblage of fauna, with dense populations of *Thalassinoides*, *Skolithus* or *Ophiomorpha* typical (Fig. 9d). The trace fossils are extremely large forms, in comparison to average in the succession, that penetrate up to 0.15 m into the underlying strata. The diameter of these trace fossils is up to 25 mm, and are also much larger than those found in related environments, that are typically no larger than 10 mm in diameter. With the exception of *Ophiomorpha*, which always occur in sand-rich facies and maintain agglutinated fecal pellet walls, the traces have unlined walls, even when preserved in fine-grained strata. Such *Glossifungites* firmground ichnofacies are commonly found in association with surfaces of sequence stratigraphic significance (e.g. MacEachern *et al.* 1992).

Abundance of gutter casts

Decimeter and smaller gutter casts with a predominant northwest to southeast orientation, occur beneath many of the basal erosion surfaces (Fig. 7c). Greensmith *et al.* (1979) and Plint (1996) interpreted the fill of such casts to be synchronous with scouring, and related their formation to the lowering of storm wavebase, during severe storms, below the offshore transition zone of the inner to middle shelf. Their NW–SE orientation in the study area is interpreted to reflect the direction of the offshore directed current which created the gutters. Though gutter casts are found throughout storm successions (e.g. Whittaker 1973), their particular abundance below sharp-based storm successions has elsewhere been interpreted to represent a so called 'pre-cursor facies', developed before the creation of the regressive surface of erosion during sea-level falls (Hadley & Elliott 1993). Similar close association of the gutter casts and sharp-based shorefaces observed in this study tends to confirm the interpretation of Hadley & Elliott (1993).

Soft sediment deformation

Soft sediment deformation is common throughout shallow marine successions. However, within the sharp-based successions of the Mesaverde Group, soft sediment deformation is especially frequent along the bases of the individual tongues, as for example along the base of high-frequency sequence 6 as shown in Fig. 5 (HOSB6). Within clastic tongues of the Virgelle Member, over one third (23 examples) of the 60 sections measured through their base, are overlain by intensely deformed storm successions. Deformation varies from large scale ball and pillow structures, with up to 6m relief, to the deformation of individual beds by shale injection (Fig. 9). Within the Claggett Formation, growth faults are developed along the sharp base of a single storm succession (Tongue C1, Fig. 9c). Fault throw ranges from 4 to 12 m, and can be traced laterally for 40 m. The fill of the hanging wall to the faults is typically sand-prone and often deformed. Internally, individual beds can be observed to thicken towards the fault plane, resulting in the development of small scale roll-over anticlines (e.g. Fig. 9c).

The abundance of soft sediment deformation features in close association with the sharp-based storm successions, suggests that their origins are intrinsically linked to the development of the regressive surface of erosion. Deformation is interpreted to have resulted from the rapid loading of large volumes of sand onto uncompacted shale-dominated and thinly bedded sandstone facies, directly after the creation of the erosion surface. The relatively rapid addition of overburden is thought to have increased pore pressure in the underlying fine-grained deposits until such a point as the rapid expulsion of pore water through the overlying succession occurred. Similar phenomenon (though at a larger scale), have been recorded in the shallow marine successions of the Niger Delta (Williams & Ugueto 1994). In the Niger system, it has been shown that growth fault movement is at its greatest during lowstands, and that the units deposited within these successions are commonly intensely deformed.

Changes in palaeocurrent orientation

Palaeocurrents within the shoreface successions were measured from basal grooves/flutes and wave to mixed current/wave ripple crestline orientations. These data, some of which are shown in Fig. 10, indicate that the general progradation direction of the Mesaverde shallow marine succession was, in the study area, to the southeast (Fig. 9). In detail, palaeocurrents within from each of the clastic tongues are slightly different, albeit with a dominant southeasterly trend, and variations of between 5° an 10° are common. The change palaeocurrent orientation is most dramatic across the basal erosion surfaces of individual tongues (e.g. Fig. 10). Changes in regional drainage patterns associated with the development of incised valleys have been widely documented (e.g. Van Wagoner *et al.* 1990; Dalrymple *et al.* 1994). In such instances, the distributary channels are pulled towards the vector of maximum base-level

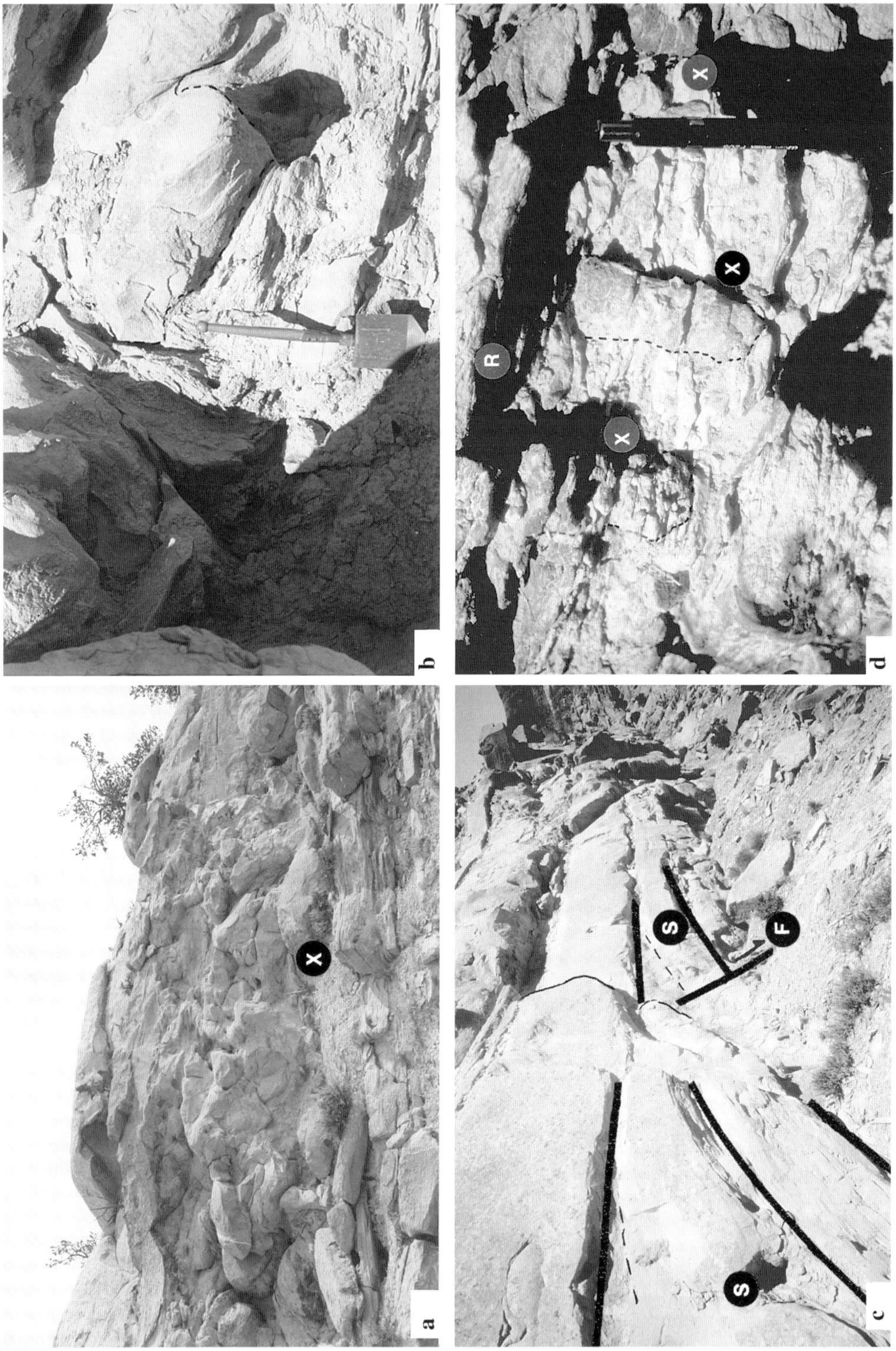
a
X
b
c
S
F
S
d
X
R
X
X

fall. It is therefore logical to assume that as distributary mouths are reoriented, the progradation direction of the associated shoreline will also be subtly realigned. Such processes are thought to be responsible for the most dramatic changes in palaeocurrent orientation occuring across sequence boundaries observed here.

Erosional truncation

At the majority of measured sections, the contact between the basal surface of each tongue and the underlying succession appears planar over the length of the individual exposure. However, when the individual sections are correlated, it is apparent that the base of the sharp-based sandstones is erosive, and truncates the underlying strata. In Fig. 5, the basal high-order sequence boundary to sequence six (SEQ 6 HOSB, base of tongue V5 labeled) is shown. This surface was traced over 20 km up-dip, from its down-dip pinch out, east of the Gebo measured section, to the point where it is truncated below the Virgelle low-order sequence boundary, west of Cottonwood Creek. When traced east to west, this surface progressively truncates (i) the underlying aggradational shale-dominated cycles, (ii) the sequence 5 maximum flooding surface (SEQ 5 HOMFS) and (iii) backstepping, heterolithic parasequences of the transgressive systems tract. A similar pattern is observed at the base of sequence 7 (Fig. 5). Thus, although the angle of truncation is less than 0.5°, making it difficult to observe at any single measured section, it is clear that the processes that created the sharp basal surfaces to the tongue were highly erosive, and that the resulting erosion is most intense of proximal areas and decreases down dip.

Regional development

It may be argued that local autocyclic processes (i.e. geometry of shoreline, shelf width, rapid progradation due to high sediment supply and freak storms), could produce a sharp-based shoreface (e.g. Van Wagoner 1995*b*). Many sharp-based shorefaces have been recognized in the stratigraphic record (e.g. Roep *et al.* 1979; Cant 1984; McCrory & Walker 1986; Eyles & Clark 1986; Rosenthal & Walker 1987; Davis & Bryers 1989; Van Wagoner 1995) and undoubtedly not all rest upon surfaces created by allocyclic processes. However, within the Mesaverde Group the features described above are developed in close association with the bases of the individual clastic tongues that are expressed on a basin wide scale. One surface at the base of tongue V3 has been tied with confidence from Syncline Draw to Oregon Basin, and tentatively correlated to Elk Basin (137 km north of Syncline Draw). Such regional persistence indicates that these surfaces are not a local phenomenon, driven solely by allocyclic processes.

Interpretation

To understand the sequence stratigraphic significance of these regional erosion surfaces, the processes responsible for their formation need to be considered and understood. The erosion surfaces within the sharp-based shoreface successions considered so far are quite unlike those at the base of incised valleys, since no subaerial exposure or channel incision has occurred. Erosion at the base of sharp-based shorefaces is considered to be caused by the lowering of relative base-level (e.g. Plint 1988), enabling storm-wave action to scour the sea floor in advance of the prograding shoreline. This submarine erosive process is referred to as 'erosional regression' (Curray 1964; Bosselini *et al.* 1989), with the erosive surface being termed the 'regressive surface of marine erosion' (RSME, Posamentier *et al.* 1992). Because these erosion surfaces are expressed regionally within the Mesaverde Group, they are interpreted to have formed in response to a decrease in accommodation space due to a fall in relative sea-level. They are not thought to be of autocyclic origin. Additional evidence for this interpretation has been gained through tracing the sharp-based shorefaces up-dip into unequivocal sequence bounding surfaces associated with subaerial exposure and fluvial channel incision as is discussed further later.

Fig. 9. (**a–c**) Outcrop photographs of the various styles of soft sediment deformation observed in association with regressive surfaces of marine erosion. (**a**) Ball and pillow structures are well developed within the massive proximal shorefaces (note: shovel for scale X), Tongue V4 Cottonwood Creek. (**b**) In the distal heterolithic successions, discrete shale injection structures are observed to penetrate the sandstones dominated by hummocky cross stratification, Tongue V5 Ronoco Mine. (**c**) Growth faults exhibiting sediment thickening (S) into the plane of the fault (F) developed in Tongue C2, Syncline Draw. (**d**) Example of *Glossifungites* ichnofabrics as are locally developed along the regressive surfaces of marine erosion (R). Here, sand filled burrows (X) penetrate the underlying fine-grained heterolithic deposits of the offshore transition zone (lithofacies association S2) in Tongue C2 Ronoco Mine.

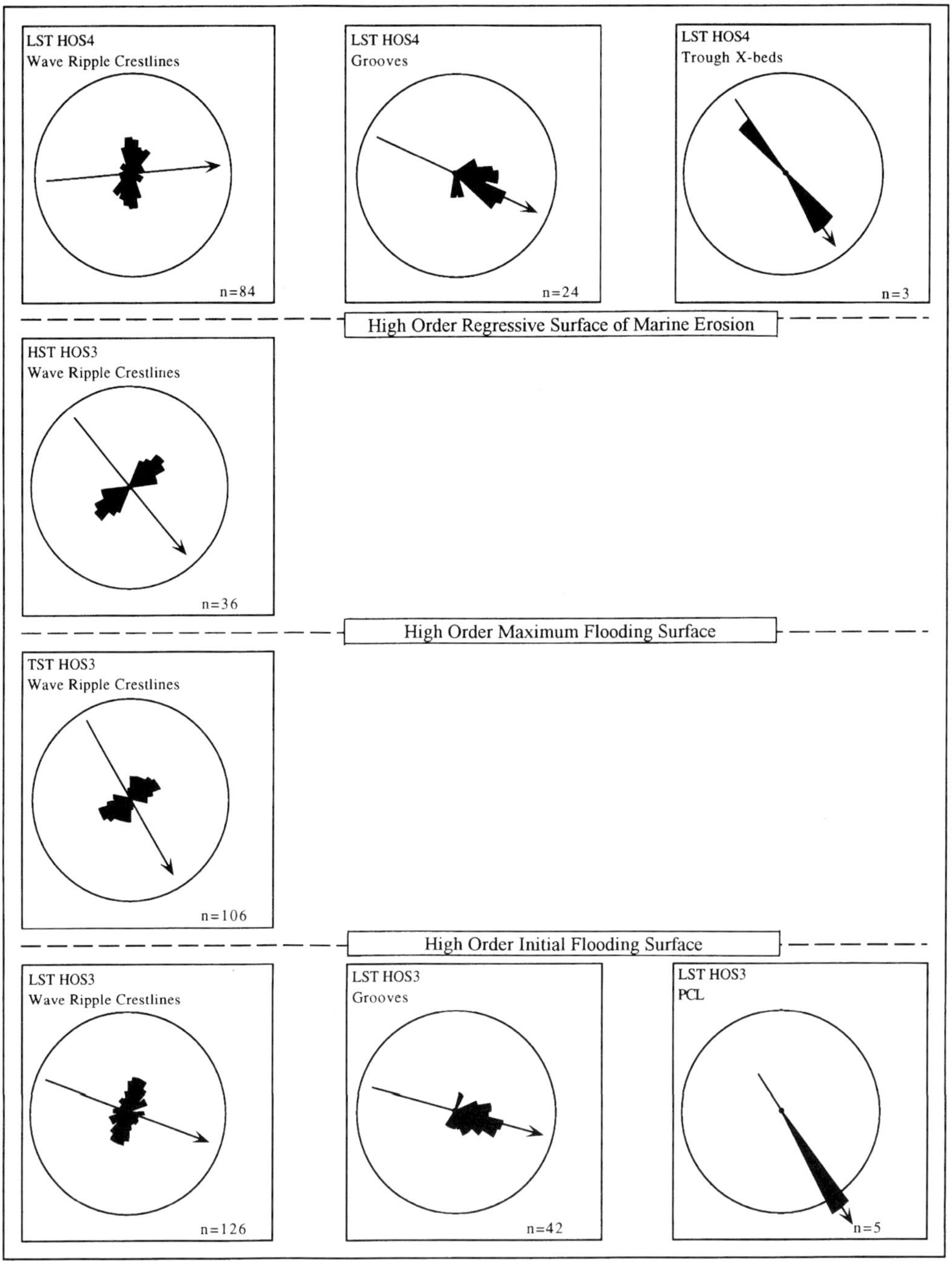

Fig. 10. Regional palaeocurrent data collected from the V3 and V4 shallow marine clastic tongues of the Virgelle Member. These data illustrate the subtle changes in sediment dispersal patterns developed within the individual high-order systems tracts. In the lowstand systems tract (LST) of tongue V3 (HOS3), both wave-crest and groove data indicate sediment transport was to the ESE. In the transgressive systems tract (TST), the transport direction changed to a SSE orientation that persisted into the highstand systems tract (HST). However, the palaeocurrents change across the regressive surface of marine erosion developed at the base of the next clastic tongue, with sediment transport returning to a predominant ESE direction. This trend is interpreted to reflect the subtle changes in shoreline orientation which would occur during periods of relative sea level fall. Key to symbols: LST, lowstand systems tract; TST, transgressive systems tract; HST, highstand systems tract; HOS, high-order sequence; PCL, primary current lineation.

Internal hetrogeneities of shoreface sandbodies

Up-dip amalgamation. Within the individual shoreface sandstone tongues, the expression of individual shoaling-upwards cycles, or parasequences, is best preserved in the distal hetrolithic facies associations (S1, S2 and SF1; e.g. Figs 5, 8). In up-dip areas the development of sand-prone facies (SF2 and SF/FS) is more areally extensive than expected from normal shoreline progradation. In these up-dip areas, flooding surfaces can be difficult to ientify. However, where parasequence flooding surfaces could be traced landward into these relatively massive shorefaces, they were expressed as intensively bioturbated zones, within which all sedimentary structures were destroyed. However in the study area, none of these surfaces could be traced with any degree of confidence far landward, due to the amalgamation of sand bodies and surfaces.

In many of the massive shorefaces, internal amalgamation/erosion surfaces are abundant. These surfaces are observed to truncate the sedimentary structures developed within the underlying succession, and are commonly overlain by an intraformational mud clast lag. The amalgamation surfaces may well result from normal storm/fair-weather processes. It is thought that local autocyclic processes which create these features overprint the expression of regional allocyclic base-level rise in these shallow water facies, resulting in the poor preservation of the parasequence flooding surface.

Hiatal surfaces. Within the massive shorefaces some internal erosion surfaces appear to have a much greater significance than that of the more localized amalgamation surfaces described above. These distinctive and relatively steeply basinward-dipping surfaces (up to 3–4°) create large scale internal hetrogeneities within the shoreface, and are well preserved within the Claggett Formation (e.g. Fig. 11). One particuar surface can be traced over 70 m downdip, as it truncates underlying units and their associated sedimentary structures (Fig. 11). Locally, small patches of *Glossifungites* and also laminated mud- and siltstones (<0.1 m thick) are preserved along the erosion surface, separating the two sandstones. The latter contain primary deposition structures and are not reworked.

These steeply dipping surfaces are interpreted to represent an erosive/hiatal phase during the overall trend of shoreface progradation. The surface shown in Fig. 11 shallows in dip basinward as it amalgamates with the basal regressive surface of marine erosion, and it is logical to assume that the same erosive processes were involved in its creation. Most authors (Plint 1988, 1996; Posamentier *et al.* 1992; Hunt & Tucker 1992, 1995; Plint & Nummedal this volume) infer the step wise progradation of the shoreface during sea-level falls, as it is widely thought that falls do not occur at a constant rate. The surface described above (e.g. Fig. 11) is interpreted as an example of how an erosive step, formed during progressive stepwise in relative sea-level fall, is manifested within the otherwise relatively massive portions of the strandplains. It is apparent from the deposition of laminated mudstones and firm ground fabrics across the surface that there was some hiatus between its development and resumption of shoreface deposition. These erosive scours, created during periods of decreasing accommodation space result in the cannibalisation of the distal, finer-grained facies so that a complete, proximal to distal suite of shoreface facies is only preserved during times when there is preservation or the addition of accommodation space. It is worthy of note that comparable steeply dipping surfaces in well exposed shoreface sandstones of similar stratigraphic origin are also described by Mellere & Steel (this volume).

Key surface development from shoreface to coastal plain; incised valleys, interfluves and ravinement surfaces

Within the Eagle and Judith River Formations, shorefaces can be traced up-dip to their correlative deposits within the non-marine succession (Fig. 2, tongues G1, G2, G3 and JR1). Within each coastal plain succession, a series of multistorey channelized sandbodies have been identified (lithofacies CHT, Table 1).

Using the criteria outlined in previous nonmarine studies (e.g. Shanley & McCabe 1991 1993, 1994; Aitken & Flint 1994, 1995; Van Wagoner 1995), the multi-storey channelized sandbodies have been identified as incised valley fills. In each case, there is a marked upward increase in the vertical and lateral connectivity of the individual channels, which are commonly, but not exclusively, of low sinuosity. Each incised valley fill can be correlated regionally in both dip and strike profiles. Where the incised valleys are not expressed as multi-storey channel bodies, they can be tied to well drained pedogenic horizons interpreted as interfluves. The regional erosion associated with each valley (10 15 m) is significantly greater than that observed at the base of the individual channels within the valley (4–6 m). These surfaces are interpreted as sequence boundaries created by high-order falls in relative base-level.

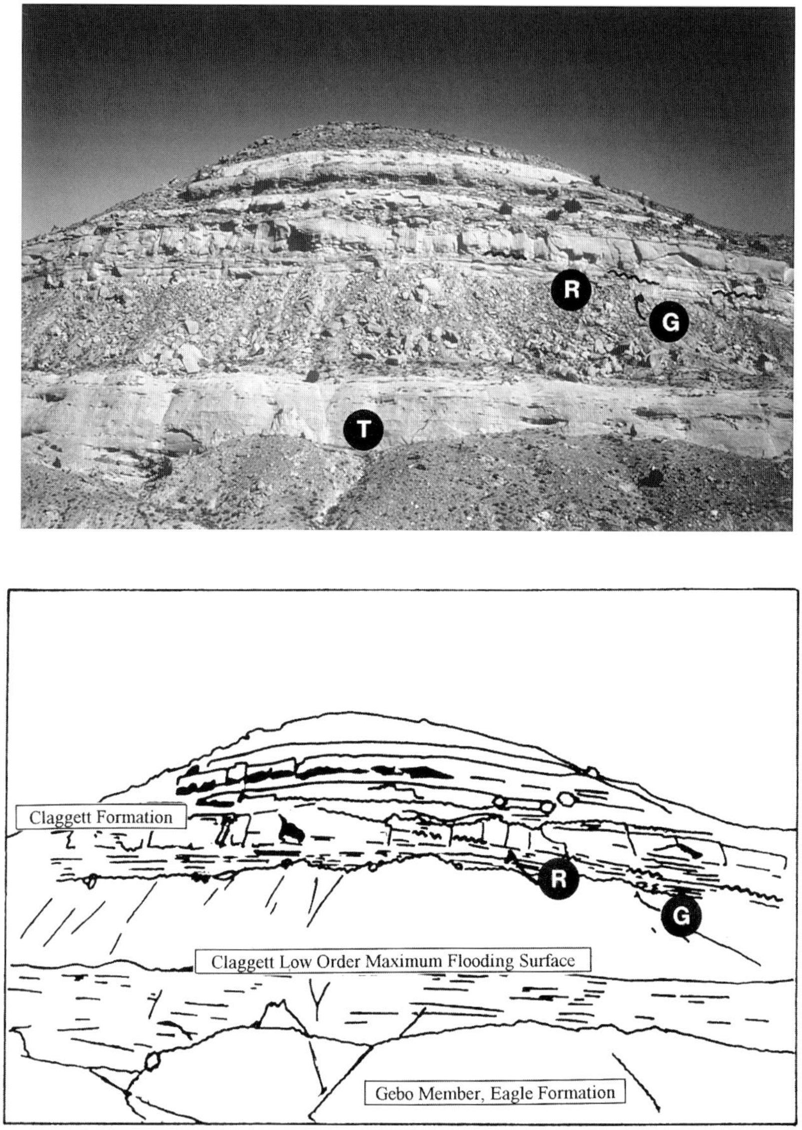

Fig. 11. Photograph and line drawing of Tongue C2 at Syncline Draw (located in Fig. 2), illustrating the basinward dipping internal hetrogeneities developed within the massive strandplains. The erosion surface R dips to the right at an angle of 1.5°, truncating the bedforms below, before it merges with the basal RSME. G marks the location of the growth faults illustrated in Figure 9c. Tongue C1 is expressed as distal facies and as such is hidden in the scree below Tongue C2. T is the tidal inlet developed in association with Tongue G3 (see Fig. 5). There is approximately 40 m of stratigraphy between the top of the Claggett low order maximum flooding surface and the top of the C2 shoreface tongue.

In all of the examples studied, the unconformities developed beneath incised valley fills can be traced laterally to interfluve areas. At the contact between the incised valley fill and sharp-based shorefaces (Fig. 12), the basal unconformity rises over the shoreface sediments and correlates with a well-developed pedogenic horizon. This subaerial exposure surface is typified by extremely well developed root systems that penetrate 0.5 to 1.5 m into the underlying strandplain. Commonly this zone has a mottled appearance with brown/purple patches developed within the matrix.

Clearly this association of incised valley and

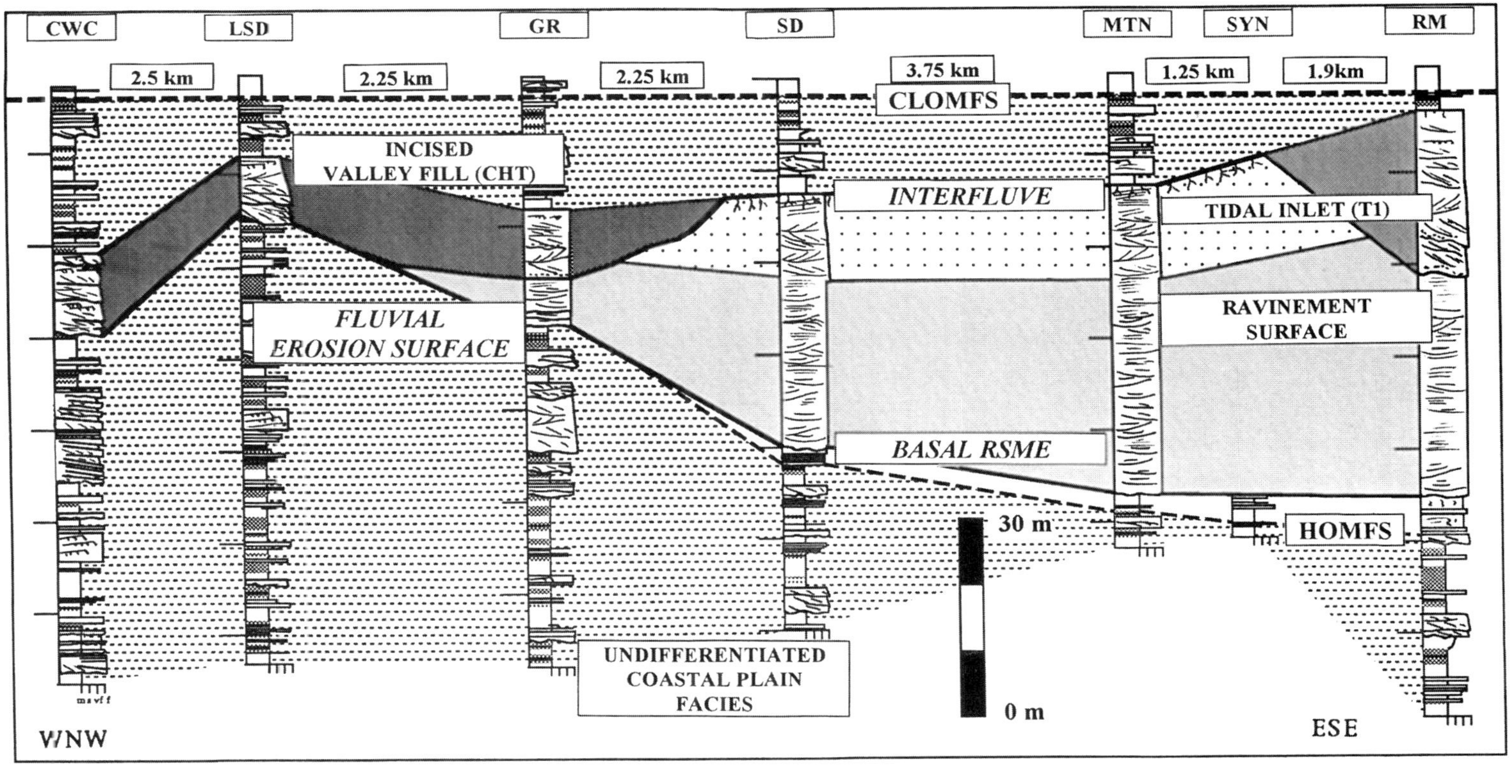

Fig. 12. When Tongue G3 of the Gebo Member in the Eagle Formation is traced up-dip it can be correlated with confidence into the correlative non-marine system. Here the basal surface of fluvial erosion (interpreted as a high order sequence boundary) can be observed to rise over the strandplain to correlate with a well developed interfluve surface. When traced down-dip the interfluve can be observed to have been destroyed through transgressive ravinement at the base of a tidal inlet (far right). It is in these localities, that the regressive surface of marine erosion (RSME) provides the principal marine indicator of a basinward shift in facies. Note the up-dip truncation of the high order maximum flooding surface (HOMFS) beneath the basal RSME. Vertical scale bar is 30 m, length of cross section is 13.9 km. The datum is the Claggett low order maximum flooding surface (CLOMFS). For key see Fig. 4.

interfluve is consistent with sequence boundary development within the depositional sequence model of Posamentier *et al.* (1988) and Posamentier & Vail (1988). However, as noted by many authors (Van Wagoner *et al.* 1988; Posamentier *et al.* 1992; Ainsworth & Pattison 1994; Plint 1996), interfluve surfaces are commonly modified by transgressive marine erosion (ravinement) resulting in the development of a composite surface. In the Mesaverde Group erosion associated with tidal inlets (lithofacies T1, Table 1 & Fig. 12, RM section) reworked the upper parts of the strandplains during transgression, removing evidence of subaerial exposure and ultimately sequence boundary development. At no locations within the Mesaverde Group have interfluves been observed to have developed upon the offshore facies of the underlying shoreface assemblage (e.g. lithofacies S1, S2, SF1 & SF2, Table 1).

Discussion

Genetic v. sequence stratigraphy?

If the genetic sequence stratigraphic model is applied to the Mesaverde succession described here, the regressional component would incorporate all of the aggradational and strongly progradational parasequences into one genetic stratal unit, with the backstepping parasequences forming the transgressive phase. However, as has been demonstrated, there is a clear and abrupt change from an aggradational to strongly progradational trend within the regressive strata, normally marked by an important regional erosional truncation surface. Progradation of the shoreline in the sandstone tongues examined in this study is far from gradual, as would occur under conditions of normal regression. As such the development of the major stratal surface within the shoreface profiles is not addressed in the genetic stratigraphic model of Galloway (1989*a*), and for these reasons is not thought appropriate to successions examined in this study.

Alternatively, the RSME that is present at the base of each tongue and associated with an abrupt change in parasequence stacking patterns, and many other features as discussed above, may be regarded as the depositional sequence boundary (*sensu* Van Wagoner *et al.* 1988). This seems to be the most logical system to follow, whereby the major stratal surface within the sandstone tongues, the RSME, is regarded as the marine expression of a sequence boundary (e.g. Plint 1988, 1996). Using this scheme, the uppermost surface of the aggradational parasequence set that is coincident with the sharp basal surface of the first strongly progradational parasequence represents a sequence boundary. Following the original definition of a parasequence (e.g. Van Wagoner *et al.* 1990) the shoaling-up cycles above the basal RSME are clearly anomalous as their lower boundaries are not marine flooding surfaces. However, it is also clear that these cycles are 'in special positions within a sequence' whereby 'parasequences may be bounded either above or below by sequence boundaries' (Van Wagoner 1995*b*).

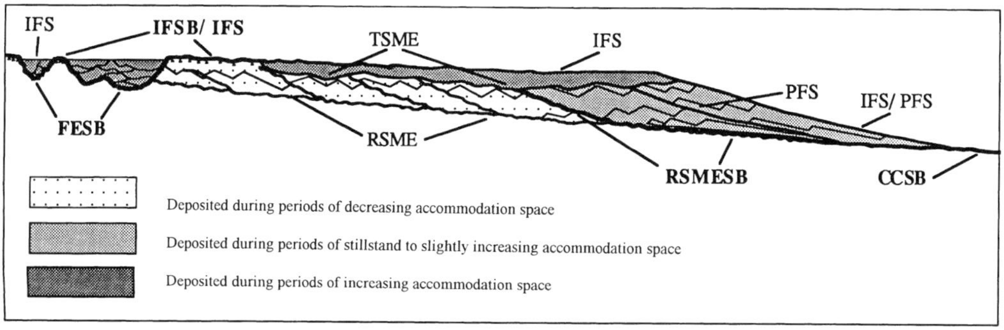

Fig. 13. Schematic model diagram illustrating the principal components of the 'lowstand' strandplains of the Mesaverde Group within the Bighorn Basin. Sediments deposited during periods of decreasing accommodation space, are interpreted as falling stage (or forced regressive) systems tract deposits (*sensu* Plint & Nummedal this volume). Those deposited under stillstand to slightly increasing accommodation are Lowstand Systems Tract. Those deposited during increasing accommodation, or as the result of transgressive reworking are assigned to the Transgressive Systems Tract. RSME, regressive surface of marine erosion; RSMESB, regressive surface of marine erosion sequence boundary; CCSB, correlative conformity sequence boundary; FESB, fluvial erosion sequence boundary; IFSB/IFS, interfluve sequence boundary/initial flooding surface; IFS, initial flooding surface; TSME, transgressive surface of marine erosion; PFS, parasequence flooding surface; IFS/PFS, initial/parasequence flooding surface.

Development of stratal surfaces

The principal features of the sharp-based shoreface systems, and equivalent non-marine strata of the Mesaverde Group in the Big Horn Basin are summarized in Fig. 13. Here the fluvial erosion/interfluve surface and the basal RSME respectively form the upper and lower bounding surfaces to the amalgamated massive shoreface deposits. When traced up-dip and down-dip, these two surfaces merge, forming an easily recognizable sequence boundary (RSMESB, Fig. 13). Internally, the sigmoidal-shaped regressive package of relatively massive amalgamated shoreface sandstones (Fig. 13, shown dotted, without tone), contains minimal evidence of parasequence preservation, as already discussed. Instead it is characterized by the development of internal hetrogeneities (erosion surfaces) that pass downward to the regional lower bounding RSME, and sandstones that tend to lack of fine-grained, fair-weather mudstones and siltstones facies. Together, these and the many other diagnostic features of the lower bounding surface indicate that the sharp-based sandstones and their lateral equivalents were deposited during periods of decreasing accommodation space, when relative sea-level was progressively falling.

It is only in the most distal portions of the tongues, basinward of the last internal erosion surface, that complete shallow marine shoaling-upward successions are preserved, as shown in Fig. 13 over and basinward of surface RSMESB. These parasequences are interpreted to have been formed when accommodation space was not being destroyed (fall/stillstand of relative sea-level). That more than one parasequence is commonly developed is taken to indicate that accommodation space was being created, indicating the onset of longer-term relative sea-level rise, and deposition of the lowstand wedge.

Placement of the sequence boundary

We interpret the different conditions of accommodation destruction and creation, under which the sharp-based amalgamated and succeeding expanded storm-dominated shoreface successions respectively formed (e.g. Fig. 13, positioned below and above surface labelled RSMESB), to be related to different parts of a relative sea-level cycle. However, the question then arises as to which of the two major surfaces (fluvial erosion/interfluve or basal RSME) is the sequence boundary, the placement of which will ultimately define to which systems tract (and sequence) the different parts of the tongue belong. We have found this to be a far from easy practical decision to make, reflecting differences apparent in the literature specific to forced regressions (e.g. Posamentier *et al.* 1992; Hunt & Tucker 1992 1995; Kolla *et al.* 1995; Plint & Nummedal this volume, Posamentier & Morris this volume).

RSME as the sequence boundary. In consideration of sharp-based shoreface profiles, many authors have favoured placing the sequence boundary along the RSME, beneath the shoreface (e.g. Plint 1988; Posamentier *et al.* 1992; Ainsworth & Pattison 1994; Fitzsimmons 1994*b*; Johnson & Fitzsimmons 1994). Indeed, when the distal deposits of the Mesaverde storm systems (Virgelle and Claggett Tongues) are studied in isolation, this is the only regionally correlatable surface across which a basinward shift in the depositional system can be observed.

Subaerial exposure/fluvial incision as the sequence boundary. However, in the proximal areas of the Mesaverde sandstone tongues, the basal RSME is itself truncated and overlain by the fluvial erosion/interfluve surface (e.g. the Gebo and Judith River Tongues; also see Fig. 12 CWC-GR sections). It is this surface which many authors would argue is the true high-order expression of the sequence boundary (e.g. Van Wagoner 1990; Hunt & Tucker 1992, 1995; Ainsworth 1994; Plint & Nummedal this volume). Within the study area, transgressive reworking of interfluve surfaces commonly removes the expression of this surface, therefore negating its use as an easily idenfiable, chronostratigraphic surface for correlation (see also argument of Posamentier & Morris this volume). Even if it were possible to trace these surfaces into the distal shallow marine successions, many authors would invoke the development of the sequence boundary as a cryptic correlative conformity, which would be difficult to recognize in the outcrop sections, rather than recognize the distal expression of the RSME (e.g. Mutti & Allen 1987; Mutti & Sgavetti 1987; Plint 1988, 1996; Ainsworth & Pattison 1994).

Surface development

It would seem that both interpretations, though valid, seem to be only partially correct and partly contradictory when applied in the successions studied here.

The RSME. As considered previously, the presence of the internal erosion surfaces within the amalgamated shoreface profiles, supports an interpretation of punctuated stepwise progradation of the shoreline. This interpretation is supported by the localized development of the *Glossifungites* type hiatal ichnofabrics, on both the basal RSME and internal erosion surfaces. Consequently, because the basal RSME is an amalgam of these events it is clearly a diachronus surface, that becomes progressively younger basinward. This diachroneity would seem to preclude interpretation of the regressive surface of marine erosion as a finite chronostratigraphic surface (sequence boundary), as for example argued by Posamentier *et al.* (1992) and Posamentier & Morris (this volume). However, when the processes which created the fluvial erosion/interfluve surface are examined, it can be argued that these surfaces are also diachronus.

Incised valleys. The bases of incised valleys are recognized to be composite surfaces created by fluvial erosion (e.g. Dalrymple *et al.* 1994). During a relative sea-level fall, the erosive locus of the primary channel is known to migrate within the valley meander belt. Abandoned fluvial systems are preserved as terrace deposits (Shanley & McCabe 1991, 1994), and are interpreted to record deposition during periods of decreasing accommodation space. As sea-level falls, the subaerially exposed interfluve surface becomes subject to pedogenesis. Clearly the areas exposed during the beginning of the fall will be exposed for a greater length of time than those subsequently exposed by continued forced regression. Interfluve maturity therefore would be expected to increase in a landward direction, a reflection of how this surface is also diachronus.

Surfaces and systems tracts

It is clear from the above discussion that within the non-marine and shorefaces settings considered here, arguements based on diachroneity of surfaces are of little practical use in chosing between which surface should represent the 'master' sequence boundary. A plethora of surfaces (erosive and/or maturing) are developed at different times in response to a single sea-level fall. At this resolution we are beyond the level at which a finite, basin-wide surface, marking an instant in time, can be identified. Furthermore, it also seems illogical that in the classic three system tract sequence stratigraphic models sea-level fall (and sequence boundary formation), is considered instantaneous whereas, and in stark contrast, the history of relative sea-level is contained within the whole of the transgressive system tract.

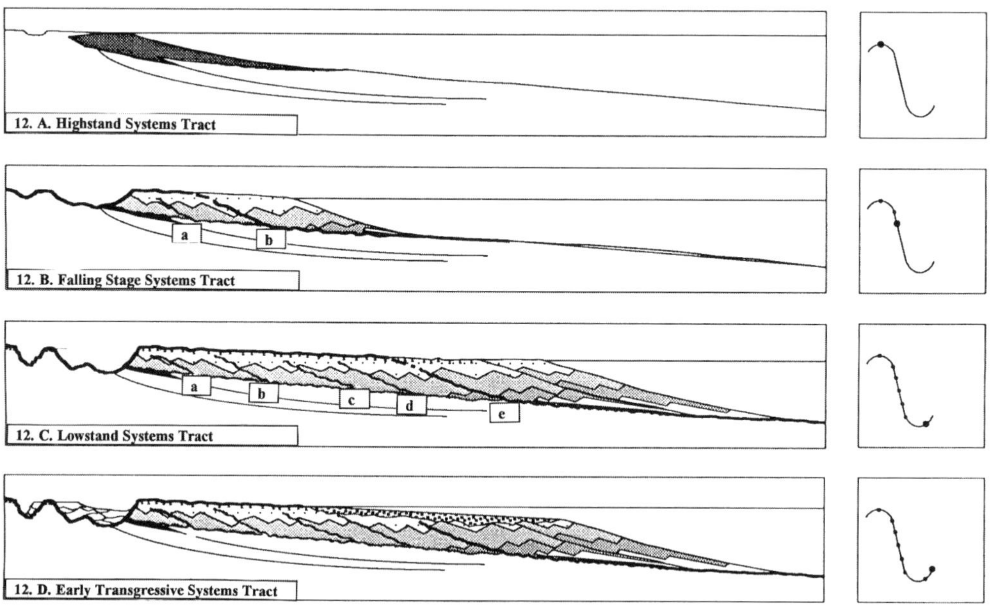

Fig. 14. Hypothetical development of falling stage (forced regressive), lowstand and transgressive systems tracts in a high order sequence of the Mesaverde Group of the Bighorn Basin (modified from Plint 1988, 1996). The highstand sand-dominated strata deposited in shoreface environments are depicted in dark tone (A).

Hunt & Tucker (1992, 1995) and also Plint & Nummedal (this volume) argued that a surface must exist, which would separate sediments deposited during times of decreasing accommodation space, from those deposited after sea-level had fell to its maximum extent. In the sediments of the Mesaverde Group, this surface must be the fluvial erosion/interfluve, which caps the amalgamated shoreface deposits. However, if this is the case, sediments of the same age in the incised valley (fluvial terrace deposits, pedogenic soils) to time equivalent shoreline deposits will be separated by a surface of supposed chronostratigraphic significance, the sequence boundary.

A dynamic model from the Mesaverde Group. Rather than considering the depositional system to be governed by abrupt, chronostratigraphically significant events, its evolution must be considered as a continum through processes of erosion, transport and deposition. Fig. 14 illustrates how the systems of the Mesaverde Group are thought to evolve, based on observations from the study area.

Figure 14a shows the position of the progradational highstand shoreline, whereby regression is interpreted to result from a sediment supply that outpaces relative sea-level rise, so filling the available accommodation space. In this situation sufficient space would be available for a complete shoaling-upward storm-dominated shoreface succession to develop (e.g. Fig. 6a). In this shoreface profile autocyclic processes may compress fair-weather wavebase (e.g. Van Wagoner 1995*b*), resulting in the localized development of sharp-based storm successions.

Figure 14b shows the onset of relative sea-level fall, the result of compression of the storm wavebase across the whole shoreface profile. The latter results in the formation of a regional sharp-based shoreface profile that is erosionally-based, the RSME truncating the pre-existing highstand shoreface deposits (e.g. Fig. 14b). In time-equivalent up-dip reaches of the profile, the fluvial system incises to adjust to the new shoreline position, and the interfluves become areas of sediment bypass. Consequently, the rate of sediment supply to the shoreline is increased, resulting in rapid progradation of the strandplain across the RSME. It is thought that a complete shoaling-upward gradational transition of depositional facies (offshore to onshore) would exist at any point during this fall. However, as long as the fall proceeds, these deposits would be successively cannibalized, and reincorporated into the massive sandy strandplain.

Internally, seaward dipping internal erosion surfaces would be developed (e.g. points a and b Fig. 14b). These relatively steep surfaces (e.g. Fig. 11) may record subtle realignments of the distributary mouth during hiatuses in the fall and/or oscillations in the sea-level fall. In updip areas, older parts of the upper strandplain become coincident with the subaerially exposed interfluve.

Figure 14c shows sea-level to have fallen through a series of steps (a, b, c, d) to its lowest position (point e). A regionally correlatable surface, marking the maximum down shift is at this time developed, and is marked by a bold line in Fig. 14c. Its expression changes down-dip from (i) a fluvial erosion surface, passing as (ii) an interfluve *above* the sharp-based shorefaces deposited during sea-level fall to (iii) the most basinward and final regressive surface of marine erosion (e.g. the bold RSMESB surface in Fig. 13, surface in Fig. 14c) before becoming (iv) a correlative conformity. Overlying this surface, the shoreline depositional systems prograde into the available accommodation space, downlapping onto the distal expression of the RSME and its correlative conformity in the distal areas (e.g. Figs 13, 14c). During this period, smaller scale, high-order flooding events create accommodation space enabling the preservation of complete parasequences.

During the ensuing transgression (Fig. 14d), the strandplain is reworked by ravinement at the base of migrating tidal inlets and in shoreface environments forming the transgressive surface of marine erosion (TSME, Fig. 13). As an important consequence the interfluve/RSME transition is destroyed. Finally, in response to the increase in base-level, the incised valley is filled so modifying its basal surface, and the development of the transgressive system tract is initiated.

Implications for existing stratigraphic models for sediments deposted during forced regression

From our observations in the Mesaverde Group, as summarized in the above depositional model discussed above (Figs 13, 14), it is apparent that two surfaces related to the same sea-level fall are contemporaneously developed within the clastic tongues; the RSME and the fluvial erosion/interfluve surface. Within a single clastic tongue three broad zones can be recognized. In up-dip areas, the fluvial erosion surface clearly separates the deposits of each sequence of fall and rise. Down-dip, it is the RSME which does this. However, between these areas, a zone exists where both surfaces are expressed (RSME and interfluve), separated by massive amalgamated shoreface deposits. It was partly to this end that Hunt & Tucker (1992, 1995) invoked the presence of a fourth forced regressive which is specific to periods of sea-level fall and decreasing accommodation space (equivalent to the falling stage systems tract of Plint & Nummedal this volume). This systems tract could be distinguished from the lowstand systems tract, typified by stillstand to slightly rising sea-level (where accommodation space is in equilibrium to slightly increasing).

Plint (1996) eloquently reiterated this argument, recognising that Posamentier *et al.* (1992) 'do not follow their observations to a logical conclusion', a theme furthered in this volume (Plint & Nummedal). For although Posamentier *et al.* (1992) recognized the importance of deposition during sea-level fall, they did not distinguish a systems tract with discrete evolutionary origins, which would discriminate it from the rest of the lowstand deposits. Nummedal & Cole (1993), Nummedal *et al.* (1995), and Plint & Nummedal (this volume) also followed a four systems tract approach, interpreting the first phase as the falling stage systems tract, specific to times of sea-level fall, followed by the lowstand systems tract, marking the onset of relative sea-level rise. This terminology enables the processes active during the fall (forced regression) to be removed from the systems tract nomenclature, as we contend that this process only describes part of the developing depositional system.

Surface continuity and signiificance. Unfortunately no single 'master' surface can been traced throughout a single tongue within the Mesaverde Group (e.g. Figs 13, 14). As noted this may be a reflection of the specific depositional system studied, with a significant component of transgressive reworking, and the cryptic expression of the last seaward dipping surface which will separate similar facies across a sand on sand erosive contact. However, by examining the sediments between the candidate surfaces, and through the recognition of the effects of different available accommodation space on lithofacies development, systems tracts can be assigned.

Within the Mesaverde Group, the massive shoreface components of each individual clastic tongue are interpreted to be the expression of the falling stage (or forced regressive) systems tract formed during relative sea-level fall (cf. Hunt & Tucker 1992, 1995; Plint & Nummedal this volume). Where complete progradational upward-shoaling parasequences are preserved, these units are interpreted to have been deposited during relative sea-level still stand and the onset of rise; they are placed with a lowstand systems tract that is considered to develop during times of lowstillstand and relative base-level rise.

Rather than focusing on the depositional response to an evolving system, recent debate has centered on the development and significance of a 'master chronostratigraphic surface' (e.g. Posamentier *et al.* 1992; Hunt & Tucker 1992, 1995; Kolla *et al.* 1995; Plint & Nummedal this volume; Posamentier & Morris this volume). However, sediments deposited at the same time (fluvial terrace sandstones, pedogenic soil horizons and parts of the massive shoreface) can be separated by a such a surface, contradicting the widely held view that such surfaces do in all places have significance as chronostratigraphic surfaces separating older strata below from younger strata above. Equally, it is clear that fluvial erosion, interfluve development and compression of storm wavebase all occur at the same time, so it is important to ask why should one expression assume more significance than another?

This study has shown that through tracing the expression of the sharp-based storm systems, into distal environments, the regional nature of the basal RSME can have as much chronostratigraphic significance as an incised valley/interfluve surface. It is also noted that in distal environments it becomes the principal expression of a surface related to base-level fall. On the whole, a chronostratigraphic sequence boundary does form, separating the majority of sediments deposited after sea-level fall from those developed before. By recognising the processes (changes in available accommodation space) which occur during the fall, a fourth falling stage (or forced regressive) system tract can be identified, constrained by two surfaces, both of which are formed in response to a single (gradual or stepwise) fall in base-level.

The sequence boundary is expressed as both a fluvial erosion/interfluve surface and a RSME. However, where both surfaces are developed in a single section, the uppermost fluvial erosion/interfluve must be treated as the ultimate expression of the depositional systems response to sea-level fall (e.g. Figs 13, 14; Hunt & Tucker 1992, 1995; Plint & Nummedal this volume). Here the RSME records the initial expression of the fall, the shoreface records deposition during that fall, and the fluvial erosion/interfluve as the final basinwide expression of that fall. However, shorelines do not disappear during falls, they only relocate, and basinward of this point, it is in these areas that the RSME becomes the ultimate expression of the fall, and must be interpreted as the sequence boundary (Figs 13, 14).

Conclusion

The regressive surface of marine erosion can be traced with confidence into distal shallow marine environments. It forms the basal bounding surface to the falling stage (forced regressive) systems tract, which is bounded above by the fluvial erosion/interfluve surface interpreted as the sequence boundary. Up-dip the RSME is truncated by the sequence boundary, whereas down-dip it merges with the sequence boundary before passing into a correlative conformity when traced basinward. The finite trace of the sequence boundaries transition from subaerial to submarine is cryptic, for not only is it a subtle sand on sand contact, but it is also prone to removal by transgressive ravinement. Thus identification of the RSME enables the expression of sea-level fall to be recognized when no sequence boundary is apparent.

Even though the RSME and sequence boundary are diachronus, by recognising the expression of sediments deposited under different conditions of accommodation space, the falling stage and lowstand systems tracts can be distinguished. It could be argued that the diachroneity involved in the creation of strandplains of the falling stage system tract will be of no consequence when these sand bodies are correlated. However, if these successions are to be successively exploited as potential reservoirs, the temporal evolution of the depositional system must be understood. The anatomy of smaller scale erosion surfaces, which dissect the forced regressive systems tract, may ultimately have a critical impact on creating successful reservoir models.

Initial field work was carried out while both authors were in receipt of a NERC Ph.D. studentships at the University of Liverpool. Additional funding was supplied by Amoco, BP and Mobil. The authors thank Dave Hunt and Lee Krystinik for their constructive reviews. Additional discussions with our colleagues within the Strat Group have benefited this paper, in particular J. F. Aitken and J. A. Howell. Dave Hunt is additionally thanked for his extreme patience during the writing of this manuscript.

References

AINSWORTH, R. B. 1994. Marginal marine sedimentology and high resolution sequence analysis; Bearpaw-Horseshoe Canyon transition, Drumheller, Alberta. *Bulletin of Canadian Petroleum Geology*, **42**, 26–54.

—— & PATTISON, S. A. J. 1994. Where have all the lowstands gone? Evidence for the attached lowstand systems tracts in the Western Interior of North America. *Geology*, **22**, 415–418.

——, BOSSCHER, H. & NEWALL, M. J. 2000. Forward modelling of forced regressions: evidence for the genesis of attached and detached lowstand systems. *This volume*.

AITKEN, J. F. & FLINT, S. S. 1994. High-frequency sequences and the nature of incised-valley fills in fluvial systems of the Breathitt Group (Pennsylvanian), Appalachian Foreland Basin, eastern Kentucky. *In*: DALRYMPLE, R. W., BOYD, R. & ZAITLIN, B. A. (eds) *Incised-Valley Systems: Origin and Sedimentary Sequences*. Society of Economic Paleontologists and Mineralogists, Special Publications, **51**, 353–368.

—— & —— 1995. The application of high resolution sequence stratigraphy to fluvial systems: a case study from the Upper Carboniferous Breathitt Group, eastern Kentucky, U.S.A. *Sedimentology*, **42**, 3–30.

ASQUITH, D. O. 1970. Depositional topography and major marine environments, Late Cretaceous Wyoming. *American Association of Petroleum Geologist Bulletin*, **54**, 1184–1224.

—— 1974. Sedimentary models, cycles and deltas, Upper Cretaceous, Wyoming. *American Association of Petroleum Geologist Bulletin*, **58**, 2274–2283.

BALSLEY, J. K. 1980. *Cretaceous wave dominated delta systems, Book Cliffs, east-central Utah*. American Association of Petroleum Geologists Continuing Education Course Field Guides.

BOSELLINI, A., MUTTI, E. & RICCI-LUCCHI, F. 1989. *Rocce e successioni sedimentarie*. Utet, Torino.

BRENCHLEY, P. J., FLINT, S. S. & STROMBERG, S. G. 1993. Quantitative facies discrimination and the application of sequence stratigraphy to bed length modeling of shallow marine hetrolithic facies. *In*: ESCHARD, R. & DOLIGEZ, B. (eds) *Subsurface reservoir characterization from outcrop observations*. Editions Technip, Paris, 1–8.

CANT, D. J. 1984. Development of shoreline-shelf sand bodies in a Cretaceous epiric sea deposit. *Journal of Sedimentary Petrology*, **54**, 541–556.

CURRAY, J. R. 1964. Transgressions and regressions. *In*: MILLER, R. L. (ed.) *Papers in Marine Geology*. MacMillan, New York, 175–203.

DALRYMPLE, R. W., BOYD, R. & ZAITLIN, B. A. (eds) 1994. *Incised-Valley Systems: Origin and Sedimentary Sequences*. Society of Economic Paleontologists and Mineralogists, Special Publications, **51**.

DAVIS, H. R. & BYERS. C. W. 1989. Shelf sandstones in the Mowry Shale: evidence for deposition during Cretaceous sea level falls. *Journal of Sedimentary Petrology*, **59**, 548–560.

DEVINE, P. E. 1991. Transgressive origin of channelised estuarine deposits in the Point Lookout Sandstone, Northwestern New Mexico: a model for the U.S. Western Interior. *American Association of Petroleum Geologists Bulletin*, **75**, 1039–1063.

ELLIOTT, T. 1986. Clastic Shorelines. *In*: READING, H. G. (ed.) *Sedimentary Environments and Facies*. Blackwell Scientific Publications, Oxford, 143–177.

EYLES, N. & CLARK, B. M. 1986. Significance of hummocky and swaley cross stratification in late Pleistocene lacustrine sediments of the Ontario basin, Canada. *Geology*, **14**, 679–682.

FITZSIMMONS, R. J. 1994*a*. *Field-based testing of high resolution sequence stratigraphic concepts; the Campanian Eagle Formation, Mesaverde Group, Bighorn Basin, Northwestern Wyoming, U.S.A.* PhD thesis, University of Liverpool.

—— 1994*b*. Identification of high order sequence boundaries and land attached forced regressions. *In*: JOHNSON, S. D. (ed.) *High resolution sequence stratigraphy: Innovations and applications Abstract volume*. Department of Earth Sciences, University of Liverpool, 332–333.

GALLOWAY, W. E. 1989*a*. Genetic stratigraphic sequences in basin analysis I: architecture and genesis of flooding surface bounded depositional units. *American Association of Petroleum Geologists Bulletin*, **73**, 125–142.

—— 1989*b*. Genetic stratigraphic sequences in basin analysis II: application to Northwest Gulf of Mexico Cenozoic basins. *American Association of Petroleum Geologists Bulletin*, **73**, 143–154.

GILL, J. R. & COBBAN, W. A. 1966*a*. *The Red Bird section of the Upper Cretaceous Pierre Shale in Wyoming*. United States Geological Survey Professional Papers, **393-A**.

—— 1966*b*. *Regional Unconformity in Late Cretaceous, Wyoming*. United States Geological Survey Professional Papers, **550-B**.

—— 1973. *Stratigraphy and geological history of the Montana Group and equivalent rocks, Montana, Wyoming and North Dakota*. United States Geological Survey Professional Papers, **776**.

GREENSMITH, J. T., RAWSON, P. F. & SHALABY, S. E. 1979. An association of minor fining-upwards cycles and aligned gutter marks in the middle Lias (Lower Jurassic) of the Yorkshire coast. *Proceedings of the Yorkshire Geological Society*, **42**, 525–538.

HADLEY, D. F. & ELLIOTT, T. 1993. The sequence stratigraphic significance of erosive based shoreface sequences in the Cretaceous Mesaverde Group of northwest Colorado. *In*: POSAMENTIER, H. W., SUMMERHAYES, C. P., HAQ, B. U. & ALLEN, G. P. (eds) *Sequence Stratigraphy and Facies Associations*. International Association of Sedimentologists, Special Publications, **18**, 521–535.

HAMBERG, L. & NIELSEN, L. H. 2000. Shingled, sharp-based sandstones and stepwise forced regression during sea-level fall in a shallow basin, Upper Triassic Gassum Formation, Denmark. *This volume*.

HICKS, J. F. 1993. *Chronostratigraphic analysis of foreland basin sediments of the latest Cretaceous, Wyoming, U.S.A.* PhD thesis, Yale University.

HUNT, D. & TUCKER, M. E. 1992. Stranded parasequences and the forced regressive wedge systems tract: deposition during base-level fall. *Sedimentary Geology*, **81**, 1–9.

—— & —— 1992. Stranded parasequences and the forced regressive wedge systems tract: deposition during base-level fall – reply. *Sedimentary Geology*, **95**, 147–160.

JOHNSON, S. D. 1995. *High resolution sequence stratigraphy of Upper Cretaceous strata, Bighorn Basin, Wyoming, U.S.A.* PhD thesis, University of Liverpool.

—— & FITZSIMMONS, R. J. 1994. High resolution sequence stratigraphy: The characterisation and genesis of key sequence stratigraphic surfaces in fluvial and shallow marine successions: Examples from the Mesaverde Group of the Bighorn Basin, Wyoming, U.S.A. *In*: JOHNSON, S. D. (ed.) *High resolution sequence stratigraphy: Innovations and applications (Abstract volume)*. Department of Earth Sciences, University of Liverpool, 86–92.

KOLLA, V., POSAMENTIER, H. W. & EICHENSEER 1995. Stranded parasequences and the forced regressive system tract: deposition during base level fall – discussion. *Sedimentary Geology*, **95**, 139–145.

MACEACHERN, J. A., RAYCHAUDHURI, I. & PEMBERTON, S. G. (1992) Stratigraphic applications of the Glossifungites ichnofacies: delineating discontinuities in the rock record. *In*: PEMBERTON, S. G. (ed.) *Applications of Ichnology to Petroleum Exploration*. Society of Economic Paleontologists and Mineralogists, Core Workshop, 17, Calgary, June 21, 1992, 169–198.

MCCRORY, V. L. C. & WALKER, R. G. 1986. A storm and tidally induced prograding shoreline – Upper Cretaceous Milk River Formation of southern Alberta, Canada. *Sedimentology*, **33**, 47–60.

MCGOOKEY, D. P., HAUN, J. D., HALE, L. A., GOODELL, H. A., MCCUBBIN, D. G., WEIMER, R. J. & WULF, G. R. 1972. Cretaceous System. *In*: *Geologic Atlas of the Rocky Mountain Region, United States of America*. Rocky Mountain Association of Geologists, Denver, Colorado, 190–228.

MELLERE, D. & STEEL, R. 2000. Style contrast between forced regressive and lowstand/transgressive wedges in the Campanian of south-central Wyoming. *This volume*.

MUTTI, E. & ALLEN, G. 1987. Storm dominated sandstone deposits: an indicator of relative sea level variations (abstract). *In*: *Society of Economic Paleontologists and Mineralogists Mid Year Meeting*, 59.

—— & SGAVETTI, M. 1987. Sequence Stratigraphy of the Upper Cretaceous Aren strata in the Orcau-Aren region, south-central Pyrenees, Spain: Distinction between eustatically and tectonically controlled depositional sequences. *Annali delliUniversita di Ferrara: Science della terra*, 1, 1–21.

NUMMEDAL, D. & COLE, R. D. 1993. Sequence stratigraphy of the Castlegate and Desert Sandstones, Utah: an alternate view (abstract). *In*: *American Association of Petroleum Geologists Annual Convention Abstracts Volume*, New Orleans.

——, GUPTA, S., PLINT., A. G. & COLE, R. D. 1995. The falling stage systems tract: definition, character and expression in several examples from the Cretaceous of the U.S. Western Interior(abstract). *In*: *Sedimentary response to forced regression: recognition, interpretation and reservoir potential*. Conference Abstracts, Burlington House, London, 7–8 September 1995.

OBRADOVITCH, J. D. 1993. A Cretaceous time scale. *In*: CALDWELL, W. G. E. & KAUFFMANN, E. G. (eds) *Evolution of the Western Interior Basin*. Geological Association of Canada, Special Papers, **39**, 379–396.

PLINT, A. G. 1988. Sharp-based shoreface sequences and ìoffshore barsî in the Cardium Formation of Alberta: their relation to relative changes in sea level. *In*: WILGUS, C. K., HASTINGS, B. S., KENDALL, C. G. ST. C., POSAMENTIER, H. W., ROSS, C. A. & VAN WAGONER, J. C. (eds) *Sea Level Changes: An Integrated Approach*. Society of Economic Paleontologists and Mineralogists Special Publications, **42**, 357–370.

—— 1996. Marine and non-marine systems tracts in fourth-order sequences in the Early-Middle Cenomanian, Dunvegan Alloformation, northeastern British Columbia, Canada. *In*: HOWELL, J. A. & AITKEN, J. F. (eds) *High Resolution Sequence Stratigraphy: Innovations and Applications*. Geological Society Special Publications, **104**, 159–191.

—— & NUMMEDAL, D. 2000. The falling stage systems tract: recognition and importance in sequence stratigraphic analysis. *This volume*.

POSAMENTIER, H. W. & MORRIS, W. S. 2000. aspects of the stratal architecture of forced regressive deposits. *This volume*.

—— & VAIL, P. R. 1988. Eustatic controls on clastic deposition II – sequence and systems tract models. *In*: WILGUS, C. K., HASTINGS, B. S., KENDALL, C. G. ST. C., POSAMENTIER, H. W., ROSS, C. A. & VAN WAGONER, J. C. (eds) *Sea Level Changes: An Integrated Approach*. Society of Economic Paleontologists and Mineralogists Special Publications, **42**, 125–154.

——, ALLEN, G. P., JAMES, D. J. & TEESON, M. 1992. Forced regressions in a sequence stratigraphic framework; concepts, examples and exploration significance. *American Association of Petroleum Geologists Bulletin*, **76**, 1687–1709.

——, JERVEY, M. T. & VAIL, P. R. 1988. Eustatic controls on clastic deposition I – conceptual framework. *In*: WILGUS, C. K., HASTINGS, B. S., KENDALL, C. G. ST. C., POSAMENTIER, H. W., ROSS, C. A. & VAN WAGONER, J. C. (eds) *Sea Level Changes: An Integrated Approach*. Society of Economic Paleontologists and Mineralogists Special Publications, **42**, 109–124.

ROEP, T. H. B., BEETS, D. J., DRONKERT, H. & PAGNIER, H. 1979. A prograding coastal sequence of wave built structures of Messinian age, Sorbas, Almeria, Spain. *Sedimentary Geology*, **22**, 135–163.

ROSENTHAL, L. R. P. & WALKER, R. G. 1987. Lateral and vertical facies sequences in the Upper Cretaceous Chungo Member, Wapiaba Formation, Southern Alberta. *Canadian Journal of Earth Sciences*, **24**, 771–783.

SEVERN, W. P. 1961. *General Stratigraphy of the Mesaverde Group, Bighorn Basin, Wyoming*. Wyoming Geological Association Guidebook, 16th Annual Field Conference.

SHANLEY, K. W. & MCCABE, P. J. 1991. Alluvial architecture in a sequence stratigraphic framework – detailed correlations between marine and non-marine strata. *In*: *1991 NUNA conference on high-resolution sequence stratigraphy. Abstracts*, 50–51.

—— & —— 1994. Alluvial Architecture in a sequence stratigraphic framework – a case study from the upper Cretaceous of southern Utah, U.S.A. *In*: FLINT, S. S. & BRYANT, I. D. (eds) *The Geological Modeling of Hydrocarbon Reservoirs*. International Association of Sedimentologists, Special Publications, **15**, 21–56.

—— & —— 1994. Perspectives on the sequence stratigraphy of continental strata. *American Association of Petroleum Geologists Bulletin*, **78**, 544–568.

VAN WAGONER, J. C. 1995*a*. Overview of sequence stratigraphy of foreland basin deposits: terminology, summary of papers and glossary of sequence stratigraphy. *In*: VAN WAGONER, J. C. & BERTRAM, G. T. (eds) *Sequence stratigraphy of foreland basin deposits: outcrop and subsurface examples from the Cretaceous of North America*. American Association of Petroleum Geologists Memoirs, **64**, ix-xxi.

—— 1995*b*. Sequence stratigraphy and marine to non-marine facies architecture of foreland basin strata, Book Cliffs, Utah, U.S.A. *In*: VAN WAGONER, J. C. & BERTRAM, G. T. (eds) *Sequence stratigraphy of foreland basin deposits: outcrop and subsurface examples from the Cretaceous of North America*. American Association of Petroleum Geologists Memoirs, **64**, 137–224.

——, CAMPION, K. M. & RAHMANIAN, V. D. 1990. *Siliciclastic sequence stratigraphy in well logs, cores and outcrops: concepts for high-resolution correlation of time and facies*. American Association of Petroleum Geologists; Methods in Exploration Series, **7**.

——, POSAMENTIER, H. W., MITCHUM, R. M., VAIL, P. R., SARG, J. F., LOUTIT, T. S. & HARDENBOL, J. 1988. An overview of the fundamentals of sequence stratigraphy and key definitions. *In*: WILGUS, C. K., HASTINGS, B. S., KENDALL, C. G. ST. C., POSAMENTIER, H. W., ROSS, C. A. & VAN WAGONER, J. C. (eds) *Sea Level Changes: An Integrated Approach*. Society of Economic Paleontologists and Mineralogists Special Publications, **42**, 39–45.

WALKER, R. G. & PLINT, A. G. 1992. Wave and storm-dominated shallow marine systems. *In*: WALKER, R. G. & JAMES, N. P. (eds) *Facies Models; response to sea level change*. Geological Association of Canada, 219–238.

WHITTAKER, J. H. MCD. 1973. Gutter Casts a new name for scour-and-fill structures: with examples from the Llandoverian of Ringerike and Malmoya, Southern Norway. *Norsk Geologisk Tidsskr*, **53**, 403–417.

WIEMER, R. J. 1984. Relations of unconformities, tectonics and sea level changes, Cretaceous of the Western Interior, U.S.A. *In*: SCHLEE, J. S. (ed.) *Interregional unconformities and hydrocarbon exploration*. American Association of Petroleum Geologists Memoirs, **36**, 7–35.

WILLIAMS, H. & UGUETO, G. 1994. High resolution tectono-sedimentary modeling of Niger delta reservoirs. *In*: *American Association of Petroleum Geologists Annual Convention Abstracts* Volume, Denver, 284.

Style contrast between forced regressive and lowstand/transgressive wedges in the Campanian of south-central Wyoming (Hatfield Member of the Haystack Mountains Formation)

DONATELLA MELLERE[1] & RONALD STEEL[2]

[1]*Statoil, Petroleum Technology, 4035-Stavanger, Norway (e-mail: Dome@Statoil.Com)*

[2]*Department of Geology and Geophysics, University of Wyoming, Laramie, Wyoming, USA*

Address for correspondence: Donatella Mellere, Statoil, Petek-Stra, 4035-Stavanger, Norway

Abstract: The Campanian Hatfield Member of the Haystack Mountains Formation is composed of two well-exposed marine sandstone tongues that extend up to 35 km basinward from their earliest shoreline position into the Western Interior Seaway. Each tongue (H1 and H2) is comprised of two parts that have characteristic architecture, external geometry and facies assemblages. Together, the tongues form a stratigraphic sequence that is formed of four systems tracts and bounded by erosional unconformities. The sequence is interpreted to have been generated over an interval of less than 1 Ma during a fall-to-rise cycle of relative sea level. The earliest and latest systems tracts of the sequence, interpreted as lowstand prograding deltaic wedge and forced regressive shoreface respectively, are distinguished on the basis of their position with respect to the sequence-bounding unconformities, reconstructed shoreline trajectories, and by their component facies that indicate the dominant depositional regime.

The mapped basinward shift of the Hatfield 1 lowstand prograding wedge from the previous shoreline deposits and erosional relief on the sequence boundary, indicates a relative sea-level fall prior to its deposition. The lowstand prograding wedge consists of parasequences that are dominated by tidally influenced cross-stratified sandstones and step for more than 30 km basinward, and are readily distinguished from the underlying highstand shoreface facies. Distal aggradational stacking of the lowstand produced a slightly rising shoreline trajectory that in combination with proximal onlap against the underlying erosional unconformity indicates accumulation under conditions of rising relative sea-level with abundant sediment supply. The domination of tidally influenced facies and an estimated relief of at least 20 m in proximal reaches of the underlying sequence boundary suggests that the lowstand wedge was a tidally dominated deltaic system localized and fed through an incised valley. This systems tract resembles other cross-stratified Mancos-type sandstone bodies of the Western Interior Seaway which have been under debate. However, unlike most of these, the Hatfield 1 has great outcrop extent and the updip relationship of the lowstand wedge with the older shoreline deposits can be traced. The overlying retrogradational Hatfield 1 transgressive systems tract has comparable facies to the lowstand wedge and also shows proximal onlap of the sequence boundary, suggesting that it developed within a tidally influenced estuary. As such, the lowstand and transgressive systems tracts form a distinctive cross-bedded tidally influenced lithosome that is readily distinguished from the wave-dominated lithosomes of the preceding Hatfield 1 highstand systems tract and the overlying Hatfield 2 highstand and forced regressive systems tracts.

The Hatfield 2 forced regressive systems tract is a wave-dominated shoreface that like the preceding Hatfield 2 highstand shoreface is strongly progradational. However, in contrast to the highstand shoreface from which it builds, the forced regressive shoreface is relatively thin, lacks shaley offshore transitional facies at its base, and displays a downstepping trajectory relative to the underlying MFS. The basal surface of the forced regressive shoreline also has an enrichment of coarse glauconitic grains derived from erosion of the underlying condensed section whereas the upper bounding surface of the systems tract is an erosional unconformity, documenting the maximum fall in relative sea level. There is a clear sedimentological distinction of the lowstand and forced regressive systems tracts because whereas the former has a tidally influenced facies association, forced regressive facies tend to be wave-dominated. Such facies partitioning and style contrast are thought to reflect the less-confined nature of the highstand and forced regressive shorelines in comparison to the incised or embayed nature of the lowstand and transgressive shorelines.

From: HUNT, D. & GAWTHORPE, R. L. (eds) *Sedimentary Responses to Forced Regressions*. Geological Society, London, Special Publications, **172**, 141–162. 1-86239-063-0/00/$15.00

The Campanian Haystack Mountains Formation of the Mesaverde Group crops out in the Rawlins Uplift and the western portion of the Hanna Basin, south-central Wyoming (Fig. 1). It consists of a series of sandstone tongues that exhibit an overall basinward progradation. As can be seen in Fig. 2, many of the tongues are comprised of two distinctive lithosomes with (i) a lower coarse-grained and cross-stratified regressive to transgressive sandbody and (ii) an overlying and downlapping shoreface sandbody.

In the Haystack Mountains Formation, the cross-bedded component of the sandstone tongues has previously been interpreted as plume-derived, shelf-ridge deposits (e.g. Tillman & Martinsen 1985; McClurg 1990; Davies 1990). Such an interpretation is suggested by their offshore position with respect to the inferred contemporaneous shoreline (Gill *et al.* 1970). More recently, some of the units within the Mesaverde Group previously interpreted as offshore ridges (e.g. Tillman & Martinsen 1984; Gaynor & Swift 1988) have been reinterpreted as forced regressive shorelines. As such they are thought to have been deposited during or following relative sea-level fall that induced a seaward shift of the depocentre (e.g. Davis & Byres 1989; Posamentier *et al.* 1992; Walker & Bergman 1993).

This study is based on the stratigraphic framework established by Mellere & Steel (1995*a*) and Mellere (1996) in the southern outcrop area of the Haystack Mountains Formation between the towns of Rawlins and Saratoga, Wyoming (Fig. 2). Here, we describe and examine the facies successions, geometry and the sequence stratigraphy of the Hatfield 1 and 2 sandstone tongues where both cross-bedded and shoreface lithosomes are present (Fig. 2). Our sequence stratigraphic and facies analysis of the cross-bedded lithosome gives no support to recent suggestions (e.g. Walker & Bergman 1993) that such sandbodies may represent forced regressive shorelines.

Shoreface lithosomes within the Hatfield tongues are characterized by hummocky cross-stratification and thick packages of swaley cross-bedded strata. As such they are quite easily distinguished from the cross-bedded lithosomes that show evidence of tidal influence and reworking (Fig. 2). In the model presented here, the distinctive cross-bedded lithosome of

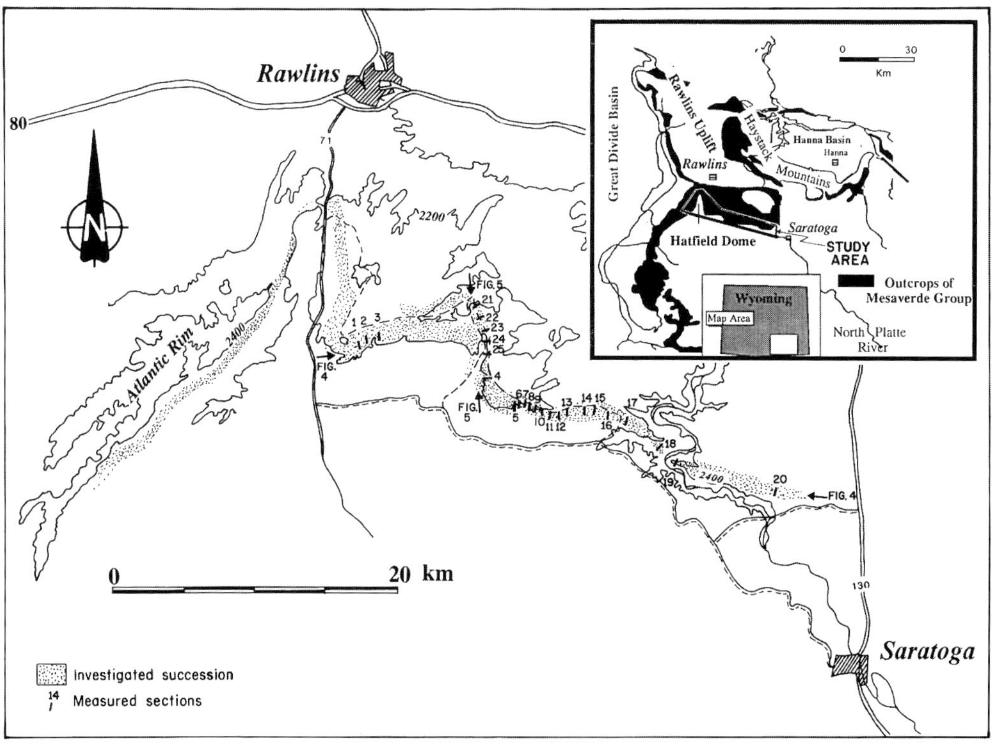

Fig. 1. General location of the study area with the outcrop of Mesaverde Group and trace of the measured sections and cross-sections used in this study. Contour interval in feet above sea-level.

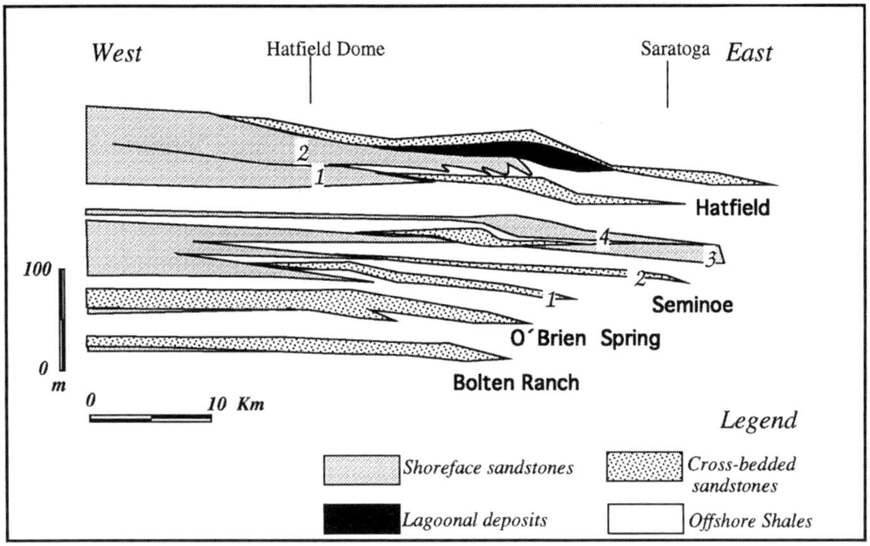

Fig. 2. Stratigraphy of the Haystack Mountains Formation between the Hatfield Dome and Saratoga (location in Fig. 1) as described by Mellere & Steel (1995*a*). Note the composite architecture of the uppermost sandstone tongues, consisting of combined shoreface and cross-bedded lithosomes. The tongues display an overall regressive stacking pattern.

Hatfield 1 is interpreted to represent lowstand and transgressive systems tracts, deposited in a tide-dominated delta (progradational parasequences) and under estuary-mouth conditions (backstepping parasequences), respectively. These systems tracts developed partially basinward and over the preceding highstand Hatfield 1 shoreface sandstone lithosome from which they are separated by a sequence boundary. The lowstand and transgressive systems tracts were partly confined within, and also developed partially beyond the mouth of an incised valley system. The succeeding highstand and forced-regressive units and clinoforms of the Hatfield 2 shoreface lithosome downlap onto the underlying maximum flooding or estuarine deposits. The highstand clinoforms evolve seaward into offlapping forced regressive shoreface clinoforms, the topsets of which drop basinward to topographically lower levels, juxtaposing shallower shoreface facies (seaward) against deeper shoreface facies (landward).

Stratigraphy and palaeogeography

The Campanian Haystack Mountains Formation, in south central Wyoming is part of the lower portion of the Mesaverde Group (Weimer 1960, 1966; Gill *et al.* 1970, Van Horn 1979; Roehler 1989). It consists of a progradational stack of siliciclastic shallow-marine sandstone tongues partially enveloped by offshore shales (Fig. 2). In the southern outcrop area, the Haystack Mountains formation has been informally divided into eight members (Mellere & Steel 1995*a*) that in order of decreasing age are: (1) Bolten Ranch Sandstone member, (2) O'Brien Spring Member, (3–7) Seminoe 1, 2, 3 and 4 members (the middle unnamed member of Gill *et al.* 1970 and Gill & Cobban 1973) and (8) Hatfield Member (Fig. 2).

The Hatfield Member is the focus of this study, and consists of two major sandstone tongues, Hatfield 1 and 2 (Fig. 2). These tongues are separated from the underlying Seminoe members by a 20 m thick marker interval of black shales. Both Hatfield tongues are widespread, laterally continuous and easily recognizable by means of their distinctive topographic expression in the region between the Hatfield Dome, south of Rawlins, and Saratoga (Figs 1 and 2).

Hatfield 1 is a wedge-shaped sandstone body up to 30 m thick that pinches out toward the east into equivalent basinal deposits. The bulk of the basinward part of the wedge is formed by a tabular cross-bedded sandstone lithosome that truncates the underlying shoreface sediments (Fig. 2). Hatfield 2 as recognized by Mellere & Steel (1995) represents the Hatfield Member as originally defined and measured by Smith *et al.* (1965) and Gill *et al.* (1970) in the Hatfield Dome. Hatfield 2 also consists of two distinct lithosomes, separated by a major surface of erosion identified as a sequence boundary (Mellere & Steel 1995*a*, *b*). The lower lithosome of Hatfield 2 is dominated by hummocky followed by swaley cross-strata. The upper lithosome represents a basinward facies shift of several kilometres and consists of a tidally influenced trough cross-bedded sandstone wedge that passes landward into lagoonal and coastal plain deposits in the southeastward part of the basin (Fig. 2). We focus attention on the facies and architectural details of the upper cross-bedded Hatfield 1 sandbody, its bounding surfaces and the downlap of the shoreface sandstones of Hatfield 2 onto it.

Palaeogeographic reconstruction for the time interval of the Hatfield Member (Gill & Cobban 1973; Roehler 1990) indicates a shoreline which extended along the margin of the Western Interior Seaway from Rock Springs to Rawlins (Fig. 3). The expansion of a large lobate delta across a platform located westwards of Rawlins separated a large embayment to the west from a strandline (the Hatfield strandline) to the east.

Methods

Twenty-five stratigraphic profiles are presented from the study area (Figs 4 and 5), as located on Fig. 1. Correlations have been walked out within a 70 m thick interval of Hatfield members 1 and 2 that can be traced down-dip for 40 km. Logged sections are spaced at 0.3–4.5 km (Figs 1, 4), enabling facies analysis and the definition of a high-resolution stratigraphic framework. Dissection of the topography provides some oblique-strike oriented outcrops perpendicular to the general west-east dip trend (sections 21–24, Fig. 1) allowing a partial 3D examination of the studied succession (see Fig. 5).

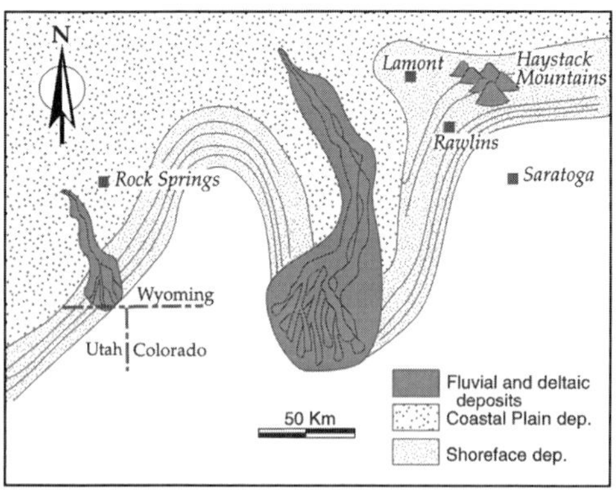

Fig. 3. Palaeogeographic reconstruction during the time of deposition of the Hatfield Sandstone Member of the Haystack Mountain Formation at the time of *Baculites asperiformis*. Note the position of the major delta system to the SW of the study area. the main outcrop belt studied runs south east from Rawlins to Saratoga representing a proximal-distal section as seen in Figs 2 and 4. Map modified from Roehler (1990).

As with all correlation panels of the type presented in Figs 4 and 5, the choice of datum is important, and may even be critical for the correct interpretation of the panel's architecture (see also Plint & Nummedal this volume). In this study a combination of two levels of maximum flooding were picked to aid the alignment of the panel, one within the Hatfield sandstone tongue, the other in the shale immediately below (Figs 4 and 5). Neither of these levels were placed exactly horizontal on the dip oriented correlation panel, but instead both were made to slope generally basinward (to the southeast) at a very gentle angle over the 40 km length of the panel, an approach also advocated and utilized by Plint & Nummedal (this volume).

Facies associations

Two main facies associations were recognized within the study succession: (I) a coarsening-upward shoreface facies association grading from offshore shales that is truncated seaward and down-dip by (II) a cross-bedded wedge. The latter in turn, is downlapped by shoreface deposits of the overlying Hatfield 2 tongue. The stratigraphic context of these two lithofacies associations are illustrated in the upper part of Figure 4.

(I) The shoreface facies association

The shoreface facies association characterizes the lower part of the Hatfield 1 and Hatfield 2 sandstone tongues (sections 1–3, Fig. 4). In the Hatfield Dome area (sections 1–3, Figs 1, 4), the Hatfield 1 sandstone tongue overlies black shales and consists of a coarsening-upward shoreface succession of heterolithic shales and sandstones passing upward into planar to hummocky cross-stratified sandstone facies. The shoreface profile in this sandbody is clearly incomplete as the middle and upper shoreface zones are missing, interpreted to have been eroded beneath the overlying unconformity. In the same area, the younger Hatfield 2 shoreface facies consists of hummocky and swaley cross bedded strata at its top. This upper shoreface profile is also incomplete, again lacking the upper shoreface and foreshore zones because of the overlying erosional unconformity (Fig. 4).

Five broad lithofacies types were distinguished within the shoreface lithosomes of the

Hatfield Member and include; (1) organic-rich black shales, (2) thin bedded siltstones and shales, (3) bioturbated and rarely slumped heterolithic units of sandstones and siltstones, (4) hummocky cross-stratified sandstones, and (5) swaley cross-stratified sandstones.

(I1) Organic-rich, black shales. Very dark to black organic-rich semi-fissile mudstones and silty mudstones that form an interval up to 20 m thick at the base of the Hatfield sandstone tongues. The shales are mostly structureless and unbioturbated, although thinly laminated intervals are also present. A few pelecypods have been recovered.

Interpretation. The fissility, lack of identifiable biogenic structures and rare body fossils together imply either anaerobic conditions on the shelf bottom or a substrate unfavourable for colonization by most benthic organisms (e.g. Driese *et al.* 1991). The facies was probably deposited in anaerobic waters by pelagic settling and from minor bottom flows. Considering the generally anoxic conditions interpreted, the whole interval is suggested to represent a zone of maximum flooding and sediment starvation prior to the deposition of the bioturbated shales and sandstones of the prograding Hatfield Member shoreface. The top of the black shale unit is used as one of two datum levels for alignment of Fig. 4.

(I2) Shales with thin sandstone intercalations. Lithofacies 2 overlies the fissile black mudstones and is characterized by siltstones and black claystones with thin layers of very fine sandstone, mainly organized into a coarsening-upward trend. The sandstone/shale ratio ranges between 30% and 70%. Sandstone layers are typically 30–70 mm thick, show planar to lenticular geometry and sometimes have superimposed wave-ripple caps. Bed contacts may be diffuse, irregularly and sharp or erosional with local scouring. The sandstone beds tend to become slightly thicker and more closely spaced upwards. Bioturbation can be intense with a typical ichnofabric of *Chondrites*, *Helmintoidae* and *Planolites* traces. The main sedimentary structures are thin beds of planar, current-ripple and occasional wave-ripple lamination.

Interpretation. Lithofacies 2 is interpreted in terms of an outer shelf environment below effective wavebase and dominated by currents of low velocity and strength that were able to rework and resediment sand on an offshore muddy substrate. Pervasive bioturbation by burrowing infauna suggests deposition under better oxygenated conditions compared to facies 1, presumably in shallower water, close to or above the top of the pycnocline. The absence of grading in the sandstone beds and the presence of escape structures and wave ripples suggest deposition by storm-generated geostrophic flows upon which oscillatory wave currents were superimposed (Duke 1990).

(I3) Heterolithic units of sandstone and siltstone. Facies 3 gradationally overlies facies 2 and underlies both the hummocky cross-bedded strata of Hatfield 1 (Fig. 4) and the strongly bioturbated facies II1 with *Teichichnus* and *Rhizocorallium* that characterizes the lower part of the cross-bedded wedge (Fig. 4). In both cases the upper contact is sharp. The heterolithic facies can reach 7–8 m in thick, and consists of fine to very fine-grained sandstone beds intercalated with siltstone and shale. Individual sandstone beds display a tabular geometry and can be followed for tens of metres. Sandstone beds range from 3 to 200 mm and average 70 mm thick, while shale interbeds are 20–30 mm thick, and the sandstone/shale ratio ranges from 70% to 90%.

The facies is strongly bioturbated and trace fossils dominated by *Chondrites*, *Helmintoidae*, *Planolites* and *Paleophycus* occur mainly on the soles and along top surfaces of beds; vertical to subvertical traces are less common. The strong degree of bioturbation tends to obscure sedimentary structures so contacts between sandstone beds and bounding strata tend to be diffuse and poorly defined. Only the thickest sandstone beds, 0.1–0.25 m thick, display sharp and flat bases, as well as very fine planar lamination and hummocky cross-stratification. The generally sharp or undulatory upper boundaries of these beds reflect common preservation of ripple forms. The facies is usually organized into coarsening- and thickening-upward 0.75–1.50 m thick packages, characterized by an upward reduction of bioturbation and the presence of *Terebellina* and *Aulichnites* traces in the uppermost layers.

Interpretation. Deposition of the sandstones and siltstones records storm-generated wave and current activity, alternating with periods of quiescence. Individual beds are mostly of storm origin. The sharp, flat bases of the thickest beds can be indicative of high flow velocities (Hunter & Clifton 1982) whereas planar lamination in the sandstone beds suggests conditions of plane bed deposition (Dott & Bourgeois 1982). This heterolithic facies is interpreted to represent deposition in a middle to outer shelf setting, possibly within a transitional to lower shoreface environment.

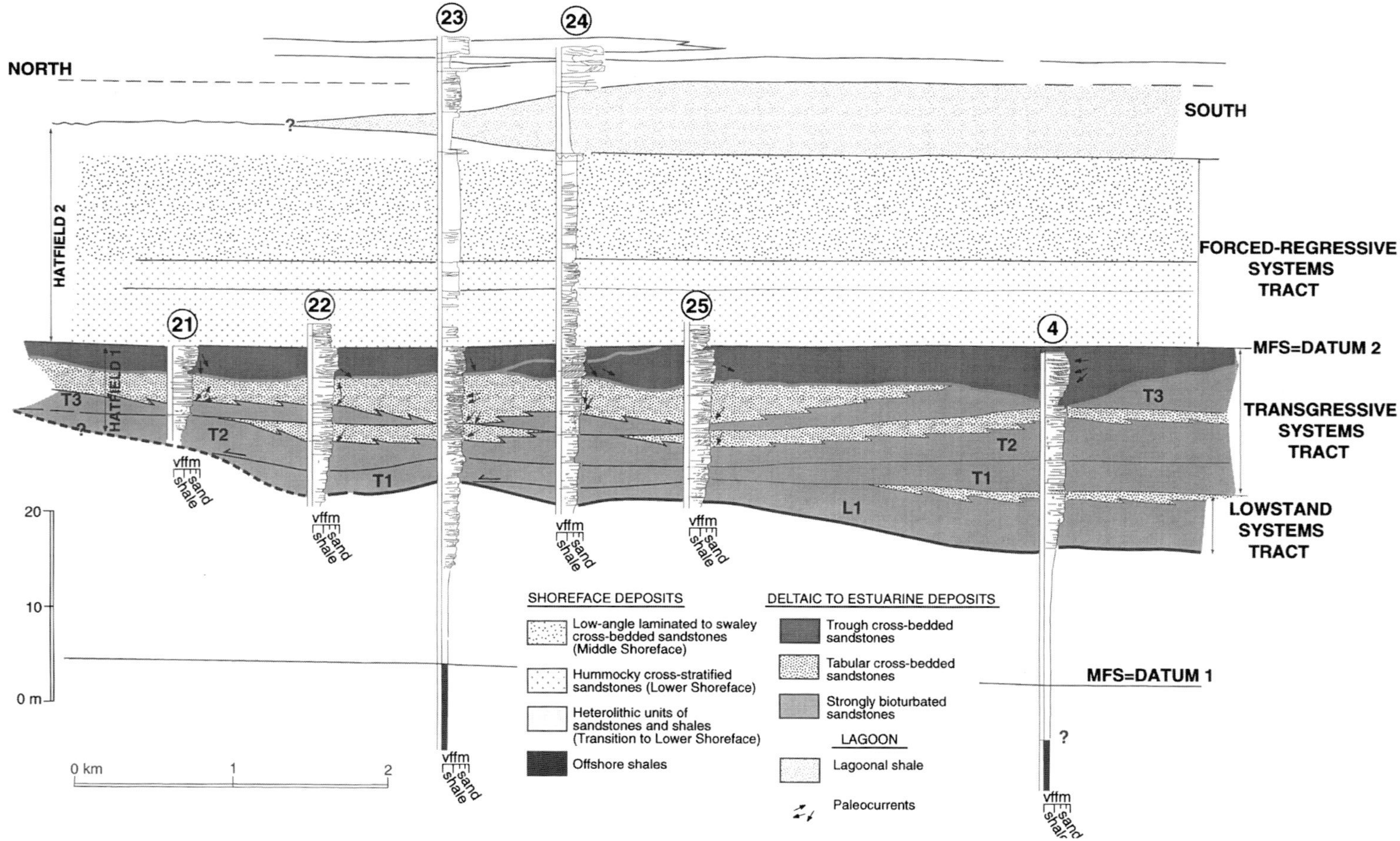

Fig. 5. Additional oblique-oriented correlation panel and architectural interpretation of the cross-bedded lithosome of Hatfield 1 sandstone tongue and the overlying shoreface sediments of Hatfield 2. The maximum flooding surface between the two lithosomes has been chosen as the main datum that in this strike-oriented perspective is taken as horizontal. The cross-section (see location in Fig. 1 and legend in Fig. 4), is broadly parallel to the Hatfield 2 shoreline. The sequence boundary at the base of the cross-bedded wedge here is inferred to be located at the boundary between grey siltstone and shales and coarser-grained, buff bioturbated sandstones. Note the onlap relationships of the lowstand and transgressive parasequences onto the sequence boundary, and the abrupt superposition of the hummocky cross-strata of Hatfield 2 onto the tidally-dominated sandbody of Hatfield 1.

(I 4) Hummocky cross-stratified sandstone facies (HCS). This facies sharply overlies units of heterolithic siltstones and sandstones, and characterizes the basal portion of the Hatfield 1 and Hatfield 2 sandstone tongues in their most proximal (i.e. westernmost) locations (Fig. 4). Downdip, hummocky cross-bedded strata are eroded in Hatfield 1, whereas in Hatfield 2 they directly overlie the Hatfield 1 cross-bedded lithosome and are in turn sharply overlain by swaley cross-stratified sandstones (sections 4 to 12, Fig. 4 and sections 21–15, Fig. 5).

The HCS facies is up to 15 m thick and consists of medium- to very thickly bedded amalgamated very fine to lower fine-grained sandstones that form both laterally continuous sheets and discontinuous lensoid bodies. The facies is typified by medium-scale sets of hummocky cross-bedded strata. Individual sandstone beds range in thickness from 80 mm to 1.3 m, the thickest beds being composed of several stacked hummocky sets each a few centimetres to 0.35 m thick. Laterally, the thick sandstone beds split into thin hummocky laminated sandstones with interlaminated siltstone. The boundaries between individual hummocky sets are accentuated by shale layers and by the presence of clay and siltstone rip-up clasts at their base. Commonly, bed bases are sharp and scoured.

The thickest hummocky bedded sandstone intervals are separated by heterolithic 30–300 mm thick units of siltstone, very fine rippled sandstone layers and black organic-rich claystones. Sandstone interbeds are usually very thin (30–50 mm) whereas shale is present mainly as 10–30 mm thick lenses and layers. Rare *Ophiomorpha* sp. traces are observed within beds, although are locally abundant at the base of sandstone beds and associated with siltstone layers. Hummocky beds have sharp flat bases, with rare flute or groove casts. In the more proximal parts of Hatfield 1, the basal hummocky facies is marked by a horizon of gutter casts oriented perpendicular to the shoreline and isolated within heterolithic sandstones and shales. The gutter casts can be 0.3–0.8 m thick, and are typically filled by strongly bioturbated fine-grained sandstones. The lack of lithological contrast makes identification of specific forms difficult, although few *Thalassinoides* and *Skolithos* have been recognized. In the more basinward position, hummocky bedded strata are separated by bioturbated sandstones with a small form of *Teichichnus* traces (e.g. sections 24 to 12, Fig. 4).

Interpretation. The hummocky facies is interpreted to represent deposition in an inner shelf setting within a storm-dominated lower shoreface environment (Dott & Bourgeois 1982). The interbedding of this facies with the thin-bedded sandstone and shale facies can be explained by the combination of storm-wave generated oscillatory flows (Southard *et al.* 1990) and a storm-generated, unidirectional mean flows (Driese *et al.* 1991). In the continuum of storm-generated structures present along a deepening shelf profile, HCS represents deposition in a more proximal position relative to wave-rippled laminations of the distal shelf (Aigner 1985). The structures and the geometry of HCS beds indicates deposition from waning flow. The sharp-based soles and scouring of beds indicate an abrupt change from mud to sand deposition (except in the case of amalgamated sequences) and are interpreted to result from initiation of a high-velocity competent flow (Hunter & Clifton 1982). The deep, steep-sided gutter casts are thought to record periods of intense localized erosion in a generally cohesive muddy substratum. As suggested by Duke (1990) and Plint & Norris (1991), their consistent shore-normal orientation indicate that they were formed largely as a result of oscillatory wave scour.

(I5) Swaley cross-stratified sandstones. This facies is developed in the Hatfield 2 tongue where it can reach 10 m thick and sharply overlies the hummocky cross-stratified sandstone facies (Fig 4, sections 1–12; Fig. 5, sections 23–24). It is characterized by upper fine-grained sandstones organized into superimposed concave-upward large broad troughs that scour into each other (Fig. 6). Single sets are commonly 30–50 mm thick, usually have sharp bases above very fine sandstones and lack siltstones and shale interbeds. The swales consist of low angle, parallel to slightly divergent gentle cross-stratification that follows their erosive basal surface. The top of the sets exhibit rare wave ripple-lamination and bioturbation is absent or moderate and characterized by a few *Ophiomorpha* sp. and *Macaronichus* traces. The facies grades upwards into parallel, unbioturbated sandstones, but it is generally truncated by a sequence boundary, as recognized at the top of Hatfield 2 sandstone tongue. The facies grades basinward into hummocky cross-bedded strata.

Interpretation. The low-angle gentle scours typical of this facies are interpreted as swaley cross-stratification (Leckie & Walker 1982) produced in a middle shoreface environment. The absence of mud is consistent with such a depositional setting where continuous wave agitation inhibits the settling of mud. Swaley cross-stratification seems to be a common feature of the shoreface and forced regressive deposits of the

Fig. 6. Swaley cross-stratification in Hatfield 2. Detail of superimposed concave-upward swales that scour into each other. Hammer for scale.

Alberta Foreland Basin (Leckie & Walker 1982; McCrory & Walker 1986; Plint & Walker 1987; Plint & Norris 1991; Ainsworth 1994; Plint & Nummedal this volume). It is interpreted as a storm-dominated structure formed above fair-weather wavebase, in water depth of about 5–10 m (Plint & Norris 1991) as also indicated by the presence of *Macaronichnus* traces (Clifton & Thompson 1978). Unstable environmental conditions prevented the settlement of organisms and only a few well-adapted forms were able to prosper in this high-stress environment.

(II) The cross-bedded facies association

The cross-bedded facies association dominates the extensive wedge-shaped lithosome developed in the most basinward (eastern) reaches of the Hatfield 1 sandstone tongue (Figs 2, 4). The cross-bedded sandbody is up to 30 m thick in the central part of the basin (Fig. 4) and it is characterized by upper fine- to medium-grained sandstones that display an overall thickening- and coarsening-upward trend. Four major lithofacies were recognized; (1) strongly bioturbated sandstones, (2) planar- to wedge cross stratified sandstones, (3) trough cross-bedded sandstones and (4) glauconitic sandstones, and all but the latter are differentiated on Figs 4 and 5.

(II1) Strongly bioturbated sandstone facies. Facies 1 consists of intensely bioturbated structureless to horizontally bedded fine sandstones up to 15 m thick. This facies are generally developed in the basal portion of the cross-bedded wedge (e.g. Figs 4 and 7a).

Bioturbation is pervasive throughout the interval, obliterating almost all of the original sedimentary structures; only a faint horizontal stratification delineating 0.2–0.3 m thick beds and occasional wedge-like 0.10–0.35 m thick cross-bedded sets can be recognized. The diversity of trace fossils is low although those present are abundant filled by siltstone and shale, and include *Teichichnus*, *Cylindrichnus, Diplocraterion, Ophiomorpha* and *Rossellia*. *Teichichnus* traces, oriented longitudinal and oblique to bedding surfaces, dominate the ichnofabric. The few laminated sandstone beds recognized display tabular-wedge cross-stratification, sharp-bases and rippled tops. In places, beds shows an apparent low-angle stratification although three-dimensional exposures reveal that these low-angle structures result from an oblique cut of higher-angle cross-beds.

Interpretation. The extreme density of burrows suggests overall conditions of moderate- to low-energy. Overall rates of bioturbation were apparently able to keep-up or outpace those of sedimentation (Howard 1978; Driese *et al.* 1991), though rare preservation of high angle cross-stratification within bioturbated beds suggests that the sand was generally deposited by megaripple migration.

(II2) Planar to wedge cross-stratified sandstone facies. This cross-stratified sandstone facies makes up the bulk of the cross-bedded wedge. The facies gradationally overlies the strongly bioturbated sandstone facies (II1), and passes upward through an erosional contact to the trough cross-bedded sandstones of Facies II3. It is up to 10 m thick, and the cross-bedded sets show a general thickening-upward trend. Seen

Fig. 7. The cross-bedded lithosome of Hatfield 1 at profile 7 (Fig. 4). Note at the base of the outcrop the strongly bioturbated sandstone facies (a) that grade upwards into tabular wedge cross-strata (b). Hammer for scale (circled).

from a distance, the cross-bedded units show gently inclined and frequently intersecting master-bedding surfaces that are usually difficult to detect near the outcrop.

The upper fine-grained sandstones consist of medium to large-scale sets of planar, tabular- and wedge-type cross-strata 50 mm to 0.6 m thick (Figs 7 and 8). Planar sets are composed of normally graded centimetre-thick foresets that are sometimes bundled where thicker sandstones alternate with mud drape couplets. Double mud drapes within the cross-bedded sets occur in outcrops bordering the North Platte River (Figs 1, 4, section 18 and see Fig. 8) but are not particularly abundant in other sections. Wave-ripple lamination, often with a complex interference pattern characterizes the top of most of the cross-sets. Shaley, sideritized rip-up clasts occur across the lower bounding surface of the sets and form in thin zones along the foresets. Shale lenses and laminae mainly in the basal portions of the facies are commonly interbedded with the trough cross-bedded strata and are up to 50 mm thick. Bioturbation varies from weak to strong and it is characterized by *Ophiomorpha*, *Teichichnus*, *Rhizocorallium* and *Thalassinoides*, with the uppermost intervals of the cross-bedded facies dominated by *Ophiomorpha*. Palaeocurrent patterns are usually unimodal, toward SSW, although a subordinate mode toward the north is also present.

Fig. 8. General view of tabular- to wedge-shaped cross-stratified sandstone beds at profile 18 (Fig. 4). The cross-bedded sets are often bundled and show double mud drapes along foresets and set boundaries.

Interpretation. As proposed by Mellere & Steel (1995*a*) and Mellere (1996), the planar cross-bedded sandstone facies is interpreted to have been deposited by the migration of megaripples and dunes driven by tidal and subordinate storm currents in an outer estuarine/tide-dominated delta setting. Although cross-bedded sandstones can characterize most of the upper shoreface, the two facies contrasts significantly in terms of their grain size, ichnofabric and sedimentary structure, with the hummocky and swaley deposits of the typical shorefaces of the Haystack Mountains Formation.

Tidal activity is strongly suggested by the erosive base of the cross-beds, decimetre-scale high-angle cross-stratification with internal bundling of foresets, the occurrence of mud-drape couplets along cross-sets (de Raaf & Boersma 1971; Reinech & Singh 1973; Visser 1980; Boersma & Terwindt 1981; Smith 1988) and an abundance of clay- and siltstone pebbles (Allen 1980). Although double mud-drapes are only occasionally preserved, the presence of abundant mud clasts suggests their original extent was greater. The general interlayering of sandstones and shales indicates fluctuation in flow velocities that is more common in tidal than upper shoreface settings. The fact that bipolar palaeocurrents are not abundant and herring-bone structures are rare is thought to indicate that the system was dominated by one prevalent (ebb) current. The gradual superposition of facies II2 over facies II1 suggests progressively shallower water conditions accompanied by an increase in the energy and strength of tidal currents. The abundant but relatively low-diversity ichnofabric assemblage supports the hypothesis of a stressed estuarine depositional setting.

In terms of bedding complexity, the gently-dipping nature of the master surfaces within the cross bedded sandstones are comparable to the class VI sand waves of Allen (1980). They also show similarity to the class V sand waves of Allen (1980) in that the cross-beds and presumed second-order erosion surfaces dip in the same general direction, with little or no preservation of herring-bone structures.

(II3) Trough cross-bedded sandstones. This facies is generally most important in the uppermost part of the cross-bedded wedge (Fig. 4). It normally displays a channelized and erosional contact (Fig. 9) onto the tabular cross-bedded sandstones of Facies II2, and is in places overlain by glauconitic sandstones (Facies II4). It consists of white-weathering and well sorted upper fine- to lower medium-grained sandstones, with large to small-scale sets of trough and planar cross-beds (Fig. 10). The facies can reach 9–10 m thick and shows a lower coarsening- and thickening-upward trend, followed by a thinning- and

Fig. 9. General view of the white weathering, trough cross-bedded sandstones in broad channels at locality 18 (Fig. 4). The basal erosive surface, just above the person's head (dotted), is marked by a shell lag and dinosaur bones. The channelized body is interpreted to possibly represent a distributary channel within the tidally influenced deltaic system. Note the internal low-angle cross beds.

Fig. 10. Large, sigmoidal bedforms that migrate within the channel system at the top of the cross-bedded sandstones as seen at profile 19 (Fig. 4). Jacob staff (1.5 m) for scale.

fining-upward tendency. Cross-stratified sets are up to 2 m thick and are characterized by trough and tabular geometries. The basal portion of the trough cross-bedded interval typically displays a rapid thickening-upward trend of the compound trough cross-sets. Thin clay layers and double mud drapes along the foresets are particularly common and are occasionally bundled. Up-section the troughs become broader and in places they mimic swaley cross-stratification.

Dip of the foresets is consistent, and towards either 140–160° or 240–270° (i.e. profiles 12–15, Fig. 4) with dips ranging from 10° to 20°, although these data come mainly from the base of the trough cross-bedded interval. Bioturbation is very slight to absent and characterized by *Ophiomorpha*, *Thalassinoides* and *Aulichnites* traces. As shown in Fig. 4a, the trough cross-bedded sandstones fill broad scours a few metres deep but up to 5 km wide eroded into the tabular cross-bedded facies. In a few places, as at profile 18 (Fig. 4), the contact is marked by aligned shells and an odd dinosaur bone. At profile 19 (Fig. 4), the contact with the lower cross-bedded sandstones is clearly channelized and relatively large sigmoidal bedforms occur within some of the channels (Fig. 10).

Interpretation. As shown in the cross-section (Fig. 4), the contact between the tabular cross-bedded facies and the white-weathering trough cross-bedded facies is scoured and channelized. The presence of lateral accretion and planar-tabular cross beds up to 1 m thick indicate the presence of strong currents within the channels. The occurrence of bi-directional palaeocurrent indicators and double mud drapes suggest a tidal origin for these deposits (de Raaf & Boersma 1971; Reinech & Singh 1973; Visser 1980; Boersma & Terwindt 1981; Smith 1988). The sandstones were likely deposited in channels that intersected the planar cross-bedded estuarine-mouth lobes, and were possibly feeders of these ebb-delta lobes. The fact that the thickest bedforms are oriented seaward (140–150°), but in a direction perpendicular to the shoreline, suggests that they were deposited within major ebb-tidal channels. The distinctive upward-shallowing and widening of the trough cross-beds in the later stage of the channel fill is accompanied by a wider range of palaeocurrent dispersal, suggesting that tidal currents had become less confined. As indicated by Yang & Nio (1989) in the outer reaches of ebb deltas tidal currents commonly exhibit rotating pattern.

(II4) Glauconitic sandstones. This facies forms discontinuous lenses up to 2 m thick at the top of the cross-bedded wedge and is particularly well developed in outcrops bordering the North Platte River in the southernmost part of the basin (Fig. 11). The lower contact is abrupt and is in places erosional and cemented, with a visible relief up to 0.2 m. The upper contact with the overlying offshore shales is sharp and flat. The abundance of glauconitic grains leads to a distinctive green-brown weathering colour that enables easy

Fig. 11. General view of the cross-bedded lithosome of Hatfield 1 along the North Platte River at the eastern margin of the study region in proximity to profile 19. Note the ridged or swaley upper surface of Hatfield 1 which may represent the original topography. The lithosome is capped by a dark resistant weathering glauconitic sandstones. Cliff is 20 m high above the river and represents most all of the cross-bedded Hatfield 1 lithosome in this area.

identification at outcrop. In addition, the coarseness and presence of lithified horizons make this facies resistant so that it commonly forms a prominent ledge over the underlying tabular and trough cross-bedded sandstones (Fig. 11).

The facies itself consists of bioturbated to tabular cross-stratified medium-grained glauconitic sandstones with abundant sideritic mudstone clasts that are mainly concentrated along the cross-laminae or on set bounding surfaces. It is strongly bioturbated with ichnofabric characterized by large traces of *Teichichnus, Ophiomorpha* and *Skolithos*. Indeed, *Teichichnus* can be up to 0.7 m long. The glauconitic and siderite content increases upward, and rare ammonites and oyster shells were found at the top of the sandstones.

Interpretation. The position of the facies at the top of the cross-bedded wedge, the presence of authigenic minerals and the strong bioturbation indicate that the glauconitic medium-grained sandstones constitute a condensed interval. Condensed sections (Van Wagoner *et al.* 1988; Loutit *et al.* 1988) are associated with marine depositional hiatuses. The presence of lithified horizons indicates surfaces of omission and non-deposition developed as a consequence of the general sediment starvation associated with the condensed interval (Loutit *et al.* 1988). Although the rate of deposition is likely to have been extremely low, the presence of planar cross-beds indicate that there were occasionally sufficient sediments and strong enough energy to generate migrating bedforms.

Sequence Stratigraphy of the Hatfield tongues

The Hatfield Member sandstone tongues form a distinctive two-tiered progradational wedge near the top of the Haystack Mountains Formation (Fig. 2). The Hatfield wedge progrades further basinward than any of the underlying sandstone tongues, and it is positioned near the culmination of an overall regressive succession deposited over an interval of approximately 3 Ma (Gill *et al.* 1970). The Hatfield wedge described herein is bounded by two unconformities (Figs 4, 5) and consists of four distinctive parts or systems tracts, each having a characteristic combination of facies and stacking architecture.

The lower two parts of the Hatfield wedge, representing the Hatfield 1 tongue are dominated by a tidally influenced facies association. The lower part of Hatfield 1 has a progradational architecture (L1-L7 in Fig. 4b) whereas a retrogradational stacking pattern is characteristic of the upper part (T1-T3; Figs 4b and 5). In contrast, the Hatfield 2 tongue, also comprised of two parts is progradational and of wave-dominated shoreface origin. The two parts of this tongue differ from each other in terms of the trajectory angle of the prograding shoreface. The older shoreface segments displays a normal regressive horizontal to basinward-rising trajectory (H1–3, Fig. 4b), whereas the younger segments (F1-F4, Fig. 4b) are thinner, and compressed or foreshortened, and show a downward-sloping trajectory of at least 15 m fall over 10 km (relative to the flooding surfaces that constrain the correlation panel). In the landward area of the panel there are some 9 m of marine shales and transitional facies between the Hatfield sandstone 1 and 2 tongues (profiles 1–3, Fig. 4) that are absent in the basinward reaches, where the F1–4 overlying regressive shoreface sands rest directly onto the retrogradational part of Hatfield 1 (e.g. profiles 4–12, Fig. 4a; profiles in Fig. 5).

The four systems tracts of the 40 km long Hatfield wedge form a genetic sequence (*sensu* Helland-Hansen & Gjelberg 1994) with a duration of less than 1 Ma, as inferred from Gill *et al.* (1970). The internal architecture of the wedge is documented in detail below.

The shoreface deposits below the Hatfield 1 unconformity

These deposits downlap onto the black organic-rich shales, representing the older condensed section used as one of the datum levels in Figs 4 and 5, and are truncated above by the Hatfield 1 unconformity. The vertical trend from offshore shales comprised of mudstones and siltstones to hummocky cross-stratified sandstones seen in profiles 1 and 2 (Fig. 4), reflects a gradual increase in wave activity associated with conditions of progressive shallowing. The shoreface facies and their downdip offshore equivalents are interpreted to have accumulated on the shelf under conditions of relative sea-level highstand. The sandy shoreface parasequences recognized in profiles 1 and 2 are missing farther basinward, presumably eroded below the more steeply dipping sequence-bounding Hatfield 1 unconformity (Fig. 4).

The unconformity below Hatfield 1 sandstone tongue

The base of Hatfield 1 sandstone tongue shows evidence of erosional truncation, and a marked basinward shift in facies whereby shallow-marine sandstones are positioned over offshore shales (Fig. 4), although there is no clear evidence of subaerial exposure in the study area. Both facies and stratigraphic relationships indicate that the erosional unconformity in Hatfield 1 is a sequence boundary (Figs 4, 5; Van Wagoner *et al.* 1990). It is interpreted to have formed when the depositional system underwent an abrupt basinward shift across a relatively shallow platform during base-level fall. As such the basal sequence bounding surface represents a non-accretionary forced regression in the studied area (*sensu* Helland-Hansen & Gjelberg 1994).

Erosion across the unconformity removed up to 10 m of lower shoreface deposits near profile 1 (Fig. 12). The amount of erosion appears to increase basinward at least as far as profile 11 (Fig. 4). Between profiles 1 and 4 the unconformity overlies transitional shoreface deposits (Figs 4 and 5). In the strike-oriented correlation panel (Fig. 5) the erosive surface displays a southerly dip across which there is an abrupt outcrop colour change between the grey heterolithic sandstone and shale units below and the buff bioturbated sandstones above. Some sideritic nodules of uncertain origin are present along the surface. This most distinctive and easily recognizable contact is taken as the formal base of the cross-bedded wedge in this part of the basin. The contact here resembles the base of the Shannon Sandstone in Salt Creek area where Walker & Bergman (1993) interpreted phosphatic pebbles aligned along the basal surface as reworked nodules.

At and beyond profile 4, the sequence boundary overlies offshore shales (Fig. 4). Analysis of the correlation panels suggests that the maximum erosion across the sequence boundary is some 25 m, so that the strata overlying the sequence boundary occupied an incised valley at least in the up-dip reaches of the cross-bedded lithosome (Figs 4 and 5). Such a semi-confined setting is consistent with the tidally-influenced facies of the overlying lowstand (L1–7) and transgressive systems tracts (T1–3), as noted below.

The Hatfield 1 sandstone wedge

In terms of facies associations and geometry, the long-distance progradation of the Hatfield 1 wedge broadly resembles the 'offshore shelf ridges' (Tillman & Martinsen 1984; Palmer & Scott 1984; Gaynor & Swift 1988) or the more recently reinterpreted 'lowstand shorelines' of the Cretaceous Western Interior Seaway (e.g.

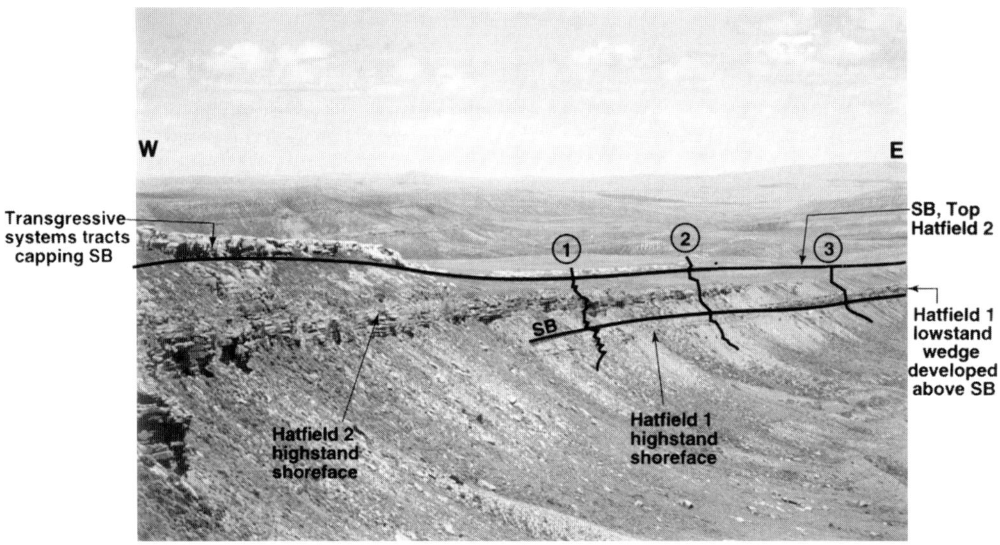

Fig. 12. Position of the unconformities above the Hatfield 2 shoreface and below the Hatfield 1 cross-bedded wedge at locality 1 (Fig. 4). Note the eastward and basinward development of Hatfield 1 lowstand wedge compared to the Hatfield 1 highstand shoreface. The older (updip) shoreface lithosome is truncated by the sequence boundary. This is associated with a basinward shift of facies and the development, above the sequence boundary, of a wedge-shaped body of cross-bedded sandstones representing the lowstand wedge and transgressive systems tracts (e.g. units L1-7 and T1-4 in Fig. 4). The dark cliff-forming Hatfield 2 highstand shoreface sandstones is approximately 25 m thick between the two sequence boundaries (locations 1 to 3; Fig. 4).

Plint 1988, 1991; Posamentier *et al.* 1992; Walker & Bergman 1993; Ainsworth & Pattison 1994; and papers by Ainsworth *et al.*, Fitzsimmons & Johnson, Plint & Nummedal and Posamentier & Morris this volume). Here, the cross-bedded sandstones comprising the lowstand and transgressive systems tracts are interpreted to have been deposited in environments dominated by tidal currents. These conditions, absent in the underlying shoreface deposits, appear to have been initiated during the relative rise of sea-level immediately after the erosion of the shelf. Abundant sediment supply, valley-filling along a deeply-eroded shoreline, and rising relative sea-level produced a lower highly progradational tract and an upper retrogradational one, each showing a significant tidal influence not seen in either the earlier or later highstand shoreline lithosomes of the Hatfield 1 and 2 sandstone tongues, respectively.

The lowstand systems tract (L1–L7 in Fig. 4b) is 5–20 m thick and displays a variably sharp-based to transitional contact with the underlying strata. The wedge can be walked out basinward over a distance of more than 20 km. The systems tract displays an overall wedge-like geometry and a general coarsening- and thickening-upward trend with the superposition of bioturbated tabular and trough cross-stratified sandstones.

Although not exposed, the basinward pinch-out of the lowstand systems tract sandstones is believed to occur just east (?1–3 km) of profile 20 (Fig. 4). Landward thinning of the systems tract occurs by onlap onto the sequence boundary between profiles 1 and 4 (Fig. 4), and laterally between profiles 21 and 24 (Fig. 5). In its landward reaches the systems tract is represented by parasequence L1, is dominated by bioturbated sandstones and has only subordinate tabular cross-stratification. Downdip, at its eastern end, the wedge is dominated by tabular cross-stratification with well-developed tidal signatures (cf. Fig. 8). The upper bounding surface of the systems tract shows a series of broad swales, probably related to the growth of successive lobate(?) parasequences (Fig. 11). Facies association and geometry suggest the seaward growth of a tide-dominated delta system (Mellere & Steel 1995*a*, *b*; Mellere 1996). More proximally, this systems tract is thought to have been confined within an incised valley cut into the underlying shoreface strata, whereas its distal end built out into a more open basin. The sequence stratigraphic context is that of a lowstand progradational wedge (*sensu* Hunt & Tucker 1992; Mellere & Steel 1995*b*). It represents a period of initial relative sea-level rise following the lowest stance of sea-level, when sediment supply was still sufficient to compensate for the increased accommodation space.

The transgressive systems tract is a set of 3 parasequences T1–T3 in Figs 4b and 5. Each parasequence shows a progradational to retrogradational trend, and as a set these display an overall landward-stepping retrogradation or transgressive stratigraphic pattern. The grain-size profile through the transgressive systems tract (T1–T3) is irregular, often displays a repeatedly blocky character (e.g. profiles 4, 7, 12, Fig. 4) and varies markedly from locality to locality (e.g. profiles 4, 5, 7, 23 and 24). Such irregularity is expected in thick transgressive systems tracts that consist of multiple flooding surfaces and parasequences. Moreover, the uppermost part of the parasequences often show stacking of thin carbonate-cemented bioturbated to cross-stratified beds (Fig. 13, cycles T2–3). Such sedimentological variability and complexity is probably induced by a high degree of sediment reworking and starvation in tracts with multiple erosional ravinement surfaces (e.g. Siggerud *et al. in press*).

Downdip correlation between localities 24–16 in Fig. 4 has been established through walking-out of the parasequence boundaries. Here the transgressive systems tract is 2–12 m thick with component parasequences varying between 2 and 8 m thick. Basinward it wedges out just landward of profile 20, and thins out by onlap of the sequence boundary landward between profiles 1 and 4 (Fig. 4), and 21–23 (Fig. 5). The facies are dominated by ripple-laminated/tabular cross-stratified sandstones alternating with strongly bioturbated sandstones in which shale laminae and large amounts of shale rip-up clasts are especially abundant. Clearly, these facies resemble those of the underlying lowstand prograding wedge (Fig. 4, compare a and b).

Figure 14 illustrates the outcrop expression of the oldest transgressive parasequence at locality 15 where it overlies deposits of the preceding lowstand systems tract. Here the transgressive surface defines a change in parasequence stacking from forward- to backstepping, although it is not accompanied by a facies change. The white-weathering medium-grained trough-cross-stratified channelized sandstone bodies at the top of each parasequence, are thought to represent a channel system intersecting the estuary-mouth cross-bedded lobes. The channels are up to 10 m thick, 1–6 km wide and locally developed as laterally continuous sheets (i.e. between profiles 2 and 8, Fig. 4). Although facies analysis

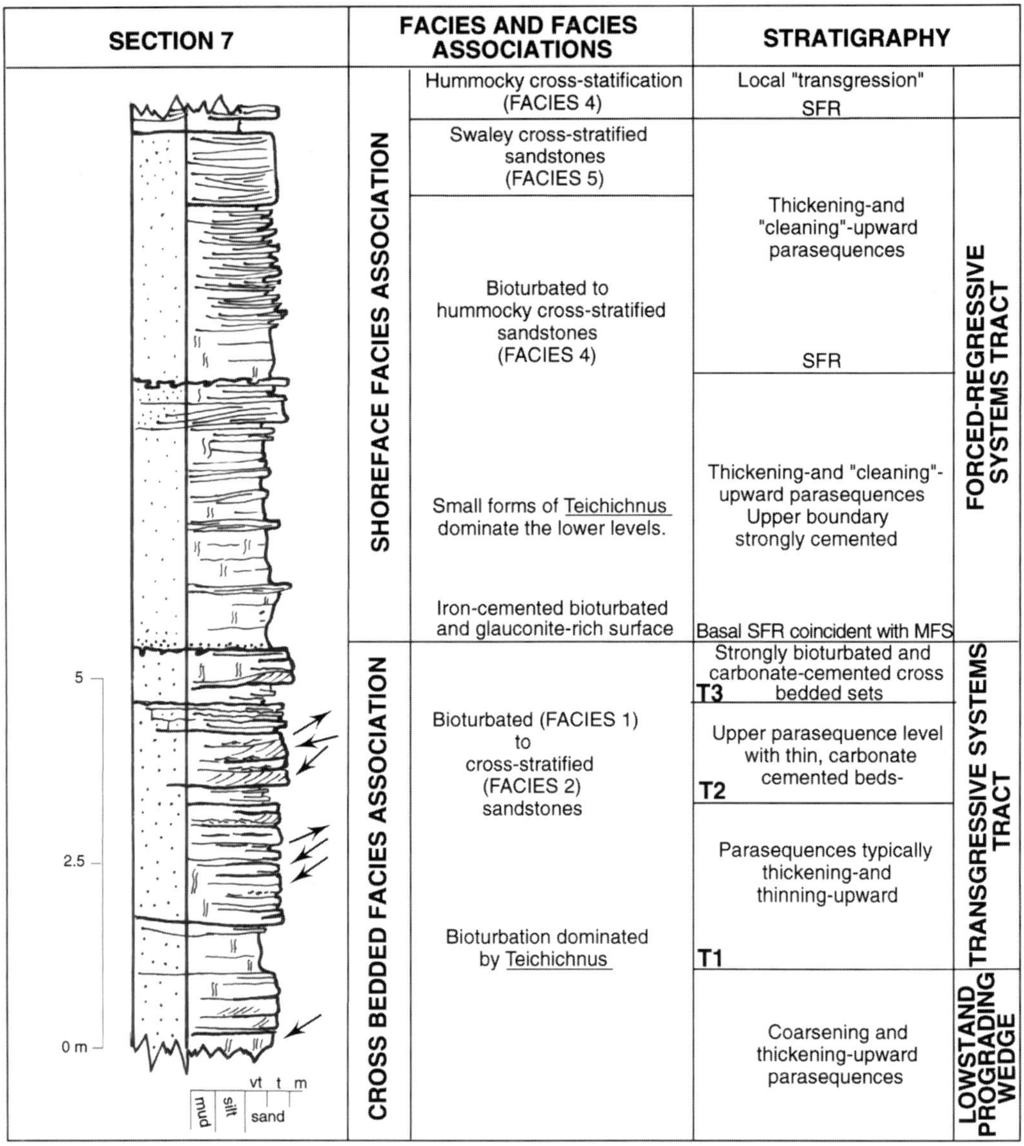

Fig. 13. Detail of the cross-bedded facies association of the LST, TST and the overlying forced-regressive shoreface at profile 7 (see Fig. 4). Note the pattern of transgressive parasequences, typically thickening and thinning-upwards, and the overlying sandy shoreface parasequences. The basal surface of forced regression, coincident with the maximum flooding surface, is bioturbated, iron-cemented and enriched in glauconite grains. Minor surfaces of forced regression (SFR) are strongly bioturbated and carbonate cemented.

indicates very shallow-water conditions, no subaerial exposure features were revealed in any of the examined sections. The occurrence of (i) mud couplets in the bedforms deposited in the lower portion of the trough cross-bedded interval and (ii) the upward increase in trough size suggests that tidal influence was greatest in the basal zones of the channel fill.

The greater tidal influence of the backstepping transgressive systems tract suggests that the lowstand tidally-influenced delta had transformed into an estuary. We use the term estuary as being the transgressive equivalent of a regressive river delta (see Dalrymple *et al.* 1990). The T1–T3 architecture of the transgressive systems tract indicates that it developed during rising

Fig. 14. Outcrop detail of the oldest transgressive parasequence at profile 15 (Fig. 4), overlying tidally influenced, bioturbated to planar wedge cross-bedded strata of the lowstand tract. The bounding surface is a transgressive surface (marked by thick line) and defines a change in parasequence trend (from basinward- to landward-stepping). The transgressive unit here is broadly channelized, and consists of trough cross-strata overlying bioturbated sandstones. View towards the east, person for scale.

relative sea-level, but with a higher accommodation to sediment supply ratio than during deposition of the underlying lowstand systems tract. The tidally influenced nature of the facies strongly suggests that parasequences T1–T3 accumulated in a valley-confined setting at least in their most proximal reaches.

The maximum flooding surface and the condensed section

The top of the cross-bedded wedge is marked by a sharp irregular surface of non-deposition, characterized by strong bioturbation, lithification, shell lags and glauconitic-rich sandstones (e.g. Fig. 11). This is interpreted as the maximum flooding surface (MFS), and is the level onto which the clinoforms of the overlying highstand and forced regressive systems tracts downlap (Fig. 4, sections 2–11 & Fig. 5). The glauconitic sandstones of the condensed section constitute only a minor portion of the maximum flooding sedimentary record, as they normally occur as patchy, thin (up to 2 m) lenses draping the topographic swales at the top of the underlying cross-bedded lithosome in distal sections (e.g. Fig. 11). In most of the profiles the maximum flooding surface is expressed by extensive burrowing and lithification (see MFS in Fig. 13). The high degree of bioturbation and enrichment in authigenic minerals suggest that a considerable period of time was associated with the zone of maximum flooding.

The Hatfield 2 prograding shoreface lithosome

The Hatfield 2 tongue consists of a 25–30 m thick shoreface succession of heterolithic mudstones and siltstones overlain by hummocky passing to swaley cross-stratified sandstones in its proximal reaches (profiles 1–3, Fig. 4). This succession coarsens-upward and is inferred to have low-angle clinoform surfaces (time lines) that downlap onto the maximum flooding zone on top of the underlying cross-stratified lithosome (H1–H3 in Fig. 4B). This shoreface can be followed as a consistently thick unit for some 8 km across the correlation panel (Fig. 4), and is interpreted in terms of a highstand systems tract, because it is overlain by a sequence boundary, downlaps onto a maximum flooding surface and it displays a sub-horizontal to slightly aggradational progradational trajectory. It is interpreted to have formed during conditions of stable or slightly rising relative sea-level, with a markedly lower accommodation to sediment supply ratio than for the underlying transgressive systems tract.

In contrast to the succeeding forced-regressive systems tract, the highstand systems tract appears to have been developed by sub-horizontal progradation of a shoreface, as judged by the laterally consistent thickness of its component facies, i.e. of swaley, hummocky, heterolithic and mudstone–siltstone facies belts respectively (Fig. 4). It should be noted that the lowermost mudstone–siltstone shoreface unit is some 9 m thick here. The basinward thinning and disappearance of this unit coincides with the later onset of forced regression. In the most proximal segment of the Hatfield 2 shoreface the highstand strata distinguished herein were mistakenly included in the forced regressive portion of the shoreface by Mellere & Steel (1995*a*).

The Hatfield 2 forced regressive shoreface lithosome

Basinward of profile 4 the shoreface lithosome continued to prograde and downlap onto the maximum flooding surface immediately above the cross-bedded lithosome (Figs 4 and 5). The evidence of continued progradation can still be seen in the form of 2 or more parasequences at any locality. The parasequence boundaries slope gently basinwards, and represent the shoreface clinoform surfaces (F1–F4 in Fig. 4b). However, the shoreface units of the forced regressive systems tract (Figs 4b, 5, 13 and 15) exhibit a number of features that distinguishes them from the older highstand shoreface.

(1) The shoreface package (clinoform height) is reduced almost by half in thickness (e.g. compare profiles 7, 8 with 2, 3 in Fig. 4b).

(2) The shales and siltstones that are thickly developed at the base of the highstand shoreface are completely absent, so that bioturbated to hummocky cross-stratified sandstones rest directly on tidal deposits of the preceding transgressive systems tract (Figs 13, 15).

(3) The more even or gradual upward-coarsening highstand shoreface profiles are replaced by more uniform or blocky shoreface profiles as at localities 6–11, and occasionally even by a slight upward-fining trend as at profiles 7–10 (Fig. 4).

(4) Restoration of the underlying maximum flooding surface to have a gentle seaward dip, in an attempt to mimics its original depositional basinward dip (cf. Plint & Nummedal this volume), shows that the uppermost level of the shoreface lithosome has a slight but significant downward and basinward shift in successive progradational segments, when traced from profile 3 of the highstand systems tract to profile 11 at the basinward end of the forced regressive systems tract (Fig. 4).

(5) The shoreface facies belts (swaley, hummocky and heterolithic respectively) are displaced downward with respect to each other within successively younger shoreface segments (localities 4–11, Fig. 4). Downward displacement of the shoreface occurs across slightly erosive and carbonate-cemented clinothem boundaries that appear as erosively-based and somewhat blocky parasequences at outcrop (e.g. locality 7, Fig. 13).

(6) Within the hummocky cross-bedded strata, the ichnofabric becomes dominated by a small form of *Teichichnus* suggesting the persistence of restricted basin conditions.

These characteristics are developed within the shoreface package between profiles 3 and 11 in the correlation panel (Fig. 4). We suggest that they reflect reduction in water depth in this segment of the shoreline during regression caused in part by falling relative sea-level. Such a fall would have caused a reduction in thickness of the accompanying shoreface package, as the shoreface clinoform would have had less accommodation space in which to develop. This type of regression occurring during sea-level fall (e.g. Plint 1988) has been termed forced regression (Plint 1991; Posamentier *et al.* 1992) and the resultant systems tract has been referred to as the forced regressive systems tract (Hunt & Tucker 1992, 1995; Helland-Hansen & Gjelberg 1994) or the falling stage systems tract (cf. Plint & Nummedal this volume). Figure 15 illustrates the typical compressed or foreshortened forced regressive wave-dominated shoreface lithosome that lacks basal shales and directly overlies the tidally influenced estuarine lithosome of the transgressive systems tract.

Between localities 6 and 10 (Fig. 4), there is a retrogradational tendency in the uppermost 8 m of the forced-regressive shoreface profile. This has been interpreted as local transgression resulting (at these localities) from relatively high accommodation with respect to sediment supply that may be related to deposition out of the plane of section.

Surface(s) of forced regression

The basal bounding surface of the forced regressive systems tract, in contrast to the updip-equivalent surface at the base of the highstand systems tract, is erosive (Fig. 15). This surface is marked by sideritic nodules and has an unusually

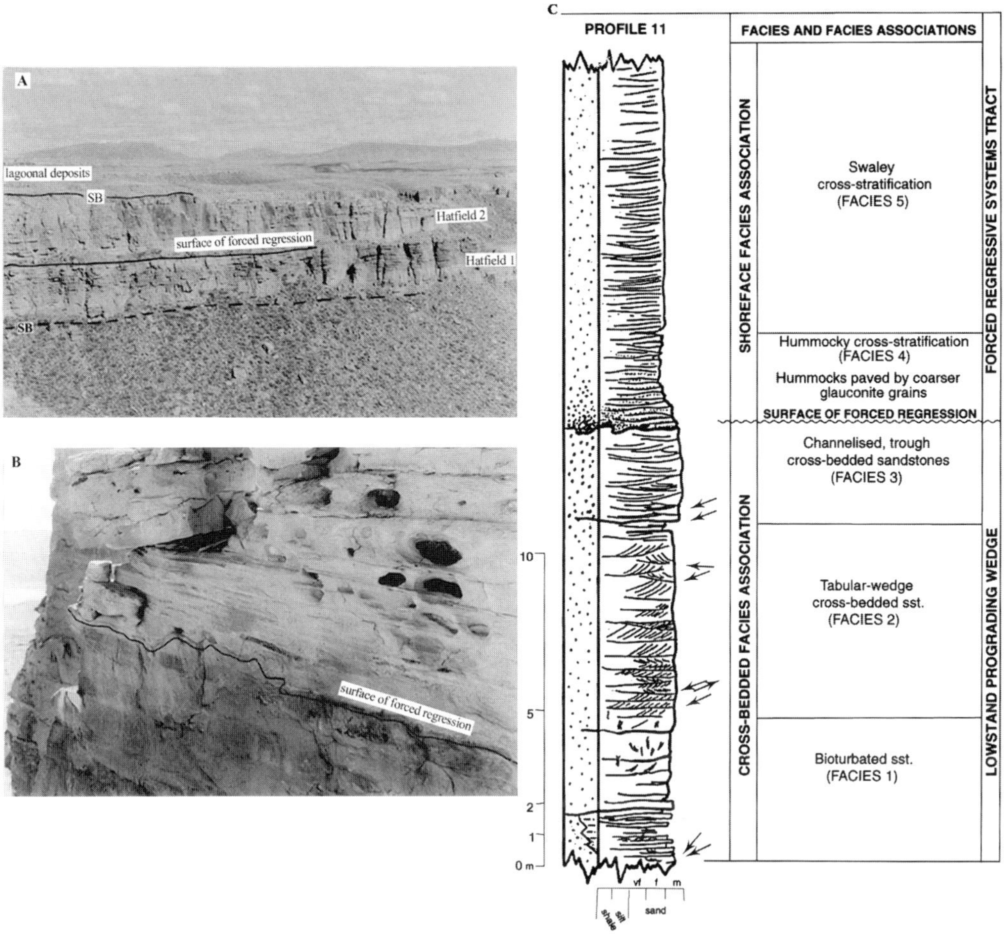

Fig. 15. (**a**) View of the abrupt superposition of the forced regressive wave-dominated shoreface of Hatfield 2 onto the tidally influenced estuarine deposits of the Hatfield 1 transgressive systems tract at locality 10 (Fig. 4). The contact between Hatfield 1 and 2 is a surface of forced regression, coincident with the maximum flooding. Height of the cliff is approximately 20 m between the two sequence boundaries. (**b**, **c**) Detail of the basal surface of forced regression at profile 11 (Fig. 4). At the contact, the hummocky cross-strata of the forced regressive shoreface are enriched in sideritic clasts and glauconite grains, documenting wave erosion into the underlying condensed section during falling sea level.

high concentration of coarse glauconitic grains. The glauconitic grains also occur in the bioturbated and hummocky cross-strata up to 2 m above the basal surface (Fig. 15), and are particularly concentrated within the gently dipping laminae of the hummocky sets. Concentration of glauconitic grains provides evidence of significant erosion along the base of the forced regressive tract, as these grains have clearly been incorporated from the underlying condensed section.

The basal surface of forced regressive sediments represents a 'regressive surface of marine erosion' or the 'surface of forced regression' (*sensu* Nummedal 1992; Helland-Hansen & Gjelberg 1994). It is a composite or master erosion surface, resulting from the amalgamation of a number of minor erosional surfaces. The latter separate updip from the basal surface and rise landwards up through the shoreface body as clinoform surfaces as shown in the correlation panel between localities 4 and 11 (Fig. 4). As already noted, these minor forced regressive surfaces tend also to have a concentration of coarse grains, and are the surfaces across which the forced regressive shoreface gradually downsteps from its former highstand position.

Upper sequence boundary

In this study the forced regressive systems tract lies entirely below the upper sequence boundary (Figs 4 and 5). The upper sequence boundary depicted in the correlation panel (Fig. 4) is overlain by lagoonal deposits, that in turn pass basinward and are overlain by a second Hatfield cross-bedded lithosome (Figs 2 and 4). These relationships indicate a more severe basinward shift of the depositional system, with the unconformity truncating the forced regressive systems tract, than seen in the Hatfield 1 tongue. The lagoonal deposits overlying the Hatfield 2 sequence boundary onlapped the unconformity as part of a younger lowstand prograding wedge developed during the ensuing relative sea-level rise (see also Mellere & Steel 1995*a*).

Conclusions

(1) The Hatfield Member of the Haystack Mountains Formation in the Hatfield Dome area of south central Wyoming, provides a well-exposed example of a depositional sequence deposited during a fall to rise cycle of relative sea-level change over an interval that was probably less than 1 Ma in duration.

(2) In the study area the sequence is bounded by erosional unconformities, is up to 40 m thick and extends as a composite tongue some 30–40 km into the basin, significantly beyond the pinch-out of the conventional shoreline. The most basinward-extended part of the sequence has characteristics typical of sandbodies earlier interpreted as offshore bars or shelf ridges.

(3) Four systems tracts can be distinguished between the sequence boundaries. The oldest systems tract extends some 20 km basinward of the pinch-out to the underlying shoreface, from which it is separated by a sequence boundary. The overlying sandstone tongue consists of tide-dominated deltaic deposits stacked in a strongly progradational pattern, although near its basinal pinch-out an aggradational stacking pattern is developed. This systems tract is interpreted as a lowstand progradational wedge, developed during early rise of relative sea level, but under conditions of relatively low accommodation to sediment supply ratio. In its proximal reaches, this systems tract probably occupies an incised valley.

(4) Above the lowstand wedge and below the level of maximum marine flooding there is a second tidally influenced lithosome. This consists of a retrogradational series of parasequences and is interpreted as a transgressive systems tract, deposited during conditions of rising relative sea-level but with higher accommodation to sediment supply ratio than in the underlying tract. The tidal influence suggests some continued topographic confinement, perhaps of an estuary in its upper reaches.

(5) Above the maximum flooding surface there is a wave-dominated shoreface lithosome with a progradational stacking pattern. This lithosome has 2 distinct parts. In its landward reaches the shoreface clinoforms have a horizontal to slightly aggradational trajectory, characteristic of a highstand systems tract developed during stillstand to slightly rising relative sea-level. Basinward of the highstand systems tract the shoreface lithosome exhibits the following change in character:

(a) it is reduced to half its thickness;
(b) the top of the shoreface body physically downsteps into the basin;
(c) the deepest shoreface facies, that is well developed updip in the highstand systems tract, is absent;
(d) the basal surface of this tract, the surface of forced regression, is associated with an enrichment in glauconitic grains – these are derived from the underlying condensed section, and document the erosive nature of the surface of forced regression.

This development, resultant from reduced space on the shoreface following a drop of relative sea-level, characterizes the forced regressive systems tract.

The work on the Haystack Mountains Formation was financed by Norske Shell, in an effort to study outcrops analogous to the Troll reservoir sandstones. We thank B. Ainsworth and an anonymous reviewer for helpful comments on an earlier version of the manuscript.

References

AIGNER, T. 1985. *Storm deposition Systems*. Springer-Verlag, New York.

AINSWORTH, R. B. & PATTISON, S. 1994. Where have all the lowstands gone? Evidence for attached lowstand systems tracts in the Western Interior of North America. *Geology*, **22**, 415–418.

——, BOSSCHER, H. & NEWELL, M.J. 2000. Forward modelling of forced regressions: evidence for the genesis of attached and detached lowstand systems. *This volume*.

ALLEN, J. R. L. 1980. Sand waves: a model of origin and internal structure. *Sedimentary Geology*, **26**, 281–328.

BOERSMA, J. R. & TERWINDT, J. H. J. 1981. Neap-spring tide sequences of intertidal shoal deposits in a mesotidal estuary. *Sedimentology*, **28**, 151–170.

CLIFTON, H. E. & THOMPSON, J. K. 1978. *Macaronichnus segregatis* – a feeding structure of shallow marine polychaetes. *Journal of Sedimentary Petrology*, **48**, 1293–1302.

DALRYMPLE, R. W., KNIGHT, R. J., ZAITLIN, B. A. & MIDDLETON, G. V. 1990. Dynamics and facies model of a macrotidal sandbar complex, Cobequid Bay – Salmon river estuary (Bay of Fundy). *Sedimentology*, **37**, 577–612.

DAVIES, H. 1990. Quantitative field studies of Cretaceous shelf-ridge sandstones of the Western Interior Basin, U.S.A. *Abstract. 13th International Sedimentological Congress*, England, 119–120.

DAVIS, H. R. & BYERS, C. W. 1989. Shelf sandstones in the Mowry Shale: evidence for deposition during Cretaceous sea level falls. *Journal of Sedimentary Petrology*, **59**, 548–560.

DE RAAF, J. F. M. & BOERSMA, J. R. 1971. Tidal deposits and their sedimentary structures. *Geologie en Mijinbow*, **50**, 479–503.

DOTT, R. H. & BOURGEOIS, J. 1982. Hummocky stratification: significance of its variable bedding sequences. *Geological Society of American Bulletin*, **93**, 663–668.

DRIESE, S. G., FISHER, M. W., EASTHOUSE, K. A., MARKS, G. T., GOGOLA, A. R. & SCHONER, A. E. 1991. Model for genesis of shoreface and shelf sandstone sequences, southern Appalachians: Palaeoenvironmental reconstruction of an Early Silurian shelf system. *In*: SWIFT, D. J. P., TILLMAN, R. W. & THORNE, J. A. (eds) *Shelf Sand and Sandstone Bodies*. International Association of Sedimentologists, Special Publications, **14**, 309–338.

DUKE, W. L. 1990. Geostrophic circulation or shallow marine turbidity currents? The dilemma of palaeoflow patterns in storm-influenced prograding shoreline systems. *Journal of Sedimentary Petrology*, **60**, 870–883.

FITZSIMMONS, R. & JOHNSON, S. 2000. Forced regressions; recognition, architecture and genesis in the Campanian of the Bighorn Basin, Wyoming. *This volume*.

GAYNOR, G. C. & SWIFT, D. J. P. 1988. Shannon Sandstone depositional model: sand ridge formation on the Campanian western interior shelf. *Journal of Sedimentary Petrology*, **58**, 868–880.

GILL, J. R. & COBBAN, W. A. 1973. Stratigraphy and geologic history of the Montana Group and equivalent rocks, Montana, Wyoming and North and South Dakota. *USGS Professional Paper* **776**, 37 p.

——, MEREWETHER, E. A. & COBBAN, W. A. 1970. *Stratigraphy and nomenclature of some Upper Cretaceous and Lower Tertiary rocks in south-central Wyoming*. USGS Professional Papers, **667**.

HELLAND-HANSEN, W. & GJELBERG, J. G. 1994. Conceptual basis and variability in sequence stratigraphy: a different perspective. *Sedimentary Geology*, **92**, 31–52.

HOWARD, J. D. 1978. Sedimentology and trace fossils. *In*: BASAN, P. B. (ed.) *Trace Fossil Concept*. Society for Economic Paleontologists and Mineralogists Short Course Notes, **5**, 13–47.

HUNT, D. & TUCKER, M. E. 1992. Stranded parasequences and the forced regressive wedge systems tract: deposition during base level fall. *Sedimentary Geology*, **81**, 1–9.

—— & —— 1995. Stranded parasequences and the forced regressive wedge systems tract: deposition during base-level fall; Reply. *Sedimentary Geology*, **95**, 139–145.

HUNTER, R. E. & CLIFTON, H. E. 1982. Cyclic deposits and hummocky cross-stratification of probable storm origin in Upper Cretaceous rocks of Cape Sebastian area, southwestern Oregon. *Journal of Sedimentary Petrology*, **52**, 127–144.

LECKIE, D. A. & WALKER, R. G. 1982. Storm and tide dominated shorelines in Cretaceous Moosebar – Lower Gates interval – outcrop equivalents of Deep Basin gas trap in western Canada. *American Association of Petroleum Geologists Bulletin*, **66**, 138–157.

LOUTIT, T. S., HARDENBOL, J., VAIL, P. & BAUM, G. R. 1988. Condensed sections: the key to age determination and correlation of continental margin sequences. *In*: WILGUS, C. K., HASTINGS, B. S., KENDALL, C. G., POSAMENTIER, H. W., ROSS, C. A. & VAN WAGONER, J. C. (eds) *Sea-level Changes: an Integrated Approach*. Society for Economic Paleontologists and Mineralogists, Special Publications, **42**, 183–213.

MCCLURG, T. A. 1990. *Lithosome relationships and depositional environment of the O'Brien Spring Sandstone (Late Cretaceous), south-central Wyoming*. Unpublished MS. Thesis, University of Wyoming.

MCCRORY, V. L. C. & WALKER, R. G. 1986. A storm and tidally-influenced prograding shoreline – Upper Cretaceous Milk River Formation of Southern Alberta, Canada. *Sedimentology*, **33**, 47–60.

MELLERE, D. 1996. Seminoe 3, a tidally influenced lowstand wedge and its relationship with subjacent highstand and overlying transgressive deposits, Haystack Mountains Formation, Cretaceous Western Interior, Wyoming (USA). *Sedimentary Geology*, **103**, 249–272.

—— & STEEL, R. J. 1995*a*. Facies architecture and sequentiality of nearshore/shelf sandbodies Haystack Mountains Formation – Wyoming U.S.A. *Sedimentology*, **42**, 551–574.

—— & STEEL, R. J. 1995*b*. Variability of lowstand wedges and their distinction from forced-regressive wedges in the Mesaverde Group, southeast Wyoming. *Geology*, **23**, 803–806.

NUMMEDAL, D. 1992. The falling sea level systems tract in ramp settings: *Abstract*, SEPM Theme Meeting, Ft.Collins, CO, August 1992.

PALMER, J. R. & SCOTT, A. J. 1984. Stacked shoreline and shelf sandstone of La Ventana Tongue (Campanian), northwestern New Mexico. *American Association of Petroleum Geologists Bulletin*, **68**, 74–91.

PLINT, A. G. 1988. Sharp-based shoreface sequences and 'offshore bars' in the Cardium Formation of Alberta: their relationship to relative changes in sea level. *In*: WILGUS, C. K., HASTINGS, B. S., KENDALL, C. G., POSAMENTIER, H. W., ROSS, C. A. & VAN WAGONER, J. C. (eds) *Sea-level Changes: an Integrated Approach*. Society for Economic Paleontologists and Mineralogists, Special Publications, **42**, 357–370.

—— 1991. High-frequency relative sea-level oscillations in Upper Cretaceous shelf clastics of the Alberta foreland basin: possible evidence for a glacio-eustatic control. *In*: MACDONALD, D. I. M. (ed.) *Sedimentation, Tectonics and Eustasy*. International Association of Sedimentologists, Special Publications, **12**, 409–428.

—— & NORRIS, B. 1991. Anatomy of a ramp margin sequence: facies successions, paleogeography, and sediment dispersal patterns in the Muskiki and Marshybank formations, Alberta Foreland Basin. *Bulletin of Canadian Petroleum Geology*, **39**, 18–42.

—— & NUMMEDAL, D. 2000. The falling stage systems tract: recognition and importance in sequence stratiraphic analysis. *This volume*.

—— & WALKER, R. G. 1987. Cardium Formation 8. Facies and environment of Cardium shoreline and coastal plain in the Kakwa Field and adjacent areas, Northwestern Alberta. *Bulletin of Canadian Petroleum Geology*, **35**, 48–64.

POSAMENTIER, H. W. & MORRIS, W. R. 2000. Aspects of the stratal architecture of forced regressive deposits. *This volume*.

——, ALLEN, G. P., JAMES, D. P. & TESSON, M. 1992. Forced regressions in a sequence stratigraphic framework: concepts, examples, and exploration significance. *American Association of Petroleum Geologists Bulletin*, **76**, 1687–1709.

REINECK, H. E. & SINGH, I. B. 1973. *Depositional Sedimentary Environments – with reference to Terrigenous Clastics*. Springer-Verlag, Berlin.

ROEHLER, H. W. 1989. *Depositional environments and correlation of the Cretaceous Mesaverde Group and associated Upper Cretaceous Formations, Rock Springs to Atlantic Rim, southwest Wyoming*. US Geological Service Miscellaneous Field Studies Map **MF-2078**.

—— 1990. *Stratigraphy of the Mesaverde Group in the central and eastern Greater Green River Basin, Wyoming, Colorado, and Utah*. US Geological Survey Professional Papers, **1508**.

SIGGERUD, E., STEEL, R. & POLLARD, J. in press. Ravinement complexes and bored pebble: example from a transgressive systems tract, SE Ebro Basin, Spain. *In*: MARZO, M. & STEEL, R. (eds) *San Llorenc del Munt Fan Delta Complex. Sedimentary Geology, Special Volume*.

SMITH, D. G. 1988. Tidal bundles and mud couplets in the McMurray Formation, northeastern Alberta, Canada. *Bulletin of Canadian Petroleum Geology*, **36**, 216–219.

SMITH, J. H., GRAS, V. B. & FABINY, E. M. 1965. Southeast Hatfield Dome (measured section). *In*: DEVOTO, R. H. & BITTER, R. K. (eds) *Sedimentation of Late Cretaceous and Tertiary Outcrops, Rock Springs Uplift*. Wyoming Geological Association, 19th Field Conference, plate III.

SOUTHARD, J. B., LAMBIE, J. M., FEDERICO, D. C., PILE, H. T. & WEIDMAN, C. R. 1990. Experiments on bed configurations in fine sands under bidirectional purely oscillatory flow, and the origin of hummocky cross-stratification. *Journal of Sedimentary Petrology*, **60**, 1–17.

TILLMAN, R. W. & MARTINSEN, R. S. 1984. The Shannon shelf-ridge sandstone complex, Salt Creek Anticline area, Powder River Basin, Wyoming. *In*: TILLMAN, R. W. & SIEMERS, C. T. (eds) *Siliciclastic Shelf Sediments*. Society for Economic Paleontologists and Mineralogists, Special Publications, **34**, 85–142.

—— & —— 1985. *Upper Cretaceous Shannon and Haystack Mountains Formation Field Trip, Wyoming*. Field trip of the Society for Economic Paleontologists and Mineralogists Mid-Year Meeting, Golden, Colorado.

VAN HORN, M. D. 1979. *Stratigraphy of the Almond Formation, east-central flank Rock Springs uplift, Sweetwater County, Wyoming – a mesotidal-shoreline model for the Late Cretaceous*. Master Thesis, Colorado School of Mines, Golden.

VAN WAGONER, J. C., MITCHUM, R. M., CAMPION, K. M. & RAHMANI, V. D. 1990. *Siliciclastic sequence stratigraphy in well logs, cores, and outcrops: techniques for high resolution correlation of time and facies*. American Association of Petroleum Geologists Bulletin Methods in Exploration Series, **7**.

——, ——, POSAMENTIER, H. W., VAIL, P. R., SARG, J. F., LOUTIT, T. S. & HANDERBOL, J. 1988. An overview of the fundamentals of sequence stratigraphy and key definitions of sequence stratigraphy. *In*: WILGUS, C. K., HASTINGS, B. S., KENDALL, C. G., POSAMENTIER, H. W., ROSS, C. A. & VAN WAGONER, J. C. (eds) *Sea-level Changes: an Integrated Approach*. Society for Economic Paleontologists and Mineralogists, Special Publications, **42**, 39–47.

VISSER, M. J. 1980. Neap-springs cycles reflected in Holocene subtidal large-scale bedform deposits: a preliminary note. *Geology*, **8**, 543–546.

WALKER, R. G. & BERGMAN, K. M. 1993. Shannon Sandstone in Wyoming: a shelf-ridge complex reinterpreted as lowstand shoreface deposits. *Journal of Sedimentary Petrology*, **63**, 839–851.

WEIMER, R. J. 1960. Upper Cretaceous stratigraphy, Rocky Mountain area. *American Association of Petroleum Geologists Bulletin*, **44**, 1–20.

—— 1966. Time-stratigraphic analysis and petroleum accumulations Patrick Draw Field, Sweetwater County, Wyoming. *American Association of Petroleum Geologists Bulletin*, **50**, 2150–2175.

YANG, C. S. & NIO, S. D. 1989. An ebb-tide delta depositional model – a comparison between the modern Eastern Scheldt tidal basin (southwest Netherlands) and the Lower Eocene Roda Sandstone in the southern Pyrenees (Spain). *Sedimentary Geology*, **64**, 175–196.

Forward stratigraphic modelling of forced regressions: evidence for the genesis of attached and detached lowstand systems

R. B. AINSWORTH[1,2], H. BOSSCHER[1,3] & M. J. NEWALL[1,4]

[1] *Shell International Exploration and Production B.V., Technology and Research Centre, PO Box 60, 2280 AB, Rijswijk, The Netherlands*

[2] *Present address: Woodside Energy Ltd, 1 Adelaide Terrace, Perth, WA 6000, Australia*

[3] *Present address: A/S Norske Shell, Risavikvegan 180, PO Box 40, 4056 Tanager, Norway*

[4] *Present address: Sarawak Shell Berhad, Locked Bag No. 1, 98009, Miri, Sarawak, Malaysia*

Abstract: A complex series of interactions between subsidence, eustasy and sediment supply determine whether a forced regressive shoreface will be physically *attached* to underlying sandy deposits, or *detached* and encased in marine mudstone. Using a Shell-developed and proprietary forward stratigraphic modelling system, these interactions of controls on clastic depositional geometries can be simulated. Upper Cretaceous subsurface and outcrop data from the Western Interior Basin of Canada form the basis of this study. The outcrop data suggest that high frequency (200 000 years), relatively low amplitude (8 m) relative sea-level changes occurred during deposition of the studied succession. When these parameters are convolved with the Haq third-order eustatic sea-level curve and a constant rate of subsidence and sediment supply, a series of attached lowstand, forced regressive shorefaces are generated by the forward stratigraphic modelling system. In order to generate a detached lowstand shoreface a relatively large magnitude, high frequency eustatic fall or a high frequency tectonic uplift is required. The modelling studies also suggest that (i) forced regressive deposits preferentially develop as attached lowstand systems, (ii) detached lowstand systems in most cases initially develop as attached lowstand systems which are subsequently detached by transgressive–regressive wave erosion, (iii) if the sequence boundary is picked below the sharp-based, forced regressive shoreface (attached and detached lowstand systems tract interpretation) or above it (falling-stage systems tract interpretation) it must remain in that position relative to the correlative down-dip, sharp-based shoreface sandbodies regardless of whether those bodies are attached or detached and (iv) the role of sediment supply as a controlling parameter in the generation of forced regressions appears to be a secondary one.

The definition of forced regression used in this paper is that of Posamentier *et al.* (1992). That is, *a forced regression is defined as a seaward translation of a shoreline occurring in response to relative sea level fall and independent of sediment flux variations.* In recent years the location of the sequence boundary with respect to forced regressive deposits has been a matter of debate (e.g. Hunt & Tucker 1992, 1995; Kolla *et al.* 1995). For the purposes of this paper, the sequence boundary is placed below the sharp-based, forced regressive shoreface (following the systematics of Posamentier *et al.* 1991, 1992; Ainsworth & Pattison 1994; Fitzsimmons 1995; Kolla *et al.* 1995; Pattison 1995). The implications of placing the sequence boundary above the sharp-based, forced regressive shoreface (cf. Ainsworth 1991, 1992, 1994; Hunt & Tucker 1992, 1995; Nummedal 1992*a*, *b*; Plint 1996) are addressed in the discussion section of this text.

Forced regressive systems can be sub-divided into two major categories, those that are physically *attached* to underlying sandy deposits (attached lowstands; Ainsworth & Pattison 1994), and those that are *detached* from underlying sandy deposits and encased in marine mudstone (detached lowstands; Ainsworth & Pattison 1994) (Fig. 1). In the subsurface the detached systems may form stratigraphic trapping configurations for hydrocarbons. In order to define play concepts and exploration strategies to target these deposits, an understanding of the processes that control their genesis is important. Forward stratigraphic modelling studies based on well-constrained datasets can aid in determining these controls and hence enhance the prediction of such potential trapping configurations in other less well constrained areas.

The main purpose of this paper is to demonstrate how forward stratigraphic modelling can provide insights into the genesis of attached and detached forced regressive systems. This paper deals exclusively with forced regressive deposits in a foreland basin setting. However, many of the concepts and ideas can be readily translated

From: HUNT, D. & GAWTHORPE, R. L. (eds) *Sedimentary Responses to Forced Regressions*. Geological Society, London, Special Publications, **172**, 163–176. 1-86239-063-0/00/$15.00

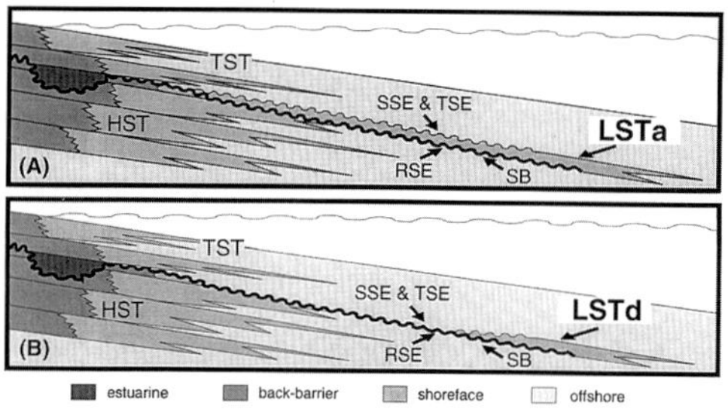

Fig. 1. (**a**) The attached lowstand systems tract (LSTa). Note that the lowstand shoreface is physically attached up-dip to underlying sandy deposits. (**b**) The detached lowstand systems tract (LSTd). Note that the lowstand shoreface is physically detached from underlying sandy deposits and encased in marine mudstone. As such it forms a potential stratigraphic trap for hydrocarbons. RSE, Regressive surface of erosion; TSE, transgressive surface of erosion; SSE, subaerial surface of exposure; SB, sequence boundary; HST, highstand systems tract; TST, transgressive systems tract.

to analogous depositional systems in other tectonic regimes.

Database

The study area lies within the Canadian Western Interior foreland basin (Fig. 2). The basin developed during the late Jurassic as a response to the accretion of terranes in the cordillera to the west. The focus of this study is the non-marine to marginal marine (Campanian–Maastrichtian) Horseshoe Canyon Formation (Fig. 2). This formation forms the base of a thick, clastic wedge – the Edmonton Group (Irish 1970). It intertongues southeastward with the marine shales of the Bearpaw Formation to form the Bearpaw–Horseshoe Canyon transition zone (Shepheard & Hills 1970).

The data set utilized in this study provides a rare opportunity to generate input data for a forward stratigraphic model on two different scales: a regional scale (regional well log correlation panels) and a high resolution scale (outcrop observations). This provides a relatively high degree of confidence over the input variables for modelling on both a regional and a local scale.

The regional, dip-oriented (west–east) well log correlation panel (Fig. 2) is modified from O'Connell *et al.* (unpublished data). This cross-section illustrates how the overall Horseshoe Canyon progradation is punctuated by one major transgression. It also shows a significant basinward shift of the shoreline subsequent to this major transgression (A in Fig. 2).

The outcrop data (Fig. 3) are derived from Saunders (1989) and Ainsworth (1991, 1992, 1994). The outcrop area lies about 12 km southeast of the town of Drumheller (Fig. 2). At this location the Horseshoe Canyon Formation is represented by stacked, attached, sharp-based shorefaces (Plint 1988), fluvio-estuarine,

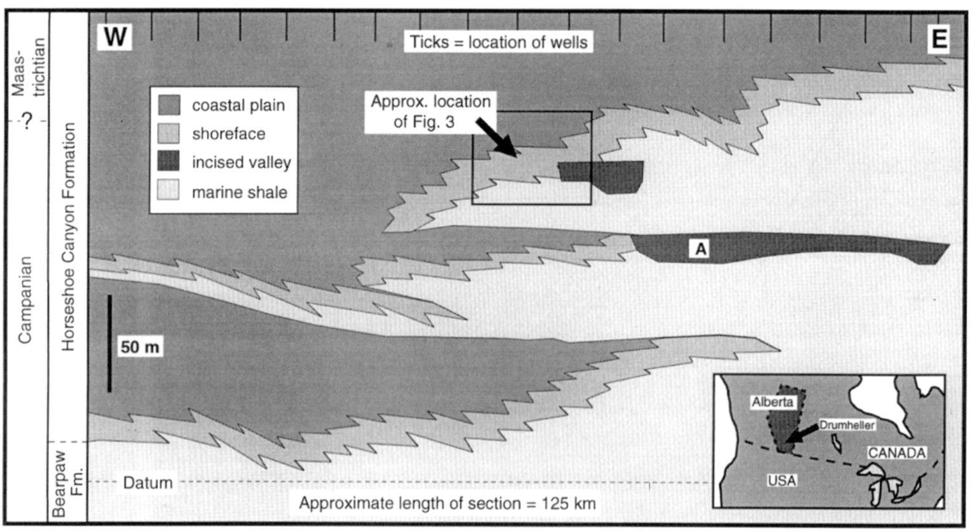

Fig. 2. Bearpaw–Horseshoe Canyon schematic well-log cross-section and location map. Letter *A* represents the tongue that is the focus of the modelling simulation efforts in this paper. Cross-section is modified from O'Connell, S. C., Rottenfusser, B. A. & Langenberg, C. W. (unpublished data).

lagoonal and coastal plain deposits. The strata are divided into seven disconformity bounded units (lettered A to G; Fig. 3). Each unit contains a transgressive and regressive phase. Incised valleys and attached sharp-based shoreface deposits are identified in five of the units (A, B, C, D and F; the incised valley associated with unit D is located further up-dip of section shown in Fig. 3; see Ainsworth 1994, fig. 3, p. 29). These features suggest that the studied deposits exhibit evidence for at least five cycles of relative sea-level change (Ainsworth 1991, 1992, 1994). Absolute dating of the deposits using K/Ar methods (Lerbekmo & Coulter 1985) indicates that the outcrop study interval represents an approximate duration of 1 Ma. Hence, each relative sea-level cycle can be estimated to have a duration of 200 000 years. As distances of shoreface progradation and marine flooding can be measured from the outcrop, approximate magnitudes of relative sea-level change can also be calculated by estimating a palaeo-slope for the shelf (Ainsworth 1994). Estimated values range from 4 m to 12 m. For this study an average value of 8 m was used.

Forward stratigraphic modelling

Numerical simulation of sedimentary basin stratigraphy provides a useful means to develop and quantify concepts of basin evolution (Lawrence *et al.* 1990). It is a valuable tool for the prediction of facies distributions and architecture, for constraining interpretations of subsurface data, for testing exploration scenarios in frontier basins, and for performing sensitivity analyses to determine the fundamental controls on observed basin stratigraphy.

Both in academia and industry, forward stratigraphic modelling applications have evolved rapidly over recent years into easy-to-use programs for evaluating different geological models (Franseen *et al.* 1991; Griffiths & Hadler-Jacobsen 1995). Forward stratigraphic modelling programs present the user with control of the most important parameters that govern the stratigraphic evolution of sedimentary basins: basin geometry, sediment supply, eustasy and subsidence. Interactive parameter editing allows for rapid comparisons of different geological hypotheses. In addition the user can define more specific parameters such as model duration, detailed stratigraphic geometry and sediment properties.

The Shell-developed and proprietary quantitative stratigraphic modelling program utilized in this paper was developed over a period of years and is based on the principles defined in numerous publications (e.g. Aigner *et al.* 1989, 1990; Lawrence *et al.* 1990; Levell & Leu 1993;

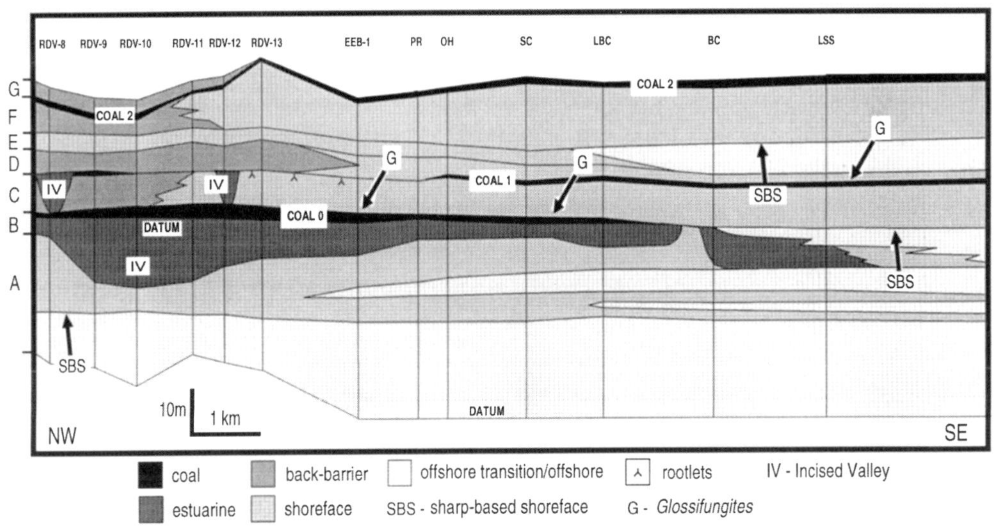

Fig. 3. Bearpaw–Horseshoe Canyon outcrop cross-section. Strata are divided into seven disconformity bounded units, A to G, defined by marine and brackish flooding surfaces. Note incised valleys (IV) in units B and C and also sharp-based shorefaces (SBS) in units A, C and F. Both features indicate evidence for relative sea level falls. Vertical lines represent locations of measured sections. Data from RDV-8 to RDV-13 are from Ainsworth (1991, 1992, 1994). Data from location EEB-1 to LSS are from Saunders (1989). See Fig. 2 for relative location in regional stratigraphy. Modified after Ainsworth & Pattison (1994).

Lawrence 1994; Shuster & Aigner 1994). It simulates depositional processes in an evolving sedimentary basin using a geometric/deterministic approach similar to that described in detail by Ross *et al.* (1995). The calculated stratigraphic geometries result from the interplay of a set of predefined primary controls: depositional profile, sediment volume and calibre, subsidence rates, eustatic variations (Fig. 4), and a series of secondary controls such as wavebase depths, shoreface profiles and erosion rates (both subaerial and submarine). The modelling program is two-dimensional and in the examples used in this paper does not consider the potential effects of sediment flux from out of the plane of section.

Modelling run 1: initial model

The objective of this run was to simulate the stratigraphic architecture derived from the regional well-log correlation panel (Fig. 2).

Input parameters for this run (Fig. 4) were derived from a number of sources. Subsidence rates were taken from the open literature (Chamberlain *et al.* 1989) and were applied as constants. Subsidence rates decreased from the thrust front in the west (80 m Ma^{-1}) towards the basin in the east (30 m Ma^{-1}). The model was run from 78 Ma to 70 Ma (the age of the studied interval) using the Haq third-order eustatic sea-level curve (Haq *et al.* 1988). High-frequency (200 000 years), relatively low-amplitude (8 m) sea-level changes (values derived from the outcrop data; Ainsworth 1994) were superimposed onto the third-order eustatic sea-level curve. Temporal variation in sediment supply was the most difficult variable to constrain. In order to keep the input parameters as simple as possible, the sedimentation rate was held at a constant value (10^{10} m^3 Ma^{-1}) throughout the modelling run. This figure was chosen to ensure that the accommodation space created during the simulation was infilled with sediment. Fair-weather wave-base was set at 10 m and storm wave-base at 15 m.

On a regional scale the modelling run (Figs 5 and 6) compares favourably with the observed stratigraphy. The model simulates the lower progradational tongue, the major overlying flooding event and the ensuing progradational wedge (compare Figs 2 and 5). However, there is one major discrepancy; the modelling run does not simulate the smaller tongue (A in Fig. 2) that progrades a large distance into the basin. The basinward portion of the tongue consists of an incised valley fill (Fig. 2). In the Western Interior basin it is common for incised valleys to feed detached forced regression shorelines situated in more basinward locations (e.g. Pattison & Walker 1994). Therefore a detached shoreface (and hence a potential stratigraphic trap) may lie further down-dip, off the line of section shown in Fig. 2. To aid in prediction of stratigraphic trapping configurations, it is therefore important to ascertain the reason for the progradation of this anomalous wedge (A in Fig. 2).

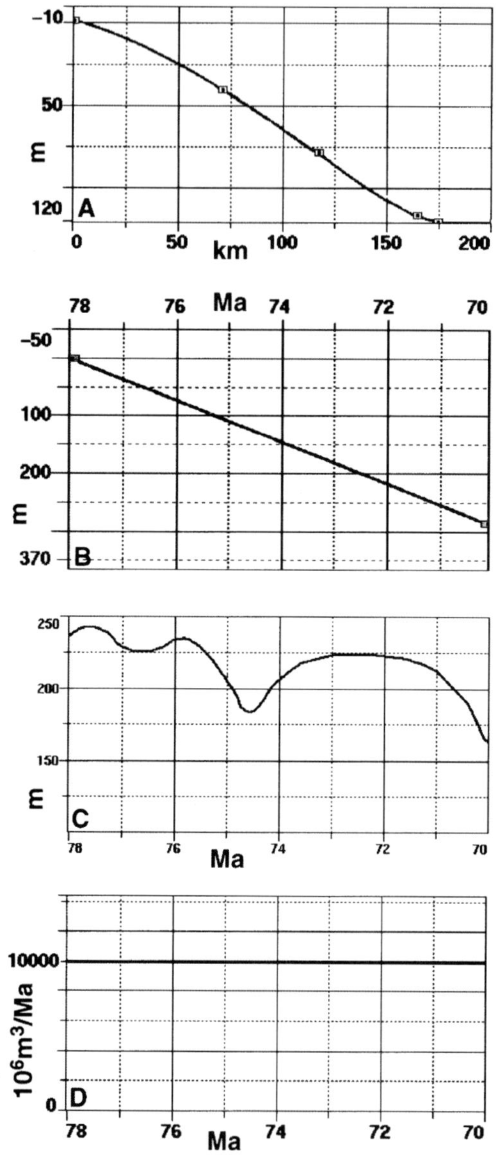

Fig. 4. Primary modelling input variables. Actual values are those used for Modelling Run 1. (**a**) initial basin geometry, (**b**) subsidence curve, (**c**) eustatic sea-level curve, (**d**) sediment supply curve.

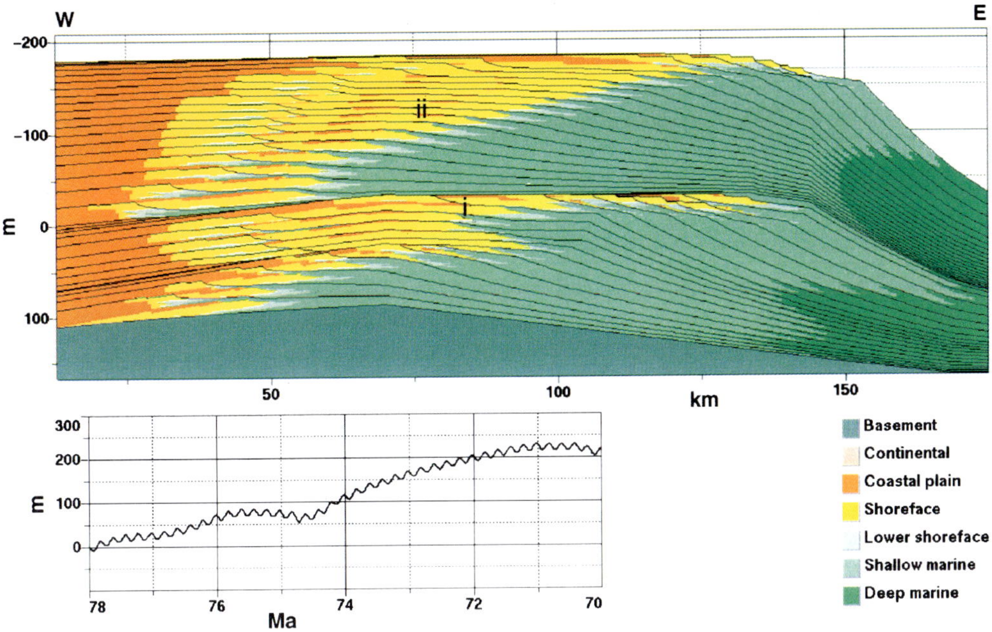

Fig. 5. Modelling run 1, initial model: resultant model and relative sea-level curve. Compare with Fig. 2. Note the favourable comparison between the data and the model. Also note that the *A* tongue in Fig. 2 is not replicated in the model suggesting that an external forcing factor is responsible for the progradation of this tongue. Note the third-order attached lowstand that develops (i) and also the higher frequency fourth-order attached lowstands (ii). Relative sea level curves for all models are derived from a point at 100 km. The term *basement* in the caption refers to the inferred shelf-slope basin topography at the initiation of the model. The overall thickening of the succession towards the west is a result of the relatively higher subsidence rates employed in the west to simulate the foreland basin setting.

The presence of this tongue suggests that it is a product of an external forcing factor not included in the modelling run. Such a factor could be (i) a previously uncharted relatively high amplitude eustatic sea level fall, (ii) a rapid increase in sediment supply rate or (iii) a period of uplift. Each of these hypotheses was tested by running the same model as the initial run but with differing eustatic, sediment supply and subsidence histories respectively.

Modelling run 2: eustatic spike model

All parameters were kept exactly the same as modelling run 1 except for the introduction of a 35 m eustatic sea-level fall at approximately 73.75 Ma (Figs 7 and 8). This resulted in a relative sea-level fall of 25 m (Fig. 7). The relative sea-level fall produced by the eustatic spike resulted in the initial progradation of an attached sharp-based shoreface, which was subsequently detached by ensuing transgressive wave erosion during the rapid return to the previous rate of accommodation space development (Figs 7 and 8).

It is apparent from these results that a previously uncharted, high-frequency (400 000 years), moderate-amplitude (35 m) eustatic sea-level fall could have resulted in the generation of a detached lowstand system in the appropriate location.

Modelling run 3: sediment spike model

All parameters were kept exactly the same as modelling run 1 except for a sediment spike which was introduced into the model at approximately 73.75 Ma (Fig. 9). As no change in the accommodation space regime was introduced, the relative sea-level curve was identical to modelling run 1 (compare Figs 5 and 9). The increase in sediment supply (Fig. 9) produced a substantial basinward shift of the shoreface but only under conditions of *normal* regression (Fig 9). Hence a gradational and not a sharp-based shoreface was generated. A coeval decrease in

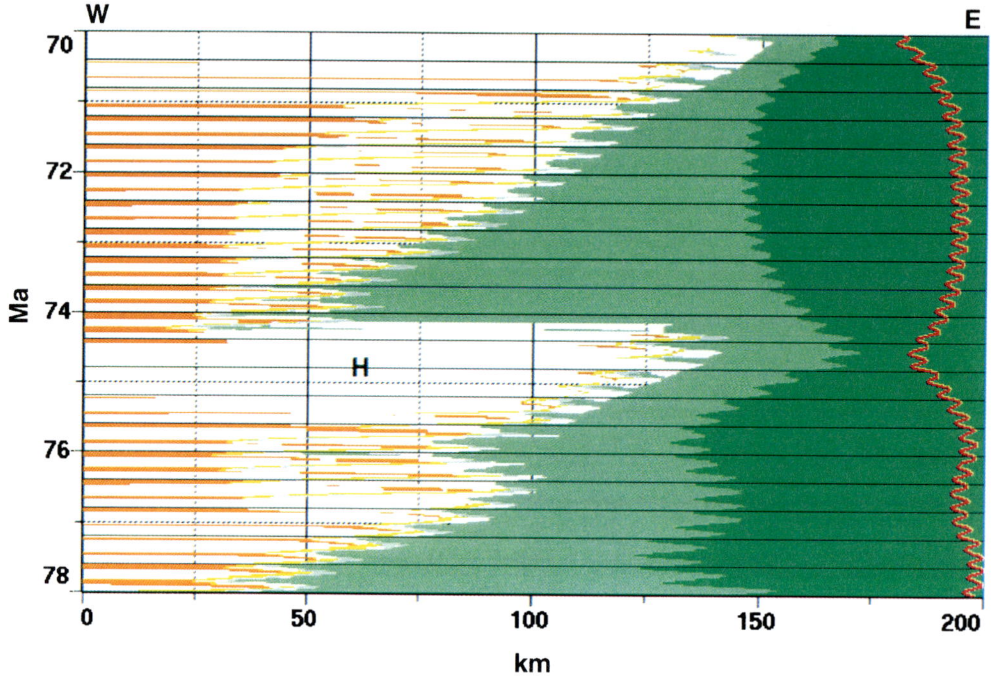

Fig. 6. Modelling run 1, initial model: chronostratigraphic diagram (see Fig. 5). Note the hiatus associated with the 3rd order attached lowstand (H) and the hiati that develop at the base of the high frequency attached lowstands. See Fig. 5 for legend. White represents hiati or erosion. Red line on right represents eustatic sea level curve.

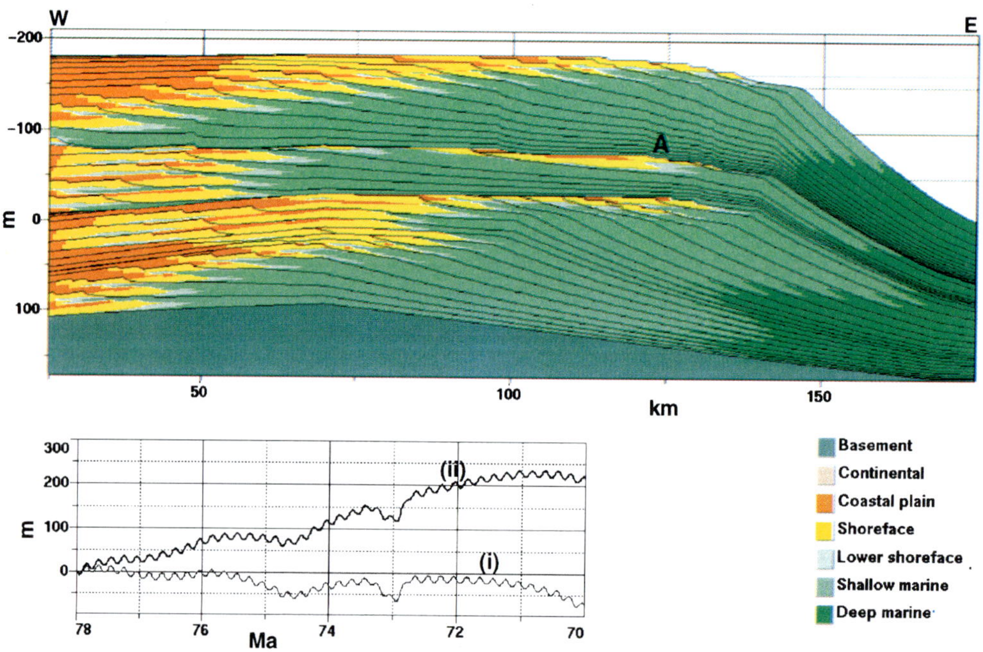

Fig. 7. Modelling run 2, eustatic spike model: resultant model, eustatic sea-level curve (i) and relative sea-level curve (ii). Compare with Fig. 5. Note the development of the fourth-order detached lowstand (A).

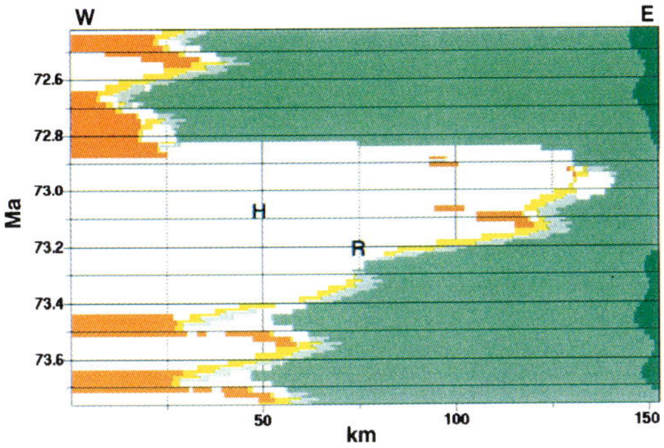

Fig. 8. Modelling run 2, eustatic spike model: chronostratigraphic diagram (see Fig. 7). Note the hiatus associated with the fourth-order detached lowstand (H) and the ravinement surface (R). Compare with Fig. 6. See Fig. 7 for legend.

accommodation space is required to produce a forced regression and hence generate a sharp-based shoreface. The role of sediment supply as a controlling parameter in the generation of forced regressions thus appears to be a secondary one.

Modelling run 4: tectonic spike model

All parameters were kept exactly the same as modelling run 1 except for the introduction of a 30 m tectonic uplift at approximately 73.75 Ma (Fig. 10). This resulted in a relative sea-level fall of 25 m (Fig. 10). The relative sea-level fall produced by the tectonic spike resulted in the initial progradation of an attached sharp-based shoreface which was subsequently detached by ensuing transgressive and regressive wave erosion during the rapid return to the previous rate of accommodation space development (Fig. 11).

It is apparent from these results that a 30 m tectonic uplift over a period of 0.8 Ma could have resulted in the generation of a detached lowstand system in the appropriate location. In a study of the Western Canada Basin, Peper (1993) suggested that such rates could probably be accomplished by isostatic uplift of the foreland during a period of tectonic quiescence in the thrust belt. He appealed to erosion rates in the order of 1 mm per year to reduce the load in the orogenic wedge. These rates of erosion are not excessive when present day erosion rates of 0.5 to 5 mm per year from active orogenic belts

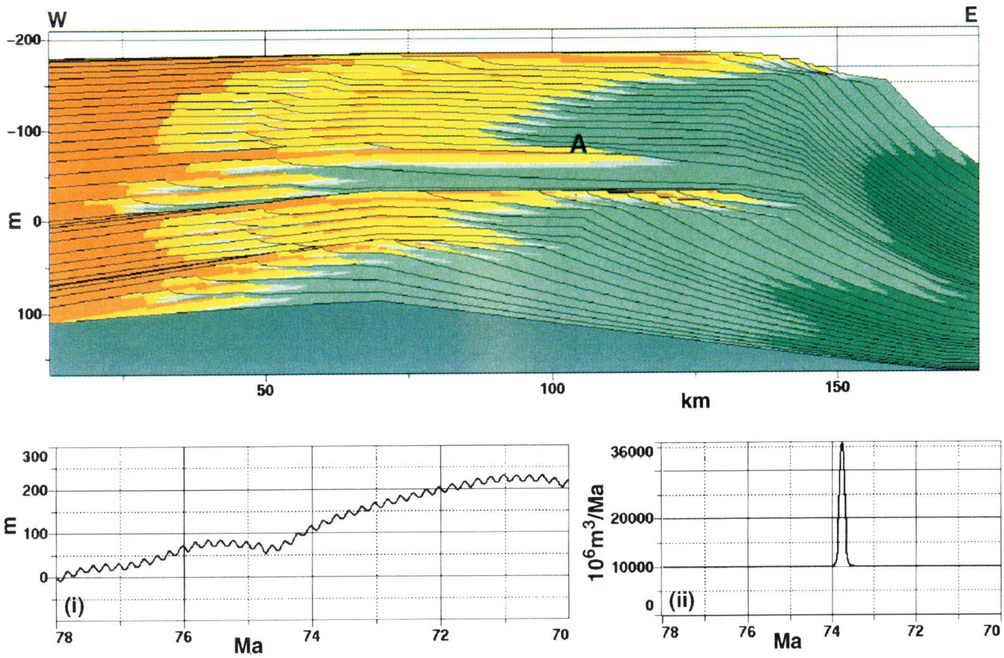

Fig. 9. Modelling run 3, sediment spike model: resultant model, relative sea-level curve (i) and sediment input curve (ii). Note that the shoreface remains attached during the progradation of the tongue (A) and that the shorefaces in this tongue are gradationally based. See Fig. 7 for legend.

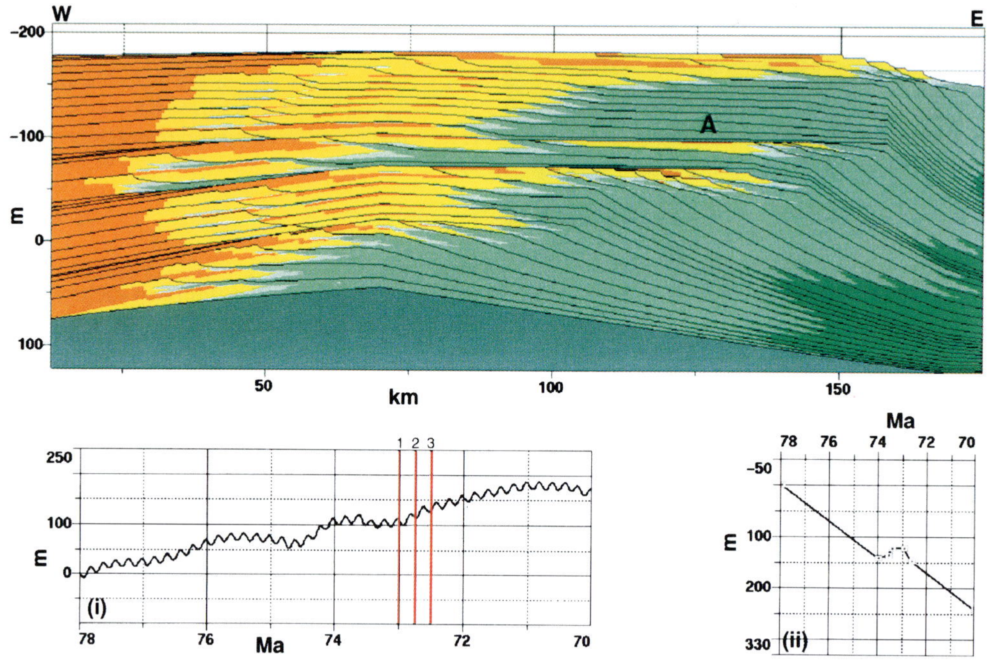

Fig. 10. Modelling run 4, tectonic spike model: resultant model, relative sea-level curve (i) and subsidence curve (ii). Compare with Fig. 5. Note the development of the fourth-order detached lowstand (A). See Fig. 11 for the detailed evolution of this system. Numbers 1, 2 and 3 on the relative sea level curve (i) refer to time steps illustrated in Fig. 11. See Fig. 7 for legend.

(Leeder 1991) are considered. Another method of reducing the load in the thrust belt and hence creating uplift in the foreland is by thrust sheet unloading mechanisms (Jamieson & Beaumont 1988). However, no time-scale for unloading is suggested by the authors. Peper (1993) and Plint *et al.* (1993) both suggest that the migration of the foreland flexural bulge could create localized uplifts on the appropriate timescale.

Discussion

Genesis of attached lowstand systems

During the modelling runs attached lowstands are generated on two temporal and physical scales. A large scale attached lowstand system develops as a product of the initial third order relative sea level fall (Figs 5, 7, 9 and 10). This third-order lowstand remains attached to the underlying deposits as the amount of sediment eroded during the ensuing third order transgression is insufficient to detach the system (Fig. 6).

The smaller-scale attached lowstand sharp-based shorefaces that are observed in outcrop (Fig. 3) are also well simulated by all the modelling runs (Figs 5, 7, 9, 10 and 12a). They are the products of the fourth-order (200 000 year), low-amplitude (8 m) sea-level falls superimposed upon the third-order eustatic sea-level signature. The substantial progradation distances of these individual forced regressive shorefaces (15–30 km) in response to only 8 m of relative sea-level fall is a product of the relatively low shelf gradient typical of this ramp type setting (approximately 1 in 1600; Ainsworth 1994). The relatively constant shelf gradient maintained in the modelling runs explains the observed stacks of attached lowstand deposits (Figs 5, 7, 9, 10 and 12a). Apart from being typical of the Horseshoe Canyon outcrop deposits, similar stacks of fourth order, attached lowstand deposits have been described from other formations in the Western Interior Basin e.g. The Claggett Formation, Wyoming (Johnson 1995), The Eagle Formation, Wyoming (Fitzsimmons 1995) and the Belly River Formation, Alberta (Power & Walker 1996).

To develop these high-frequency attached

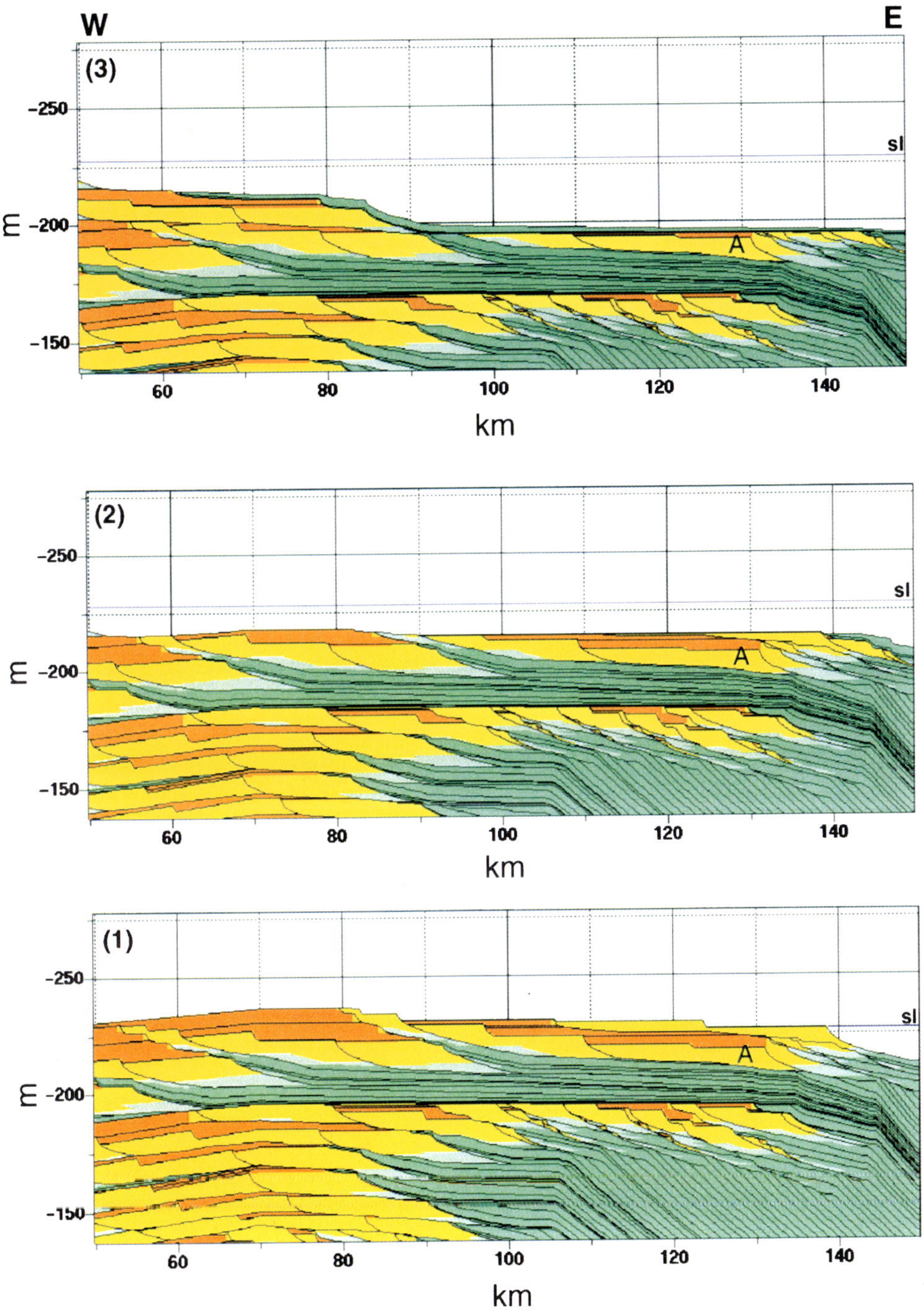

Fig. 11. Modelling run 4, tectonic spike model: time steps through the model illustrating the development of the fourth-order detached lowstand (A). (1) Note initial attachment of lowstand shoreface. (2) Note detachment of shoreface by subsequent transgressive wave erosion processes. (3) Note enhanced erosion by subsequent transgressive and regressive wave erosion. See Fig. 7 for legend and Fig. 10 for location of time steps on relative sea-level curve; sl, sea level.

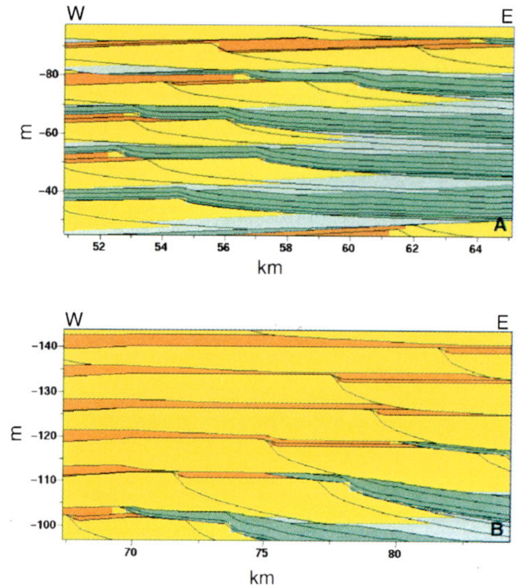

Fig. 12. Detail of sharp-based shoreface deposits. (**a**) Modelled sharp-based shorefaces overlying marine shale. Compare with Fig. 3. Note the sharp-based shorefaces become gradational towards the east. (**b**) Modelled sharp-based shorefaces overlying coastal-plain deposits. The bases of these shorefaces represent composite transgressive/regressive wave erosion surfaces. Compare with Figs 3 and 13. See Fig. 7 for legend.

lowstand deposits, the rates of relative sea-level fall (8 m in 100 000 years) were such for each shoreface that a wave-ravinement surface was cut in front of the prograding system (the sharp-base to the shoreface) whilst sufficient accommodation space was maintained to deposit the shoreface sediment. The shorefaces become gradationally based in a basinward direction. This change from sharp-based to gradationally based occurs when accommodation space increases in response to the ensuing high frequency rises in relative sea level (Fig. 12a).

In the palaeo-landward outcrops, individual sharp-based shorefaces are observed to directly overlie composite transgressive–regressive wave-ravinement surfaces cut into coastal plain deposits (Figs 3 and 13). This relationship is also well modelled in each of the runs (Figs 5, 7, 9, 10 and 12b). These correspondences between the models and the detailed outcrop data suggest a reasonable confidence level in the employed modelling algorithms.

Genesis of detached lowstand systems

A detached lowstand system was generated by both a moderate amplitude (35 m), high frequency eustatic sea-level fall (modelling run 2) and a period of rapid tectonic uplift (modelling run 4). In both cases the forced regressive shoreface was initially attached to the underlying sandy deposits before being subsequently detached by ensuing transgressive and regressive wave erosion (Fig. 11). This result suggests that if sediment is supplied to a shoreline during a relative sea-level fall, then that shoreline will initially be attached to the underlying sandy deposits (Figs 11 and 14a). If detachment of the shoreline occurs then it occurs as a consequence of subsequent transgressive and regressive wave erosion (Figs 11 and 14a). This conclusion seems logical when one considers that in real time there will always be a coastline during a relative sea-level fall and that if sediment is available, then it will always be deposited at that coastline. A sediment bypass zone cannot develop *during* a relative sea-level fall unless sediment supply is cut-off during the fall and there is no sediment available to be deposited at the coastline. Most sediment bypass zones will be *apparent* sediment bypass zones (Ainsworth & Pattison 1994) generated by wave erosion during ensuing transgression and regression (Figs 11 and 14a).

Placement of sequence boundary

The above conclusion regarding the genesis of detached sandbodies has implications for the positioning of the sequence boundary.

If the systematics of Posamentier *et al.* (1991), Ainsworth & Pattison (1994), Fitzsimmons (1995), Kolla *et al.* (1995), Pattison (1995) and this paper are followed then the regressive surface of erosion below the sharp-based shoreface that forms during relative sea-level fall represents the sequence boundary (Fig. 14a). This then suggests that the detached sandbody produced by ensuing wave erosion is a detached *lowstand* sandbody (Fig. 14a).

If, however, the systematics of Ainsworth (1991, 1992, 1994), Hunt & Tucker (1992, 1995), Nummedal (1992*a*, *b*) and Plint (1996) are followed then the sub-aerial exposure surface above the sharp-based shoreface that forms during sea-level fall and lowstand represents the sequence boundary (Fig. 14b). This placement then suggests that the detached sandbody produced by ensuing wave erosion is *not* a detached lowstand sandbody, but that it is part of the falling-stage systems tract (Ainsworth 1991, 1992, 1994; Nummedal 1992*a*, *b*) (Fig. 14b). If this placement of the sequence boundary is favoured then it infers that the majority of detached sandbodies formed during relative sea-level fall and lowstand are actually members of the falling-stage systems tract (Fig. 14b). If the placement of the sequence boundary above the

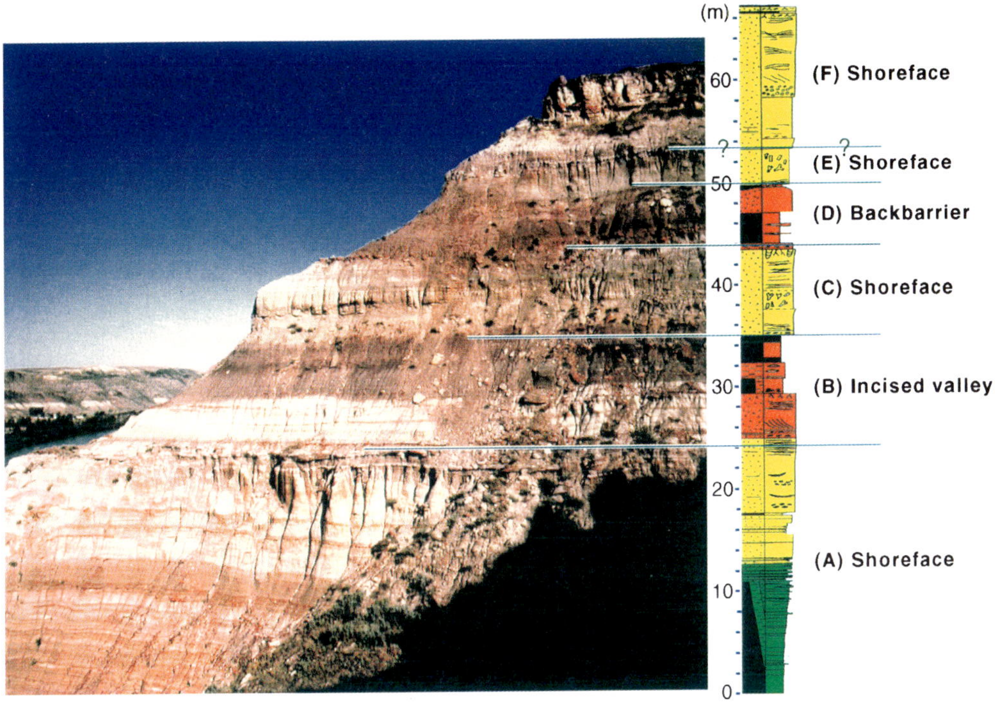

Fig. 13. Sharp-based shorefaces overlying coastal-plain deposits (near RDV–13, see Fig. 3). Compare with detailed model results in Fig. 12b. Modified after Ainsworth (1994).

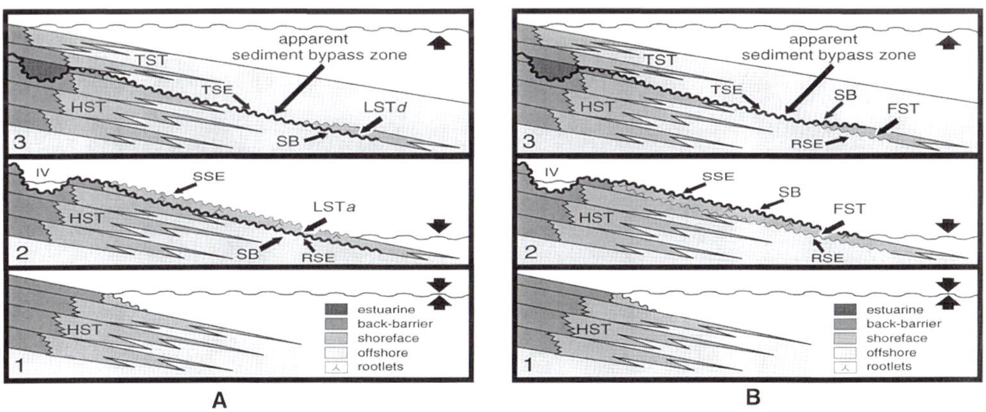

Fig. 14. Generalized model for the generation of detached lowstands and implications of placement of the sequence boundary. Note in A2 that the lowstand shoreface is initially attached during the relative sea level fall and lowstand (LSTa). Detachment of the shoreface (LSTd) occurs during the ensuing transgression via wave erosion processes (A3) to develop an *apparent* sediment bypass zone. In A2 and A3 sequence boundary (SB) is placed at the regressive surface of erosion (RSE). Therefore detached sandbody in A3 is placed within the LSTd. In (B) the SB is placed at the subaerial surface of exposure (SSE). Therefore detached sandbody in B3 is placed within the Falling-stage Systems Tract (FST). Note in B that the SB is substantially modified by wave erosion during transgression and any evidence of previous sub-aerial exposure (rootlets in B2) is removed. TSE, transgressive surface of erosion; IV, incised valley. Modified after Ainsworth & Pattison (1994).

sharp-based shoreface is favoured, then we may have to revise the naming convention for the isolated detached systems (currently termed lowstand systems tracts) and re-name them as falling-stage systems tracts.

These results also suggest that figures that

depict the sequence boundary passing from above an attached sharp-based, forced regressive shoreface (the falling-stage systems tract) to below a sharp-based *detached* lowstand systems tract shoreface (e.g. Plint 1996, fig. 25) are confusing the timing and genesis of these surfaces. If due regard is given to the processes that govern its timing and genesis, the sequence boundary must remain in the location initially picked i.e. it must remain either below the sharp-based shoreface (Fig. 14a) or above it (Fig. 14b) regardless of whether the correlative down-dip, sharp-based shoreface sandbodies concerned are attached or detached.

Conclusions

The following represent the major conclusions derived from this study.

(1) Forced regressive deposits preferentially develop as *attached* lowstand systems.

(2) *Detached* lowstands representing potential stratigraphic traps in most cases develop as *attached* lowstand systems that are subsequently detached by transgressive–regressive wave erosion.

(3) If the sequence boundary is picked below the sharp-based, forced regressive shoreface (attached and detached lowstand systems tract interpretation) or above it (falling-stage systems tract interpretation) it must remain in that position relative to the correlative down-dip, sharp-based shoreface sandbodies regardless of whether those bodies are attached or detached.

(4) In this example, the detached lowstand system generated appears to be a product of either eustatic or tectonic forcing mechanisms.

(5) The role of sediment supply as a controlling parameter in the generation of forced regressions appears to be a secondary one. It may be important in the generation of *detached* lowstand systems where an *apparent* sediment bypass zone can develop if sediment supply is cut-off during a relative sea-level fall and is then resumed at sea-level lowstand.

The authors wish to thank J. Barwis, J. Keating and reviewers J. Howell, K. Taylor and D. Waltham for the numerous constructive comments made on earlier versions of this manuscript. We also wish to acknowledge the contributions of numerous Shell colleagues towards the development of the forward stratigraphic modelling system. In particular, K. de Waal and H. Lammers are thanked for their invaluable assistance. We are grateful to Shell International Exploration and Production B.V. for permission to publish this work.

References

AIGNER, T., DOYLE, M., LAWRENCE, D., EPTING, M. & VAN VLIET, A. 1989. Quantitative modeling of carbonate platforms: Some examples. *In*: CREVELLO, P. D., WILSON, J. L., SARG, J. F. & READ, J. F. (eds) *Controls on carbonate platform and basin development*. Society of Economic Paleontologists and Mineralogists, Special Publications, **44**, 27–37.

——, BRANDENBURG, A., VAN VLIET, A., DOYLE, M., LAWRENCE, D. & WESTRICH, J. 1990. Stratigraphic modelling of epicontinental basins; two applications. *Sedimentary Geology*, **69**, 167–190.

AINSWORTH, R. B. 1991. *Sedimentology and high resolution sequence stratigraphy of the Bearpaw–Horseshoe Canyon transition (Upper Cretaceous), Drumheller, Alberta, Canada*. MSc Thesis, McMaster University, Hamilton, Ontario, Canada.

—— 1992. *Sedimentology and sequence stratigraphy of the Upper Cretaceous, Bearpaw–Horseshoe Canyon transition, Drumheller, Alberta*. American Association of Petroleum Geologists Annual Convention, Calgary, Field Trip Guidebook **#7**.

—— 1994. Marginal marine sedimentology and high resolution sequence analysis; Bearpaw–Horseshoe Canyon transition, Drumheller, Alberta. *Bulletin of Canadian Petroleum Geology*, **42**, 26–54.

—— & PATTISON, S. A. J. 1994. Where have all the lowstands gone? Evidence for attached lowstand systems tracts in the Western Interior of North America. *Geology*, **22**, 415–418.

CHAMBERLAIN, V. E., LAMBERT, R. ST J. & MCKERROW, W. S. 1989. Mesozoic sedimentation rates in the Western Canada Basin as indicators of the time and place of tectonic activity. *Basin Research*, **2**, 189–202.

FITZSIMMONS, R. 1995. *High resolution sequence stratigraphy of the Upper Cretaceous Eagle Formation, Wyoming, USA*. Society of Economic Paleontologists and Mineralogists, Research Conference, Tongues, ridges and wedges. Highstand versus lowstand architecture in marine basins, Casper, Wyoming. Field trip guidebook.

FRANSEEN, E. K., WATNEY, W. L., KENDALL, C. G. ST. C. & ROSS, W. 1991. Sedimentary Modelling: computer simulations and methods for improved parameter definition. *Kansas Geological Survey Bulletin*, **233**, 63–99.

GRIFFITHS, C. M. & HADLER-JACOBSEN, F. 1995. Practical dynamic modelling of clastic basin fill. *In*: STEEL, R. J., FELT, V. L., JOHANNESSEN, E. P. & MATHIEU, C. (eds) *Sequence Stratigraphy on the Northwest European Margins*. Norsk Petroleumforening, Special Publications, **5**, 31–49.

HAQ, B. U., HARDENBOL, J. & VAIL, P. R. 1988. Mesozoic and Cenozoic chronostratigraphy and cycles of sea-level change. *In*: WILGUS, C. K., HASTINGS, B. S., KENDALL, C. G. ST. C., POSAMENTIER, H. W., ROSS, C. A. & VAN WAGONER, C. (eds) *Sea-level changes: an integrated approach*. Society of Economic Paleontologists and Mineralogists, Special Publications, **42**, 71–108.

HUNT, D. & TUCKER, M. E. 1992. Stranded parasequences and the forced regressive wedge systems tract: deposition during base-level fall. *Sedimentary Geology*, **81**, 1–9.

—— & —— 1995. Reply: Stranded parasequences and the forced regressive wedge systems tract: deposition during base-level fall. *Sedimentary Geology*, **95**, 147–160.

IRISH, E. J. W. 1970. The Edmonton Group of south-central Alberta. *Bulletin of Canadian Petroleum Geology*, **18**, 125–155.

JAMIESON, R. A. & BEAUMONT, C. 1988. Orogeny and metamorphism: A model for deformation and pressure-temperature-time paths with applications to the central and southern Appalachians. *Tectonics*, **7**, 417–445.

JOHNSON, S. J. 1995. Forced regressive shallow marine sandbodies: Characterization, formation and heterogeneity. *In*: *Society of Economic Paleontologists and Mineralogists Research Conference, Tongues, ridges and wedges. Highstand versus lowstand architecture in marine basins*, Casper, Wyoming (abstracts).

KOLLA, V., POSAMENTIER, H. W. & EICHENSEER, H. 1995. Discussion: Stranded parasequences and the forced regressive wedge systems tract: deposition during base-level fall. *Sedimentary Geology*, **95**, 139–145.

LAWRENCE, D. T. 1994. Evaluation of eustasy, subsidence and sediment input as controls on depositional sequence boundaries. *In*: WEIMER, P. & POSAMENTIER, H. W. (eds) *Siliciclastic sequence stratigraphy: Recent developments and applications.* American Association of Petroleum Geologists, Memoirs, **58**, 337–367.

——, DOYLE, M., & AIGNER, T. 1990. Stratigraphic Simulation of sedimentary basins: Concepts and calibration. *American Association of Petroleum Geologists Bulletin*, **74**, 273–295.

LEEDER, M. R. 1991. Denudation, vertical crustal movements and sedimentary basin infill. *Geologische Rundschau*, **80**, 441–458.

LERBEKMO, J. F. & COULTER, K. C. 1985. Late Cretaceous to Early Tertiary magnetostratigraphy of a continental sequence: Red Deer Valley, Alberta, Canada. *Canadian Journal of Earth Sciences*, **22**, 567–583.

LEVELL, B. K. & LEU, W. 1993. Stratigraphic basin modelling – recent advances. *In*: DORE, A. G., AUGUSTSON, J. H., HERMANRUD, C., STEWART, D. J. & SYLTA, O. (eds) *Basin modelling: Advances and applications*. Norsk Petroleum-forening, Special Publications, **3**, 71–83.

NUMMEDAL, D. 1992*a*. The falling sea-level systems tract in ramp settings. *In*: *SEPM Theme Meeting*, Fort Collins, Colorado (abstracts), 50.

—— 1992*b*. Sequence stratigraphy in ramp settings – with application to Upper Cretaceous rocks in the San Juan basin (keynote speech). *In*: *New Mexico Geological Society Guidebook, 43rd Field Conference, San Juan Basin*, **IV**, 185–186.

PATTISON, S. A. J. 1995. Sequence stratigraphic significance of sharp-based lowstand shoreface deposits, Kenilworth Member, Book Cliffs, Utah. *American Association of Petroleum Geologists Bulletin*, **79**, 444–462.

—— & WALKER, R. G. 1994. Incision and filling of a lowstand incised valley: Late Albian Viking Formation at Crystal, Alberta, Canada. *Journal of Sedimentary Research*, **B64**, 365–379.

PEPER, T. 1993. *Tectonic control on the sedimentary record in foreland basins. Inferences from quantitative subsidence analyses and stratigraphic modelling.* PhD Thesis, Vrije Universiteit, Amsterdam, The Netherlands.

PLINT, A. G. 1988. Sharp-based shoreface sequences and 'offshore bars' in the Cardium Formation of Alberta: their relationship to relative changes in sea level. *In*: WILGUS, C. K., HASTINGS, B. S., KENDALL, C. G. ST. C, POSAMENTIER, H. W., ROSS, C. A. & VAN WAGONER, J. C. (eds) *Sea-level Changes- an intergrated approach.* Society of Economic Paleontologists and Mineralogists, Special Publications, **42**, 357–370.

—— 1991. High-frequency relative sea-level oscillations in Upper Cretaceous shelf clastics of the Alberta foreland basin: possible evidence for a glacio-eustatic control? *In*: MACDONALD, D. I. M. (ed.) *Sedimentation, tectonics and eustasy: Sea level changes at active margins.* International Association of Sedimentologists Special Publications, **12**, 409–428.

—— 1996. Marine and nonmarine systems tracts in fourth-order sequences in the Early–Middle Cenomanian, Dunvegan Alloformation, northeastern British Columbia, Canada. *In*: HOWELL, J. A. & AITKEN, J. F. (eds) *High Resolution Sequence Stratigraphy: Innovations and Applications*. Geological Society, London, Special Publications, **104**, 159–191.

——, HART, B. S. & DONALDSON, W. S. 1993. Lithospheric flexure as a control on stratal geometry and facies distribution in Upper Cretaceous rocks of the Alberta foreland basin. *Basin Research*, **5**, 69–77.

POSAMENTIER, H. W., ALLEN, G. P., JAMES, D. P. & TESSON, M. 1992. Forced regressions in a sequence stratigraphic framework: concepts, examples, and exploration significance. *American Association of Petroleum Geologists Bulletin*, **76**, 1687–1709.

——, ERSKINE, R. D. & MITCHUM, R. M. JR. 1991. Models for submarine fan deposition within a sequence stratigraphic framework. *In:* WEIMER, P. & LINK, M. H. (eds) *Seismic facies and sedimentary processes of submarine fans and turbidite systems*. Springer Verlag, New York, 127–136.

POWER, B. A. & WALKER, R. G. 1996. Allostratigraphy of the Upper Cretaceous Lea Park – Belly River transition in central Alberta, Canada. *Bulletin of Canadian Petroleum Geology*, **44**, 14–38.

ROSS, W. C., WATTS, D. E. & MAY, J. A. 1995. Insights from stratigraphic modelling: mud-limited versus sand limited depositional systems. *American Association of Petroleum Geologists Bulletin*, **79**, 231–258.

SAUNDERS, T. D. A. 1989. *Trace fossils and sedimentology of the Appaloosa Sandstone:*

Bearpaw–Horseshoe Canyon transition, Dorothy, Alberta. MSc Thesis, University of Alberta, Edmonton, Canada.

SHEPHEARD, W. W. & HILLS, L. V. 1970. Depositional environments Bearpaw–Horseshoe Canyon (Upper Cretaceous) transition zone Drumheller 'Badlands', Alberta. *Bulletin of Canadian Petroleum Geology*, **18**, 166–215.

SHUSTER, M. W. & AIGNER, T. 1994. Two-dimensional synthetic seismic and log cross-sections from forward stratigraphic models. *American Association of Petroleum Geologists Bulletin*, **78**, 409–431.

Tectonically enhanced forced regressions: examples from growth folds in extensional and compressional settings, the Miocene of the Suez rift and the Eocene of the Pyrenees

ROBERT L. GAWTHORPE[1], MATT HALL[1,2], IAN SHARP[1,3], TOM DREYER[4]

[1]*Basin & Stratigraphic Studies Group, Department of Earth Sciences, University of Manchester, Manchester M13 9PL, UK (e-mail: Rob.Gawthorpe@man.ac.uk)*

[2]*Present address: Statoil, Exploration & Production Division, N-4035 Stavanger, Norway*

[3]*Present address: Norsk Hydro Research Centre, Sandsliveien 90, N-5020 Bergen, Norway*

[4]*Norsk Hydro Research Centre, Sandsliveien 90, N-5020 Bergen, Norway*

Abstract: This paper examines the stratal geometries and facies stacking patterns associated with forced regressions around fault-propagation folds in extensional and compressional settings. Case studies are documented from: (i) the Miocene of the Suez rift and (ii) the Eocene of the Ainsa piggyback basin, Pyrenees. Despite the different tectonic settings, the stratal geometries and facies stacking patterns are remarkably similar. Distinctive sharp-based shoreface sandstones, formed as a result of forced regression, were deposited around growth anticlines. The forced regressive shoreface sandstones 'shale-out' rapidly basinward away from the growth anticlines and sit abruptly within offshore mudstones of highstand (HST) and transgressive (TST) systems tracts along the flanks of the growth anticlines.

As fold amplification proceeded, older sandbodies were rotated to dip more steeply, and there is commonly a 2–5° angular difference between successive forced regressive sandbodies. This progressive tilting, coupled with marine erosion during relative sea-level fall has completely removed HST and TST deposits near anticline crests, and led to vertical amalgamation of individual forced regressive sandbodies. The resulting stratal geometries clearly result from the tectonic enhancement of forced regression.

In recent years the original Exxon sequence stratigraphic model, consisting of three systems tracts (e.g. Posamentier & Vail 1988; Van Wagoner *et al.* 1990), has been modified to take into account the nature of shallow-marine sedimentation and stratal surface development during relative sea-level fall (e.g. Plint 1988; Hunt & Tucker 1992; Posamentier *et al.* 1992; Ainsworth & Pattison 1994; Helland-Hansen & Gjelberg 1994; papers in this volume). In particular, several research groups advocate a four-fold systems tract scheme, with the introduction of the forced regressive or the falling stage systems tract to distinguish stratigraphy formed during falling relative sea level from the three systems tracts deposited during relative sea level rise (e.g. Hunt & Tucker 1992, 1995; Plint & Nummedal this volume) (Fig. 1).

Many of the detailed case studies documenting the key stratal surfaces, facies stacking patterns and stratal geometry associated with forced regressive deposits are based on areas dominated by relatively simple regional subsidence, for example the Cretaceous Western Interior Seaway, USA and Canada (e.g. Plint 1988; Posamentier *et al.* 1992; Ainsworth & Pattison 1994; Pattison 1995; papers in this volume by Ainsworth *et al.*, Fitzsimmons & Johnson, Mellere & Steel and Plint & Nummedal). Forced regressive deposits from these areas are commonly bounded below by a surface of regressive marine erosion and above by a surface of subaerial exposure/incision that may be locally modified by subsequent marine transgressive erosion. They form an overall offlapping wedge, with younger parts of the wedge deposited at stratigraphically lower elevations and in more basinward positions (Fig. 1). Proximal areas of the forced regressive wedge are characteristically incomplete in terms of facies stacking patterns and thickness.

To date, there have been few studies of the geometry and architecture of forced regressive deposits from tectonically active areas, where uplift and tilting of the depositional surface around faults and folds would be expected to exert a major influence on sequence development (e.g. Gawthorpe *et al.* 1994; 1997). This paper presents two examples of shallow-marine, forced regressive deposits developed around growing fault-propagation folds: one from an extensional setting and one from a compressional setting. The extensional example comes from the Miocene Nukhul Formation, Suez rift,

From: HUNT, D. & GAWTHORPE, R. L. (eds) *Sedimentary Responses to Forced Regressions*. Geological Society, London, Special Publications, **172**, 177–191. 1-86239-063-0/00/$15.00

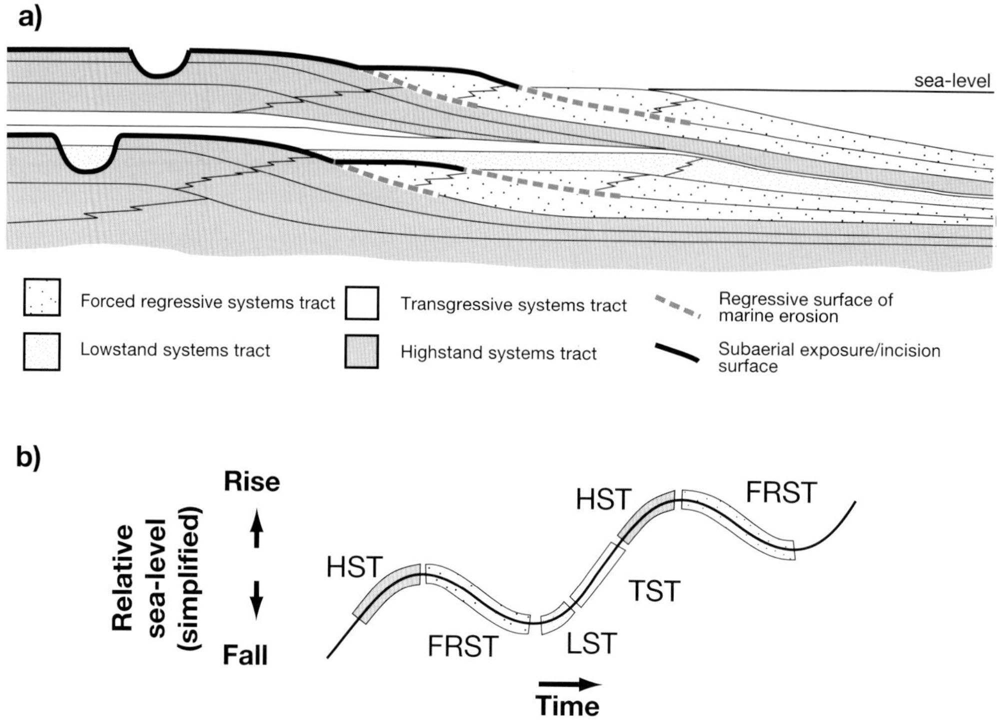

Fig. 1. Idealized sequence stratigraphic model showing the position of the forced regressive systems tract: (**a**) dip-section highlighting systems tracts and key stratal surfaces developed during relative sea-level fall, (**b**) simplified relative sea-level interpretation. This is a four systems tract model based on an Exxon depositional sequence in a ramp setting such as in Van Wagoner *et al.* (1990) and Hunt & Tucker (1992).

Sinai, Egypt and the compressional case study comes from the Eocene Sobrarbe Formation, Ainsa basin, South Pyrenean foreland basin, Spain. For both examples, the facies, facies-stacking patterns and stratal geometries are documented and the influence of structural evolution on forced regression is discussed.

Forced regressions around propagating normal faults: the Nukhul Formation, El Qaa fault block, Suez rift

Geological setting

This example comes from the Miocene early syn-rift succession deposited in the immediate hanging wall of the Baba–Sidri fault in the Suez rift (Fig. 2). This fault defines the NE margin of the El Qaa fault block (Fig. 2), one of two large-scale fault blocks, 10–20 km wide and over 30 km long, that constitute the central dip province of the Suez rift (e.g. Patton *et al.* 1995).

The Baba–Sidri fault is a steeply dipping (60–80°) down-to-the-west normal fault with up to 2 km of throw. Bedding around the fault describes a faulted monocline geometry, with a broad, upward-widening asymmetric syncline, the El Qaa syncline, developed in its hanging wall (Moustafa 1987; Sharp *et al.* in press) (Fig. 2). The El Qaa syncline has a short, steeply dipping eastern limb adjacent to the fault zone and a broad western limb, dipping shallowly back towards the fault (Fig. 2).

The early syn-rift succession displays marked thickness and facies variability around the rift topography and is well exposed in the El Qaa syncline. Where the succession is complete, it consists of a lower fluvial 'red bed' facies association (Abu Zenima Formation), overlain by tidally-influenced and shallow marine sediments of the Nukhul Formation (e.g. Patton *et al.* 1995; Sharp *et al.* in press). Where fully developed, this deepening upward succession may be over 200 m thick. Locally, however, the lower units may be absent, resulting in a substantially

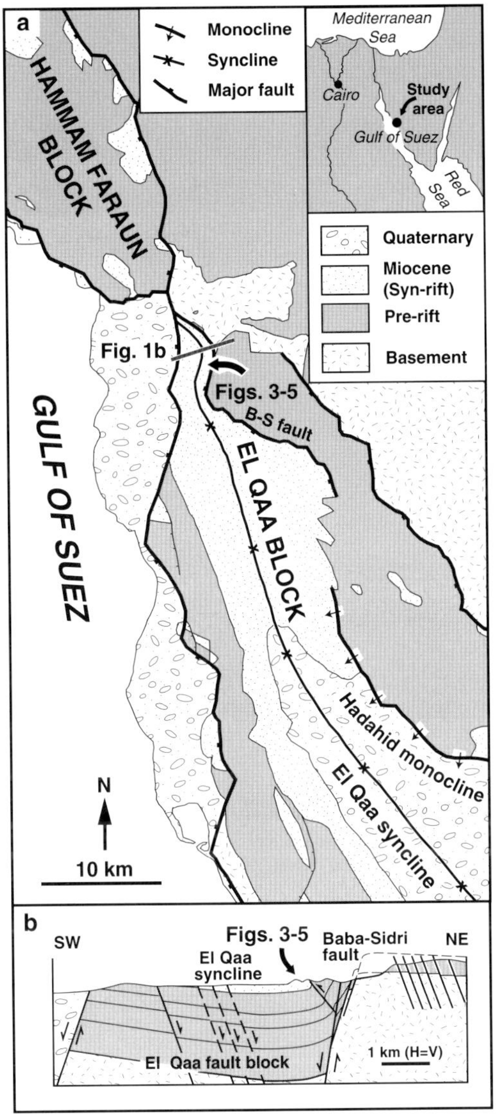

Fig. 2. (**a**) Simplified location of the study area in the northern part of the El Qaa fault block, Suez rift. Inset shows location within the Suez rift; B–S fault, Baba–Sidri fault. (**b**) Cross-section from south Wadi Baba showing the structural style of the Baba–Sidri fault. Note the general faulted monocline geometry of bedding around the fault zone and the asymmetric synclinal form of the syn-rift in the immediate hanging wall of the fault. Also note small displacement faults, some with reverse displacement splaying off the fault zone. Location of Figs 3–5 is shown.

thinner succession, only a few tens of metres in thickness. This case study concentrates on the marine part of the Nukhul Formation, exposed immediately south of Wadi Baba in the northern part of the El Qaa fault block (Figs. 2, 3), where the effects of normal fault growth on sequence development can be clearly identified (Gawthorpe *et al.* 1997).

Facies associations and sequence stratigraphy

In the study area the Nukhul Formation can be subdivided into a number of high-frequency sequences and systems tracts based on downshift and major marine flooding surfaces (Fig. 3). The formation is thickest (>50 m) along the axis of the El Qaa syncline and thins to *c.* 35 m towards the Baba–Sidri fault (Figs 4 & 5). Whereas both the tidal and shallow marine deposits are present in the centre of the El Qaa syncline, where the formation is thickest, only the younger, shallow marine deposits are found on the fold limbs where the formation is much thinner (Fig. 4). This study focuses on the marine part of the Nukhul Formation, and in particular a lower wedge-shaped unit, bounded above and below by regionally correlatable marine flooding surfaces (Figs 3, 4, 5). These key stratal surfaces have been mapped at 1: 10 000 on aerial photographs throughout the northern part of the El Qaa syncline. Internally, the interval is dominated by two main facies associations: (i) grey/green mudstones and (ii) highly bioturbated sandstones. These two facies associations are intercalated and together comprise 95% of the succession; the other 5% consists of large-scale trough cross-bedded, fine-medium grained sandstones (Figs 3, 5). The volumetrically significant facies are described in more detail below.

Grey/green mudstones. These form recessive weathering intervals up to 20 m thick and are composed of laminated to massive mudstones, with subordinate decimetre-thick coarsening-upward units. Bioturbation within the mudstones is commonly indistinct, but locally *Planolites* and *Chondrites* can be identified. The coarsening-upward units grade from mudstone into planar laminated siltstone/fine sandstone, with *Ophiomorpha* and *Thalassinoides* in the uppermost 0.50 m. The upper surfaces are sharp and commonly strongly cemented by ferroan carbonate. When traced towards the axis of the El Qaa syncline, the coarsening-upward units 'shale-out' and the top of the units are marked by carbonate concretions, 50 mm to 0.5 m in diameter.

The mudstone-dominated nature of the succession and the style of bioturbation together

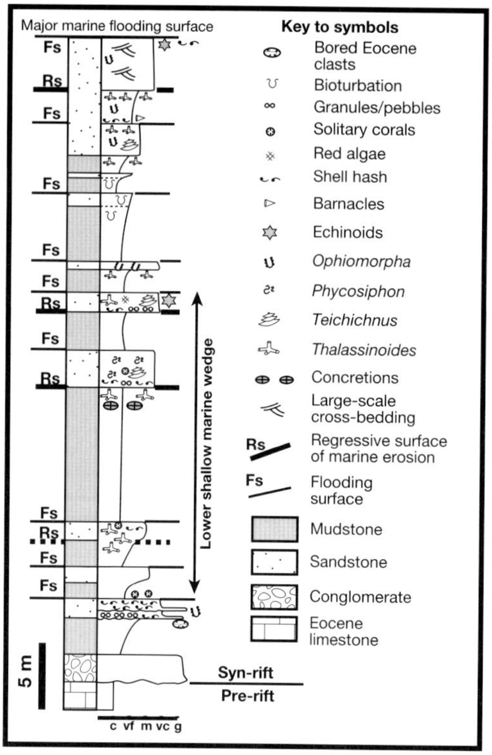

Fig. 3. Graphic sedimentological log for Location C on the correlation panel, Fig. 5 (location of section shown in Fig. 2). The section of interest, the lower shallow marine wedge is indicated. Note the two sharp-based sandbodies near the top of this interval. Major regressive surfaces and marine flooding surfaces are shown.

suggest a low energy marine setting, generally below fair-weather and storm wavebase. As such, the grey/green mudstones are interpreted to have been deposited in an offshore to offshore transition zone environment. Within this general depositional setting, the sharp-topped coarsening-upward units are thought to represent phases of progradation, with the sharp tops representing marine flooding surfaces. Thus the coarsening-upward units are interpreted as distal parasequences (e.g. Van Wagoner *et al.* 1990). The occurrence of both *Thalassinoides* colonization and carbonate cementation, in the uppermost few tens of centimetres of parasequences, reflects early lithification associated with reduced sediment accumulation rates during marine flooding (cf. Taylor & Gawthorpe 1993; Taylor *et al.* 1995).

Highly bioturbated sandstones. These typically form 1–10 m thick packages that have sharp contacts with underlying and overlying units. The sandstones are fine- to medium-grained and are intensely bioturbated by *Ophiomorpha*, *Thalassinoides*, *Teichichnus* and *Planolites* (Fig. 3). Internally, bedding is rarely evident, but locally traces of parallel lamination and hummocky cross-stratification are preserved. The ichnofabrics and trace fossils present are reminiscent of the highly bioturbated lower shoreface sandstones such as the Upper Jurassic Fulmar and Ula formations, North Sea (e.g. Taylor & Gawthorpe 1993; Martin & Pollard 1996). A lower shoreface interpretation for the highly bioturbated sandstones of the Nukhul Formation is also supported by the rare hummocky cross-stratification, which suggests that the sandstones were affected by storm wave activity. The bioturbation reflects colonization during lower energy, fair weather conditions and thus the highly bioturbated nature of the sandbodies suggests prolonged periods of fair weather and/or low-energy storms.

Sequence stratigraphy. There is clear evidence for erosional truncation below many of the sharp-based highly bioturbated sandbodies, and lag deposits, including bored and encrusted concretions derived from underlying mudstones may be present at their base (Figs 3–5). In addition to the lag deposits, the sharp bases to the sandbodies are commonly marked by 10–20 mm diameter *Thalassinoides* burrows extending up to 1.5 m into the underlying lithologies. These burrows are filled with granules and shell hash similar in composition to the lag at the base of the sandbodies and are interpreted as a firmground omission colonization (e.g. Taylor & Gawthorpe 1993). The sharp bases and the intraformational lags, together with the lack of coarsening-upward trends within the underlying mudstones, indicate marked regressive marine erosion at the base of the shoreface sandbodies. This is consistent with sandbody formation as a result of forced regression during times of relative sea-level fall (e.g. Plint 1988; Hunt & Tucker 1992; Posamentier *et al.* 1992).

The tops of the sandbodies are also sharp, marked by an abrupt facies shift into offshore mudstones, and are commonly associated with omission colonization by *Thalassinoides*, early diagenetic cementation and firmground oyster colonization (Figs 3 & 5). As with the tops of the distal parasequences in the mudstone-dominated intervals, biogenic colonization and cementation are interpreted to reflect reduced sediment accumulation rates during development of marine flooding surfaces. The marked landward shift in facies from forced regressive shoreface

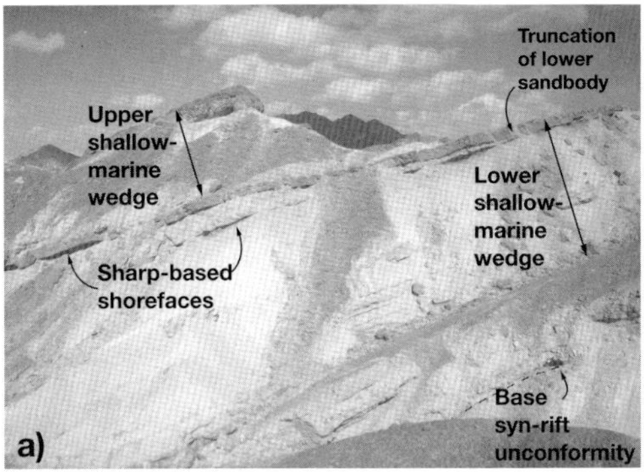

Fig. 4. (**a**) Oblique view of the area to the SW of Location B, Fig. 5, illustrating the truncation of the lower sharp-based sandbody show in Fig. 3 by the overlying sharp-based shoreface sandbody that caps the top of the lower shallow marine wedge. View looking to NW, thickness of the upper shallow marine wedge is approximately 25 m. (**b**) Detail of the base of the sharp-based shoreface that caps the top of the lower shallow marine wedge from N of Wadi Baba. Note the lack of coarsening upward from the mudstone into the shoreface sandbody and the large concretions reworked from the underlying mudstones; the concretions are often heavily bored.

sandstones below to offshore mudstones above suggests that the tops of the sandbodies are major marine flooding surfaces marking the end of maximum regression of the shoreline. Hence they are interpreted as transgressive surfaces (cf. Fig. 1).

Based on the sequence stratigraphic interpretation of the sandbodies, the grey-green mudstone units are bounded below by a transgressive surface and above by a regressive surface of marine erosion. Hence they are interpreted as the deposits of transgressive (TST) and highstand (HST) systems tracts. In the field it is not possible to identify a maximum flooding surface, or facies stacking patterns within the mudstone units which would permit subdivision into the two systems tracts. The up-dip shoreface deposits equivalent to these offshore mudstones are not preserved in the study area. In fact, the transgressive and highstand shorelines were located in what is now the footwall of the Baba–Sidri fault. The effects of footwall uplift mean that no syn-rift deposits are now preserved in the footwall.

Stratal geometry and controls on forced regression

The marine wedge that is the focus of this study dips to the west, away from the Baba–Sidri fault, and thins from >30 m in the core of the El Qaa syncline to <10 m adjacent to the fault over a distance of 2 to 2.5 km (Fig. 5). This thinning is achieved by a combination of erosional truncation below sharp-based, forced regressive sandbodies and onlap onto underlying units (Fig. 5). Erosion at the base of individual sharp-based shorefaces is most pronounced near to the Baba–Sidri fault and there is 2° to 5° angular difference between individual sandbodies, with older sandbodies dipping more steeply than younger ones. Individual shoreface sandbodies tend to be vertically amalgamated adjacent to the Baba–Sidri fault and undergo rapid lateral facies transition into offshore mudstones away from the fault. For example sandbodies of 3–5 m in thickness 'shale-out' into offshore mudstones over a horizontal distance of 200–400 m and sandbodies rarely extend more than 1 km away from the Baba–Sidri fault.

The systematic variations in thickness and facies, together with the marked angularity between successive stratal units, suggest that tectonics played an important role in sequence stratigraphic evolution and the development of forced regressions. In particular, the combination of stratigraphic thinning towards the

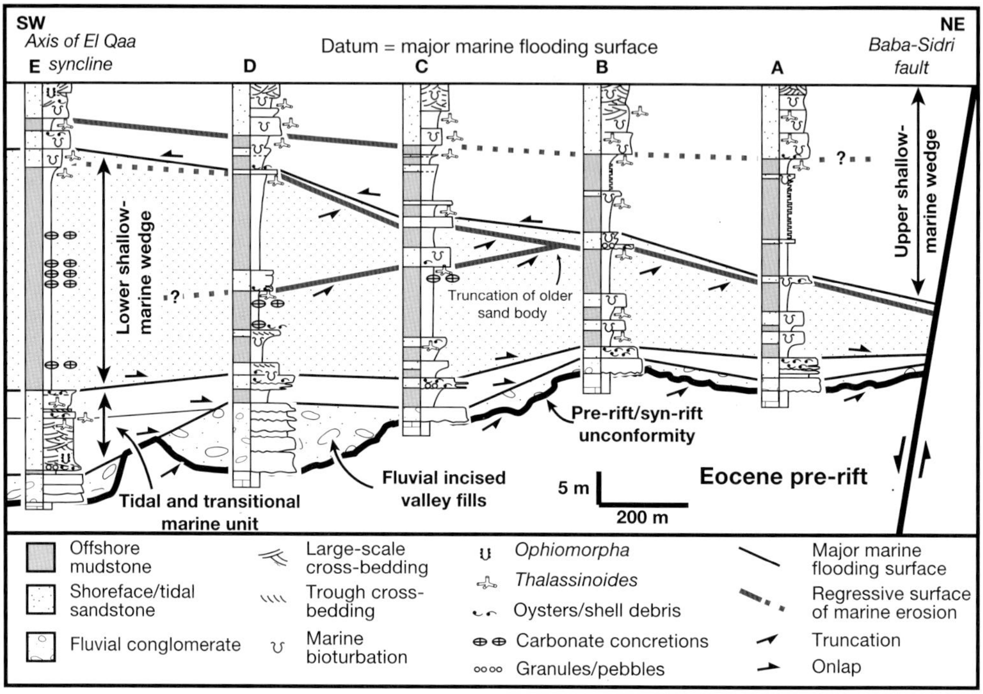

Fig. 5. Correlation of the Nukhul Formation between the Baba–Sidri fault and the El Qaa syncline, Wadi Baba (see Fig. 2 for location). The lower shallow marine wedge is highlighted. Below this, tidal sediments are restricted to the axis of the El Qaa syncline and localized fluvial deposits form broad incised valley fills. Above this, the upper shallow marine wedge clearly thickens towards the Baba–Sidri fault. Only major stratal surfaces are illustrated (after Gawthorpe *et al.* 1997).

Baba–Sidri fault and progressive tilting of syn-rift stratigraphy away from the fault zone indicate that the Baba–Sidri fault was blind during the deposition of the lower marine wedge (Fig. 6) (Gawthorpe *et al.* 1997). Surface deformation above the blind fault tip was associated with growth folds; a west-facing monocline above the fault zone and, to the west, the El Qaa syncline (Fig. 6) (Sharp *et al.* in press).

The growth and tightening of these folds had a major impact on erosion and deposition during forced regression and on sequence development in general. During phases of relative sea-level rise, TST and HST shorelines were located to the east of the study area, in what is now the footwall of the Baba–Sidri fault. Only mud-dominated offshore facies accumulated to the west of the blind fault in the El Qaa growth syncline. During relative sea-level fall, the combination of marine regressive erosion and bed rotation around the growth monocline resulted in complete removal of the up-dip portions of mudstone-dominated highstand and transgressive systems tracts and deposition of sharp-based shoreface sandbodies to the west of the blind fault (Fig. 6). One consequence of this *tectonically enhanced forced regression* is that the forced regressive sandbodies either become amalgamated around the flanks of the growth monocline, or are completely removed by regressive erosion at the base of the next forced regression. The key stratal surfaces, when traced towards the growth monocline, also tend to be either truncated below younger regressive marine erosion surfaces or become amalgamated into composite surfaces recording one or more cycles of relative sea-level change.

Forced regressions around propagating thrust faults: the Sobrarbe Formation, Ainsa basin, South Pyrenean foreland basin

Geological setting

The Ainsa basin is a 20 × 50 km piggyback basin that forms the central element of the South Pyrenean foreland basin, flanked by the

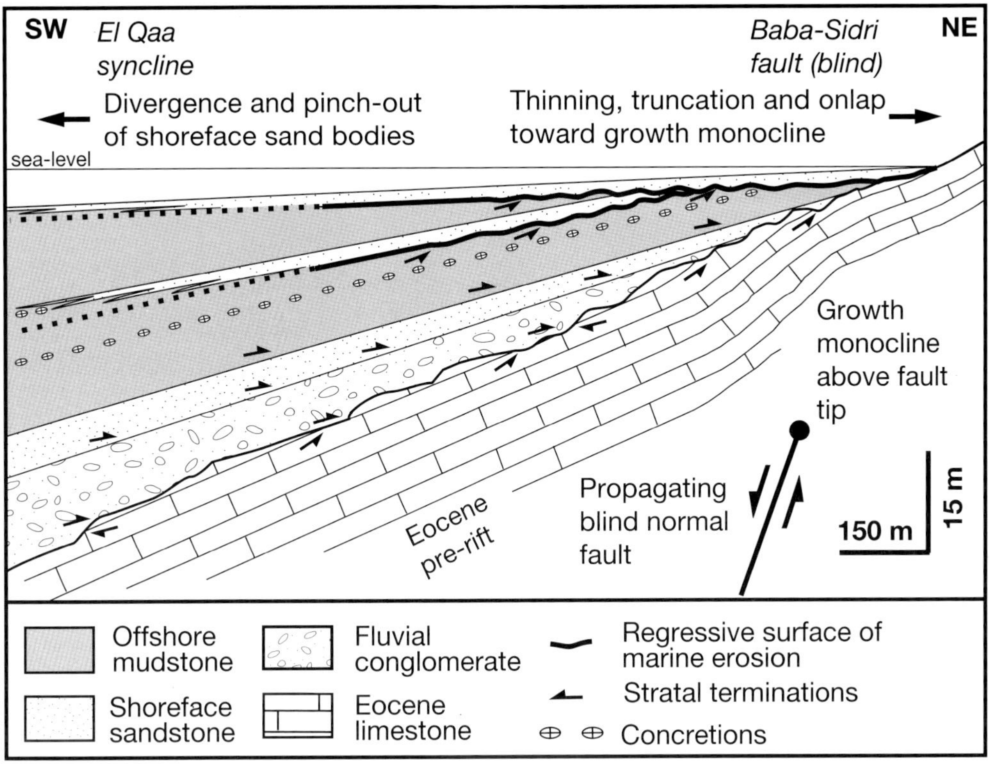

Fig. 6. Tectono-stratigraphic interpretation of the lower shallow marine wedge. The wedge is interpreted to have developed in response to growth folding above a blind propagating normal fault, the Baba–Sidri fault. Note progressive basinward rotation and enhancement of marine regressive erosion surfaces due to fold amplification and up-dip amalgamation of key stratal surfaces.

Tremp–Graus basin to the east and the Jaca–Pamplona basin to the west (Puigdefàbregas 1975; Brunet 1986) (Fig. 7). The Ainsa basin has a synclinal form, its east and west margins defined by the Mediano and Boltaña anticlines respectively. In the central and southern parts of the basin, two intra-basinal anticlines, the Olson and Arcusa anticlines, deform the basin fill (Fig. 7a). The northern margin of the basin is marked by the Cotiella thrust complex; the southern margin is associated with the frontal ramp of the Gavarnie nappe (Sierras Marginales thrust) (Muñoz 1992) (Fig. 7b). Although displacement on the frontal ramp of the Sierras Marginales thrust affected stratal architecture, the syn-depositional stratal geometry is also thought to have been influenced by the growth of the intra-basinal Arcusa anticline (Dreyer *et al.* 1999).

The Sobrarbe Formation is Upper Lutetian in age and forms the central part of an approximately 3000 m thick package of Lutetian to Priabonian sediments that comprise the Campodarbe Group (Bentham & Burbank 1996) (Fig. 7). The Campodarbe Group forms an overall regressive succession with thick lower slope mudstones and turbidite channels at the base (San Vicente Formation), passing upward into the shallow-marine Sobrarbe Formation, which in turn passes into coastal plain and fluvial sediments (Escanilla Formation) (Mutti *et al.* 1988). Regional studies (e.g. de Federico 1981; Puigdefàbregas *et al.* 1991) suggest that, during deposition of the Sobrarbe Formation, sediment was transported into the basin from the southeast across the growing Mediano anticline, with the flanking anticlines and frontal thrust ramp also probably contributing sediment to the basin. The examples of forced regression presented here focus on the highstand sequence set of the Las Gorgas composite sequence, one of four composite sequences that form the Sobrarbe Formation (Dreyer *et al.* 1999).

The Sobrarbe Formation is superbly exposed on the western limb of the basin, and this allows stratal surfaces to be traced unequivocally over

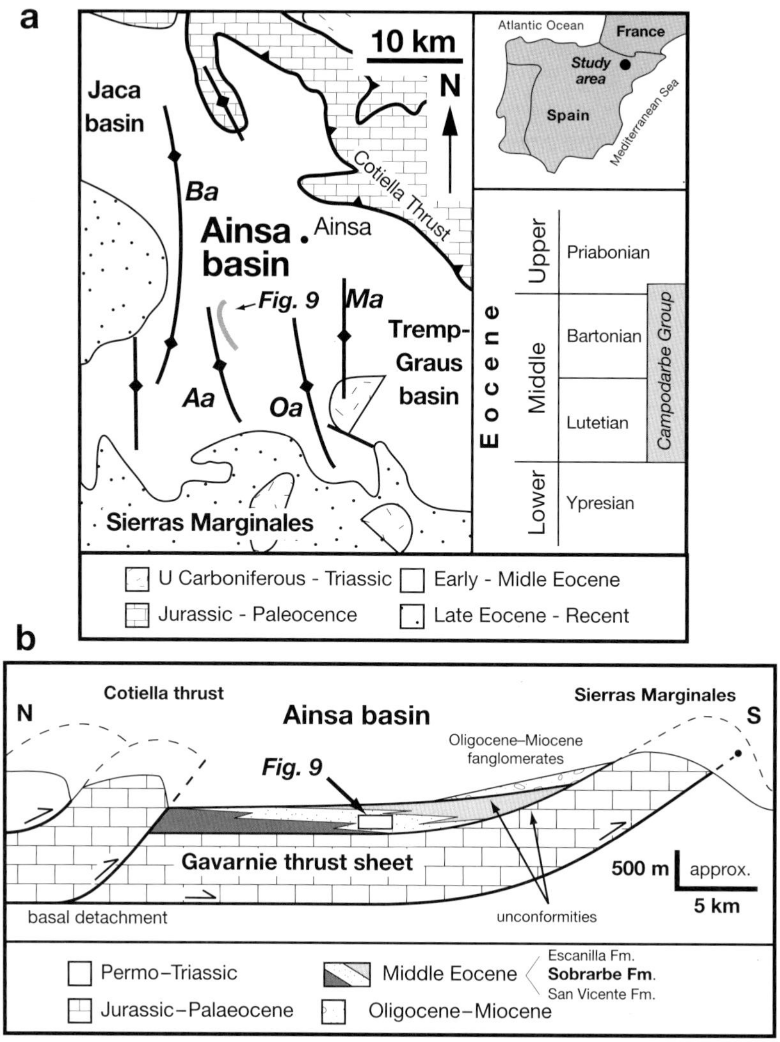

Fig. 7. (**a**) Simplified location and stratigraphy of the Ainsa Basin. The Campodarbe Group is split into the San Vicente Formation, the Sobrarbe Formation (the subject of this case study) and the Escanilla Formation. The location of Fig. 9 is shown. Aa, Arcusa anticline; Ba, Boltaña anticline; Ma, Mediano anticline; Oa, Olson anticline. (**b**) Schematic N–S cross-section along the axis of the Ainsa basin, showing the overall northward progradation of the shallow marine Sobrarbe Formation and the structural position of the studied section.

several kilometres. As a result, the forced regressive components of the stratigraphy can be clearly identified and the influence of compressional growth folds on their development can be determined.

Facies associations and sequence stratigraphy

The part of the highstand sequence set of the Las Gorgas composite sequence documented here consists of three main facies associations: (i) blue-grey mudstones, (ii) highly bioturbated silty sandstones and (iii) trough cross-bedded medium-grained sandstones (Figs 8 & 9). Although a detailed facies analysis is beyond the scope of this paper, the general characteristics of the facies associations are outlined below.

Blue-grey mudstones. This facies association accounts for about 40% of the Las Gorgas composite sequence and occurs in units up to 30 m thick (Figs 8, 9, 10). Although dominated by

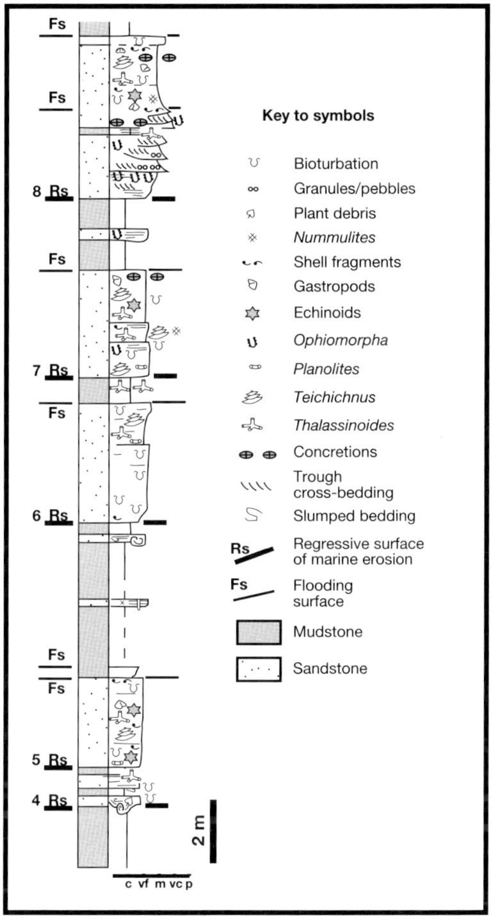

Fig. 8. Representative graphic sedimentological log for the Sobrarbe Formation (log L12, Fig. 9). The log illustrates the overall regression of the Sobrarbe Formation, with slope/offshore mudstones at the base passing upward into highly bioturbated lower shoreface sandstones, which are overlain by trough cross-bedded upper shoreface deposits. Note the sharp-bases and abrupt tops to the shoreface sandbodies. Numbers refer to the main regressive surfaces; major marine flooding surfaces are also indicated.

mudstones, the facies association also contains rare, fining-upward sandstones that are generally a few tens of centimetres thick. These are erosively based and show an upward transition from planar lamination to ripple cross-lamination with sparse bioturbation. The mudstones contain a diverse assemblage of marine fauna, including *Nummulites*, foraminifers, gastropods and bivalves. Bioturbation is locally intense and characterized by *Thalassinoides, Planolites* and *Teichichnus*. The faunal assemblage clearly indicates that the blue-grey mudstones were deposited in a marine environment, and the dominance of mudstone and the trace fossil assemblage are both consistent with a generally low energy environment, below storm wavebase (e.g. Pemberton *et al.* 1992). The thin sandstones correspond to episodes of higher energy and are interpreted as either storm event beds or turbidites. Thus this facies association represents deposition in an offshore to slope environment.

Highly bioturbated silty sandstones. These are grey-brown in colour and make up approximately 35% of the study interval (Figs 8 & 10). The sandstones generally form units 1–8 m thick and are highly bioclastic, containing abundant *Nummulites* and subordinate *Velates* gastropods, *Dentalium* scaphopods and infaunal echinoids (Fig. 10). Bioturbation is generally pervasive, resulting in a lack of preserved sedimentary structures. *Thalassinoides, Planolites* and *Teichichnus* are the most common ichnotaxa, with *Ophiomorpha nodosa* and *Diplocraterion parallelum* also present. This facies association is very similar to the highly bioturbated sandstones of the Nukhul Formation, described in the first case study. The diverse trace- and body-fossil assemblages suggest a shallow marine environment, below fair-weather wavebase, and the sandstones are therefore interpreted as lower shoreface deposits (e.g. Pemberton *et al.* 1992; Taylor & Gawthorpe 1993).

Trough cross-bedded medium-grained sandstones. This facies association accounts for approximately 20% of the study interval, and occurs in units 2–5 m thick that overlie and pass down-dip into the highly bioturbated silty sandstones. Trough cross-bedding typically occurs in sets 0.3–1 m thick, that commonly have a mud-chip or granule lag along set boundaries. Body fossils are restricted to rare *Nummulites* and shell debris, and the sandstones are sparsely bioturbated by *Ophiomorpha*. The sedimentary structures, and body- and trace-fossil assemblages together indicate a high energy shallow marine environment. Taking into account the vertical and lateral relationships with respect to the highly bioturbated silty sandstones, an upper shoreface setting is suggested.

Sequence stratigraphy. Figure 8 illustrates the abrupt nature of many of the facies transitions in the Sobrarbe Formation. In particular, in the proximal parts of the depositional system, the shoreface sandbodies overlie offshore/slope deposits and have sharp, angular basal contacts

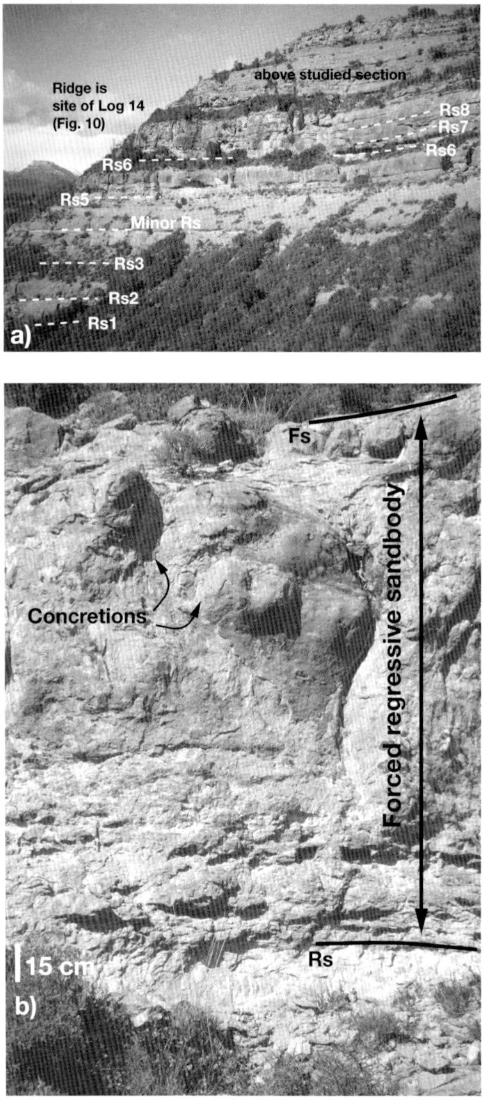

Fig. 9. (**a**) General view of the studied interval of the Sobrarbe Formation around Log 14 (Fig. 10). The main regressive surfaces of marine erosion are labelled as in Fig. 10. (**b**) Detail of forced regressive sandbody composed of highly bioturbated silty sandstone facies association; Sequence 6. The base of the sandbody is a sharp, erosional contact (Rs) with omission colonization by *Thalassinoides* penetrating down into the underlying offshore mudstones. The top of the sandstone, a marine flooding surface (Fs) is associated with bioclastic accumulation and the nodular weathering pattern is related to early carbonate cementation.

(Fig. 9b). Omission colonisation, indicated by preferentially cemented *Thalassinoides*, commonly penetrates as much as 1 m below the base of the sharp-based shoreface sandstones into the underlying mudstones (Figs 8 & 9b). Granule lags and mud-chips are also common in the lowermost 0.2–0.5 m of the sharp-based shoreface sandbodies. These relationships suggest that the sharp-based shorefaces are forced regressive deposits that formed during relative sea-level fall (e.g. Plint 1988; Hunt & Tucker 1992; Posamentier *et al.* 1992; Ainsworth & Pattison 1994).

The tops of the forced regressive shoreface sandbodies, unless truncated by an overlying regressive erosion surface, are marked by an abrupt landward shift in facies that typically places offshore/slope mudstones directly on top of the shoreface sandstones (Figs 8, 9 and 10). In addition, the tops of the sandbodies are associated with bioclastic accumulations, *Thalassinoides* omission colonization and early diagenetic carbonate cementation (Fig. 8). Such colonization and cementation are features commonly associated with reduced sediment accumulation (e.g. Van Wagoner *et al.* 1990; Taylor & Gawthorpe 1993; Taylor *et al.* 1995) and suggest that the tops of the shoreface sandbodies represent marine flooding surfaces. The pronounced landward shift in facies and the change in facies stacking patterns on top of forced regressive sandbodies suggest that the overlying mudstone-dominated units represent transgressive and highstand systems tracts.

Stratal geometry and controls on forced regression

Figure 10 summarizes the stratal geometry and facies stacking patterns in a N–S section that approximates to a depositional dip section. A marked basinward (northward) expansion characterizes the succession and is associated with a northward divergence of stratal surfaces into the basin. For example, in log L18 at the southern (proximal) end of the section, the sequence above regressive surface 3 has a preserved thickness of 10 m, whereas in log L7 it is 55 m thick (Fig. 10). If we consider the whole of the studied interval, the total down-dip expansion is about 400% over a distance of 4 km. This northward expansion in sequence thickness coincides with an increase in the mudstone: sandstone ratio as the shoreface sandstones 'shale-out' basinward into offshore/slope mudstones (Fig. 10).

The shoreface sandbodies, when traced updip (to the south), become sharp-based and show increasing amounts of erosional truncation

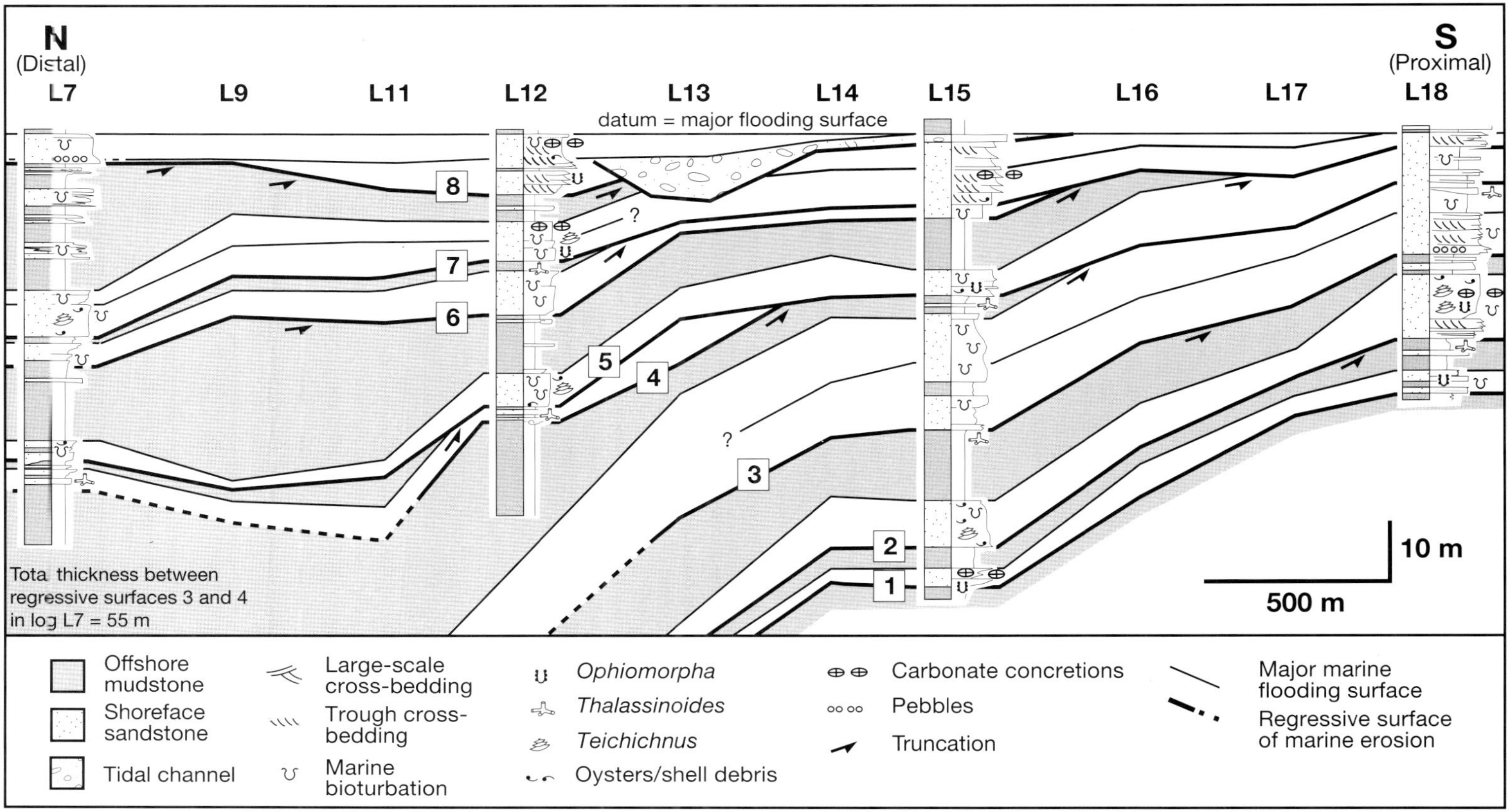

Fig. 10. Correlation panel from the western side of the Ainsa Basin (see Fig. 7 for location). Stratal geometry is divergent from right to left (S to N) and this is associated with an increase in mudstone-dominated offshore mudstones. Note the increase in truncation of underlying regressive surfaces of marine erosion to the S, leading to amalgamation and even complete truncation of several stratal packages. Numbers on the correlation refer to the main regressive surfaces of marine erosion; numbers prefixed by L along the top of the correlation refer to graphic logs used in its construction.

at their bases (Fig. 10). In most cases this results in complete removal of the mudstone-dominated TST and HST intervals and amalgamation of individual sharp-based shoreface sandbodies (e.g. L18, Fig. 10). In addition, the sharp bases of the forced regressive shoreface sandbodies form angular unconformities with up to 5° difference in dip above and below the surfaces.

The marked angularity between successive forced regressive sandbodies, coupled with a progressive decrease in dip of younger sandbodies suggest that the blind thrusts beneath the Sierras Marginales thrust and the structures beneath the Arcusa anticline were active during deposition of the Sobrarbe Formation (Dreyer *et al.* 1999). The propagation of these faults caused progressive rotation and folding of the depositional surface, that in turn led to marked variations in accommodation development over horizontal distances of only a few kilometres. This accounts for the overall wedge-shaped geometry of the sequences and the southward thinning toward the fold/thrust (Fig. 11). Thus it would appear that sufficient accommodation space was generated down the flanks of the anticline to allow the development of relatively deep water mudstone-dominated environments, whereas along the crest of the anticline accommodation was limited.

Variation in accommodation development and associated tilting of the depositional surface had a major impact on the geometry of the sequences and systems tracts, particularly the connectivity of the forced regressive sandbodies. It is not clear whether the relative sea-level falls that generated the forced regressive sandbodies were mainly due to local tectonic uplift or regional/global sea-level falls. However, the combination of relative sea-level fall and tectonic rotation around the developing growth anticline led to the complete removal of the up-dip portions of transgressive and highstand systems tracts of older sequences (Fig. 11). Thus, in the southern, up-dip part of the succession (e.g. log L18, Fig. 10), regressive surfaces of marine erosion cut down into older forced regressive deposits, producing multi-storey forced regressive shoreface sandbodies 10–30 m thick (Fig. 10). As these composite sandbodies are traced down-dip away from the growth anticlines they split into a number of sharp-based, forced regressive shoreface sandbodies separated by transgressive and highstand mudstones. Thus many of the key stratal surfaces in up-dip locations are composite surfaces that record one or more episodes of relative sea-level fall and rise. These composite surfaces are clearly indicated in Fig. 10, for example, regressive surfaces 5 and 7.

Summary

Although the two case studies of forced regression come from markedly contrasting tectonic settings, their stratigraphy displays strong similarities in terms of stratal geometry and facies stacking patterns. Perhaps this is not surprising, as deformation of the depositional surface in both cases was dominated by progressive rotation of the limbs of growth folds. This rotation and progressive basinward tilting of the depositional surface led to an overall stratal geometry characterized by a basinward divergent wedge within which progressively younger forced regressive wedges display shallower dips. When combined with relative sea-level fall, fold limb rotation led to the development of tectonically-enhanced forced regressive wedges, the basal surfaces of which are marked by angular unconformities. Amplification of growth anticlines also led to a rather unusual spatial partitioning of systems tracts. In both examples, highstand and transgressive systems tract deposits are only preserved in basinal settings and coarse-grained deposits of the forced regressive systems tract are preferentially preserved up-dip (around the crests of growth anticlines). This is the opposite of what would normally be predicted in sequence stratigraphic models (cf. Fig. 1), but is explained by the complete removal of earlier-formed transgressive and highstand deposits around the crests of growth anticlines during relative sea-level fall.

Furthermore, due to enhanced regressive marine erosion, many key stratal surfaces are truncated around the crests of growth anticlines. Thus, although transgressive surfaces, maximum flooding surfaces and regressive surfaces of marine erosion may be identified down the flanks of growth folds, only composite surfaces marked by a basinward shift in facies are preserved around the crests. If there is limited aggradation around fold crests, as is the case in the two examples presented here, then these composite surfaces may represent several depositional sequences preserved down-dip within the basin. Although these so-called 'progressive' or 'growth' unconformities have been described from growth strata associated with compressional settings (e.g. Riba 1976; Ford *et al.* 1997), they are poorly documented from around normal fault zones.

The characteristics of the bounding stratal surfaces and internal facies stacking patterns of the forced regressive wedges presented in this paper are similar to previously documented examples of forced regressions in shallow marine settings (e.g. Plint 1988; Posamentier *et al.* 1992;

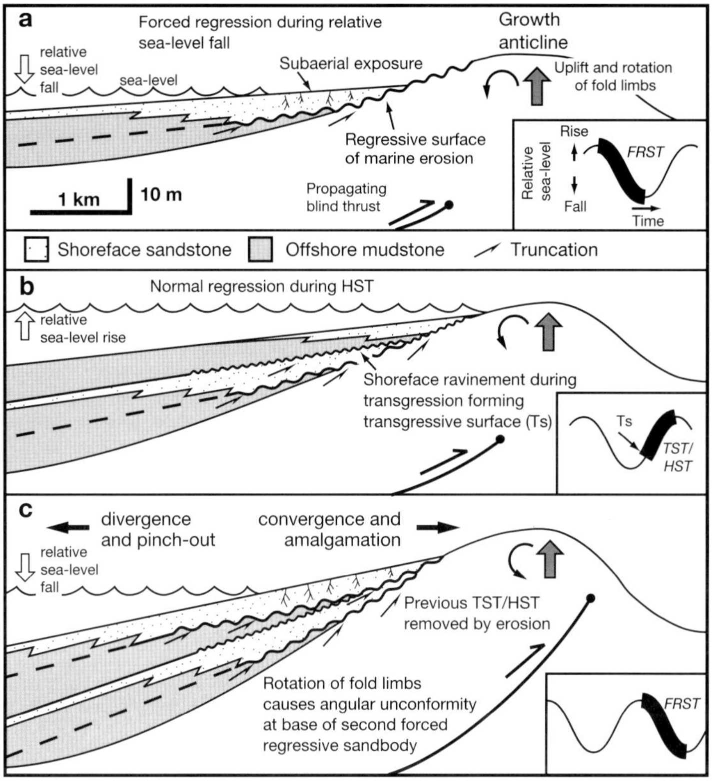

Fig. 11. Tectono-stratigraphic interpretation of the development of the high frequency sequences in the Las Gorgas composite sequence of Dreyer *et al.* (1999). Scales are approximate; insets show relative sea-level interpretation. (**a**) Relative sea-level fall cuts a regressive surface of marine erosion and is associated with deposition of a forced regressive shoreface sandbody. The top of the shoreface sandbody becomes a subaerial exposure surface as sea-level continues to fall. (**b**) Relative sea-level rise results in an abrupt landward shift in facies and is associated with transgressive erosion. Near the end of relative sea-level rise, progradation of a highstand shoreface occurs by normal regression. (**c**) As a result of ongoing tectonic uplift and rotation around the growth anticline, erosion during the next relative sea-level fall forms an angular unconformity with respect to the previous high frequency sequence. The previous TST/HST are partly removed and a second sharp-based forced regressive sandbody is deposited.

Ainsworth & Pattison 1994; Pattison 1995; and papers in this volume by Ainsworth *et al.*, Fitzsimmons & Johnson, Mellere & Steel and Plint & Nummedal), but the overall stacking patterns and stratal geometries are markedly different. The differences are largely due to the contrasting length-scales of deformation. In the two case studies presented here, variations in subsidence/uplift and rotation of bedding are due to local fault propagation folding and occur on a kilometre-scale. In contrast, the examples cited from the literature largely come from areas dominated by flexural subsidence, where variations in subsidence and tilting occur over horizontal distances of many tens of kilometres. These differences in the tectonic regime not only affected the overall stacking patterns and stratal geometries, but also had a marked impact on the length scales of the forced regressive sandbodies. In the examples presented here, from around growth anticlines, the forced regressive sandbodies are typically only a few hundred metres to a few kilometres in length (Figs 5 & 10). In contrast, the examples from the Cretaceous Western Interior Seaway tend to be longer, with those presented by in this volume by Ainsworth *et al.*, Fitzsimmons & Johnson, Mellere & Steel and Plint & Nummedal commonly several tens of kilometres in length.

R.L.G. and I.S. acknowledge financial support from a 1995 Realising Our Potential Award (GR3/R95/27), from Norsk Hydro and Amoco. M.H. would like to thank the Natural Environmental Research Council for supporting his PhD research in the Ainsa Basin (GT4/93/219/G). The authors would like to thank J. R. Underhill, C. Puigdefàbregas, S. Gupta, A. M. Taylor, J. Dolson and S. Gooda who helped with work in both the Suez rift and the Ainsa basin.

References

AINSWORTH, R. B. & PATTISON, S. A. J. 1994. Where have all the lowstands gone? Evidence for attached lowstand systems tracts in the Western Interior of North America. *Geology*, **22**, 415–418.

——, BOSSCHER, H. & NEWALL, M. J. 2000. Forward modelling of forced regressions: evidence for the genesis of attached and detached lowstand systems. *This volume*.

BENTHAM, P. & BURBANK, D. W. 1996. Chronology of Eocene foreland basin evolution along the western oblique margin of the south-central Pyrenees. *In*: FRIEND, P. F. & DABRIO, C. J. (eds) *World & Regional Geology Series*, 6: *Tertiary Basins of Spain: The Stratigraphic Record of Crustal Kinematics*. Cambridge University Press, Cambridge, 144–152.

BRUNET, M. F. 1986. The influence of the Pyrenees on the development of adjacent basins. *Tectonophysics*, **129**, 343–354.

DE FEDERICO, A. 1981. *La sedimentacion de talud en el sector occidental de la cuenca Palaeogena de Ainsa*. PhD dissertation, Universitat Autonoma de Barcelona, Spain.

DREYER, T., CORREGIDOR, J., ARBUES, P. & PUIGDEFÀBREGAS, C. 1999. Architecture of the tectonically-influenced Sobrarbe delta complex in the Ainsa Basin, northern Spain. *Sedimentary Geology*, **127**, 127–169.

FITZSIMMONS, R. & JOHNSON, S. 2000. Forced regressions: recognition, architecture and genesis in the Campanian of the Bighorn Basin, Wyoming. *This volume*.

FORD, M., WILLIAMS, E. A., ARTONI, A., VERGÈS, J. & HARDY, S. 1997. Progressive evolution of fault-related fold pairs from growth strata geometries, Sant Llorenc de Morunys, SE Pyrenees. *Journal of Structural Geology*, **19**, 413–441.

GAWTHORPE, R. L., FRASER, A. J. & COLLIER, R. E. LL. 1994. Sequence stratigraphy in active extensional basins: implications for the interpretation of ancient basin-fills. *Marine and Petroleum Geology*, **11**, 642–58.

——, SHARP, I., UNDERHILL, J. R. & GUPTA, S. 1997. Linked sequence stratigraphic and structural evolution of propagating normal faults. *Geology*, **25**, 795–798.

HELLAND-HANSEN, W. & GJELBERG, J. G. 1994. Conceptual basis and variability in sequence stratigraphy: a different perspective. *Sedimentary Geology*, **92**, 31–52.

HUNT, D. W. & TUCKER, M. E. 1992. Stranded parasequences and the forced regressive wedge systems tract: deposition during base level fall. *Sedimentary Geology*, **81**, 1–9.

—— & —— 1995. Reply: Stranded parasequences and the forced regressive wedge systems tract: deposition during base level fall. *Sedimentary Geology*, **95**, 147–160.

MARTIN, M. & POLLARD, J. E. 1996. The role of trace fossil (ichnofabric) analysis in the development of depositional models of the Upper Jurassic Fulmar Formation of the Kittiwake Field (Quadrant 21 UKCS). *In*: HURST, A., JOHNSON, H. D., BURLEY, S. D., CANHAM, A. C. & MACKERTICH, D. S. (eds) *Geology of the Humber Group: Central Graben and Moray Firth, UKCS*. Geological Society, London, Special Publications, **114**, 163–185.

MELLERE, D. & STEEL, R. 2000. Style contrast between forced regressive and lowstand/transgressive wedges in the Campanian of south-central Wyoming. *This volume*.

MOUSTAFA, A. R. 1987. Drape folding in the Baba–Sidri area, eastern side of the Suez rift. *Egypt Journal of Geology*, **31**, 15–27.

MUÑOZ, J. A. 1992. Evolution of a continental collision belt: ECORS-Pyrenees crustal balanced cross-section. *In:* MCCLAY, K. R. (ed.) *Thrust Tectonics*. Chapman & Hall, London, 235–246.

MUTTI, E., SEGURET, M. & SGAVETTI, M. 1988. *Sedimentation and deformation in Tertiary sequences in the southern Pyrenees*. American Association of Petroleum Geologists Mediterranean Basins Conference Field Guide, **7**.

PATTISON, S. A. J. 1995. Sequence stratigraphic significance of sharp-based lowstand shoreface deposits, Kenilworth Member, Book Cliffs, Utah. *American Association of Petroleum Geologists Bulletin*, **79**, 444–462.

PATTON, T. L., MOUSTAFA, A. R., NELSON, R. A. & ABDINE, S. A. 1995. Tectonic evolution and structural setting of the Suez Rift. *In:* LANDON, S. M. (ed.) *Interior Rift Basins*. American Association of Petroleum Geologists Memoirs, **59**, 7–55.

PEMBERTON, S. G., MACEACHERN, J. A. & FREY, R. W. 1992. Trace fossil facies models: environmental and allostratigraphic significance. *In*: WALKER, R. G. & JAMES, N. P. (eds) *Facies Models: Response to Sea-Level Change*. Geological Association of Canada, 47–72.

PLINT, A. G. 1988. Sharp-based shoreface sequences and 'offshore bars' in the Cardium Formation of Alberta; their relationship to relative changes in sea-level. *In*: WILGUS, C. K., HASTINGS, B. S., KENDALL, C. G. ST. C., POSAMENTIER, H. W., ROSS, C. A. & VAN WAGONER, J. C. (eds) *Sea-level changes: An integrated approach*. SEPM Special Publications, **42**, 357–370.

PLINT, A. G. & NUMMEDAL, D. 2000. The falling stage systems tract: recognition and importance in sequence analysis. *This volume*.

POSAMENTIER, H.W. & VAIL, P. R. 1988. Eustatic controls on clastic deposition II – Sequence and systems tracts models. *In*: WILGUS, C. K., HASTINGS, B. S., KENDALL, C. G. ST. C., POSAMENTIER, H. W., ROSS, C. A. & VAN WAGONER, J. C. (eds) *Sea level changes: An integrated approach*. SEPM Special Publications, **42**, 125–154.

——, ALLEN, G. P., JAMES, D. P. & TESSON, M. 1992. Forced regressions in a sequence stratigraphic framework: concepts, examples and exploration significance. *American Association of Petroleum Geologists Bulletin*, **76**, 1687–1709.

PUIGDEFÀBREGAS, C. 1975. La sedimentacion molasica en la Cuenca de Jaca. *Monografia del Instituto de Estudios Pirineos*, **104,** 1–88.

——, NIJMAN, W. & MUÑOZ, J. A. 1991. *Alluvial deposits of the successive foreland basin stages and their relation to the Pyrenean thrust sequences*. 4th International Conference on Fluvial Sedimentology, Guidebook Series, Servei Geologic de Catalunya.

RIBA, O. 1976. Syntectonic uncomformities of the Alto Cardener, Spanish Pyrenees: a genetic interpretation. *Sedimentary Geology*, **15**, 213–233.

SHARP, I. R., GAWTHORPE, R. L., UNDERHILL, J. R., GUPTA, S. in press. Fault-propagation folding in extensional settings. Examples of structural style and sedimentary response from the Suez Rift, Sinai, Egypt. *Geological Society of America Bulletin*, in press.

TAYLOR, A. M. & GAWTHORPE, R. L. 1993. Application of sequence stratigraphy and trace fossil analysis to reservoir description: examples from the Jurassic of the North Sea. *In*: PARKER, J. R. (ed.) *Petroleum Geology of Northwest Europe: Proceedings of the 4th Conference.* Geological Society, London, 317–335.

TAYLOR, K. G., GAWTHORPE, R. L. & VAN WAGONER, J. C. 1995. Stratigraphic control on laterally persistent cementation, Book Cliffs, Utah. *Journal of the Geological Society, London,* **152**, 225–228.

VAN WAGONER, J. C., MITCHUM, R. M., CAMPION, K. M. & RAHMANIAN, V. D. 1990. *Siliciclastic Sequence Stratigraphy in Well Logs, Cores and Outcrops: Concepts for High Resolution Correlation of Time and Facies*. American Association of Petroleum Geologists Methods in Exploration Series, **7**.

Recognition and distinction of normal and forced regression in cyclothemic strata: a Plio-Pleistocene case study from eastern North Island, New Zealand

DOUGLAS W. HAYWICK

Department of Geology, Geography and Meteorology, University of South Alabama, Mobile, AL 36688-0002, USA

Abstract: The Tangoio block of eastern North Island, New Zealand contains an exceptionally well exposed, 550 m thick sequence of Plio-Pleistocene cyclothemic sedimentary rocks, the Petane Group, that were deposited along the western margin of a shallow (<200 m) pericontinental seaway. The cyclicity that characterizes the Petane Group in this area was generated through recurring 40 ka (sixth-order) sea-level changes of c. 75–150 m. Data from more than 100 sections indicates that the strata consist of five distinct cyclothems, each consisting of a lower formation of mudstone deposited in midshelf environments (highstand system tract, HST), and an upper formation dominated by coarse-grained facies deposited in shallow marine and occasionally non-marine environments. Western exposures of HST mudstones in the lower portion of the Petane Group shoal up into shallow marine siliciclastic sandstones, fluvial gravels and siltstones. The transition is gradational and is consistent with progradation of a siliciclastic shoreline during late sea-level highstand. The shoaling upward siliciclastic sandstone interval represents a regressive system tract (RST) whereas the fluvial beds represent a lowstand system tract (LST). LST gravel beds are sharply overlain by a transgressive surface of erosion and fining-upward siliciclastic sandstone facies of a transgressive systems tract (TST).

Carbonate sediments replaced siliciclastic sandstones as the dominant coarse-grained lithology in the upper Petane Group, probably due to bypass of siliciclastic sediment during sea-level fall. Transitions from HST mudstone into carbonate sand and bioclastic limestone (coquina) are generally sharp everywhere except the downdip eastern portions of the Tangoio block. The contacts represent regressive surfaces of erosion produced during falling stages of sea-level and/or ravinement surfaces formed during rising phases of sea-level (transgressive surfaces of erosion). Carbonate facies atop these erosional surfaces are generally interpreted as TST deposits, except for the uppermost limestones in the Petane Group that occur below a sequence boundary and therefore represent a forced regression systems tract (FRST). Stratal position is the only distinguishing characteristic of FRST limestones in the Petane Group, otherwise they are identical to TST deposits. In eastern portions of the Tangoio block, HST mudstones pass gradationally into RST calcareous sandstones and LST limestones. Systems tracts here are bounded by correlative conformities rather than erosive surfaces indicating continuous sedimentation during sea-level fall and lowstand. Differences in sedimentation and subsidence rates between carbonate and siliciclastic sedimentary systems, are responsible for the different stratal architecture of systems tracts in the Petane Group.

Forced regressions and their impact on sedimentary successions have generated significant interest among sedimentologists and petroleum geologists. Since the earliest studies (e.g., Vail *et al.* 1977; Posamentier & Vail 1988; Van Wagoner *et al.* 1988), numerous researchers have recognized forced regression deposits in siliciclastic and carbonate strata (e.g., Plint 1988; Trincardi & Field 1991; Dam & Surlyk 1992; Hunt & Tucker 1992; Walker & Wiseman 1995; Hart & Long 1996; Naish & Kamp 1997). The response of sedimentation to sea-level fall is frequently a sharp lower bounding surface and an abrupt change from shelfal to shallow marine or shoreface facies (e.g., Suter *et al.* 1987; Tesson *et al.* 1990; Hunt & Tucker 1992; Posamentier *et al.* 1992). This transition records wave scouring in advance of seaward migration of the shoreline (Posamentier *et al.* 1992; Posamentier 1995) and the subsequent loss of accommodation space during falling sea level (Walker 1995*a*).

Posamentier & Allen (1993) noted that the stratal architecture induced by forced regressions was a function of several factors. They and others (e.g., Plint 1991; Hunt & Tucker 1992; Ainsworth & Crowley 1994; Naish & Kamp 1997; Gawthorpe *et al.* this volume) have suggested that sea-floor physiography, sediment

From: Hunt, D. & Gawthorpe, R. L. (eds) *Sedimentary Responses to Forced Regressions*. Geological Society, London, Special Publications, **172**, 193–215. 1-86239-063-0/00/$15.00

flux, tectonics, rate of sea-level fall and changes in sedimentation as depocenters are forced seaward all play some role in controlling the distribution of sedimentary facies. This gives rise to considerable variability in facies and in stratigraphic disposition (see Posamentier *et al.* 1992).

In order to resolve systems tracts in outcrop, including those produced during forced regressions (FRST), it is necessary to have good vertical and spatial exposure over large areas. One place where outcrop of this quality occurs is the Tangoio block, a 450 km^2 coastal region of eastern North Island, New Zealand (Haywick *et al.* 1991; Fig. 1). The strata within the Tangoio block are Plio-Pleistocene in age and are over 550 m thick. Haywick *et al.* (1991) assigned the sedimentary succession to the Petane Group and recognized 11 distinct lithostratigraphic units (Fig. 2), most of which could be traced over the entire Tangoio block. For the middle portion of the sedimentary succession, formational contacts that are exposed along incised river valleys can be walked along up to 15 km inland from the coast. With the exception of a slight eastward dip (<10°) associated with monoclinal folding, the Tangoio block is tectonically undeformed. There are no major structural breaks or faults anywhere in the study area that might complicate lateral correlations or assessment of along strike variability of systems tracts.

The Petane Group is cyclothemic in disposition and the 11 formations which comprise it can be grouped into five mudstone–sandstone\limestone formational couplets (Fig. 2). Siliciclastic sand is the dominant coarse-grained lithology in the lower portion of the Petane Group. It is, however, largely replaced by temperate carbonate sand and bioclastic limestone (coquina) in the upper Petane Group (Fig. 2). Fine-grained mudstone-dominated formations contain micro- and macrofaunal assemblages indicative of mid- to outer shelf deposition (*c.* 50–200 m water depth; Haywick & Henderson 1991). Siliciclastic sand- and limestone-dominated formations

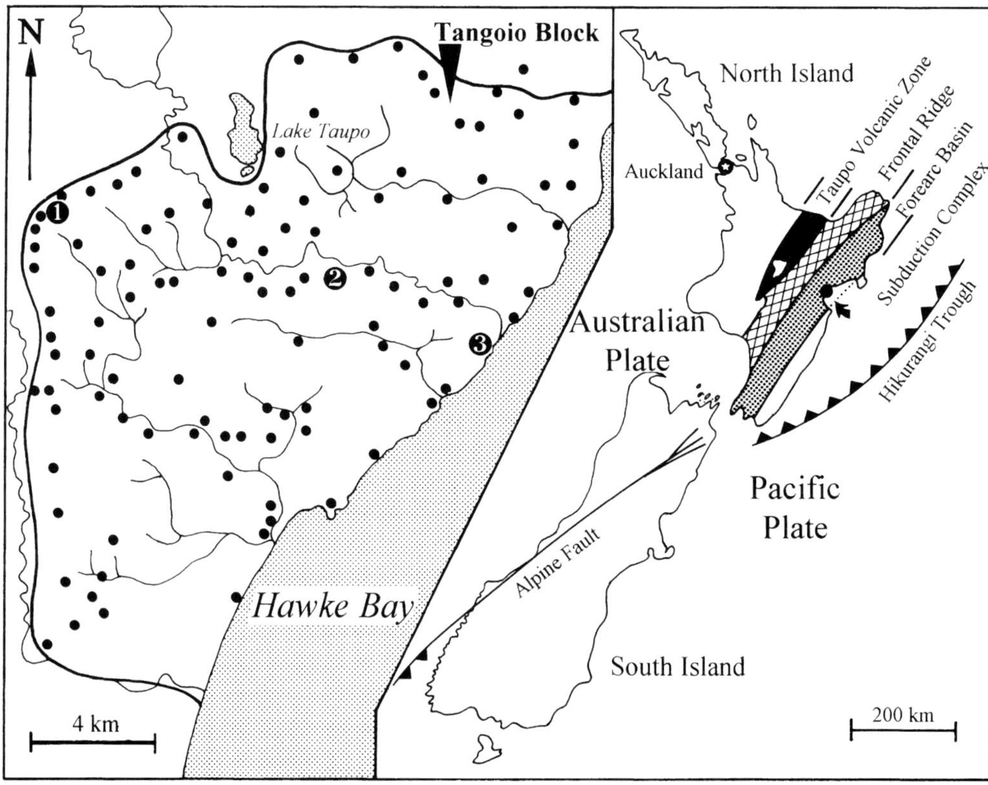

Fig. 1. Location of sedimentary sections examined in the Tangoio block in this study (closed circles) and the location of the Tangoio block relative to major tectonic features of the eastern North Island, New Zealand. The position of the Tangoio block is indicated by the arrow. Numbers refer to the locations of sections 1, 2 and 3 in Fig. 8.

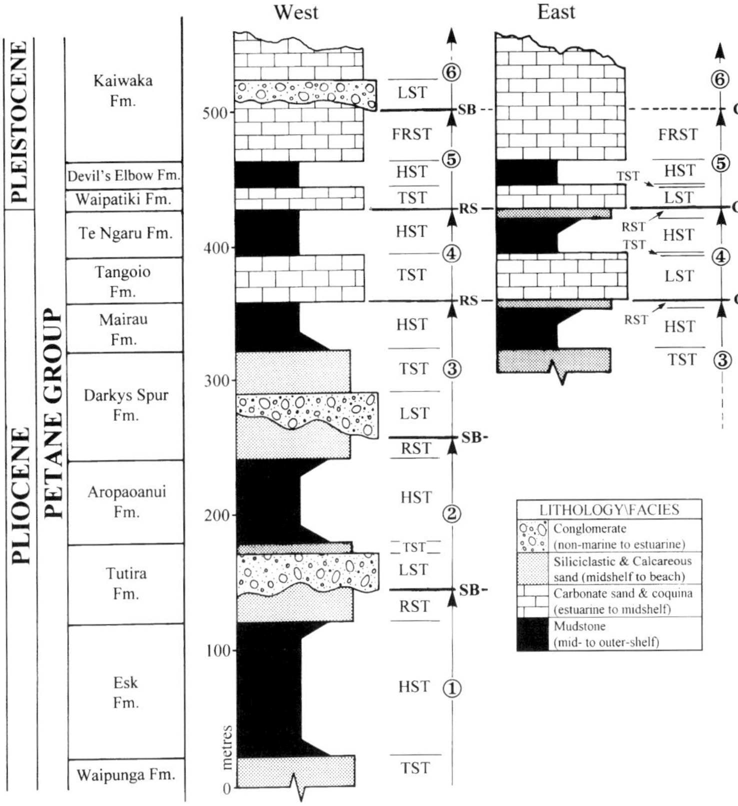

Fig. 2. Composite stratigraphic column for the Petane Group showing the general stratigraphy, lithology, interpreted depositional environments and systems tracts of fine-grained and coarse-grained portions of the stratigraphy. Numbers refer to sequences. LST, lowstand systems tract; HST, highstand systems tract; RST, regressive systems tract (see discussion section for full explanation of the RST); TST, transgressive systems tract; FRST, forced regressive systems tract; SB, sequence boundary; RS, ravinement surface; CC, correlative conformity.

contain fauna, ichnofossils and/or sedimentary structures consistent with fluvial, estuarine and shallow-marine depositional environments (*c.* 0–50 m water depth). The cyclothemic nature of the Petane Group is attributed to recurring sea-level fluctuations (Vella 1963; Beu & Edwards 1984), most probably associated with 40 ka orbicular oscillations (Haywick *et al.* 1992).

Bounding surfaces marking abrupt changes between facies can be found in strata throughout the Tangoio block; however, they are neither ubiquitous, nor consistent everywhere in the study area. The purpose of this paper is to summarize the sedimentological character of systems tracts (particularly those associated with regressions) and their bounding surfaces in the Tangoio block, and to outline the factors responsible for the overall stratal architecture of the Plio-Pleistocene strata in this region of New Zealand. The mixed lithology of coarse-grained formations in the Petane Group also allows the role of sediment type and sedimentation rate on overall systems tract architecture in a shallow shelf setting to be assessed.

Geological and tectonic setting

The North Island of New Zealand lies along a convergent plate boundary, the Hikurangi Trough, where oceanic crust of the Pacific Plate is obliquely subducted beneath oceanic crust of the Australian Plate (Walcott 1978; Cole & Lewis 1981; Cole 1984; Fig. 1). Eastern North Island is structurally complex and contains many classic elements of a collisional plate boundary. These include a highly faulted subduction complex, a forearc basin and a frontal ridge, the Ruahine Range. To the east of the Hikurangi

Trough lies a 150 km wide thrust fault-controlled accretionary borderland consisting of Neogene anticlinal ridges, and linear, slope-parallel accretionary basins. The sedimentary fill within these basins ranges from 200 to 2000 m in thickness (van der Lingen & Pettinga 1980; Pettinga 1982). The highest accretionary basin is broader than those nearer the Hikurangi Trough and is filled by thick Neogene sediment. It represents the main forearc basin of the subduction complex.

The forearc basin began to shallow in the Miocene eventually constricting into a narrow, northeast-southwest-trending pericontinental seaway (Beu *et al.* 1980; Kamp *et al.* 1988). Water depths along the western margin of the seaway, where the Tangoio area lay, decreased to mid- to outer shelf depths (<200 m) by Pliocene time. During deposition of the Petane Group, sedimentation in the seaway was strongly influenced by tidal currents and waves, and less so by storms (Haywick *et al.* 1992). Sediments of the Petane Group are the uppermost strata of the forearc basin sedimentary wedge in eastern North Island (Beu *et al.* 1980), and are underlain by Pliocene shallow marine/fluvial sedimentary rocks and by thick deposits of outer shelf and abyssal Miocene mudstones (Grindley 1966).

Continued convergence eventually began to directly uplift the Tangoio area during Pliocene and into the Early Pleistocene (Kamp & Nelson 1987). The presence of frontal ridge-derived pebbles within carbonate sand and coquina, and discrete conglomerate beds in the uppermost portion of the sedimentary succession in the western Tangoio block, suggests rapid uplift culminating in subaerial exposure during deposition of the Kaiwaka Formation.

The highest elevations occur in the western portion of the Tangoio block where the eastward dip is the greatest (approximately 10°). The dip shallows further east measuring approximately 2° at coastal exposures. River incisions and landslide scours have created a rugged upland topography that is characterized by flat-topped hills bounded by steep scarps and cliffs. Terraces produced by sea-level changes as described by Pillans (1983) in the Wanganui region of western North Island, have not been recognized anywhere in the Tangoio block.

Sedimentology of the Petane Group

The sedimentological framework of the Petane Group is based upon data collected from over 100 sedimentological sections (Fig. 1) and the detailed mapping of formations across the Tangoio block (Haywick *et al.* 1992). Mapping was accomplished primarily by walking along formational contacts. Due to the eastward dip of the strata, the lowermost lithostratigraphic units (Waipunga, Esk and Tutira Formations) were only well exposed in the western Tangoio block. The remaining formations, particularly the Darkys Spur – Devil's Elbow interval, could be mapped out across the Tangoio block permitting 3-dimensional fence post modelling of sedimentary facies (Haywick *et al.* 1992) and systems tracts. Only the lower Kaiwaka Formation is laterally correlative across the Tangoio block. Upper portions of this formation are extensively incised making lateral correlation of facies very difficult. The Kaiwaka Formation is best exposed in the western Tangoio block.

Most Petane Group formations are remarkably continuous across the Tangoio block and only display relatively minor changes in thickness (Haywick *et al.* 1991). Offshore drilling has revealed that Plio-Pleistocene strata below the Hawke Bay seafloor further east of the Tangoio block (Fig. 1) are dominated by marine sandy-mudstones rather than cyclothems of siliciclastic sandstones/limestones and mudstones (Heffer & Milne 1976). This demonstrates that Petane Group cyclothems pinch-out or pass laterally into more offshore facies eastward of the Tangoio block.

Haywick *et al.* (1992) distinguished five distinct facies associations within Petane Group strata in the Tangoio block. They were; (1) *non-marine association* (three component facies), (2) *estuarine association* (three component facies), (3) *siliciclastic shoreline association* (six component facies), (4) *carbonate shelf association* (four component facies) and (5) *offshore association* (four component facies). Table 1 summarizes the major sedimentological and palaeontological characteristics of the facies in each association.

The non-marine association is restricted to portions of the Tutira, Darkys Spur and Kaiwaka Formations in the western portion of the Tangoio block (Fig 2). It is dominated by thick lenticular beds (1–10 m) of trough cross-stratified coarse siliciclastic sand and pebbles up to 0.2 m in diameter. Gravel beds commonly scour up to 3 m into underlying marine sand and gravel beds (shingle beach deposits), and frequently interdigitate with thin beds (0.1–0.5 m) of laminated to massive mudstone (Fig. 3). The siltstones contain rootlets, comminuted carbonaceous fragments and plant imprints. Clasts of siltstone are common basal constituents in the gravel beds (Fig. 3). Haywick *et al.* (1992) regarded the gravel-dominated facies as channel deposits and gravel bars located in braided river environments. Limited palaeocurrent data

Table 1. *Summary of facies and facies associations recognized within Petane Group sedimentary rocks in the Tangoio block (modified from Haywick* et al. *1992)*

Facies association (constituent facies)	Lithology	Fauna\flora	Sedimentary\biogenic structures	Environment of deposition
(A) Non-marine				
(1) poorly sorted gravel	Gravel, siliciclastic sand	None	Trough cross-stratification	Braided river
(2) laminated siltstone	Siliciclastic siltstone	Rootlets\plant imprints	Parallel laminations, ripples	Overbank
(3) ash beds	Pumiceous sand and silt	Rootlets\plant imprints	Trough cross-stratification	Overbank
(B) Estuarine				
(1) massive mudstone	Mudstone, pumice	Restricted bivalves, rootlets	Common ichnofossils, rootlets	Muddy estuary
(2) pebbly mudstone	Mudstone, gravel	Restricted bivalves, rootlets	Rare ichnofossils, rootlets	Estuary\beach
(3) flaser bedded sand and silt	Mudstone, sand, silt	None	Rare ichnofossils	Tidal flat
(C) Siliciclastic shoreline				
(1) laminated sandy gravel	Siliciclastic sand	None	Parallel lamination/bedding,	Shingle beach
(2) low angle cross-stratified sand	Siliciclastic sand	Diverse shoreface molluscs	Low angle cross-stratification	Upper shoreface
(3) laminated sand	Siliciclastic sand	Diverse shoreface molluscs	Parallel lamination, ripples	Lower shoreface
(4) bioturbated silty sand	Siliciclastic sand	Shoreface\inner shelf molluscs	Parallel lamination, abundant burrows	Transitional
(5) hummocky cross-stratified sand	Siliciclastic sand	Rare molluscs	Hummocky cross-stratification	Inner shelf
(6) massive silty sand	Siliciclastic sand\silt	Inner to midshelf molluscs	Abundant bioturbation	Inner-midshelf
(D) Carbonate shelf				
(1) cross-stratified sand	Carbonate sand, coquina	Diverse near shore molluscs	Planar tabular cross-stratification	Tidal sea floor
(2) bidirectional cross-stratified sand	Carbonate sand, coquina	Diverse near shore molluscs	Bidirectional cross-stratification	Tidal sea-floor
(3) laminated bioclastic limestone	Coquina	Diverse near shore molluscs	Parallel\low angle cross-stratification	Tidal shell bar
(4) cross-stratified bioclastic limestone	Coquina	Near shore\estuarine molluscs	Mega cross-stratification	Tidal channel
(E) Offshore				
(1) massive mudstone	Siliciclastic mudstone	Diverse midshelf fauna	Abundant bioturbation	Mid-outer shelf
(2) weakly laminated mudstone	Siliciclastic mudstone	Diverse midshelf fauna	Parallel lamination, abundant burrows	Mid-outer shelf
(3) graded silt\sand mudstone	Silt, sand, pumice	Rare mid-shelf fauna	Graded bedding, parallel lamination	Mid-outer shelf
(4) shelly mudstone	Siliciclastic mudstone	Midshelf fauna (oysters)	Abundant bioturbation	Midshelf shell bed

Fig. 3. Interbedded siliciclastic sand and gravel comprising LST deposits (LST) in the Darkys Spur Formation at section 1 (Fig. 1). The gravel bed pictured is lenticular (contacts are highlighted) and scours approximately 0.50 m downward into siliciclastic shingle beach deposits. The inset shows rotated blocks of rootlet-bearing siltstone clasts (arrow) contained within the gravel beds.

obtained from trough cross-stratification in the gravel beds suggested that the rivers flowed eastward toward the seaway from the Ruahine Range. The siltstone facies was considered to be primarily floodplain deposits. Similar environments occur in this area of eastern North Island today (Suggate 1978).

A significant amount of pyroclastic material (tephra and pumiceous sand) also occurs within the non-marine association and was derived from plinian-style andesitic eruptions in the Taupo Volcanic Zone (Walker 1973). The vast majority of the pyroclastic deposits are cross-stratified suggesting reworking rather than primary airfall. Much of it was probably transported into the Tangoio area by rivers flowing eastward from the Taupo Volcanic Zone and Ruahine Range (Fig. 1). The Ruahine Range (frontal ridge in Fig. 1) consists of thick Permian–Triassic siliciclastic ('greywacke') sandstone and was probably the source of all Neogene-aged siliciclastic sediment deposited into the forearc basin.

The estuarine association is a relatively rare component of siliciclastic sand and carbonate-dominated portions of cyclothems. It consists of either mudstone, pumiceous mudstone or pebbly mudstone beds up to 3 m thick and is generally massive to bioturbated. All three component facies contain a diagnostic restricted marine fauna including the estuarine bivalve *Austrovenus stuchburyi* (Beu & Maxwell 1990) and the brackish water foraminifera *Ammonium beccarri aoeteana* (Hayward 1986). The presence of rootlets and ichnofossils in some outcrops confirms the estuarine origin of the facies within this association.

The siliciclastic shoreline association predominantly consists of coarse-grained components in cyclothems of the lower Petane Group, but individual facies are found throughout most formations in the Petane Group

(Haywick *et al.* 1992). This association consists of 6 facies. Shingle (pebble) beach facies which forms association 1, is comprised of parallel-bedded coarse sandstones and discoidal gravel beds with rare abraded shell fragments and is restricted to the Darkys Spur and Tutira Formations. Facies association 2 on the other hand is widespread. It is dominated by low angle cross-stratified fine to medium siliciclastic sandstones with pebble-lined troughs and a diverse nearshore macrofauna including the sand dollar *Fellaster zelandiae*. The widespread lower shoreface sediments of facies association 3 consist of parallel-laminated to bioturbated fine siliciclastic sandstones containing nearshore macrofossils and shoreface ichnofossils including *Ophiomorpha nodosa* (Fig. 4). Transitional shoreface–inner shelf sediments of facies association 4 are also widespread, and represented by massive to strongly bioturbated fine siliciclastic sand containing shallow marine macrofossils. In contrast, facies association 5 is restricted to Darkys Spur and Tutira Formations. It consists of massive to hummocky cross-stratified, poorly sorted siliciclastic sandstones. Finally, inner to midshelf, massive very silty sand containing both nearshore and offshore macro- and microfossils make up the widespread facies association 6. Overall, this association of facies is interpreted as a high energy, wave-dominated, but storm-influenced, shoreline environment very typical to that found along eastern New Zealand today (Haywick *et al.* 1992). The present coastline adjacent to the Tangoio block is an excellent modern analogue for the siliciclastic sand assemblage as it contains shingle beach and upper and lower shoreface environments that are identical to those facies identified in the Petane Group. These sediments grade laterally into mudstone further into Hawke Bay.

The carbonate shelf association is restricted to coarse-grained formations in the upper portion of the Petane Group (Tangoio, Waipatiki, Kaiwaka Formations; Fig. 2). It consists of four facies that are dominated by either medium to coarse, well-rounded carbonate sand, or by bioclastic limestone (coquina) comprised mostly of abraded shells. Differentiation of facies was made primarily on the basis of sedimentary structures and fauna (Table 1), but they were all attributed to current-swept and tidally-influenced, shallow marine depositional environments (*c.* 0–50 m; Haywick *et al.* 1992). All four facies contain diverse sedimentary structures (Fig. 5) that vary from flaser bedding to parallel lamination, and bidirectional ripple cross-stratification to mega-cross stratification (up to 10 m thick cosets). Palaeocurrent directions obtained from cross-beds indicated that tidal currents were primarily orientated northeast-southwest, generally parallel to the trend of the Plio-Pleistocene pericontineltal seaway (Haywick *et al.* 1992).

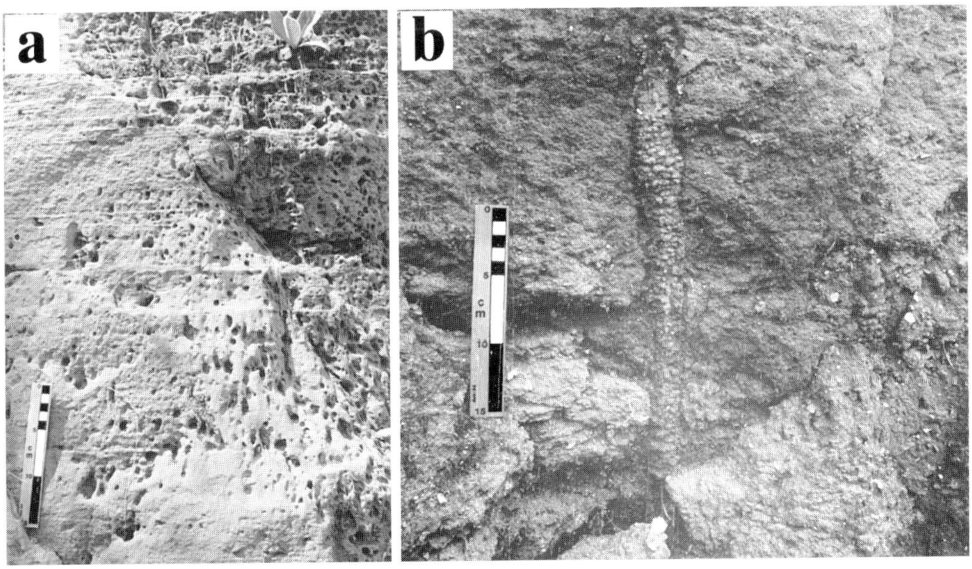

Fig. 4. (**a**) Low-angle cross-stratification and (**b**) the nearshore ichnofossil *Ophiomorpha nodosa* within shoreface facies of the siliciclastic shoreline association. Darkys Spur Formation, Section 1 (Fig. 1), western Tangoio block.

Fig. 5. Diverse sedimentary structures consisting of planar tabular cross-stratification, bi-directional cross-stratification and parallel lamination within carbonate sand facies in the Waipatiki Formation, Section 1 (Fig. 1), western Tangoio block.

Bioclastic limestone and carbonate sand beds display significant lateral variation in sedimentary structures suggesting rapid interdigitation of facies and possibly overlapping bars and shell beds. Haywick *et al.* (1992) explained the large-scale cross-stratification in this facies association (Fig. 6) as resulting from tidal bar migration and the fill of tidal channels. This interpretation was based, in part, on the presence of estuarine macrofossils in some mega-cross-stratified limestones.

The offshore assemblage consists primarily of mudstone and siltstone and contains a diagnostic mid- to outer shelf (*c.* 50–200 m) faunal assemblage of bivalves (Haywick 1990) and foraminifera (Hayward 1986; Haywick & Henderson 1991). Of the four facies in this assemblage, the most abundant and pervasive are massive silt-rich mudstones (Fig. 7). Weakly laminated mudstone and graded bedding containing thin (< 5 mm thick) siliciclastic sandstones or volcanic ash beds occurs in some formations, but is rare. The ash may represent primary airfall deposits derived from the Taupo Volcanic Zone. Graded siliciclastic sandstone beds were spatially associated with hummocky cross-stratified sand facies and were attributed to re-suspension of sediment during storm events similar to that observed by Gagan *et al.* (1988) on the Queensland, Australia shelf.

Petane Group systems tracts

Sedimentological and faunal criteria, in addition to the age, thickness and basin setting of the strata, leave little doubt that the characteristic cyclicity of the Petane Group was generated through recurring sea-level fluctuations of ±100 m (Haywick & Henderson 1991; Haywick *et al.* 1992). The sedimentary succession can therefore be divided into cyclothems containing distinct lowstand, transgressive and highstand systems tracts (LST, TST and HST, respectively), bounded by important stratal surfaces (e.g., sequence boundary, SB; maximum flooding surface, MFS; downlap surface, DLS; etc.; Vail *et al.* 1977). Figure 2 summarizes the systems tract nomenclature for Petane Group formations in the eastern and western portions of the Tangoio block. The introduction of a regressive systems tract (RST) is utilized in strata that are thought to originate during times of relative sea-level fall, but contain no diagnostic features that allow them to be definitively assigned to the forced regressive systems tract (following Naish & Kamp 1997). Figure 8 details along-strike variability of systems tracts within the coarse-grained formations in the lower Petane Group (siliciclastic sand-dominated) and the upper Petane Group (carbonate-dominated).

Fig. 6. Large scale cross-stratification in a bioclastic limestone (1) down-cutting into carbonate sand (2). Both facies contain shallow marine bivalve faunas and sedimentary structures consistent with sedimentation in nearshore and estuarine environments. Kaiwaka Formation, Section 2 (Fig. 1), central Tangoio block.

Fig. 7. Typical exposure of offshore mudstone facies in the Esk Formation in the western portion of the Tangoio block. The mudstone facies is typically massive and contains midshelf molluscan and foraminiferal faunas. Width of railroad tunnel: approximately 4 m. Section located 2 km south of Section 1 in Fig. 1.

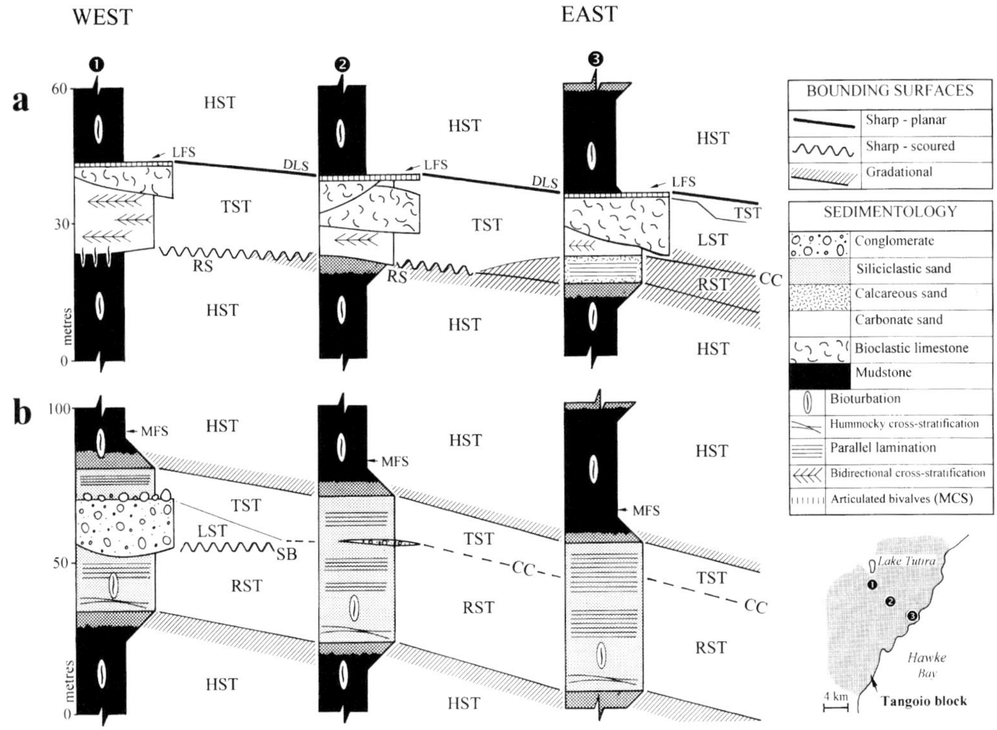

Fig 8. Summary of bounding surfaces and systems tracts within cyclothems in (**a**) the upper Petane Group and (**b**) the lower Petane Group (see Fig. 1 for location of sections). HST, highstand systems tract; RST, regressive systems tract (see discussion section for full explanation of the RST); TST, transgressive systems tract; FRST, forced regressive systems tract; SB, sequence boundary; RS, ravinement surface; CC, correlative conformity; LFS, local flooding surface; MFS, maximum flooding surface. Note change in vertical scale between A and B.

The Tutira, Darkys Spur and Kaiwaka Formations in the western portion of the Tangoio block contain non-marine facies (which represent non-marine LST deposits; Figs 2, 8), and can be fitted to the sequence stratigraphic model of Vail *et al.* (1977) reasonably easily. The lack of non-marine facies elsewhere in the study area, and the change from siliciclastic sand to temperate carbonate sedimentation during lowstands, complicates systems tract interpretations in the upper portion of the Petane Group. Beu (1995) identified an important (transgressive) ravinement surface between offshore mudstone and overlying carbonate shelf facies in a road cut in the western portion of the Tangoio block (Fig. 9). He regarded the contact as a sequence boundary and the overlying carbonate facies as a TST. Moreover, he suggested that all carbonate facies throughout the Petane Group and across the Tangoio block represent TST deposits. Careful examination during this study demonstrated that ravinement surfaces, such as the one identified by Beu (1995) commonly separate carbonate and offshore mudstone facies (Fig. 8a), but only in western and central portions of the Tangoio block. These ravinement surfaces are absent from all but the uppermost portion of the Petane Group in sections in the east of the Tangoio block (Fig. 8a). This variation in vertical facies transitions and their relationships with stratal surfaces in the Petane Group require that two distinct systems tract patterns (motifs) be proposed for carbonate-bearing cyclothems in eastern and western portions of the Tangoio block as is illustrated in Fig. 2. Whereas in the west, the carbonate facies are transgressive in origin and rest on composite sequence boundary/flooding surfaces (Fig. 2, left), in the east the carbonate facies are interpreted to develop over the correlative conformities (to up dip sequence boundaries) and be lowstand in origin (Fig. 2, right). Figure 2 illustrates that similar carbonate facies appear in quite different stratigraphic positions in the up- and downdip portions of the Tangoio block. The following sections discuss the characteristics of

Fig. 9. Sharp contact along a ravinement surface (interpreted as a transgressive surface of erosion) between mudstone (recessive interval) and overlying carbonate sand. Waipatiki Formation, Section 1 (Fig. 1), western Tangoio block.

the lowstand, transgressive and highstand systems tracts and their bounding surfaces, all of which must be outlined before discussing the forced regressive systems tract.

Lowstand systems tracts

Siliciclastic-dominated cycles. As previously stated, non-marine facies in the Tutira, Darkys Spur and Kaiwaka Formations in the western Tangoio block are interpreted as LST fluvial gravel beds (Figs 2, 8a). They vary in total thickness from 15 to 35 m. The erosive basal contacts of these non-marine facies with the underlying marine shoreface sediments represent sequence boundaries (Figs 2 left, 6, 8b). Non-marine gravel beds in the Darkys Spur Formation grade eastward into marine siliciclastic shoreface facies (Fig. 8b). Here it can be demonstrated that the sequence boundary passes eastward into a correlative conformity (cf. Walker & Wiseman 1995) within approximately 5 km of the western most sections. The lowstand shoreline must therefore have lay between sections 1 and 2 on Fig. 8b, and would have been aligned northeast-southwest (parallel to the trend of the periconti-nental seaway).

In the east of the study area, the differentiation of forced regressive and lowstand systems tract sedimentation is difficult within regressive shoreface sandstones (e.g. Fig. 8b, Sections 2, 3). The regressive sandstones coarsen from upward-shoaling offshore facies below (HST), and are bounded above by a correlative conformity overlain by deepening-upward sediments of clear transgressive (TST) affinity. These problematic shoreface sediments are assigned to neither highstand nor lowstand systems tracts (that are respectively developed below or absent), are bounded above by a correlative conformity and a retrogradational stacking pattern of the TST (Fig. 8b). Instead, these shoreface sediments are placed within a non-diagnostic regressive systems tract, the development of which is discussed further below.

Mixed carbonate siliciclastic cycles. Interpretation of the carbonate facies as TST deposits (e.g. Fig. 8a, section 1) was made on the basis of the abrupt contact developed between the carbonate sediments and offshore mudstones in the west of the study area (Beu 1995). In the eastern portion of the Tangoio block passage between carbonate and clastic sediments is very different. Here there are no abrupt contacts, such as sequence boundaries or transgressive ravinement surfaces (e.g. see Fig. 2, right and

Fig. 8a, section 3). Instead, massive siliciclastic sandstones (inner to midshelf facies) grade upward into bi-directional cross-bedded carbonate sands (tidal facies) that in turn grade into nearshore to estuarine bioclastic limestones (Fig. 8a, Section 3).

Haywick (1990) concluded that the entire 20–35 m thick shoaling-upward succession in the eastern part of the Tangoio block was deposited during the falling and lowstand segments of a eustatic cycle (Fig. 8a, Section 3). In the absence of evidence for subaerial exposure (including diagenetic overprinting; cf. Steinhauff & Walker 1995), and the position of flooding surfaces within these offshore cyclothems, Haywick (1990) interpreted the carbonate facies in this area of the Tangoio block as marine LST deposits. This lowstand wedge developed on a correlative conformity, downdip of the area of the shelf undergoing subaerial exposure (e.g. Fig. 8a, Sections 2 to 3). Of practical concern, the exact position of the correlative conformity is difficult to pin-point due to the gradational nature of the transition from siliciclastic sands into carbonates (Fig. 8a, Section 3). The most practical choice of surface is that separating the carbonate and siliciclastic sediments. It is thought that as incised valley development is most important during lowstands, siliciclastic supply to the shoreline would have been localized, therefore allowing more widespread development of carbonate sediments. However, it is realized that there is likely to be some diachroneity in this facies transition and it is practical rather than an ideal choice of surface.

Transgressive systems tracts

Siliciclastic-dominated cycles. In the Tutira and Darkys Spur Formations (western portion of the Tangoio block), upper contacts of LST fluvial gravel beds are commonly marked by a prominent ravinement surface and a 0.5 m thick marine conglomerate consisting of shoreface bivalves, pebbles and boulders, some of which exceed 0.3 m in diameter (e.g. Fig. 8b, Section 1). The basal ravinement surface of this systems tract is equivalent to a transgressive surface of erosion (e.g. Nummedal & Swift 1987; Walker & Plint 1992). Such lag deposits progressively grade upward into sandstone facies of the siliciclastic shoreline association and then into mudstones of the offshore association. This fining- and deepening-upward succession varies from 10 to 30 m in thickness, and is interpreted as an accretionary TST deposited during sea-level rise following deposition of the LST. In eastern portions of the Tangoio block the facies and development of the transgressive systems tract is similar. However, in these areas identifying the base of the systems tract can be more problematic as the basal transgressive lag is absent.

Mixed carbonate–siliciclastic cycles. Beu's (1995) suggestion that carbonate facies in sharp contact with offshore mudstone facies represents a TST is the most reasonable interpretation for carbonate facies in the western and central portions of the Tangoio block (Fig. 8a, Sections 1, 2; with the exception of the Kaiwaka Formation; discussed shortly). The majority of the TST in each cyclothem consists of a coarsening upward succession of carbonate sands and bioclastic limestones that are indistinguishable from LST limestones further east. The critical difference is the ravinement surface that places the carbonate sediment in direct and erosive contact with offshore mudstone (Fig. 9). Below the ravinement surface, vertical ichnofossils 10–20 mm in diameter penetrate downward into mudstone (Fig. 10). They are filled with carbonate sand and shell hash from the overlying TST and are similar to ichnofossils left by boring pholads such as *Barnea (Anchmasa) similis* (Beu & Maxwell 1990) in the Wanganui area of western North Island (Abbott 1994). *Barnea similis* is not found *in situ* within borings or within carbonate facies in the Petane Group as it is in the Wanganui area, but the ichnofossils it leaves behind are indicative of a hard or firm substrate. The significance of these borings, including those found in the Petane Group, is that they generally mark disconformities including wave-planed surfaces, ravinement surfaces (Beu & Maxwell 1990; Abbott & Carter 1994) and as in this case, transgressive surfaces of erosion.

The upper contact between bioclastic limestone and overlying offshore mudstone is always sharp within the study area (Figs 8 & 11). In central and western portions of the Tangoio block, it is marked by a 0.1 to 0.5 m thick shellbed consisting of articulated bivalves (genus *Glycymeris*) and oysters in an offshore mudstone matrix. The shellbed is somewhat thicker atop limestones in the eastern Tangoio block (up to 1 m), and contains abundant *Eumarcia plana* bivalves (sometimes in place of *Glycymeris* sp.). The matrix contains more siliciclastic sand-grade sediment than mudstone. The beds of articulated bivlaves are interpreted as condensed sediments deposited atop a local flooding surface (LFS) during rapid sea-level rise. Across the study area it is important to realise the quite different significance of the

Fig. 10. Vertical borings (*Barnea (Anchmasa) similis*) filled with shell hash penetrating into HST mudstone from carbonate sand below a transgressive surface of erosion. Borings are filled with shell hash from the overlying sand. Scale bar divisions: 1 cm. Waipatiki Formation, Section 1 (Fig. 1), western Tangoio block.

Fig. 11. Sharp planar contact at the top of the Tangoio Formation between bioclastic limestone (TST) and overlying mudstone (HST). The contact is characterised by an accumulation of articulated bivalves and corresponds to a midcycle shellbed. The outcrop is well exposed in cliff sections due to pervasive cementation of the coarse grained carbonate facies. Section 2 (Fig. 1), central Tangoio block.

shellbeds. In the western and central portions of the Tangoio block, these sediments represent only the upper most portion of the TST. However, in the east, where bioclastic carbonate facies are interpreted as marine LST deposits, the shellbed may represent the entire TST (Fig. 8).

In all locations, the shellbeds grade upward into offshore mudstones, generally within 1 m of the top of the bioclastic limestone. The shellbed is inferred to be overlain by a downlap surface at the base of the mudstones which are interpreted as part of the overlying HST (Haywick 1990). In fact because of their condensed nature, regional significance and position, and the inferred downlap of siliciclastic mudstones, these shellbeds are considered to represent maximum flooding surfaces (equivalent to mid-cycle shellbeds (MCS) of Abbott & Carter 1994). No other stratal surfaces are recognized in these TST limestones.

Early highstand systems tracts

Siliciclastic and mixed cycles. Fine-grained formations contain mid- to outer-shelf microfossil and macrofossil assemblages and doubtless represent HST deposits. Haywick & Henderson (1991) examined the foraminiferal palaeoecology of the Mairau Formation (at Section 1 in Fig. 1) and found that the deepest assemblages (*c*. 150–200 m) occurred near the base of the formation, within 3 m of the top of the underlying Darkys Spur Formation. Similar observations were made for two sections through the Te Ngaru Formation. This implies that the maximum flooding surface (MFS) of each eustatic cycle *lies just above* the top of the underlying TST, so that the main body of the offshore deposits represent highstand sedimentation.

Foraminiferal data also indicate that water depths shallowed gradually from a maximum of 150–200 m at the MFS to 50 m or less at the top of the HST (Haywick & Henderson 1991). From this interval, midshelf mudstone grades upwards, over 5–15 m, into massive to horizontally bedded sandy mudstone and silty siliciclastic sand.

Distinction of regressive and forced regressive systems tracts

In the Tangoio block, two styles of regressive deposition can be distinguished. In western exposures of the lower Petane Group, intervals between HST mudstones and LST conglomerates are characterized by a shoaling-upward succession consisting of hummocky cross-stratified sand at the base, passing through parallel to low-angle cross-stratified sand (shoreface facies), into laminated sandy gravel (beach facies; Fig. 8b). Transitions between facies are gradational throughout this interval which is consistent with shoreline progradation during late highstand (i.e. 'normal regression'; Plint 1988) rather than forced regression. For this reason, these shoaling intervals clearly represent deposition as a part of the HST, as opposed to FRST deposits (Figs 2, left and 8b, Section 1). However, for reasons discussed in the following section it is in some cases preferred to maintain these deposits within the RST.

Comparable regressive sediments assigned to the RST within the mixed carbonate-siliciclastic cycles in the upper Petane Group occur between HST mudstone and LST limestones in the eastern Tangoio block (Fig. 8a, section 3). They are similar to the RST deposits described from the lower Petane Group, but with three important differences: (1) they are significantly thinner (average 5 m); (2) they lack shoreface and shingle beach facies; (3) they contain much more calcareous detritus (up to 50% carbonate grains versus < 5% carbonate grains in the regressive lower intervals of the Petane Group); (4) they are characterized by gradational lower and upper surfaces (Fig. 12). In addition, these regressive deposits have a quite different position in the basin and occur beneath a correlative conformity over which lowstand carbonate cross-bedded and bioclastic facies occur (Fig. 12; LST gravel beds do not occur in this portion of the Tangoio block). These RST sediments are interpreted to have been deposited during late rise *and* during relative sea-level fall due to their stratigraphic and basinal position, and the rapid, but apparently gradual, shoaling of palaeobathymetry from offshore highstand mudstones. Because of the lack of development of bounding surfaces within these strata they are neither appropriately assigned to the HST or the FRST. This is why they are assigned to a less diagnostic RST.

Fig. 12. Gradational contacts between HST mudstone and overlying calcareous sand (RST) and carbonate sand (LST) in eastern portions of the Tangoio block. No surfaces of erosion occur in this sedimentary package indicating uninterrupted deposition during late rise, falling and lowstand of sea level. Mairau–Tangoio Formation, Section 3 (Fig. 1), eastern Tangoio block. It is the lack of distinctive stratal surfaces in such sections that demands the usage of a less diagnostic regressive systems tract as discussed in detail in text.

The Pleistocene Kaiwaka Formation, at the top of the Petane Group in the western Tangoio block, also contains shoaling-upward (regressive) deposits, albeit of carbonate shelf facies rather than siliciclastic shoreline facies (Fig. 2).

However, here there *is* a significant difference in the transition from HST to regressive deposits. The basal contact between the carbonate shelf facies and underlying mudstone is sharp and erosive. In fact, with the exception that no *Barnea (Anchmasa) similis* borings occur below it, this marine erosion surface is practically identical to the transgressive ravinement surfaces at the base of TST carbonate facies lower in the Petane Group (Fig. 8). However, this interval of the Kaiwaka Formation cannot be interpreted as a TST because LST fluvial gravel beds lie above it (Fig. 2). Consequently, the carbonate facies are considered a FRST (Hunt & Tucker 1992) and the surface on which they lay a regressive surface of marine erosion produced through wave planning during sea-level fall. Lowstand systems tract gravel beds cannot be traced very far toward the east because of poor exposure of outcrop. It would, however, appear that the gravel beds pinch out or grade into thick bioclastic limestone beds in the central portion of the Tangoio block. This implies that the sequence boundary at the base of the LST passes eastward into a correlative conformity (Fig. 2).

Discussion

The cyclic strata of the Tangoio block were deposited during Plio-Pleistocene times that were characterized by high-amplitude, high-frequency glacioeustatic sea-level changes with long-term falls and short, rapid rises. However, in contrast to sedimentation on other time equivalent margins (e.g. as discussed in this volume by Chiocci; Hernández-Molina *et al.*; Kolla *et al*; McMurray & Gawthorpe; Trincardi & Corregarri), only a small fraction of the stratigraphy preserved in the Tangoio block can be unequivocally attributed to deposition during relative sea-level fall, and the forced regressive systems tract. The following section examines some of the reasons why this is so.

Sedimentation and subsidence rates

The cyclicity which characterizes the Petane Group in eastern North Island was generated through eustatic sea-level changes (Haywick *et al.* 1992), yet only the carbonate-dominated cycles near the top of the sedimentary succession unequivocally fulfil the criteria of a FRST, and marine lowstand deposits. Regressive deposits lower in the Petane Group that consist of siliciclastic sand do not apparently record forced regressions, and lowstand systems tracts have not been differentiated; they are characterized by gradual shallowing upward cycles from unequivocal offshore marine highstand mudstones to regressive shoreface sandstones (e.g. Figs 2, 8b). One of the main differences between carbonate and siliciclastic sedimentary systems in the Petane Group would appear to be controlled by contrasting rates of deposition and subsidence. A combination of different rates of sediment flux/production and subsidence is thought to be the most reasonable explanation for why forced regressive and lowstand systems tract cannot be differentiated and/or apparently did not develop in marine environments in siliciclastic cycles, whereas they can in mixed carbonate–clastic cycles.

Siliciclastic settings. In modern coastal areas offshore New Zealand, siliciclastic sand is currently being deposited at rates in excess of 1 m ka^{-1} (Kamp & Nelson 1987). In the Plio-Pleistocene, it appears that rates of siliciclastic sandstone and mudstone were probably as high or even higher. For example, depositional rates of 4 m ka^{-1} are suggested from the lower Petane Group (Haywick *et al.* 1997). Here, cyclothems deposited during a single 40 ka sea-level oscillation can exceed 150 m in thickness (Haywick 1990). Such rates suggest at least a local combination of high rates of subsidence and sedimentation; subsidence rates that may have been in excess of the 1.25 m ka^{-1} that would cancel out most fourth-order Pleistocene sea-level falls (e.g. Hunt & Gawthorpe this volume).

Subsidence rates across the Tangoio area are thought to have been relatively rapid and uniform during deposition of the lower Petane Group (Haywick *et al.* 1992). This is because of the lateral continuity of facies observed within the Darkys Spur Formation, suggesting a shallow depositional gradient and close balance between rates of sediment supply and subsidence (cf. Cant & Hein 1986). In addition, modern subsidence rates in the forearc basin adjacent to the Hikurangi Trough although highly variable, can exceed several m ka^{-1} (Lewis 1980; Wells 1989). Overall, Haywick (1990) concluded that thick cyclothems in the Lower Petane Group resulted from a combination of rapid sedimentation filling accommodation space created through high rates of tectonic subsidence. Thus, in the lower Petane Group it is possible that significant shoreface progradation occurred (i) during the late highstand, before sea-level fall, or in absence of sea-level fall due to rapid subsidence cancelling the glacioeustatic falls, (ii) during falling sea-level (e.g. forced regression) or (iii) during lowstands.

As summarized in Fig. 13, evidence from the plethora of sedimentary sections examined on the Tangoio block (>100) suggest relative continuity of deposition from the highstand, through falling sea-level to lowstand conditions in the siliciclastic-dominated successions. In the study area (schematically shown by sections 1–3 in Figs 8 and 13) highstand conditions were associated with deposition of mudstones in offshore environments, with upper shoreface environments most likely positioned westward (Fig. 13a, c). The onset of relative sea-level fall resulted in rapid eastward progradation of the sand-prone shoreface environments, and accounts for the rapid progradation of shoreface facies across a wide shelf area. No distinct lowstand deposits are differentiated. Therefore, the change from highstand, through sea-level fall to lowstand was apparently relatively seamless, and resulted in rapid shallowing and progradation successions in the absence of major erosional/stratal surfaces. This is why the regressive strata are placed within the non-diagnostic RST; becuase sediments deposited during highstand, fall and lowstand conditions cannot easily be differentiated.

Mixed carbonate-siliciclastic settings. Sediment production and deposition rates for temperate shelf carbonates are much lower than those discussed for the siliciclastic settings above. Modern temperate shelf carbonates such as those that occur in open marine, current-scoured environments around New Zealand (Nelson 1978), southern Australia (James & Bone 1991; Boreen & James 1995) and elsewhere, are typically deposited at 0.02 to 0.05 m ka^{-1} (e.g. Barbera *et al.* 1980; Nelson 1988; Collins 1988; Smith 1988). Plio-Pleistocene carbonate facies in the Tangoio block were deposited in a shallow, tide-dominated, embayed coastline setting, where carbonate sedimentation rates may have been higher, between 0.1

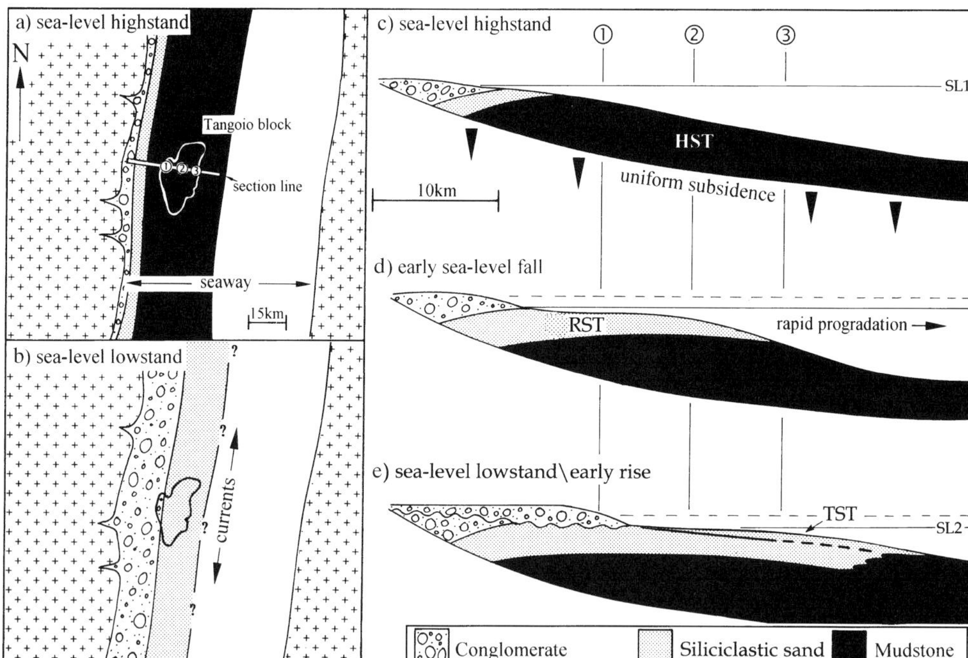

Fig. 13. Schematic palaeogeographic maps and cross sections summarising deposition of the lower Petane Group in the western portion of the pericontinental seaway. (**a**) Palaeogeography during sea-level highstands. (**b**) Palaeogeography during sea-level lowstands. (**c**) Schematic cross sections (not to scale) of sedimentation during periods of sea-level highstand – note that the study area is characterized by widespread development of offshore facies; (**d**) early sea-level fall, and the main phase of progradation; (**e**) sea-level lowstand and the onset of relative rise. The approximate positions of sections 1, 2 and 3 (Figs 1, 8) during the Plio-Pleistocene are indicated on both the palaeogeographic maps and the cross-sections. Sea-level highstands were characterised by high rates of sedimentation, abundant accommodation space. See text for detailed discussion of the depositional systems evolution in response to relative sea-level changes.

and 0.5 m ka^{-1} (Haywick 1990; Haywick *et al.* 1992), based upon comparisons with similar environments elsewhere (Farrow *et al.* 1984). However, even at this higher rate, temperate carbonate sedimentation in the Tangoio area would still have been almost an order of magnitude lower than the rate of siliciclastic sand sedimentation. The thickness of the cycles in this upper part of the Petane Group succession is also reduced. This leads to an inescapable conclusion; both the input of siliciclastic sand into the Tangoio area during deposition of the upper Petane Group, and subsidence rates must have been greatly reduced in order for carbonate sand and bioclastic limestone to form TST, LST and FRST deposits.

Bypass of siliciclastic sand. Plio-Pleistocene sedimentary rocks to the east (basinward) of the Tangoio block consist of siliciclastic sand and silt (Fig. 14b, d; Heffer & Milne 1976). The occurrence of these sands indicates that siliciclastic sand was being deposited concurrently with Petane Group carbonates, but in deeper water environments than those of the pericontinental seaway. Haywick *et al.* (1992) therefore concluded that siliciclastic sand must have bypassed the Tangoio area during falling and lowstand stages of sea-level. Sediment bypassing is typical of shelves with shallow gradients (Posamentier 1995) as are interpreted for the Tangoio area. Haywick *et al.* (1992) argued that as sea-level fell, the Tangoio area became more and more isolated from the main fairways for siliciclastic sand, interpreted to have been localized within incised valleys. This localization of siliciclastic supply is thought to have allowed carbonate facies to accumulate as LST, FRST and TST deposits (e.g. Fig. 14). High concentrations of siliciclastic sand within RST, TST and mid-cuycle shell bed deposits in eastern sections in the Tangoio block are consistent with siliciclastic sand bypassing, and suggests an eastern source of non-carbonate sediment at these times (e.g. see Fig. 14b).

The initiation of siliciclastic sediment bypass is though to have been driven by tectonism elsewhere in the forearc basin/seaway. Regional uplift directly affected the Tangoio area during deposition of the Kaiwaka Formation, although Haywick *et al.* (1992) argued that uplift elsewhere in the forearc basin began to affect siliciclastic sand transport pathways much earlier than this (i.e., during deposition of the Mairau Formation). Kamp & Nelson (1987) discussed the role of tectonism in the evolution and distribution of Neogene limestones in eastern North Island. They concluded that the tectonic regime controlled the rate of terrigenous sediment input to shelf areas and produced ecologically favourable environments for carbonate production (e.g., isolated basins with shallow marine depositional conditions). Overall, extensive carbonate deposition only occurred where siliciclastic sand influx was restricted.

Systems Tract Architecture

Lower Petane siliciclastic cycles. The development of systems tracts during sea-level highstand, early sea-level fall and sea-level lowstand for cyclothems in the lower portion of the Petane Group is summarized in Fig. 13. The high rates of sedimentation in siliciclastic shoreface environments resulted in rapid progradation of shoreline facies outward across the Tangoio shelf during late highstand (Fig. 13c) and early sea-level fall (Fig. 13d). As a consequence, mudstones of the HST grade into shoaling upward RST sand facies everywhere in the Tangoio area (Fig. 13c to d, Sections 1–2). Sea-level lowstand sedimentation is thought to have been characterized by continued eastward progradation of shoreface facies, and contemporaneous deposition of LST fluvial gravel beds in the west where the Tangoio area was subaerially exposed. However distinction of the lowstand and forced regression deposits has not been possible due to the gradational nature of contacts and therefore the lack of obvious stratigraphic surfaces. Accordingly, the sequence boundary is considered to develop at the end of the lowstand, occurring below non-marine LST in updip areas (Fig. 13e, Section 1) and passing into a correlative conformity in central and eastern portions of the study area (Fig. 13e, Sections 2–3).

The placing of the sequence boundary in these clastic-dominated cycles differs from that suggested in most idealized stratigraphic models discussing forced regressive and lowstand sedimentation (e.g., Posamentier *et al.* 1992; Hunt & Tucker 1992, 1995; Kolla *et al.* 1995). There has been considerable debate concerning the positioning of forced regressive systems tracts (FRST) in the sequence stratigraphic model (e.g. see Kolla *et al.* 1995 and Hunt & Tucker 1995) with some authors placing them above sequence boundaries (e.g., Posamentier *et al.* 1992; Posamentier & Allen 1993), and others below sequence boundaries (e.g., Van Wagoner *et al.* 1990; Hunt & Tucker 1992). Hunt & Tucker (1995) argued convincingly for placing FRST deposits below sequence boundaries (one reason being that it would position the sequence boundary at the lowest point of sea-level during

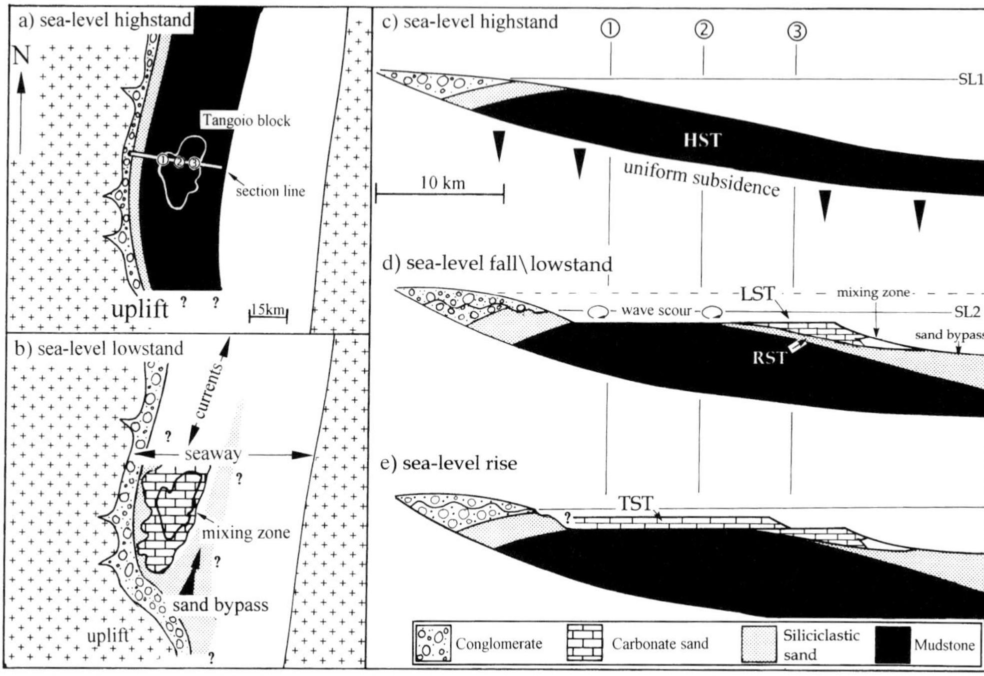

Fig. 14. Schematic palaeogeographic maps and cross-sections summarising deposition of the upper Petane Group. (**a**, **b**) Palaeogeography in the western portion of the pericontinental seaway during sea-level highstands and sea-level lowstands, respectively. Schematic and cross-sections (not to scale) of sedimentation during (**c**) periods of sea-level highstand, (**d**) sea-level fall\lowstand and (**e**) sea-level rise. Siliciclastic sand must have bypassed the Tangoio area during sea-level lowstands in order for carbonate facies to accumulate. Uplift may have altered distribution pathways of siliciclastic sand forcing it to be deposited into deeper areas of the seaway and possibly developing a sediment mixing zone east of the Tangoio area. A marine erosion surface (regressive surface of erosion) is though to have been produced in central and western portions of the Tangoio area during falling sea-level. In the east (Section 3), RST sand was deposited during this time. There is no evidence in either the Tangoio Formation or the Waipatiki Formation to support deposition of FRST during this phase of sea-level change, however, subsequent sea-level rise may have removed them through marine processes giving rise to development of the subsequent transgressive ravinement surface. FRST limestone is only present in the Kaiwaka Formation (not shown). Limestone was deposited during sea-level lowstands in eastern portions of the Tangoio area. Sea-level rise caused the carbonate facies to backstep over the Tangoio area and resulted in TST deposits resting directly on the ravinement surface.

a eustatic cycle). However, in this study the most practical and indeed only readily identifiable surface within the extensive shoreface profiles is that associated with the onset of retrogradation (e.g. the transgressive surface, coincident with the correlative conformity in Fig. 8), marking the most basinward regression of the shoreline (e.g. Figs 8b, 13e).

As rates of relative sea-level rise outpaced those of clastic supply, TST deposits encroached shoreward westward across the Tangoio area (Fig. 13e). In the east, where TST deposits overlay a correlative conformity, the transition from RST facies into TST facies was gradational. In the west, rising sea-level first drowned the LST gravel beds depositing a transgressive lag deposit (Arnott 1995) atop of a wave-planned ravinement surface. TST deposits consisting of fining upward sand facies were initially deposited atop the ravinement surface but were eventually replaced by HST mudstone at the onset of sea-level highstand, as recorded by microfauna.

Upper Petane mixed carbonate–siliciclastic cycles. The systems tract architecture for the upper Petane Group is pictured in Fig. 14. As in the lower Petane Group, sea-level highstands were characterized by widespread deposition of offshore mudstones (Fig. 14c), with clastic shoreface sandstone deposition occurring west of the study area. The understanding of deposition during late highstand and onset of relative sea-level fall is poor, due to lack of widespread preservation (e.g. Fig. 14d, Sections 1–2), but appears to have been markedly different. Regressive sediments attributed to the RST, represent this interval of time and are only preserved in the eastern portion of the Tangoio block. It would appears that either RST sand (comparable to that seen in Fig. 13d) was never deposited over much of the Tangoio area, or it was removed through wave planning during sea-level fall (e.g. Fig. 14d), and the ensuing rise. Indeed, a similar fate may explain the absence of FRST limestones in all but the Kaiwaka Formation, as Haywick (1990) concluded that the composite sequence boundary/transgressive erosion surface atop HST mudstones probably recorded many metres of erosion. TST deposits comprising the Tangoio and Waipatiki Formations were deposited atop a transgressive surface of erosion that most likely overprints an earlier regressive surface of erosion. If this scenario is correct, ravinement surfaces that separate HST mudstone from overlying TST limestone represent significant periods of time in which both RST and FRST deposits might have been removed.

Carbonate sand and bioclastic limestone facies were deposited atop RST sand facies in the Tangoio and Waipatiki Formations in the eastern Tangoio block during sea-level lowstands (e.g. Fig. 14d). The presence of a zone of mixed siliciclastic-carbonate facies to the east of the Tangoio area is inferred given the calcareous composition of RST facies in this area, and concurrent deposition of siliciclastic sand in deeper portions of the seaway (Fig. 14d). The mixed carbonate-siliciclastic sediments of the RST (e.g. Figs 8a – Section 3, 14d) are therefore thought to represent deposition during relative sea-level fall or forced regression. The overlying transition into carbonate-dominated tidal facies is thought to represent the correlative conformity to the updip subaerial exposure surface (e.g. Fig. 8a). A similar change from storm/wave-dominated processes to tide-dominated processes has been observed elsewhere between forced regressive and lowstand deposits (e.g. Mellere & Steel this volume). However, since in the study area wavebase erosion has removed evidence for updip subaerial exposure, and the change from siliciclastic to carbonate-dominated facies is somewhat arbitrarily used as a criterion for differentiating systems tracts, it is considered better to place these sediments within a RST, rather than the FRST.

Relative sea-level rise resulted in the westward shift of carbonate facies, and deposition of a TST atop the transgressive surface of erosion (Fig. 14e). Eventually sea-level rise outpaced carbonate sedimentation drowning the carbonate facies and allowing the development of a so called 'mid-cycle' shellbed atop limestone formations (e.g. Fig. 8a). Sharp planar contacts between the Tangoio and Waipatiki Formations and overlying mudstones record the migration of a downlap surface across the Tangoio area. HST mudstones were deposited directly atop this surface.

Conclusions

Several key points about regressive sedimentation and the distinction of the regressive highstand, forced regressive and lowstand systems tracts can be made from this study of Plio-Pleistocene deposits in the Tangoio block, New Zealand.

(1) Sediments were deposited during times characterized by dominant 4th-order glacio-eustatic signals with prolonged periods of fall, and rapid rises. However, unlike in many other areas of the World, the recognition of distinctive stratal units attributed solely to forced regressive sedimentation is surprisingly uncommon.

(2) In conditions of high siliciclastic sediment supply and ample accommodation space, as occurred during deposition of the clastic-dominated lower Petane Group, the only obvious physical bounding surfaces are: (i) at the base of LST gravel beds deposited onto shoreface deposits in updip areas, which corresponds to the sequence boundary and (ii) atop transgressive ravinement surfaces atop LST gravel beds. Both surfaces are restricted to western portions of the Tangoio block. Elsewhere, boundaries consist of correlative conformities which are not manifested by discrete surfaces outcrop; boundaries are instead placed between deposits with progradational (e.g. RST) and retrogradational stacking patterns (e.g. the TST).

(3) In the lower Petane Group rapid progradation during late sea-level highstands, falls and sea-level lowstands led to uninterrupted deposition of regressive sedimentary packages. No distinctive regressive surfaces of marine erosion were produced, and because of the gradational nature of the successions deposited, and absence of diagnostic surfaces, differentiation of HST, FRST and LST is not practical. Instead the regressive strata are attributed to a broad regressive systems tract (RST) encompassing

deposition under conditions of late rise, falling and lowstand of relative sea-level.

(4) In the mixed carbonate–siliciclastic cycles of the upper Petane Group, temperate carbonate sediment was deposited at lower rates than siliciclastic sand and probably under conditions of lower subsidence rates and reduced creation of accommodation space. RST attributed to deposition during relative sea-level fall were only preserved/produced in the easternmost portion of the study area where they grade up into marine LST carbonate deposits. Again here are no physical bounding surfaces within this HST–RST–LST sedimentary package.

(5) Sea-level fall produced marine erosion surfaces in the updip central and western portions of the Tangoio area over which FRST limestone beds may have been deposited. However, only the Kaiwaka Formation preserves such deposits. In this region, the other two carbonate formations consist of TST deposits, an interpretation that is made on the basis of the bounding surfaces that lay below them (transgressive surfaces of erosion) and above them (maximum flooding surfaces). FRST limestones within the Kaiwaka Formation are essentially identical in terms of their sedimentology to TST and LST limestones elsewhere. Only the presence of a physical bounding surface (sequence boundary below LST gravel beds) above limestones in the Kaiwaka Formation permits their interpretation as FRST deposits.

(6) Within the Kaiwaka Formation, the sequence boundary appears to pass eastward into a correlative conformity. Although poor exposure limits this conclusion, it appears that FRST deposits in this area of the Tangoio block would occur below a less than distinctive physical surface. It is quite evident that it is often practically impossible to distinguish FRST carbonates from overlying TST deposits produced during the subsequent sea-level rise. The entire carbonate package would probably be interpreted as a TST (as was done here for the limestones that comprise the Tangoio and Waipatiki Formations in western and central portions of the Tangoio block).

(7) This study clearly highlights the practical difficulties that can exist in attempting to identify systems tracts in outcrop, even where the coverage of a study area with large amounts of pin-point data (e.g. >100 sections), and the existence of forced regressions is well known. The study also illustrates the need for exceptional lateral continuity of strata in outcrop studies, a logical inference of which is that considerable caution needs to be applied in subsurface interpretations based on few widely spaced boreholes/log data.

This paper resulted from my attendance at a Geological Society of London workshop on forced regressions. I thank R. Gawthorpe and D. Hunt for the opportunity to attend that workshop and also thank the participants for several valuable discussions on forced regressions and sea-level change. B. Carter, D. Hunt, T. Naish and L. Quinn provided comments on an earlier version of this manuscript. Mike Young proof read and suggested corrections on the final version of the paper. I am particularly indebted to B. Carter and T. Naish for those long discussions on top of Petane Group outcrops in New Zealand. Funding for the early portion of this study was provided in part by the Australian Research Grants Scheme (ARGS), and in part by the Commonwealth Scholarship and Fellowship Plan (CSFP). Additional funding was provided by the Geological Research Fund, the College of Arts and Science, the Graduate School and the President's Travel Fund (all courtesy of the University of South Alabama).

References

ABBOTT, S. T. 1994. *Sedimentology of Mid-Pleistocene Cyclothems from Wanganui, New Zealand.* PhD thesis, James Cook University of North Queensland, Townsville, Australia.

—— & CARTER, R. M. 1994. The sequence architecture of mid-Pleistocene (c. 1.1–0.4 Ma) cyclothems from New Zealand: facies development during a period of orbital control on sea-level cyclicity. *In*: DE BOER, P. L. & SMITH, D. G. (eds) *Orbital Forcing and Cyclic Sequences.* International Association of Sedimentologists Special Publications, **19**, 367–394.

AINSWORTH, R. B & CROWLEY, S. F. 1994. Wave-dominated nearshore sedimentation and 'forced' regression: post abandonment facies, Great Limestone Cyclothem, Stainmore, UK. *Journal of the Geological Society, London*, **151**, 681–695.

ARNOTT, R. W. C. 1995. The parasequence definition – are transgressive deposits inadequately addressed? *Journal of Sedimentary Research*, **B65**, 1–6.

BARBERA, C., CARANNATE, G., D'ARGENIO, B. & SIMONE, L. 1980. Miocene calcareo dell'Appennino meridionale: contributo della paleoecologia alla construnzione di un modello ambientale. *Annali della Università delle Ferrara (N.S.)* **6**, 281–299.

BEU, A. G. 1995. Pliocene limestones and their scallops. *Lithostratigraphy, Pectinid Biostratigraphy and Paleogeography of eastern North Island Late Neogene Limestone.* New Zealand Institute of Geological and Nuclear Sciences Monograph, **10**.

—— & EDWARDS, A. R. 1984. New Zealand Pleistocene and Late Pliocene glacio-eustatic cycles. *Palaeogeography, Palaeoclimatology, Palaeoecology*, **46**, 119–142.

—— & MAXWELL, P. A. 1990. *Cenozoic Mollusca of New Zealand*. New Zealand Geological Survey Palaeonological Bulletin, Lower Hutt, **58**.

——, GRANT-TAYLOR, T. L. & HORNBROOK, N. DE B. 1980. *The Te Aute Limestone facies, Poverty Bay to Northern Wairarapa. 1:250,000*. New Zealand Geological Survey Miscellaneous Series Map, New Zealand Department of Science and Industrial Research, Wellington, New Zealand, **13**.

BOREEN, T. D. & JAMES, N. P. 1995. Stratigraphic sedimentology of Tertiary cool-water limestones. *Journal of Sedimentary Research*, **B65**, 142–159.

CANT, D. J. & HEIN, F. J. 1986. Depositional sequences in ancient shelf sediments: some contrasts in style. *Canadian Society of Petroleum Geologists Memoir*, **11**, 303–312.

COLE, J. W. 1984. Taupo-Rotorua Depression: an ensialic marginal basin of the North Island, New Zealand. *In:* KOKELAAR, B. P. & HOWELLS, M. F. (eds) *Marginal Basin Geology. Volcanic and Associated Sedimentary and Tectonic Processes in Modern and Ancient Marginal Basins*. Geological Society, London, Special Publications, **16**, 109–120.

—— & LEWIS, K. B. 1981. Evolution of the Taupo-Hikurangi subduction system. *Tectonophysics*, **72**, 1–21.

COLLINS, L. B. 1988. Sediments and history of the Rottnest Shelf, southwest Australia: a well documented, non-tropical continental margin. *Sedimentary Geology*, **60**, 15–50.

DAM, G. & SURLYK, F. 1992. Forced regressions in a large wave- and storm-dominated anoxic lake, Rheatian-Sinemurian Kap Stewart Formation, East Greenland. *Geology*, **20**, 749–752.

FARROW, G. E., ALLEN, N. H. & AKPAN, E. B. 1984. Bioclastic carbonate sedimentation on a high-latitude, tide-dominated shelf: northeast Orkney Islands, Scotland. *Journal of Sedimentary Petrology*, **54**, 373–393.

GAGAN, M. K., JOHNSTON, D. P. & CARTER, R. M. 1988. The Cyclone Winifred storm bed, central Great Barrier Reef shelf, Australia. *Journal of Sedimentary Petrology*, **58**, 845–856.

GAWTHORPE, R. L., MCMURRAY, L., HALL, M. & HUNT, D. 1995. The variability of forced regressive deposits in tectonically active settings. *In*: *Sedimentary Response to Forced Regression Congress, Geological Society, September 7 & 8*, London, UK

GRINDLEY, G. W. 1960. Sheet 8: Taupo, 1st edition. Geological Map of New Zealand 1:250,000, *Department of Scientific and Industrial Research*, Wellington.

HART, B. S. & LONG, B. F. 1996. Forced regressions and lowstand deltas: Holocene Canadian examples. *Journal of Sedimentary Research*, **B66**, 820–829.

HAYWARD, B. W. 1986. *A guide to palaeoenvironmental assessment using New Zealand Cenozoic foraminiferal faunas*. New Zealand Geological Survey Report PAL, **109**.

HAYWICK, D. W. 1990. *Stratigraphy, Sedimentology, Palaeoecology and Diagenesis of the Petane Group (Plio-Pleistocene) in the Tangoio Block, Central Hawke's Bay, New Zealand*. PhD thesis, James Cook University of North Queensland, Townsville, Australia.

—— & HENDERSON, R. A. 1991. Foraminiferal paleobathymetry of Plio-Pleistocene cyclothemic sequences, Petane Group, New Zealand. *Palaios*, **6**, 586–599.

——, CARTER, R. M. & HENDERSON, R. A. 1992. Sedimentology of 40,000 year Milankovitch-controlled cyclothems from central Hawke's Bay, New Zealand. *Sedimentology*, **39**, 675–696.

——, LOWE, D. A., BEU, A. G. HENDERSON, R. A. & CARTER, R. M. 1991. Plio-Pleistocene (Nukumaruan) lithostratigraphy of the Tangoio block, and origin of sedimentary cyclicity, central Hawke's Bay, New Zealand. *New Zealand Journal of Geology & Geophysics*, **34**, 213–225.

HEFFER, K. & MILNE, A. D. 1976. *Well completion report, Hawke Bay-1. Volume 2 – geology.* BP, Shell, Aquitaine and Todd Petroleum Ltd.

HUNT, D. & TUCKER, M. E. 1992. Stranded parasequences and the forced regressive wedge systems tract: deposition during base level fall. *Sedimentary Geology*, **92**, 1–9.

—— & —— 1995. Stranded parasequences and the forced regressive wedge systems tract: deposition during base level fall – reply. *Sedimentary Geology*, **95**, 147–160.

JAMES, N. P. & BONE, Y. 1991. Origin of a cool water, Oligo-Miocene deep shelf limestone, Eucla Platform, southern Australia. *Sedimentology*, **38**, 323–341.

KAMP, P. J. J. & NELSON, C. S. 1987. Tectonic and sea-level control on non-tropical Neogene limestones in New Zealand. *Geology*, **15**, 610–613.

——, HARMSEN, F. J., NELSON, C. S. & BOYLE, S. F. 1988. Barnacle-dominated limestone with giant cross-beds in a non-tropical, tide-swept Pliocene forearc seaway, Hawke's Bay, New Zealand. *Sedimentary Geology*, **60**, 173–195.

KOLLA, V., POSAMENTIER, H. W. & EICHENSEER, H. 1995. Stranded parasequences and the forced regressive wedge systems tract: deposition during base level fall – discussion. *Sedimentary Geology*, **95**, 139–145.

LEWIS, K. B. 1979. *Foraminifera on the continental shelf and slope off southern Hawkes Bay, New Zealand*. New Zealand Oceanographic Institute Memoir, **84**.

—— 1980. Quaternary sedimentation on the Hikurangi oblique-subduction and transform margin, New Zealand. *In*: BALLANCE, P. F. & READING, H. G. (eds) *Sedimentation in Oblique-slip Mobile Belts.* International Association of Sedimentologists, Special Publications, **4**, 171–189.

NAISH, T. R. & KAMP, P. J. J. 1997. Sequence stratigraphy of sixth-order (41 k.y.) Pliocene–Pleistocene cyclothems, Wanganui Basin, New Zealand: a case for the regressive systems trace. *Geological Society of America Bulletin*, in press.

NELSON, C. S. 1978. Temperate shelf carbonate sediments in the Cenozoic of New Zealand. *Sedimentology*, **25**, 737–771.

—— 1988. An introductory perspective on non-tropical shelf carbonates. *Sedimentary Geology*, **60**, 3–12.

NUMMEDAL, D. & SWIFT, D. J. P. 1987. Transgressive stratigraphy at sequence-bounding unconformities: some principles derived from Holocene and Cretaceous examples. *In*: NUMMEDAL, D., PILKEY, O. H. & HOWARD, J. D. (eds) *Sea-level Fluctuation and Coastal Evolution*. Society of Economic Paleontogists and Mineralogists, Special Publications, **42**, 358–370.

PETTINGA, J. A. 1982. Upper Cenozoic structural history, coastal southern Hawke's Bay, New Zealand. *New Zealand Journal of Geology and Geophysics*, **25**, 149–191.

PILLANS, B. 1983. Upper Quaternary marine terrace chronology and deformation, south Teranaki, New Zealand. *Geology*, **11**, 292–297.

PLINT, A. G. 1988. Sharp-based shoreface sequences and 'offshore bars' in the Cardium Formation of Alberta: their relationship to relative changes in sea-level. *In*: WILGUS, C. K., HASTINGS, B. S., KENDALL, C. G. ST. C., POSAMENTIER, H. W., ROSS, C. A. & VAN WAGONER, J. C. (eds) *Sea-level Changes: An Integrated Approach*. Society of Economic Paleontogists and Mineralogists, Special Publications, **42**, 358–370.

—— 1991. High Frequency relative sea-level oscillations in Upper Cretaceous shelf clastics of the Alberta foreland basin. Possible evidence for a glacio-eustatic control? *In*: MACDONALD, D. I. M. (ed.) *Sedimentation, Tectonics and Eustasy: Sealevel Changes at Active Margins*. International Association of Sedimentologists, Special Publications, **12**, 409–428.

POSAMENTIER, H. W. 1995. Controls on forced regression stratal architecture. *In*: *Sedimentary Response to Forced Regression Congress, Geological Society, September 7 & 8*, London, UK.

—— & ALLEN, G. P. 1993. Variability of the sequence stratigraphic model: effect of local basin factors. *Sedimentary Geology*, **86**, 91–109.

—— & VAIL, P. R. 1988. Eustatic controls on clastic deposition II – sequence and systems tract models. *In*: WILGUS, C. K., HASTINGS, B. S., KENDALL, C. G. ST. C., POSAMENTIER, H. W., ROSS, C. A. & VAN WAGONER, J. C. (eds) *Sea-level Changes: An Integrated Approach*. Society of Economic Paleontogists and Mineralogists, Special Publications, **42**, 125–154.

—— & ALLEN, G. P., JAMES, D. P. & TESSON, M. 1992. Forced regressions in a sequence stratigraphic framework: concepts, examples and exploration significance. *American Association of Petroleum Geologists Bulletin*, **76**, 1687–1709.

SMITH, A. M. 1988. Preliminary steps toward formation of a generalized budget for cold-water carbonates. *Sedimentary Geology*, **60**, 323–332.

STEINHAUFF, D. M. & WALKER, K. R. 1995. Recognizing exposure, drowning, and 'missed beats': platform-interior to platform-margin sequence stratigraphy of Middle Ordovician limestones, east Tennessee. *Journal of Sedimentary Research*, **B65**, 183–208.

SUGGATE, R. P., STEVENS, G. R. & TE PUNGA, M. T. (eds) 1978. *The Geology of New Zealand*. Government Printer, Wellington, New Zealand, 2 vols, 820p.

SUTER, J. R., BERRYHILL, H. L. & PENLAND, S. 1987. Late Quaternary sea level fluctuations and depositional sequences, southwest Louisiana continental shelf. *In*: NUMMEDAL, D., PILKEY, O. L. & HOWARD, J. D. (eds) *Sea-level Fluctuation and Coastal Evolution*, Society of Economic Paleontogists and Mineralogists Special Publications, **41**, 199–219.

TESSON, M., GENSOUS, B., ALLEN, P. & RAVENNE, C. 1990. Late Quaternary deltaic lowstand wedges on the Rhône continental shelf, France. *Marine Geology*, **91**, 325–332.

TRINCARDI, F. & FIELD, M. E. 1991. Geometry, lateral variation and preservation of downlapping regressive shelf deposits: eastern Tyrrhenian Sea margin, Italy. *Journal of Sedimentary Petrology*, **61**, 775–790.

VAIL, P. R., MITCHUM, R. M. JR., TODD, R. G., WIDMIER, J. M., THOMPSON, S., SANGREE, J. B., BUBB., J. N. & HATLELID, W. G. 1977. Seismic stratigraphy and global changes of sea level. *In*: CLAYTON, C. E. (ed.) *Seismic Stratigraphy – Applications to Hydrocarbon Exploration*. American Association of Petroleum Geologist Memoirs, **26**, 49–212.

VAN DER LINGEN, G. J. & PETTINGA, J. R. 1980. The Makara Basin: a Miocene slope-basin along the New Zealand sector of the Australian-Pacific obliquely convergent plate boundary. *In*: BALLANCE, P. F. & READING, H. G. (eds) *Sedimentation in Oblique-slip Mobile Zones*. Special Publication, International Association of Sedimentologists, **4**, 191–215.

VAN WAGONER, J. C., POSAMENTIER, H. W., MITCHUM, R. M., VAIL, P. R., SARG, J. F., LOUTIT, T. S. & HARDENBOL, J. 1988. An overview of sequence stratigraphy and key definitions. *In:* WILGUS, C. K., HASTINGS, B. S., KENDALL, C. G. ST. C., POSAMENTIER, H. W., ROSS, C. A. & VAN WAGONER, J. C. (eds) *Sea-level Changes: An Integrated Approach*. Society of Economic Paleontogists and Mineralogists, Special Publications, **42**, 39–40.

VELLA, P. 1963. Plio-Pleistocene cyclothems, Wairarapa, New Zealand. *Transactions of the Royal Society of New Zealand, Geology*, **2**, 15–50.

WALCOTT, R. I. 1978. Present tectonics and Cenozoic evolution of New Zealand. *Geophysical Journal of the Royal Astronomical Society*, **52**, 137–317.

WALKER, G. P. L. 1973. Explosive volcanic eruptions – A new classification. *Geologische Rundschau*, 431–446.

WALKER, R. G. 1995*a*. Falling stage and lowstand shorefaces resulting from forced regressions. *In*: *Sedimentary Response to Forced Regression Congress, Geological Society, September 7 & 8*, London, UK.

—— 1995*b*. Sedimentary and tectonic origin of a transgressive surface of erosion: Viking Formation, Alberta, Canada. *Journal of Sedimentary Research*, **B65**, 209–221.

—— & PLINT, A. G. 1992. Wave – and storm-dominated shallow marine systems. *In*: WALKER, R. G.

& JAMES, N. P. (eds) *Facies Models: Response to Sea Level Change.* Geological Association of Canada, 219–238.

—— & WISEMAN, T. R. 1995. Lowstand shorefaces, transgressive incised shorefaces, and forced regressions: examples from the Viking Formation, Joarcam Area, Alberta. *Journal of Sedimentary Research*, **B65**, 132–141.

WELLS, P. 1989. Burial History of Late Neogene sedimentary basins on part of the New Zealand convergent plate marbin. *Basin Research*, **2**, 145–160.

Response of Plio-Pleistocene mixed bioclastic–lithoclastic temperate-water carbonate systems to forced regressions: the Calcarenite di Gravina Formation, Puglia, SE Italy

MARCELLO TROPEANO[1] & LUISA SABATO[2]

[1]*Centro di Geodinamica, Università della Basilicata, via Anzio 10, 85100 Potenza, Italy (e-mail: tropeano@unibas.it)*

[2]*Dipartimento di Geologia e Geofisica, Università di Bari. Campus Universitario, via Orabona 4, 70125 Bari, Italy (e-mail: l.sabato@geo.uniba.it)*

Abstract: Upper Pliocene–lower Pleistocene shallow-marine temperate-water carbonates of the Calcarenite di Gravina Formation crop out in the Murge area of Puglia, SE Italy, and record a regional subsidence-driven transgression that was punctuated by higher-frequency forced regressions. Sedimentation occurred during the drowning of a complexly faulted island archipelago whose bedrock was exclusively composed of deformed Cretaceous platform carbonates. High-energy temperate-water bioclastic carbonate systems dominated marine environments, but bioclasts were locally mixed with carbonate lithoclasts derived from the Cretaceous limestones bedrock and supplied to the shoreline via ephemeral rivers. This setting allows us to compare the depositional response of bioclastic-dominated and mixed bioclastic–lithoclastic temperate-water carbonate systems to relative sea-level changes, and in particular to forced regressions within a long-term transgressive sequence set.

Bioclastic dominated temperate-water carbonate systems are comprised of a nearshore non-depositional abrasion zone and an offshore accumulation zone; long-term subsidence led to erosional transgression through nearshore abrasion and bioerosion of the drowning archipelago. The bioclastic-dominated carbonate system was best developed during relative sea-level rises and highstands, with offshore cyclic subtidal carbonate successions interpreted to record higher-frequency relative sea-level fluctuations. Forced regressions and lowstands were associated with basinward migration of the abrasion zone and development of a subaerial exposure surface that passed basinward into marine rock- and softgrounds on the shelf; little additional sediment was supplied from updip karstic areas of the island archipelago where superficial drainage was limited.

In contrast, mixed bioclastic–lithoclastic carbonate systems are characterized by reciprocal sedimentation, developed where ephemeral rivers supplied carbonate lithoclasts to the shoreline. In these systems, bioclastic sedimentation typified relative sea-level rises and highstands whereas forced regressions and lowstands were associated with the development of coarse lithoclastic deposits. Forced regressive–lowstand deposits are represented by narrow progradational gravel beaches in ramp settings whereas small coarse-grained deltas formed against steep fault-bounded coastlines; both lack an aggradational component. Lower surfaces of the forced regressive–lowstand units are sharp and record abrupt basinward facies shifts. However, these basal surfaces were largely inherited, formed in the nearshore abrasion zone of the preceding transgressive–highstand bioclastic-dominated carbonate system. Rockgrounds formed in this way were not substantially modified by marine shoreface erosion during sea-level fall. The upper bounding surfaces of the forced regressive/lowstand deposits are also marine in origin and developed in response to rapid sea-level rise and landward translation of the shoreline. These surfaces were associated with nearshore abrasion and ravinement so that subaerial exposure surfaces were reworked in the marine environment and have very low preservation potential. Accordingly, the forced regression/lowstand sediment bodies are bounded by marine erosion surfaces and enclosed within sediments and/or surfaces formed in offshore environments.

The aim of this paper is to present two case studies of representative coastal terrigenous deposits that are enclosed within deeper-shelf limestones belonging to the transgressive Plio-Pleistocene Calcarenite di Gravina Formation that crops-out in Puglia, Southern Italy (Fig. 1). We consider development of coeval bioclastic and mixed bioclastic–lithoclastic temperate-water carbonate systems in terms of their facies, sequence stratigraphy and examine the nature of surface development. The terrigenous deposits are exclusively composed of rounded fragments of Cretaceous limestone (lithoclastic carbonate sands and gravels) deposited in beach–shoreface

From: HUNT, D. & GAWTHORPE, R. L. (eds) *Sedimentary Responses to Forced Regressions*. Geological Society, London, Special Publications, **172**, 217–243. 1-86239-063-0/00/$15.00

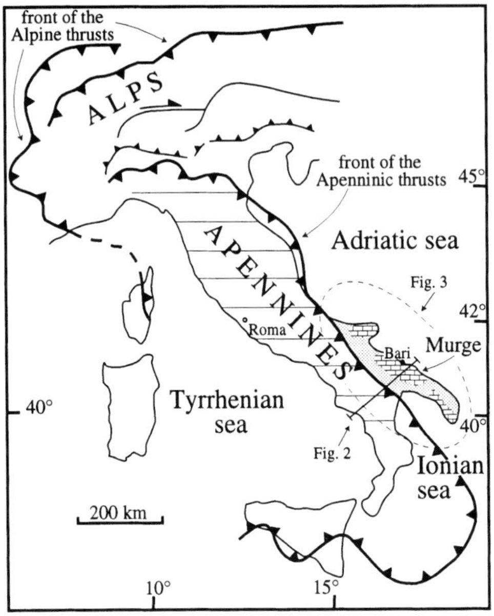

Fig. 1. Schematic structural map of Italy. The shaded area delimits the Puglia region in which the Murge area is located. Modified from Doglioni (1994).

and deltaic environments during relative sea-level falls and lowstands. These lithoclastic rocks are enclosed in surfaces and/or strata formed in mixed bioclastic–lithoclastic shallow-marine temperate-water carbonate systems. Sedimentation was reciprocal, with bioclastic-dominated transgressive and highstand deposits and lithoclastic forced regressive (FRST, *sensu*, Hunt & Tucker 1993) and lowstand (LST) deposition. Because the lithoclastic deposits sit abruptly on marine erosion surfaces and record basinward facies shifts they have direct similarities to the forced regressive 'sharp-based shoreface sequences' enclosed in offshore mudstones observed in siliciclastic shallow-marine successions (e.g. Plint 1988 and papers by Ainsworth *et al.*, Fitzsimmons & Johnson, Mellere & Steel and Plint & Nummedal this volume).

Background

Carbonates produced on open shelves, ramps and in non-tropical (*sensu* Nelson 1988) shallow-marine systems, are subjected to many of the same physical processes (i.e. hydrodynamic) that affect sediments in siliciclastic shelf/ramp systems (James 1990; Tucker & Wright 1990). Such carbonate systems are not bordered by protective shallow-water barrier reefs or shoal-rims, and therefore can display facies similar to those characteristic of shallow-marine siliciclastic systems (Burchette & Wright 1992). Along high-energy wave-dominated coasts, as considered here, three distinctive zones are commonly distinguished: beach (backshore and foreshore), shoreface and offshore transition. The beach zone occurs in an emerged and intertidal coastal setting, the shoreface extends from mean low-water level to mean fair-weather wavebase, and the offshore-transition zone extends from mean fair-weather wavebase to mean storm wavebase (e.g. Reading & Collinson 1996). Offshore facies are deposited in areas below mean storm wavebase (mid- and outer shelf/ramp).

By way of contrast, a rather different environmental zonation has been proposed from the study of modern and fossil cool-water carbonate systems. The base of wave abrasion and base of swells divides these systems into (i) a nearshore abrasion zone that corresponds to the inner shelf/ramp where rates of erosion and offshore sediment transport are higher than those of carbonate production, (ii) a swell zone that represents the mid- shelf/ramp where the sediments produced *in situ*, derived from the nearshore zone, and relict ones are frequently reworked by waves and bioturbated and (iii) a deeper zone of the outer shelf/ramp where non-phototrophic carbonate production occurs (James *et al.* 1992; Boreen & James 1995; Wright & Burchette 1996; James & Clarke 1997).

The sequence stratigraphy of carbonate systems can show a very different response to relative sea-level changes in comparison to siliciclastic counterparts (Kendall & Schlager 1981; Schlager 1992; James & Kendall 1992; Handford & Loucks 1993; Hunt & Tucker 1993; Wright & Burchette 1996). The main difference is that *in situ* production is greatest during relative rise and highstands in carbonate systems (Schlager 1991; Schlager *et al.* 1994; Pomar & Ward 1995), whereas in siliciclastic systems sediment supply is augmented during times of relative fall and lowstand. During relative sea-level fall two different types of carbonate sedimentation are distinguished; autochthonous material derived from *in situ* production and allochthonous debris, calciclastic sediments mechanically derived from the preceding highstand (Sarg 1988). However, sediment production on carbonate shelves is often reduced during falls and lowstands because the area for shallow water carbonate production is reduced (Schlager 1992; James & Kendall 1992; Handford & Loucks 1993; Hunt & Tucker 1993). During these times little sediment is derived from the platform top which undergoes subaerial diagenesis and karstification rather than mechanical reworking and cannibalization of older deposits to augment sediment supply (e.g. Hunt & Tucker 1993). Nevertheless, an increase in siliciclastic sediment supply during sea-level falls and lowstands is typical of mixed siliciclastic–carbonate depositional systems where sedimentation is characteristically reciprocal (e.g. Wilson 1967, 1975).

In response to relative sea-level falls or a forced regression (*sensu* Plint 1988; Posamentier *et al.* 1990, 1992*a*, 1992*b*) cool-water carbonate and siliciclastic systems will likely react quite differently. In shallow marine storm-dominated siliciclastic systems the abrupt seaward shifting of the shoreline in response to a forced regression is often recorded by a sharp-based shoreface sequence disconformably developed on deeper muddy facies (e.g. Bergman & Walker 1987; Plint 1988; Walker & Plint 1992; Plint & Nummedal this volume). Rather differently, in cool-water carbonate successions either marine condensed sections or hardgrounds form in middle- outer-shelf/ramp environments in response to a relative sea-level fall as the extensive non-depositional nearshore zone shifts offshore (e.g. Boreen & James 1995). Accordingly, along the margins of the Murge archipelago a quite different sedimentary expression is expected in areas of bioclastic and mixed bioclastic–lithoclastic temperate-water carbonate sedimentation to falling relative sea-level.

Geological setting

The studied deposits crop out in the Murge area of Puglia, SE Italy (Figs 1, 2), and belong to the upper Pliocene-lower Pleistocene carbonate dominated Calcarenite di Gravina Formation (Azzaroli 1968; Iannone & Pieri 1979). The formation unconformably overlies faulted Cretaceous strata of the Apulian intraoceanic Tethyan platform. The Apulian platform was a relic of Mesozoic rifting and passive margin development across the Adria African lithospheric promontory (D'Argenio 1974; Channel *et al.* 1979; Ricchetti 1980), and became an emerged continental region at the end of the Mesozoic (Ricchetti *et al.* 1988).

During the Neogene, the Apulian platform became part of the foreland to the southern Apennine mountain chain (Selli 1962; D'Argenio *et al.* 1973) (Fig. 2). The Apulian foreland became divided by the Gargano, Murge and Salento structural highs (Ricchetti *et al.* 1988) (Fig. 3). From mid- Pliocene times, the Apulian foreland underwent a relatively rapid increase in regional subsidence (Ciaranfi *et al.* 1979; Iannone & Pieri 1982), as a consequence of eastward migration of the south Apennines orogenic system and rollback of the subducting Adria plate (Malinverno & Ryan 1986; Royden *et al.* 1987; Doglioni 1991). It was in response to this subsidence, in the order of >1 km Ma^{-1} in the foredeep depocentre (Doglioni 1993, 1994), that regional transgression resulted in the progressive drowning of the Murge structural high. As it was transgressed, this high became a large island archipelago composed exclusively of a Cretaceous limestone bedrock (Pieri 1980).

High subsidence rates and low rates of sediment accumulation led to the deposition of a thin (no more than a few tens of metres thick) upper Pliocene–lower Pleistocene mantle of bioclastic and/or lithoclastic carbonates on the faulted Cretaceous rocks of the Murge high, as shown in Fig. 4 (Iannone & Pieri 1979, 1983). The vertical separation of comparable shallow

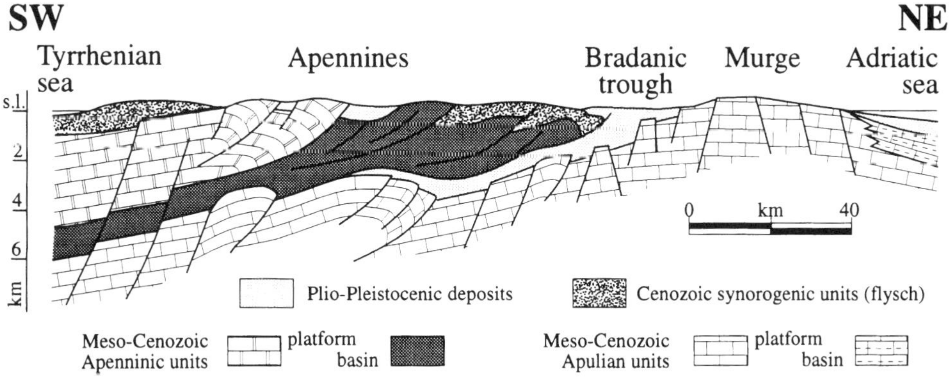

Fig. 2. Geological cross-section showing the main structural features of the southern Apennines orogenic system (modified from Sella *et al.* 1988). For location see Fig. 1.

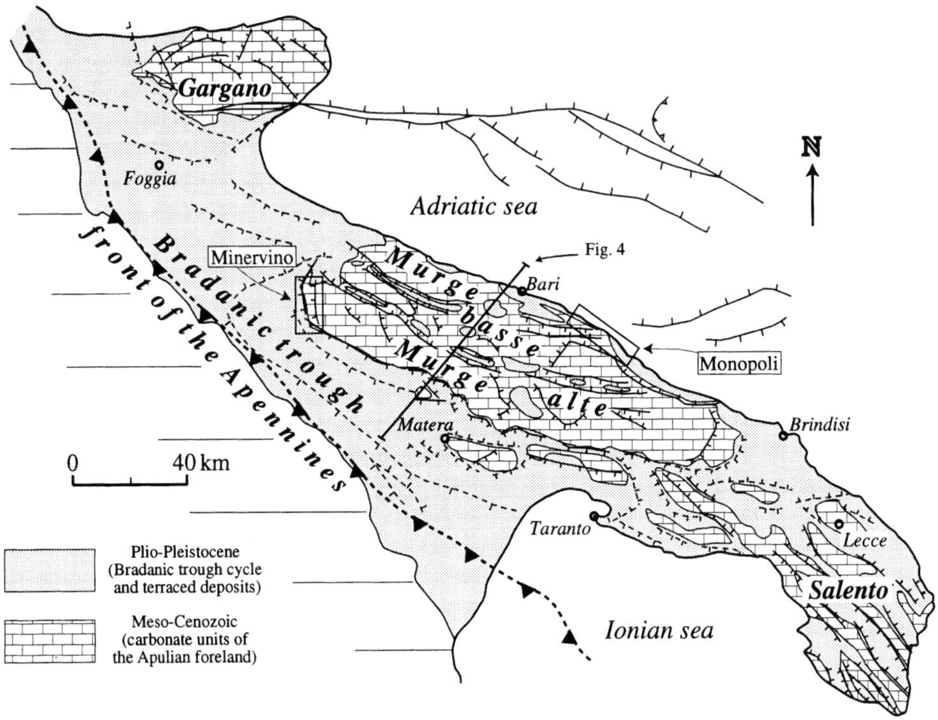

Fig. 3. Schematic geological map of the Puglia region showing the position of the Gargano, Murge and Salento structural highs of the Apulian foreland. The insets show location of the study areas (modified from Pieri *et al.* 1997). For location see Fig. 1.

marine Plio-Pleistocene carbonate deposits on the flanks of the high indicates a local long-term relative sea-level rise with a minimum amplitude of 350–400m (e.g. Fig. 4). The carbonate-dominated system was subsequently drowned by clays of the Argille subappenine Formation derived from the Apennines thrust belt during the Pleistocene (Pieri *et al.* 1996; Fig. 4).

The antecedent topography of the Murge high played an important control on the development of the Calcarenite di Gravina Formation (Pieri 1975; Iannone & Pieri 1979), as is typical of accommodation-dominated settings (Swift & Thorne 1991; Swift *et al.* 1991). Antecedent relief of the Murge high during the Plio-Pleistocene was that of a large island characterized by a large central NW–SE-trending 15–20 by 60–80 km plateau, today represented by the Murge alte, some 500–600 m above sea-level (Figs 3, 4). This central plateau area was flanked by fault-bounded NE dipping-displaced blocks (up to 15–20 by 60–80 km), with smaller and narrower 3–5 by 10–20 km blocks that variably dip SW, NW and SE; these today comprise the Murge basse plateau and the Apulian Adriatic shelf (Figs 3 and 4). This simple structure was itself cut by NW–SE-trending narrow grabens (Iannone & Pieri 1982).

Faulting of the Murge high mostly occurred prior to deposition of Plio-Pleistocene transgression and deposition of the Calcarenite di Gravina Formation (Iannone & Pieri 1983). In a regional sense deposits of this formation progressively onlap the Murge high by (i) flooding narrow shore platforms around palaeoislands (horsts) or their tops (Tropeano 1994*a*, *b*), (ii) drowning narrow straits (grabens) (Iannone & Pieri 1983) or (iii) onlapping degraded fault

Fig. 4. Schematic geological section across the Murge high. Note the stratigraphic relationships between bedrock (Cretaceous limestone) and overlying Plio-Pleistocene units. The Calcarenite di Gravina Formation is a few tens of metres thick and onlaps the flanks of the Murge high. The Calcarenite di Gravina Formation is bounded by a long-term ravinement surface below and by a drowning unconformity above.

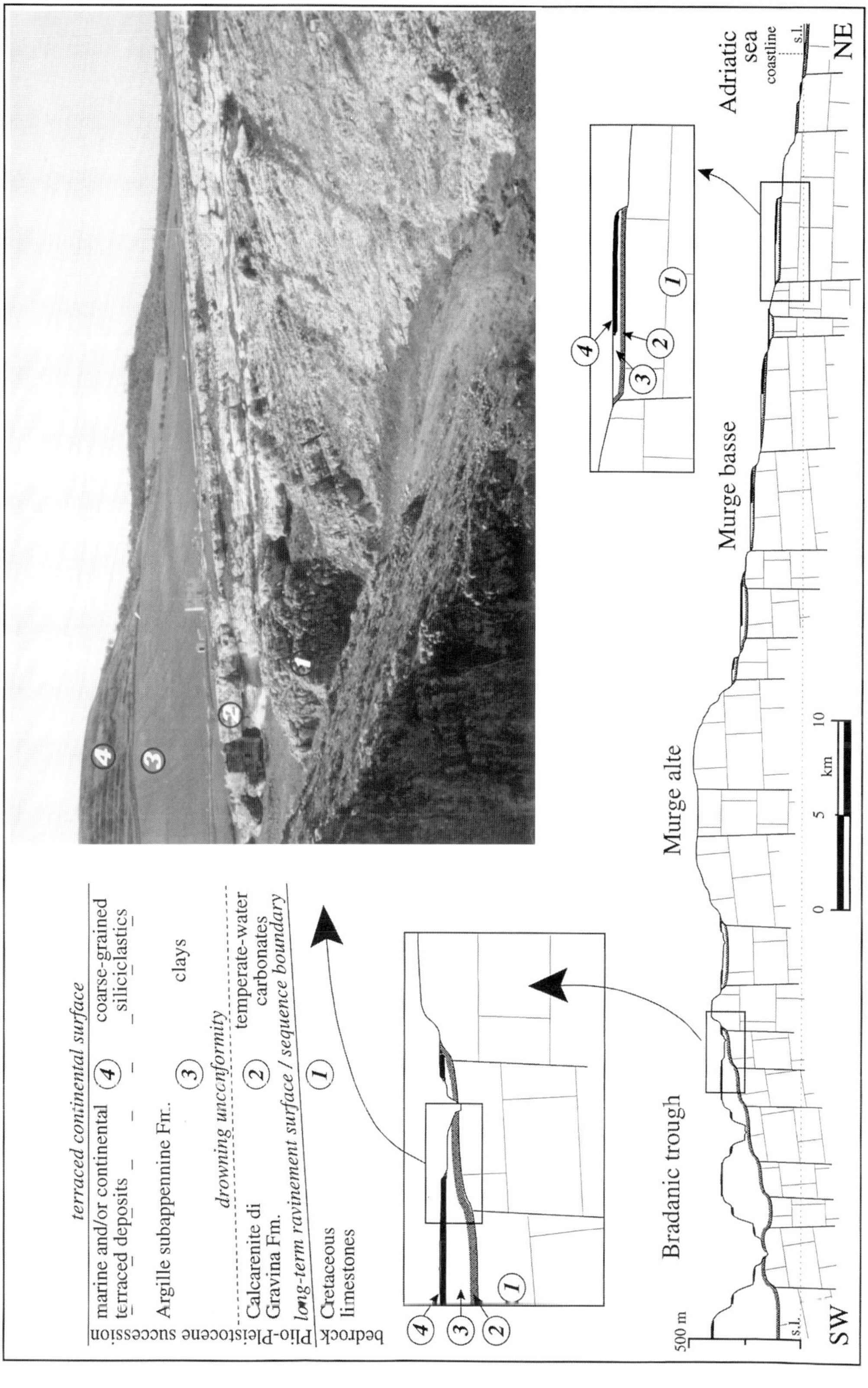
terraced continental surface
marine and/or continental terraced deposits
4
coarse-grained siliciclastics
Argille subappennine Fm.
3
clays
drowning unconformity
Calcarenite di Gravina Fm.
2
temperate-water carbonates
long-term ravinement surface / sequence boundary
Cretaceous limestones
1
Plio-Pleistocene succession
bedrock
Bradanic trough
Murge alte
Murge basse
Adriatic sea
coastline
s.l.
500 m
0
5
10
km
SW
NE

scarps (Iannone & Pieri 1979; Sabato 1996). Locally, small syndepositional extensional faults with maximum displacements of a few metres are observed in upper Pliocene–lower Pleistocene deposits. However in general active faulting did not significantly influence Plio-Pleistocene deposition (Tropeano *et al.* 1994).

Since the mid-Pleistocene the Murge region has undergone regional uplift (Ciaranfi *et al.* 1983; Doglioni *et al.* 1994), recorded by terraced deposits that disconformably overlie upper Pliocene–lower Pleistocene strata and the Cretaceous bedrock (Fig. 4; Ciaranfi *et al.* 1994; Doglioni *et al.* 1996; and references therein). It is as a consequence of this uplift and subaerial erosion that the Calcarenite di Gravina Formation is exposed today, and its relationship with the underlying bedrock and internal sedimentology and architecture revealed.

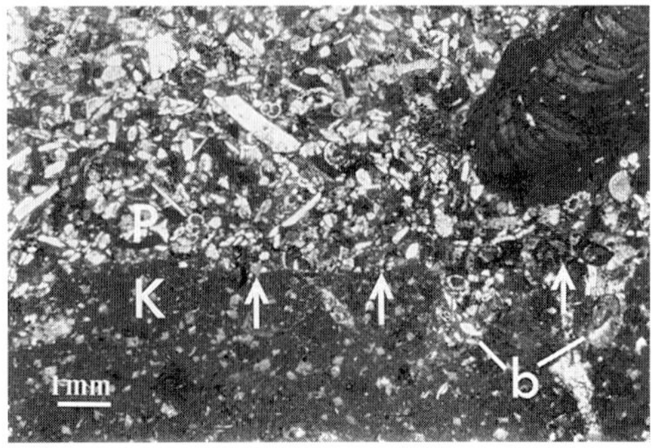

Fig. 5. Thin section showing the sharp contact between bedrock (K, Cretaceous limestone) and bioclastic packstone (P, Plio-Pleistocene = Calcarenite di Gravina Formation). Note the filled borings (b) in the bedrock, and mixture of shallow water benthic and pelagic fauna in the succeeding Plio-Pleistocene sediments. The bioclast in the top right is a red algae.

The Calcarenite di Gravina Formation

In the last 20 years interest in the Calcarenite di Gravina Formation has focused on its sedimentology and stratigraphy (e.g. see review of Tropeano 1994*a*), building on a long history of research initiated by Di Stefano & Viola (1892). A major limitation on detailed correlation within the formation is that biostratigraphic studies have only been carried out at a relatively few locations (D'Onofrio 1960; Ricchetti 1970; D'Alessandro & Iannone 1982; Bromley & D'Alessandro 1987; Caldara 1987; Tropeano *et al.* 1994; Taddei Ruggiero 1996). These studies confirm a regional late Pliocene–early Pleistocene age (Ciaranfi *et al.* 1988) although the lack of a more detailed and precise chronostratigraphic framework makes exact correlations of isolated outcrops difficult. However, contiguous sections along sea cliffs, incised river valleys and quarried areas permit physical tracing and correlation of many important stratigraphic surfaces and bodies.

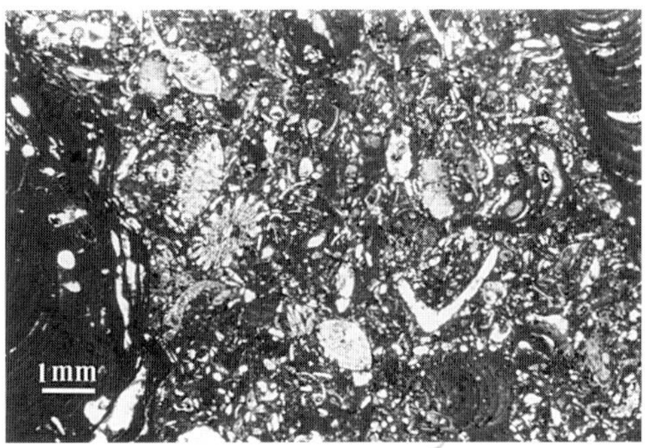

Fig. 6. Thin section of typical bioclastic facies of the Calcarenite di Gravina Formation showing a benthic-dominated open marine fauna, that lacks a shallow warm water (photozoan) fauna.

Composition

The Calcarenite di Gravina Formation is exclusively comprised of carbonate sediments that are both autochthonous and terrigenous in origin (Azzaroli 1968; Dell'Anna *et al.* 1968). The autochthonous component is dominated by bioclasts created on the shelf (Figs 5, 6) whereas the terrigenous grains are composed of rounded fragments of Cretaceous limestones (Figs 7, 8). Both carbonate grain types generally characterize coarse-grained facies that lack a significant mudstone component (Tropeano 1994*a*, *b*).

Skeletal grains are the basic components of the Calcarenite di Gravina Formation and consist of abundant bivalves, echinoids, red algae, serpulids and benthic forams with fragments of barnacles, brachiopods, gastropods, bryozoans and rare planktonic foraminifera (e.g. Figs 5, 6). This long-recognized carbonate assemblage (e.g. Di Geronimo 1969; Ricchetti 1970; Iannone & Pieri 1979; D'Alessandro & Iannone 1982; Bromley & D'Alessandro 1987; Caldara 1987) can be reinterpreted as a temperate-water deposit (Tropeano 1994*a*, *b*), following the work of Pieri (1975). The skeletal assemblage is comparable to the molechfor facies of Carannante *et al.* (1988), typical of a temperate-water open shelf or ramp carbonate factory. Significantly, the modern shelf offshore

Fig. 7. Pebble lag at the base of the Calcarenite di Gravina Fm (P) on the abraded and bored bedrock (K, Cretaceous limestone). Pebbles are rounded fragments of Cretaceous limestone. Arrows indicate the boundary. Pen for scale.

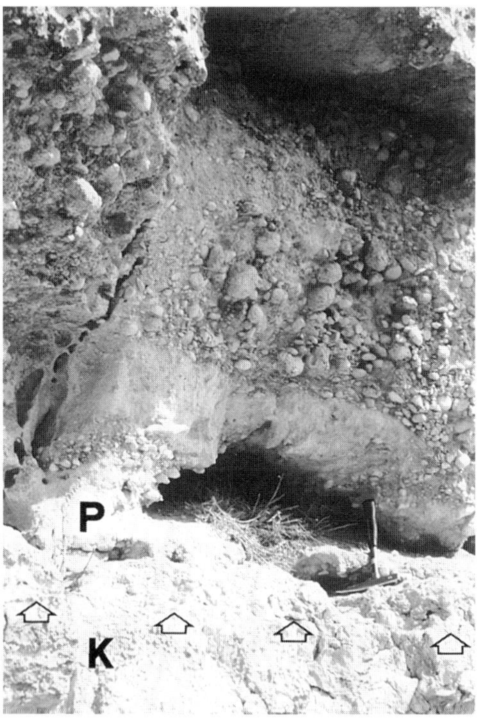

Fig. 8. Thick calclithite-calcirudite (sandy-conglomeratic) beds at the base of the Calcarenite di Gravina Fm (P) on the abraded and bored bedrock (K, Cretaceous limestone). Arrows indicate the boundary. Hammer for scale.

of Puglia, located between 40° and 42°N in the Mediterranean-temperate zone (Fig. 1), is characterized by a comparable temperate foramol carbonate assemblage (Viel & Zurlini 1986; Aiello *et al.* 1995); the late Pliocene–early Pleistocene setting of the Apulian foreland was not dis-similar to that of Puglia today, particularly to that of the Salento peninsula.

Terrigenous carbonates (calclithites and calcirudites) of the Calcarenite di Gravina Formation are commonly composed of coarse sand- and gravel-sized carbonate lithoclasts eroded from the Cretaceous bedrock of the Murge islands during transgression. Terrigenous sediment occurs as either (i) minor components of mixed bioclastic-lithoclastic carbonate deposits or (ii) the main components of lithoclastic facies. Sedimentary structures are indicative of deposition in wave- and/or storm-dominated environments. Presence of the Cretaceous limestone clasts in nearshore deposits indicates that the coastline was locally fed by terrigenous sediment via ephemeral-rivers (Sabato 1993). However, these sediment sources appear to have been small due to (i) the relatively small area and drainage basins of Murge archipelago being transgressed and (ii) the carbonate nature of the bedrock that lead to karst type drainage networks. Locally, the terrigenous carbonate deposits comprise the whole Plio-Pleistocene succession, as at Matera to the west of the Murge alte (Fig. 3; Tropeano 1994*a*, *b*), but these settings and their depositional systems differ from those described in this paper (Pomar & Tropeano 1998; 2000).

Lower bounding surface

The lower boundary of the Plio-Pleistocene deposits is a complex erosional surface developed on the Cretaceous bedrock (Perrella 1964; Figs 4, 5, 7, 8) that represents a composite sequence boundary/transgressive surface. In detail this surface is a composite one, formed of a series of marine erosional surfaces that become progressively younger higher on the Murge uplift (Fig. 4). Marine erosion has mostly removed evidence of subaerial exposure. In a regional sense the lower boundary is interpreted as a ravinement surface (*sensu* Swift 1968; Nummedal & Swift 1987), developed in response to a subsidence driven relative sea-level rise (Tropeano 1994*a*, *c*). It represents a long-term ravinement surface in the sense of Liu & Gastaldo (1992). Ravinement occurred through wave abrasion in the upper shoreface/nearshore abrasion zone of the shelf as indicated by a high density of molluscs and sponges borings in the bedrock that are characteristic of infralittoral and/or upper circalittoral environments (D'Alessandro & Iannone 1982; Bromley & D'Alessandro 1987) (Figs 5, 7). Comparable erosion surfaces are formed in coastal settings

Fig. 9. Amalgamation of mixed bioclastic and lithoclastic carbonates due to intense burrowing causing reorientation of rounded clastic grains. The latter are composed entirely of granule-grade Cretaceous limestone bedrock. Scale bar in 10 mm increments.

where rates of sediment supply are much less than those of accommodation development (Demarest & Kraft 1987), and are typical of storm/wave-dominated temperate water carbonate shelves where the upper shelf is generally current swept (James *et al.* 1994; Boreen & James 1995).

Facies

Typically, the Calcarenite di Gravina Formation is only a few tens of metres thick, and consists of a basal lithoclastic carbonate pebble lag (Fig. 7). This is overlain by amalgamated coarse-grained facies, characterized either by bioclastic-dominated packstone–grainstones or mixed bioclastic–lithoclastic calcarenites and calcirudites (Fig. 9). These amalgamated facies vertically stack in either (i) metre-scale allostratigraphic units bounded by subhorizontal diastems (*sensu* Walker 1990, 1995; Fig. 10) or (ii) subhorizontal decimetre-scale rhythmic beds (Figs 11, 12). The amalgamation of shallow-marine coarse-grained skeletal and terrigenous sediments appears to result from intense bioturbation (e.g. Fig. 9; Reinech & Sing 1980) and the influence of multiple superimposed storm events (Specht & Brenner 1979; Swift & Thorne 1991), as is typical in areas of low sediment supply (Swift *et al.* 1991). This interpretation is consistent with the temperate water setting indicated by the fauna, where carbonate production rates are generally low (Simone & Carannante 1985, 1988; James & Clarke 1997).

Carbonate facies and cyclicity

The facies and surfaces of the allostratigraphic units and characteristic rhythmically bedded strata of the Calcarenite di Gravina Formation are comparable in terms of their lithostratigraphic, sedimentological and textural characteristics to the burrowed grainstone–packstone cycles and bioturbated packstone–wackestone rhythmic beds described from cool-water Cenozoic carbonates in southern Australia (James & Bone 1991, 1994; Boreen & James 1995). Accordingly, analogous facies of the Calcarenite di Gravina Formation are interpreted to have been deposited in mid–deep open shelf/ramp settings below the level of effective abrasion down-dip from a non-depositional abrasion zone. Subtidal cycles within these facies are thought to record high-frequency relative sea-level fluctuations (e.g. Goldhammer *et al.* 1987; Collins 1988; Osleger 1991; James & Bone 1991; Jones & Desrochers 1992; Soreghan & Dickinson 1994; Boreen & James 1995; and references therein).

The diastems bounding the carbonate cycles are softground omission surfaces interpreted to form in response to the lowering of relative

Fig. 10. Angular unconformity (white arrows) between the Cretaceous limestone (K) and the overlying Plio-Pleistocene Calcarenite di Gravina Fm (P). Note the vertical stacking of metre-scale units bounded by subhorizontal diastems. Hammer (black arrow) for scale.

Fig. 11. Spectacular canyons ('*gravine*') exposure of a complete 15–20 m thick section through the Calcarenite di Gravina Formation along the Bradano River. Here the formation is comprised of bioclastic amalgamated offshore facies stacked in decimetre-scale rhythmic subhorizontal bed sets. White arrows indicate the boundary between Cretaceous limestone below (K) and Calcarenite di Gravina Fm above (P). Circle indicates the location of Fig. 12.

Fig. 12. Decimetre-scale rhythmic subhorizontal beds are separated by bioturbated softground surfaces that weather prominently (detail of Fig. 11). Hammer for scale.

sea-level, and the scouring and erosion of the sea floor as the nearshore non-depositional zone is superimposed across sub-wavebase environments of the preceding highstand. Softground development reflects the absence of pervasive early marine cements that generally characterize tropical carbonates; loose grains are prone to recycling during sea-level falls and lowstands in the temperate-water environments. The resulting omission surface formed is comparable to that at the base of sharp-based shorefaces formed during sea-level fall through downshift of storm wavebase across offshore environments in siliciclastic systems (e.g. Plint 1988; Plint & Nummedal this volume). However, an important difference is that because of the constant sweeping of currents across the sea floor no sediments are deposited in the nearshore zone during times of falling sea level and lowstand. Instead deposition may be expected further downdip.

To surmise, relative sea-level falls in areas of carbonate deposition are characterized by an expansion of subaerial exposure and a downshift of the non-depositional zone where rockgrounds (*sensu* Clari *et al.* 1995) are formed updip and pass downdip to the softground bounding omission surfaces of subtidal cyclic carbonates.

Lithoclastic carbonate bodies

In areas of lithoclastic sediment supply to the coastline, the Calcarenite di Gravina Formation contains localized and isolated sigmoidal bodies almost exclusively composed of terrigenous carbonates that are bounded by erosional surfaces and/or enclosed in offshore carbonate facies. Rather than present detailed stratigraphic correlations for the whole formation that may be undermined by problems related to dating, we present examples of the representative facies, surfaces and processes associated with lithoclastic forced regression deposits within the Calcarenite di Gravina Formation. In particular we examine (i) a coarse-grained beach package deposited in a ramp setting and (ii) a coarse-grained delta sequence deposited against a faulted and cliff-bounded coastline. We adopt the terminology developed for siliciclastic or clastic rocks in general, for the textural classification of these lithoclastic carbonates. However, it should be noted that in this case: (i) sands and gravels indicate rounded limestone fragments (carbonate extraclasts) and (ii) sandstones and conglomerates indicate lithoclastic-dominated calcarenites and calcirudites (calclithites and calcrudites).

Ramp setting

The Calcarenite di Gravina Formation crops-out semi-continuously for some 35 km between Bari and Monopoli along the seacliffs of the Apulian Adriatic coast, and is particularly well exposed along a 10 km section of coast between Polignano and Monopoli (e.g. Fig. 13). Here, the Cretaceous bedrock structurally belongs to the lowermost

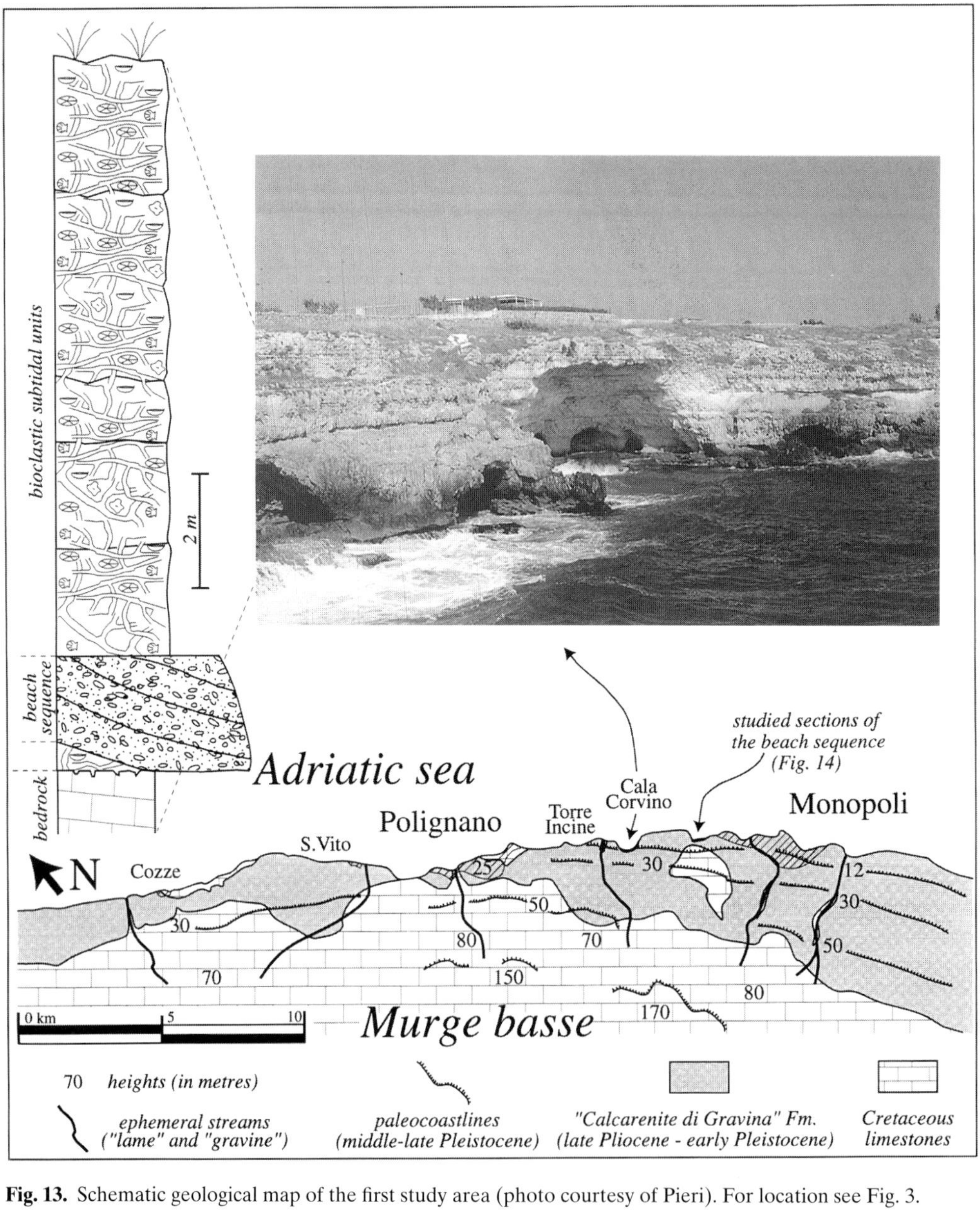

Fig. 13. Schematic geological map of the first study area (photo courtesy of Pieri). For location see Fig. 3.

plateau of the Murge basse, and dips gently basinward (Figs 3 and 4). The modern coastline is cut by short karstic canyons locally referred to as 'lame'. These are oriented perpendicular to shoreline and in combination with seacliff exposures allow for three-dimensional control of the facies and surface architecture. Representative and well-exposed stratal relationships are particularly well developed approximately 2 km NE of Monopoli (Figs 13, 14).

Surfaces and stratigraphic features. The Plio-Pleistocene succession sits unconformably on the Cretaceous bedrock, the upper surface of which dips 2–3° northeast, is wave-abraded and bioeroded by sponge and *Lithophaga* borings. The bedrock has a terraced morphology as it is cut by 1–2 m high steps. The overlying Plio-Pleistocene succession has a maximum thickness of 15 m and is comprised of (i) a lower discontinuous thin package of bioclastic facies, (ii) an erosionally-based sigmoid unit of conglomerates and (iii) an upper tabular package of bioclastic facies (e.g. see log, Fig. 13). Both the lower and upper packages of bioclastic sediments consist of burrowed and amalgamated bioclastic facies deposited in offshore open-shelf/ramp environments (Fig. 13). The eroded lower package is up to 1.5 m thick (Fig. 14). The upper package consist of metre-scale units bounded by soft-ground hiatal surfaces (D'Alessandro & Iannone 1982) interpreted to develop during short-term relative sea-level falls and lowstands in response to the lowering of wavebase. The upper package of bioclastic facies overlying the conglomerate forms tabular allostratigraphic units that are continuous for several kilometres along the coastline (e.g. at Cala Corvino; Fig. 13).

The sigmoidal unit is composed of sand-rich conglomerates (Figs 14–16) that are relatively thin (<2 m). They pinch-out landward over approximately 150 m from their maximum thickness (Fig. 14), and can be traced basinward for <200 m before exposure is lost due to orientation of the modern seacliffs. The lower bounding surface to the sigmoidal unit is erosional and the conglomerates sit on both Cretaceous bedrock and an eroded remnant of the preceding sequence composed of offshore bioclastic facies (Figs 14, 15). The upper boundary of the conglomerate unit is a sharp planar surface with a gentle basinward dip, marking an abrupt deepening from beach facies to bioclastic offshore burrowed and amalgamated facies (Figs 14, 16). Internally the conglomerates are typically organized into high-angle (<20°) seaward (NE)

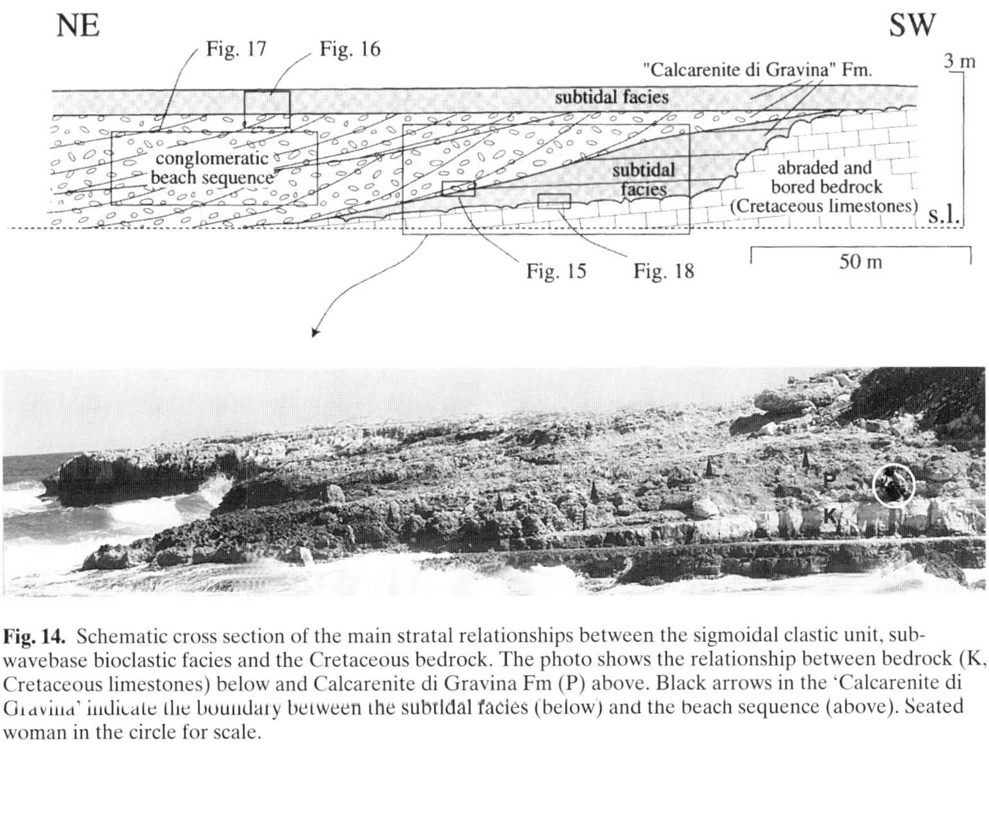

Fig. 14. Schematic cross section of the main stratal relationships between the sigmoidal clastic unit, sub-wavebase bioclastic facies and the Cretaceous bedrock. The photo shows the relationship between bedrock (K, Cretaceous limestones) below and Calcarenite di Gravina Fm (P) above. Black arrows in the 'Calcarenite di Gravina' indicate the boundary between the subtidal facies (below) and the beach sequence (above). Seated woman in the circle for scale.

Fig. 15. Detail of the sharp contact between offshore carbonates (below) and conglomerates of the beach sequence (above). For location see Fig. 14.

Fig. 16. Close-up view of the erosional contact (ravinement surface) between conglomerates of the beach sequence (below) and intense burrowed offshore carbonates (above). Bar is one metre long. For location see Fig. 14.

Fig. 17. Conglomeratic offshore dipping clinostratified beachface facies. Note the rich-oysters layers gently dipping seaward, and intensely bioeroded limestone clasts. Hammers for scale. For location see Fig. 14.

dipping cross beds that pass seaward into less-steeply dipping (1–2°) alternating beds of sandstones and conglomerates rich in oysters (Fig. 17).

Facies. In the vicinity of Monopoli, the eroded and bored upper surface of the Cretaceous carbonates is overlain by a thin (<1.5 m thick) unit of mixed bioclastic and lithoclastic carbonate strata (Figs 14, 18). These are bioturbated, fine upward, have a mixed bioclastic–lithoclastic sandy matrix and contain (20–60 mm) pebble-sized clasts. Larger clasts are rounded, bored and encrusted by bryozoans and sponges. Other constituents are abundant nodular bryozoans, pectinaceans and benthic foraminifers that are concentrated just above the Cretaceous bedrock.

The overlying erosionally-based coarse-grained body is composed of two sandy-conglomeratic bedsets, respectively 0.8 m and 1.2 m thick, each having a basal lag of bored rounded cobbles (0.1–0.15 m) and a fauna of rare pectinaceans fragments, oysters and barnacles. These normal- to reverse-graded strata display seaward dips of up to 20° in the lower bedset and 10° in the upper bed set. Individual beds are 0.1–0.2 m thick and dominated by moderately sorted clast-supported conglomerate with a scarce coarse mixed bioclastic–lithoclastic sandy matrix. Conglomerate beds are composed mostly of rounded to well-rounded, bladed and bored pebble grade clasts (average 60 mm diameter) with subangular bladed pebbles (average 10 mm diameter); cobbles (<0.15 m) and random disarticulated convex-up oyster shells are also observed. Conglomerate beds alternate with layers of large robust oysters (<0.1 m) in growth position with a sand-rich matrix. Depositional structures include imbricated disc-shaped pebbles and oyster shells with a dominant onshore dip (Fig. 19).

Offshore the conglomerates pass into less inclined (<1°–2°) alternating beds of sandstone and conglomerate. Sandstones are composed of well-sorted coarse sand and skeletal grains, and generally lack depositional structures.

Fig. 18. Sharp contact between the abraded and bored bedrock (K, Cretaceous limestones) and the subtidal carbonates at the base of Calcarenite di Gravina Fm (P). For location see Fig. 14.

Fig. 19. Detail of onshore pebble imbrication in the conglomeratic facies exposed at the Monopoli locality (land to right). Close-up view of Fig. 17, hammer for scale.

Conglomeratic beds are erosionally based, normal- to reverse-graded, matrix-supported and contain small bored pebbles within a sand grade matrix with rare oysters. Crests of the upper bedset are cut alongshore by 0.3–0.4 m thick lenses of cross-laminated coarse sands with scattered pebbles that dip alongshore. Overall the amount of sandy matrix and intensity of bioturbation increases and the size/abundance of oysters decreases so that the lithoclastic carbonate body fines up.

The coarse-grained lithoclastic package has structures and textures analogous to the modern gravel beaches of Creta (Postma & Nemec 1990), and to siliciclastic and carbonate beachface deposits (Rainone *et al.* 1981; Dabrio *et al.* 1985; Massari *et al.* 1986; Massari & Parea 1988; Pomar & Tropeano 2000). The sedimentary structures and seaward alternation of sands and conglomerates is also comparable to those of coarse-grained shorefaces (e.g. Leithold & Bourgeois 1984; Massari & Parea 1988; Hart & Plint 1995). Thus, the unconformity-bounded conglomeratic package at Monopoli is interpreted to have been deposited in beach-upper shoreface environments subject to a microtidal tidal regime.

Sequence stratigraphy. The lower surface of the conglomerate package truncates the underlying sub-wave/swellbase facies and represents a major basinward facies shift and forced regression; it is the most proximal marine expression of a high-frequency (?fourth-order) sequence boundary (Figs 13–15). The progradational beachface conglomerate is relatively thin (<2 m) and fines upward. The lack of aggradation indicates that deposition occurred during a period of sea-level stillstand following relative sea-level fall; as such the package is interpreted as a FRST–LST deposit. The sharp and erosional upper boundary of this package is a transgressive surface with offshore sub-wavebase facies recording a major landward facies shift (Figs 13, 14 and 19). Updip of the FRST–LST deposits the sequence boundary and transgressive surfaces become a coincident composite surface across

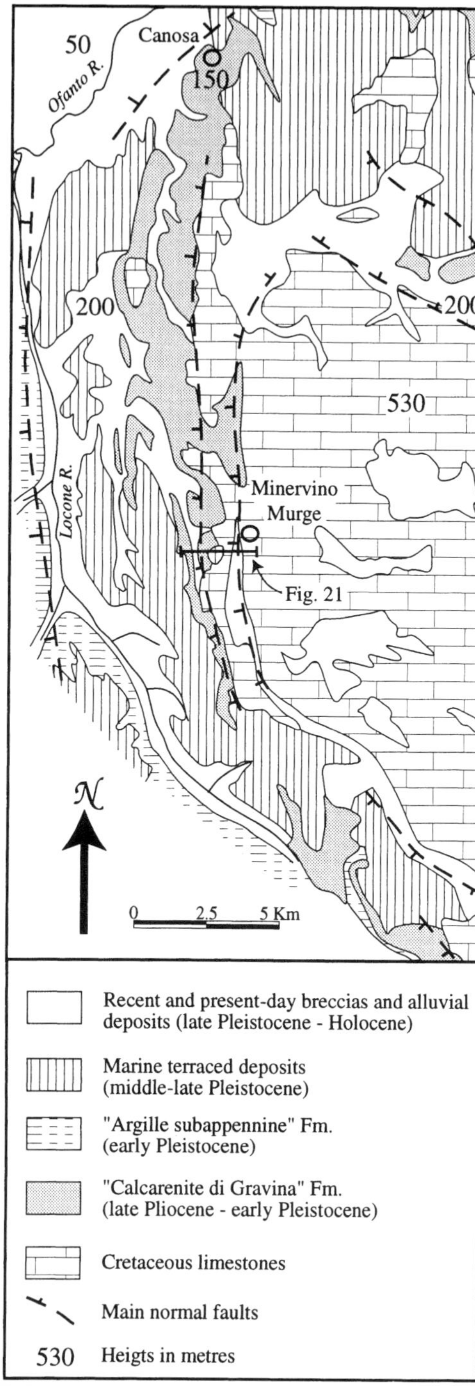

Fig. 20. Schematic geological map of the second study area, close to the village of Minervino. For location see Fig. 3.

Fig. 21. Below, a geological cross-section showing the main stratigraphic relationships developed between the Cretaceous substratum and the overlying Plio-Quaternary deposits close to Minervino locality. In the middle, detail is shown of the geological relationships of the Calcarenite di Gravina Formation within which a coarse-grained delta is developed. The two units that comprise the delta are visible along the wall of a quarry (photo). Dashed lines separate bioclastic-dominated offshore facies (a and d) from lithoclastic ones (b, foreset delta; c, shoreface/offshore-transition deposits).

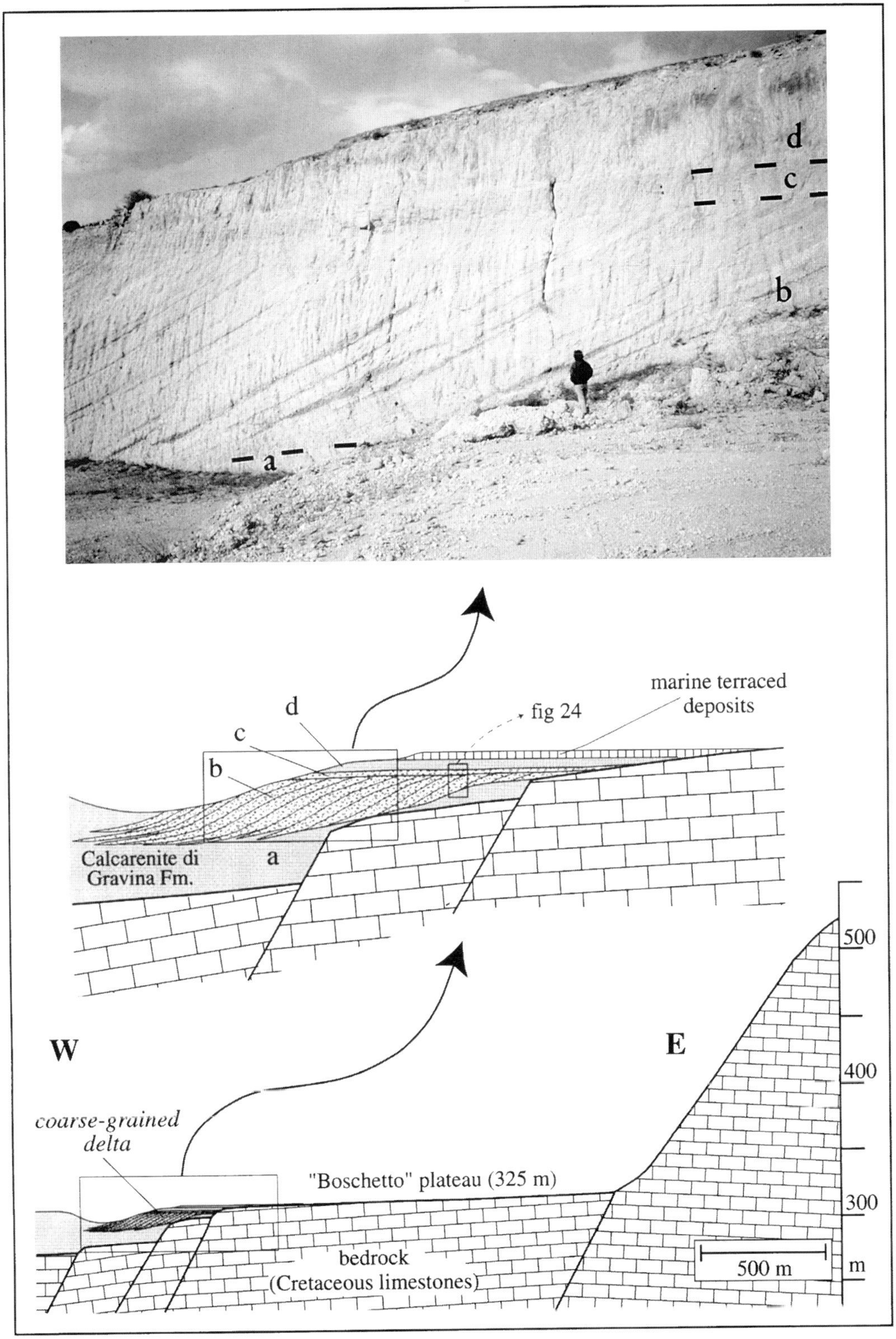
d
c
b
a
d
c
b
fig 24
marine terraced deposits
Calcarenite di Gravina Fm.
a
W
E
coarse-grained delta
"Boschetto" plateau (325 m)
bedrock (Cretaceous limestones)
500 m
500
400
300
m

which evidence for subaerial exposure has been removed by transgressive ravinement related to the passage of the non-depositional abrasion zone across the upper shelf.

Faulted seacliff setting

The western margin of the Murge alte plateau in the Minervino area displays one of the best examples of steep, fault-bounded coastal relief (Figs 3 and 20). Here the Plio-Pleistocene carbonates onlap the local substratum against either narrow and gentle basinward dipping shore platforms or against palaeocliffs. The succession is up to 40 m thick and includes a conglomerate body (<12 m thick), interpreted as a delta, deposited against a complex bedrock relief (Sabato 1993, 1996; Fig. 21). The modern relief is cut by steep karstic canyons (locally termed gravine), and comparable systems in the Plio-Pleistocene would likely have formed local sediment sources to the shoreline.

Surfaces and stratigraphic features. The conglomerate body is situated adjacent to a terrace feature near a degraded major fault scarp (Martinis 1961; Pieri 1980). This N–S-trending scarp is up to 150 m high and represents the structural boundary between the Murge alte (*c.* 500 m in height) and the so called Boschetto lower terraced surface (*c.* 325 m in height) (Fig. 21). Smaller and subparallel buried fault scarps define a WSW-dipping stepped structure (Fig. 21).

The conglomerate body is encased within offshore facies (Fig. 21) that are organized in metre-scale beds. The lowermost package of these facies (Fig. 21, unit a) represents both (i) an eroded remnant of a preceding sequence updip and (ii) toe-of-slope facies to the delta conglomerates downdip (Fig. 21, unit a).

The delta erosionally truncates the underlying older Plio-Pleistocene carbonate sequence and onlaps the Cretaceous bedrock updip (Fig. 21). The lower erosion surface of the delta dips 10° westward and is overlain by sponge, echinoid, and *Lithophaga*-bored pebbles and cobbles (Fig. 22), and is locally cut by scours and deformed by load casts (e.g. Fig. 23). Downdip the delta downlaps on the underlying Pleistocene subtidal carbonates, the Cretaceous bedrock, and also interfingers with time equivalent toe-of-slope carbonates. Updip, the delta foresets (Fig. 21, unit b) toplap against the subhorizontal lower surface of unit c, on which there is a lag deposit of outsized *Lithophaga*-bored cobbles (<0.2 m). Unit c has a sharp parallel subhorizontal contact with the overlying unit d. Progressively, units a-d onlap the basement relief further updip (Fig. 21).

Fig. 22. Marine borings in pebbles and cobbles located along the erosional surface separating the coarse-grained delta and the underlying deposits. Some clasts are composed of subrounded fragments of granules layers.

Fig. 23. Detail of a load cast at the contact between the subtidal (offshore) carbonates and the overlying conglomeratic deposits.

Facies. The subtidal carbonate facies (i) predate (Fig. 21, unit a updip), (ii) are downdip equivalents (Fig. 21, unit a downdip) and (iii) also postdate the Minervino delta sequence (Fig. 21, unit d). These deposits consist of 1–2 m thick bedsets of coarse bioclastic sediments that are occasionally lithoclastic. Their base is commonly an erosional surface covered by a thin veneer (<0.3 m) of bored pebbles with disarticulated

Fig. 24. Walls of the quarry showing the relationship between the conglomeratic body (units b and c) and the underlying and overlying offshore deposits (units a and d), in the proximal area (note the reduced thickness of the foreset beds unit). Units a–d are indicated as shown in Fig. 21.

valves of large (<0.2 m) pectinaceans and oysters. These sediments are typically intensely bioturbated and contain abundant skeletal grains including bivalves, echinoids, bryozoans, corals, barnacles, benthic and rare planktonic foraminifers within a sand grade matrix. Locally, beds of conglomerates with granules, pebbles (<6 mm) and a rare fauna of oysters, red algae, pectinaceans occur. Despite extensive bioturbation, low-angle cross-stratification is widely observed in these facies. Height of the delta foresets indicates that these facies were deposited in water depths in excess of 10 m, even in areas protected from the open ocean waves and currents (e.g the west side of the Murge uplift).

The Minervino delta has a length of approximately 200 m, foresets up to 10 m in height and interfingers with toe-of-slope subtidal carbonate facies. The most proximal part of the delta sits unconformably on the Cretaceous bedrock and is comprised of *Lithophaga*-bored boulders (<0.7 m) indicating shallow marine conditions. Relief of the west-dipping delta foresets increases basinward from a few tens of centimetres to 10 m, thickening that is accompanied by a decrease of foreset dips from an angle of 30° to a few degrees (compare Figs 21, 24). Foreset beds are wedge-shaped up to 1 m thick and have an angular base that is often erosive and truncates older foreset strata. Beds are exclusively comprised of pebble-boulder grade limestone lithoclasts derived from the Murge Cretaceous bedrock with large oysters (<0.2 m) colonizing bedding surfaces. The colonization is thought to indicate alternating periods of no fluvial supply (allowing oysters to colonize and grow) and periods of resedimentation on the slopes and/or reactivation of the fluvial system. The larger clasts and macrofossils appear to be concentrated at the foreset toe by subaqueous sediment avalanche processes.

Coarse-grained foreset beds are mostly normal graded, although massive and reverse graded beds also occur. Beds are both clast- and matrix-supported and contain well-rounded pebbles with an average diameter of 80 mm and a maximum of 0.2 m; matrix-supported beds are composed of sub-angular 30–50 mm pebbles with abundant coarse sandy-calcarenitic matrix. Many well-rounded clasts are spherical, but both blade- and disc-shaped clasts are observed; the disc-shaped clasts often show a fabric of a(p)b(t) type (*sensu* Walker 1975). The dip of foresets decreases asymptotically downdip as the clastic facies pass into bottomsets composed of matrix-rich coarse sands and interbedded conglomeratic beds. The bottomsets have smaller clasts and more abundant macrofossils in comparison with the foreset beds.

The tabular unit c that rests with erosional unconformity on the truncated delta foresets and Cretaceous bedrock (Fig. 20, 24) has a thickness of about 2 m. It fines upward and its upper surface is intensely bioturbated by echinoids (Fig. 25). The unit is comprised of conglomerates and coarse sands with a fine sand-grade matrix, and displays very low-angle (1–2°) clinostratified beds up to 0.3 m thick. Beds are normally graded and pebble grade in the lower part (40–50 mm), fining to small pebbles (average 10 mm) in the upper part with an occasional fauna of small molluscs (10–20 mm). Imbricated clasts dip both land- and seaward.

The delta is interpreted to be a classic Gilbert-type delta (Gilbert 1885), that has been wave cut, terraced and overlain by a package that records upward deepening through shoreface to offshore environments (e.g. units c-d, Figs 21 and 24; Sabato 1993, 1996). The facies and dimensions of the delta are comparable to those of the Gilbert delta described from Dalmatia by Babic *et al.* (1985) and Postma *et al.* (1988), while

Fig. 25. Characteristic biogenic structures resulting from echinoid bioturbation in the upper conglomeratic unit (c).

its sedimentary and geometric features are similar to those of the 'terraced delta' described by Postma & Cruickshank (1988). The supply of clastic sediment to the delta was probably via an ephemeral stream localized within a karstic canyon, similar to the '*gravine*' (canyons) observed today in Murge alte karstic area. Indeed, a remnant of this feeder canyon is preserved in the delta apex where large boulders set in a sandy conglomeratic matrix, fill and fossilize an erosional feature that cuts through the original steep coastal relief. Since the stream that fed the delta was apparently confined within canyon, avulsion would have been impossible, so that the cessation of clastic supply was unlikely related to autocyclic processes.

Sequence stratigraphy. The lower surface of the deltaic conglomerates passes from an updip erosional and locally angular unconformity to a downdip conformity (e.g. Figs 21, 24). Downlap onto the underlying eroded sequence of sub-wave/swellbase facies represents a major basinward facies shift and forced regression (Figs 21, 24). The lack of preserved topsets means that it cannot be determined if the delta was deposited during the fall or low-stillstand conditions, and so it is interpreted as a FRST–LST package. The sharp and erosional upper boundary of the forced regressive–lowstand delta is a transgressive surface (Figs 21, 24). Updip of the FRST–LST delta the sequence boundary and the transgressive surface become coincident. Transgressive ravinement has eroded both the delta top (e.g. topsets if any were deposited) and, with the exception of the fossilized canyon, any evidence of subaerial exposure updip of the palaeoshoreline.

Discussion

Late Pliocene–early Pleistocene high-amplitude (>350–400 m) subsidence-driven transgression of the Murge structural high, allows comparison of the deposits and surfaces formed in response to higher-order (? fourth-order) forced regressions in bioclastic-dominated and mixed bioclastic–lithoclastic temperate-water carbonate systems, preserved within a transgressive sequence set. The main controls on the depositional system were (i) antecedent physiography and the exclusively carbonate composition of the bedrock, (ii) the temperate climate that led to production of lithoclasts in the emerged karstic region and temperate-water carbonate production in marine environments and (iii) along strike variation in clastic supply. In the following section we illustrate the depositional systems characterizing the Plio-Pleistocene transgression and discuss the significance of forced regression deposits on the Murge high in terms of their enclosing surfaces and deposits, their preservation potential, their recognition, nature of their bounding surfaces, implications for interpretation of the lithoclastic carbonate strata and along-strike variability.

Depositional systems and ravinement

Transgression and ravinement occurred as the Plio-Pleistocene shorelines migrated landward and upward across the flanks of the Murge high on a complex surface, the relief and gradient of which were controlled by the antecedent uplift, tilting and erosional degradation of faulted blocks (e.g. Figs 4, 14, 21, 26, 27). Some of the

Fig. 26. Block diagrams showing the development of the shallow-marine systems on the Murge flanks. (**I**) Temperate-water carbonate systems mainly developed during relative sea-level rises and highstands, and were composed of a nearshore abrasion (non-depositional) zone and an offshore accumulation zone. (**II**) Landward shifting of these systems allowed the abrasion and bioerosion of the drowning Cretaceous substrata and produced an erosional transgression. (**III**) Sea-level falls led to the sedimentation of lithoclastic forced-regression/lowstand deposits in sharp and/or erosional contact on preceding offshore carbonates. In open shelf/ramp coastal settings (on the right) narrow gravelly beaches prograded; in fault-related cliff settings (on the left) small coarse-grained deltas were built up. In offshore environments softground omission surfaces were developed at these times.

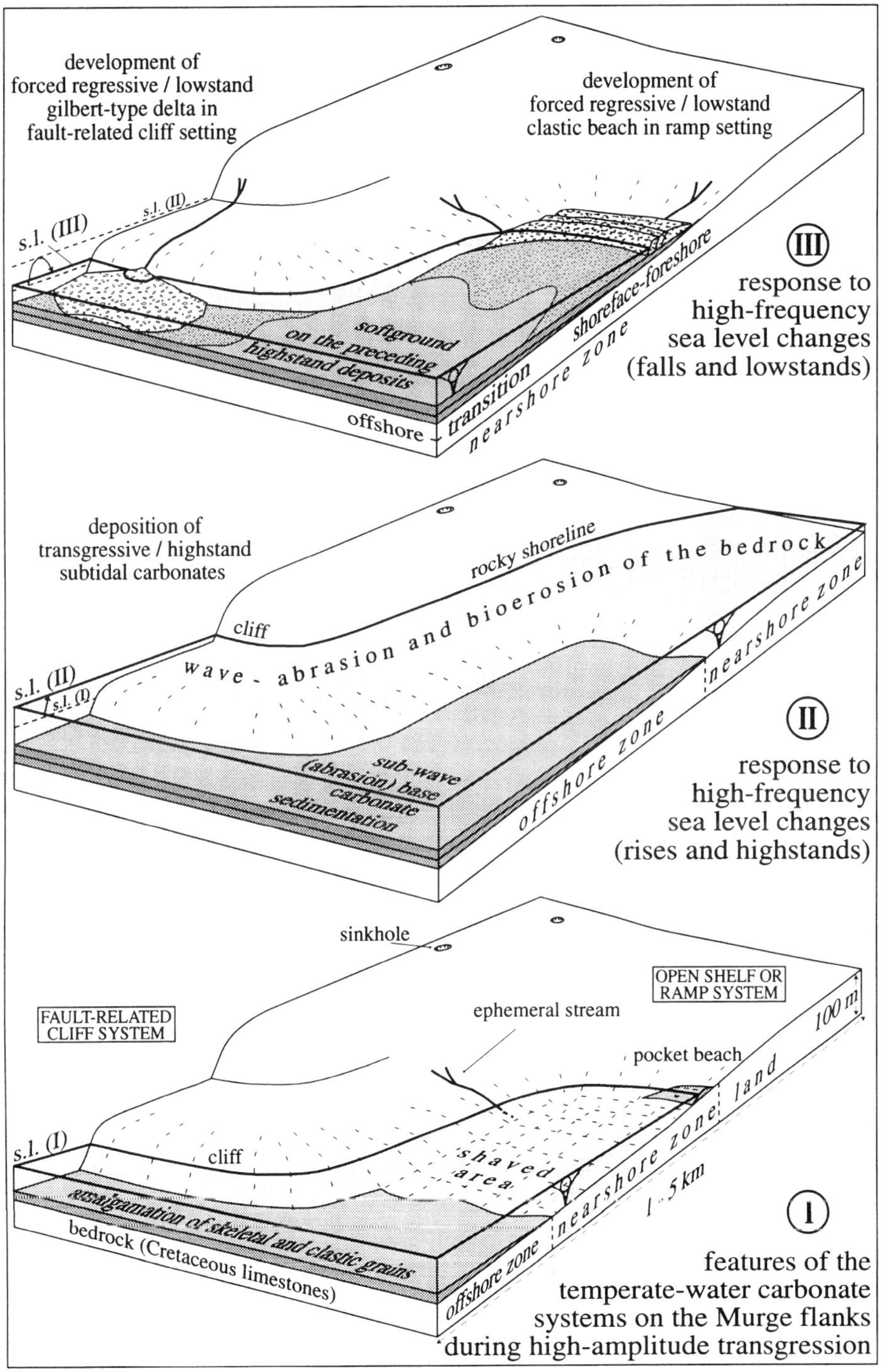
development of
forced regressive / lowstand
gilbert-type delta in
fault-related cliff setting
development of
forced regressive / lowstand
clastic beach in ramp setting
s.l. (III)
s.l. (II)
softground
on the preceding
highstand deposits
shoreface-foreshore
transition
offshore
nearshore zone
III
response to
high-frequency
sea level changes
(falls and lowstands)
deposition of
transgressive / highstand
subtidal carbonates
rocky shoreline
cliff
wave - abrasion and bioerosion of the bedrock
s.l. (II)
s.l. (I)
sub-wave
(abrasion) base
carbonate
sedimentation
offshore zone
nearshore zone
II
response to
high-frequency
sea level changes
(rises and highstands)
sinkhole
OPEN SHELF OR
RAMP SYSTEM
FAULT-RELATED
CLIFF SYSTEM
ephemeral stream
pocket beach
100 m
land
s.l. (I)
cliff
shaved
area
amalgamation of skeletal and clastic grains
bedrock (Cretaceous limestones)
nearshore zone
offshore zone
1 - 5 km
I
features of the
temperate-water carbonate
systems on the Murge flanks
during high-amplitude transgression

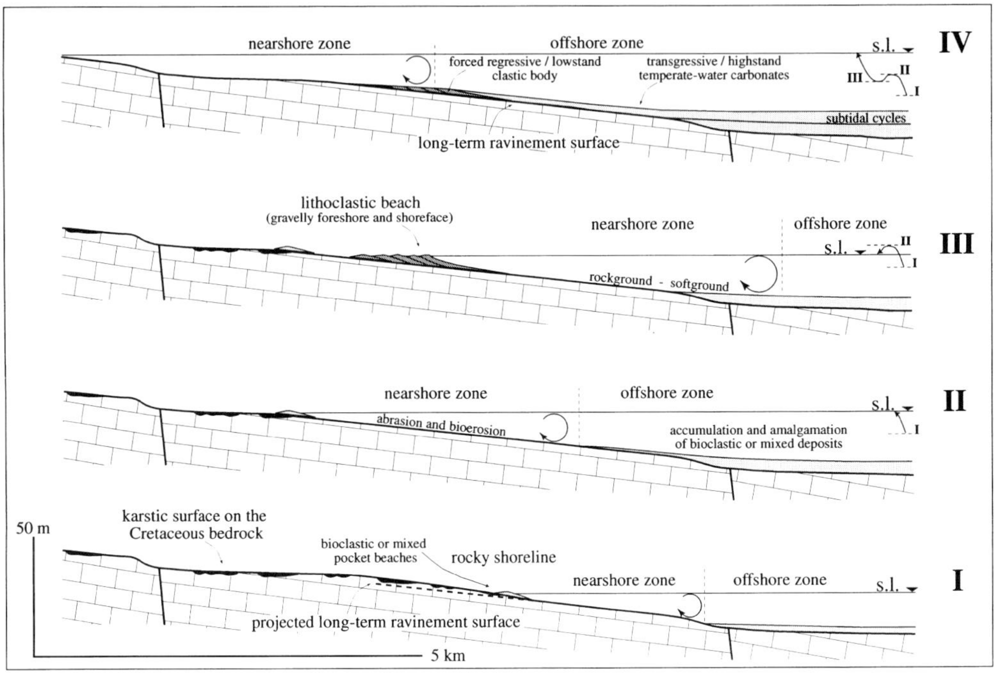

Fig. 27. Development of reciprocal sedimentation in the Calcarenite di Gravina Fm. During relative sea-level rise temperate-water bioclastic-dominated carbonate systems dominated on the Murge flanks and erosional transgression occurred (**I** and **II**). Relative sea-level changes led to the development either of carbonate subtidal cycles or, in terrigenous fed coasts, of reciprocal sedimentation. (**III**) During lowering of sea level, lowstand lithoclastic coastal deposits frequently occupied the non-depositional nearshore zone of the preceding highstand system, as indicated by the presence of conglomeratic beds on the abraded and bored bedrock. (**IV**) During the following relative sea-level rise the terrigenous supply drastically reduced, bioclastic-dominated carbonate systems restored, and the erosional transgression resumed producing a long-term ravinement surface. Modified from Tropeano & Sabato (1998).

transgressed settings were likely complex because of the presence of degraded major fault scarps between cross-cutting displaced blocks. It is likely that such areas displayed sea-floor profiles similar to those of the fault-related cliff systems observed by Hernández-Molina *et al.* (1994) on the flanks of the Cabrera Island, Spain (Fig. 26/I and II, on the left).

Formation of ravinement surfaces on the bedrock of the Murge flanks occurred as during transgression the Plio-Pleistocene depositional systems were similar to the 'classical' cool-water carbonate systems, having a nearshore profile characterized by (i) a narrow inner zone, dominated by bedrock shorelines that varied from low and gently-inclined to vertical cliffs and (ii) a more extensive zone dominated by marine erosion and the boring of the substrata (Figs 26, 27; Tropeano 1994*a*, *c*, 1995). Downdip of this abrasion zone, offshore carbonate sediment accumulation occurred. Back stepping of these depositional systems led to migration of the nearshore abrasion zone, producing a long-term ravinement surface on the bedrock. This eroded, bored and 'transgressively bypassed' surface was onlapped by aggradational offshore bioclastic-dominated carbonate facies (e.g. Fig. 26/II). The nearshore abrasion zone of these depositional systems occupied settings comparable to the shoreface/offshore-transition zones in siliciclastic shelf-ramp systems and is directly comparable to the shaved shelf of James *et al.* (1994) reported from southern Australia. However, in enclosed Mediterranean-type seas the wavebase interfaces that delimit these high-energy coastal zones are shallower than those of the oceanic seas due to limited wave fetch (Pomar & Tropeano 2000). The Calcarenite di Gravina depositional systems have a modern counterpart along the western Salento coast of the Puglia region (Fig. 3), which is characterized by an extremely irregular series of bioclastic pocket-beaches and rocky coasts. Beach bodies are <2–4 km in length, <10–20 m wide, from few hundred millimetres to few metres thick, and rest on a rocky substratum. Sediment distribution drastically reduces nearshore in 3–7 m water depth, with deeper environments characterized by a rocky seafloor (Dal Cin & Simeoni 1987). Offshore, shelf is again characterized by the presence of carbonate sediments (Viel & Zurlini 1986; Aiello *et al.* 1995) (e.g. Figs 26/I, 27/II).

Preservation. As a consequence of transgressive erosion, the preservation potential of accretionary Plio-Pleistocene deposits on the flanks of the Murge high was poor. The upper bounding surfaces of FRST–LST deposits of the Calcarenite di Gravina Formation are typically erosional transgressive surfaces (e.g. Figs 13, 14, 16, 21, 24, 27). Updip of the FRST–LST deposits, the underlying sequence boundary becomes coincident with the transgressive surface to form a composite sequence boundary/transgressive surface; expression of the higher-frequency sequence boundaries is almost always lost on the long-term ravinement surface. Across this composite surface all macroscopic evidence of subaerial exposure, and any older Plio-Pleistocene deposits have been removed; it is essentially a marine rockground representing a major erosional unconformity between Cretaceous and Plio-Pleistocene successions. Transgressive ravinement means that only sediments and surfaces formed (i) during the latest increments of a fall and (ii) during low-stillstand conditions downdip of the shoreline are preserved (e.g. Fig. 27). Consequently, any forced-regressive sediments updip of the lowest position of the shoreline were removed during transgression so that evidence of the gradual or punctuated nature of the sea-level falls was lost. If it was not for the higher-order relative sea-level falls that punctuated transgression of the Murge uplift it seems unlikely that there would be any accretionary deposits on its flanks, but instead a single enclosing long-term transgressive surface of erosion.

Forced regressions and lowstands: recognition

Recognition of forced regressions within the Calcarenite di Gravina Formation is critically dependant on (i) the preservation of offshore Plio-Pleistocene facies of older sequences below lithoclastic dominated FRST–LST deposits in mixed bioclastic–lithoclastic settings and (ii) interpretation of hiatal surfaces in offshore sub-wavebase carbonate environments.

Lower bounding surfaces. In areas of mixed bioclastic–lithoclastic deposition, the sharp and erosive contact that juxtaposes shallow-water beach and delta deposits above offshore carbonate facies indicates relative sea-level fall with a minimum amplitude of 10 m. This estimate is based on comparison with the modern Salento nearshore profile, where a sea-level fall of more than 10 m is needed to superimpose beach sediments on offshore ones.

The lower contact of the conglomeratic beachface and deltaic FRST–LST deposits is normally erosive. It is observed to truncate underlying offshore bioclastic facies so that the lower regressive erosion surface (*sensu* Nummedal *et al.* 1993) developed on both these strata, and exhumed and reworked the underlying basal transgressive surface. During sea-level falls, such wave-driven erosion occurs at the base of the foreshore–upper shoreface in response to the 'Bruun rule' (e.g. Dominguez & Wanless 1991); similarly, in deltaic settings this kind of erosion is though to be related to the rapid introduction of coarser sediment during an abrupt seaward shift of the shoreline (Posamentier & Vail 1988; Posamentier *et al.* 1992*a*).

In the areas characterized by bioclastic-dominated systems, forced regressions and lowstands are thought to be represented by the development of hiatal surfaces within subtidal cyclic carbonate deposits (Fig. 27). In such settings, deposition occurs below wave-abrasion base in offshore environments, as updip areas are characterized by current-sweeping of sediment and bioerosion. During the fall and lowstand of sea level the abrasion zone shifts downward and basinward to winnow and rework the subtidal deposits of the preceding highstand, forming hiatal softground surfaces. These pass updip to rockgrounds on the Cretaceous bedrock that are indistinguishable from those formed at any other stance of sea level (Fig. 27/III and IV).

Upper bounding surfaces. The upper bounding surface of the FRST–LST deposit is a transgressive ravinement surface that marks the abrupt upward deepening of the facies. In particular, at the top of the delta unit the transgressive ravinement surface is marked by a lag comprised of the larger clasts of the eroded delta facies. In this setting, the large amount of lithoclasts from the substratum and/or a surviving terrigenous input allowed clastic sedimentation in the initial stages of the relative sea-level rise (e.g. unit c, Figs 21, 24). The latter passes vertically into offshore facies, recording the gradual restoration of the bioclastic-dominated carbonate system.

Transgressive v. forced regressive–lowstand deposition

In the ramp and fault-cliff bound depositional settings on the flanks of the Murge uplift the preservation of an eroded remnant of bioclastic offshore facies below the overlying coarse clastic facies is critical to the recognition of forced regressive deposits. For without the presence of these offshore facies no evidence for relative sea-level fall would be preserved; the FRST–LST deposits would sit directly on the Cretaceous bedrock. In this latter case, many of the conglomeratic beds regionally observed at the base of the formation, that mostly rest on abraded and bored bedrock (i.e. Fig. 8; Azzaroli *et al.* 1968*a*, *b*; Boenzi *et al.* 1971*a*, *b*; Martinis & Robba 1971), may well require quite drastic re-interpretation as they are commonly attributed solely to transgressive processes. Using the criteria presented herein, we have also been able to reinterpret many of these clastic strata as being forced regressive and lowstand deposits. Accepting that the abraded and bored surface of the bedrock might be assumed as an offshore-transition/proximal offshore facies, such a reinterpretation has a precedent in the situation described by Plint (1988) and Walker & Plint (1992) where the sharp-based 'offshore bars' of the Cardium Formation in Canada that are enclosed in offshore facies were reinterpreted as forced regression-lowstand deposits.

Along-strike variability

Variability in the depositional response to forced regressions can be attributed to (i) changes in relief and gradient of the Cretaceous bedrock and (ii) localized supply of clastic sediment from ephemeral streams, as summarized in Figs 26 and 27. In ramp-like areas thin shallow-water FRST–LST deposits are preserved as accommodation space is limited, whereas relatively thick deposits are preserved when the low position of sea level encounters steep antecedent basement relief. In areas without lithoclastic supply forced regressions–lowstands are associated with sediment starvation, bypass and scouring to form soft- and rockground surfaces, whereas in areas with lithoclastic supply thin coastal deposits are associated with the last increments of the fall and the lowstand.

Conclusions

This study has been able to demonstrate the development of forced-regressive and lowstand deposits on the flanks of the Murge high during the Plio-Pleistocene transgression. Forced regressive-lowstand deposits (?fourth-order) are preserved within a transgressive sequence set. In localized areas of lithoclastic supply reciprocal sedimentation occurred, with shallow-water lithoclastic-dominated forced regressive–lowstand sedimentation and bioclastic-dominated transgressive–highstand deposits. Previously, many of the lithoclastic units preserved within the Calcarenite di Gravina were interpreted as transgressive lags, however we have been able to re-interpret these as forced regressive-lowstand deposits on the basis of the abrupt basinward facies shifts that occur at their base. Mainly however, Plio-Pleistocene sedimentation is dominated by deposition of a thin veneer of bioclastic-dominated temperate-water carbonate deposits that onlap deeply eroded Cretaceous bedrock. In these areas times of falling and lowstand are though to be associated with softground formation within sub-wavebase cyclic carbonate successions.

Transgressive ravinement on the wave-dominated and abraded flanks of the Murge high is the dominant process during the Plio-Pleistocene so that it appears that the only significant lithoclastic deposits preserved are forced regressive–lowstand systems tracts; for these units escaped the effects of significant nearshore transgressive erosion. The lower bounding surfaces of these lithoclastic sequences are erosional and pass downdip from softgrounds developed on eroded remnants of older offshore carbonate facies to rockgrounds formed on the Cretaceous bedrock. Without the presence of these latter strata, criteria for relative sea-level falls would be lost, because updip of the forced regressive–lowstand strata all evidence of subaerial exposure is removed during transgression. Therefore, in the accommodation-dominated setting of the Murge uplift during the Plio-Pleistocene it was only (i) the last instance of the fall, (ii) the low-stillstand condition and occasionally (iii) the onset of the rise that is recorded by Plio-Pleistocene accretion on the bored Cretaceous bedrock and multiple surface development.

This paper is partly based on the M. Tropeano doctoral dissertation and benefits by many ideas of our teacher P. Pieri. We are grateful to him and to L. Pomar which passionately discussed our field data and elaborations, and provided helpful suggestions to an earlier version of the manuscript. The manuscript has also benefited by comments of R. Gawthorpe and an anonymous reviewer. Thanks are due to E. Franseen for his comments and to B. Ward for his vigorous and useful suggestions to an advanced version of the manuscript. Special thanks are due to D. Hunt, for his corrections and fine suggestions, which have strongly improved our paper, and for his editorial patience. Encouragements of C. Doglioni to present and write this paper and discussions with him, J. Fornos and F. J. Hernández-Molina were appreciated. We are indebted with our colleagues, particularly G. Coppola, about the pataphysic approach to forced-regression concepts. Finally we would like to thank our parents and our friends who have tolerated our bad humour.

This research was supported by CNR (grants 96.00827.CT13, 98.00273.CT05 and 99.00700.CT05 to P. Pieri), ASI (grants ARS 96–47 to P. Pieri) and MURST (grants 60% 98 and 99 to L. Sabato).

References

AIELLO, G., BRAVI, S., BUDILLON, F., CRISTOFALO, G. C., D'ARGENIO, B., DE LAURO, M., FERRARO, L., MARSELLA, E., MOLISSO, F., PELOSI, N., SACCHI, M. & TRAMONTANO, M. A. 1995. Marine geology of the Salento shelf (Apulia, south Italy). Preliminary results of a multidisciplinary study. *Giornale di Geologia*, **57** (1/2), 17–40.

AINSWORTH, R. B., BOSSCHER, H. & NEWALL, M. J. 2000. Forward modelling of forced regressions: evidence for the genesis of attached and detached lowstand systems. *This volume*.

AZZAROLI, A. 1968. Studi illustrativi della Carta Geologica d'Italia – Formazioni Geologiche. *Servizio Geologico d'Italia*, **I**, 183–185.

——, PERNO, U. & RADINA, B. 1968*a*. *Note illustrative della carta Geologica d'Italia, F°188 'Gravina di Puglia'*. Servizio Geologico d'Italia.

——, RADINA, B., RICCHETTI, G. & VALDUGA, A. 1968*b*. *Note illustrative della carta Geologica d'Italia, F°189 'Altamura'*. Servizio Geologico d'Italia.

BABIC, L., POSTMA, G., ROE, S. L. & ZUPANIC, J. 1985. Gravelly Gilbert delta sedimentation in the Paleogene Promina Formation, Dalmatia, Yugoslavia (abstract) *In*: *Proceedings 6th Regional Meeting (Lerida), International Association of Sedimentologists*, 26–29.

BERGMAN, K. M. & WALKER, R. G. 1987. The importance of sea-level fluctuations in the formation of linear conglomerate bodies; Carrot Creek Member of Cardium Formation, Cretaceous Western Interior Seaway, Alberta, Canada. *Journal of Sedimentary Petrology*, **57**, 651–665.

BOENZI, F., PALMENTOLA, G., PIERI, P. & VALDUGA, A. 1971*a*. *Note illustrative della carta Geologica d'Italia, F°165 e F°176 'Trinitapoli' e 'Barletta'*. Servizio Geologico d'Italia.

——, RADINA, B., RICCHETTI, G. & VALDUGA, A. 1971*b*. *Note illustrative della carta Geologica d'Italia, F°201 'Matera'* Servizio Geologico d'Italia.

BOREEN, T. D. & JAMES, N. P. 1995. Stratigraphic sedimentology of tertiary cool-water limestones, SE Australia. *Journal of Sedimentary Research*, **B65**, 142–159.

BROMLEY, R. G. & D'ALESSANDRO, A. 1987. Bioerosion of the Plio-Pleistocene transgression of Southern Italy. *Rivista Italiana di Paleontologia e Stratigrafia*, **93**, 379–442.

BURCHETTE, T. P. & WRIGHT, V. P. 1992. Carbonate ramp depositional systems. *Sedimentary Geology*, **79**, 3–57.

CALDARA, M. 1987. Segnalazione di Pliocene medio e superiore nelle Murge settentrionali (Puglia). *Bollettino della Società Geologica Italiana*, **106**, 153–162.

CARANNANTE, G., ESTEBAN, M., MILLIMAN, J. D. & SIMONE, L. 1988. Carbonate lithofacies as paleolatitude indicators: problems and limitations. *Sedimentary Geology*, **60**, 333–346.

CHANNEL, J. E. T., D'ARGENIO, B. & HORWATH, F. 1979. Adria, the African Promontory, in Mesozoic Mediterranean Palaeogeography. *Earth Science Review*, **15**, 213–292.

CIARANFI, N., GHISETTI, F., GUIDA, M., IACCARINO, G., LAMBIASE, S., PIERI, P., RAPISARDI, L., RICCHETTI, G., TORRE, M., TORTORICI, L. & VEZZANI, L. 1983. *Carta Neotettonica dell'Italia meridionale*. Progetto Finalizzato Geodinamica del CNR, Pubblicazione **515**.

——, MAGGIORE, M., PIERI, P., RAPISARDI, L. RICCHETTI, G. & WALSH, N. 1979. *Considerazioni sulla neotettonica della Fossa bradanica*. Progetto Finalizzato Geodinamica del CNR, Pubblicazione **251**.

——, PIERI, P. & RICCHETTI, G. 1988. Note illustrative alla carta geologica delle Murge e del Salento (Puglia centromeridionale). *Memorie della Società Geologica Italiana*, **41**, 449–460.

——, —— & —— 1994. Linee di costa e terrazzi marini pleistocenici nelle Murge e nel Salento: implicazioni neotettoniche ed eustatiche (abstract) *In*: *Proceedings 77th Congress, 'Società Geologica Italiana', Bari, Italy*, 170–172.

CLARI, P. A., DELA PIERRE, F. & MARTIRE, L. 1995. Discontinuities in carbonate succession: identification, interpretation and classification of some Italian examples. *Sedimentary Geology*, **100**, 97–121.

COLLINS, L. B. 1988. Sediments and history of the Rottnest shelf, southwest Australia: a swell-dominated, non-tropical carbonate margin. *Sedimentary Geology*, **60**, 15–49.

DABRIO, C. J., GOY, J. L. & ZAZO, C. 1985. A model of conglomeratic beaches in tectonically-active areas (Late Pleistocene-Actual, Almeria, Spain) (abstract) *In*: *Proceedings 6th European Regional Meeting International Association of Sedimentologists, Lleida, Spain*, 104–105.

DAL CIN, R. & SIMEONI, U. 1987. Processi erosivi e trasporto dei sedimenti nelle spiagge pugliesi fra S. Maria di Leuca e Taranto (Mare Ionio). Possibili strategie di intervento. *Bollettino della Società Geologica Italiana*, **106**, 767–783.

D'ALESSANDRO, A. & IANNONE, A. 1982. Pleistocene

carbonate deposits in the area of Monopoli (Bari province): sedimentology and palaeoecology. *Geologica Romana*, **21**, 603–653.

D'ARGENIO, B. 1974. Le piattaforme carbonatiche periadriatiche. Una rassegna di problemi nel quadro geodinamico mesozoico dell'area mediterranea. *Memorie della Società Geologica Italiana*, **13**, 137–159.

——, PESCATORE, T. & SCANDONE, P. 1973. Schema geologico dell'Appennino meridionale (Campania e Lucania). *Accademia Nazionale dei Lincei*, **182**, 49–72.

DELL'ANNA, L., GARAVELLI, C. L. & NUOVO, G. 1968. Sui cosiddetti 'tufi calcarei' della Regione Pugliese-Lucana. *Periodico di Mineralogia*, **37**, 657–715.

DEMAREST, J. M. II & KRAFT, J. C. 1987. Stratigraphic record of Quaternary sea levels: implication for more ancient strata. *In*: NUMMEDAL, D., PILKEY, O. H. & HOWARD, J. D. (eds) *Sea-level fluctuation and coastal evolution*. Society of Economic Paleontologists and Mineralogists, Special Publications, **41**, 223–239.

DI GERONIMO, I. 1969. I depositi quaternari della costa fra Brindisi e Torre Canne. *Atti dell'Accademia Gioenia di Scienze Naturali in Catania*, serie VI, **20**, (Supplemento di Scienze Geologiche), 195–224.

DI STEFANO, G. & VIOLA, C. 1892. L'età dei tufi calcarei di Matera e di Gravina e il sottopiano 'Materino' M. E. *Bollettino del Regio Comitato Geologico d'Italia*, **23**, 125–149.

DOGLIONI, C. 1991. A proposal for the kinematic modelling of W-dipping subduction – possible application to the Tyrrenian-Apennines system. *Terra Nova*, **3**, 423 –434.

—— 1993. Some remarks on the origin of foredeeps. *Tectonophysics*, **228**, 1–20.

—— 1994. Foredeeps versus subduction zones. *Geology*, **22**, 271–274.

——, MONGELLI, F. & PIERI, P. 1994. The Puglia uplift (SE-Italy): an anomaly in the foreland of the Apenninic subduction due to buckling of a thick continental litosphere. *Tectonics*, **13**, 1309–1321.

——, TROPEANO, M., MONGELLI, F. & PIERI, P. 1996. Middle-Late Pleistocene uplift of Puglia: an 'anomaly' in the Apenninic foreland. *Memorie della Società Geologica Italiana*, **51**, 101–117.

DOMINGUEZ, J. M. L. & WANLESS, H. R. 1991. Facies architecture of a falling sea-level strandplain, Doce River coast, Brazil. *In*: SWIFT, D. J. P., OERTEL, G. F., TILLMAN, R. W. & THORNE, J. A. (eds) *Shelf sand and sandstone bodies*. International Association of Sedimentologists, Special Publications, **14**, 259–281.

D'ONOFRIO, S. 1960. Sui foraminiferi di alcuni campioni raccolti presso Gravina di Puglia. *Giornale di Geologia*, **28**, (1958/59), 203–210.

FITZSIMMONS, R. & JOHNSON, S. 2000. Forced regressions: rcognition, architecture and genesis in the Campanian of the Bighorn Basin, Wyoming. *This volume*.

GILBERT, G. K. 1885. The topographic features of lake shores. *U.S. Geological Survey, 5th Annual Report*, 69–123.

GOLDHAMMER, R. K., DUNN, P. A. & HARDIE, L. A. 1987. High-frequency glacio-eustatic sea level oscillation with Milankovitch characteristics recorded in Middle Triassic cycle platform carbonates, northern Italy. *American Journal of Science*, **287**, 853–892.

HANDFORD, C. R. & LOUCKS, R. G. 1993. Carbonate depositional sequences and systems tracts-response of carbonate platform to relative sea-level changes. *In*: LOUCKS, R. G. & SARG, J. F. (eds) *Carbonate sequence stratigraphy: recent development and applications*. American Association of Petroleum Geologists Memoirs, **57**, 3–41.

HART, B. S. & PLINT, A. G. 1995. Gravelly shoreface and beachface deposits. *In*: PLINT, A. G. (ed.) *Sedimentary facies analysis*. International Association of Sedimentologists, Special Publications, **22**, 75–99.

HERNÁNDEZ-MOLINA, F. J., SOMOZA, L., REY, J. & POMAR, L. 1994. Late Pleistocene-Holocene sediments on the Spanish continental shelves: Model for very high resolution sequence stratigraphy. *Marine Geology*, **120**, 129–174.

HUNT, D. & TUCKER, M. E. 1993. Sequence stratigrahy of carbonate shelves with an example from the mid-Cretaceous (Urgonian) of southeast France. *In*: POSAMENTIER, H. W., SUMMERHAYES, C. P., HAQ, B. U. & ALLEN, G. P. (eds) *Sequence stratigraphy and facies association*. International Association of Sedimentologists, Special Publications, **18**, 307–341.

IANNONE, A. & PIERI, P. 1979. Considerazioni critiche sui 'Tufi calcarei' delle Murge. Nuovi dati litostratigrafici e paleoambientali. *Geografia Fisica e Dinamica Quaternaria*, **2**, 173–186.

—— & —— 1982. Caratteri neotettonici delle Murge. *Geologia Applicata e Idrogeologia*, **18**, 147–159.

—— & —— 1983. Rapporti fra i prodotti residuali del carsismo e la sedimentazione quaternaria nell'area delle Murge. *Rivista Italiana di Paleontologia e Stratigrafia*, **88**, 319–330.

JAMES, N. P. 1990. Cool water carbonates: viable analogues for paleozoic limestones? (abstract) *In*: *Proceedings 13th International Sedimentological Congress, Nottingham, England*, 245–246.

—— & BONE, Y. 1991. Origin of a cool-water, Oligo-Miocene deep shelf limestone, Eucla Platform, southern Australia. *Sedimentology*, **38**, 323–341.

—— & —— 1994. Paleoecology of cool-water, subtidal cycles in mid-Cenozoic limestones, Eucla Platform, southern Australia. *Palaios*, **9**, 457–476.

—— & CLARKE. (eds) 1997. *Cool-Water Carbonates*. Society for Sedimentary Geology (SEPM), Special Publications **56**.

—— & KENDALL, A. C. 1992. Introduction to carbonate and evaporite facies models. *In*: WALKER, R. G. & JAMES, N. P. (eds) *Facies Models – response to sea level changes*. Geological Association of Canada, 265–275.

——, BONE, Y., VON DER BORCH, C. C. & GOSTIN, V. A. 1992. Modern carbonate and terrigenous sediments on a cool-water, high energy, mid-latitude shelf: Lapedece, southern Australia. *Sedimentology*, **39**, 877–903.

——, BOREEN, T. D, BONE, Y. & FEARY, D. A. 1994.

Holocene carbonate sedimentation on the west Eucla Shelf, Great Australian Bight: a shaved shelf. *Sedimentary Geology*, **90**, 161–177.

JONES, B. & DESROCHERS, A. 1992. Shallow platform carbonates. *In*: WALKER R. G. & JAMES N. P. (eds) *Facies Models – response to sea level changes*. Geological Association of Canada, 277–301.

KENDALL, C. G. ST. G. & SCHLAGER, W. 1981. Carbonates and relative changes in sea level. *Marine Geology*, **44**, 181–212.

LEITHOLD, E. L. & BOURGEOIS, J. 1984. Characteristics of coarse-grained sequences deposited in nearshore, wave-dominated environments – examples from the Miocene of south-west Oregon. *Sedimentology*, **31**, 749–775.

LIU, Y. & GASTALDO, R. A. 1992. Characteristics of a Pennsylvanian ravinement surface. *Sedimentary Geology*, **77**, 197–213.

LOMANDO, A. J. & HARRIS, P. M. 1991. Mixed cabonate – siliciclastic sequences. Preface. *In*: LOMANDO, A. J. & HARRIS, P. M. (eds.) *Mixed carbonate–siliciclastic sequences*. SEPM Core Workshop, **15**, iii–vi.

MALINVERNO, A. & RYAN, W. B. F. 1986. Extension of the Tyrrhenian Sea and shortening in the Apennines as result of arc migration driven by sinking lithosphere. *Tectonics*, **5**, 227–245.

MARTINIS, B. 1961. Sulla tettonica delle Murge nord-occidentali. *Rendiconti Accademia Nazionale dei Lincei*, serie VIII, **31**, 299–306.

—— & ROBBA, E. 1971. *Note illustrative della carta Geologica d'Italia, F°202 'Taranto'*. Servizio Geologico d'Italia.

MASSARI, F. & PAREA, G. C. 1988. Progradational gravel beach sequences in a moderate- to high-energy, microtidal marine environment. *Sedimentology*, **35**, 881–913.

——, ——, RAINONE, M. L., VEDOVATO, L. & VIVALDA, P. 1986. Elementi di sedimentologia delle paleospiagge pleistoceniche marchigiane. *In*: *Proceedings 'Riunione Gruppo Sedimentologia CNR', Ancona, Italy*. 81–103.

MELLERE, D. & STEEL, R. 2000. Style contrast between forced regressive and lowstand/transgressive wedges in the Campanian of south-central Wyoming. *This volume*.

NELSON, C. S. 1988. An introductory perspective on non-tropical shelf carbonates. *Sedimentary Geology*, **60**, 3–12.

NUMMEDAL, D. & SWIFT, D. J. P. 1987. Transgressive stratigraphy at sequence-bounding unconformities: some principles derived from Holocene and Cretaceous examples. *In*: NUMMEDAL, D., PILKEY, O. H. & HOWARD, J. D. (eds.) *Sea-level fluctuation and coastal evolution*. Society of Economic Paleontologists and Mineralogists, Special Publications, **41**, 241–260.

——, RILEY, G. W. & TEMPLET, P. L. 1993. High-resolution sequence architecture: a chronostratigraphic model based on equilibrium profile studies. *In*: POSAMENTIER, H. W., SUMMERHAYES, C. P., HAQ, B. U. & ALLEN, G. P. (eds.) *Sequence stratigraphy and facies association*. International Association of Sedimentologists, Special Publications, **18**, 55–68.

OSLEGER, D. 1991. Subtidal carbonate cycles: implication for allocyclic vs. autocyclic controls. *Geology*, **19**, 917–920.

PERRELLA, G. 1964. Aspetti della trasgressione plio-calabriana nel tratto Ponte Impiso – Oppido Lucano (Puglia – Lucania). *Bollettino della Società Geologica Italiana*, **83**, 73–76.

PIERI, P. 1975. Geologia della città di Bari. *Memorie della Società Geologica Italiana*, **14**, 379–407.

—— 1980. Principali caratteri geologici e morfologici delle Murge. *Murgia sotterranea, Bollettino Gruppo Speleo Martinense*, Martina Franca, **2**, 13–19.

—— 1991. Evoluzione geomorfologica e arretramenti costieri olocenici nei dintorni di Monopoli. *Itinerari Speleologici*, II, **5**, 121–130.

——, FESTA, V., MORETTI, M. & TROPEANO, M. 1997. Quaternary tectonic activity of the Murge area (Apulian foreland – southern Italy). *Annali di Geofisica*, **40**, 1395–1404.

——, SABATO, L. & TROPEANO, M. 1996. Significato geodinamico dei caratteri deposizionali e strutturali della Fossa bradanica nel Pleistocene. *Memorie della Società Geologica Italiana*, **51**, 501–515.

PLINT, A. G. 1988. Sharp-based shoreface sequences and 'offshore bars' in the Cardium Formation of Alberta: their relationship to relative changes in sea level. *In*: WILGUS, C. K., HASTING, B. S., KENDALL, C. G. ST.C., POSAMENTIER, H. W., ROSS, C. A. & VAN WAGONER, J. C. (eds) *Sea-level changes: an integrated approach*. Society of Economic Paleontologists and Mineralogists, Special Publications, **42**, 357–370.

—— & NUMMEDAL, D. 2000. The falling stage systems tract: recognition and importance in sequence stratigraphic analyisis. *This volume*.

POMAR, L. & TROPEANO, M. 1998. Insights for large-scale cross-bedded sandbodies encased in offshore deposits: the Calcarenite di Gravina in Matera (Southern Italy) (abstract) *In*: *Proceedings 15th International Sedimentological Congress, Alicante, Spain*, 631–633.

—— & —— 2000. The Calcarenite di Gravina Formation in Matera (southern Italy): new insights for coarse-grained, large-scale, crossbedded bodies encased in offshore deposits. *American Association of Petroleum Geologists Bulletin*, in press.

—— & WARD, W. C. 1995. Sea-level changes, carbonate production and platform architecture: the Llucmajor platform, Mallorca, Spain. *In*: HAQ (ed.) *Sequence stratigraphy and depositional response to eustatic, tectonic and climating forcing*. Kluwer Academic Publishers, Netherlands, 87–112.

POSAMENTIER, H. W. & VAIL, P. R. 1988. Eustatic controls on clastic deposition II – Sequence and systems tract models. *In*: WILGUS, C. K., HASTINGS, B. S., KENDALL, C. G. ST. C., POSAMENTIER, H. W., ROSS, C. A. & VAN WAGONER, J. C. (eds) *Sea-level changes: an integrated approach*. Society of Economic Paleontologists and Mineralogists, Special Publications, **42**, 125–154.

——, ALLEN, G. P. & JAMES, D. P. 1992a. High resolution sequence stratigraphy – The East Coulee Delta, Alberta. *Journal of Sedimentary Petrology*, **62**, 310–317.

——, ——, —— & TESSON, M. 1992b. Forced regressions in a sequence stratigraphic framework: concepts, examples, and exploration significance. *American Association of Petroleum Geologists Bulletin*, **76**, 1687–1709.

——, JAMES, D. P. & ALLEN, G. P. 1990. Aspects of sequence stratigraphy: recent and ancient examples of forced regressions (abstact) *American Association of Petroleum Geologists Bulletin*, **74**, 742.

POSTMA, G. & CRUICKSHANK, C. 1988. Sedimentology of a late Weichselian to Holocene, terraced fan delta, Varangerfjord, northern Norway. *In*: NEMEC, W. & STEEL, R. J. (eds) *Fan deltas: sedimentology and tectonic settings*. Blackie and son, London, 144–157.

—— & NEMEC, W. 1990. Regressive and transgressive sequences in raised Holocene gravelly beach, southwestern Crete. *Sedimentology*, **37**, 907–920.

——, BABIC, L., ZUPANIC, J. & ROE, S. L. 1988. Delta-front failure and associated bottomset deformation in marine, gravelly Gilbert-type fan delta. *In*: NEMEC, W. & STEEL, R. J. (eds) *Fan deltas: sedimentology and tectonic settings*. Blackie and son, London, 91–102.

RAINONE, M., NANNI, T., ORI, G. C. & RICCI LUCCHI, F. 1981. A prograding gravel beach in Pleistocene fan delta deposits south of Ancona, Italy (abstact) *In*: *Proceedings 2nd European meeting, International Association of Sedimentologists, Bologna, Italy*, 155–156.

READING, H. G. & COLLINSON, J. D. 1996. Clastic coasts. *In*: READING, H. G. (ed.) *Sedimentary environments: processes, facies and stratigraphy*, Blackwell, Oxford, 154–231.

REINECK, H. E. & SINGH, I. B. 1980. *Depositional sedimentary environments*. Springer-Verlag, New York.

RICCHETTI, G. 1970. Nuove osservazioni sui depositi plio-pleistocenici nei dintorni di Taranto. I 'Tufi' calcarei con 'Arctica islandica'. *Bollettino della Società Geologica Italiana*, **89**, 3–10.

—— 1980. Contributo alla conoscenza strutturale della Fossa bradanica e delle Murge. *Bollettino della Società Geologica Italiana*, **99**, 421–430.

——, CIARANFI, N., LUPERTO SINNI, E., MONGELLI, F. & PIERI, P. 1988. Geodinamica ed evoluzione sedimentaria e tettonica dell'avampaese apulo. *Memorie della Società Geologica Italiana*, **41**, 57–82.

ROYDEN, L., PATACCA, E. & SCANDONE, P. 1987. Segmentation and configuration of subducted lithosphere in Italy: An important control on thrust-belt and foredeep-basin evolution. *Geology*, **15**, 714–717.

SABATO, L. 1993. Gravelly terraced delta in the Lower Pleistocene 'Calcarenite di Gravina' Formation (Southern Italy) (abstact) *In*: *Proceedings 3th International Workshop on Fan Deltas, Korea*, 32–33.

—— 1996. Delta calcareo terrazzato nella Calcarenite di Gravina (Pleistocene inferiore). Minervino, Murge nord-occidentali. *Memorie della Società Geologica Italiana*, **51**, 517–526.

SARG, J. F. 1988. Carbonate-sequence stratigraphy. *In*: WILGUS, C. K., HASTINGS, B. S., KENDALL, C. G. ST.C., POSAMENTIER, H. W., ROSS, C. A. & VAN WAGONER, J. C. (eds.) *Sea-level changes: an integrated approach*. Society of Economic Paleontologists and Mineralogists, Special Publications, **42**, 155–181.

SCHLAGER, W. 1991. Depositional bias and environmental change – important factors in sequence stratigraphy. *Sedimentary Geology*, **70**, 109–130.

—— 1992. *Sedimentology and sequence stratigraphy of reefs and carbonate platforms*. American Association of Petroleum Geologists Course Notes, **34**.

——, REIJMER, J. J. G. & DROXLER, A. 1994. Higstand shedding of carbonate platforms. *Journal of Sedimentary Research*, **B64**, 270–281.

SELLA, M., TURCI, C. & RIVA, A. 1988. Sintesi geopetrolifera della Fossa bradanica (avanfossa della catena appenninica meridionale). *Memorie della Società Geologica Italiana*, **41**, 87–107.

SELLI, R. 1962. Il Paleogene nel quadro della geologia dell'Italia meridionale. *Memorie della Società Geologica Italiana*, **3**, 737–789.

SHANMUGAM, G. 1988. Origin, recognition, and importance of erosional unconformities in sedimentary basins. *In*: KLEINSPEHN, K. L. & PAOLA, C. (eds) *New perspective in basin analysis*. Springer-Verlag, New York, 83–108.

SIMONE, L. & CARANNANTE, G. 1985. Evolution of a carbonate open shelf up to its drowning: the case of the Southern Apennines. *Rendiconti dell'Accademia delle Scienze Fisiche e Matematiche in Napoli*, IV, **53**, 1–43.

—— & —— 1988. The fate of foramol ('temperate-type') carbonate platforms. *Sedimentary Geology*, **60**, 347–354.

SOREGHAN, G. S. & DICKINSON, W. R. 1994. Generic types of stratigraphic cycles controlled by eustasy. *Geology*, **22**, 759–761.

SPECHT, R. W. & BRENNER, R. L. 1979. Storm-Wave genesis of bioclastic carbonates in Upper Jurassic epicontinetal mudstones, East-Central Wyoming. *Journal of Sedimentary Petrology*, **49**, 1307–1322.

SWIFT, D. J. P. 1968. Coastal erosion and transgressive stratigraphy. *Journal of Geology*, **76**, 444–456.

—— & THORNE, J. A. 1991. Sedimentation on continental margins, I: a general model for shelf sedimentation. *In*: SWIFT, D. J. P., OERTEL, G. F., TILLMAN, R. W. & THORNE, J. A. (eds) *Shelf sand and sandstone bodies*. International Association of Sedimentologists, Special Publications, **14**, 3–31.

——, PHILLIPS, S. & THORNE, J. A. 1991. Sedimentation on continental margins, VI: lithofacies and depositional systems. *In*: SWIFT, D. J. P., OERTEL, G. F., TILLMAN, R. W. & THORNE, J. A. (eds) *Shelf sand and sandstone bodies*. International Association of Sedimentologists, Special Publications, **14**, 89–152.

TADDEI RUGGIERO, E. 1996. Biostratigrafia e paleoecologia delle Calcareniti di Gravina nei dintorni di Cerignola (brachiopodi e foraminiferi). *Memorie della società Geologica Italiana*, **51**, 197–207.

TROPEANO, M. 1993. Strutture sedimentarie deformative in calcareniti di spiaggia plio-pleistoceniche delle Murge. *Giornale di Geologia*, **55**, 201–212.

—— 1994*a*. *Sistemi costieri carbonatici nella Calcarenite di Gravina (Pliocene superiore – Pleistocene inferiore) nell'area delle Murge e della Fossa bradanica*. PhD Thesis, University of Bari.

—— 1994*b*. Caratteri deposizionali della Calcarenite di Gravina (Pliocene superiore – Pleistocene inferiore) sul bordo orientale della Fossa bradanica nell'area di Matera. *In: Guida alle escursioni del 77° Congresso della Società Geologica Italiana Bari*. Quaderni della Biblioteca Provinciale di Matera, **15**, 67–86.

—— 1994*c*. Superfici di 'ravinement' su un substrato roccioso carbonatico. Esempi nelle Murge nord-orientali (abstract) *In*: *Proceedings 77th Congress, 'Società Geologica Italiana', Bari, Italy*, 234–235.

—— 1995. Forced regression in a trasnsgressive non-tropical carbonate unit: examples from the Calcarenite di Gravina Fm. (Plio-Pleistocene) Puglia, Southern Italy (abstract) *In*: *Proceedings 'Sedimentary Responses to Forced Regression: Recognition, Interpretation and Reservoir Potential', Geological Society, London*, 78–80.

—— & SABATO, L. 1998. Reciprocal sedimentation and forced regressions in the Calcarenite di Gravina Fm (late Pliocene – early Pleistocene, Puglia, SE Italy) (abstract) *In*: *Proceedings 15th International Sedimentological Congress, Alicante, Spain*, 771–772.

——, MARINO, M. & PIERI, P. 1994. Evidenze di tettonica distensiva plio-pleistocenica al margine orientale della Fossa bradanica: l'Horst di Zagarella. *Il Quaternario*, **7**, 597–606.

——, SABATO, L., PIERI, P. & DOGLIONI, C. 1998. Geodynamic and stratigraphic history of the bradanic trough (south-Apennines foredeep – Italy) (abstract) *In*: *Proceedings 15th International Sedimentological Congress, Alicante, Spain*, 772–773.

TUCKER, M. E. & WRIGHT, V. P. 1990. *Carbonate sedimentology*. Blackwell, Oxford.

WALKER, R. G. 1975. Conglomerate: sedimentary structures and facies models. *In*: HARMS, J. C., SOUTHERD, J., SPEARING, D. R. & WALKER, R. G. (eds) *Depositional environments as interpreted from primary sedimentary structures and stratification sequences*. Society of Economic Paleontologists and Mineralogists, Shourt Course Lectures Notes, **2**, 133–158.

—— 1990. Facies modeling and sequence stratigraphy. *Journal of Sedimentary Geology*, **60**, 777–786.

—— 1995. An incised valley in the Cardium Formation at Ricinus, Alberta: reinterpretation as an estuary fill. *In*: PLINT, A. G. (ed.) *Sedimentary facies analysis*. International Association of Sedimentologists, Special Publications, **22**, 47–74.

—— & PLINT, A. G. 1992. Wave- and storm-dominated shallow marine systems. *In*: WALKER, R. G. & JAMES, N. P. (eds.) *Facies Models – response to sea level changes*. Geological Association of Canada, 219–238.

VIEL, M. & ZURLINI, G. (eds) 1986. *Indagine ambientale del sistema marino costiero della Regione Puglia*. ENEA.

WILSON, J. L. 1967. Cyclic and reciprocal sedimentation in Virgilian Strata of Southern New Mexico. *Geological Society of American Bullettin*, **78**, 805–818.

—— 1975. *Carbonate facies in geologic history*. Springer-Verlag, Berlin.

WRIGHT, V. P. & BURCHETTE T. P. 1996. Shallow-water carbonate environments. *In*: READING, H. G. (ed.) *Sedimentary environments: processes, facies and stratigraphy*. Blackwell, Oxford, 325–394.

Quaternary forced regression deposits in the Adriatic basin and the record of composite sea-level cycles

FABIO TRINCARDI & ANNAMARIA CORREGGIARI
Istituto per la Geologia Marina, via Gobetti 101, 40129 Bologna, Italy
(e-mail: fabio@igm.bo.cmr.it)

Abstract: High-resolution seismic and sedimentological data from the Central Adriatic basin reveals a Quaternary shelf-perched wedge that is comprised of four prograding units, the top surfaces of which are truncated by regional erosional surfaces. Internal reflector geometry indicates that each unit developed during highstand to falling sea-level conditions. Falling sea level resulted in successive downward and seaward shifts of the shoreline (forced regression). Because progradation occurs as a depositional continuum from highstand to lowstand conditions, sequence boundaries are difficult to recognize and correlate on a regional scale. In fact, continued sea-level fall produces several downshift surfaces of limited lateral extent. The regional erosional surfaces that truncate and hence bound the prograding units are composite in origin. The erosion surfaces formed during times of shelf subaerial exposure and were modified by shoreface and marine erosion during each subsequent rapid sea-level rise. All of these composite erosional surfaces become conformable seaward of the youngest depositional shoreline break formed at the end of each phase of progradation. These composite sequence-bounding erosional surfaces are draped by mud that is the distal equivalent of the overlying progradational unit deposited following rapid landward shifts of the shoreline.

Facies analysis and stratigraphic data in the youngest progradational unit indicate that each of the four progradational units formed in response to fourth-order (100–120 ka) cyclicity. The four progradational units stack to form an aggradational–retrogradational sequence set that records a longer-term relative sea-level rise. Such a trend can reflect regional subsidence and/or a longer-term component of rise in the Quaternary eustatic signal. In the Adriatic basin, the key factors that favour the preservation of the forced regression deposits within the Quaternary shelf-perched wedge are: (1) the composite nature of relative sea-level cycles where a longer-term relative sea-level rise interacts with shorter-term (fourth to fifth-order) sea-level cycles; (2) the asymmetry of the eustatic signal that, reinforced by local subsidence, yields relative rises of large magnitude and short duration; (3) the high amplitude of the higher-frequency fifth-order signal that drives relative sea-level falls of short duration. This kind of composite cyclicity also controlled the formation and preservation of forced regression deposits on other Quaternary continental margins as well as in ancient sedimentary successions where these deposits may occur in backstepping or aggradational multistorey sequence sets.

The stratigraphic importance of progradational units deposited during episodes of relative sea-level fall has long been recognized (e.g. Curray 1964; Vail *et al.* 1977; Plint 1988). Forced regressions deposits are important because they record significant portions of a relative sea-level fall that in updip locations are represented by a shelf-wide erosional surface. Moreover, forced regression deposits can provide a more precise measure of the actual amount of relative sea-level fall on a continental margin where downstepping offlap breaks are preserved (e.g. Pomar & Ward 1994; Franseen *et al.* 1998; papers in this volume). The factors controlling the timing of deposition and the internal facies architecture of forced regressive deposits have been reviewed recently and placed in a sequence-stratigraphic framework (Posamentier *et al.* 1992; Hunt & Tucker 1992, 1995; Helland-Hansen & Gjelberg 1994; Helland-Hansen & Martinsen 1996; Plint & Nummedal this volume; Posamentier & Morris this volume). While the mechanism of formation of forced-regression deposits is relatively well understood (Posamentier *et al.* 1992; Boyd *et al.* 1992), the factors controlling their preservation are debatable (Posamentier & Allen 1993; Trincardi 1994; Helland-Hansen & Gjelberg 1994; Plint & Nummedal this volume). For example, low rates of subsidence are envisaged to explain the formation of forced-regression deposits (Posamentier & Allen 1993; Gawthorpe *et al.* 1994), whereas in contrast high-rates of relative sea-level rise seem necessary to explain the preservation of these deposits during subsequent transgressions (Tesson *et al.* 1993; Piper & Aksu 1992).

During the Quaternary, high-frequency and high-amplitude changes in glacioeustatic sea

From: HUNT, D. & GAWTHORPE, R. L. (eds) *Sedimentary Responses to Forced Regressions*. Geological Society, London, Special Publications, **172**, 245–269. 1-86239-063-0/00/$15.00

level were related primarily to changes in the volume of water stored in the ice caps of the Northern Hemisphere. These changes were controlled by variations in the astronomic parameters such as eccentricity, obliquity and precession that drive gradual changes in the motion of the earth around the sun and control mean annual insolation of the northern hemisphere (Chappell & Shackleton 1986; Martinson *et al.* 1987). While these parameters change gradually, ice caps grow slowly and then retreat rapidly (Shackleton 1987; Anderson & Thomas 1991; Blanchon & Shaw 1995). Standard oxygen isotope records in pelagic sediment provide a proxy for the cyclic variations in eustatic sea-level during the Quaternary. These variations are characterized by eustatic gradual falls of sea-level punctuated by more rapid rises (Aharon 1983; Chappell & Shackleton 1986; Martinson *et al.* 1987). Major sea-level rises occur every *c.* 100 ka, exceed magnitudes of 100 m and show maximum rates on the order of 10 m ka^{-1} (Fairbanks 1989). Highstand conditions, associated with warm climatic intervals, represent less than 10% of the last 800 ka, whereas the remaining time is dominated by falling sea level and lowstand (Trincardi 1994; Sydow & Roberts 1994; Morton & Suter 1996; Chiocci this volume; Hernández-Molina *et al.* this volume).

The Adriatic epicontinental basin represents a foreland basin related to the Cenozoic emplacement of the Apennine chain, and provides a good example of a Quaternary shelf-perched wedge (SPW) composed of high-frequency progradational units in which an important component of sedimentation can be attributed to sedimentation during forced regression (Fig. 1). The four regressive units forming the SPW stack to form an aggradational–retrogradational sequence set (Fig. 2). Each regressive unit includes both highstand and forced regression deposits and consists of fine-grained sediment (mud to fine sand). In this paper we present new data illustrating the internal geometry of this multi-storey shelf-perched wedge and discuss the factors that control the preservation of forced-regression deposits.

Forced regressions on Quaternary Mediterranean margins

The Mediterranean basin comprises a variety of young episutural basins formed in response to the relative movements that occurred between Africa, Eurasia and other smaller plates during the Cenozoic. Because of uplift and erosion of adjacent fold-and-thrust belts sediment accumulation rates are high on several Mediterranean margins. The prevailing tectonic and sediment supply regime, in concert with a Quaternary glacioeustatic signal characterized by slow falls, rapid high-amplitude rises and short intervals in the high and low stillstand condition, favoured the deposition and preservation of Quaternary forced-regression deposits within stacks of regressive units. Depending on the rates of sediment supply and tectonic subsidence, the regressive units can stack to form backstepping or aggrading sequence sets. On the Rhone shelf, four backstepping regressive units compose a shelf-perched wedge (Tesson *et al.* 1990, 1993); each unit is composed of several prograding wedges that record a progressive seaward and downward displacement of the shoreline (Posamentier *et al.* 1992). Seaward tilting of the margin by differential compaction and tectonic subsidence are interpreted to have played a key role in preservation of these regressive units by providing seaward-increasing accommodation space during the Quaternary (Tesson *et al.* 1993). Similar interplay between these mechanisms were envisaged for other regressive Quaternary deposits delimited by transgressive erosional surfaces on margins like the northestern Alboran Sea (Hernández-Molina *et al.* 1994, 1996; Ercilla & Alonso 1996), the Gulf of Cadiz (Hernández-Molina *et al.* this volume), sectors of the Eastern Tyrrhenian margin (Marani *et al.* 1988; Chiocci this volume) and also in the Gulf of Mexico (e.g. Sydow & Roberts 1994; Kolla *et al.* this volume).

Additional examples of Quaternary forced regression in Mediterranean basins come from both extensional and collisional settings. On a shelf sector of Western Turkey, deltaic wedges separated by regional transgressive surfaces record a 100-ka eustatic cyclicity of the fourth-order (Piper & Aksu 1992). In places however, tectonic deformation may hamper detailed reconstruction of the older regressive units (Brooks & Ferentinos 1984; Trincardi & Field 1991; Field & Trincardi 1991; McMurray & Gawthorpe this volume). In general terms, literature review reveals that forced regression deposits in the Quaternary relate to multiple high-frequency cycles of sea-level change that punctuate longer-term relative sea-level rises.

Data and methods

High-resolution seismic profiles were collected during a cruise of the R/V *Urania* in August 1993 (Fig. 3). The shot interval was set at 0.25 seconds using a 300 joule UNIBOOM sound source. Recording length typically

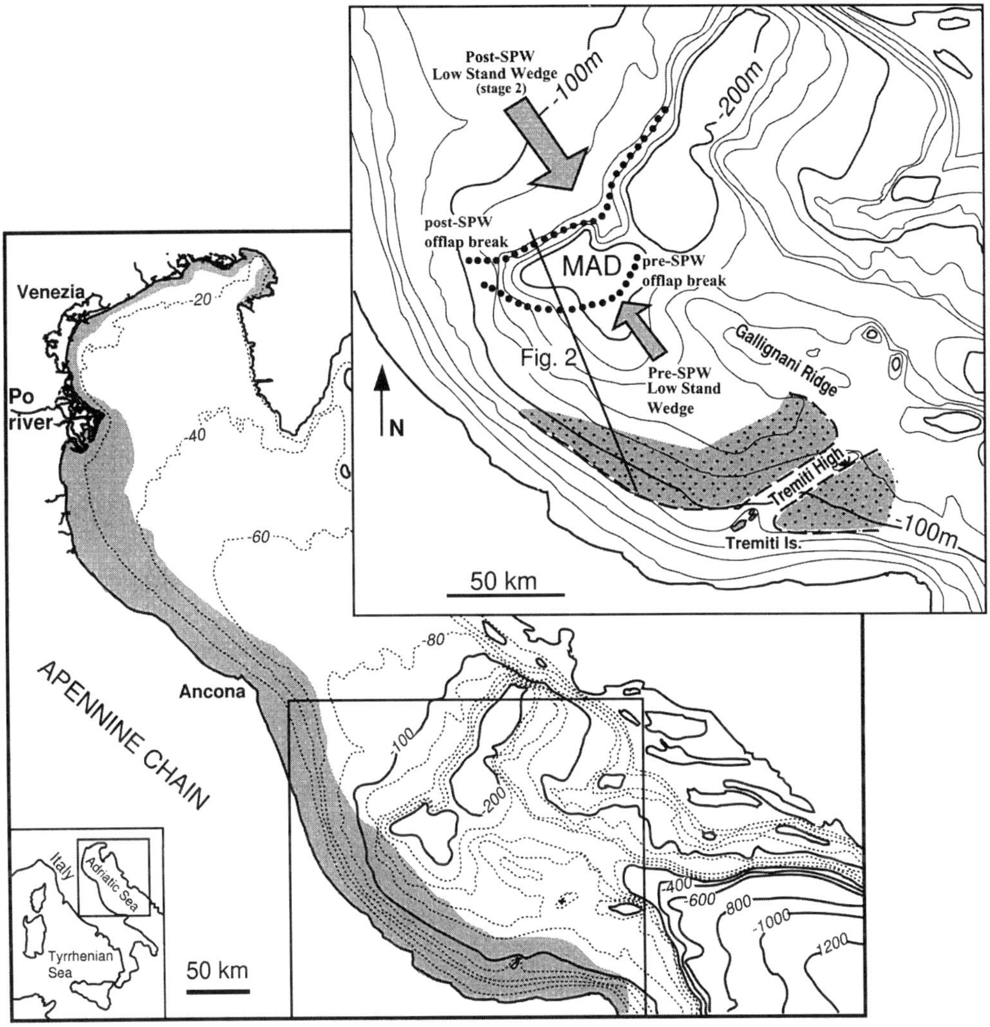

Fig. 1. The lower part of the figure shows a simplified bathymetric map of the Central Adriatic basin showing the depocenter of combined late-Quaternary transgressive (TST) and highstand (HST) systems tracts (grey tone), and the extent of the study area (box). An enlargement of the figure shows late Quaternary shelf-perched wedge (dotted) and the depositional shoreline breaks of the low-stand wedges that stratigraphically overlay and underlay the shelf-perched wedge (see also Fig. 2). Tectonic uplift during upper Quaternary occurred along the Tremiti High and the Gallignani Ridge, see text for further discussion. MAD is the Mid Adriatic Deep.

was 0.25 or 0.5 s. Additional information about the uppermost 10–50 m comes from a 3.5 kHz pinger firing at 0.125 s intervals. Piston cores were collected using either 6 or 12 m long barrels. The maximum recovery was 10 m on the slope, whereas cores on the shelf typically recovered 4–8 m. The land-based DelNorte positioning system and GPS provided an accuracy within 10 m and allowed us to place confidently the core stations on specific features detected on seismic profiles. The selection of the best targets for coring also took into account the reinterpretation of the unpublished seismic data available in the area from former IGM cruises (Correggiari *et al.* 1992; Trincardi *et al.* 1994, 1996*a*). About 40 accelerator mass spectrometry (AMS) ^{14}C dates on planktonic foraminifers provide good control on the ages of all the systems tracts composing the Late Quaternary depositional sequence; these data are reported and discussed elsewhere (Asioli 1996; Langone *et al.* 1996; Trincardi *et al.* 1996*b*). In this paper when reporting dates in years BP we implicitly refer to uncalibrated ^{14}C years before present.

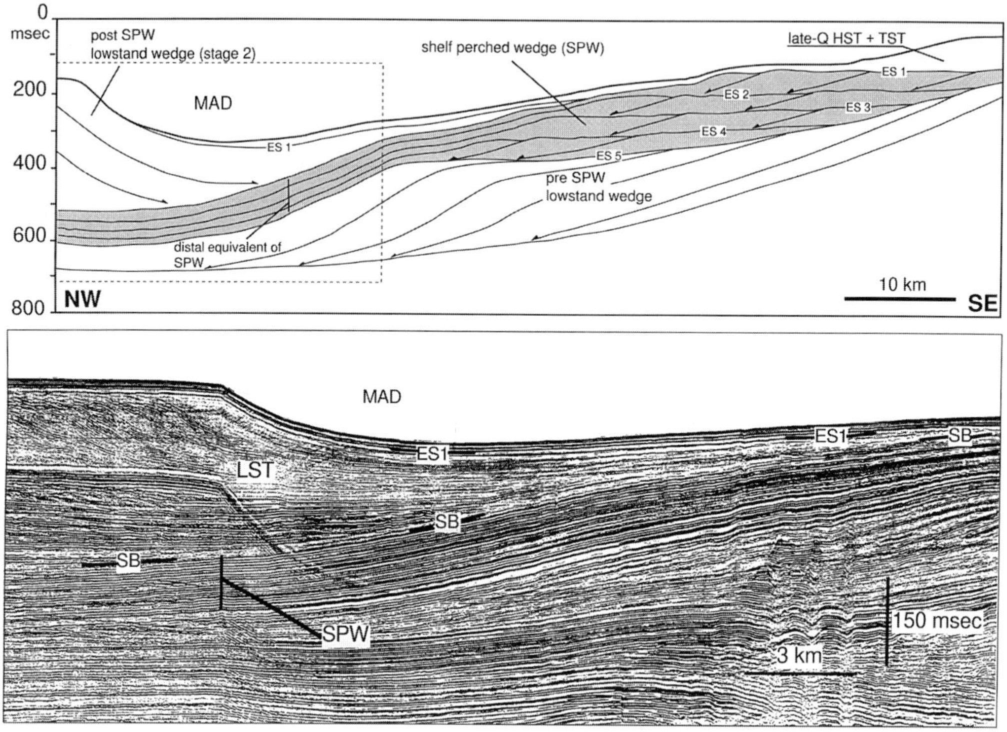

Fig. 2. Schematic combined line drawing based on high-resolution profiles across the Central Adriatic basin in the upper part shows the stratigraphic relation between the shelf-perched wedge (SPW) to the south and the overlaying lowstand progradational wedge from the northwest. Note also that: (1) the former rests below the sequence boundary (SB) while the latter lays immediately above it and (2) the late Quaternary transgressive systems tract (TST) and highstand systems tract (HST) are expanded in the south and reduced in thickness across the Mid-Adriatic Deep (MAD). The lower seismic data shows detail of intermediate-resolution common-depth-point seismic profile (recorded using a 24-channel streamer and a 20-cubic inch water-gun sound source) across the MAD and along the NW-SE basin axis (dashed box in the schematic line drawing). The profile clearly documents the 200 m thick late Quaternary LST above the sequence boundary (SB); the LST thins southward where several cores penetrated the SB and the forced regressive system tract (FRST) below it (details in Figs 5 and 12). The combined thickness of the TST and HST in the MAD is about 10 m and surface ES1 remains obscured below the outgoing pulse in this kind of record (see texts for details).

The Adriatic basin

The Adriatic epicontinental sea corresponds to the Plio-Quaternary foreland basin of the Apennine chain (Ori *et al.* 1986; Royden *et al.* 1987; Argnani *et al.* 1993). The basin is an elongated and narrow (800 × 200 km) semi-enclosed epicontinental shelf that surrounds the Mesoadriatic Deep (MAD); the MAD is a small remnant basin about 250 m deep (Figs 1 and 2). Seismic-stratigraphic studies show that the concentric fill of the MAD consists of progradational wedges of variable size deposited during the Pliocene and Quaternary (Ciabatti *et al.* 1987). In this small central basin, a continuous section of marine mud was deposited during the Quaternary. All the unconformities that punctuate the shelf-perched wedge on the surrounding shelf have a correlative conformable surface within this marine section (Fig. 2).

Seismic-stratigraphic studies document major changes in sediment flux, depositional style and direction of progradation during the eastward translation of the Apennine chain in Pliocene and Quaternary times (Ori *et al.* 1986). During the Late Pleistocene, the largest progradational unit was fed from the north by the combined supplies of the Po Plain (the major drainage area feeding the basin) and several smaller coalescing drainage basins of the uplifting Apennine chain (Trincardi *et al.* 1996*b*).

The Adriatic foreland region is segmented in

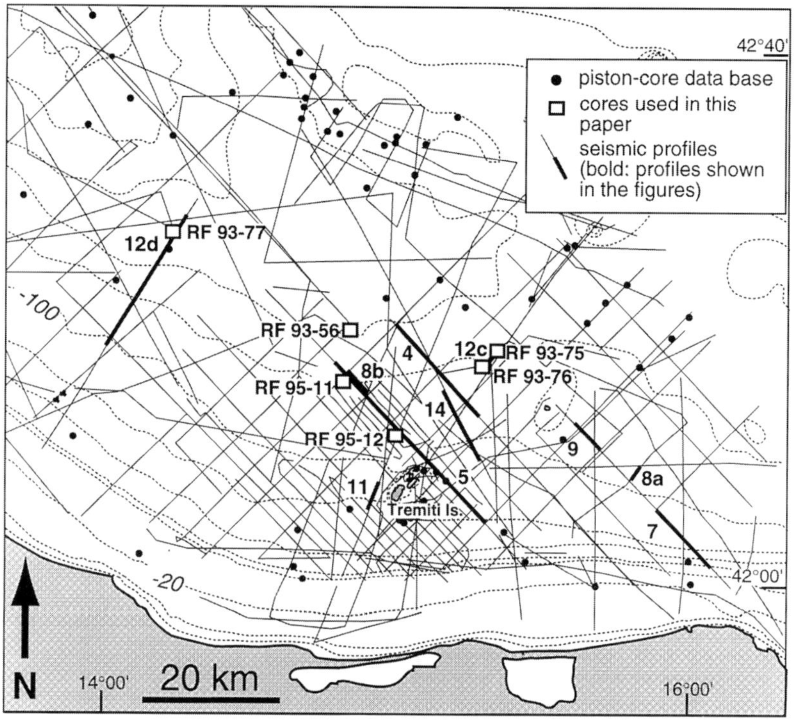

Fig. 3. Map showing the database of seismic profiles and piston cores in the study area. Schematic bathymetry is given every 20 m. The modern – 140 m bathymetric contour approximates the position of the shoreline break at maximum lowstand of sea level during the Last Glacial Maximum. The seismic profiles shown in this paper (bold lines) refer to figure numbers. The cores discussed in Figs 5 and 12 are indicated by open squares.

distinct domains that show different lithospheric thickness, state of deformation and rates of subduction (Royden *et al.* 1987). The study area lies in a region of intense tectonic deformation during the Quaternary, as indicated by historical seismicity (Anderson & Jackson 1987; Tinti *et al.* 1995). Tectonic deformation and uplift occurred during the Late Quaternary within two distinct domains in the southern area, the Tremiti High and the Gallignani Ridge (Figs 1 and 5). The former is a NE-SW ridge affected by uplift before and during the deposition of the shelf-perched wedge (Figs 4, 5 and 6). The Tremiti High acted as a right lateral transfer zone separating two segments of the Adriatic plate that underwent different rollback rates (Doglioni *et al.* 1994). Within this geodynamic setting, the Central Adriatic area underwent high subsidence rates through the Quaternary while the Puglia foreland, to the south, underwent uplift since mid Quaternary times and buckling of the lithosphere (Doglioni *et al.* 1994). The area north of the Tremiti transfer zone underwent tilting and differential subsidence during the Quaternary whereas in contrast the marine area immediately south of this feature was subject to lower rates of subsidence (Fig. 1). This differential subsidence is indicated by (i) structural maps of the base of the Pliocene (Correggiari *et al.* 1992) and (ii) by the reduced thickness of the SPW to the south indicating a lower rate of accommodation development (Figs 5, 6). The structural map of the base of the shelf-perched wedge confirms the separation of a subsiding area to the north from a more stable area to the south (Fig. 5); the main depocentre of the SPW (exceeding 150 ms) occurs north of the Tremiti transfer zone (Fig. 6). The Gallignani Ridge is a more complex structural relief trending NW–SE where Pliocene and Quaternary beds are folded and faulted. Deposits that represent the distal equivalent of the shelf-perched wedge thin toward the ridge and can be sampled selectively as well as the underlying older Pleistocene beds (Pasini *et al.* 1993).

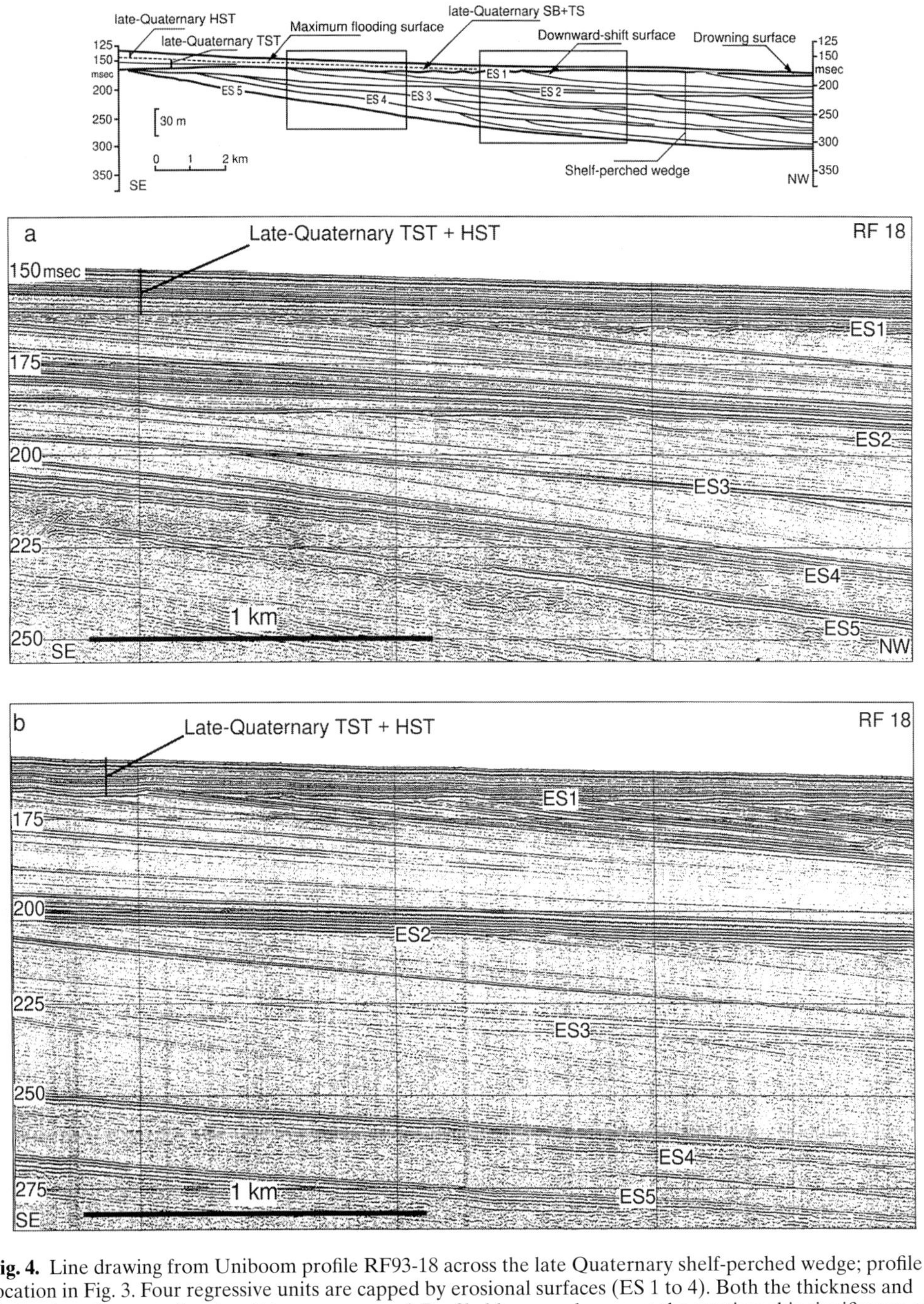

Fig. 4. Line drawing from Uniboom profile RF93-18 across the late Quaternary shelf-perched wedge; profile location in Fig. 3. Four regressive units are capped by erosional surfaces (ES 1 to 4). Both the thickness and the landward extent of each unit increase upward. Profile blow-ups document the stratigraphic significance of the regional erosional surfaces that separate progradational units. The relief associated with each erosion surface increases from the oldest (ES 4) to the youngest (ES 1, see also Fig. 8).

Seismic stratigraphy

In the Adriatic basin, high-resolution seismic data allow the definition and regional correlation of four Late Quaternary systems tracts: the highstand systems tract (HST), resting above the maximum flooding surface (Mfs; Correggiari *et al.* 1992); the transgressive systems tract (TST), positioned between the Mfs, above, and the Transgressive surface (Ts), below (Trincardi *et al.* 1994); the lowstand systems tract (LST) resting above a regional sequence boundary; and the Forced-Regressive Systems Tract (FRST). This fourth systems tract lies below the three other systems tracts (the HST, the TST and the distal correlative of the LST), and is part of a shelf-perched wedge (defined following Posamentier & Vail 1988 and Tesson *et al.* 1990, 1993). The shelf-perched wedge is a composite aggradational–retrogradational sequence set (*sensu* Mitchum & Van Wagoner 1991) that includes three other units of forced regression separated by regional erosional surfaces that are not associated with any detectable (> 1 m) TST deposit with a backstepping configuration (Figs 2 and 4).

Our core control on depositional facies and ages is extended to the three upper systems tracts (HST, TST and LST). Core control is limited in the FRST, where only the most seaward wedges of forced regression could be sampled, as well as in the rest of the shelf-perched wedge. The interpretation of the origin of the shelf-perched wedge and related key surfaces, therefore, is based on our understanding of the overlying Late Quaternary systems tracts that are accordingly discussed first.

Deposits above the shelf-perched wedge

Sediment fluxes from the north, the west and the south toward the MAD were markedly different during the Late Pleistocene and caused significant lateral changes in thickness within each systems tract. These changes are particularly evident along the axis of the basin (e.g. Fig. 2). The following section summarizes the main characteristics of the systems tracts and surfaces formed above the shelf-perched wedge (HST, TST and LST) and presents some sedimentological and chronostratigraphic data from cores. In the following section we examine the stratigraphy downward from the modern depositional surface to the unconformity associated with the pervious glacial maxima (surface ES1, Fig. 2).

Highstand systems tract. The Holocene HST rests on the Mfs dated to about 5 ka (Trincardi *et al.* 1994, 1996*b*). The HST is up to 30 m thick and consists primarily of a mud wedge (Fig. 7) and a drape of mud less than 2 m thick that represents its distal equivalent (Fig. 4). The depocentre of the HST is confined to the western side of the basin (Figs 1 and 3, inset). Two factors control this sediment distribution: (1) the western side of the basin is rimmed by rivers that drain and supply sediment from the rising Apennine chain and the Po Plain to the north and (2) the modern oceanographic circulation that distributes the sediment along the western basin margin toward the southeast (Malanotte-Rizzoli & Bergamasco 1983; Trincardi *et al.* 1994, 1996*b*).

Transgressive systems tract. The Late Quaternary TST in the Adriatic basin contains coastal-plain and marine components separated by a Ravinement surface (Rs; Trincardi *et al.* 1994). Across large areas of the outer shelf the Rs coincides with the transgressive surface (ES 1; Figs 8 and 9). In these locations no backstepping coastal-plain deposits occur so that a composite marine erosional surface represents the seaward extent of the Rs and the TST. Backstepping parasequences composed of offshore mud occur between the erosion surface (ES 1) and the Mfs (Figs 7, 8). In contrast, southwest of the Tremiti Islands, the TST includes a thick coastal plain wedge composed of landward onlapping back-barrier muds capped by landward-prograding flood-tidal-delta deposits (Correggiari *et al.* 1992; Trincardi *et al.* 1994, 1996*a*; Cattaneo & Trincardi 1999).

Lowstand systems tract. The LST was deposited during the last glacial maximum and includes two components: a lower chaotic unit and a progradational wedge filling the MAD from the northwest (Trincardi *et al.* 1996*b*; Fig. 2). The lower unit is extensive and thick (as much as 30 m); it shows a chaotic internal geometry that is consistent with the emplacement of several depositional bodies by mass-failure processes that became ponded within the MAD. The prograding unit originated from the northwest and derived from sediment supplied by the ancient Po and other rivers that drained the Apennine chain and the Southern Alps (Trincardi *et al.* 1996*a*).

The thickness of the LST is about 250 m in its depocentre, immediately north of the MAD, where it followed 150 km of shelf progradation. The LST thins to a minimum of only a few metres on the southern flank of the MAD (Fig. 2). Farther to the south, the erosional surface ES 1 represents the distal equivalent of the northern lowstand wedge on the shelf

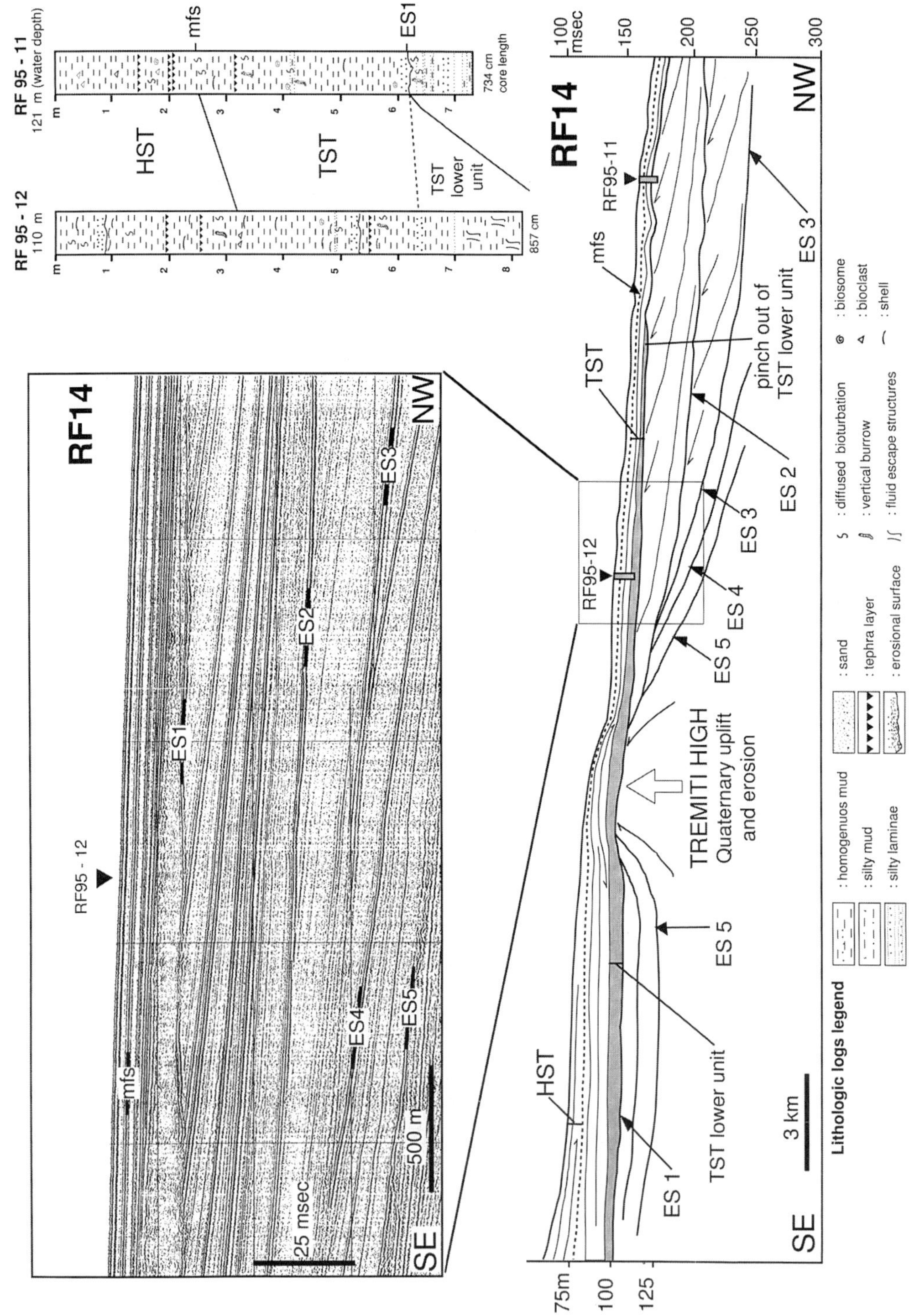
RF 95 - 12
110 m
857 cm
RF 95 - 11
121 m (water depth)
734 cm
core length
HST
TST
TST
lower
unit
mfs
ES1
RF14
RF95 - 12
mfs
ES1
ES2
ES3
ES4
ES5
25 msec
500 m
SE
NW
RF14
RF95-12
RF95-11
mfs
TST
HST
pinch out of
TST lower unit
ES 1
ES 2
ES 3
ES 4
ES 5
TST lower unit
TREMITI HIGH
Quaternary uplift
and erosion
3 km
75m
100
125
100
msec
150
200
250
300
SE
NW
Lithologic logs legend
: homogenuos mud
: silty mud
: silty laminae
: sand
: tephra layer
: erosional surface
: diffused bioturbation
: vertical burrow
: fluid escape structures
: biosome
: bioclast
: shell

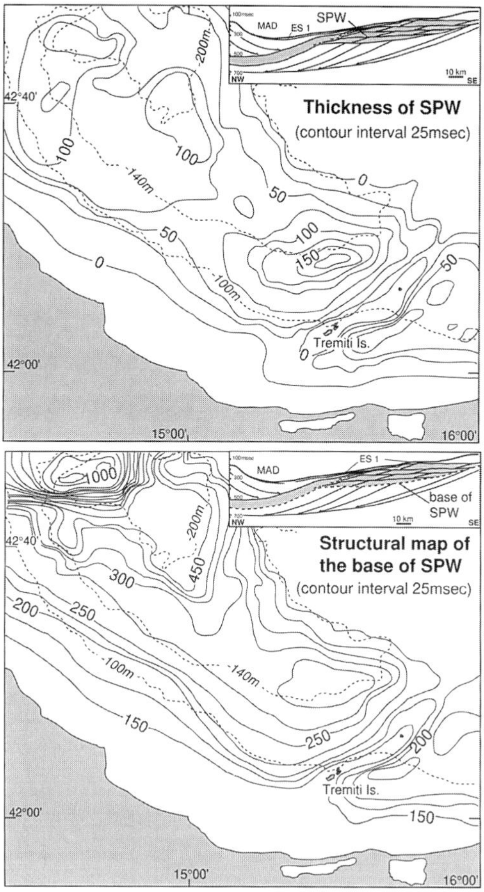

Fig. 6. (**a**) Isopach map of the late Quaternary shelf-perched wedge (SPW). The main depocenter of the SPW is north of the Tremiti High; the SPW has a relatively thick and draped 'distal' equivalent farther north and in the MAD. (**b**) Structural map of the base of the SPW showing the uplifting Tremiti High in the south. Inset profiles highlight stratigraphic relationship across the MAD (see also Fig. 2); grey pattern: SPW and its distal equivalent; dashed horizon: base of the SPW.

above the SPW. Cores obtained in this area are characterized by lower rates of lowstand accumulation and yield age data proving the entire progradational unit (lowstand wedge) to have originated during Isotope Stage 2 and the glacial maxima that ended at approximately 18 ka BP (Trincardi *et al.* 1996*b*).

Erosional surface 1 (ES 1). Figure 10 is a structure contour map of the erosional surface ES 1 formed at the top of the most recent regressive unit of forced regression within the shelf-perched wedge. The surface shows an erosional ridge and swale morphology at depths greater than 145 ms, approximately 100–110 m. Ridges and swales are elongated parallel to the regional contour around the NE–SW-trending Tremiti High. The relief between the ridge tops and adjacent lows is typically 5–6 msecs., or approximately 4 m, and corresponds primarily to changes in lithology (grain size and composition) and cementation in the underlying regressive unit (Fig. 8). Comparable transgressive erosional surfaces occur on other margins below the marine component of the TST (Trincardi & Field 1991; Saito 1994). The time interval during which the Late Quaternary erosional bevelling took place is not fully constrained and likely shows spatial and temporal variability related to physiography of the underlying surface, hydrodynamic conditions on the shelf and sediment supply (Trincardi *et al.* 1994).

The Pleistocene shelf-perched wedge

The shelf-perched wedge in the central Adriatic is up to 100 m thick and is composed of four regressive units separated by regional transgressive erosional surfaces. The shelf-perched wedge is fed from the south (Tremiti High) and thins to the north into a draped unit composed of plane-parallel reflectors; this unit is approximately 70 m thick and extends to the MAD and north of it (Figs 2, 6).

Seismic stratigraphy. The line drawing of a seismic profile shot along the basin axis defines the stratigraphic position and internal architecture of the shelf-perched wedge (Figs 2, 4). The base of the shelf-perched wedge is a basin-wide

Fig. 5. Profile line drawing of line RF93-14 (lower) showing the erosional truncation of strata flanking the uplifting Tremiti High (vertical arrow, below); this high separates a more stable southern area from a more subsiding northern area. The progressive tilting and erosion of stratigraphy demonstrates that tectonic deformation and uplift occurred during the deposition of the shelf-perched wedge (SPW) between surfaces ES 1 and ES 5. Note also the stratigraphic arrangement of units above surface ES 1 on the shelf. Immediately above ES 1 a seismic unit characterized by laterally-discontinuous or chaotic reflections (as shown in detail of seismic line RF14 above) was cored in site RF95-12 where marine muds show benthic faunas indicative of vicinity of fresh-water input (Asioli pers. comm.). This unit is the lower unit of the late Quaternary TST. Based on the available data the presence of fluvial deposits in this unit seems unlikely; if present, these deposits would have to be restricted to the lowermost portion of the unit.

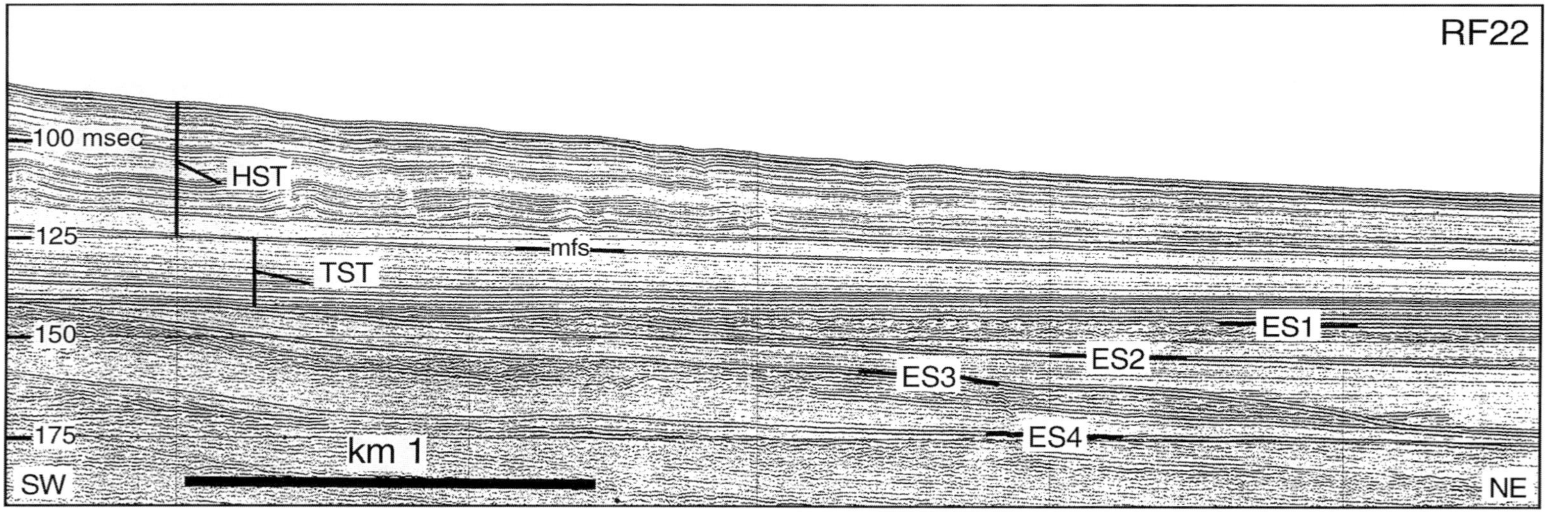

Fig. 7. Uniboom profile RF93-22 showing the late Quaternary TST and HST on the inner shelf separated by the maximum flooding surface (Mfs). Load structures in the HST deposits relate to increased sediment accumulation rates above the Mfs (Correggiari *et al.* 1992). The late Quaternary TST in this area consists of laterally continuous offshore mud (Trincardi *et al.* 1994). See Fig. 3 for profile location.

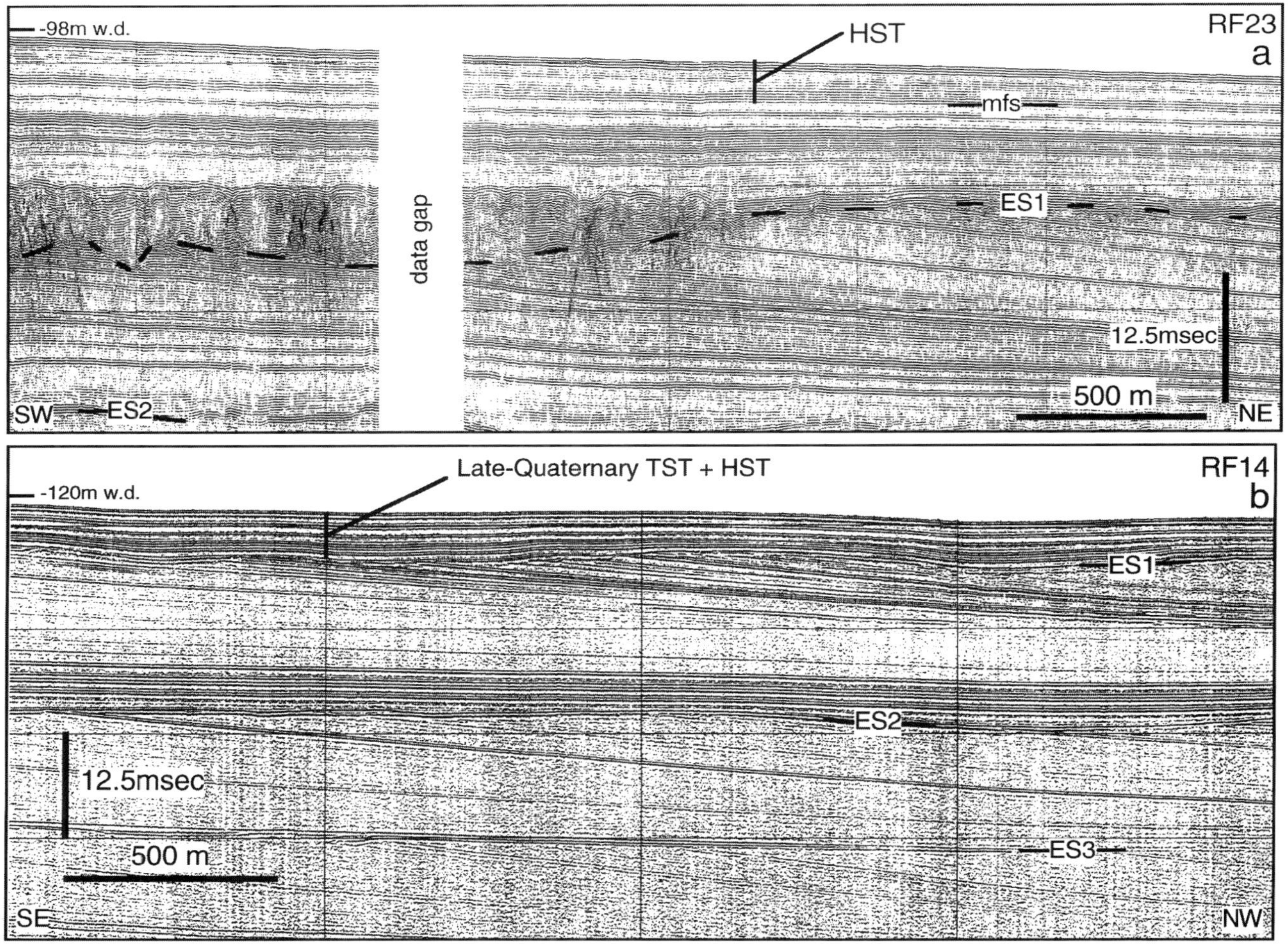

Fig. 3. (**a**) Uniboom profile RF93-23 illustrates the nature of the Transgressive erosional surface ES 1. This surface originated during the subaerial exposure of the margin during lowered sea level and was subsequently modified during the late-Quaternary sea-level rise. (**b**) Uniboom profile RF93-14 from the outer shelf north of the Tremiti High. Note the systematic increase in the amount of erosion associated to the ES surfaces from the oldest (ES 3) to the youngest (ES 1). See Figs 3 and 10 for profiles location.

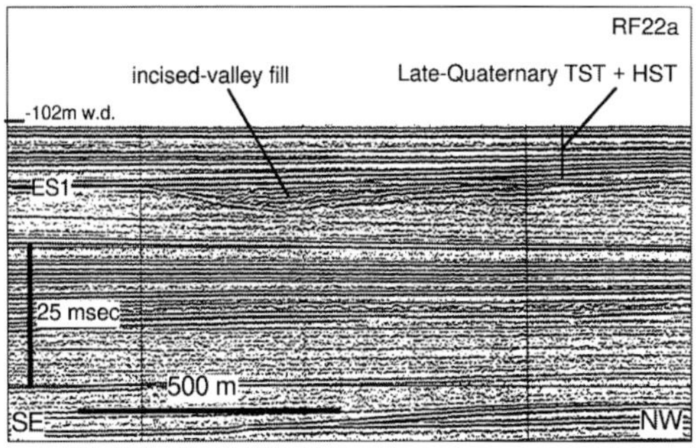

Fig. 9. Detail of Uniboom profile RF93-22a south of the Tremiti High that shows an example of incised-valley-fill deposit of reduced thickness developed across surface ES 1. No information on facies is available from direct coring in this deposit although comparable seismic units are pentrated elsewhere as discussed in text.

surface that marks the drowning of an older progradational wedge (ES 5, Figs 2, 4); the offlap break of this older wedge is now about 300 m below present day sea-level (Figs 1, 2). The shelf-perched wedge is composed of four regressive units separated by regional transgressive erosional surfaces (named ES 1 to ES 4, from top to bottom). These transgressive surfaces mark rapid landward shifts of the shoreline that occurred at the end of each regressive phase, analogous to the surfaces developed between the Holocene HST and preceding TST. Each regressive unit consists of several progradational sub-units that progressively shift seaward and downward (Fig. 11); the sub-units are sharp-based in their proximal parts and show a conformable base seaward (Figs 4 and 11).

The four regressive units show a systematic and regionally-consistent increase in thickness from the oldest (typically 10–15 ms thick) to the youngest (typically 30–40 ms thick). Each regressive unit reaches its maximum thickness at the most seaward depositional shoreline break, where accommodation space was greatest, and time for transgressive reworking minimal. All the wedges composing each regressive unit shale out into distal mud drapes that veneer the underlying transgressive erosional surfaces or their correlative conformities (Figs 4, 8 and 11). This character is derived through the combination of a large, mud-dominated sediment supply and relatively low energy oceanographic conditions relative to other modern continental margins (e.g.: the Rhone Shelf, Tesson *et al.* 1990; Cadiz Margin, Hernández-Molina *et al.* this volume).

Previous interpretations of the origin of the four regressive units in the Adriatic basin called for highstand progradation punctuated by erosional surfaces that originated during glacial lowstands (Savelli *et al.* 1990). Two observations suggest, however, that the regressive units were deposited during times of falling sea level: (1)

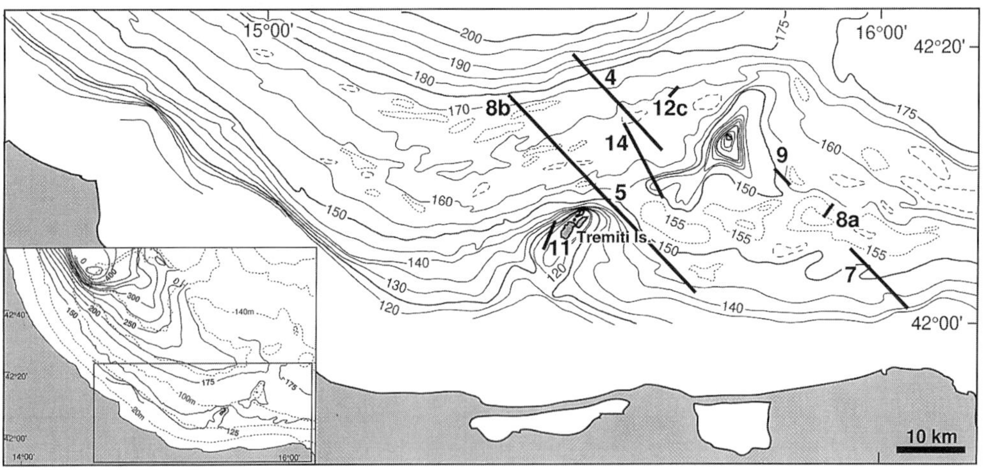

Fig. 10. Structural map of the surface ES 1 showing a ridge-and-swale morphology across the outer shelf between 130 and 110 m water depth (170 and 150 ms contours, respectively). Contour interval is every 5 ms, with heavy lines every 25 ms. Dashed and dotted lines are in outer-shelf areas (135–165 ms) characterized by uneven topography, are at 2.5 ms intervals, and indicate local highs and lows, respectively. Inset shows the smoothed regional contour of the same horizon (continuous lines, contours in msec) superimposed on a simplified bathymetric map with sea-floor depth in metres.

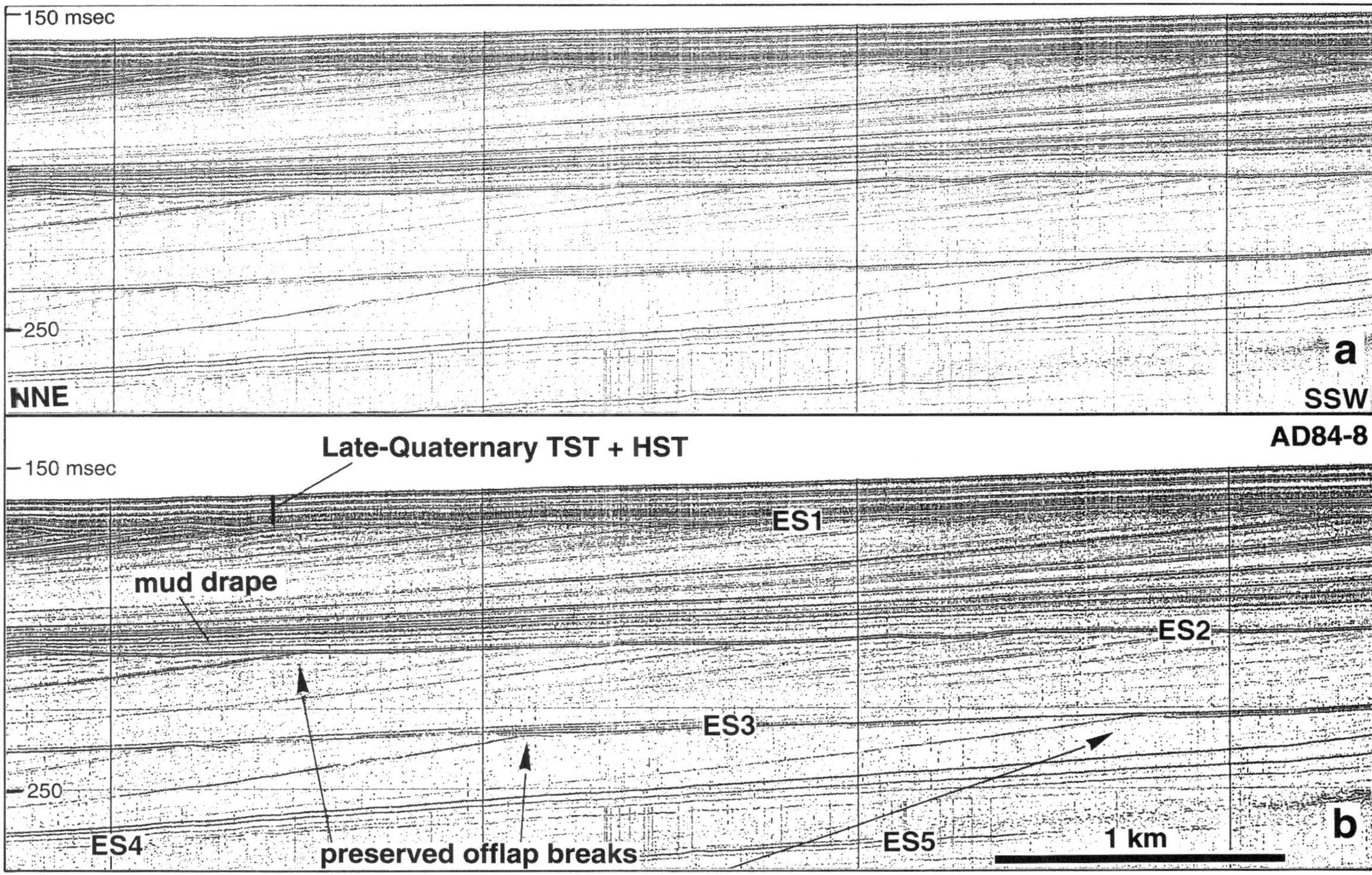

Fig. 11. Uniboom profile AD84-8 showing downstepping offlap breaks below surface ES 3 and ES 2 (arrows). This is a key piece of evidence in the interpretion of the prograding units as forced-regression deposits. These downward-shifted offlap breaks are better preserved where transgressive erosional reworking was negligible (ES 4 and ES 3). Note also that thickness of prograding units increases from bottom to top, and that the low-angle wedge above surface ES 1 encompasses the distal equ valent of late-Quaternary TST and HST. Profile location in Fig. 10.

the occasional preservation of successive offlap breaks that shift downward and seaward (Fig. 11); this key feature, although not commonly preserved in other siliciclastic falling-sea-level deposits, is not consistent with a regime of highstand deposition and (2) during the late Pleistocene, highstand conditions characterize only a small portion of each sea-level cycle; it would be surprising to have deposition only during these intervals and not during more-prolonged times of sea-level fall accompanied by river rejuvenation and downcutting (see also discussions in Trincardi 1994; Sydow & Roberts 1994; Morton & Suter 1996; Chiocci, Hernández-Molina *et al.* and Kolla *et al.* this volume).

Older erosional surfaces. The top of each regressive unit is a shelf-wide erosional surface that began to form during lowstand subaerial exposure of the shelf and continued to develop during shoreface retreat related to the ensuing relative sea-level rise. The four transgressive surfaces merge landward and become conformable seaward of each corresponding depositional shoreline break (Fig. 2). On the shelf, where these transgressive surfaces have an erosional character, mud drapes represent the distal equivalent of the overlying progradational wedges. The erosional relief associated with the transgressive surfaces increases consistently from the oldest to the most recent surface. The two lowest surfaces (ES 4 and ES 3) are interpreted to record more rapid drowning of the margin and landward shifts of the shoreline during which negligible erosion of the underlying offlap breaks took place (Figs 8, 11). On the contrary, surfaces ES 2 and ES 1 are characterized by enhanced erosional relief (on the order of several metres; Correggiari *et al.* 1992; Trincardi *et al.* 1994) and the toplap terminations beneath these surfaces are less commonly preserved. Backstepping TST deposits, which are well developed above surface ES 1, are thin and limited in extent above surface ES 2 (Fig. 8, lower), and do not occur above the older ES surfaces. This observation reinforces the interpretation that the older ES 3 and ES 4 of the SPW represent more rapid transgressions. A more rapid transgression may reflect high rates of relative sea-level rise and/or comparatively lower rates of sediment supply.

Incised valley fills. South of the Tremiti High, all the surfaces appear more erosive and all the progradational units are thinner than their equivalents to the north (Figs 5, 8 and 9). In this southern area, remnants of incised valleys on surface ES 1 are few metres deep and several hundreds of metres wide (Fig. 9). Beside these more localized incised valley features seismic profiles perpendicular to the regional bathymetric contour show a negative relief of approximately 6 m with a landward dipping component similar to that observed on erosional surfaces in ancient deposits at the base of fluvial fills (e.g. Van Wagoner 1995, figs 5, 6). The seismic unit that typically fills these broader lows on surface ES 1 (Fig. 10) consists of variable-amplitude laterally discontinuous reflections or is acoustically transparent. Profile RF93-23, shot at the reduced speed of 2 knots (to increase data coverage and minimize the effects of vertical exaggeration on reflector geometries), documents that the incision is filled by a tabular unit with diffused small-scale discontinuities, diffractions and gently-inclined reflectors (Fig. 8, upper). The top of the infilling unit is irregular; the lateral discontinuities of this unit can alternatively be attributed to the occurrence of small scale channels and creeks or to deformation induced by fluid-escape structures (Correggiari *et al.* 1992). Direct facies information for this unit is limited because of the thick overlying section; however, core RF95-12 recovered 2 m of homogenous mud from this unit (Fig. 5). Within this unit benthic Foraminifera like *Ammonia beccarii* and *A. perlucida* indicate the dominance of fresh-water influx in a marine environment (Asioli pers. comm. 1996). This implies that at least the cored upper part of this fill is not fluvial but may have deposited in an open lagoon or embayed coastal area under the influence of fresh water.

Core data and facies. Several cores provide complementary shelf and basin stratigraphic records that constrain ages and facies of the Late Quaternary HST and TST above surface ES 1 in the Adriatic basin (Asioli *et al.* 1996, 1999; Lowe *et al.* 1996; Langone *et al.* 1996; Trincardi *et al.* 1996*a*, *b*). Below ES 1 piston cores reached only the most seaward wedge within the most recent forced regression unit (FRST). Very fine sand to silt comprises the most seaward unit of forced regression (Fig. 12). Scattered pebbles and granule–pebble-grade bioclastic sediment marks the transgressive/ravinement surface ES 1 on the shelf (Cattaneo & Trincardi 1999). Beyond the shelf edge, surface ES 1 occurs within a continuous marine succession of mudstones that represent the distal equivalent of the forced regressive wedge, below, and the Late Quaternary TST, above (Trincardi *et al.* 1994, 1996*a*; Asioli 1996). We did not reach older subunits of forced regression in a more landward position because of the prohibitive thickness of

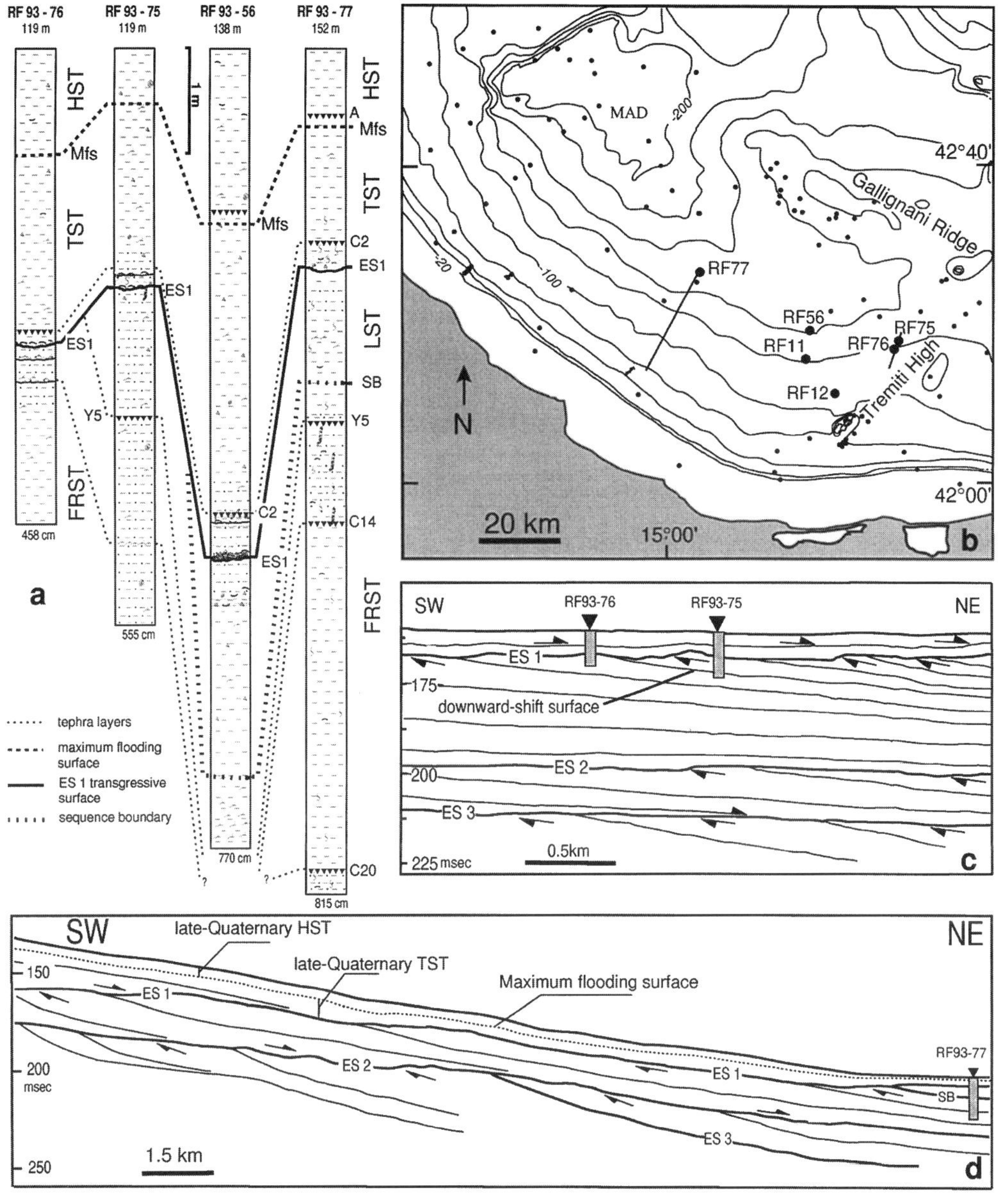

Fig. 12. (**a**) Correlation of four cores that pentrated units below surface ES 1. Lithologic symbols as in Fig. 5. Core RF93-77 reached deposits of Isotope Stage 4 and possibly 5a, according to tephra recognition (Calanchi *et al.*, 1996) and biostratigraphy (Asioli, 1996). RF93-75 and -76, on the outer shelf, penetrated two adjacent subunits of the FRST that are respectively landward and seaward of the most recent downward shift surface. In both cores the regressive deposits are muddy and rich in organic matter. Note that the Y5 tephra layer of the Campanian Ignimbrite is eroded between core RF93-75 and core RF93-76 by the ES 1 surface. The Sequence boundary (SB) at the top of the FRST merges with surface ES1 on the shelf. (**b**) Map showing locations of the cores (including, for reference, cores RF95-11 and -12, discussed in Fig. 5) and sea-floor bathymetry. (**c**) Line drawing of 3.5 kHz profile RF93-56 showing the location of cores RF93-75 and -76 relative to surface ES1 and the most recent downward-shift surface in the FRST recognized from seismic data. (**d**) Line drawing of 3.5 kHz profile Palmas 94-2 showing the location of core RF93-77 on the upper slope.

the overlying Late Quaternary TST and HST to piston cores. However, we surmise that sediment here is even finer-grained than in the cored unit because these slightly older wedges have more gentle and lower amplitude foresets (Figs 4 and 8, lower).

Interpretation of the observed cyclicity. The four regressive units that compose the shelf-perched wedge in the Central Adriatic record four episodes of relative sea-level highstand and fall and provide an expanded record of high-frequency relative sea-level cycles during Quaternary times. The four regressive units stack to form composite aggradational-retrogradational sequence set. This stacking pattern implies a long-term sea-level rise of either tectonic (subsidence-driven) or eustatic origin.

Several cores were collected to constrain the age of the deposits below and above the shallowest erosional surface (ES 1) and its correlative conformable surface in deeper waters (Figs 12, 13 and 14). Preliminary stratigraphic analyses in cored sediment indicate that the uppermost erosional surface (ES 1) records the Late Quaternary sea-level rise at the end of Isotope Stage 2 (Asioli 1996; Trincardi *et al.* 1996). Below ES 1, Core RF77 (Figs 12, 13) encountered deposits that record Stages 2, 3 and 4 based on independent and consistent lines of evidence including; (i) Foraminifera (Asioli 1996), (ii) pollen spectra (Lowe *et al.* 1996), (iii) tephrachronology (Calanchi *et al.* 1996) and (iv) AMS ^{14}C dates in the upper part of the core (Trincardi *et al.* 1996). The identification of tephra layer C20 near the base of the core (Calanchi *et al.* 1996, 1998), in particular, suggests a correlation of the lowermost part of this section to the transition between Stages 5a and 4 (Paterne *et al.* 1988).

By interpolating our data down section, the next erosional surface (ES 2) may represent a record of the sea-level rise from Isotope Stage 6 to Stage 5. If this assumption is correct, the observed units record a fourth-order Quaternary cyclicity in which transgressive erosional surfaces punctuate the stratigraphic record every *c.* 100 ka (Figs 13, 14). This interpretation is consistent with other data and interpretations of Quaternary deposits on other margins like the

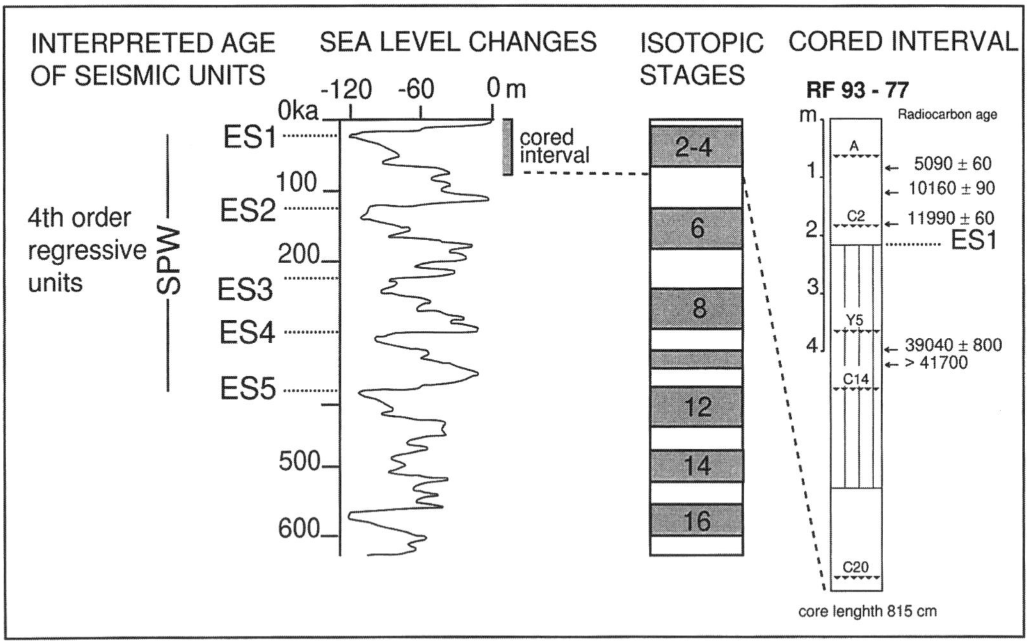

Fig. 13. Mean sea-level changes during the Upper Quaternary inferred from Oxygen Isotope record from SPECMAP (Martinson *et al.*, 1987), with inferred position of ES surfaces in the Adriatic shelf-perched wedge. The sedimentary section below ES 1 was penetrated in core RF93-77. A schematic diagram based on planktonic forams from core RF93-77 (Asioli, 1996) is shown on the right with main tephra layers (Calanchi *et al.* 1996, 1998) and AMS ^{14}C uncalibrated ages (Trincardi *et al.* 1996*b*). Vertically ruled area indicates lack of planktonic forams during intervals of falling sea-level and lowstand. Surface ES 1 marks the top of the interval without planktonic foraminifera and the onset of the sea level rise.

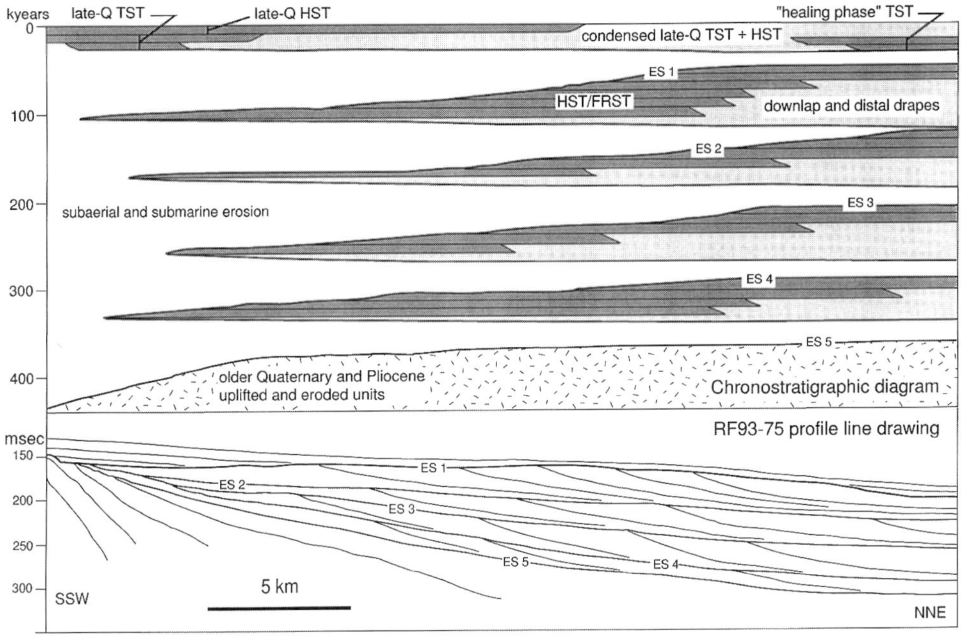

Fig. 14. Chronostratigraphic 'Wheeler' diagram based on profile RF93-75 (below) and all the available core data. The diagram highlights the relationship between HST/FRST progradational deposits (dark grey) and the erosional surfaces that punctuate the shelf-perched wedge. Note the landward increase in magnitude of erosional hiatuses (white area) associated with all the ES surfaces and the occurrence of condensed deposition (light grey) below and seaward of downlap terminations. Speckled patterned area below ES 5 refers to older units of lower Quaternary to Pliocene age.

Gulf of Mexico (Coleman & Roberts 1988; Sydow & Roberts 1994; Kolla *et al.* this volume) and the Mediterranean (Piper & Aksu 1992; van Andel *et al.* 1990; Chiocci this volume; Hernández-Molina *et al.* this volume). In this view, the erosional surfaces ES 1 to 5 originated during Stage 2/1, 6/5, 8/7, 10/9 and 12/11 boundaries, respectively (Figs 12, 13 and 14).

Discussion

In shallow-marine settings down to wavebase, forced regressions are recorded by sharp-based shoreface deposits that may be basinally isolated by a zone of sedimentary bypass (Plint 1988, 1991; Posamentier *et al.* 1992; Helland Hansen & Martinsen 1996; Plint & Nummedal this volume). Where preserved, progressively younger offlap breaks are displaced seaward and downward, and provide a measure of the relative sea-level fall. The shoreface deposits that originated through the process of forced regression typically are narrow and elongated parallel to the regional contour. They are not necessarily related to incised valleys (Tesson *et al.* 1990; Trincardi & Field 1991). The lower bounding surface is a wave-scoured erosional surface that becomes a gradational contact distally, where it corresponds to a low-angle downlap surface (Plint 1988; Posamentier *et al.* 1992; Plint & Nummedal this volume; Posamentier & Morris this volume). The upper bounding surface is a combined sequence boundary and transgressive erosional surface in the area landward of the lowstand shoreline. This surface becomes conformable seaward, where it marks a sudden deepening in the palaeobathymetry (drowning surface, Posamentier *et al.* 1992).

Our data show the architecture of fine-grained prograding shoreface deposits with evidence of downward and seaward shifting offlap breaks (Figs 11, 14 and 15). It is of primary importance to point out that each regressive unit represents a time interval characterized by highstand to falling sea-level, and is capped by a widespread transgressive erosional surface that is draped by mud (Fig. 14). The set of four progradational units composing the SPW have an overall aggradational stacking pattern and pinchout progressively landward (Figs 2, 4, and 14). Based on the evidence from the Adriatic basin, we discuss the current sequence-stratigraphic definitions of forced regression deposits, the position and the sedimentologic expression of the sequence boundary, the factors that control the preservation and the volumetric importance relative to other deposits originated during different phases of a cycle of relative sea-level change.

Recognition of forced-regressive deposits

The Adriatic example shows that the concept of forced regressive systems tract (Hunt & Tucker 1995), falling sea-level systems tract (Helland-Hansen & Gjelberg 1994), or falling stage systems tract (Plint & Nummedal this volume) is applicable in the case of Quaternary high-order sequences controlled largely by sea-level cycles that are characterized by relatively low rates of fall, rapid rises and short lowstand and highstand intervals. In the Adriatic, shelf-perched wedge deposition occurred as a continuum from highstand to falling sea-level conditions, as discussed by Hunt & Tucker (1995); in particular, in our study area each regressive unit represents a suite of attached forced regressions (Ainsworth & Pattison 1994). Because HST and FRST can not easily and unequivocally be separated they must be considered and placed together in the same sequence below the master sequence bounding surface (Figs 14 and 15). This is quite different to the scheme advocated by Posamentier *et al.* (1992) and Posamentier & Morris (this volume).

Our data and recent work on Late Quaternary margins shows that a distinction between HST and sediments deposited during subsequent falling sea-level conditions is not obvious, based only on geometrical evidence (Field & Trincardi 1991; Trincardi 1994; Sydow & Roberts 1994; Morton & Suter 1996; Chiocci this volume; Kolla *et al.* this volume). Indeed, in the case of the four regressive units examined in the study area it appears exceedingly difficult to consistently place the transition from HST and FRST across the entire study area on a particular downward shift surface (Figs 14, 15). Studies in ancient successions also documented that, in many cases, there is no practical way of defining a physical surface that records the initiation of a relative sea-level fall (Hunt & Tucker 1995). Forced regressions are naturally 'cannibalistic' and sediments that deposited during the early phases of this process have the least possibility of being preserved. This applies in particular to relative sea-level falls in the order of 100 m, as in the case of Quaternary deposits. Given these difficulties and the fact that progradation occurs on very-low-angle surfaces, typically less than 0.5–1° (Figs 11, 15), in the absence of very high-resolution seismic data, the composite shelf-perched wedge in the Adriatic could be overlooked and attributed to only one cycle of highstand mud progradation.

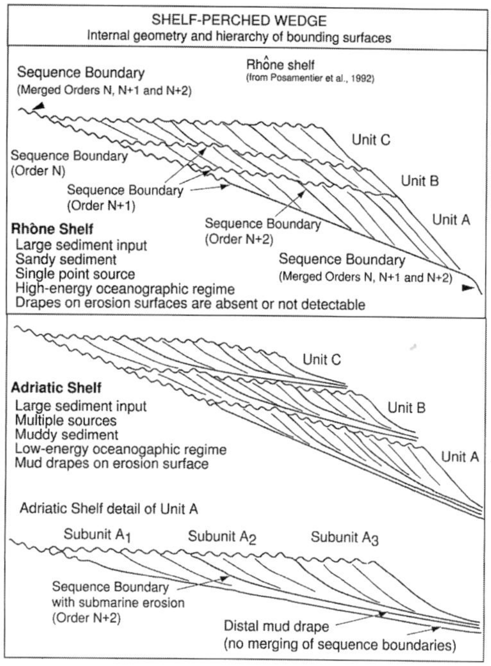

Fig. 15. Stratigraphic relationships between progradational deposits and bounding surfaces in the Adriatic basin in comparison with the model of Posamentier *et al.* (1992) based on data from the Rhône shelf. In the Adriatic basin the lower-energy oceanographic regime and finer-grained sediment input favour the deposition of mud drapes on former erosional surfaces. The occurrence of such drapes makes it impossible to identify 'master' sequence boundaries as interpreted in the case of the Rhône margin. All downward-shift surfaces in the Adriatic basin are equally important and quite subtle features; the best master surfaces here are provided by the transgressive erosional surfaces that separate the regressive units.

Evidence from seismic stratigraphic data and isopach maps of the shelf-perched wedge document the importance of structural control on deposition and preservation of forced-regression deposits in the study area (Figs 4, 5). In particular, changes in the rates of local tectonic subsidence north of the Tremiti transfer zone may explain the change in toplap geometry from lower units with preserved offlap breaks and upper units with more erosional tops (Fig. 11). Along-strike variability in deposit thickness is also a consequence of shelf morphology and location of uplifted areas; in the northernmost part of the study area the line drawing in Fig. 11d shows that surface ES 3 is truncated by the overlying ES 2 in a more seaward position, compared to the area north of the Tremiti High (Figs 4, 5 and 14).

Stratigraphic position of sequence boundaries

Two alternative sequence stratigraphic definitions and paradigms place the sequence boundary on the lower bounding surface below the shoreline wedge of a forced regression (e.g. Posamentier *et al.* 1992; Posamentier & Allen 1993; Posamentier & Morris this volume), or on the upper surface above the forced-regression deposits (Hunt & Tucker 1992, 1995; Trincardi 1994; Helland-Hansen & Gjelberg 1994; Mellere & Steel 1995; Plint & Nummedal this volume). In both definitions the sequence boundary is expressed by a bypass surface landward of the wedge of forced regression. The second definition of the sequence boundary includes forced-regression deposits in a fourth systems tract (FRST/FSST) within the sequence below it (Hunt & Tucker 1992; Plint & Nummedal this volume). This scheme applies to the Adriatic basin, where the forced-regression deposits clearly rest below the major regional erosion surface and sequence boundary, and the northward prograding lowstand systems tract (Figs 2, 14 and 15).

When a sea-level fall is punctuated by multiple high-frequency sea-level cycles, Posamentier *et al.* (1992) and Posamentier & Morris (this volume) argued that the *master sequence boundary* should be placed at the base of these higher-order component sequences (Fig. 15). However, in the case of the Adriatic basin, downward-shift surfaces are erosive only in their uppermost reach and become conformable within a few hundred metres in the seaward direction. Furthermore, no correlation to the underlying ES surfaces is possible because of the occurrence of a detectable distal mud drape immediately above the ES surfaces. Similar asymptotic clinoform reflections were described on high-resolution seismic profiles in the Gulf of Mexico (Morton & Suter 1996). In this respect the forced-regression deposits in the Adriatic basin differ substantially from those in coarser-grained and/or higher-energy depositional systems (Posamentier *et al.* 1992). For this reason, none of the multiple downward-shift surfaces is a key surface of regional extent suitable for long-distance correlation (Figs 14, 15). Therefore, we agree with Hunt & Tucker (1992, 1995) and Plint & Nummedal (this volume) that the erosional surface above the FRST is regionally more extensive and a more practical surface for correlation, compared to more restricted downward-shift surfaces (Fig. 14).

Expression of bounding surfaces

The erosional surfaces ES1-4 in the Adriatic shelf-perched wedge are the only key surfaces that are regionally extensive and provide the easiest way to subdivide our stratigraphic record; erosional surfaces of this kind have been observed on other Quaternary margins (e.g. Field & Trincardi 1991; Chiocci this volume; Hernández-Molina *et al.* this volume; Kolla *et al.* this volume). Each of the ES surfaces originated at times of maximum lowstand and most extensive subaerial exposure of the shelf. The preservation of the offlap-break terminations of the units immediately below some of these surfaces may have resulted from cementation during subaerial exposure that prevented their erosion; a similar process was suggested to explain the formation of 'hardpans' at the top of similar regressive Quaternary deposits in Greece (van Andel *et al.* 1990; McMurray & Gawthorpe this volume). In the case of the Adriatic, however, the youngest and more erosional of these surfaces (ES 1), records a phase of submarine erosion and reworking. Submarine erosion is supported by two lines of evidence: (1) the ridge-and-swale morphology that characterizes this surface is parallel to the modern bathymetry and to local structural relief (Fig. 10) and (2) where cored, the sedimentologic expression of surface ES 1 is given by a lag of polymictic to bioclastic coarse sand with nearshore and shallow-marine faunas (Fig. 12).

It is worth pointing out that in the study area, incised valley systems are encountered only in areas with low rates of subsidence in water depths close to stance of sea-level during maximum lowstand conditions (Figs 8, 9). In contrast to other Quaternary margins (e.g. Sydow & Roberts 1994; Chiocci this volume; Hernández-Molina *et al.* this volume; Kolla *et al.* this volume), farther landward and updip incision related to incised valleys is not present; it is unclear if this is due to a lack of development or preservation. This lack of preservation during transgression implies that fluvial erosion at the end of each lowstand was probably less incised than on many other modern continental margins (Kindinger 1988; Knebel & Circè 1988; Anderson *et al.* 1996; Morton & Suter 1996).

In the study area, the structural map of surface ES 1 shows a broad low that parallels the regional bathymetric contour on the outer shelf (Fig. 10); on profiles perpendicular to the margin this low shows a landward dipping component similar to that observed on erosional surfaces at the base of ancient fluvial-fill deposits (e.g. Van Wagoner 1995). Similar geometric relationship between regressive deposits, below, and fluvial facies, above, were documented in the northeast Gulf of Mexico (Suter & Penland 1987; Sydow & Roberts 1994; Kolla *et al.* this volume). Available core control in the study area, however, documents that the tabular unit filling this low consists of homogeneous mud deposited in a shallow-marine environment during the early stages of the Late Quaternary sea-level rise (Fig. 5).

Factors that control the preservation of forced regressions deposits

While the origin of forced-regression deposits has been explained in detail, the factors that can favour the preservation of several stacks of such deposits are not as well understood (Posamentier *et al.* 1992; Posamentier & Allen 1993). Based on the new evidence presented here from the Adriatic Sea, favourable conditions must be met during both the subsequent intervals of lowstand and sea-level rise. During lowstand intervals, forced-regression deposits are best preserved (i) in the absence of widespread downcutting related to stream rejuvenation (Posamentier *et al.* 1992) and (ii) in the lack of major shelf-edge mass failure. In some cases, progradation across a former irregular outer-shelf physiography may have favoured the preservation of forced regression deposits that are located seaward of major steps (Field & Trincardi 1991; Trincardi & Field 1991). During shoreline transgression, preservation of forced-regression deposits is greatest if shoreface erosion is reduced. Factors limiting the amount of transgressive shoreface erosion include: (1) the presence of a low wave energy with negligible wave erosion and microtidal regime; in the case of the Adriatic Sea, the very small extent of the basin at the onset of the Late Quaternary sea-level rise may imply shallow wavebase and reduced wind-driven currents (Trincardi *et al.* 1994); (2) the low gradient of the transgressed shelf that results in a larger horizontal translation for any given amount of sea-level rise; (3) the high rates of shoreline transgression that limit the time that shoreface erosion is active in any given area (Posamentier *et al.* 1992; Correggiari *et al.* 1996); (4) the amount of time elapsed between successive highstands that controls the amount of accommodation space that can be filled during each successive lowstand during a high-frequency cycle (Piper & Aksu 1992). This last factor explains why highstand systems tract deposits have the lowest preservation potential in times of high-frequency, high-amplitude of sea-level cyclicity (Thorne & Swift 1991); this limitation also applies to the early wedges of each forced regression unit, in particular in a regime of 'attached' forced regression.

In many cases, forced regression deposits occur within composite aggradational-retrogradational sequence sets; this observation applies to Quaternary deposits, like those on the Mediterranean margin and older successions too (e.g. Pomar & Ward 1994; Posamentier *et al.* 1992; Butler *et al.* 1995; Tropeano & Sabato this volume). This association implies that these deposits are more readily preserved when the rising limb of a longer-term component of sea-level change is punctuated by higher-frequency cycles. In this case, the equilibrium point, defined as the point where subsidence rates equal the rate of eustatic sea-level change (Jervey 1988; Posamentier & Allen 1993), would move seaward for a relatively short time interval and rapidly return to a more landward position.

Quaternary Mediterranean margins provide several examples of forced regression deposits that record closely spaced long periods of sea-level fall separated by short intervals of rapid rises associated with flooding of the margin. The overall aggradational–retrogradational stacking pattern of the longer-term composite depositional sequences record the interplay between distinct frequency components of the eustatic signal, subsidence and sediment flux (e.g. Mitchum & Van Wagoner 1991). In their model of composite sequences Mitchum & Van Wagoner (1991) assumed (1) constant subsidence rate, (2) constant sediment supply, (3) that the higher the frequency component, the lower the amplitude of the related signal and (4) that all frequency components of the composite eustatic signal are sinusoidal. In the Quaternary cycles considered here, the high-frequency component of the eustatic signal may have a relatively higher amplitude compared to (i) the longer term signal and (ii) hence results in substantial short-term long-distance seaward shifts of the shoreline (Figs 13 and 16). In terms of sediment supply the MAD is relatively complex, receiving sediment supply from the north and the south and this makes for major temporal and spatial variability in sequence development.

As observed on other margins, the asymmetric shape of the relative sea-level curve with relatively-slow and stepped falls and high-rate rises

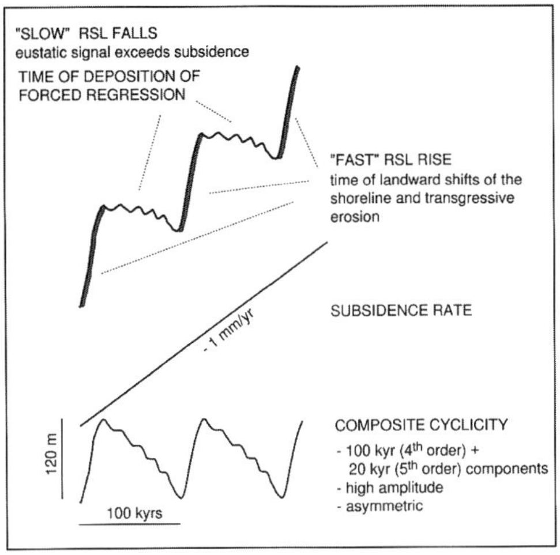

Fig. 16. Cartoon showing possible mechanism to explain the preservation of forced-regression deposits in the geological record. The lowest part of the diagram shows a composite and asymmetric fourth-order eustatic curve with stepped falls and faster rises. The middle shows the addition of an assumed constant subsidence rate of 1 mma^{-1} to provide the upper relative sea-level curve. On the relative curve, the combination of the composite eustatic curve and constant subsidence result in reduced amplitude of prolonged relative sea-level falls and acceleration of relative sea-level rises; the latter plays a key role in favouring the preservation of forced regression deposits.

of high amplitude and short duration is of primary importance in favouring the preservation of forced regression deposits (Trincardi & Field 1991; Hernández-Molina *et al.* this volume). Furthermore, on rapidly subsiding margins, this asymmetric pattern of the eustatic signal would be enhanced because subsidence would counteract sea-level falls but would interfere constructively with sea-level rises, therefore increasing their rates (Fig. 16). On the other hand, higher subsidence rates result in a destructive interference with the eustatic falls and prevent complete cannibalization of deposits originated during the early phases of sea-level fall. The prolonged intervals of relative sea-level fall are more conducive to the formation of a suite of highstand to forced regression deposit (HST/FRST) while the high-rate and high-amplitude rise component favours their preservation (Figs 14 and 16).

Volumetric importance of forced regressive systems tracts

In a depositional regime controlled by a high-amplitude and asymmetric sea-level cyclicity, different systems tracts represent substantially different amounts of time (see chronostratigraphic diagram in Fig. 14), although volume of a systems tract does not necessarily relate directly to its duration. On the basis of the downward extrapolation of our new stratigraphic data, the amount of time represented by each of the highstand–forced regression units (HST/FRST) in the SPW is in the order of 80%, the rest being occupied by much shorter intervals of maximum lowstands and rising sea-level (Figs 13 and 14). Williams (1988), Chiocci (this volume) and Hernández-Molina *et al.* (this volume) suggest a similar figure. In the study area around the Tremiti High, HST/FRST deposits represent virtually all of the depositional stratigraphic record while ES surfaces record times of sea-level lowstand and rise (Figs 4 and 14).

Comparison of the relative importance of the most recent regressive unit (HST-FRST) between surfaces ES 1 and ES 2 and the Late Quaternary depositional sequence above it, shows that the relative importance of the HST/FRST decreases to the north of the study area as a function of contrasting supply regimes. This trend mainly reflects the major stratigraphic expansion of the overlying Late Quaternary lowstand systems tract (LST, cored in site RF93-77, Figure 12) toward the area influenced by the Po and other smaller Apennine rivers (Fig. 2). Similar Late Quaternary LST are reported from other margins that received large sediment fluxes (Boyd *et al.* 1989). To the north of the MAD the Late Quaternary LST is more than 200 m thick and represents a translation of the shoreline at least 150 km along the basin axis. Although it was deposited during less than 10–15 ka (Trincardi *et al.* 1996*b*), the LST only represents approximately 10% of the Late Quaternary sea-level cycle.

In general, the relationship between the time encompassed and the relative volume of each systems tract is not straightforward because of their different preservation potential during subsequent cycles and because sediment flux is far from constant in time or space (Langone *et al.* 1996; Trincardi *et al.* 1996*a*). In particular, significant short-term variations in supply regime in the study area reflect climatic changes that occurred at a frequency higher than the eustatic signal (Asioli *et al.* 1996; Trincardi *et al.* 1996*a*). Such short-term supply pulses are also well documented on other margins (Kolla & Perlmutter 1993; Blum *et al.* 1994).

Conclusions

Within the Adriatic basin, as on several other Mediterranean margins, a shelf-perched wedge developed during discontinuous relative sea-level falls and predates the Late Quaternary lowstand of Isotope Stage 2. This shelf-perched wedge records several fourth- (100 ka) and fifth-order (20 ka) cycles of sea-level change in the Pleistocene. These fourth-order cycles are of high amplitude (*c*. 100 m) and appear to punctuate a longer term trend of sea-level rise (tectonic and/or eustatic in origin); for this reason they represent a particular case of composite cyclicity in which the fourth-order lower-frequency component is markedly asymmetric and the fifth-order higher-frequency component carries a relatively high-amplitude signal.

The superimposition of several different frequencies of cyclicity may explain (1) the overall aggradational–retrogradational composite sequence set in which the erosionally bounded regressive units occur, (2) the occurrence of regional transgressive surfaces that record times of constructive interference between the rising components of relatively lower and higher frequency curves, (3) the preservation of high-frequency sequences of forced regression and, in particular, the recurrent preservation of the offlap breaks in their topset terminations and (4) the erosional relief on the transgressive surfaces as a function of the short time elapsed during each interval of transgressive shoreface retreat.

In a depositional regime controlled by a high-amplitude and asymmetric sea-level cyclicity, each of the fourth-order units in the shelf-perched wedge is exclusively composed of highstand to forced regression systems tracts (HST/FRST), that account for 80% of the time, the rest being occupied by much shorter intervals of maximum lowstand and rising sea level (Figs 13 and 14).

The shelf-perched wedge in the Adriatic consists dominantly of muddy sediment and progradation occurs on very low-angle surfaces; very fine sand occurs only in the topset termination of the most seaward units of forced regression. This fact implies that, in the absence of very high-resolution seismic images or exceptional outcrop data, other composite shelf-perched wedges of this kind could be overlooked and misleadingly attributed solely to one phase of highstand progradation.

We thank IGM colleagues and technicians and Captains M. Gentile and V. Lubrano and the crew of R/V *Urania* for helping during cruises RF93 and RF95. We also thank A. Asioli, A. Cattaneo and E. Dinelli for sharing the results of their stratigraphic reconstructions based on foraminifera, magnetic susceptibility correlations and tephrochronology. Critical reading of the manuscript by D. Hunt, J. Anderson, G. Plint, M. Field, F. Massari and an anonymous reviewer is gratefully acknowledged. This is IGM contribution no. 1071.

References

AHARON, P. 1983. 140,000 year isotope climatic record from raised coral reefs in New Guinea. *Nature*, **304**, 720–723.

AINSWORTH, R. B. & PATTISON, S. A. 1994. Where have all the lowstands gone? Evidence for attached lowstand systems tracts in the Western Interior of North America. *Geology*, **22**, 415–418.

ANDERSON, H. & JACKSON, J. 1987. Active tectonics of the Adriatic region. *Geophysical Journal Royal Astronomical Society*, **36**, 201–214.

ANDERSON, J. B. & THOMAS, M. A. 1991. Marine ice-sheet decoupling as a mechanism for rapid, episodic sea-level change: the record of such events and their influence on sedimentation. *Sedimentary Geology*, **70**, 131–160.

——, ABDULAH, K., SARZALEJIO, S., SIRINGAN, F. & THOMAS, M. A. 1996. Late Quaternary sedimentation and high-resolution sequence stratigraphy of the east Texas shelf. *In*: DE BATIST, M. & JACOBS, P. (eds) *Geology of Siliciclastic Shelf Seas*, Geological Society, London, Special Publications, **117**, 95–124.

ARGNANI, A., FAVALI, P., FRUGONI, F., GASPERINI, M., LIGI, M., MARANI, M. MATTIETTI, G. & MELE, G. 1993. Foreland deformation pattern in the Southern Adriatic Sea. *Annali di Geofisica*, **36**, 229–247.

ASIOLI, A. 1996. High-resolution foraminifera biostratigraphy in the Central Adriatic basin during the last deglaciation: a contribution to the PALICLAS project. *Memorie Istituto Italiano di Idrobiologia,* **55**, in press.

ASIOLI, A., TRINCARDI, F., CORREGGIARI, A., LANGONE, L., VIGLIOTTI, L., VAN DER KAARS, S. & LOWE, J. J. 1996. The late-Quaternary deglaciation in the Central Adriatic basin. *Il Quaternario, Italian Journal of Quaternary Sciences*, **9**, in press.

——, ——, LOWE, J. J. & OLDFIELD, F. 1999. Short-term climate changes from the Central Adriatic basin during the Late-Glacial: comparison with other Mediterranean records and GRIP event stratigraphy. *Journal of Quaternary Science*, **14**, 373–381.

BLANCHON, P. & SHAW, J. 1995. Reef drowning during the last deglaciation: evidence for catastrophic sea-level rise and ice-sheet collapse. *Geology* **23**, 4–8.

BLUM, M. D., TOOMEY, R. S. & VALASTRO, S. JR. 1994. Fluvial response to Late Quaternary climatic and environmental change, Edward Plateau, Texas. *Palaeoceanography, Palaeoclimatology, Palaeoecology*, **108**, 1–21.

BOYD, R., DALRYMPLE, R. & ZAITLIN, B. A. 1992. Classification of clastic coastal depositional environments. *Sedimentary Geology*, **80**, 139–150.

BOYD, R., SUTER, J. & PENLAND, S. 1989. Relation of sequence stratigraphy to modern sedimentary environments. *Geology*, **17**, 926–929.

BROOKS, M. & FERENTINOS, G. 1984. Tectonics and sedimentation in the Gulf of Corinth and the Zakynthos and Kefalinia Channels, Western Greece. *Tectonophysics*, **101**, 25–54.

BUTLER, R. W. H., GRASSO, M. & LICKORISH, H. 1995. Plio-Quaternary megasequence geometry and its tectonic controls within the Maghrebian thrust belt of south-central Sicily. *Terra Nova*, **7**, 171–178.

CALANCHI, N., CATTANEO, A., DINELLI, E., GASPAROTTO, G. & LUCCHINI, F. 1998. Tephra layers in Late Quaternary sediments of the central Adriatic Sea. *Marine Geology*, **149**, 191–209.

——, DINELLI, E., LUCCHINI, F. & MORDENTI, A. 1996. 'Paliclas' multi-proxy investigation of paleoenvironmental history of Central Italy: chemostratigraphy of late-Quaternary sediments from Lago Albano and Central Adriatic Sea cores. *Memorie Istituto Italiano di Idrobiologia*, **55**, 247–263.

CATTANEO, A. & TRINCARDI, F. 1999. The late-Quaternary transgressive record in the Adriatic epicontinental sea: basin widening and facies partitioning. *In*: BERGMAN, K. & SNEDDEN, J. (eds) *Isolated Shallow Marine Sand Bodies: Sequence Stratigraphic Analysis and sedimentologic Interpretation*. SEPM Special Publications, **63**, 127–146.

CHAPPEL, J. & SHACKLETON, N. J. 1986. Oxygen isotopes and sea level. *Nature*, **324**, 137–140.

CHIOCCI, F. 2000. Depositional response to Quaternary fourth-order sea-level falls on the northern Latium margin (Tyrrhenian Sea, Italy). *This volume*.

CIABATTI, M., CURZI, P. V. & RICCI LUCCHI, F. 1987. Quaternary sedimentation in the Central Adriatic Sea. *Giornale di Geologia*, **49**, 113–125.

COLEMAN, J. M. & ROBERTS, H. H. 1988. Sedimentary development of the Louisiana continental shelf related to sea level cycles. *Geo-Marine Letters*, **8**, 63–119.

CORREGGIARI, A., TRINCARDI, F. & ROVERI, M. 1992. Regressioni 'forzate', regressioni 'deposizionali' e fenomeni di instabilita' in unita' progradazionali tardo-quaternarie (Adriatico Centrale). *Giornale di Geologia*, **54**, 19–36.

CORREGGIARI, A., FIELD, M. E. AND TRINCARDI, F. 1996. Late Quaternary transgressive large dunes on the sediment-starved Adriatic shelf. *In*: DE BATIST, M. & JACOBS, P. (eds) *Geology of Siliciclastic Shelf Seas*, Geological Society, London, Special Publications, **117**, 155–169.

CURRAY, J. R. 1964. Transgressions and regressions. *In*: MILLER, R. L. (ed.) *Papers in Marine Geology*. Macmillan, 175–203.

DOGLIONI, C., MONGELLI, F. & PIERI, P. 1994. The Puglia uplift (SE Italy): an anomaly in the foreland of the Apennine subduction due to buckling of a thick continental lithosphere. *Tectonics*, **13**, 1309–1321.

ERCILLA, G. & ALONSO, B. 1996. Quaternary siliciclastic sequence stratigraphy of western Mediterranean passive and tectonically active margins: the role of global versus local controlling factors. *In*: DE BATIST, M. & JACOBS, P. (eds) *Geology of Siliciclastic Shelf Seas*. Geological Society, London, Special Publications, **117**, 95–124.

FAIRBANKS, R. G. 1989. A 17,000 year glacio-eustatic sea level record: influence of glacial melting rates on the younger Dryas event and deep-ocean circulation. *Nature*, **342**, 637–642.

FIELD, M. E. & TRINCARDI, F. 1991. Regressive coastal deposits on Quaternary continental shelves: preservation and legacy. *In*: OSBORNE, R. H. (ed.) *From shoreline to abyss*. SEPM Special Publications, **46**, 107–122.

FRANSEEN, E. K., GOLDSTEIN, R. H. & FARR, M. R. 1998. Quantitative controls on location and architecture of carbonate depositional sequences: upper Miocene, Cabo de Gata region, SE Spain. *Journal of Sedimentary Research*, **68**, 283–298.

GAWTHORPE, R. L., FRASER, A. J. & COLLIER, R. E. LL. 1994. Sequence stratigraphy in active extensional basins: implications for the interpretation of ancient basin-fills. *Marine and Petroleum Geology*, **11**, 642–658.

HELLAND-HANSEN, W. & GJELBERG, J. G. 1994. Conceptual basis and variability in sequence stratigraphy: a different perspective. *Sedimentary Geology*, **92**, 31–52.

—— & MARTINSEN, O. J. 1996. Shoreline trajectories and sequences: description of various depositional-dip scenarios. *Journal of Sedimentary Research*, **66B**, 670–688.

HERNÁNDEZ-MOLINA, F. J., SOMOZA, I. & LOBO, F. 2000. Seismic stratigraphy of the Gulf of Cádiz continental shelf: a model for late Quaternary very high-resolution sequence stratigraphy and response to sea-level fall. *This volume*.

——, —— & REY, J. 1996. Late Pleistocene-Holocene high-resolution sequence analysis on the Alboran Sea continental shelf. *In*: DE BATIST, M. & JACOBS, P. (eds) *Geology of Siliciclastic Shelf Seas*. Geological Society, London, Special Publication, **117**, 95–124.

——, ——, —— & POMAR, L. 1994. Late Pleistocene-Holocene sediments on the Spanish continental shelves: model for very high resolution sequence stratigraphy. *Marine Geology*, **120**, 129–174.

HUNT, D. & TUCKER, M. E. 1992. Stranded parasequences and forced regressive wedge systems tract: deposition during base-level fall. *Sedimentary Geology*, **81**, 1–9.

—— & —— 1995. Stranded parasequences and forced regressive wedge systems tract: deposition during base-level fall – reply. *Sedimentary Geology*, **95**, 147–160.

JERVEY, M. 1988. Quantitative geological modelling of siliciclastic rock sequences and their seismic expression. *In*: WILGUS, C. K., HASTINGS, B. S., KENDALL, C. G. St. C., POSAMENTIER, H. W., ROSS, C. A. & VAN WAGONER, J. C. (eds) *Sea-level Changes – An Integrated Approach*. SEPM Special Publications, **42**, 47–69.

KINDINGER, J. L. 1988. Seismic stratigraphy of the Mississippi-Alabama shelf and upper continental slope. *Marine Geology*, **83**, 79–94.

KNEBEL, H. J. & CIRCÈ, M. 1988. Late Pleistocene drainage systems beneath Delaware Bay. *Marine Geology*, **78**, 285–302.

KOLLA, V. & PERLMUTTER, M. A. 1993. Timing of turbidite sedimentation on the Mississippi fan: *American Association of Petroleum Geologists Bulletin*, **77**, 1129–1141.

——, BIONDI, P., LONG, B. & FILLON, R. 2000. Sequence stratigraphy and architecture of the late Pleistocene Lagniappe delta complex, northeast Gulf of Mexico. *This volume.*

LANGONE L., ASIOLI, A., CORREGGIARI, A. & TRINCARDI, F. 1996. Age-depth modelling through the late-Quaternary deposits of the Central Adriatic Basin. *Memorie Istituto Italiano di Idrobiologia*, **55**, 177–196.

LOUTIT, T. S., HARDENBOL, J., VAIL, P. & BAUM, G. R. 1988. Condensed sections, the key to age determination and correlation of continental margin sequences. *In*: WILGUS, C. K., HASTINGS, B. S., KENDALL, C. G. ST. C, POSAMENTIER, H. W., ROSS, C. A. & VAN WAGONER, J. C. (eds) *Sea-level Changes – An Integrated Approach.* SEPM Special Publications, **42**, 183–213.

LOWE J. J., ACCORSI C. A., ASIOLI A., VAN DER KAARS S., TRINCARDI F., 1996. Pollen-stratigraphical records of the last glacial-interglacial transition (ca. 14–9 ka BP) from Italy: a contribution to the PALICLAS project. *In*: Late-Glacial and early Holocene climatic and environmental changes in Italy. *Il Quaternario, Italian Journal of Quaternary Sciences*, **9**, 627–642.

MALANOTTE RIZZOLI, P. & BERGAMASCO, A. 1983. The dynamics of the coastal region of the North Adriatic Sea. *Journal of Physical Oceanography*, **13**, 1105–1130.

MARANI, M., TAVIANI, M., TRINCARDI, F., ARGNANI A., BORSETTI, A. M., & ZITELLINI, N. 1988. Pleistocene progradation and postglacial events of the NE Tyrrhenian continental shelf between the Tiber river delta and Capo Circeo. *Memorie Societá Geologica Italiana*, **36**, 67–89.

MARTINSON, D. G., PISIAS, N. G., HAYS, J. D., IMBRIE, J., MORE, T. G. JR. & SHACKLETON, N. J. 1987. Age dating and the orbital theory of the ice ages: development of a high-resolution 0 to 300.000-year chronostratigraphy. *Quaternary Research*, **27**, 1–30.

MCMURRAY, L. S. & GAWTHORPE, R. L. G. 2000. Along-strike variability of forced regressive deposits: late Quaternary, northern Peloponnesos, Greece. *This volume.*

MELLERE, D. & STEEL, R. 1995. Variability of lowstand wedges and their distinction from forced-regressive wedges in the Mesaverde Group, southeast Wyoming. *Geology*, **23**, 803–806.

MITCHUM, R. M. & VAN WAGONER, J. C. 1991. High-frequency sequences and their stacking patterns: sequence-stratigraphic evidence of high-frequency eustatic cycles. *Sedimentary Geology*, **70**, 131–160.

MORTON, R. A. & SUTER, J. R. 1996. Sequence stratigraphy and composition of late-Quaternary shelf-margin deltas, Northwest Gulf of Mexico. *American Association of Petroleoum Geologists Bulletin*, **80**, 505–530.

ORI, G. G., ROVERI, M. & VANNONI, F. 1986. Plio-Pleistocene sedimentation in the Apenninic-Adriatic foredeep (Central Adriatic Sea, Italy). *In*: ALLEN, P. A. & HOMEWOOD, P. (eds) *Foreland basins*. International Association of Sedimentologists, Special Publications, **8**, 183–198.

PASINI, G., COLALONGO, M. L., CURZI, P. V. & TAVIANI, M. 1993. Analisi paleoecologica degli ambienti tardo-quaternari del Dosso Gallignani (Adriatico centro-meridionale) mediante lo studio paleontologico di carote. *Bollettino della Societá Paleontologica Italiana*, **32**, 1–18.

PATERNE, M., GUICHARD, R. & LABEYRIE, L. 1988. Explosive activity of the South Italian volcanoes during the past 80,000 years as determined by marine tephrochronology. *Journal of Volcanology and Geothermal Research*, **34**, 153–172.

PIPER, D. J. W. & AKSU, A. E. 1992. Architecture of stacked Quaternary deltas correlated with global oxygen isotopic curve. *Geology*, **20**, 415–418.

PLINT, A. G. 1988. Sharp-based shoreface parasequences and 'offshore bars' in the Cardium formation of Alberta; their relationship to relative changes in sea level. *In*: WILGUS, C. K., HASTINGS, B.S., KENDALL, C. G. St. C., POSAMENTIER, H. W., ROSS, C. A. & VAN WAGONER, J. C. (eds) *Sea-level Changes – An Integrated Approach.* SEPM Special Publications, **42**, 357–370.

—— 1991. High-frequency relative sea-level oscillations in the Upper Cretaceous shelf clastics of the Alberta foreland basin: possible evidence for glacio-eustatic control? *In*: MACDONALD, D. I. M. (ed.) *Sedimentation, Tectonics and eustasy*. International Association of Sedimentologists, Special Publications, **12**, 409–428.

—— & NUMMEDAL, D. 2000. The falling stage systems tract: recognition and importance in sequence stratigraphic analysis. *This volume.*

POMAR, L. & WARD, W. C. 1994. Response of a late Miocene Mediterranean reef platform to high-frequency eustasy. *Geology*, **22**, 131–134.

POSAMENTIER, H. W. & ALLEN, G. P. 1993. Siliciclastic sequence stratigraphic patterns in foreland ramp-type basins. *Geology*, **21**, 455–458.

—— & MORRIS, W. S. 2000. Aspects of the stratal architecture of forced regressive deposits. *This volume.*

—— & VAIL, P. R. 1988. Eustatic control on clastic deposition, II-sequence and systems tract models, *In*: WILGUS, C. K., HASTINGS, B. S., KENDALL, C. G. St. C., POSAMENTIER, H. W., ROSS, C. A. & VAN WAGONER, J. C. (eds) *Sea-level Changes – An Integrated Approach*. SEPM Special Publications, **42**, 125–154.

——, JAMES, D. P., ALLEN, G. P. & TESSON, M. 1992. Forced Regressions in a sequence stratigraphic framework: concepts, examples, and exploration significance. *American Association of Petroleoum Geologists Bulletin*, **76**, 1687–1709.

ROYDEN, L. E., PATACCA, E. & SCANDONE, P. 1987. Segmentation and configuration of subducted lithosphere in Italy: an important control on thrust-belt and foredeep-basin evolution. *Geology*, **15**, 714–717.

SAITO, Y. 1994. Shelf sequence and characteristic bounding surfaces in a wave-dominated setting: Latest Pleistocene-Holocene examples from the Northeast Japan. *Marine Geology*, **120**, 105–127.

SAVELLI, D., TRAMONTANA, M. & WEZEL, F. C. 1990. Cyclic sedimentation and erosion of Quaternary sedimentary wedges off the Gargano promontory (Southern Adriatic Sea). *Bollettino di Geofisica Teorica e Applicata*, **3**, 163–175.

SHACKLETON, N. J. 1987. Oxygen-isotopes, ice volumes and sea level. *Quaternary Science Reviews*, **6**, 183–190.

SUTER, J. R. & PENLAND, S. 1987. Late Quaternary sea-level fluctuations and depositional sequences, Southwest Louisiana continental shelf. *In*: NUMMEDAL, D., PILKEY, O. H. & HOWARD, J. D. (eds) *Sea-Level Fluctuation and Coastal Evolution*. SEPM Special Publications, **41**, 199–219.

SYDOW, J. & ROBERTS, H. H. 1994. Stratigraphic framework of a late-Pleistocene shelf-edge delta, Northeast Gulf of Mexico. *American Association of Petroleoum Geologists Bulletin*, **78**, 1276–1312.

TESSON, M., ALLEN, G. P. & RAVENNE, C. 1993. Late Pleistocene shelf-perched lowstand wedges on the Rhône continental shelf. *In*: POSAMENTIER, H. W., SUMMERHAYES, C. P., HAQ, B. U. & ALLEN, G. P. *Sequence stratigraphy and facies associations*. International Association of Sedimentologists, Special Publications, **18**, 183–196.

——, GENSOUS, B., ALLEN, G. P. & RAVENNE, C. 1990. Late Quaternary deltaic lowstand wedges on the Rhone continental shelf, France. *Marine Geology*, **91**, 325–332.

THORNE, J. A. & SWIFT, D. J. P. 1991. Sedimentation on continental margins, VI: a regime model for depositional sequences, their component systems tracts, and bounding surfaces. *In*: SWIFT, D. J. P., OERTEL, G. F., TILLMAN, R. W. & THORNE, J. A. (eds) *Shelf Sand and Sandstone Bodies.* International Association of Sedimentologists, Special Publications, **14**, 189–255.

TINTI, S., MARAMAI, A. & FAVALI, P. 1995. The Gargano promontory: An important Italian seismogenic-tsunamogenic area. *Marine Geology,* **122**, 227–241.

TRINCARDI, F. 1994. High-frequency forced regressions in the Adriatic basin: the composite nature of Quaternary eustatic cyclicity. *In*: *Second High-Resolution Sequence Stratigraphy Conference*, Tremp, Spain, 20–27 June 1994.

—— & FIELD, M. E. 1991. Geometry, lateral variability, and preservation of downlapped regressive shelf deposits: eastern Tyrrhenian margin, Italy. *Journal of Sedimentary Petrology*, **61**, 75–90.

——, ASIOLI, A., CATTANEO, A., CORREGGIARI, A., VIGLIOTTI, L. & ACCORSI, C. A. 1996*a*. Transgressive offshore deposits on the Central Adriatic shelf: architecture complexity and the record of the Younger Dryas short-term event. *Il Quaternario Italian Journal of Quaternary Sciences*, **9**, 753–762.

——, ——, ——, —— & LANGONE, L. 1996*b*. Stratigraphy of the late-Quaternary deposits in the Central Adriatic basin and the record of short-term climatic events. *Memorie Istituto Italiano di Idrobiologia*, **55**, 39–70.

——, CORREGGIARI, A. & ROVERI M. 1994. Late-Quaternary transgressive erosion and deposition in a morden epicontinental shelf the Adriatic semienclosed basin. *Geo-Marine Letters*, **14**, 41–51.

TROPEANO, M. & SABATO, L. 2000/ Response of late Pliocene–Early Pleistocene mixed carbonate–clastic temperate-water systems to forced regressions: the Calcarenite di Gravina Formation, Puglia, SE Italy. *This volume*.

VAIL, P. R., MITCHUM, R. M. JR., THOMPSON, S. 1977. Seismic stratigraphy and global changes of sea level, part 3: Relative changes of sea level from coastal onlap. *In*: PAYTON, C. E. (ed.) *Seismic stratigraphy – applications to hydrocarbon exploration*. American Association of Petroleoum Geologists Memoirs, **26**, 63–81.

VAN ANDEL, T. H., ZANGGER, E. & PERISSORATIS, C. 1990. Quaternary transgressive/ regressive cycles in the Gulf of Argos, Greece. *Quaternary Research*, **34**, 317–329.

VAN WAGONER, J. C. 1995. Sequence Stratigraphy and marine to nonmarine facies architecture of foreland basin strata, Book Cliffs, Utah, U.S.A. *In*: VAN WAGONER, J. C. & BERTRAM, G. T. (eds) *Sequence stratigraphy of foreland basin deposits*. American Association of Petroleoum Geologists Memoirs, **54**, 137–223.

WILLIAMS, D. F. 1988. Evidence for and against sea-level changes from the stable isotopic record of the Cenozoic. *In*: WILGUS, C. K., HASTINGS, B. S., KENDALL, C. G. St. C., POSAMENTIER, H. W., ROSS, C. A. & VAN WAGONER, J. C. (eds) *Sea-level Changes – An Integrated Approach*. SEPM Special Publications, **42**, 31–36.

Depositional response to Quaternary fourth-order sea-level fluctuations on the Latium margin (Tyrrhenian Sea, Italy)

FRANCESCO L. CHIOCCI[1]

Consiglio Nazionale delle Ricerche – Centro di Studio per il Quaternario e l'Evoluzione Ambientale.

[1]*Current Address: Earth Science Department, University of Rome 'La Sapienza', P.le Aldo Moro, 5, 00185, Roma, Italy (chiocci@gea.geo.uniroma1.it)*

Abstract: More than 10 000 km of high-resolution seismic profiles permit detailed study of six fourth-order seismic stratigraphic sequences deposited during the last million years on the Latium continental shelf, Tyrrhenian Sea, Italy. Sedimentation occurred on a relatively young passive margin characterized by a narrow and relatively steep shelf where sediment storage capacity in adjacent subaerial basins was limited. The Late Pleistocene deposits are locally tilted and eroded to different levels along the sea-floor so that carefully placed seafloor gravity cores help constrain the age of the seismic sequences by the dating of microfauna. Correlation with a deep well located on the coast constrains the basal third-order sequence boundary on the shelf. The seismic and a limited core database make it possible to: (1) detail variability in the architecture, stratal patterns and bounding surfaces of the sequences across the shelf and adjacent continental slope; (2) define a hierarchy of seismic units and their bounding surfaces; (3) make a correlation with the published oxygen-isotope curves; (4) develop a detailed stratigraphic framework and model for the fourth-order sequences deposited during last 0.8 Ma; (5) define the effects of the long-lasting eustatic falls on margin sedimentation; (6) recognize volumetric partitioning of sedimentation between systems tracts.

The seismo-stratigraphic expression of the third- and fourth-order sequence boundaries varies greatly from the inner to the outer part of the margin. Where subsidence allowed the preservation of lowstand systems tract (LST) deposits on the shelf, they are bounded by erosional unconformities interpreted to reflect fourth-order glacioeustatically-driven cycles. Relatively thin (<10 m) lens-shaped bodies mark the transition from the unconformities to their correlative conformities and are interpreted to have been deposited during the eustatic minimum. The deposits bounded between correlative conformities show an upward loss in acoustic transparency thought to indicate upward-coarsening and regression within the sequences. Downdip on the continental slope, sequence boundaries are concordant surfaces correlative with unconformities on the shelf. However, these surfaces are locally scoured by channellized features, interpreted to record slope erosion related to the discharge of river bedload during lowstands.

There is marked asymmetry and volumetric partitioning between systems tracts; most of the Late Quaternary deposits that comprise the Latium continental margin are interpreted to have formed during forced regression and lowstand. Offshore of the northern and central Latium shelves forced regressive and lowstand deposits account for some 1000 km^3 of shelf and slope deposition during the last eustatic cycle. In contrast, sediments attributed to the transgressive and highstand systems tracts account for approximately 37 km^3. High-frequency, high-amplitude asymmetric sea-level changes driven mainly by glacioeustasy are interpreted to have controlled deposition. Following classic three-fold sequence stratigraphic models, the unconformities created by shelf subaerial exposure and erosion represent sequence boundaries at the base of depositional sequences. However, if as is the case of the Latium margin during the Late Pleistocene, where a continental margin is formed almost exclusively of forced regressive deposits, each sequence basal boundary will paradoxically be situated above the forced regressive deposits that are deposited as the subaerial exposure surface forms, i.e. above the whole of its own depositional sequence. In this respect, the incorporation of a fourth forced regressive or falling stage systems tracts specific to times of base-level fall would help avoid this inconsistency.

This paper attempts to interpret the complex Late Pleistocene stratigraphy of the Latium margin (Fig. 1) in terms of its sequence stratigraphy, placing particular emphasis on the role of forced regressive sedimentation. Seismic-stratigraphy of the Latium margin is relatively well-constrained by an extensive grid of high-resolution seismic profiles with the dating of well and core samples offering some chronostratigraphic constraint. The seismic and

From: HUNT, D. & GAWTHORPE, R. L. (eds) *Sedimentary Responses to Forced Regressions.* Geological Society, London, Special Publications, **172**, 271–289. 1-86239-063-0/00/$15.00

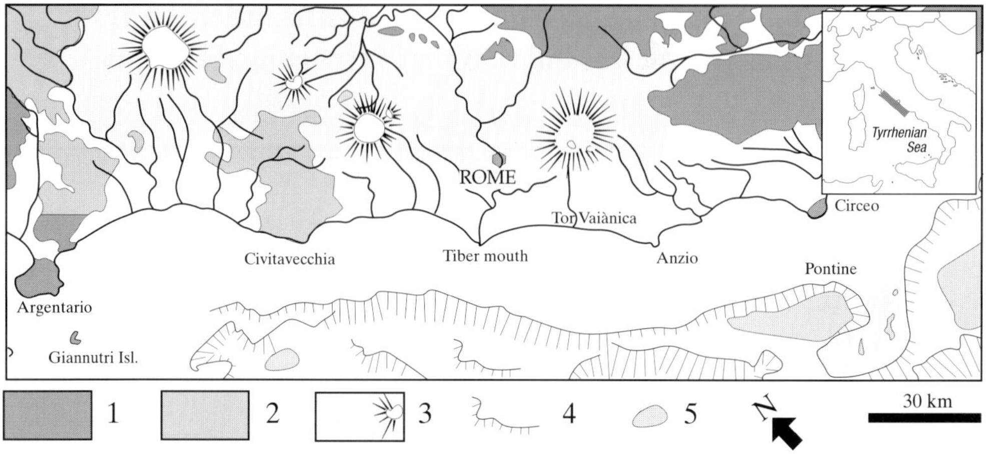

Fig. 1. Main geological and physiographical features of the Latium continental margin. (1) Lower Liassic and Upper Triassic shallow water limestones; (2) Eocene–Upper Cretaceous marly–calcareous arenaceous turbidites; (3) Quaternary volcanites; (4) shelf-break and continental slope; (5) intraslope basin.

chronostratigraphic data reveal a picture of Late Quaternary continental margin sedimentation driven by glacioeustacy. For the Late Quaternary Period was characterized by high-frequency, high-amplitude and markedly asymmetric eustatic sea-level changes, characterized by short highstands and lowstands, very rapid rises and prolonged sea-level falls. The latter played an important part in evolution of the Latium continental margin, where sedimentation appears to have been dominated by forced regression deposits, providing important insights for the principles and models of sequence stratigraphy, especially forced regressions.

Data base and geological framework

This study is based on the collation and interpretation of a large number of high-resolution seismic profiles, with total length of more than 10 000 km, shot for a variety purposes on the Eastern Tyrrhenian Margin off the Latium region of Italy (Fig. 1). This paper describes seismic-stratigraphic features of the whole Latium margin; although particular attention is given to its northern part because of the large quantity and variety of data available there (Fig. 2). In particular there are: (a) 400 km of low-resolution, deep penetration, exploration seismic (aquapulse, 5 s sweep) that provided the structural framework and constraints for the shallower part of the margin; (b) 3000 km of high-resolution single-channel profiles (sparker, uniboom, bubble pulser, 3.5 kHz pinger, 0.25–0.5 s sweep) extending from 15 m below sea-level to the shelf break, spaced about 1000 m apart that provide a very detailed threedimensional picture of the shallow shelf stratigraphy; (c) 130 km of high-resolution multichannel profiles (uniboom, 0.4 s sweep) shot in very shallow water that proved important in correlating of the shelf seismic sequences with the stratigraphy of wells drilled on the coast; (d) an exploration well on the shelf and several wells on the coast (more than 50 m deep); (e) several gravity cores on the shelf taken on seismically identified stratigraphic features.

The Latium margin is a young passive margin, characterized by microtidal siliciclastic marine sedimentation. Its evolution is linked to the post-orogenic collapse that originated the rift of the Tyrrhenian Sea. The latter may be considered as a back-arc basin with active volcanoes (Aeolian Archipelago), thin oceanic crust and a high heat flow. This small oceanic basin has developed since the Tortonian behind the eastward-propagating Apennine orogenic belt (Patacca *et al.* 1990), so that extension and the creation of ocean floor was diachronous in both a temporal and spatial sense (Kastens *et al.* 1988). During the Quaternary, extensional processes had almost ceased in the western and northern Tyrrhenian Sea, at which time the southern Tyrrhenian Sea became very active. Consequently, the eastern Tyrrhenian shelves are relatively narrow and steep (on average 14 km wide, dip 0.6°) because of their young age and active tectonic setting, characteristics that become more pronounced from north to south (Selli 1970; Savelli & Wezel 1980). The adjacent slopes are dissected by numerous intraslope

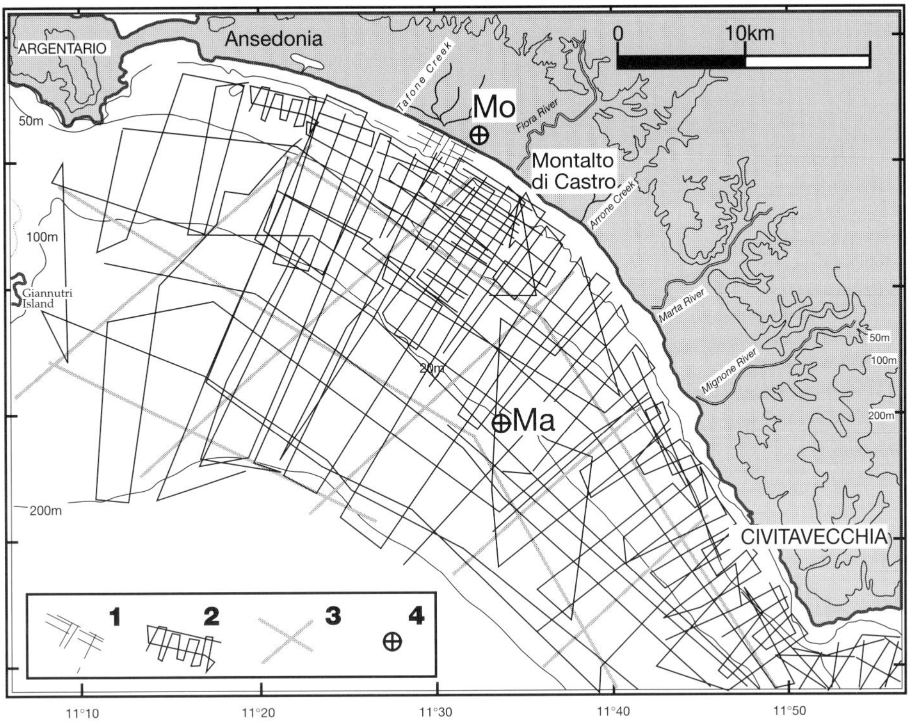

Fig. 2. Data available for the study in the northern part of the Latium margin. (1) Shallow-water high-resolution multichannel profiles; (2) high-resolution single channel profiles; (3) low-resolution, deep penetration multichannel profiles; (4) borehole (Ma, exploration well; Mo, research wells on the coast and in shallow water).

marginal basins some of which are larger than 1000 km^2. During Plio-Pleistocene times these basins trapped up to 1000–2000 metres of sediment (Fabbri *et al.* 1981).

Latium margin characters and its significance to this study

The Latium margin, despite its limited extent (less than 250 km) shows a complex stratigraphic and structural pattern that makes this area suitable for the study of the sedimentary processes and response to sea-level changes. Tectonic complexity of the margin can be related to the interaction between the high thermal and loading subsidence of its external part, and the inner shelf/coastal region characterized by different structural units with contrasting deformation style, lithologies and sediment thickness. In addition, on the central Latium shelf vertical movements related to Quaternary volcanoes are locally superimposed on to these structural features (Fig. 1).

In the northern part of the Latium shelf, exploration seismic profiles and wells (Fig. 3c) reveal the presence of thrusted Cretaceous to Oligocene arenaceous turbidites (Pietraforte Formation) overlying Upper Triassic to Early Liassic shallow-water limestones (Bartole 1990). Oligocene and older strata are carried and deformed over thrust faults that may be as shallow as 100 m below the present-day sea floor. In a regional sense, the depth of the thrusts generally increases from south to north, attributed to tilting of the margin parallel to the modern coastline (Chiocci 1991). The tectonized basement underlying the shelf apparently exerted an important control on the localization of the Late Cenozoic–Quaternary extensional basins, the border faults to which probably reactivated antecedent compressional structures. Two distinctive post-orogenic megasequences can be differentiated on regional seismic lines: (i) a postorogenic Messinian to Early Pliocene discontinuous syn-rift succession, present between the thrusts in the structural lows, (ii) a succeeding Middle Pliocene to

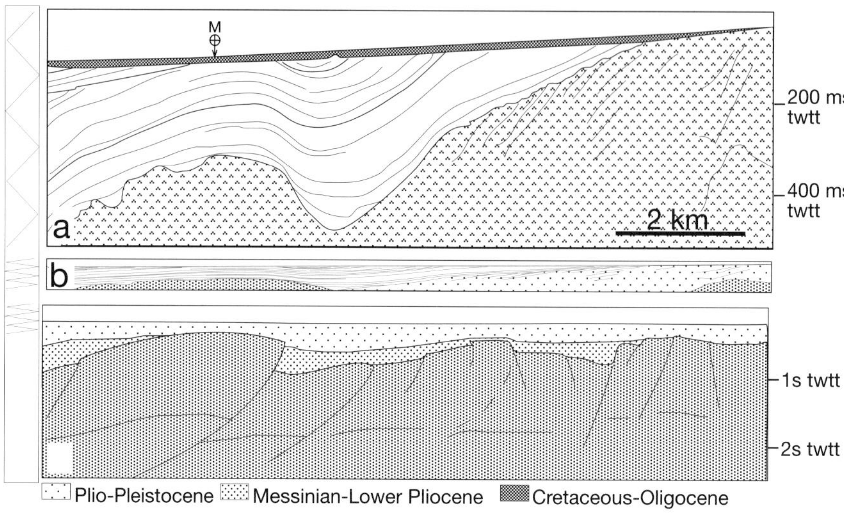

Fig. 3. (**a**) Line-drawing of a high-resolution single-channel profile in the northern Latium shelf (see Fig. 4 for location); M indicates approximate location of exploration well (see Fig. 2 for location); (**b**) same line-drawing of (a) without vertical exaggeration (using a sound speed 1500 m s^{-1}); (**c**) line-drawing of a multichannel exploration profile coinciding with the high-resolution one (after Bartole 1990). This line drawing shows the deep structure of the margin at a similar scale to that shown in (b). The left hand column schematically shows the vertical exaggeration of the line drawings. Same horizontal scale in all parts of the figure.

Pleistocene post-rift succession that rests on the syn-rift package and the thrust sheets (Fig. 3). The latter is characterized by an overall prograding architecture and its upper part is Quaternary in age.

Only occasionally does the basement structure to the passive margin form coastal relief (Quaternary volcanoes, Mesozoic–Cenozoic limestones and arenaceous turbidites) at the shoreline; the majority of the coast is rather flat and low-lying, with a coastal plain only several km in width (Fig. 1). On the northern Latium coast, Pleistocene coastal terraces outline the present shoreline in interfluve areas. Passing offshore, beaches show a complex submarine topography with bars, shoals and extensive *Phanerogama* meadows from 15 to 30 m depth (Fig. 4). Shelf morphology is quite regular and the seafloor dips 0.2 to 0.4° basinward. Downdip of the shelf-slope break, no canyons affect the slope, which dips basinward (SW) at about 2–4°.

The shelf ranges in width from 15 km in the south to 30 km to the north, with a shelf break 120–150 m deep. Erosional features related to the last glacioeustatic minimum extend to depths of 135–147 m. If a glacioeustatic minimum of sea level of 120 m below present day at 18 ka before present is assumed and a maximum depth of marine erosion is taken at 15–20 below sea level (present day value in the Tyrrhenian Sea), then estimates for shelf margin subsidence rates fall in the range of 0–0.5 m per ka.

Sediment supply to the modern coastline of the Latium margin is dominated by the Tiber, the largest river of the whole Tyrrhenian Sea, characterized by a 17 000 km^2 drainage basin and a 10^7 kg/a^{-1} suspended load, prior to the construction of dams. This sediment load is spread over the shelf and more than 150 km along the coast by a geostrophic coastal current of 0.2 to 0.4 $m^{-1}s$ (Istituto Idrografico della Marina 1982). Other rivers are relatively short, with small drainage basins less than 1000 km^2 generally oriented perpendicular to the Apennine chain. These smaller rivers supply sandy sediment to the present beaches but their contribution to shelf sedimentation is negligible.

Because the shelf is relatively narrow and the Apennine chain located adjacent to the coast, the distance between the shelf-break and the orogenic belt is rather short. Consequently, the coastal plain as well as the other subaerial basins are unable to stock a large quantity of sediment during climatic or base-level changes. This is quite different from many of the larger sedimentary systems on older, more mature passive margins from where sequence stratigraphic principles were developed. As a result, changes in base-level or climate are likely to be almost immediately recorded in the adjacent marine

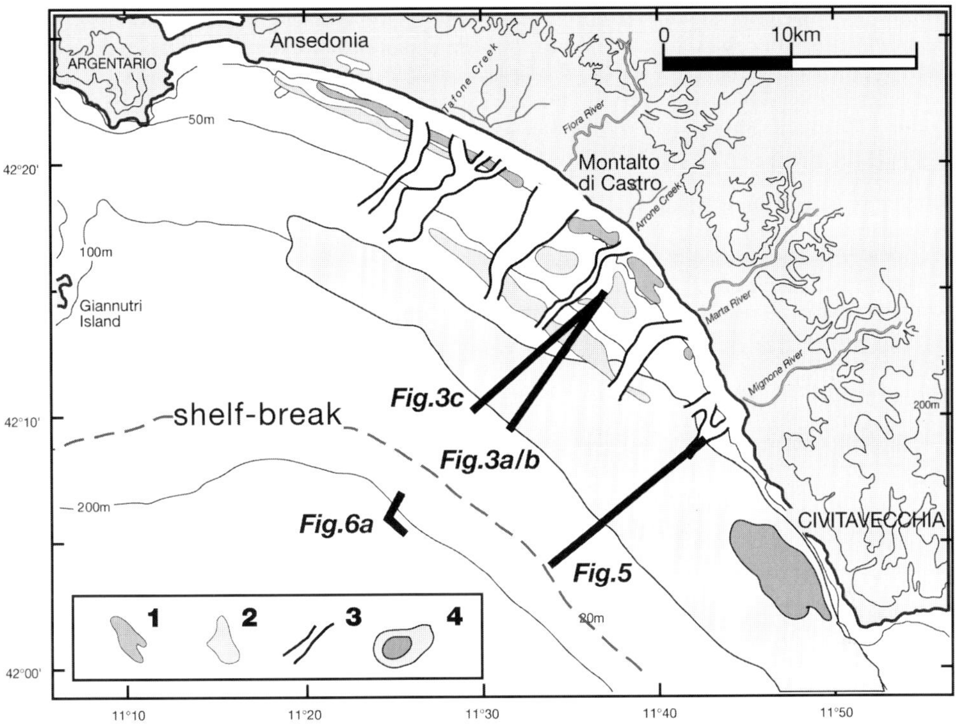

Fig. 4. Shallow subsurface features of Northern Latium shelf. (1) *Posidonia oceanica* meadow; (2) transgressive beaches; (3) buried incised valleys; (4) thickness of Holocene shelf mud (isopach 5 and 30 ms twtt). Note that the thickness of Holocene muds is distributed parallel to the coastline by northerly directed longshore currents.

basins. Because the shelf is narrow, the shift of the depositional systems in response to sea-level changes does not alter significantly the borderland morphology, except for few small volcanic islands (Pontine Archipelago) that were connected to mainland during low sea-level. Therefore a significant increase of sediment supply to the marine environments during times of falling sea-level and lowstand may not be expected. Finally, there are no canyons bypassing sediments to basin-floor and no slope-base fans have been detected.

The tectonic setting of the northern Latium margin, with coast-parallel tilting in addition to subsidence of the outer shelf has led to the variable preservation of Pleistocene sequences. In the southernmost sector of the area the Pleistocene deposits were eroded on the inner and middle shelf areas during glacial lowstands. Thus they are only present on the external part of the shelf and on the continental slope. On the contrary, to the north, subsidence allowed the preservation of the Pleistocene sequences in the middle and inner shelf (see Fig. 10). Due to differential tilting, the outcrop of different sequences in different sectors of the shelf facilitates sampling by targeted gravity coring. Analysis of the cores allows dating of the seismic sequences and the development of chronostratigraphically constrained seismic sequence stratigraphic framework for the Late Quaternary where glacioeustatic sea-level changes are well known. This framework allows the concepts and models of sequence stratigraphy, and in particular the importance of sedimentation during forced regression, to be examined and evaluated.

High-resolution seismic stratigraphy of the margin

The Late Quaternary clinostratified deposits that make up the Latium margin are truncated by the uppermost and youngest erosional unconformity (W in Fig. 5). This erosional surface, herein is referred as the Würm (i.e. last glacial) unconformity, is a composite one. It was formed by a combination processes, first related to the subaerial exposure of the shelf during the last Würm eustatic lowstand, and then by marine reworking during the subsequent Versilian sea-level rise (Aiello *et al.* 1978). The Würm unconformity loses its erosional character basinward, in proximity of the shelf break (*c.* 120–150 m deep). The basinward limit of the erosional unconformity is represented by a sharp step on the sea floor indicating that the maximum depth reached by (marine?) erosion during last glacial maximum was between 135 and 147 m below present day sea level. In the inner shelf the Würm unconformity surface is cut by incised valleys that are present in front of even the smallest rivers (Fig. 4). The incised valleys may be several tens of metres deep in proximity of the modern shoreline. The valleys are filled with multi-story acoustically transparent deposits and their relief gradually decreases offshore.

Overlying the Würm unconformity, the Holocene depositional sequence is comprised of a discontinuous veneer of sediments attributed to the transgressive and highstand systems tracts. The transgressive deposits were generally deposited in paralic environments but are rare, volumetrically subordinate, always detached, and are often found in protected areas, such as near palaeo-headlands or in palaeotopographic lows. Drowned transgressive beach deposits up to 10 m thick occur parallel to the coastline for several kilometres offshore and occur at depths of 55–65, 40 and 30 m. Possibly, these beach deposits were deposited during minor stillstands during the last eustatic rise (Tortora 1996) as has been proposed for analogous deposits in age and shelf setting found submerged in the Gulf of Cadiz, Spain (Hernández-Molina *et al.* this volume).

Holocene highstand deposits are directly related to present day depositional systems, the main deposits of which are: (i) coastal prisms that die out at a depth of about 40 m in front of river mouths and 15 m elsewhere (Tortora 1989) and (ii) a lens of shelf mud interpreted to have been deposited from the settling of the Tiber Rivers suspended load where the geostrophic alongshore currents interacts with structural offset of the coastline (Chiocci & La Monica 1993). The thickness and distribution of transgressive and highstand deposits show considerable variability along the shelf as is common to other Tyrrhenian margins (Chiocci 1994). In areas lacking sedimentary supply, the Würm erosional unconformity crops out on the sea floor.

Below the Würm unconformity, the bulk of the continental margin sedimentation is comprised of prograding clinostratified deposits that dip 2° to 4° offshore; a value similar to that of the present continental slope. The acoustic facies of these strata, characterized by a tabular external form, parallel internal configuration, concordant baselap, and high continuity of medium- to low-amplitude internal reflectors is consistent with sedimentation in continental slope environments. Within the stratal units below the Würm unconformity gullies have been found on seismic horizon C1, C2 and C3 (Figs 5, 6a). The gully features are comparable in shape and dimension to those present on the upper slope in front of the Tiber River mouth (Fig. 6b). Below gullied horizons C1–3 are two prominent erosional surfaces, horizons U1 and U2 (Fig. 5). These truncation surfaces are similar in terms of their character, extent and relief to the Würm unconformity, and pass basinward to concordant surfaces across a sharp break in slope (from 0.5 to 1°), which is characterized by the deposition of wedge-shaped prograding body (Fig. 6c). The progradational wedges are up to 10–15 m thick and may be regarded as lowstand deltas that have been suggested by Posamentier & Vail (1988) to form during 'rapid eustatic fall of short duration'. The increase of subsidence towards the external part of the margin (0.1–0.5 m ka^{-1}) generates a general offlapping stratal architecture to the Quaternary stratigraphy on the Latium shelf. The sequence bounding unconformities gradually offlap and diverge basinwards across shelf areas, whereas updip they truncate each other so that the sequences are represented by complex composite omission surfaces (Fig. 5). On the outer slope the sequence bounding unconformities delimit depositional sequences that can be several tens of metres thick.

On the southern Latium shelf, relict depositional terraces crop out on the sea floor at depths of 100–150 m flanking the relatively steep slopes of the volcanic Pontine Islands (Fig. 7a). The terraces range from 10 to 30 m thick and typically have a nearly flat top, a steep frontal slope of 10° or more and are made up of intrabasinal sediments, dominantly bioclastic sands. The terraces show an internal structure that reveals a multicyclic origin. Chiocci & Orlando (1996) demonstrated that the terraces are marine, and formed below sea-level during glacioeustatic sea-level lowstands and are related to erosion and deposition during long-lasting late Pleistocene sea-level falls. The terraces are developed parallel to the coast as far as steep slopes persist. Such deposits are best developed around the Pontine Islands but are also locally found within the

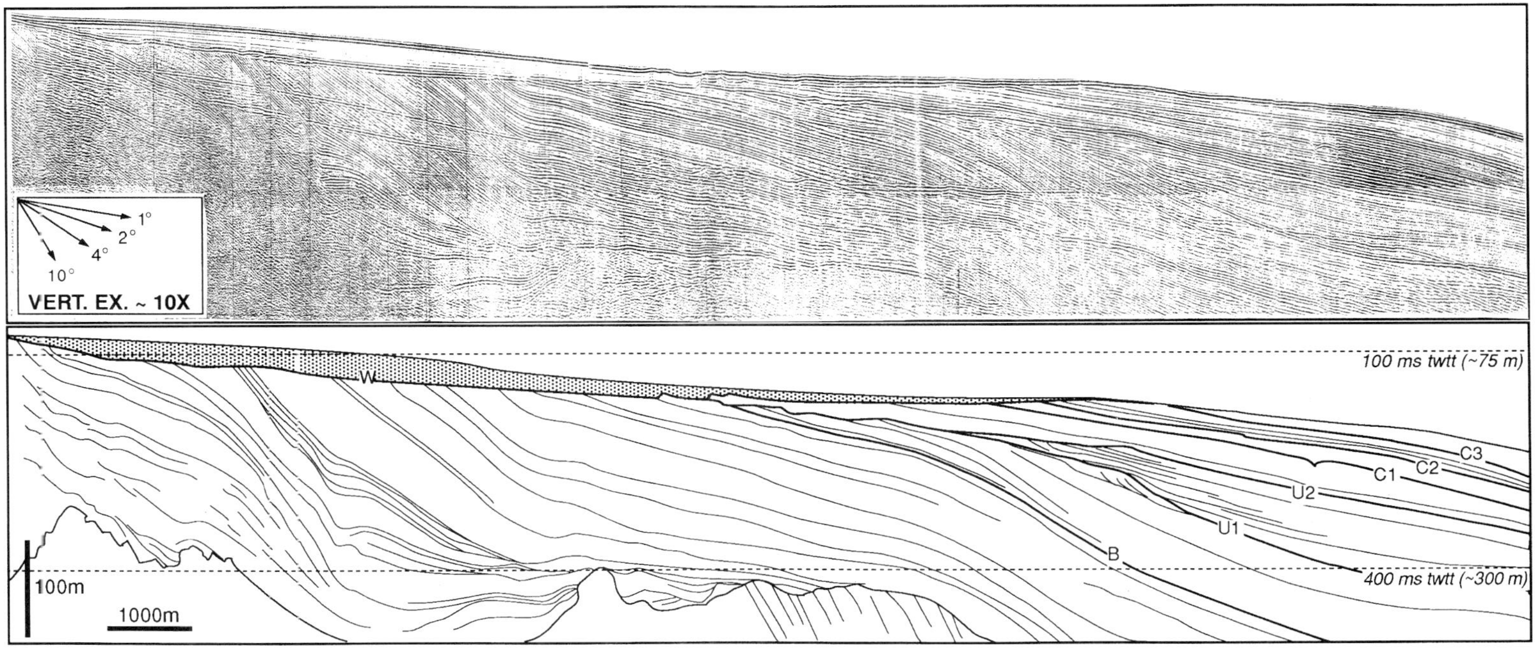

Fig. 5. High-resolution profile and line-drawing perpendicular to the coast and shelf margin, as located on Fig. 4. The Late Pleistocene sequence boundaries, B, U1, U2, C1, C2, C3 and W, are highlighted by bold lines. Note the offlapping stratal geometries resulting in updip convergence of surfaces as a result of differential subsidence and tilting of the margin.

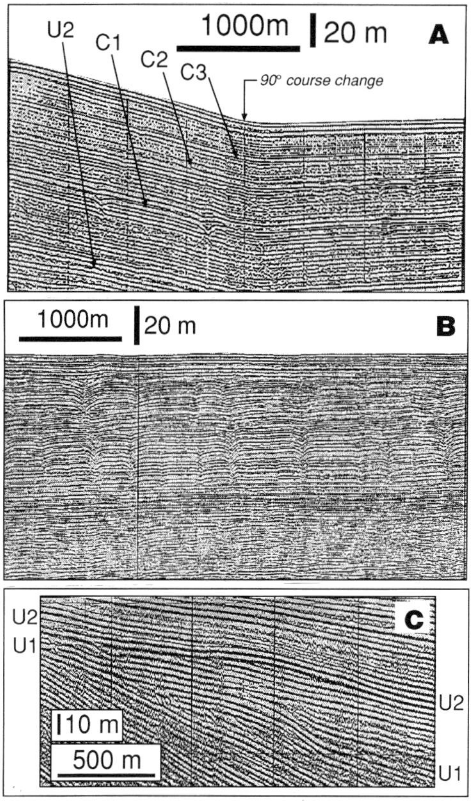

Fig. 6. (**a**) High-resolution profiles both normal (to the left) and parallel (to the right) to the continental slope (see Fig. 4 for location). Gullied surfaces are thought to represent sequence boundaries on slope environment; (**b**) buried gullies in a high-resolution profile parallel to the outer shelf offshore of Tiber River mouth; (**c**) enlargement of Fig. 5 showing depositional terraces located at the transition between erosional unconformities U1 and U2 and their correlative conformities.

other clinostratified units deposited during the early stages of margin outbuilding when the slope morphology was substantially steeper (Fig. 7b).

Interpretation of seismic stratigraphy

The interpretation of a large number of narrow-spaced high-resolution seismic profiles has allowed an extremely detailed and near 3D reconstruction of the seismic sequence stratigraphy of the Latium margin. Most of the seismic sequences appear to have been deposited during eustatic conditions very different from the present, an interpretation consistent with observations of Late Quaternary deposits on continental margins elsewhere in Italy (e.g. Chiocci 1994; Trincardi & Corregari this volume), the Mediterranean (e.g. Piper & Asku 1987; Chiocci *et al.* 1997; Hernández-Molina *et al.* this volume; McMurray & Gawthorpe this volume) and world-wide (e.g. Kolla *et al.* this volume). The evidence from along the Latium for the seismic sequence stratigraphic interpretation of margin sedimentation is as follows.

The dip and the acoustic facies of clinostratified units. Seismic facies interpretation suggests the bulk of the margin is made up of sedimentary units deposited in a continental slope environment. This observation indicates that outbuilding of the margin occurred during eustatic conditions and trends different from the present, as in today's highstand condition almost no sedimentation occurs on the slope. A similar interpretation has been suggested for the southern Latium shelf by Marani *et al.* (1986).

The absence of highstand and transgressive deposits older than last eustatic cycle. Highstand and transgressive deposits of last eustatic cycle are characterized by marked along strike variability and so have an irregular external seismic form that reflects their three-dimensional thickness distribution. Highstand systems tract deposits are localized by sediment sources, and to palaeotopography inherited from the TST (Chiocci 1994). Preserved highstand deposits of previous cycles crop out on the Latium coastline in interfluve areas where they form coastal terraces. Transgressive deposits fill the incised valleys (Milli 1992); however both these transgressive and highstand deposits are areally and volumetrically limited – they do not significantly contribute to the aggradation and progradation of the margin. Because the rates, maximum values and duration of Würm to Holocene sea-level changes are average and representative of the Pleistocene in general, it is logical to infer that highstand and transgressive deposits of previous cycles are similar in terms of thickness and distribution within the Latium margin stratigraphy. In this case, it seems likely that the transgressive and highstand deposits were eroded during the subsequent period of subaerial exposure and ensuing sea-level rise, and so do not contribute significantly to the outbuilding of the margin.

Erosional unconformities. The Würm unconformity at the base of the Holocene depositional sequence is a composite type-one sequence boundary, transgressive surface and/or maximum flooding surface. This surface was

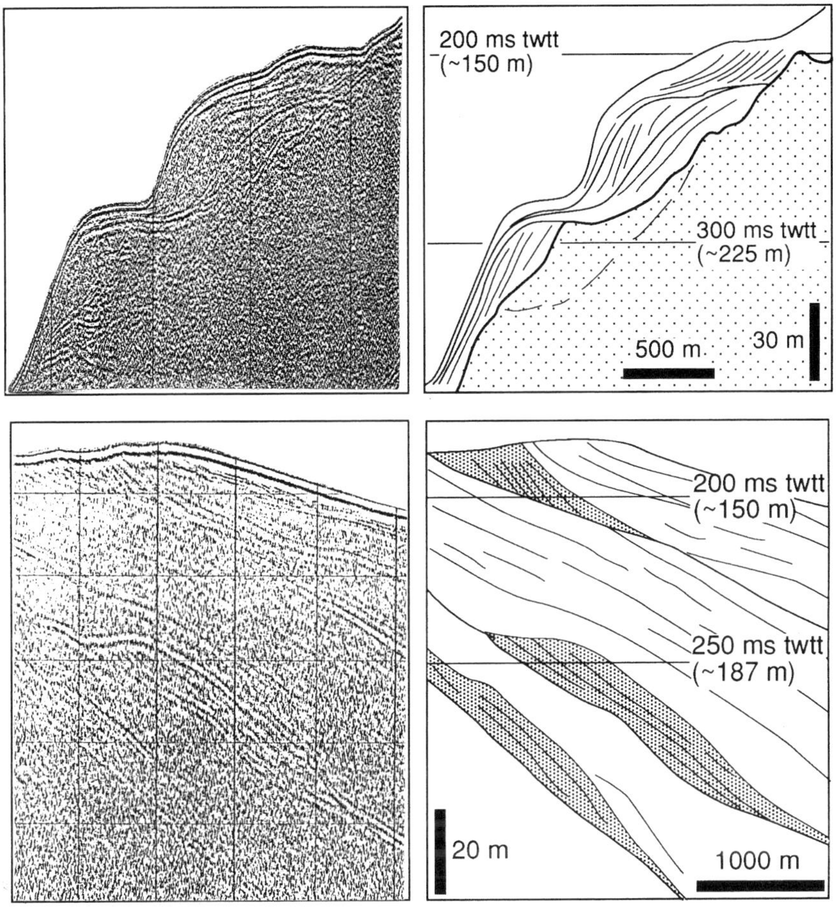

Fig. 7. (**a**) Detail of relict depositional terraces preserved on the flanks of the Pontine Islands; (**b**) depositional terraces relict on the sea floor and buried at shallow depth offshore Circeo, on the southern Latium margin. See Fig. 1 for location of the areas, and text for discussion of the spatial distribution of these features on the Latium margin.

formed by (i) marine and subaerial erosion during a sea-level fall (e.g. is a regressive surface and sequence boundary), (ii) marine ravinement during transgression where transgressive deposits are absent (e.g. is a transgressive surface) and (iii) it also represents a maximum flooding surface. Likewise the U1 and U2 erosional unconformities are interpreted to have a similar nature.

Depositional gullies. These are present on the sea floor on the upper continental slope off Tiber River mouth. They are not connected at all with present-day deltaic depositional system, dying out just inshore of the shelf beak (Bellotti *et al.* 1994). Thus they are thought to be relict from the previous Würm lowstand, when the Tiber mouth was near the shelf break and the rivers bedload was discharged directly on to the continental slope as subaqueous channelized flows (Chiocci & Normark 1992). Gullies of similar morphology, spacing and dimension are found within the seismic units making up the Latium margin, both off the Tiber mouth and in front of the other small rivers in the northern Latium coast, where they display less relief. Because of their origin, these gullied horizons are considered to mark the eustatic minimum.

Incised valleys. The incised valleys are present on the northern Latium inner shelf immediately in front of modern rivers mouths (Fig. 4). They recordthe subaerial exposure and the bypass of sediment across the shelf during times of

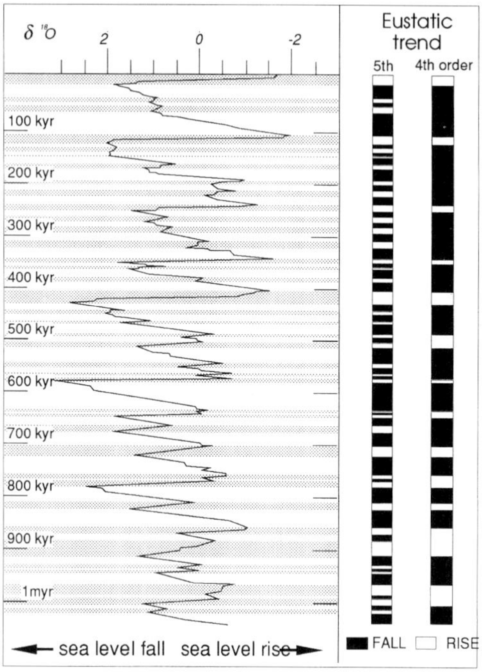

Fig. 8. Late Pleistocene $\delta^{18}O$ record (after Williams *et al.* 1988). Grey strips indicate sea-level rise, first column on the left plots sea-level falls and rises, leftmost column indicates sea-level falls and rises greater than 0.75 $\delta^{18}O$.

sea-level fall and lowstand. It is likely that such systems were the main pathway for the sediments fed to gullies on the slope.

Vertical trend of acoustic facies. Clinostratified units show high-continuity acoustic facies with a trend of diminishing acoustic transparency upsection. A similar seismic facies change has been observed within other Western Mediterranean margins such as the Rhône shelf (Torres *et al.* 1995). The trend in acoustic facies is thought to indicate coarsening-up profiles within each prograding unit caused by the offshore migration of the coastline, the narrowing of the shelf and the approach of sediment source to the upper slope environment.

Mid–Late Pleistocene eustatic sea-level changes

A reliable indication of the global ice volume during the Pleistocene is given by the $\delta^{18}O$ record from foraminifera recovered from deep water sediments. These data provide an indirect record of sea-level variations caused by glacial cycles and indicate that the Pleistocene was characterized by high-amplitude/high-frequency glacioeustatic sea-level changes (Milankovitch 1938). Spectral analysis of the $\delta^{18}O/^{16}O$ Pleistocene curve for the last million years indicates a dominant 100 ka frequency signal that is due to variation in earth orbital eccentricity, whereas the 41 and 23 ka frequencies that are related to variation in obliquity and precession were subordinate (Berger 1989). The dominant fourth-order 100 ka cyclicity of the Late Quaternary eustatic sea-level curve was extremely asymmetric, with gradual and punctuated sea-level falls, very fast and relatively constant sea-level rises and short intervals spent in the high- and low-stillstand condition. Rates of sea-level rises were typically an order of magnitude greater than those of sea-level falls (e.g. Williams 1988).

As far as the study of forced regression is concerned, if a representative composite eustatic curve for the Late Quaternary is considered (e.g. Williams 1988), it is apparent that during last million years the sea level was falling during 65% of the time (first column on the right of Fig. 8; stillstands were not considered). A similar value (60%) rises from SPECMAP curve (Pisias *et al.* 1994). If small-amplitude (less than 0.75 $\delta^{18}O$) variations are neglected, the time spent in eustatic fall increases to 75%, and to 84% for the last 500 ka (rightmost column in Fig. 8). Therefore it is reasonable to assume that during Late Quaternary, sedimentation on continental margins undergoing low rates of subsidence (e.g. <1m ka^{-1}) took place essentially under conditions of falling sea-level and basinward shifts of the shoreline (see Gawthorpe *et al.* 1994).

The exact relationship between values of $\delta^{18}O$ and the position of sea level is difficult to ascertain as several factors, in particular water temperature, determine $\delta^{18}O$ values. Matthews (1984) gives a 0.11‰ $\delta^{18}O$ value for each 10 m of sea water removed and Shackleton (1987) suggested a similar figure (0.1‰) based on comparison with the $\delta^{18}O$ record of the last glacial cycle. No matter what the exact relationship between values of $\delta^{18}O$ and sea-level is, it is important in the context of this study to notice that the $\delta^{18}O$ values of glacial extremes in stage 12 and 16 (450 ka and 600 ka respectively) were significantly greater than during any other glacial extremes in the last million years (Fig. 8). In his review of $\delta^{18}O$ excursions recorded from deep sea cores, Shackleton (1987) reported that the difference between the last glacial extreme and stage 12 and 16 extremes is as high as 0.2‰ $\delta^{18}O$, suggesting that the glacioeustatic sea level some 20 m lower than during Würm glacial period.

As subaerial erosional unconformities form during sea-level falls and extend basinward as far as the maximum lowering of sea-level, it can be predicted that erosional unconformities created at glacial maximum of 450 and 600 ka (isotopic stages 12 and 16) will be more extensive than those of previous and following cycles. Such a relationship has been reported from the Eastern Mediterranean by Piper & Aksu (1992).

Sequence boundaries within the northern Latium margin and their dating through correlation with well logs, cores and the glacioeustatic curve

Because the biostratigraphy of the Latium margin is not well-established, research efforts have focused on dating key seismic horizons interpreted to be formed during the eustatic minimum. These surfaces were dated by correlation with the data available from borehole logs and gravity cores, and placed in a chronological framework tied to Late Pleistocene sea-level curve, as discussed below.

Seismic stratigraphic column. Key horizons created at eustatic minimum and interpreted as sequence boundaries are: (i) erosional unconformities developed across shelf environment that pass basinward to correlative conformities and (ii) gullied horizons developed in the slope environment, interpreted to have formed across correlative conformities to shelf unconformities that are no longer preserved due to subaerial exposure and erosion across the shelf during the last (an possibly previous) sea-level falls and lowstand associated with times of glacial maxima.

In the study area a seismic-stratigraphic column derived from the sea floor downward can be defined as follows (Fig. 5): (1) Würm erosional unconformity (W), (2–4) three correlative conformities (gullied horizons C3, C2 and C1) and (5–6) two erosional unconformities that pass basinward to correlative conformities across depositional terraces (U2 and U1). Basinward of the U2 unconformity, depositional gullies are present on its correlative conformity. Below U1 a further key seismic surface, horizon B, is locally developed having an erosional character only in the southernmost sector of the area, passing to a concordant surface elsewhere on the margin.

Correlation borehole log – seismic stratigraphy. On the middle shelf (Ma in Fig. 2), a deep oil exploration well gives a generic Pleistocene age, based on cuttings for the upper 300 m of stratigraphy developed atop of a thrust sheet, the upper part of which is comprised of the Oligocene Pietraforte Formation (Mo in Fig. 2; Fig. 3). These data demonstrate the absence of syn-rift deposits on the structural highs and confirm the Pleistocene age of the shallowest part of the Latium margin, even in the middle shelf.

On the coast, several continuously cored wells were drilled by the Italian National Electric Board (ENEL 1993). The sedimentary succession present above the Pietraforte Formation can be synthesized as follows: (a) Miocene shale in structural lows; (b) Middle Pliocene–Lower Pleistocene marine shale; (c) Lower–Middle Pleistocene shale (Agille grigio-azzurre Formation) interpreted to have been deposited in shelf environments, the top of which shows oxidation phenomena probably related to subaerial exposure; (d) a few metres of Middle Pleistocene cemented sands and gravels deposited in foreshore to backshore environments; (e) a complex of dominantly marine sands representing multiple amalgamated sequences, the uppermost related to modern depositional systems. In the latter, the presence of several well-sorted sand layers and oxidized horizons indicates multiple subaerial exposure events (Fig. 9).

The preservation of Pleistocene sediments displays considerable variation along northern Latium margin due to coast-parallel tilting that induced relative uplift of the southern sectors compared to the north, superimposed on the subsidence of the outer shelf (Chiocci 1991). As a result, it is not possible to trace all of the key horizons across the northern Latium shelf because in the south the younger sequences erode precursors across the outer shelf, whereas to the north the older surfaces are too deep to be detected by high-resolution seismics (Fig. 10). Accordingly, it is only in the central sector of the margin offshore Montalto that the deeper 'B' horizon can be followed back to the coast (Fig. 10).

Offshore Montalto a shallow-water multichannel high-resolution seismic survey allows correlation of the seismic stratigraphy of marine areas with the stratigraphy of wells on the coast (rightmost column in Fig. 9). Essentially it has been possible to correlate the seismic 'B' horizon with the top of the Argille grigio-azzurre Formation, and eventually to the gravel and sand layer above it. Microfaunal analysis indicates an Early to Mid- Pleistocene age of the formation, consistent with the age of the Brunhes/Matuyama 780 ka magnetic reversal that has been identified at its top (Cande & Kent 1992; ENEL pers. comm. 1997). From these

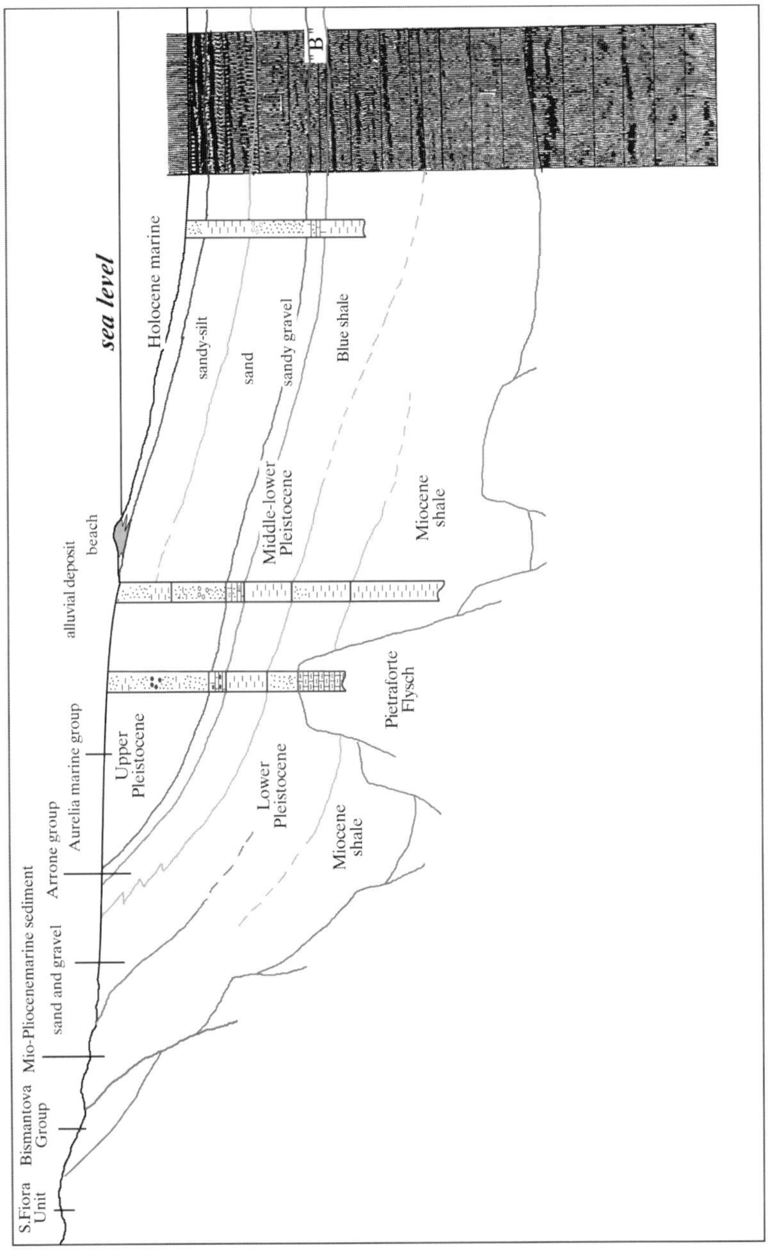

Fig. 9. Schematic sketch showing lithostratigraphy on the coast and shallow water (based on boreholes) and shallow water high-resolution seismics (after ENEL 1993).

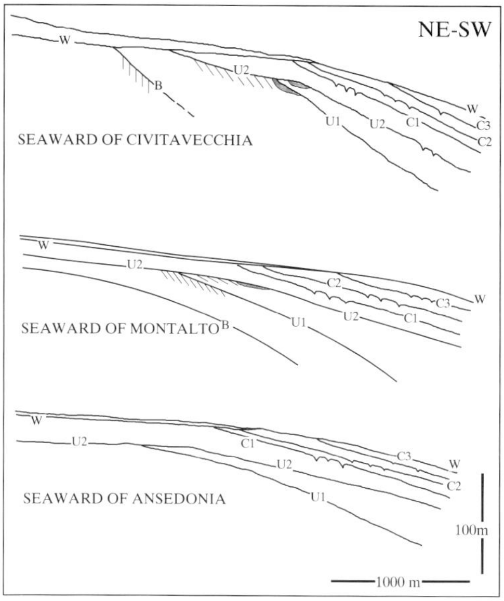

Fig. 10. Line drawings of the stratigraphic relationships developed offshore Civitavecchia (top), Montalto (centre) and Ansedonia (bottom), respectively corresponding to the southeastern, central and northwestern areas of the northern Latium continental margin depicted in Figs 2 & 4. Note the updip convergence of surfaces, relatively constant thickness of the units and widespread development of gullies across surfaces C1, C2 and C3.

chronostratigraphic constraints, the interpreted shallowing-up of the depositional environment toward the surface and the downward shift of facies associated with the seismic 'B' horizon, it can be considered the unconformity at the base of a third-order sequence having an age of about 800 ka BP (Haq *et al.* 1987). However, it is important to note that this third-order sequence boundary is not associated with an especially large excursion in the oxygen isotope record (i.e. the amplitude of stage 22 is not greater than following stages).

Correlation of gravity cores and seismic stratigraphy. Based on the results of previous seismic surveys, four gravity cores were collected to constrain the age of important seismic stratigraphic surfaces. The cores were taken where there is a minimal cover of Holocene highstand mud, as shown in Fig. 11. The 3–4 m long cores were sampled for microfauna, especially diagnostic nannoplankton and analysed by I. Raffi (National Geological Survey). The presence of *Pseudomemiliania lacunosa* in the innermost core, interpreted to have the sampled U2 unconformity, indicates an age of greater than 450 ka. The abundance of *Emiliania huxley* in the most basinward core, located 1.3 km offshore, indicates an age between oxygen isotope stages 4 and 5 of approximately 100 ka BP.

Chronostratigraphy and correlation of the margin. Once three chronostratigraphic tie points for the middle Upper Pleistocene had been established, the corresponding ages of the seismic surfaces to glacioeustatic lowstands were constrained at: W = 20 ka, U2 ≥ 450 ka and B = 800 ka. Once the age of these surfaces was known, a correlation of the erosional unconformities U1 and U2 with glacial stages 16 and 12 and the C1, C2 and C3 seismic surfaces with stages 10, 8 and 6 respectively was relatively straightforward (Fig. 12). According to this interpretation the unconformities U1 and U2 coincide with the most extreme glacioeustatic lowstands at 600 ka and 450 ka, or isotopic stages 16 and 12, respectively. The gullied conformities are interpreted to correspond to the lower amplitude glacioeustatic sea-level falls, the shelfward unconformities to which are interpreted to have been eroded during last (and likely preceding) eustatic lowstands. The position of a potential third-order sequence is provisionally identified on the basis of these correlations as illustrated in Fig. 12. Similarities of this stratigraphic framework are observed with the sequence stratigraphy of Pleistocene deposits outcropping on the coastal plain west of Rome (Chiocci & Milli 1994).

Important differences are apparent between the third- and the fourth-order sequence boundaries. Seismic surface 'B' represents the third-order sequence boundary and is marked by an abrupt lithological contrast on the coast, but is not associated with any contrast in acoustic facies on the shelf. Across the shelf seismic surface 'B' is for the most part conformable and only locally displays erosional relief. In contrast, on the coast the fourth-order sequence boundaries are quite difficult to distinguish. This is because of the up dip amalgamation of many surfaces and sequences within only few metres of stratigraphy. However, the fourth-order sequence boundaries are easily recognized on the shelf where they are characterized by erosional unconformities that pass basinward to correlative conformities.

Volumetric partitioning of sedimentation

In order to ascertain the volumetric significance and any partitioning of sediment between

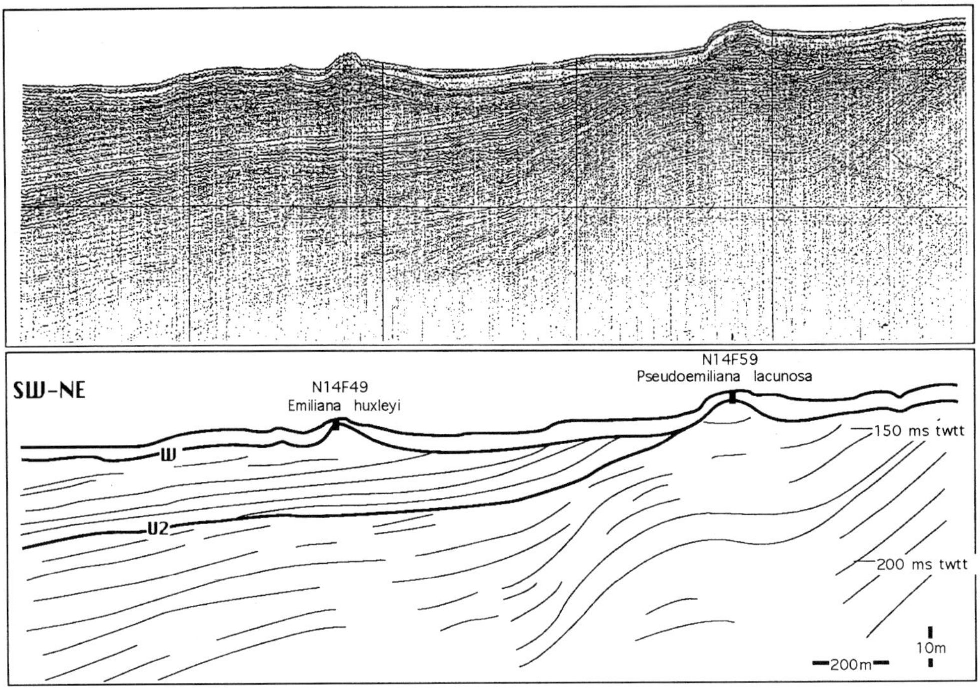

Fig. 11. High-resolution seismic profile on the outer shelf and an accompanying line drawing. The figure shows the position of two gravity cores, strategically positioned to sample and hence date surfaces W and U2 where they subcrop on the sea floor beneath a thin veneer of Holocene muds. See text for further discussion of the fauna recovered and age of these surfaces.

systems tracts, two sectors of the margin have been considered, the northern Latium margin from Tor Vaiànica to Civitavecchia, and central Latium margin, from Civitavecchia to M. Argentario (Fig. 13). Across these areas the close spacing of the seismic grid allows a rough computation of volumes of the different systems tracts deposited from 120 ka BP (i.e. last eustatic highstand) to the present day.

While the computation of highstand and transgressive deposit is quite precise, the volume of forced regression/lowstand deposits related to last eustatic cycle (i.e. above C3 horizon) is difficult to define, as the seismic network is rather poor and inhomogeneous across the continental slope. Accordingly, a homogeneous thickness of 30 m across the entire slope has been used, as appears to be consistent with the parallel reflector configuration and constant thickness observed in central Latium margin. Such computation is likely to underestimate the real volume of lowstand/forced regression deposits, as they probably extend far beyond the base of the slope, despite the fact there are no canyons feeding deep sea fans.

On the central Latium margin adjacent to the Tiber River Mouth, highstand systems tract deposits account for about 27 km^3 of the margin and transgressive deposits are lacking, although a 3.5 km^3 transgressive deposit is present just south of Tor Vaiànica (Fig. 13). The lowstand/forced regression deposits have a volume of approximately 500 km^3, although taking into account the considerations made above this figure is likely to be conservative. In northern Latium the influence of Tiber River is still felt because of longshore currents so that highstand deposits account for 9 km^3 and transgressive deposits for 1 km^3 of the margin (Fig. 13). The forced regression/lowstand deposits have about the same volume of about 500 km^3 as to the south. In consideration of these figures, it should be taken into account that the highstand and transgressive deposits currently present on the shelf would in all likelihood be eroded during next sea-level fall and lowstand, having the effect of exaggerating further the volumetric partitioning of sediment between systems tracts. It is very clear from these calculations that during the last eustatic cycle (last 120 ka) the highstand and transgressive deposits are volumetrically insignificant in comparison to the

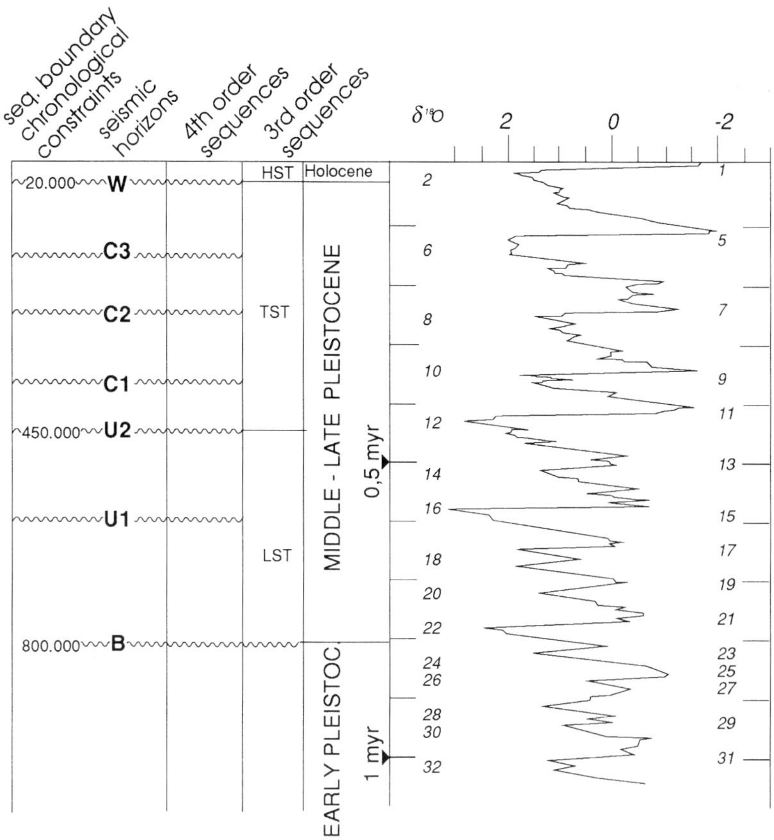

Fig.12. Proposed stratigraphy of the margin, based on the correlation of the seismo-stratigraphic features with borehole, gravity cores and the $\delta^{18}O$ record.

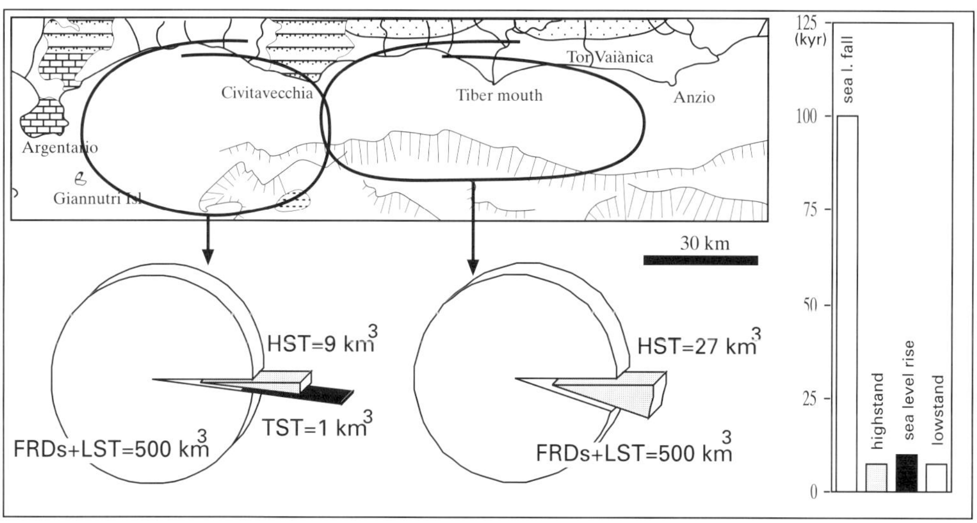

Fig. 13. Computation of volumes of highstand, transgressive and forced regressive + lowstand systems tracts in the northern and central Latium margin for the last eustatic cycle. The right-hand bar chart graphically shows duration of eustatic sea-level trends during the same period.

forced regression and lowstand deposits, even where the Tiber, the largest river and most important sediment point source feeding into the Tyrrhenian Sea is present.

Effects of high-frequency, high-amplitude sea-level changes on shelf/slope sedimentation

Having developed a detailed seismic stratigraphy of the northern Latium margin and placed it within a the chronostratigraphic framework, the succession can be viewed in terms of its dynamic response to repetitive high-frequency, high-amplitude glacioeustatically-driven sea-level changes.

(1) *During* sea-level *fall.* Forced regression is interpreted to have led to significant clinoform progradation of the shelf. Major outbuilding of the Latium margin was a consequence of sedimentation on the slope, creating depositional sequences almost exclusively formed by forced regression deposits. A similar interpretation has been demonstrated by Sydow & Roberts (1994) and Kolla *et al.* (this volume) in the Gulf of Mexico and by Hernández-Molina *et al.* (this volume) in the Gulf of Cadiz, Spain. The volumetric importance of this systems tract is attributed to the long interval of time associated with glacioeustatic sea-level fall during the Late Quaternary.

On the Latium margin sediment supply was likely augmented through the subaerial exposure, erosion and dismantling of pre-existing highstand, transgressive and of just-formed forced regression shelf deposits. However, because of the relatively narrow width and steep gradient of the shelf, and the low capacity for sediment storage in adjacent coastal plain and non-marine basins, the increase of sediment supply during times of fall and lowstand is not likely to be as significant as on more mature passive margins or systems with a greater hinterland. In the Latium margin, slope sedimentation occurred via subaqueous channelized flows (creating gullies in front of rivers mouths) and shows a coarsening-up trend that reflects the progressive shallowing of depositional environment during sea-level fall. Significantly the position of point sources (indicated by incised valleys and gullies) does not influence thickness distribution that is relatively constant along the margin. This is quite different to more sediment dominated systems such as discussed by Kolla *et al.* (this volume) and attests to the re-distribution of sediment along the lowstand shoreline by longshore currents perhaps similar in direction to those observed today. Intensity of the longshore currents may have been greater during times of forced regression and lowstand, related to more extreme climatic conditions and/or to the absence of wide shelves able to reduce the energy of the approaching waves during the eustatic minimum.

(2) *During lowstand.* The maximum seaward position reached by wavebase is interpreted to mark the transition from the erosional unconformity diachronously created during sea-level fall, to the correlative conformity formed in the marine areas below the level of effective wave base erosion (e.g. see Plint & Nummedal this volume). At eustatic minimum, depositional terraces may form on the palaeo shelf-break, and these are typically some 20 m thick and developed parallel to the coast for several tens of kilometres. In case of absence of a strong extrabasinal input and of steep bedrock (as along coast of volcanic islands), such terraces may also represent the whole of the clastic cover. The lowstand deposits may be relatively thin because of the short time spent in the low-stillstand condition before rapid sea-level rise begins and rates of sediment supply to the coastline are soon outpaced during the ensuing rise.

(3) *During sea-level rise.* The depositional systems and depocentre migrate onshore. As the shoreline backsteps, foreshore deposits created during short stillstands in the rise may be drowned and abandoned. However, in general transgressive sedimentation is volumetrically rather insignificant as the rate of sediment supply to the shoreline is drastically reduced. On the shelf a ravinement surface is often associated with reworking of the subaerial surface and this surface is generally not overlaid by any transgressive deposits, apart from in a few exceptional cases. In updip areas composite sequence boundary/transgressive/maximum flooding surfaces are formed as the sequence boundary is commonly reworked and transgressive sedimentation is minimal.

(4) *During highstand.* Once subaerial and coastal basins are filled, sedimentation on the shelf produces prograding depositional systems that are restricted to the shelf and controlled by position of point sources. This reflects the short duration of time spent in the high-stillstand condition.

Conclusions

Careful mapping of seismic stratigraphy from a dense seismic grid on the relatively young, narrow and steep Latium passive margin allows the definition of the six youngest depositional sequences. Integration of the seismic sequences with well data on the coast and gravity cores recovered from different seismic units outcropping on the seafloor indicates that the sequences were deposited during the last 0.8 Ma. As such, each sequence is fourth-order, being of a 100–120 ka duration. The age of the sequences indicates that fourth-order glacioeustacy related to orbital eccentricity was the primary driver of the stratigraphy deposited on the margin.

The six Late Quaternary stratigraphic sequences are dominated by sediments deposited during relative sea-level falls or forced regression. Highstand and transgressive deposits of cycles older than 120 ka appear to be absent. This is because during rapid glacioeustatically-driven sea-level rises and short-lived highstands (e.g. since 18 ka BP), volumetrically subordinate transgressive and highstand deposits were (and still are) being deposited on the shelf and along the coast. Such deposits have a relatively low preservation potential as they are likely to be eroded during subsequent sea-level falls and lowstands. In contrast during forced regression and sea-level fall, relatively high rates of sedimentation occurred in continental slope environments and led to outbuilding of the margin. Sediment was fed to the shelf and down the slope via incised valleys and slope channels; however, rather than deposition of discrete lobes in front of incised valleys, sheet-like outbuilding of the margin occurred because of sediment redistribution along the shoreline. Long-shore currents are thought to have been invigorated during times of sea-level fall and lowstand due to reduced shelf width and other oceanographic factors. The resulting depositional packages have a relatively constant thickness and acoustic facies that suggest a coarsening-up lithological trend. The latter is interpreted to reflect increasing proximity of the shoreline to the outer-shelf/shelf break area during forced regression.

It is generally only the sediments deposited below the glacial eustatic maximum in proximity of the shelf-slope break that are preserved and unmodified by erosional processes on the shelf. Even forced regressive sediments deposited on the shelf during early stages of forced regression were removed during further sea-level fall. In areas where extrabasinal sediment supply is minimal or lacking, wedge-shaped depositional bodies were produced at minimum sea-level. These often show a multi-story internal structure thought to record a stacking of depositional wedges related to several successive eustatic cycles.

On the shelf, forced regressive deposits are truncated by erosional unconformities that extend basinward to the shelf break. Such surfaces represent type-one sequence boundaries. In the classical Exxon model, such sequence boundaries would be at the base of the depositional sequence (e.g. Posamentier & Vail 1988). However, on the Latium shelf each of these sequence bounding surfaces is paradoxically situated *above* the whole of its own sequence; those forced regressive sediments deposited contemporaneously as sea-level fell and the subaerial exposure surface expanded basinward. To avoid this inconsistency of the classic Exxon model, it is useful to apply the fourfold subdivision of depositional sequences to Quaternary deposits (lowstand, transgressive, highstand and forced regressive systems tracts) proposed by Hunt & Tucker (1992). The forced regressive systems tract (FRST) is equivalent to the falling stage systems tract of Plint & Nummedal (this volume). According these systematics, the FRST is thought to be deposited before and during the formation of sequence boundary, i.e. erosional unconformity on the shelf and correlative conformity on the slope, at it is in the Latium margin. Kolla *et al.* (1994) were opposed to this scheme because the transition from FRST to LST would not be synchronous, so that ‘one physiographic setting would be characterized by forced regressive deposits, whereas the other setting would be characterized by deeper water turbidite systems’. During the Late Quaternary slope-base fans were not however developed against the Latium margin because (i) of the slope physiography, (ii) the fact that the lowest point of sea level only approached the shelf-slope break and (iii) lowstands represent such a short time interval; for instance sea level is at the low stillstand condition for less than 10 ka on the eustatic curve of Fig. 13.

As discussed by Chiocci (1994), because the Late Quaternary glacioeustatic sea-level signal is characterized by high-frequency and high-amplitude, asymmetric, fluctuations of similar amplitude, which repetitively dropped sea level to a similar depth range on continental margins and had lowstands and highstands of short duration, modifications to the classic three-fold sequence stratigraphic model are necessary, best accomplished by adopting a fourth forced regressive wedge systems tract, bounded above by the main or master sequence boundary (see also Hernández-Molina *et al.*, and Trincardi & Corregari and Kolla *et al.* this volume). In terms of the rock record the Quaternary is an ice-house interval that is quite different from greenhouse times that typify much of the geological history. In this regard, the late Quaternary stratigraphy of the Latium margin, dominated by forced regression deposits that are bounded by condensed sections formed during sea-level rises and highstands, can be considered an end-member situation, but may provide an important paradigm for strata deposited during other ice-house intervals (e.g. the late Carboniferous, late Ordovician and late Precambrian).

I am grateful to Emiliano Mutti for its encouragement in this study; Isabella Raffi is kindly acknowledged for nannoplankton analysis that gave an important contribution to the stratigraphic interpretation; Roberta Fiorini and Stefano Ciolli interpreted seismic data and gravity cores in their Thesis, Laura infa Guarnier revised the English language. Thanks are given to ENEL for permission to present unpublished data, to G. Ercilla, P. Tortora and A. Sposato for the critical reading of the manuscript, that was sharply revised by D. Piper and F. J. Hernández-Molina. All the data were collected with R/V *Urania* of National Research Council of Italy, whose crew is deeply acknowledged. This work is part of 396 IGCP Project 'Continental Shelves in Quaternary'.

References

AIELLO, E., BARTOLINI, C., GABBANI, G., ROSSI, S., VALLERI, G., CERTINI, L., CLERICI, C. & LENAZ, R. 1978. Studio della piattaforma continentale medio-tirrenica per la ricerca di sabbie metallifere: 1) da Capo Linaro a Monte Argentario. *Bollettino Società Geologica Italiana*, **97**, 495–525.

BARD, E., HAMELIN, B. & FAIRBANKS, R. G. 1990 U-Th ages obtained by mass spectrometry in corals from Barbados: sea level during the past 130,000 years. *Nature*, **346**, 456–458.

BARTOLE, R. 1990. Caratteri sismostratigrafici, strutturali e paleogeografici della piattaforma continentale tosco-laziale; suoi rapporti con l'Appennino settentrionale. *Bollettino Società Geologica Italiana*, **109**, 599–622.

BELLOTTI, P., CHIOCCI, F. L., MILLI, S., TORTORA, P. & VALERI, P. 1994. Sequence stratigraphy and depositional setting of the Tiber Delta: Integration of high-resolution seismics, well logs, and archeological data. *Journal of Sedimentary Research*, **B64**, 416–432.

BERGER, A. 1989**.** Pleistocene climatic variability at astronomical frequencies. *Quaternary International*, **2**, 1–14.

CANDE, S. C. & KENT, D. V. 1992 A new geomagnetic polarity time scale for the Late Cretaceous and Cenozoic. *Journal of Geophysical Research*, **97**, 13917–13951.

CHIOCCI, F. L. 1991. Evidenze di un basculamento alto-pleistocenico della piattaforma continentale del Lazio centro-settentrionale. *Studi Geologici Camerti*, **91/2**, 271–281.

— 1994. Very high-resolution seismics as a tool for sequence stratigraphy applied to outcrop scale – Examples from eastern Tyrrhenian margin Holocene/Pleistocene deposits. *American Association of Petroleum Geologists Bulletin*, **78**, 378–395.

—— & ORLANDO, L. 1996. Lowstand terraces on Tyrrhenian Sea steep continental slopes. *Marine Geology*, **134**, 127–143.

—— & LA MONICA, G. B. 1993. Shelf sedimentation and morphology as controlled by fluvial sedimentary supply. Examples from Latium coast (Tyrrhenian Sea, Italy) (abstract) *In*: *International symposium on fluvial and coastal system and environmental changes*, **5**.

—— & MILLI, S. 1994. Middle–Late Pleistocene high-resolution sequence stratigraphy of the Latium Continental Margin; Tentative Integration of Outcrop (inland) and High-resolution Seismic (Marine) Data (abstract) *In*: *Second high-resolution Sequence Stratigraphy Conference*, Tremp, Spain. 27–28.

—— & NORMARK, W. R. 1992. Effect of sea-level variation on upper-slope depositional processes offshore of Tiber Delta, Thyrrhenian Sea, Italy. *Marine Geology*, **104**, 109–122.

——, ERCILLA, G., TORRES, J. 1997. Middle–Late Pleistocene stratal architecture of Western Mediterranean Margins as the result of the stacking of lowstand deposits. In press in *Sedimentary Geology*, **112**, 195–217.

CIOLLI, S. 1995. *Analisi comparata di carote di sedimento marino e di profili sismici a riflessione ad altissima risoluzione*. Thesis University of Rome 'La Sapienza'.

ENEL 1993. *Impianto policombustibile di Montalto di Castro – Opere martittime del terminale GNL*, Relazione Generale, **5**.

FABBRI, A., GALLIGNANI, P. & ZITELLINI, N. 1981. Geologic evolution of the pery-tyrrhenian sedimentary basins. *In*: WEZEL, F. C. (ed.), *Sedimentary Basins of Mediterranean Margins*, 101–126.

FIORINI, R. 1992. *Sismostratigrafia dei depositi recenti di piattaforma tra Civitavecchia e l'Argentario*. Thesis University of Rome 'La Sapienza'.

GAWTHORPE, R. L., FRASER, A. J. & COLLIER, R. E. LL. 1994. Sequence stratigraphy in active extensional basins: implications for the interpretation of ancient basin-fills. *Marine and Petroleum Geology*, **11**, 642–658.

HAQ, B. U., HARDENBOL, J. & VAIL, P. R. 1987. Chronology of Fluctuating Sea Levels Since the Triassic. *Science*, **235**, 1156–1167.

HERNÁNDEZ-MOLINA, F. J., SOMOZA, I. & LOBO, F. 2000. Seismic stratigraphy of the Gulf of Cádiz continental shelf: a model for late Quaternary very high-resolution sequence stratigraphy and response to sea-level fall. *This volume.*

HUNT, D. & TUCKER, M. E. 1992. Stranded parasequences and the forced regressive wedge systems tract: deposition during base-level fall. *Sedimentary Geology*, **81**, 1–9.

ISTITUTO IDROGRAFICO DELLA MARINA 1982. *Atlante delle correnti superficiali dei mari italiani*. Genova.

KASTENS, K., MASCLE, J., AUROUX, C., BONATTI, E., BROGLIA, C., CHANNELL, J., CURZI, P., EMEIS, K., GLAÇON, G., HASEGAWA, S., HEIKE, W., MASCLE, G., MCCOY, F., MCKENZIE, J., MENDELSON, J., MÜLLER, C., RÉHAULT, J. P., ROBERTSON, A., SARTORI, R., SPROVIERI, R. & TORII, M. 1988. ODP Leg 107 in the Tyrrhenian Sea: Insights into passive margin and back-arc basin evolution. *Geological Society of America Bullettin*, **100**, 1140–1156.

KOLLA, V., BIONDI, P., LONG, B. & FILLON, R. 2000. Sequence stratigraphy and architecture of the late Pleistocene Lagniappe delta complex, northeast gulf of Mexico. *This volume.*

——, POSAMENTIER, H. W. & EICHENSEER, H. 1995. Stranded parasequences and the forced regressive wedge systems tract: deposition during base-level fall–discussion. *Sedimentary Geology*, **95**, 139–145.

MCMURRAY, L. S. & GAWTHORPE, R. L. G. 2000. Along-strike variability of forced regressive deposits: late Quaternary, northern Peloponnesos, Greece. *This volume.*

MARANI, M., TAVIANI, M., TRINCARDI, F., ARGNANI, A., BORSETTI, A. M. & ZITELLINI, N. 1986. Pleistocene progradation and postglacial events of the NE Tyrrhenian continental shelf between the Tiber river delta and Capo Circeo. *Memorie della Società Geologica Italiana*, **36**, 67–89.

MATTHEWS, R. K. 1984. Oxygen isotope record of ice-volume history: 100 million years of glacio-eustatic sea-level fluctuations, *In*: SCHLEE, J. S. (ed.) *Interregional Unconformities and Hydrocarbon Accumulation.* AAPG Memoirs, **36**, 97–107.

MILANKOVITCH, M. 1938. Astronomische Mittel Zur Erforschung der Erdgeschichtlichen Klimate. *Handbuch der Geophysik*, **9**, 593–698.

MILLI, S. 1992. *Analisi di facies e ciclostratigrafia in depositi di piana costiera e marino marginali – Un esempio nel Pleistocene del Bacino Romano*. PhD Thesis, University of Rome 'La Sapienza'.

PATACCA, E., SARTORI, R. & SCANDONE, P. 1990. Tyrrhenian Basins and Apenninc Arcs: Kinematic relations since Late Tortonian times. *Memorie della Società Geologica Italiana*, **45**, 425–451.

PIPER, D. J. W. & AKSU A. E. 1992. Architecture of stacked Quaternary deltas correlated with global oxyge isotopic curve, *Geology*, **20**, 415–418.

PISIAS, N. G., MARTINSON, D. G., MOORE JR., T. C., SHACKLETON, N. J., PRELL, W., HAYS, J. et BODEN, G. 1984. High resolution stratigraphic correlation of benthic oxygen isotopic records spanning the last 300,000 years. *Marine Geology*, **59**, 217–233.

PLINT, A. G. & NUMMEDAL, D. 2000. The falling stage systems tract: recognition and importance in sequence stratigraphic analysis. *This volume.*

POSAMENTIER, H. W. & VAIL, P. R. 1988. Eustatic controls on clastic deposition II – sequence and systems tract models. *In*: WILGUS, C. K. *et al.* (eds) *Sea-Level Changes: An Integrated Approach* SEPM Special Pubblications, **42**, 125–154.

——, ALLEN, G. P., JAMES, D. P., & TESSON, M. 1992. Forced Regression in a Sequence Stratigraphic Framework: Concepts, Examples, and Exploration Significance. *American Association of Petroleum Geologists Bulletin*, **76**, 1687–1709.

SAVELLI, D. & WEZEL, F. C. 1980. *Morfologic map of the Tyrrhenian Sea, scale 1: 1.250.000*: P. F. CNR 'Oceanografia e fondi marini-Bacini sedimentari' Lit. Art. Cart., Firenze.

SELLI, R. 1970. Cenni morfologici generali sul Mar Tirreno. *Giornale di Geologia*, **37**, 5–24.

SHACKLETON, N. J. 1987. Oxygen isotopes, ice volume and sea level. *Quaternary Science Reviews*, **6**, 183–190.

SYDOW, J. & ROBERTS, H. H. 1994. Stratigraphic Framework of a Late Pleistocene Shelf-Edge Delta, Northeast Gulf of Mexico. *American Association of Petroleum Geologists Bulletin*, **78**, No. 8, 1276–1312.

TORRES, J., SAVOYE, B. & COCHONAT, P. 1995. The effects of Late Quaternary sea-level changes on the Rhône slope sedimentation (north-western Mediterranean), as indicated by seismic stratigraphy. *Journal of Sedimentary Research*, **B65**, 368–387.

TORTORA, P. 1989. La sedimentazione olocenica nella piattaforma continentale interna tra il promontorio di Monte Argentario e la foce del Fiume Mignone (Tirreno centrale). *Giornale di Geologia*, **3**, 51/1, 93–117.

—— 1996. Depositional and erosional coastal processes during the last postglacial sea-level rise: an example from the central Tyrrhenian continental shelf (Italy). *Journal of Sedimentary Research,* **66**, 2, 391–405.

TRINCARDI, F. & CORREGGIARI, A. 2000. Quaternary forced-regression deposits in the Adriatic basin and the record of composite sea-level cycles. *This volume.*

WILLIAMS, D. F. 1988. Evidence for and against sea-level changes from the stable isotopic record of the Cenozoic. *In*: WILGUS, C. W., HASTINGS, B. S., KENDALL, C. G., POSAMENTIER, H. W., ROSS, C. A., & VAN WAGONER, J. C. (eds) *Sea-level changes. An integrated approach.* SEPM Special Publications, **42**, 31–36.

——, THUNELL, R. C., TAPPA, E. & RAFFI, I. 1988. Chronology of the Pleistocene oxygen isotope record 0–1.88 M.y. B. P. *Paleogeography, Paleoclimatology, Paleoecology*, **64**, 221–240.

Sequence stratigraphy and architecture of the Late Pleistocene Lagniappe delta complex, northeast Gulf of Mexico

V. KOLLA[1], P. BIONDI[1], B. LONG[2] & R. FILLON[3]

[1]*Elf Exploration-Production, CSTJF, Pau, France*

[2]*INRS-Oceanologie, Rimouski, Canada*

[3]*Texaco Inc., New Orleans, USA*

Abstract: During the last fourth-order glacial–interglacial cycle (e.g. post oxygen isotope stage 5) the Lagniappe Delta system located in the northeast Gulf of Mexico, prograded several tens of miles seaward along two main NE–SW and N–S trending fairways. The delta complex is underlain by a well-developed calcareous shale-rich condensed section that was deposited during isotope stage 5. The delta complex comprises many progradational lobes that were deposited during fifth-order sea-level falls and during the fourth-order maximum lowstand and early rise of sea level associated with isotope stages 4 to 2. Each significant fifth-order sea level fall developed a sequence boundary with an unconformity updip and a correlative conformity downdip on which a delta lobe was deposited. Autocyclic processes also lead to deposition of numerous lobes through lobe switching. During the maximum sea-level lowstand, deep erosion related to the development of an extensive incised valley system occurred across the top of the prograding wedges that were deposited during relative sea-level fall (the falling stage systems tract) and modified all of the previous updip unconformities. The base of the incised valley system, and its correlative downdip conformity, form the main fourth-order sequence boundary. It is on this surface that the last part of the delta complex was deposited during the maximum lowstand and early rise of sea-level. Thus, during the last fourth-order cycle, several fifth-order 'initial' sequence boundaries and one fourth-order 'final' sequence boundary were formed. Infilling of the incised valley system occurred mainly during the early and late rise of sea-level (isotope stages 2 and younger), prior to a major landward shift of deltaic sedimentation in response to the rapid eustatic rise in sea level during isotope stage 1.

The Late Pleistocene Lagniappe Delta system is today located on the middle to outer Mississippi–Alabama shelf in the northeast of the Gulf of Mexico, just east of the modern Mississippi delta (Fig. 1a). The first detailed study the delta utilizing high-resolution seismic data (Fig. 1a, b) was by Kindinger (1988, 1989*a*, *b*), but lacked ground truth in the form of core or borehole data. Coleman & Roberts (1988*a*, *b*) reported on two sites from the Lagniappe Delta as part of a more extensive synthesis of borings made on the Texas–Louisiana shelf and slope. However, a lack of cores, biostratigraphic and isotope data precluded precise correlation of the borehole data to major seismic stratigraphic surfaces and events evident in the seismic data existing at that time.

To remedy this problem, an industrial research consortium funded by ten oil companies (Amoco, Arco, BP, Chevron, Elf, Exxon, Marathon, Mobil, Texaco and Union Pacific) was established in 1988 to develop an integrated and detailed seismic, litho-, bio-, and chronostratigraphic framework for the Lagniappe Delta complex between the Main Pass (MP) and Viosca Knoll (VK) block areas (Fig. 1). The consortium drilled four borings from which more or less continuous cores were recovered. These borehole data were integrated and connected with the existing seismic data utilized by Kindinger (1988, 1989*a*, *b*) through the acquisition of a supplementary high-resolution seismic grid in 1991 (Fig. 1). Previously, Sydow & Roberts (1994) and Winn *et al.* (1995) reported on the results of a detailed study of the MP303 core, integrated with the consortium seismic adjacent to this site and all of the previously available seismic data.

In the Lagniappe delta area the Mississippi–Alabama shelf is about 140 km wide and has gradients of less than 0.05°, with shelf breaks in water depths of 75–110 m (Sydow & Roberts 1994; Fig. 1). The upper-slope gradients vary from 0.6° to as much as 3°. Most of the Texas–Louisiana shelf margin is considered to be an unstable passive continental margin (e.g. Winker 1982), whereas the northeast Gulf of Mexico area is an essentially stable passive continental margin (Greenlee & Moore 1988; Sydow & Roberts 1994). The Lagniappe Delta complex is overlain to the west by the modern Belize and St Bernard lobes of the Mississippi Delta.

From: HUNT, D. & GAWTHORPE, R. L. (eds) *Sedimentary Responses to Forced Regressions*. Geological Society, London, Special Publications, **172**, 291–327. 1-86239-063-0/00/$15.00

The objectives of this paper are (i) to provide a summary of stratigraphy and architecture of the Lagniappe Delta, (ii) to detail the characteristics of the key stratigraphic surfaces and develop a comprehensive sequence-stratigraphic framework of the Lagniappe delta complex and (iii) to evaluate the significance of deposition in response to forced regression (e.g. Plint 1988; Posamentier *et al.* 1992; Hunt & Tucker 1992) in building the delta.

Types of data utilized and methods of study

The database used in this paper includes all of the available high-resolution seismic and data from the four boreholes (Fig. 1b). The high-resolution seismic data comes from both the Consortium 1991 survey and earlier surveys (Fig. 1b). The sound sources used in these surveys were a triple-plate boomer and a sleeve gun or a mini-sparker, with a resolution of 1–2 m and a penetration of 100–200 ms (Sydow & Roberts 1994). In some areas of the shelf water-bottom multiple and effects of gas on seismic signal mask the geological information.

The four boreholes were drilled in the two study blocks (MP242, MP303, MP288 and VK774) in water depths of 56 m (184 ft), 72 m (236 ft), 77 m (253 ft) and 185 m (607 ft), respectively (Fig. 1). These were continuously cored with an average recovery of about 86% (Sydow & Roberts 1994; Winn *et al.* 1995). Subsequently, the cores were radiographed, photographed in black and white, and described sedimentologically. Sediment samples taken from the cores at one metre or closer spacing, were analysed for palaeontology/biostratigraphy, oxygen and carbon isotopes and coarse fraction (>0.062 mm) composition (Sydow & Roberts 1994; Winn *et al.* 1995). Foraminifera and nannofossils retrieved were analyzed in terms of their abundance, and were used to develop a biostratigraphic framework and to make palaeobathymetric estimations.

Palaeobathymetric estimates from foraminiferal studies provided relative rather than absolute water depths, but nevertheless, gave important constraints on deepening - and shallow-upward trends within the delta succession (Winn *et al.* 1995). Tests of the planktonic foraminiferal species *Globigerinoides ruber* were separated from the samples of the MP 303 and VK 774 boreholes, and analysed for oxygen and carbon isotopes. In addition, borehole gamma logs were utilized,

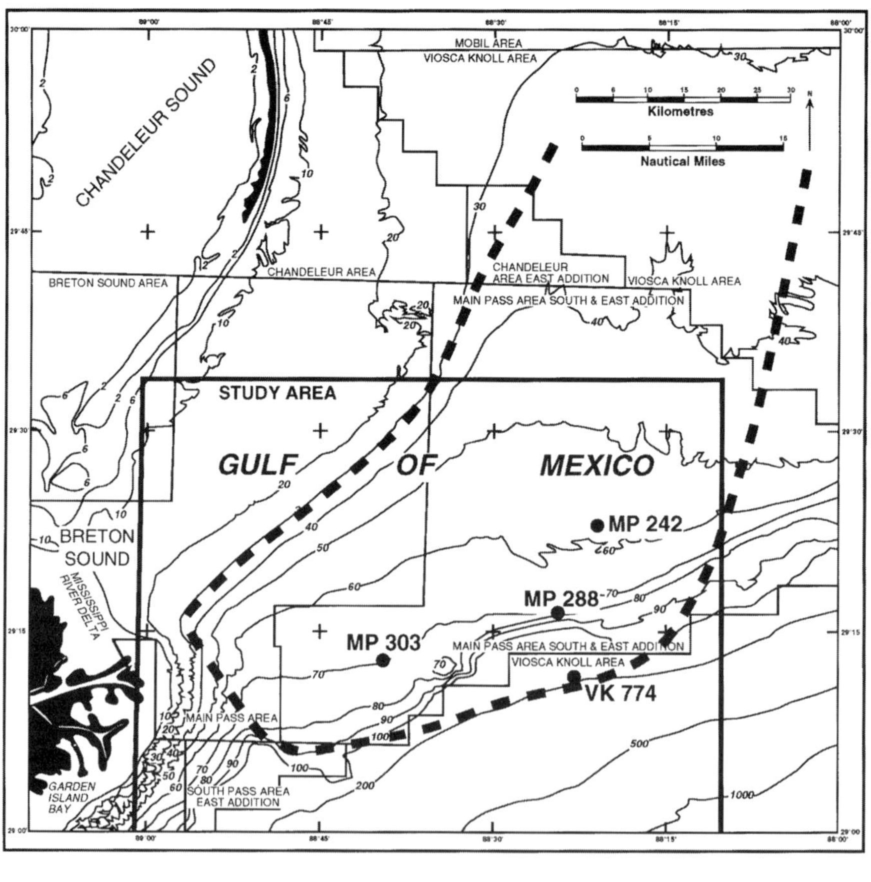

(a)

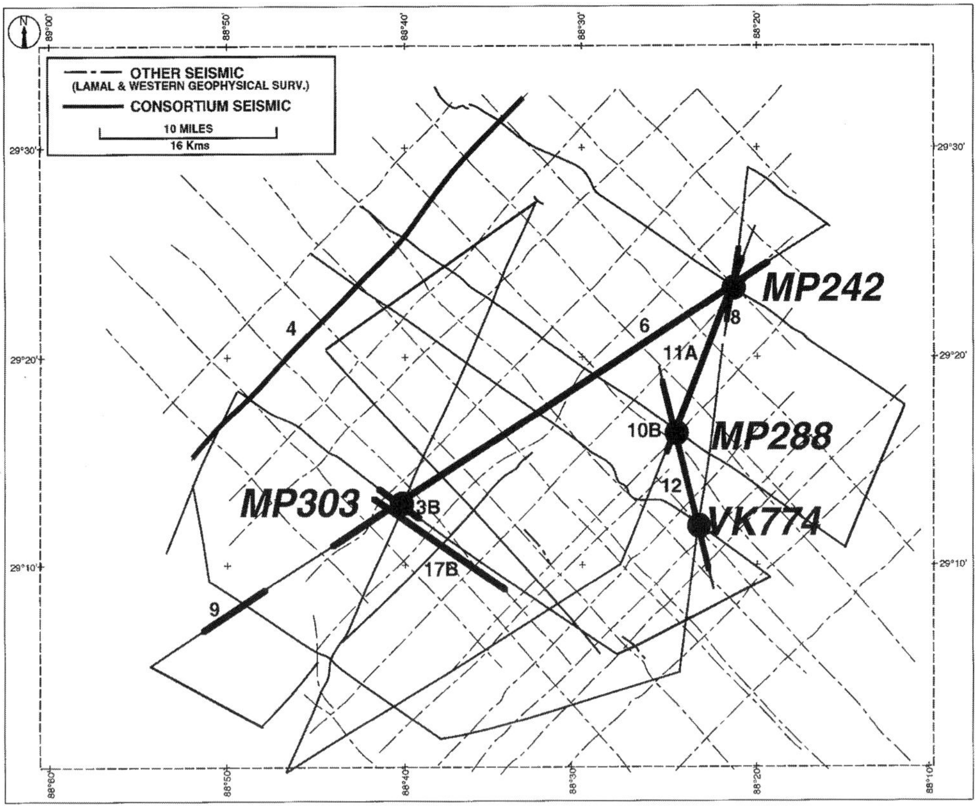

(b)

Fig. 1. (a) Location map of the study area showing the position of the boreholes drilled by the consortium and the extent of the Lagniappe Delta complex (dashed lines). Bathymetric contours are given in metres. (b) Location map showing the position of the seismic grids utilized in the study with respect to the four consortium boreholes. Portions of the consortium seismic lines illustrated herein are shown in bold, adjacent to the appropriate figure numbers.

as well as ^{14}C dates obtained on selected shell and particulate organic matter (Fig. 2). Because the lithological logs and the analytical datasets were obtained by different workers at different times, the precise equivalence of lithologies with other datasets is to an accuracy of ± 1.5 m (5 ft).

In this study the ranges of oxygen isotope values used by Emiliani (1966, 1971) is taken to discriminate between glacial, interglacial and intermediate conditions. The specific values used here are: (i) > $\delta^{18}O$ –0.5‰ for glacial intervals, (ii) between $\delta^{18}O$ –0.5 and –0.76‰ for transitional periods, (iii) $\delta^{18}O$ –0.76 to –1.6‰ for inter-glacial or melt-water during the early rise of sea-level and (iv) < $\delta^{18}O$ –1.6‰ for melt-water in the late rise or from diagenesis. However, the effects of melt-water and diagenesis do complicate the use of oxygen isotope interpretations in identifying the glacial and interglacial stages in the Gulf of Mexico (e.g. Winn *et al.* 1995). Therefore, the oxygen isotope trends, rather than the absolute values were used in combination with analysis of the biostratigraphy, sedimentology and seismic facies to develop the isotope-stage framework and to constrain the age of the key-stratigraphic surfaces identified in the boreholes. This integrated dataset allows the stratigraphy and development of the Lagniappe delta to be placed within the framework of the Late Pleistocene oxygen isotope stages shown in Fig. 2.

As a final constraint to the data and interpretations made here, an average velocity of 1.62 km s^{-1} was used in correlating the borehole data to the seismic. Using this velocity it is possible to match the maximum flooding surfaces and transgressive intervals in the MP303, MP288 and MP242 borings, and in much of the VK774 boring with the strong continuous reflectors that most probably correspond to these intervals. However, it is not possible to make a precise match of individual seismic reflectors with specific beds in the cores.

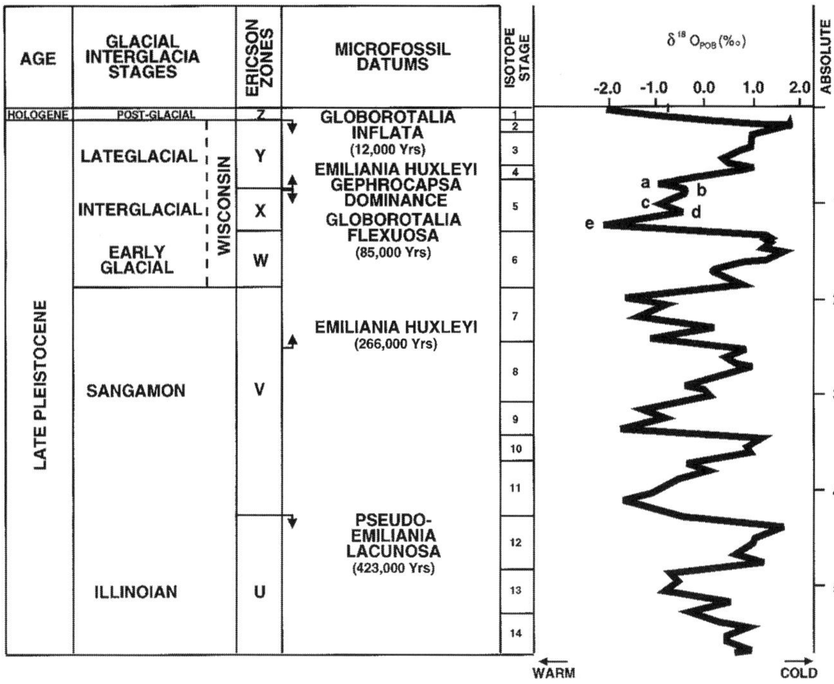

Fig. 2. The Late Pleistocene stratigraphic geologic time scale and nomenclature shown in comparison with respect to glacial and interglacial stages, $\delta^{18}O$ excursions that are used to determine eustatic sea-level changes during the Late Pleistocene (e.g. see reviews and discussion in Chiocci and Hernández-Molina *et al.* this volume), and the position of the various oxygen isotope stages referred to herein. The position of the formainiferal first and last appearance datums utilized as biostratigraphic constraints in this study, and of the Ericson zonation based on the appearance of warm and cold water formainifera are also shown. Data compiled from Imbrie *et al.* (1984) and Kohl (1986).

Key sequence stratigraphic surfaces within a chronological framework

Time framework

The last appearance datum (LAD) of the planktonic foraminifera *Globorotalia flexuosa* at approximately 85 ka BP, and the lowest level of *Emiliania huxeleyi* dominance between 82–84 ka BP in the cores provide a chronological marker signalling the end of Ericson's zone X, as shown in Fig. 2 (Kohl 1986; Kennett & Huddleston 1972; Ericson & Wollin 1968). Two additional biostratigraphic markers for older stratigraphic intervals have been provided by the first appearance datum (FAD) of *E. huxeleyi* at 266 ka BP or less (that is no older than oxygen isotope stage 8; Fig. 2), and the LAD of *Pseudoemiliania (P.) lacunosa* at 423 ka BP within isotope stages 11 and 12 (Gartner & Emiliani 1976). Within the confines of these biostratigraphic controls further stratigraphic subdivision was based on Ericson's zonation that results from alternation between the presence of the warm-water planktonic foraminifera *Globorotalia menardii* and cold-water foraminifera *G. inflata* (Fig. 2; Ericson & Wollin 1968; Kennet & Huddleston 1972). These fauna are respectively interpreted to reflect the interglacial and glacial climatic cycles. As such they aided the recognition of condensed sections and sequence boundaries. However, suppression or elevation of the fossil datums due to lithological variations and reworking did, at times, limit the use of these marker fossils. Accordingly, the full suite of data, including oxygen-isotope stages, fossil abundances and palaeobathymetric changes, combined with the succession of lithological sequences and stacking patterns, and correlation with the seismic, were utilized to refine the chronostratigraphic framework in the boreholes. In the upper part of the cores ^{14}C dating of particulate organic matter also provided some absolute age dates.

Stratigraphic surfaces

Within the bio- and chronostratigraphic framework, and also as an integral part of developing it, key sequence stratigraphic surfaces were identified utilizing essentially the same methodology as that used in the Gulf of Mexico subsurface exploration studies (e.g. Schafer 1987; Mitchum *et al.* 1993). However, some modifications were required to standard methods because of the very detailed nature of the analytical data used in this study. Calibration in cores and logs of the key sequence stratigraphic surfaces (and the stratigraphy between them) to both the seismic stratigraphic surfaces and the seismic facies provided a robust sequence stratigraphic framework for the Lagniappe Delta complex as they were traced and extended through the study area.

Sequence boundaries (e.g. Mitchum 1977; Vail 1987) are characterized in seismic sections by erosional unconformity surfaces (e.g. bases of incised valleys) and their downdip correlative conformities. In the cores these surfaces were characterized by abrupt or subtle breaks in gamma logs and lithology with an increase of grainsize, decrease in fossil content and the shallowing of palaeobathymetry in the overlying sediments. Maximum flooding surfaces, and corresponding condensed sections (e.g. Loutit *et al.* 1988), are indicated by high fossil abundances, maximum palaeobathymetry, high to moderate gamma activity and light isotope values. In the seismic data the maximum flooding surfaces and the transgressive surfaces appear to correspond to intervals of distinctive parallel and continuous reflectors or reflection doublets. However, the strength of these reflections depends on the frequency content of the high-resolution seismic. Bay-flooding and marine transgressive surfaces (e.g. Plint 1988; Mitchum *et al.* 1993; Nummedal *et al.* 1993) in cores were generally characterized by slight to significant deepening in palaeobathymetry, lithological changes (increase or decrease of grain size), presence of glauconite, increasing abundance of fossils and shells, and the occurrence of lighter isotope values above the surfaces. They were often found to correspond to distinct seismic reflectors.

In the Late Pleistocene, relative sea-level changes were primarily driven by a high-amplitude, high-frequency glacioeustatic signal (e.g. see reviews by Chiocci this volume and Hernández-Molina *et al.* this volume). Glacioeustatic falls of 100–120 m in the Late Pleistocene were not instantaneous, but were gradual, and occurred in a series of several steps within each major 100 ka fourth-order cycle (e.g. Fig. 2). Each significant sea-level fall to or below the existing shoreline (or offlap) break, located in water depths of 10–20 m (Kolla *et al.* 1995), is likely to have resulted in formation of a type 1 sequence boundary. However, it is the final and maximum fall within an individual 100 ka cycle that probably resulted in the greatest incision and the formation of a major or 'final' sequence boundary (see also Chiocci, Hernández-Molina *et al.*, McMurray & Gawthorpe and Trincardi & Corregairi this volume). Consequently, all the preceding higher-order sequence boundaries formed from the onset of a long-term (fourth-order) fall are considered to be of lesser importance, having resulted from lower-amplitude, higher-frequency sea-level falls within the more significant fourth-order cycle.

In the data from Lagniappe delta complex it is often not possible to distinguish between the allocyclic and autocyclic processes leading to formation of the individual delta lobes. As a result, within each major fourth-order sequence it is only possible to distinguish in the cores the final (main) sequence boundary and some of the higher-frequency 'initial' sequence boundaries formed during the period of sea-level fall. Here an 'initial' sequence boundary is defined as a surface formed at the onset and/or during a longer term sea-level fall, prior to sea-level having reached its lowest position. In this study such surfaces are recognized on the basis of a shallowing in palaeobathymetry, seismic onlaps/downlaps, facies shifts and changes in gamma ray log motifs.

Sequence stratigraphic framework and evolution of Lagniappe Delta during isotope stages 5 to 1

Six major fourth-order sequence boundaries, and many higher-frequency sequence boundaries and maximum flooding surfaces are identified within Late Pleistocene strata of the Lagniappe Delta complex deposited between oxygen isotope stages 14 and the present day (i.e. the last 500 ka BP; Fig. 2). However, deposits emcompassing isotope stages 5–1 (e.g. 120 ka BP to the present day) form the nucleus of this paper.

The identification of isotope stages 5 and 1 intervals and the base of an incised valley, interpreted to have resulted from fluvial erosion during the lowest position of sea-level, within the isotope stage 2 interval (Fig. 2), forms the foundation of the sequence stratigraphic framework in this study. Precise distinction between isotope stages 4, 3 and 2 within deposits of the 5

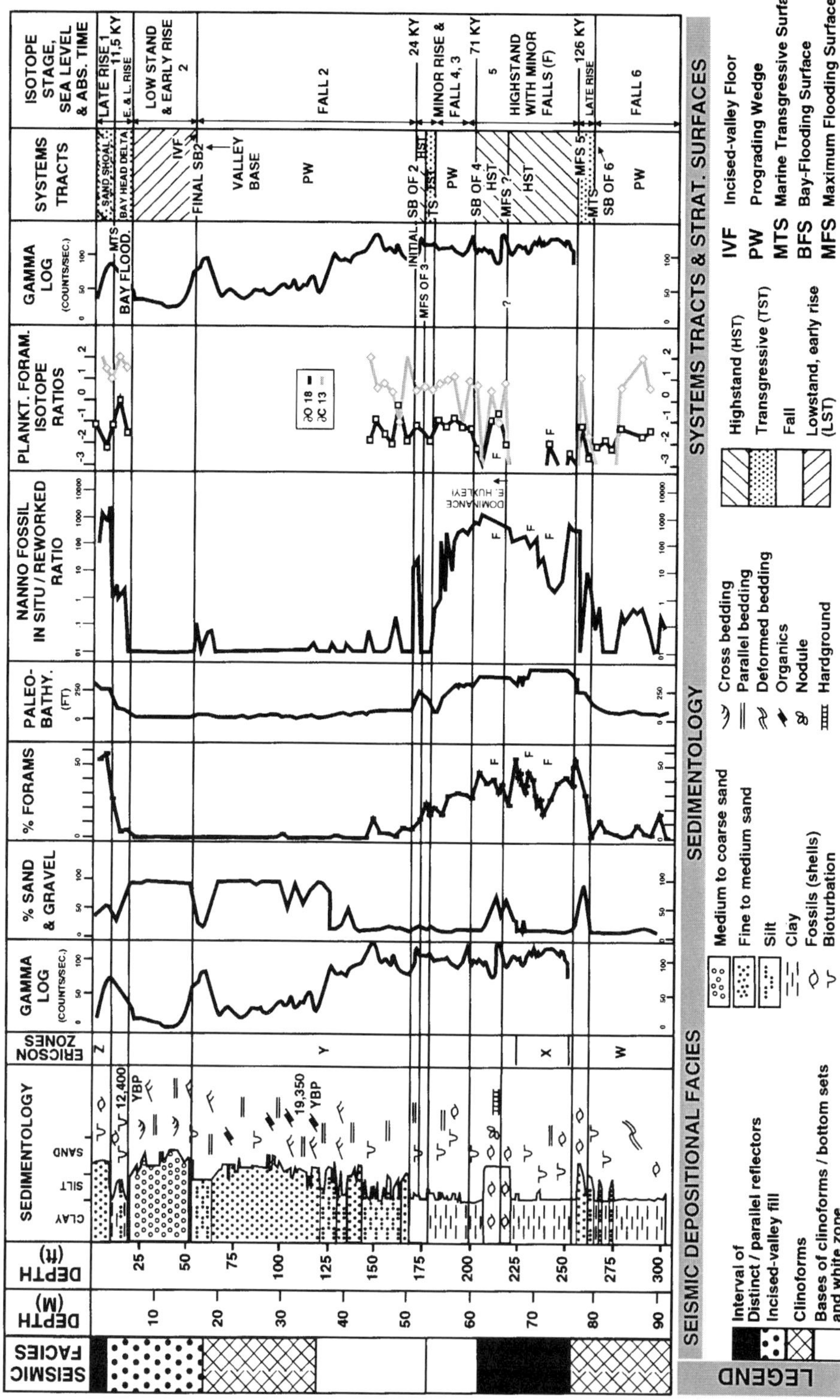

(a)

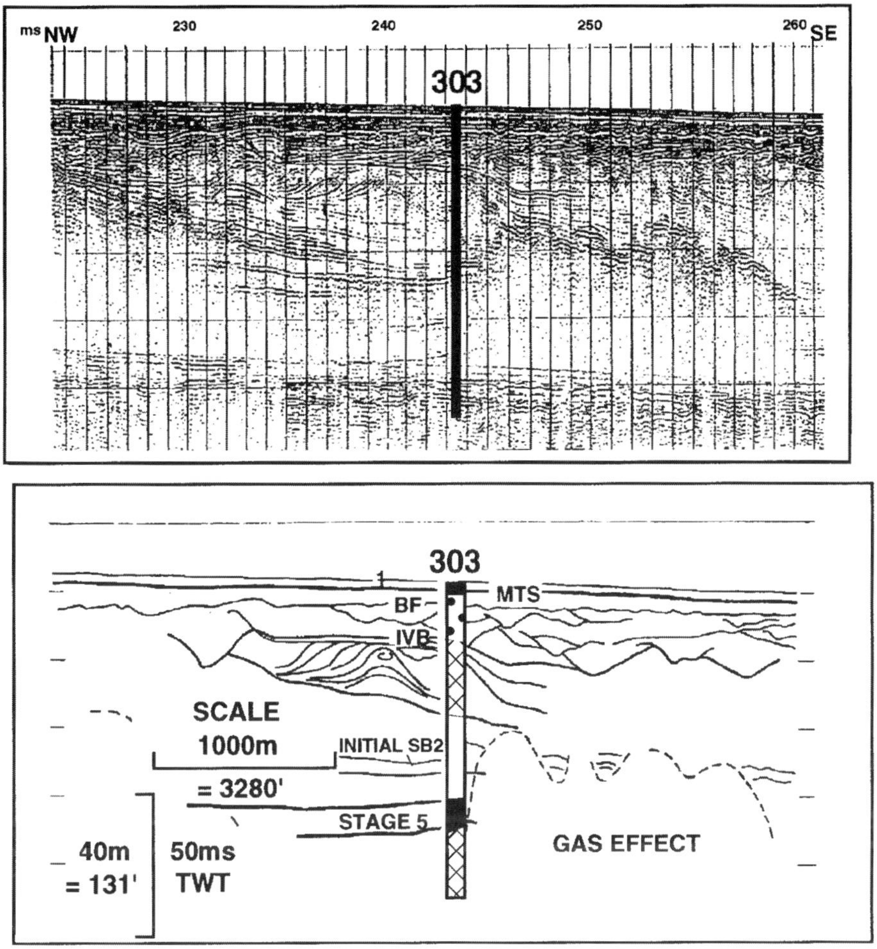

(b)

Fig. 3. (a) Summary of sedimentological, biostratigraphic, palaeoenvironmental and isotopic analytical data from the MP303 borehole, in relation to seismic facies, sequence stratigraphic interpretation, chronostratigraphy and an interpretation of relative sea-level changes. In the graphic log column, absolute age dates derived from ^{14}C dating of organic particles are also shown. The numbers 1–6 in the right-hand column refer to oxygen isotope stages shown in Fig. 2; F indicates the position of minor sea-level falls. (b) NW–SE seismic profile and summary line drawing through the MP303 borehole. These data were used to calibrate the seismic stratigraphy of the Lagniappe Delta (the position of the seismic facies interpreted in the line drawing are also shown on the left-hand column of a). Some key stratigraphic surfaces shown are: MTS, marine transgressive surface; BF, bay-flooding surface; IVB, incised valley base, and isotope stage 5 interval are calibrated. See (a) for legend to seismic facies and Fig. 1b for location of the seismic line.

to 1 stage interval is mostly uncertain except within the VK774 boring.

In the following four sections, the detailed stratigraphy encountered within each of the shelf and slope borings is discussed in turn, and related to the appropriate seismic lines. These data form the basic framework for studying the evolution of the delta, in particular its response to relative sea-level fall.

MP 303 site sequence stratigraphy

A transgressive shell-rich sand layer exists at a depth of 78.7 m (258 ft) in the MP 303 core. This layer is thought to have been formed soon after deposition of a slumped, silt-dominated delta-front section of the isotope stage 6 lowstand below (Fig. 3a). The overlying transgressive sands are succeeded by an association of

Table 1. *Biostratigraphic summary from boreholes at Consortium Sites MP303, MP242, MP288 and UK774*

Ericson Zones	Depth (ft)	Foraminifera			Nannofossils	
		Stratigraphic markers	Abundance	Environment	Stratigraphic markers	Abundance
Site MP 303						
Z	0.42–7.9	*G. menardii* etc. warm-water fauna dominate	High	Middle neritic	–	High
Y	7.9–227	*G. inflata* common to abundant but absent above (shallower than) 125 ft depth	Below 125 ft low to moderate; Above 125 ft sparse or absent	Above 125 ft, brakish to fluvial below 125 ft inner to middle neritic	Occurrence of dominant *E. huxleyi* at 219.8 ft	Low to absent
X	227.0 259.6	*G. inflata* increases above but *G. menardii* increases below 227 ft; rare occurrence of *G. flexuosa* within interval	High	Outer neritic		Generally high
W	259.6–300	–	Low to sparse	Inner neritic deepen-ing at the top	FAD of *E. huxleyi* not reached	Low to moderate
Site MP 242						
Z	0–8	Rare occurrence of *G. menardii*	Fairly high	Deepens to middle neritic	–	High
Y	8–204 ?	*G. inflata* dominant with uppermost occurrence at 64.8; the interval 102.4–106.4 ft it has only *G. inflata* with no *G. menardii*	10–100 ft, poor to moderate; 106–107.5 ft, high; 125–204; low	10–90 ft inner neritic or shallower; 90–175 ft, middle neritic 131.8–193 ft, fluvial	Dominant *E. huxleyi* at 106.4–108.6 ft; below this depth down to 214.2 ft Dominated by *Gephrocapsa* spp., *G. aperta* > 266 ka or less (up to oxygen Isotope 8)	100–121 ft, moderate to high, low in the remaining interval
X?	206.1–208	No specimens of *G. flexuosa*; *G. menardii* and *Pulleniatina obliquiloculata* occur; *G. inflata* absent within interval *G. inflata* dominant above 206 ft	High	Middle to outer neritic	–	202–214, high
?	208–257.2	No diagnostic markers	Low	Fluvial to shallow inner neritic	–	Low

Table 1. *Continued*

Ericson Zones	Depth (ft)	Foraminifera			Nannofossils	
		Stratigraphic markers	Abundance	Environment	Stratigraphic markers	Abundance
Site MP 288						
Z	0–6	*G. menardii*		Middle neritic	–	High
Y	6–287		Barren, low to moderate	Shallow inner to middle neritic	–	Barren to low to moderate
X	287–303	LAD of *G. menardii flexuosa* at 295 ft	High	Outer neritic	Dominant *E. huxleyi* at 289–291 ft (82–84 ka); below *Gephrocapsa* spp. *G. apperta* (266 ka or less)	High
Site VK 774						
Z	0–32	*G. menardii* and *G. tumida* common above & *G. inflata* common below 32 ft	High	Upper bathyal	–	High
Y	32–118.5	LAD *G. flexuosa* at 118.5 ft; *Ginflata* dominant in the interval	Low–moderate (?)	Shallow outer neritic to upper bathyal; 102–117.5 ft max. water depth	–	Low–moderate
?	118.5–165.6	Abrupt increase of *G. inflata* and decrease of *G. menardii* above 118.5 ft; below 118.5 ft, *G. menardii* & *G. ilnflata* in equal numbers	Fairly high; max. at 150–155 ft	Outer neritic to upper bathyal (137–157 ft maximum water depths)	The lowest level of dominant *E. huxeleyi* at 115–117 ft	High
?	165.6–373.4	Absent	Sparse or absent	Fluvial to shallow middle neritic	FAD *E. huxeleyi* at 167.4 ft	Sparse or absent
V	373.4–777	*G. menardii* increases in abundance at 777.7–758.8; 699–694.2; 453.2–447.8 ft	High at 778–759.8 & 453–448 ft	Fluvial to upper bathyal; 442.5–452 & 695–720 ft max. water depths	LAD *P. lacunosa* at 734.8 ft	High at 758–780; 695–700 & 450–455 ft

features indicating maximum deepening, with light oxygen isotope values at 77.4 m (254 ft) that may well represent the stage 5e maximum flooding surface (e.g. Figs 2, 3). Other candidates for younger maximum flooding surfaces within sediments deposited during isotope stage 5 occur at depths of 65.9 m (216 ft) and 60.9 m (200 ft). These surfaces are characterized by relatively high fossil abundance, deep palaeobathymetric indicators, and somewhat lighter isotope values. The precise levels of these maximum flooding surfaces are not, however, certain.

The time interval represented by isotope stage 5 (including sub stages 5a, 5b, 5c, 5d and 5e) generally represents interglacial conditions associated with highstand of eustatic sea-level, punctuated by minor falls and rises (Fig. 2). The MP303 stratigraphy from 77.4 m to 28.9 m (254–198 ft) is interpreted to represent the stage 5 interval, and partly includes Ericson's X zone (see Table 1; Fig. 3). The sediments in this interval of the core are characterized by a high abundance of fossils and carbonate, and lighter isotope values, albeit with some exceptions (Fig. 3a). The microfaunal data indicate deposition in outer neritic water depths of *c.* 90–185 m (300–600 ft).

Above the top of zone X in the borehole, there is a 3 m (10 ft) thick zone composed of nodular, fossiliferous carbonates with interstitial clay and hardground layers that contain mixed neritic and abraded reefal and clear-water fauna (Sydow & Roberts 1994; Fig. 3a). Here, precise dating is hampered by inadequate core recovery and sampling, so limiting the appearance levels of marker fossils. However, part of the nodular carbonate interval has relatively heavy oxygen isotope values. These heavy isotope values are interpreted to be indicative of relative sea-level fall during isotope stage 5, rather than in stage 4, because of the deep palaeobathymetry of the interval (Fig. 3a). The position of two or three minor sea-level falls (marked 'F' in Fig. 3a) are tentatively identified within the stage 5 interval. These stratigraphic intervals, potentially formed in response to relative sea-level fall, are suggested by relative decrease of fossil abundances, slight shallowing of palaeobathymetry and/or relatively heavy isotope values in the core data. However, not all these characteristics are present in every case to conclusively identify each of the proposed intervals of sea-level fall.

The top of isotope stage 5 is correlated to a relatively distinct seismic reflector (Fig. 3b). It is overlain by a seismically transparent or weakly reflective zone, interpreted to correspond to the very distal laminated silts and clays that are mainly attributed to deposition during isotope stage 4. The surface at the base of the silts (at a core depth of 60.3 m (198 ft)) is characterized by a shallowing of palaeobathymetry, decreasing fossil abundance and breaks in the gamma log (Fig. 3a). This surface is interpreted to represent the stage 4 sequence boundary (Fig. 3). Above it are laminated silts and clays with heavier isotope values than below. Towards a core depth of 52.7 m (173 ft) increasing fossil content, deepening of palaeobathymetry, high gamma values and light(?) isotope values are taken to represent a maximum flooding surface of oxygen isotope stage 3 (Fig. 3a).

A sharp break in lithology and the gamma log is encountered at 51.2 m (168 ft) (Fig. 3a). This surface is associated with decreasing fossil content, indications of palaeobathymetric shallowing above and the downlap of clinoforms in the seismic data (Fig. 3a, b). This surface is interpreted as a sequence boundary, and it is inferred to represent an 'initial' sequence boundary of isotope stage 2 (or of stage 3 transitional to stage 2). As is apparent in Fig. 3b, the clinoforms downlapping this surface are more steeply inclined than the progradation clinoforms of stages 4 and 3. In the core, the equivalent interval (51.2–18.9 m; 168–62 ft) is largely barren of fossils and is characterized by an upward-coarsening and thickening succession of fine-grained sands with massive, parallel bedding or cross lamination. Most of these sediments were deposited in shallow water depths close to shoreline, indicating a very significant shallowing resulting from lowering of sea-level. The interval is thus interpreted to have been deposited during the relative sea-level fall associated with isotope stage 2 (Figs 2, 3a).

Highly laminated and rippled sands and silts with marginal marine fauna occur between 18.9 and 15.9 m (62 and 52 ft), and appear to correspond with a thin landward onlapping and slightly seaward prograding seismic unit just above the main clinoform unit (Fig. 3b). This thin package is interpreted to represent the distant lateral equivalent of a delta lobe following switching of deposition away from the MP303 site. This delta switching is thought to have created a bay-like environment in which the sediments at site MP303 were deposited following slight deepening, as suggested by palaeobathymetric indicators.

A major erosion surface of significant areal extent, interpreted as the base of an incised valley on the seismic data, occurs at a depth of 15.9 m (52 ft) in the borehole (Fig. 3). This surface is interpreted to be the main (and final) sequence boundary associated with the isotope stage 2 lowstand of sea-level at approximately

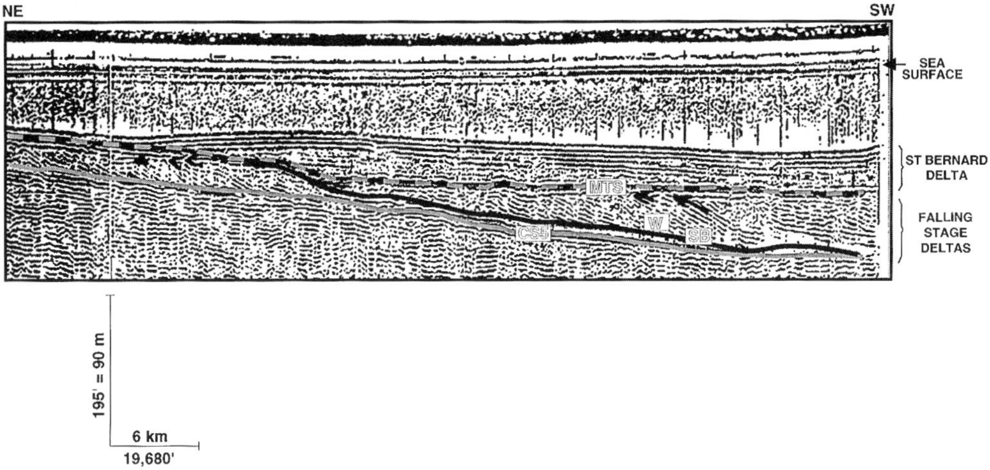

Fig. 4. A NE–SW-oriented seismic line located west of the MP303 site and outside of the incised valley complex showing well-preserved geometry of forced- regressive wedges within the Lagniappe delta complex (see Fig. 1b for location). The wedge labelled 'w' pinches out updip below the offlap break of a previously deposited clinoformed unit. The offlap break of the wedge 'w' is also down-stepped with respect to the updip unit. The top of this wedge is flat suggesting deposition during a stable sea-level after the initial step-wise fall. The top of the updip unit is sloping, suggesting deposition during the gradually falling sea-level. MTS: marine transgressive surface. CS5: condensed section of isotope stage 5.

18 ka BP. The erosional surface at the base of the incised valley is thought to have formed mainly during the lowest stand of sea-level, but may be a composite surface, the result of multiple sea-level falls during isotope stage 2. Just west of the MP303 site the updip onlap and pinchout of a delta lobe is observed in a NE–SW-oriented seismic line ('W' in Fig. 4), the eastern extension of which has been drilled at the MP303 site. This updip stratal pinchout is positioned just below the offlap break of the preceding delta (Fig. 4), and is a clear evidence of the stepwise nature of the sea-level fall. It is thought that such diagnostic stratal pinchouts and relationships were destroyed or greatly modified in the areas of intense erosion and incised valley formation close to the MP303 site, as discussed later.

The base of the incised valley in the core is very sharp and corresponds to an interval of inclined, migrating and chaotic seismic reflections (Fig. 3b). In agreement with Sydow & Roberts (1994), this interval is interpreted as a channel fill deposited within an incised valley during the early sea-level rise following the isotope stage 2 sea-level lowstand. The sediments between the base of the incised valley at 18.9 m (52 ft) and 5.5 m (18 ft) consist of coarse to medium grained, amalgamated, pebbly sand beds with frequent cross-bedding, and are barren of fossils except for two thin shell beds. Winn *et al.* (1995) interpreted the shell beds to have been reworked from older deposits and inferred the palaeobathymetry to be very shallow.

The sharp, upper boundary to the incised valley fill at 5.5 m is penetrated by root casts suggesting subaerial exposure (Winn *et al.* 1995), and apparently corresponds to a distinctive seismic reflector (Fig. 3b). Above this surface of undoubted subaerial origin, there is a distinct increase in fossil content and a deepening of palaeobathymetry (Fig. 3a). This surface at 5.5 m is interpreted to represent a bay-flooding surface formed at 12 400 ka BP. The overlying laminated silts and clays that occur between 5.5 m and 3 m (18–10 ft) contain burrows and rootlets in the lower part, and fine-grained shelly and burrowed muds in the upper part. They are associated with light to heavy isotope values and high gamma activity, and are interpreted as a fluvial to estuarine bay fill deposited during sea-level rise after 12 400 ka BP.

A significant increase in fossil content and a deepening of the litho - and biofacies suggesting a marine transgressive surface, occurs in the core at a depth of 3 m. Above this marine transgressive surface are burrowed silts that coarsen-upward into shelly medium-grained sands, with an open marine faunal assemblage indicative of middle neritic (45–90 m (150–300 ft)) to outer neritic

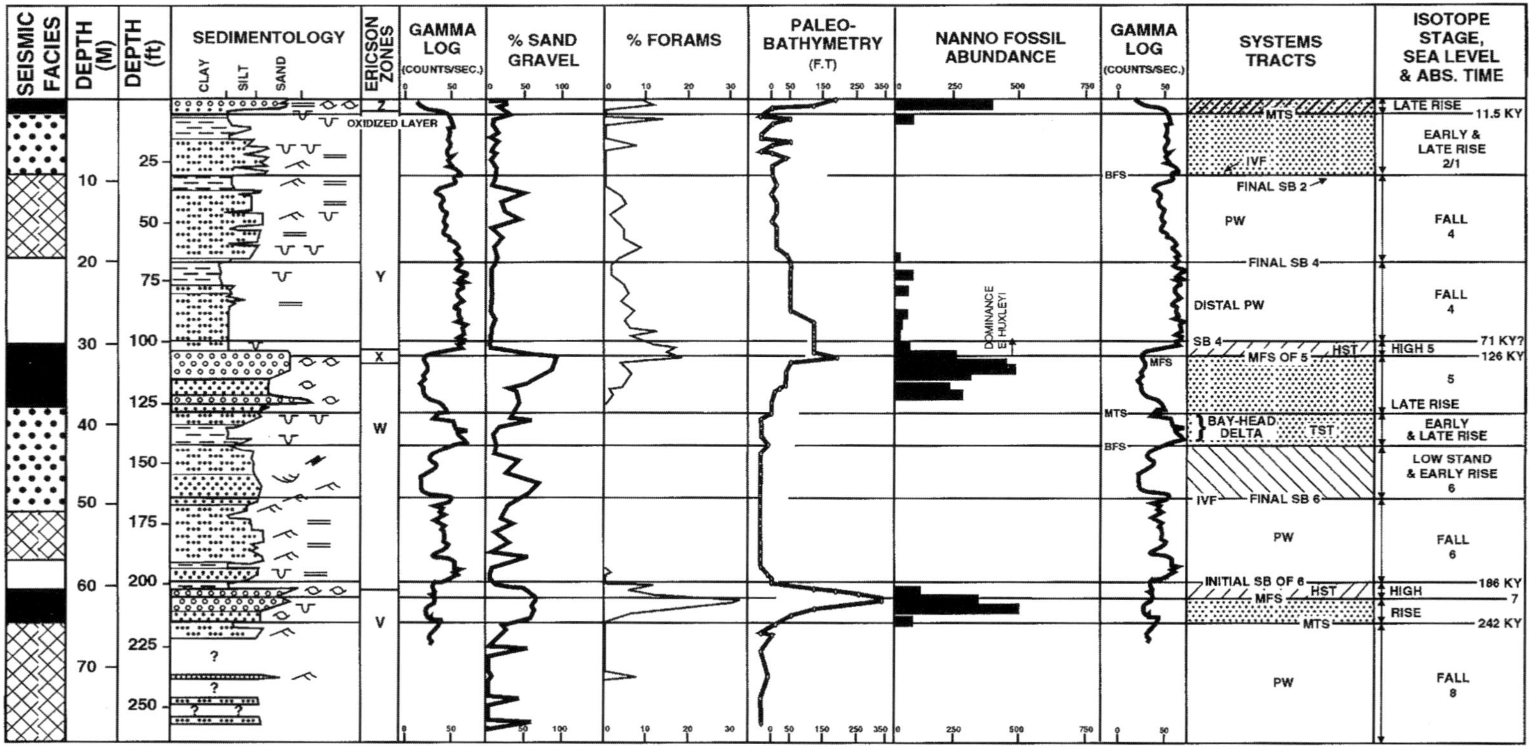

Fig. 5. Summary of sedimentological, biostratigraphic, palaeoenvironmental and isotopic analytical data from the MP242 borehole, in relation to seismic facies, sequence stratigraphic interpretation, chronostratigraphy and an interpretation of relative sea-level changes. The complimentary seismic line through this borehole is shown in Fig. 6, for comparison to the seismic facies summarized in the left hand column. The numbers 1–6 in the right-hand column refer to oxygen isotope stages. See Fig. 3a for key to symbols.

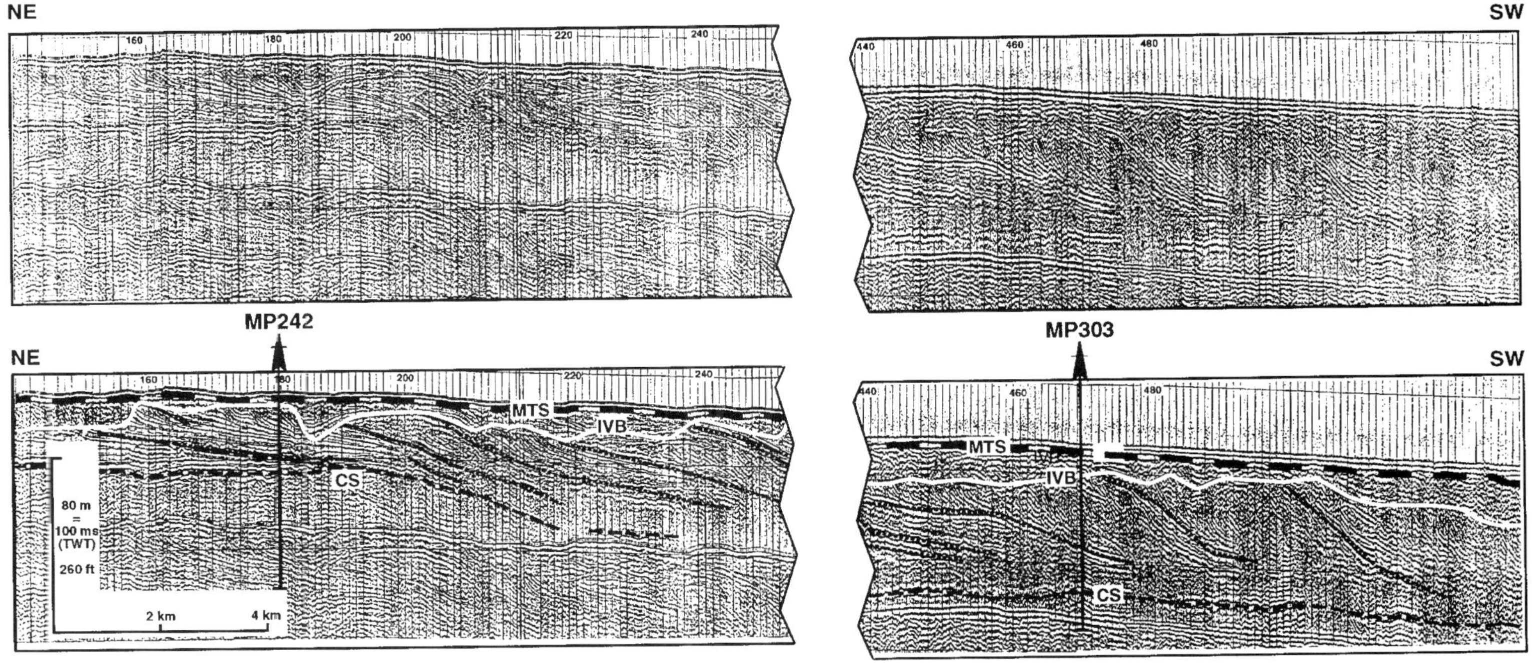

Fig. 6. NE–SW-oriented seismic line between MP 242 and 303 sites, showing the deltaic units and incised valley fill developed between sea-floor and the stage 5 condensed section (CS, dotted) (See Fig. 1b for location). Here the updip deltaic lobes tend to become thin, and pass downdip into successively younger and thicker deltaic lobes between MP 242 and MP 303. The preserved tops of delta lobes and offlap breaks are stepped down from MP 242 to MP 303 site. MTS, marine transgressive surface; IVB, incised valley base.

(90–185 m (300–600 ft)) water depths. Characteristically these sediments have light isotope values (Fig. 3a). The base of Ericson's Z zone, with the appearance of warm-water forams above it, corresponds to the marine transgressive surface observed at a depth of 3 m in the core. The overlying sediments are interpreted to be sand shoals deposited and reworked during the Holocene sea-level rise within isotope stage 1. In the seismic data, the marine transgressive surface and the maximum flooding surface formed during isotope stage 1 are indistinguishable and correspond to strong continuous reflectors (Fig. 3b).

Stratigraphy at the MP242 site and correlation along MP242-MP303 seismic section

The MP242 borehole is located updip and northeast of the MP303 site (Fig. 1). In the MP242 core (Fig. 5), the use of foraminiferal biostratigraphy proved not to be straightforward. Initial analysis indicated that Ericson's X zone occurred at a depth of 63.4–62.8 m (208–206 ft), an interpretation based on the occurrence of the warm-water *G. menardii* within this interval, and the dominance of *G. inflata* above it (Fig. 5; Table 1). However, no stratigraphic marker fossils occur either within, or close to this interval. Several lines of evidence suggested that this first interpretation was incorrect. First is the dominance of *E. huxeleyi* marker down to a depth of 32.3 m (106 ft) in the MP242 core (Table 1). In addition, the succession of lithological sequences, and the number and significance of deepening and shallowing palaeobathymetric events between the top of the core and a depth of 32.9 m (108 ft) in MP242 compares favourably with the position of Ericson's Z to X zone in MP303. Together, these data are taken to suggest that the Ericson's X zone actually occurs at a depth of 31.4–32.9 m (103–108 ft) in the MP242 borehole, and not at 63.4–62.8 m (208–206 ft) as originally thought (Fig. 5; Table 1). However, in making this interpretation, the occurrence of the cold-water species *G. inflata* between 31.4 m and 32.9 m, and the absence of the warm-water *G. menardii* must be accounted for (Table 1). The occurrence of shelly, quartz-rich sands between 39 m and 31.4 m (128–103 ft) is thought to suggest that the unusual distribution of fauna resulted from the reworking of underlying stage 6 sediments below, which contained *G. inflata*, during subsequent transgression.

Core stratigraphy and delta development. The shell-rich sand bed at a depth of 39 m (128 ft) in the MP 242 core is indicative of the first marine transgressive surface over sediments attributed to the stage 6 isotope interval. Sediments between 39 m and 31.4 m (128–103 ft) are interpreted to represent the stage 5 interval because of generally deep bathymetry and high fossil abundances within this stratigraphic interval (Fig. 5). The inferred position of Ericson's X zone is part of this interval. The seismic reflector that marks the stage 5 interval at the MP303 site is compatible with the corresponding stage 5 interval between 39 m and 31.4 m in the MP242 site (Figs 5, 6), taking into account the physiography of the shelf. Within the stage 5 interval at the MP242 site, at least one maximum flooding surface at a depth of 32.3–31.4 m (106–103 ft) is inferred from the core data. It is characterized by the highest fossil abundances and deepest palaeobathymetry attributed to sediments deposited during isotope stage 5 (Fig. 5). Sediments between this maximum flooding surface and 39 m (128 ft) consist of transgressive sandy shoals that are sedimentologically comparable to the shell-rich sandy shoals in the MP303 borehole deposited during isotope stage 1 (Figs 3a, 5).

The MP303 and MP242 sites are connected by a seismic line (Fig. 6) that is mainly positioned within a major progradational fairway of the Lagniappe Delta (see later discussion). The location of this seismic line allows the major erosive surface, interpreted to be the base of an incised valley at the MP303 site, to be directly traced to the MP242 site where it occurs at a depth of about 9.1 m (30 ft) (Figs 1b, 5, 6). Bounded above by the base of the incised valley, and below by the stage 5 reflector (CS; Fig. 6), several southwest-dipping deltaic wedges are developed between the MP242 and MP303 sites. The preserved deltaic topsets and their offlap breaks generally step downward between MP242 and MP303 (Fig. 6). The seismic stratigraphy shows that the updip, thick deltaic wedges progressively thin downdip and are replaced by thick, younger deltas. Thus, downdip equivalents of the updip delta drilled at site MP242 are likely to be represented by thin fine-grained deposits at a greater depth in the MP303 site, if they are present at all.

In the MP242 borehole there appear to be at least two significant surfaces developed between the base of the incised valley and the stage 5 reflector. The lower of these is located at 31.4–30.4 m (103–100 ft), and is characterized by sharp lithological break, decreased fossil abundances and a reduction in palaeobathymetry (Fig. 5). Burrowed clays and shelly sediments are present below this surface, and laminated silts and clays above (Fig. 5). In the seismic data

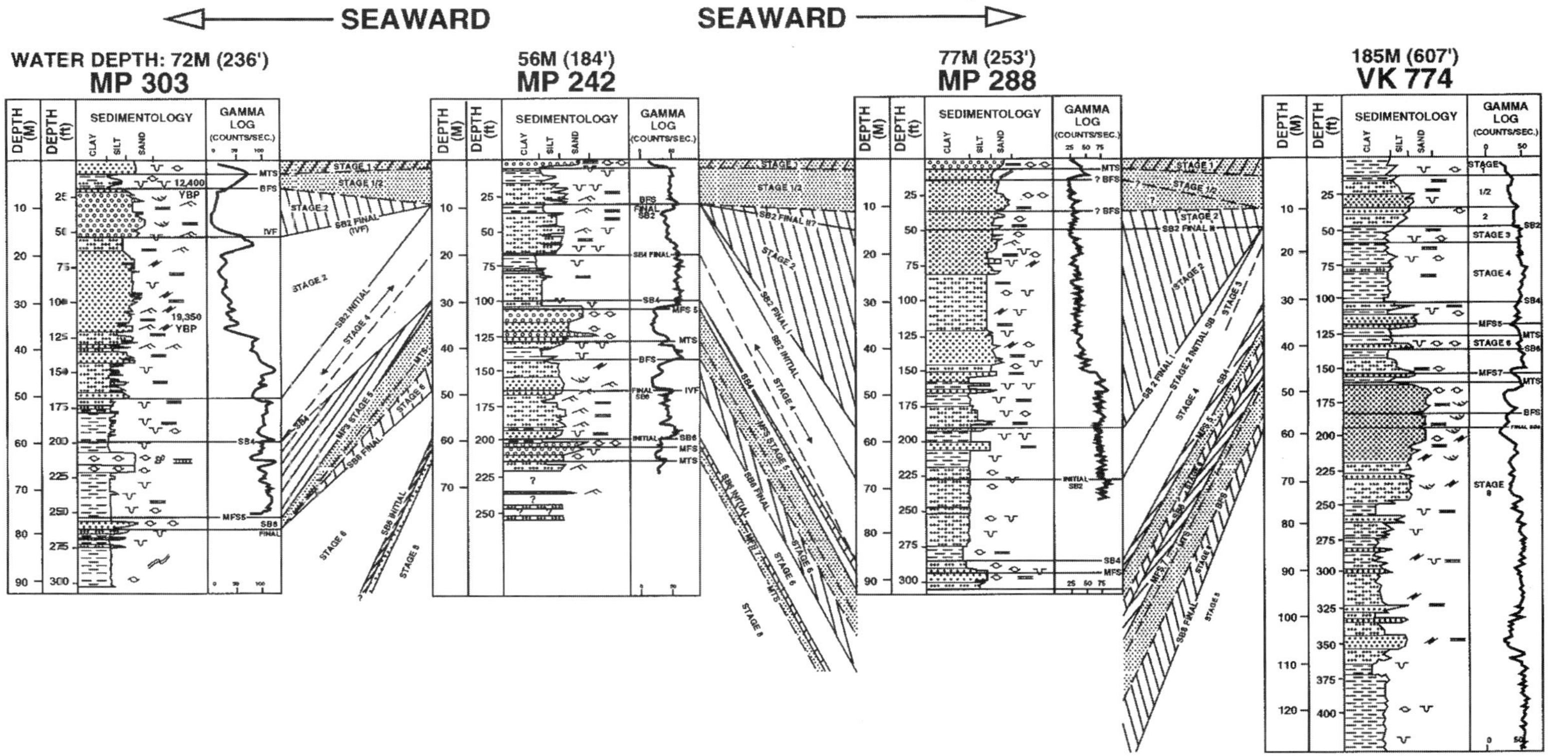

Fig. 7. Correlation panel of key stratigraphic surfaces and systems tracts between the four consortium boreholes (see Fig. 1b for location). Identificaton of stages and intervals, and the base of the stage 2 incised valley and correlative sequence boundary forms the foundation of this sequence stratigraphic study. Note the expansion of stage 2 deposits from the updip MP242 borehole (interpreted to have undergone subaerial exposure) seaward into MP303 and MP288, although this interval is relatively thin in the most basinward borehole (VK774) (see Fig. 3a for legend).

gently-inclined clinoforms are seen to downlap this surface. This downlap surface is interpreted to be a stage 4 (initial) sequence boundary, representing the onset of relative sea-level fall as recorded in the delta, and is tentatively dated at 71 ka BP (Fig. 5).

The uppermost of the two significant surfaces is located at 20.73 m (68 ft) in the MP242 borehole, and corresponds to a sharp lithological break, decreased fossil abundance and a reduction in palaeobathymetry (Fig. 5). More steeply-dipping clinoforms downlap this surface (Figs 5, 6) in comparison to those downlapping onto the surface at 31.4–30.4 m (103–100 ft). The laminated silts to fine sands with parallel and ripple laminations overlying this surface coarsen - and shallow-upward, and are associated with a reduction in fauna (20.7–11 m (68–36 ft), Fig. 5). In comparison to the main delta interval at the MP303 site this interval has a significant, although variable, foraminiferal content, is generally finer-grained, and was apparently deposited in deeper palaeobathymetry and at greater distance from fluvial input (see Figs 3a, 5).

The surface at 20.7 m (68 ft) depth can be interpreted to represent either an initial sequence boundary formed during isotope stage 2 (3 to 2 stage transition) or a boundary resulting from a further sea-level fall within stage 4 (e.g. SB 4 final). The deeper environment indicated by palaeobathymetric indicators within the strata overlying this boundary at the updip MP242 site, compared to that of the main delta interval at the MP303 site, tends to support the latter interpretation. This is because it is thought that a stage 4 sea-level fall would have resulted in less significant shallowing at the updip site than the subsequent sea-level falls during stage 2 would. Figure 7 shows this interpretation, which is consistent with the seismic data, where deltaic strata attributed to isotope stage 4 at the MP242 site thin downdip towards the MP303 site. However, because these stratal surfaces cannot be continuously traced on the seismic between the MP242 and MP303 sites, the precise

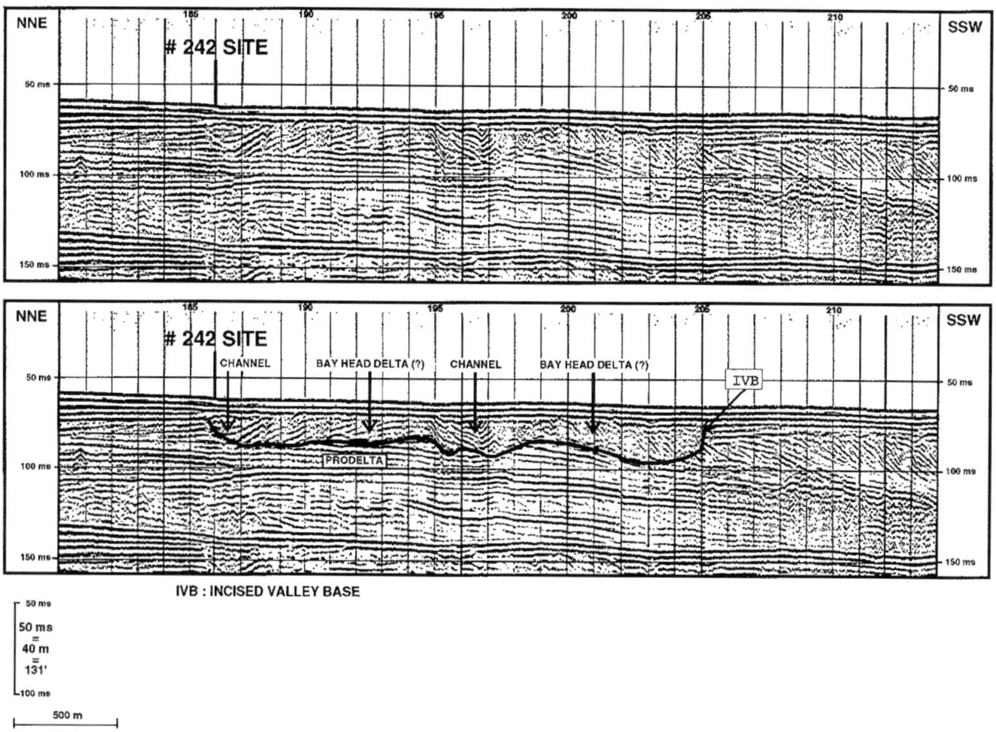

Fig. 8. Uninterpreted and interpreted seismic line showing the details of incised valley-fill showing accretionary and laterally migrating bedforms interpreted to represent bay-head deltas, channel-point or tidal bar migration. Note the location of the MP242 site at the margin of the incised valley. IVB = incised valley base. Seismic line location is shown in Fig. 1b.

timing of the delta lobe deposition during the isotope stage 4 is uncertain at these two locations. The thick stage 2 delta recorded in the MP303 borehole is thought to be absent at the MP242 site (Fig. 7), an interpretation made on the assumption that during the initial stage 2 sea-level fall, the sea floor at the latter site became subaerially exposed. Direct evidence for this exposure event is missing, although it is likely that erosion associated with major incised valley formation during the fall and lowest stand of sea-level during isotope stage 2 modified and/or removed all of the previously-formed erosive surfaces at the MP242 site.

Sediments in the MP242 borehole between 11 m and 9.1 m (36–30 ft) consist of fine-grained ripple and parallel laminated silts (Fig. 5). These strata are, in analogy to the MP303 core, thought to have been deposited in laterally distal deltaic or bay environments prior to erosion of a major incised valley in response to the maximum fall and lowstand of sea level during isotope stage 2. The palaeobathymetry at 9.1 m in the core is close to shoreline, shallower than the environments below. In contrast to the MP303 site, palaeowater depths are more variable above the base of the incised valley (9.1 m), and begin to deepen slightly (see Figs 3a, 5). Such variable deepening events could be interpreted as a bay-flooding surface, and may result from the fact that the MP242 site appears to be located at the margin of an incised valley (Figs 6, 8). From 9.1 to 1.8 m (30–6 ft) the fill of the incised valley contains a variable, but generally low content of planktonic foraminifera, and consists of largely fining-upward laminated silts in the lower part and burrowed shales in the upper part. The seismic facies of this valley-fill consist of onlapping reflections at the MP242 site and migratory (accretionary) and inclined clinoforms just away from the site (Fig. 8), that may be indicative of estuarine or bay-head deltaic fill, and fluvial or tidal point-bar deposits (Fig. 5).

At 1.8–1.5 m (6–5 ft) a marine transgressive surface is inferred at the base of Ericson's zone Z (Fig. 5). Upward-coarsening, shelly laminated sands with fauna indicative of middle neritic (45–90 m (150–300 ft)) water depths overlie this surface, and are interpreted to be shoals reworked during the isotope stage 1 sea-level rise, similar to those encountered at the MP303

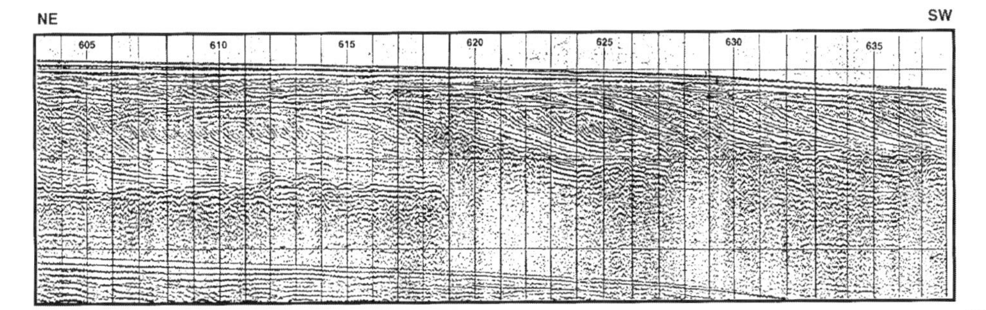

Fig. 9. An uninterpreted and interpreted portion of a seismic line located downdip of MP303 site (see Figure 1B for location). Between the stage 5 condensed section (CS5) and the main sequence boundary formed during stage 2 (IVB, solid line on interpreted line) several deltaic wedges deposited at progressively deeper levels seaward are located between shot points SP603 and 609. Downdip, between SP610 and SP635, several delta lobes (prograding complex) with offlap breaks at progressively shallower depths occur above the sequence boundary (IVB).

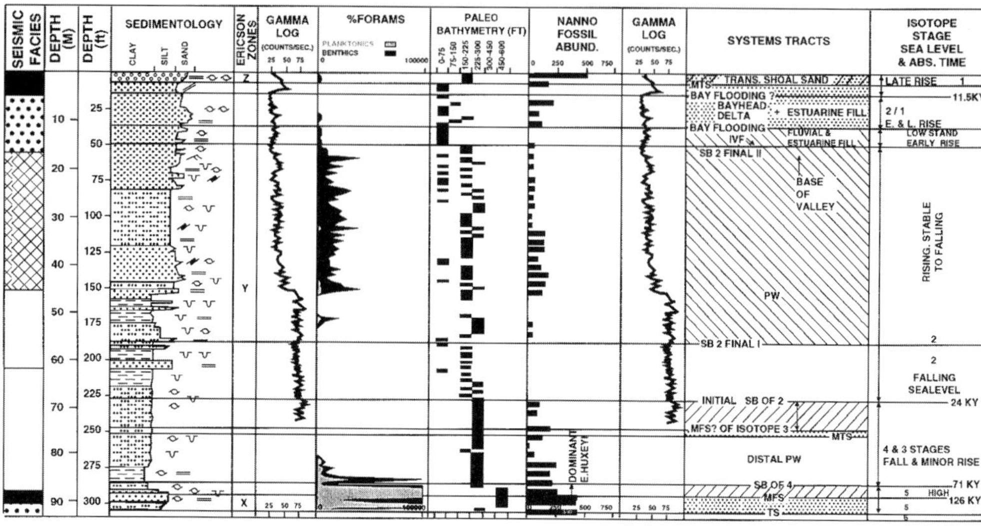

(a)

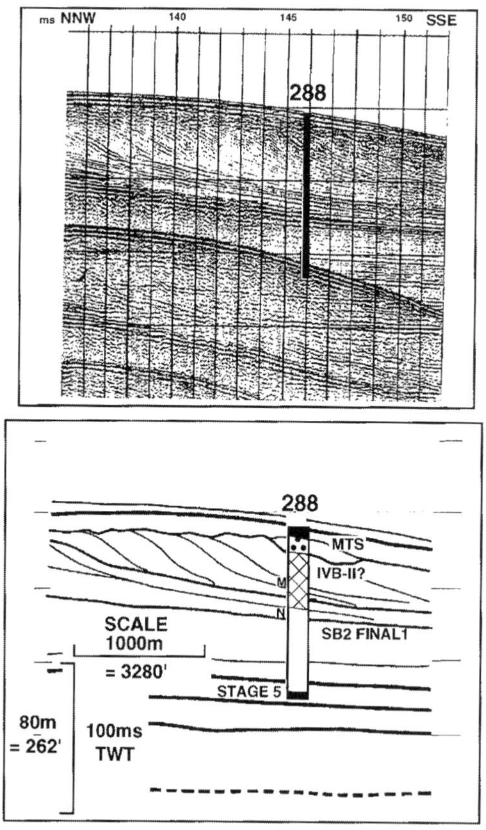

(b)

Fig. 10. (a) Summary of sedimentological, biostratigraphic, palaeoenvironmental and isotopic analytical data from the MP288 borehole, in relation to seismic facies, sequence stratigraphic interpretation, chronostratigraphy and an interpretation of relative sea-level changes based on these data and their relationship with the Late Pleistocene eustatic sea-level changes. The numbers 1–6 in the right-hand column refer to oxygen isotope stages shown in Fig. 2. See Fig. 3a for key to symbols. (b) Detail of the consortium seismic line passing through the MP288 borehole. The main seismic facies are shown and summarized in the interpreted line drawing for comparison to the borehole stratigraphy shown in (a). The prograding delta complex is here comprised of two wedges separated by a downlap surface (N and M). MTS, marine transgressive surface; IVB-II?, incised valley base (floor) equivalent to SB2 Final II; SB2 Final I, sequence boundary of isotope stage 2 (also see Figs 11 and 12) and stage 5 interval.

site (see Figs 3a, 5). Both the bay-flooding and marine transgressive surfaces are time-transgressive between the MP303 and MP242 sites.

Sequence stratigraphy downdip of MP303 site. Several younger delta lobes than the uppermost one drilled at the MP303 site exist further downdip. These lobes have offlap breaks and/or topsets that occur at progressively greater depths between SP603 and 609 (below the solid line and IVB in Fig. 9), and appear to have been deposited during relative sea-level fall. These lobes are apparently topped by the same major incised valley complex as that drilled at MP303 site. Downdip, overlying the sequence boundary (IVB in Fig. 9), there is a prograding delta complex between shot points 610 and 635. An overall aggradational component to this delta complex is apparent, and topsets to some of the lobes appear to pass updip into the fill of the incised valley complex (Fig. 9). As such, this most seaward and aggradational component of the Lagniappe Delta complex is interpreted to have been deposited during the early relative sea-level rise following the maximum fall of sea-level and related major valley incision (during isotope stage 2). Although this outermost component of the delta is generally aggradational, the presence of some downstepped lobes (between shot points 616–625) indicate that minor higher-frequency relative sea-level fall(s) and forced regression also occurred during deposition of the lowstand delta complex during isotope stage 2.

Sequence stratigraphy at the MP288 site and along MP242- MP288 site section

MP288 site is located downdip and to the SW of the MP242 borehole, in a similar water depth but to ENE of the MP303 site. The stratigraphy of the MP288 site and its relationship to the main seismic stratigraphic surfaces is shown in Fig. 10. The site is directly connected by consortium seismic lines to boreholes VK774 and MP242, and to the MP303 core via doglegs.

Position of major stratal surfaces. In the MP288 core the LAD of *G. flexuosa* occurs at a depth of 89.9 m (295 ft) (84–85 ka BP), with the lowest level of *E. huxeleyi* dominance at depths of 88.7–88.1 m (291–289 ft) (82–84 ka BP). Accordingly, the silty, shell-rich sediments associated with the deepest palaeobathymetries that occur from 87.2 m (286 ft) to the total depth of the borehole are attributed to deposition during isotope stage 5 (Fig. 10a, b; Table 1).

On the seismic line through the MP288 site, a major erosion surface is observed (Fig. 10b; IVB-II?) that corresponds to a depth of 15.2 m (50 ft) in the borehole, where it is overlain by sediments with the shallowest palaeobathymetries and a paucity of fossils (Fig. 10a). The erosion surface at 15.2 m may be attributed to either fluvial-channel erosion or a slump scar formed during stable sea-level, or alternatively interpreted as the base of an incised-valley formed in response to a further sea-level fall following deposition of the main deltaic wedge at the MP288 site. In either case, this erosional feature may well have been further enhanced during later transgression.

In seismic lines passing though the MP288 site, the erosion surface at 15.2 m in the core can be seen to merge updip with the base of the incised valley towards the MP242 borehole (Figs 11a and 12). Bounded by this erosional surface at the top, and the stage 5 reflector at the base, several deltaic wedges are apparent in the seismic line between the MP242 and MP288 sites (Fig. 11a). On the whole, updip stratal pinchouts, toplap relationships, and the position ofofflap breaks of the deltaic wedges, successively downstep basinward between the MP242 and MP288 sites (Fig. 11a). From the stratal geometries preserved, it would appear that after downstepping, each deltaic wedge downdip to about shotpoint 238 was deposited under (A) conditions of slow relative rise, (B) stable or (C) gradual sea-level fall as seen in Fig. 11a and summarized in Fig. 11b.

The main wedge of the delta (labelled 'M' in Figs 10b, 11a, 12) drilled at the MP288 site is interpreted to have been deposited during conditions of relative rise to stable sea level. However, the basal surface (labelled SB2 final I in Figs 11a, 12) of an older deltaic wedge 'N' that immediately underlies the main deltaic wedge 'M' (Figs 11a, 12), appears to be the downdip extension of the updip major incised valley. At the MP242 and MP303 sites, this incised valley is interpreted to have resulted from erosion that occurred during sea-level fall and while sea level was at its lowest position. Thus, the delta wedges M and N (Figs 11, 12) are interpreted to comprise a fourth-order lowstand prograding complex deposited during the maximum lowstand–early rise of relative sea-level of isotope stage 2.

Other important stratal surfaces. Other significant stratigraphic surfaces occur in the MP288 core between sediments attributed to deposition during stage 5, and the main erosive surface at 15.2 m (50 ft). These are located at 87.2 m (286 ft), 68.9 m (226 ft), 55.5 m (182 ft) and 47 m (154 ft), as determined from analysis of palaeobathymetric, fossil and lithologic data (Fig. 10a, b). At 87.2 m the palaeobathymetric indicators indicate dramatic shallowing from outer to deep middle neritic, followed by decreasing fossil abundances and deposition of burrowed and

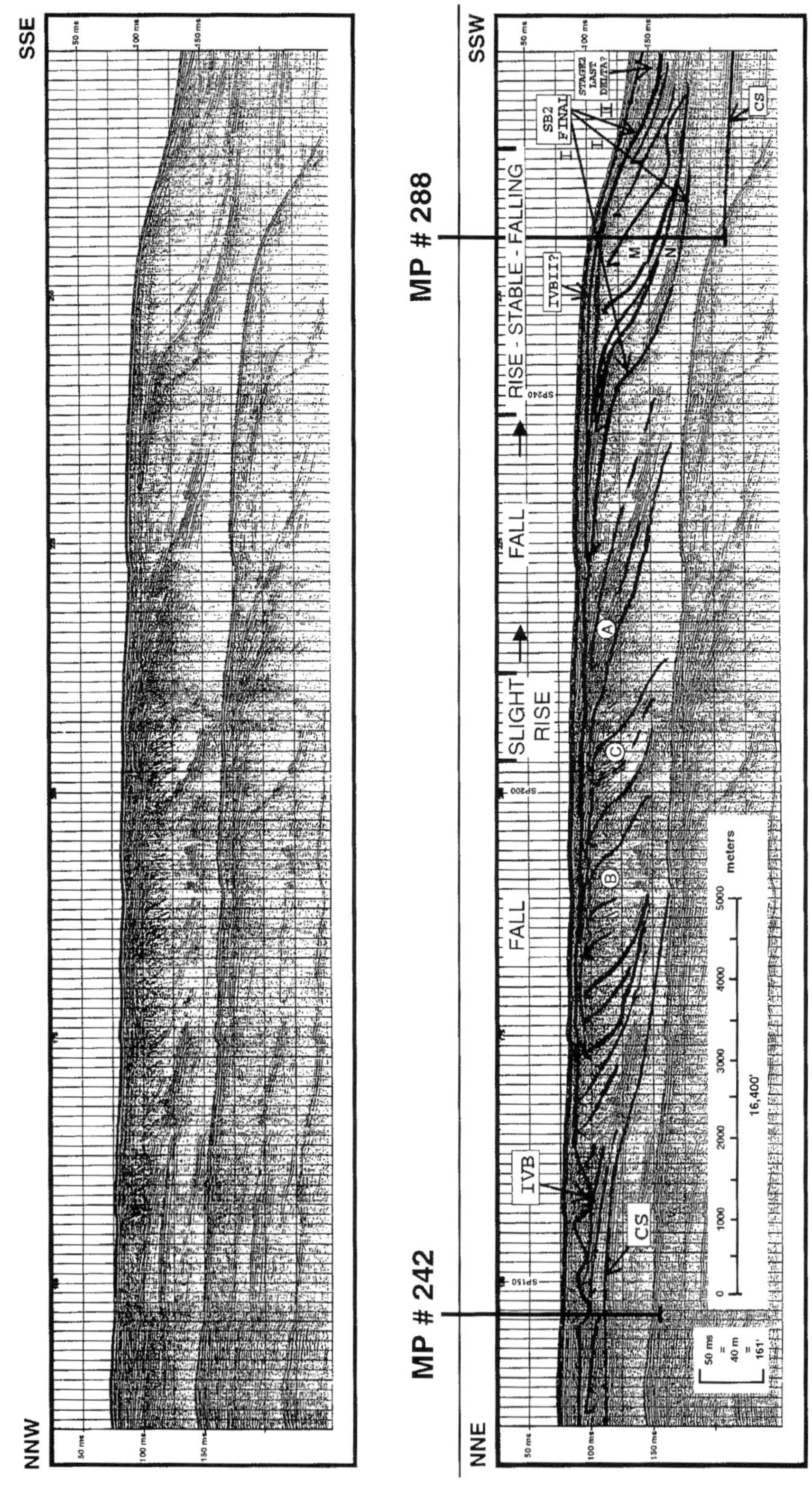
SSE
NNW
SSW
NNE
MP # 288
MP # 242
RISE - STABLE - FALLING
FALL
SLIGHT RISE
FALL
SB2 FINAL
STAGE2 LAST DELTA?
IVBII?
IVB
CS
CS
A
B
C
M
N
SP240
SP200
SP150
meters
16,400'
50 ms = 40 m = 161'

(a)

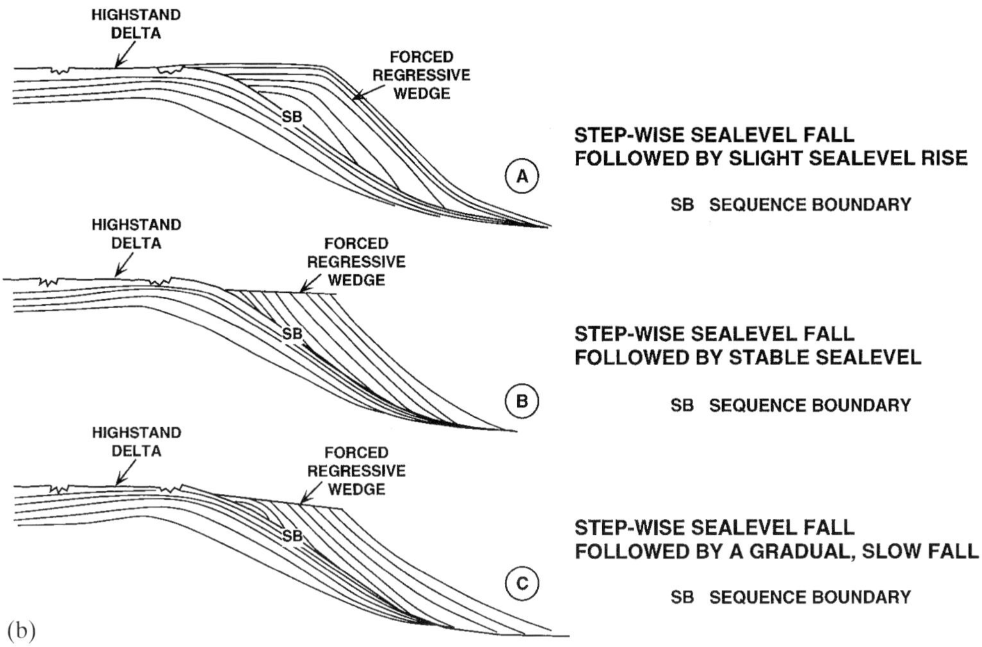

Fig. 11. (**a**) Uninterpreted and interpreted seismic line between the MP242 and MP288 boreholes (see Fig. 1b for location). The stage 5 condensed section can be seen to increase in dip and depth basinward (CS) over which the delta foresets steepen and increase in relief. The age of deltaic lobes can be clearly seen to young between the MP242 and MP288 sites. The deltaic wedges progressively step downward and basinward largely between site MP242 and shotpoint 238. Basinward of shotpoint 238 the deltas as a whole have a aggradational component. However, downdip of MP288, the 'stage 2 last delta?' may have downstepped. An interpretation of these stratal geometries in terms of relative sea-level changes is shown at the top of the lower diagram. Schematics of wedges A, B, and C are shown and interpreted in Fig. 11b. The two wedges M and N within the delta are also shown in Figs 10b, 11a and 12. IVB, incised valley base; IVBII?, incised valley base at end of stage 2 associated with SB2 final II, sequence boundary. (**b**) Idealized deltaic lobes resulting from downward shifts of relative sea level and subsequent deposition during a slow rise (A), stable (B) or gradually falling (C) sea-level. Seismic examples of wedges A, B and C are shown in (a).

laminated clays. The surface at 87.2 m is interpreted to be a sequence boundary formed during isotope stage 4, and is overlain by very distal prodelta clays.

At a depth of 68.9 m (226 ft) fossils become absent in the MP288 core, and palaeobathymetric indicators indicate abrupt shallowing from deep middle neritic (90–140 m (300–450 ft)) to shallow middle-neritic water depths (70–90 m (225–300 ft)). It is thought that this break may well correspond to an initial sequence boundary of isotope stage 2 (or stages 3 to 2 transition) as shown in Fig. 10a. Deposition occurred in largely shallow middle-neritic conditions (e.g. water depths of 45–90 m (150–225 ft)) in the succeeding sediments (between 68.9 m and 15.2 m (226 50 ft)). These strata consist of a generally upward-coarsening succession with parallel-laminated silts in the lower part and parallel to rippled sands in the upper part. At 55.5 m (182 ft) there is a surface with subtle lithological change that is downlapped by gently dipping clinoforms and associated with a palaeobathymetric shallowing (Fig. 10). This surface apparently corresponds to the base of base of wedge 'N' (labelled 'SB2 final I' in Figs 10b, 11a and 12). This sequence boundary is interpreted to have formed at the end of sea-level fall during isotope stage 2 (Figs 10b, 11A, 12). Just above this sequence boundary increased fossil content (foraminifera and nannofossils) is associated with a slight deepening, suggesting onset of relative sea-level rise (Fig. 10a).

A significant change in lithology occurs within the prograding wedge deposited during stage 2 at depth of 47 m (154 ft) in the MP288 core (Fig. 10a), with abrupt coarsening within strata deposited in middle-neritic water depths. The

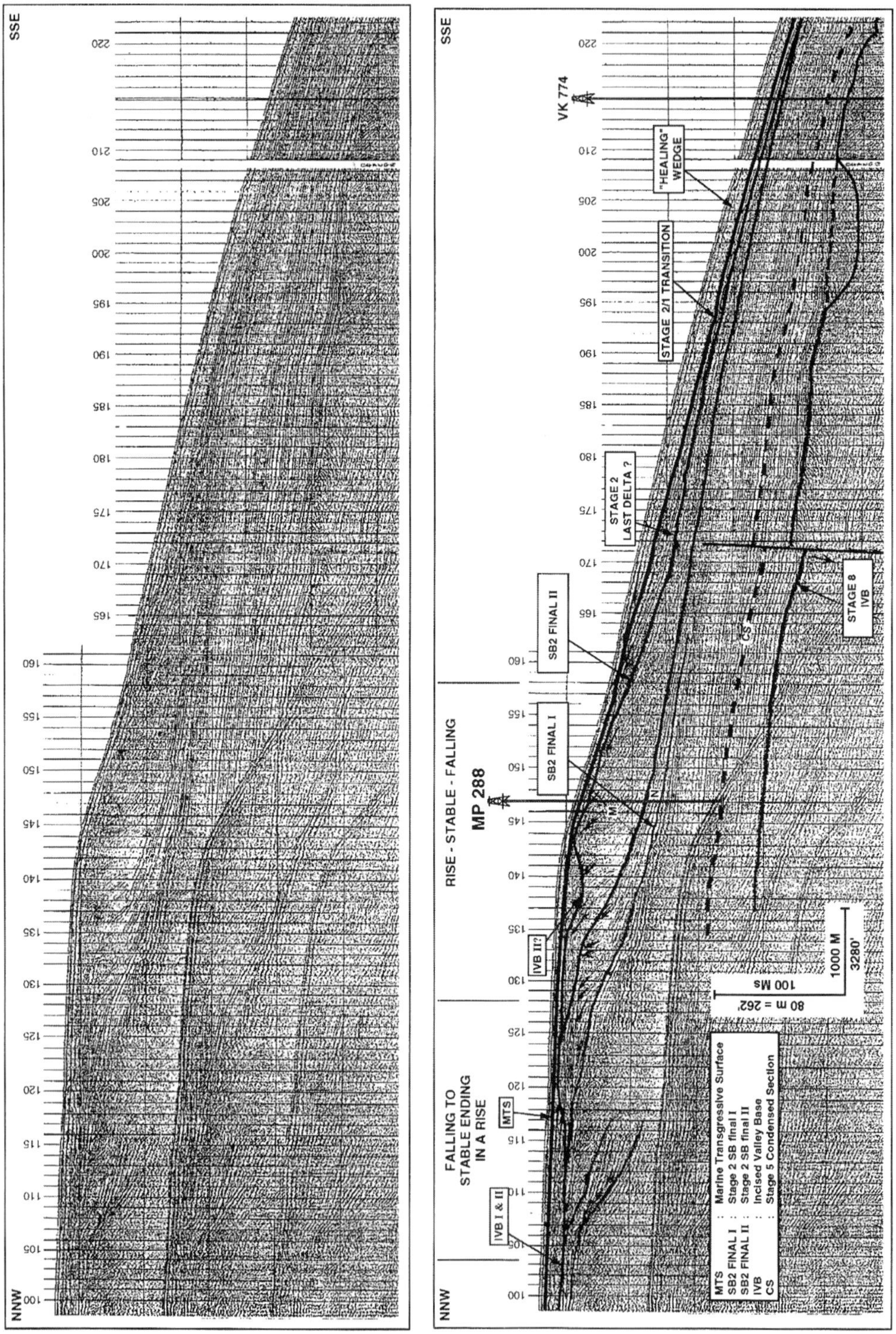
SSE
NNW
VK 774
"HEALING" WEDGE
STAGE 2/1 TRANSITION
STAGE 2 LAST DELTA ?
SB2 FINAL II
SB2 FINAL I
RISE - STABLE - FALLING
MP 288
FALLING TO STABLE ENDING IN A RISE
MTS
IVB I & II
IVB II?
STAGE 8 IVB
CS
100 Ms
80 m = 262'
1000 M
3280'
MTS : Marine Transgressive Surface
SB2 FINAL I : Stage 2 SB final I
SB2 FINAL II : Stage 2 SB final II
IVB : Incised Valley Base
CS : Stage 5 Condensed Section

corresponding surface on the seismic appears to be a downlap surface. The downlapping clinoforms show steeper dips than those that downlap the SB2 final I surface at 55.5 m (182 ft) (Fig. 10a, b, 11a, 12). Sediments within the clinoformed package, between 47 m and the major erosion surface at 15.2 m (50 ft), have a relatively high content of foraminifera (especially benthics), nannofossils and shells. This stratigraphic interval corresponds to the main deltaic wedge in the MP288 borehole (wedge M, Figs 11, 12), interpreted to have been deposited under conditions of early sea-level rise, following the lowest position of relative sea-level, on the basis of its seismic stratigraphy (wedge M, Figs 11, 12).

In comparison to the middle-neritic conditions at the MP288 site between 47 m and 15.2 m (154–50 ft), the main delta body at the MP303 site has little or no fossil content and was deposited close to the shoreline in shallower waters (Figs 5, 10). The downlap surface at 47 m in the MP288 borehole is interpreted here to mark the onset of deposition during conditions of early relative sea-level rise during isotope stage 2. The deltaic clinoform interval between 47 m and 15.2 m at the MP288 site is thought to have been deposited in greater water depths, with the delta-front located in a more landward position than at the MP303 site on the basis of sedimentology and fauna. The more downdip setting of the MP288 borehole is interpreted to have resulted in a lower sedimentation rate and a less turbid water column, leading to the higher fossil abundances in this location. The greater abundance and variety of benthic foraminifera may also have been favoured by the relatively warm-water conditions inferred by Roberts *et al.* (1997) to have existed during the early sea-level rise.

Sediments in the MP288 core between the major erosion surface at 15.2 m (50 ft) and 1.8 m (6 ft) consist of parallel and inclined laminated sands with fossils and shells that were apparently deposited in shallow water conditions, and have a variable abundance of foraminifera and nannofossils (Fig. 10a). This succession is interpreted as a largely estuarine and bay-head deltaic fill with some fluvial components, within which there are some indications of palaeobathymetric deepening that may represent flooding events. This largely estuarine succession is overlain by a major surface at 1.8 m that is associated with a significant increase in water depth, fossil content and the base of Ericson's Z zone (Fig. 10a). The overlying succession is somewhat upward-coarsening and comprises laminated shell-rich sands. The basal surface of these sediments is interpreted as a marine transgressive surface that marks the base of isotope stage 1. The overlying sands are similar to stage 1 deposits of the MP303 and 242 cores, deposited as sandy shoals reworked during the stage 1 transgression.

Origin of the MP288 erosion surface at 15.2 m. The occurrence of a marine transgressive surface at only 1.8 m (6 ft), a mixed estuarine/fluvial fill between 1.8 m and 15.2 m (6–50 ft), and the middle neritic environment of the main deltaic interval (47–15.2 m (154–50 ft)) are taken to suggest that the erosion surface at 15.2 m was not the result of fluvial and/or marine transgressive erosion. An interpretation of this erosion feature as a slump scar seems unlikely because marine transgression would then be expected at 15.2 m, but is actually observed at 1.8 m. Consequently, it is interpreted here that the erosion surface at 15.2 m resulted mainly from incised valley erosion, the result of a further fall of relative sea-level (e.g. IVB II?, Figs 11a, 12). This sea-level fall within the early rise of isotope stage 2 followed deposition of the main lowstand delta body at the MP288 site (wedges N and M, Figs 11a, 12) as the base of this incised valley clearly caps the uppermost lobe of the main progradational delta (surface IVB ii?/SB2 final II; Figs 10, 11 and 12).

The uppermost sequence boundary of the delta (SB2, final II, downdip extension of IVBII?; Figs 11, 12) is overlain by a distinctive clinoformed stratal unit with an upper erosional surface interpreted to have been formed during subsequent transgression (labelled the stage 2 last delta in Figs 11a, 12). The downdip extension of this unit was drilled by VK774, isotopic data from which indicate that it was deposited during isotope stage 2 (Figs 12, 13). Thus the deposition of this last delta probably occurred subsequent to the latest sea-level fall superimposed on the longer-term fourth-order early sea-level rise. This latest sea-level fall may have

Fig. 12. Uninterpreted and uninterpreted NNW–SSE-oriented seismic line showing the stratal patterns developed between the MP288 and VK774 sites (see Fig. 1b for location). Whereas site MP288 cored through delta clinoforms deposited during late isotope stage 2, the VK774 site can be seen to pass through their distal correlatives. Well developed at the MP288 site are wedges N and M. Interpretation of the delta stratal geometries in terms of sea-level changes is shown at the top of the lower diagram. Also shown is the upper faulted surface of the stage 8 (falling stage) delta cored at the VK774 site that is overlain and locally cut by an interpreted incised valley complex (see Fig. 13).

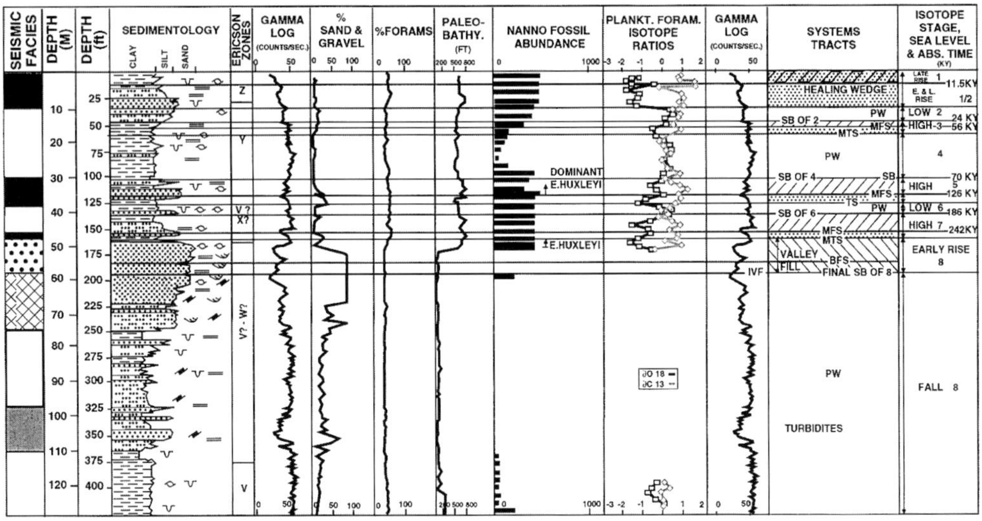

Fig. 13. Summary of sedimentological, biostratigraphic, palaeoenvironmental and isotopic analytical data from the VK774 borehole, in relation to seismic facies, sequence stratigraphic interpretation, chronostratigraphy and an interpretation of relative sea-level changes. The complimentary seismic line through this borehole is shown in Fig. 12, for comparison to the seismic facies summarized in the left hand column. The numbers 1–6 in the right-hand column refer to oxygen isotope stages. See Fig. 3a for key to symbols.

modified the updip incised valley, IVB, formed a little earlier during the maximum sea-level lowstand within stage 2. Accordingly, the surface labelled IVB II? at the MP288 site is shown in to merge with the updip incised valley (IVB) in Fig. 11a and 12. Both the bayhead flooding surface and marine transgressive surface are time-transgressive between the MP242 and MP288 sites (Fig. 7), although they can be well correlated lithologically.

Sequence stratigraphy at VK774 site and along VK774-MP288 site section

The VK774 site is located in the deepest waters (185 m) and most seaward position of all the cores, so that the sediments deposited during isotope stages 5 to 1 are represented only in the uppermost 38.4 m (126 ft) of the core. Seismic data show that stages 5–1 in the vicinity of the VK774 core are represented mainly by subparallel continuous inclined reflectors representing distal clinoforms.

Major stratal surfaces. In the VK774 core sediments deposited during the isotope stage 5 interval are located between 38.4 m and 31 m (126–102 ft), as constrained by occurrence of the planktonic foraminifera *G. flexuosa*, the lowest level of *E. huxeleyi* dominance at 35–35.7 m (115–117 ft), palaeobathymetric changes and the light-heavy trends of oxygen isotope values from the top of the core to a depth of about 38.4 m (126 ft) (Table 1; Fig. 13). Sediments between 38.4 m and 33 m are characterized by a deep palaeobathymetry (>185 m; >600 ft), light oxygen isotope values and shell-rich lithofacies (Fig. 13). The seismic reflector corresponding to the top of stage 5 interval at the VK774 site is compatible with that at the MP288 site, as shown by in Fig. 12. The position of a maximum flooding surface within the stage 5 interval is identified on the basis of a deepening in palaeobathymetry, changes in isotope values, a reduction in grain size and a kick on the gamma log (Fig. 13). Correlation of this stage 5 maximum flooding surface with the other cores is shown in Fig. 7.

Laminated silts with only subtle upward-coarsening successions, if any, heavy oxygen isotope values and relatively shallow palaeobathymetry are characteristic of sediments representing stages 4 (31–18.3 m; 102–60 ft) and 2 (14.6–10.4 m; 48–34 ft). However, whereas the stage 4 interval has a low to moderate fossil content, moderate to high abundance is typical of the stage 2 sediments. These intervals are represented by very distal clinoforms on the seismic, accounting for both the high abundance of fossils and very fine grain sizes (Figs 12, 13).

Sediments of stages 4 and 2 are separated by a distinctive 3.7 m (12 ft) thick interval that is typified by light oxygen isotope values and rather deeper palaeowater depths between 18.3 m and 14.6 m (60–48 ft; Fig. 13). This short stratigraphic interval is attributed to deposition during isotope stage 3 (Fig. 13). The main deltaic wedge sampled at the MP288 site (wedge M, Figs 11, 12) thins significantly towards the VK774 borehole as is clear in both the seismic data and correlation panel (Figs 7, 12). From the seismic data it is also apparent that the stage 2 interval drilled at the VK774 site is in part the downdip extension of the updip 'stage 2 last delta' shown in Figs 11a and 12.

A significant deepening in palaeobathymetry, associated with increased fossil content and lighter isotope values, occurs above 10.4 m (34 ft) in the VK774 core. This level also separates strata that contain *G. inflata* below from those with abundant *G. menardii* above, and is close to the base of Ericson's Z zone (Fig. 13). These changes at the VK774 site are indicative of updip transgression on the shelf margin, in the form of marine - or bay-flooding, depending on the distance landward from maximum lowstand shorelines. The gamma log between 10.4 m and the top of the VK774 core allows distinction of two subtle upward-coarsening units that consist of laminated silts with shells. The lower unit is somewhat coarser and isotopic data suggest that it was deposited during the early transgression (2/1 stage transition), with the upper unit formed during the Holocene (stage 1; Fig. 13). On the seismic lines these two coarsening-up units are represented by an onlapping wedge between the VK774 and MP288 sites (Figs 12,13). Similar seismic wedges, interpreted to have been deposited during early transgression, have been reported elsewhere in the Gulf of Mexico by Suter & Berryhill (1985), and appear similar to the healing phase wedges described by Posamentier & Allen (1993a).

It is clear that deposition at the VK774 site occurred in outer neritic water depths even during times of sea-level fall and lowstand associated with the minor and major glacials represented by isotope stages 4 and 2, respectively. Following the most recent sea-level fall during isotope stage 2, the VK774 site

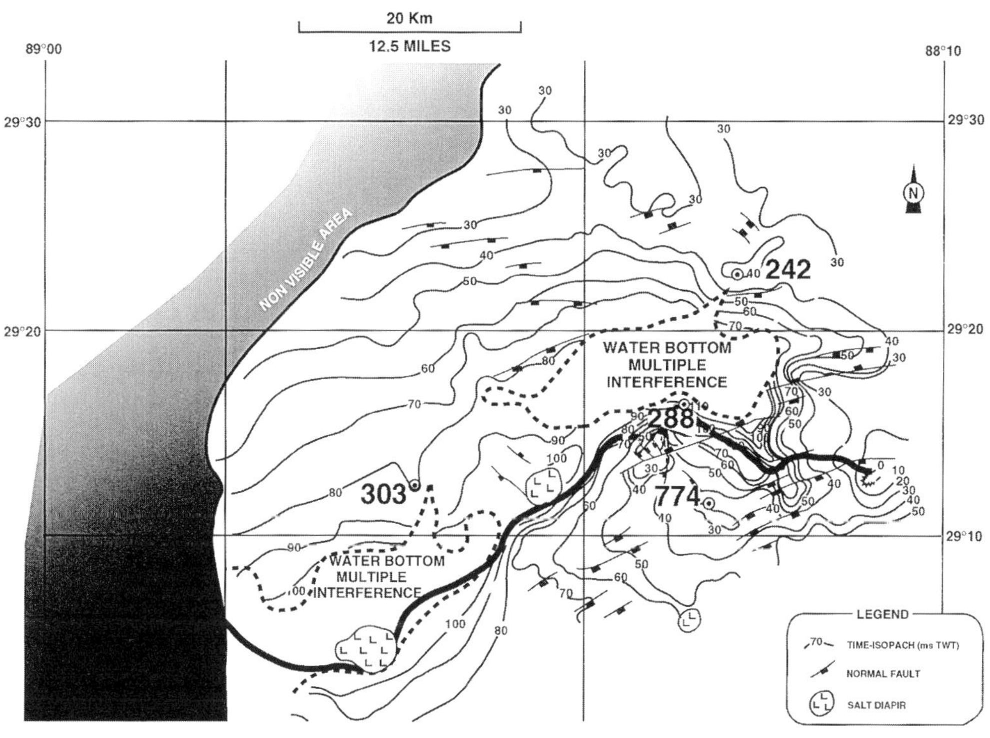

Fig. 14. Time-isopach map, in milliseconds of two-way travel time, of the total deltaic interval and incised valley fill between isotope stages 5 and 1. Although the water-bottom multiple masks some critical areas, two thick depocentres (I & II in Fig. 16) can be inferred.

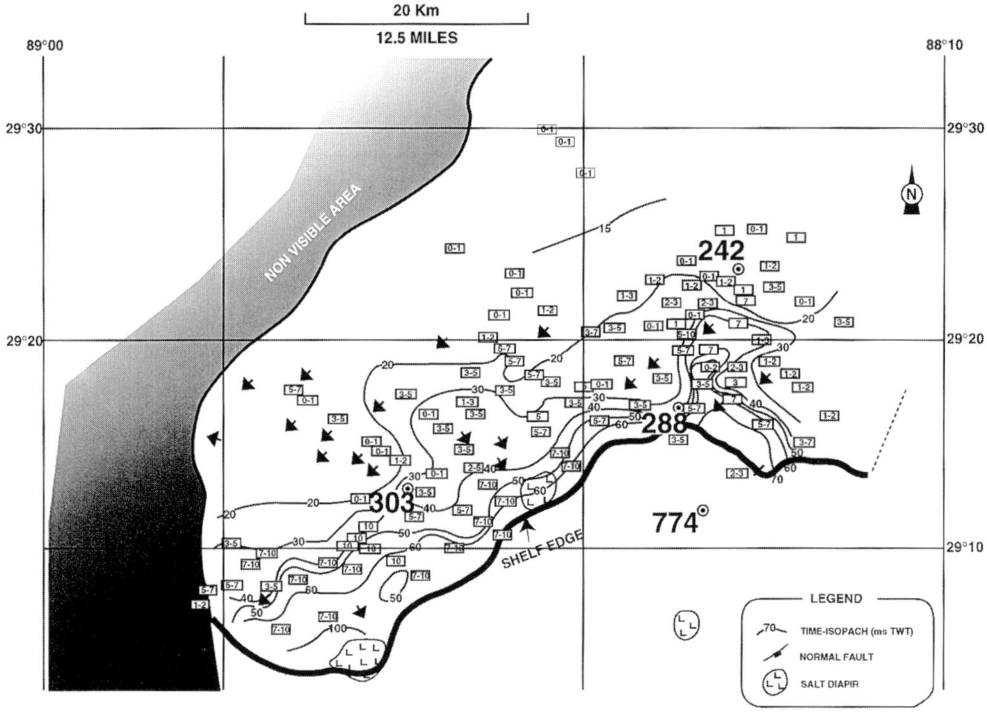

Fig. 15. Time-isopach, in milliseconds of two-way travel time, of the topmost, deltaic lobe at each location and the dips of clinoforms. The lobe thickness and the clinoform angles generally increase towards shelf edge. Two depocentres as in Fig. 14 can be inferred (see Fig. 16).

experienced deepening that, at a core depth of 10.4 m (34 ft), is the expression of the initial rise following sea-level lowstand (Fig. 13). It is interpreted that the sites on the shelf (e.g. MP242, MP288, MP303) were subaerially exposed during the lowest points of sea-level during stage 2, but probably experienced bay-flooding during the initial sea-level rise. The marine transgressive surfaces in the updip boreholes are marked by major increases in palaeobathymetry and the occurrence of warm-water foraminifera (*Globorotalia menardii*). The sediments that overlie the marine transgressive surfaces in the updip sites are interpreted as reworked sand shoals formed during isotope stage 1 (Figs 3a, 5, 10a). In the VK774 core warm-water foraminifera begin to occur in significant numbers at the base of the lower unit of the onlapping wedge at 10.4 m. This lower unit corresponds to the beginning of the 2/1 stage transition (Fig. 13). The earlier appearance of warm water foraminifera in the VK774 core than on the shelf, at the MP242, MP288 and MP303 sites, is interpreted to reflect the time-transgressive nature of the marine transgression. The upper upward-coarsening unit in the VK774 core between 10.4 m and 0 m represents stage 1 deposition, and so is equivalent to the stage 1 interval at the updip sites (Fig. 7).

Spatial and temporal evolution of the Lagniappe Delta complex

Thickness distribution of sediments and incised valley architecture

A time–isopach map of the sediments between the top of isotope stage 5 and the base of stage 1 is shown in Fig. 14, and includes all of the deltaic wedges and the incised valley fill. Unfortunately the seafloor multiple in places masks the lower part of the deltaic interval and the top of stage 5 leading to gaps in the contouring. Figure 15 is an isopach map of the uppermost deltaic interval below the incised valley fill, and shows the dip of clinoforms within this deltaic package. The distribution patterns displayed in Figs 14 and 15

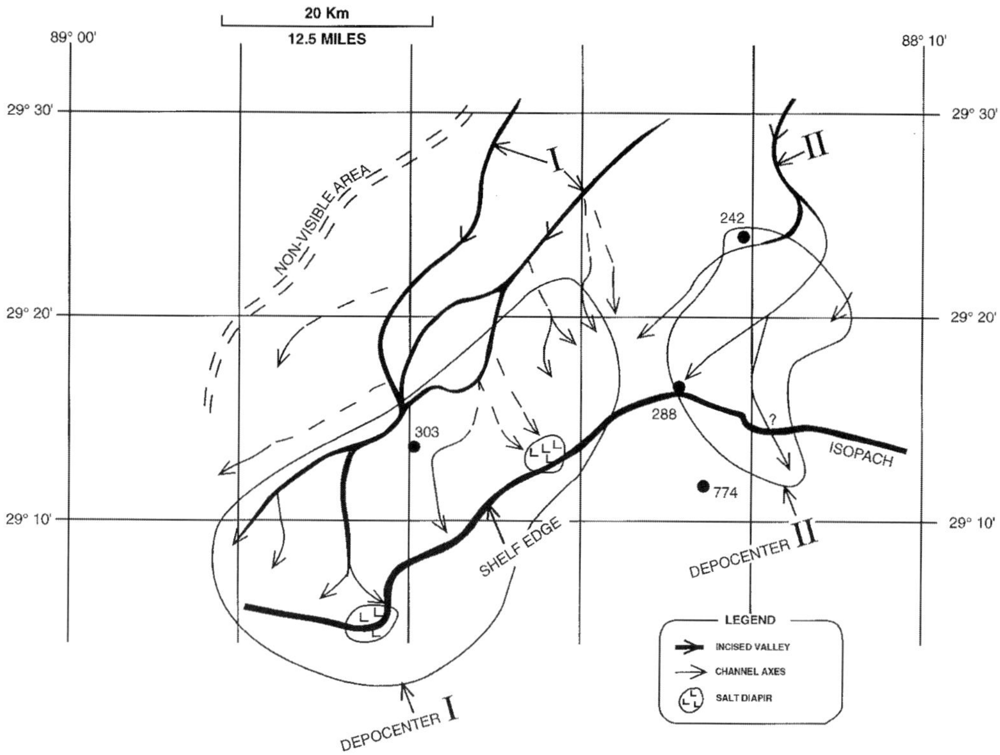

Fig. 16. Depocentres (simplified from Figs 14 and 15) and axes of incised valley-channel systems (simplified from Fig. 18) of the Lagniappe Delta complex. The main depocentre I trends NE–SW and the corresponding incised valley-channel system I is apparently displaced just to the west of it. Note the distributary channel-like features branching from the main trunk valley I. The depocentre II trends mainly N–S.

distinguish two distinct depocentres: (i) a major margin-parallel trending NE–SW (I) and (ii) a minor depocentre (II) trending NE–SW to N–S that is margin-perpendicular to the north and east of sites MP288 and VK774 (Fig. 16). From updip to downdip, the thickness and dip of clinoforms within each depocentre increase (to the SW or SE, e.g. Fig. 15), and it is clear that these two depocentres built a significant part of the present shelf margin in the study area.

A map-view of the uppermost delta lobe preserved at each location, and the axial directions of the lobes is illustrated in Fig. 17a. This map suggests that the main direction of delta progradation was from the NE to the SW or SE (Fig. 17a), although most of the individual lobes strike NW–SE or E–W within the gross progradation direction(s). A seismic line to the east of the MP303 site illustrates the eastward migration of delta lobes that is consistent with these strike trends (Fig. 17b). Clearly, not all the delta lobes shown in Fig. 17 result from changes in relative sea-level, autocyclic processes played an important role in lobe switching. It is thought that following each sea-level fall and downshift of sedimentation, subsequent delta lobe switching resulted in more than one lobe being deposited during the ensuing period of stable sea-level.

The fill of the incised valleys capping the stage 5–2 delta is shown in Fig. 18, with the contoured interval taken from the base of the incised valley to the base of stage 1. It is clear that channelisation and incised valley development is widespread across the study area. The thickest sediments are interpreted to fill the axes of the incised valleys and channels, and from the map pattern two major incised valley complexes, I and II are distinguished (Figs 16,18). The major incised valley complex (I) is associated with depocentre I and appears to show a similar NE–SW trend. This valley complex has many channel-like distributary features that fan-out downdip towards the shelf margin (Figs 16, 18). Incised valley system II also displays distributary channel-like features that fan-out towards the shelf margin, but is of lesser importance in comparison to system I.

It is thought that the axes of the incised valley/channel systems (Figs 16, 18) mainly reflect their maximum extent reached during the lowest positions of sea-level, within isotope stages 4 and 2 (e.g. Fig. 2). However, development and erosion of the incised valley systems was probably progressive, so that they reached their maximum extent at the time of lowest sea-level, within stages 4 and 2. Development of the incised valley systems was probably episodic as a result of higher-frequency falls and rises (e.g. Fig. 2) that were superimposed on the longer term (fourth-order) eustatically-driven fall between isotope stages 4 and 2. The axes of the final incised valley systems are themselves displaced from the main depocentres (Figs 16, 18). Although depocentre I and incised valley system I have the same trend, the incised valley system is displaced to the west of the main depocentre (e.g. compare Figs 16,18). It would appear that the river erosion occurred preferentially in the palaeolows that were located just to the west of the depocentre I during successive sea-level falls until the sea-level dropped to its lowest position. In a similar way, the axes of incised valley system II may have followed the palaeolows in and around depocentre II (Figs 16, 18).

Stratigraphic evolution of the Lagniappe Delta

The position, extent, stratigraphic architecture and progradation of the Lagniappe Delta was mainly controlled by physiographic setting of the shelf and slope areas, the location and direction of the fluvial inputs, relative sea-level changes and to some extent the prevailing oceanic circulation (e.g. westward longshore drift in the area). Within the study area, it is inferred that the two fluvial inputs (systems I and II, Figs 16, 18) contributed almost all of the sediment that resulted in the outbuilding of the delta, and that these correspond to the two incised valley complexes. Fluvial inputs I and II is thought to have fed depocentres I and II, respectively, during isotope stages 4 and 2.

Deposition during sea-level fall and lowstand: With each sea-level fall to or below the offlap break of the preceding deltaic wedge, incised valley erosion is interpreted to have occurred and a type 1 sequence boundary was formed. During the ensuing period of either (A) stable, (B) slow rise or (C) slow sea-level fall, a deltaic wedge was deposited on this boundary with a characteristic internal stacking pattern (e.g. Figs 11, 19). Some mass-flow deposits were likely laid down on the bottom sets of these wedges (e.g.

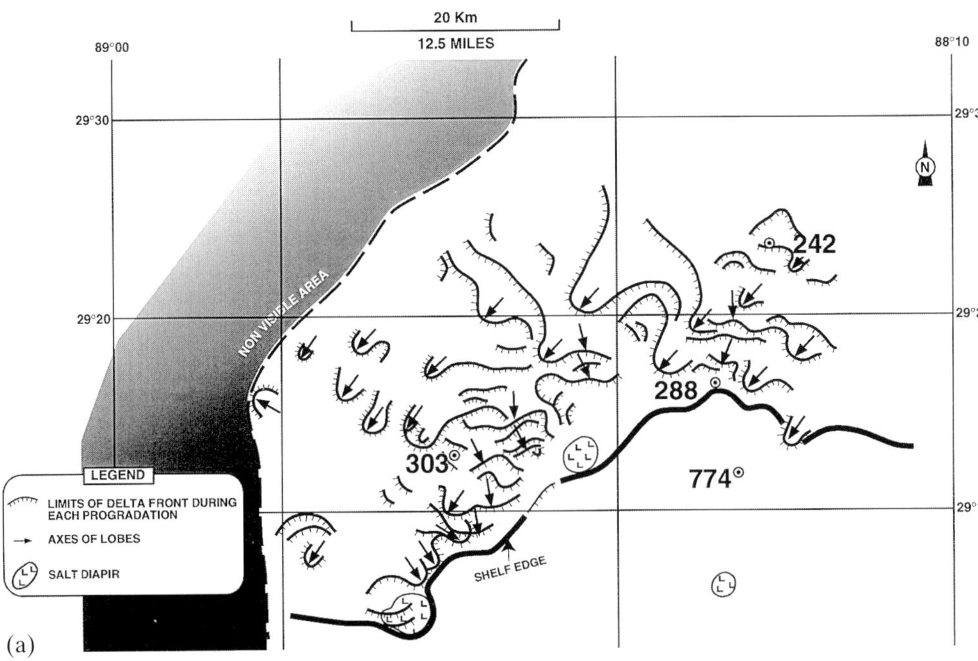

(a)

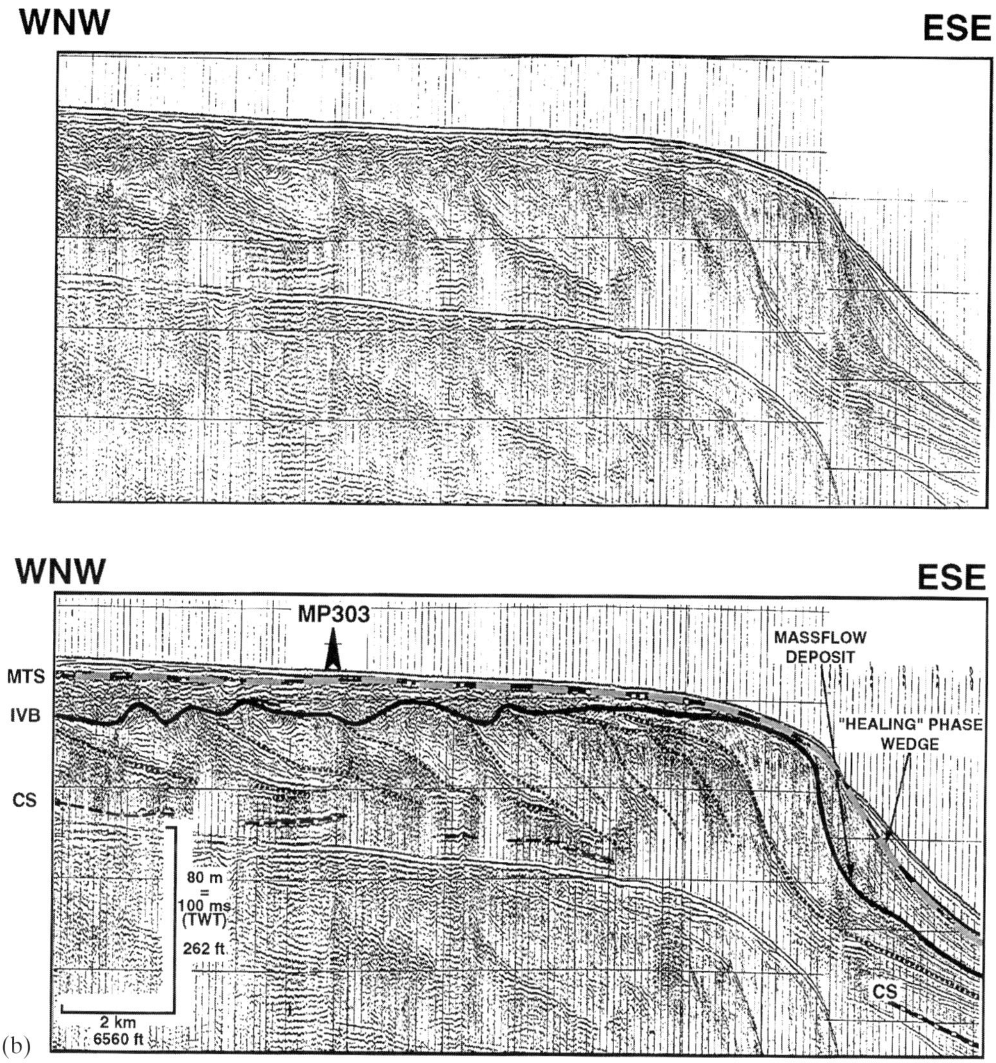

Fig. 17 (a) Extent and plan-view correlation of the topmost deltaic lobes. Although the main axis of progradation is from NE to SW and SE, the individual lobes appear to show NW–SE and E–W strike trends. (b) A seismic line displaying eastward migration of deltaic lobes east of the MP303 site consistent with the strike trends shown in (a) (see Fig. 1b for location). MTS, marine transgressive surface; IVB, incised valley base; CS, condensed section of stage 5 interval.

Fig. 19). Following the sea-level fall, and during conditions of stable or slow sea-level rise, deltas might have migrated or autocyclically shifted along strike. It is also possible that at each stand of sea level several distributary channels branched from the main trunk channel (e.g. Figs 16, 18), each of which was responsible for the deposition of at least one delta lobe. In response to each successive sea level fall, the same processes were initiated (Fig. 19), with the axis of the incised valley system successively displaced towards the palaeolows just west of the preceding lobe complex. At the same time river-input extended progressively farther seaward. As the combined magnitude of the net sea level falls increased (and hence sediment supply was augmented?), thicker lobe complexes were deposited farther seaward. By the time of the lowest positions of sea level, major incised valley erosion occurred across most of the previously highstand and falling stage delta lobes (Fig. 19), with large distributary channel-like features

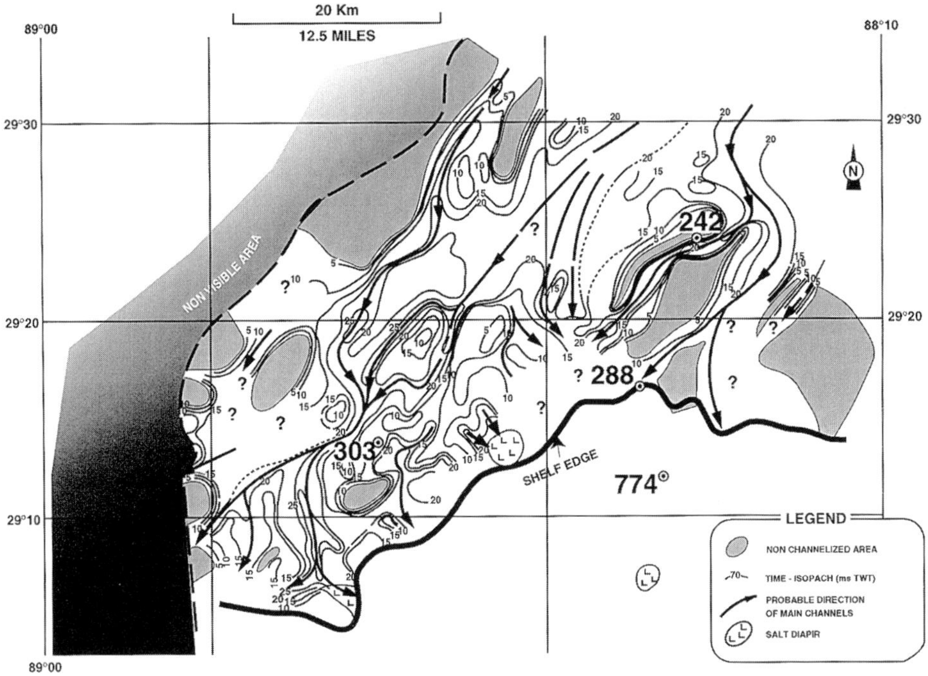

Fig. 18. Isopach map of incised valley-fill (ms TWT) and axes of valley and channel fairways inferred on the basis of maximum thickness. The two incised valleys and related channel systems inferred here are shown in Fig. 16 in relation to the depocentres.

formed in downdip areas across the outer shelf (Figs 16,18). Thus, during the maximum fall and subsequent early rise of sea-level, thick delta lobe complexes were deposited in the farthest seaward areas.

The main Lagniappe depocentre (I; Figs 16, 18) was built from the coalescence of lobe complexes deposited both during the long-term fall and during the maximum fall-early rise within isotope stages 4–2 (Fig. 19). The fluvial input II may also have similarly deposited deltaic wedges from updip (MP242) to downdip (MP288) areas. The fanning of apparently larger channels in the downdip areas close to the shelf margin probably recorded the pattern of distributary channels formed during the maximum sea-level fall. It is after this maximum fall and during the subsequent early rise that the delta at the MP288 site was deposited. However, it is here noted that in the areas downdip of both the MP288 and MP303 sites, forced regressive wedges were also deposited within the early rise during phases of high-frequency (fifth-order) sea-level fall.

Deposition during sea-level lowstand and rapid rise. During the early rise of sea-level associated with the isotope 2/1 stage transition, downdip areas of the incised valley became filled with distributary channel and bayhead delta deposits. At the same time much of the onlapping 'healing-phase' wedge (Figs 12, 19) was deposited in relatively deep waters. Further deposition in the updip areas of the incised valley occurred in the form of bayhead deltas and estuarine deposits mostly during the later sea-level rise. Above the marine transgressive surface, the stage 1 interval recovered in the borehole cores consist of mainly reworked sandy shoals and muds deposited during the transgression in the early Holocene, prior to drowning and major backstepping of the delta system to its present day site (e.g. Fig. 19).

Evaluation of systems tracts in the Lagniappe Delta

The deltaic lobes widely deposited on the inner, middle and outer shelf areas in response to long term sea-level fall during stages 4 and 2 comprise part of a falling-stage or forced regressive systems tract (*sensu* Nummedal *et al.* 1992; Plint & Nummedal this volume; Hunt & Tucker 1992, 1995). These deposits form the most important

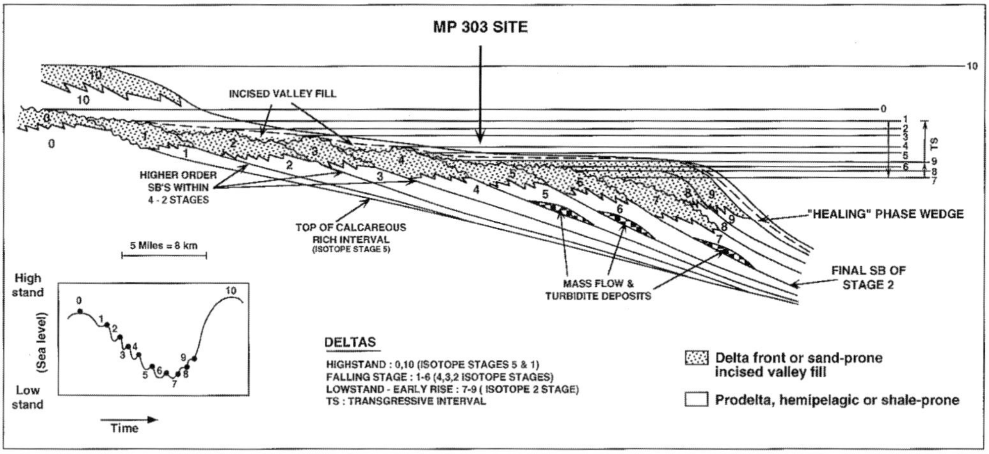

Fig. 19. Depositional model for Lagniappe Delta during the isotope stage 5 to 1 interval. Starting from a sea-level highstand at step 0, corresponding to isotope stage 5, relative sea-level is shown to fall in several steps (idealized) during isotope stages 4, 3 and 2, until the maximum drop (step 7, stage 2) and then to begin to rise (steps 8, 9) also during isotope stage 2. The present position of sea-level is shown by step 10. During each stepwise sea-level fall, a sequence boundary forms and a delta lobe is subsequently deposited over it. During the maximum sea-level drop (step 7), a surface of maximum subaerial erosion (incised valley floor) results with subsequent lobe deposition during the early sea-level rise. The incised valley is filled mainly during the early and late sea-level rise. Finally, step 10 high sea-level corresponds to HST of isotope stage 1 interval deposition.

component of the Lagniappe Delta complex, as shown schematically in Fig. 19 (time steps 1–6 and their corresponding lobes). The second most important component of the delta complex are lobes deposited on the outer shelf areas during the maximum lowstand-early rise of sea level, together with distributary channel fills, bayhead deltas and estuarine sediments deposited in the downdip areas of the incised valley during early sea-level rise (e.g. time steps 7–9, Fig. 19). The bayhead deltas and the estuarine sediments deposited during the later rise comprise the transgressive systems tract, and form a significant portion of the incised valley fill, especially in the updip areas. Even isotope stage 1 deposits consist only of a thin veneer of reworked sandy shoals and muds deposited during the latest Holocene transgression in the study area. Because of coring disturbance towards the top of the borehole, the present highstand systems tract sediments, as such, have not been recognized at the borehole sites, and their thickness is insufficient to be resolved in the seismic data. The highstand systems tract of the stage 5 interval recovered is relatively thick only in the MP303 borehole and relatively thin in the other boreholes. As is clear from Figures 3, 5, 10 and 13, sediments attributed to the highstand systems tract are only a minor component of the Lagniappe Delta complex, and is probably best developed on the inner shelf (e.g. times 0 and 10, Fig. 19).

Evolution of the Lagniappe Delta during the older oxygen isotope stage intervals

At the MP242 site, the isotope stage 6 delta interval overlies the stage 8 delta is and is itself overlain by a thick incised valley fill. At the MP303 site, part of the stage 6 delta overlain by thick quartz-rich sands inferred to be transgressive in origin (Figs 3a, 5). At the VK774 site, the deltaic strata of isotope stages 6 to 8 (Fig. 13) and older isotope stages (not shown here) have been inferred. The stage 8 delta is thick, coarse grained and is interpreted to be overlain by incised valley fill at this site (Figs 12, 13). Thus the consortium data suggest that at least during isotope stages 6 and 8, significant deposition occurred in the Lagniappe Delta complex during the falling sea level, and probably during the maximum lowstand–early rise, similar to delta development during stages 4 and 2 discussed above. Subsequent to isotope stage 8, the VK774 site remained in the upper slope-outer neritic water depths (during stages 7 to 1) so that sedimentation is characterized by distal deltaic clays, probably because of higher subsidence rates in this area.

Sequence boundaries and distinguishing characteristics of systems tracts

Sequence boundaries

The major unconformity at the base of the incised valley in the stage 5–1 Lagniappe Delta complex is interpreted to have resulted from erosion related to the maximum relative sea-level fall(s). The base of this incised valley, and its downdip correlative conformity, are considered to form the main sequence boundary (e.g. Vail 1987) of the youngest fourth-order Late Pleistocene sequence (e.g. Fig. 19). This surface is likely to be a regional onlap surface in the sense of Mitchum (1977). The major sequence boundary is underlain by the highstand and falling-stage wedges, and is overlain and onlapped by a wedge complex formed during maximum lowstand-early rise of sea-level, and also by the transgressive systems tract (e.g. Fig. 19).

Unconformity surfaces and their correlative conformities at the base and within the falling stage wedges, herein referred to as initial sequence boundaries, are interpreted as higher-order (fifth-order) type 1 sequence boundaries within the fourth-order sequence. These initial sequence boundaries resulted from high-frequency sea-level falls during the long-term fall within isotope stages 4 and 2 (Fig. 19). Traced updip, the minor sequence boundaries amalgamate into the major fourth-order sequence boundary (e.g. Fig. 19), and as such were probably modified at the time of maximum sea-level fall and major incised-valley erosion. Depending on the physiographic setting and sediment supply, it is thought that large fan systems are more likely to develop on the major sequence boundary than the preceding initial sequence boundaries.

In most data sets used in this study, the major sequence boundary forms a more regionally isochronous and more readily recognizable surface than the initial higher-order sequence boundaries. This boundary is also likely to be recognizable in multichannel seismic data used in the industry. If the very first high-order sequence boundary is taken as the master or the main sequence boundary (*sensu* Posamentier *et al.* 1992; Posamentier & Morris this volume), then the latter, more pronounced unconformity that is a more regionally isochronous surface developed during the maximum relative lowstand(s), would occur within the sequence. It is apparent from the isotope curve shown for the Late Pleistocene in Fig. 2 that it is the sequence boundary resulting from the maximum sea-level drop that gives a consistent notion of fourth-order cyclicity for the sequence development, rather than the surfaces resulting from the minor, higher-order sea-level falls. This choice of the major sequence boundary is most consistent with the arguments forwarded by Hunt & Tucker (1992, 1993, 1995), Nummedal *et al.* (1992) and Plint & Nummedal (this volume).

However, although the main sequence boundary resulting from the maximum relative sea-level drop is more likely to be regionally isochronous than the initial sequence boundaries corresponding to the higher-order sea-level falls, there could be limitations on the timing of its development. The timing of the maximum relative sea-level drop may vary from area to area within a broad region depending on the rate of eustatic fall and on the rate of subsidence (e.g. Gawthorpe *et al.* 1994). In the areas of low subsidence, the relative sea-level may continue to fall while in the areas of high subsidence the relative sea-level may drop only during the rapid eustatic falls and may rise at other times (e.g. Gawthorpe *et al.* 1994). Ideally, the timing of the maximum rate of sea-level fall resulting in the maximum net relative sea-level drop and the corresponding sequence boundary should be regionally synchronous.

In addition to the magnitude of the sea-level fall, the duration of the drop and the physiography of the depositional setting also influence the degree of the sequence boundary development. The type and viewing window of the data sets are important in recognising the sequence boundaries (e.g. Posamentier & Morris this volume). Thus, it is our preference to identify the multiple sequence boundaries as well as the most pronounced erosional surface, map them regionally and then determine and understand the causes of their origin during a particular sea-level cycle, before a surface is distinguished as the main sequence boundary.

Systems tracts

Prograding deltaic lobes or regressive wedges deposited in a shelf or ramp setting during the sea-level falls have been variously named by different authors: the lowstand-shelf phase deposits (Suter & Berryhill 1985), and in terms of systems tracts, the lowstand, perched (Posamentier & Vail 1988); the early lowstand (Posamentier *et al.* 1992; Kolla *et al.* 1995); the late highstand (Van Wagoner 1995); the falling stage (Nummedal *et al.* 1992; Plint & Nummedal *this volume*); and the forced regressive (Hunt & Tucker 1995; Helland-Hansen & Gjelberg 1994). Based on the data presented in this paper, it is

Table 2. *Distinguishing characteristics of deltas of falling stage, maximum lowstand and highstand systems tract*

Characteristics	Falling stage deltas	Maximum lowstand–early rise deltas	Highstand deltas
1 Paleobathymetric environment of the encasing shales	Encased in inner to outer shelf muds	Encased in outer shelf to upper slope muds	Encased in inner shelf muds
2 Favorable physiographic setting	Wide shelf (ramp) gradients	Wide outer shelf–upper slope	Wide inner shelf gradients
3 Position within sequence	Intermediate levels (intermediate stand)	Mainly lowermost levels with some extending into intermediate levels (lowstand)	Uppermost levels (highstand)
4 Thickness andextent of fluvial strata	Minor	More than in falling	Could be thick and extensive
5 Deltac stratal pinchout relationships and zones of sediment bypassing	Wedging of strata landward to previous offlap break or below	Same as in falling stage	Landward stratal wedging much beyond preceding offlap break.
	Downward shifts of delta tops and offlap breaks relative to preceding wedge	Same as in falling stage	No downward shifts of deltaic wedges
	Prograding unit with coarsegrained deposits; may be separated from the preceding unit by a zone of bypassing	Same as in falling stage	No zone of sedimentary bypass between the prograding unit & the preceding unit
6 Localization of coarsegrained deposits	Proximal sharp-based shorface sediments and distal gradiational shoreface sediments	Same as in falling stage	Grainsize tends to decrease gradually from proximal to distal areas
	Coarse-grained deposits in distal clinoforms or encased in distal offshore muds	Same as in falling stage	
7 Nature of lower bounding surface	Updip, the surface is an unconformity that grades basinward into an erosional conformable surface under the delta ('SB' of the higher order)	Updip, the surface is a major unconformity that grades basinward into a conformable surface under the delta (major SB)	Deltas overlie and downlap the maximum flooding surface
	Updip: restricted or no landward onlaps of strata on the surface; downdip: deltaic clinoforms downlap on the surface	Updip: landward, regional onlapping of overlying strata on major 'SB'; downdip: clinoforms mayor may not downlap on major SB.	
8 Nature of upper bounding surface	Major unconformity (incised valley floor – major SB) bounds the top	Major transgressive surface limits the top	Major unconformity overlies the highstand delta similar to falling stage

Table 2. *Continued*

Characteristics	Falling stage deltas	Maximum lowstand–early rise deltas	Highstand deltas
9 Degree of progradation/aggradation	Extensive, wide progradation	Prograde and aggrade; thick; less areal extent (?)	Could be extensive, but likely to be less than falling stage; aggrade & prograde
10 Relationship of transgressive deposits	Overlie the major SB that in turn overlies the deltas; tend to be thick in incised valley	Overlie the prograding wedge that in turn overlies the major SB	Overlies major SB that in turn overlies the delta
11 Dip of progradational unit	Gentle to steep foresets Top of regressive complex may have seaward dips relative to the more horizontally bedded underlying units	Steep foresets	Gently dipping foresets
12 Growth faulting and depocentres	May be localized by growth faults	More commonly localized by growth faults	May not be commonly confined by growth faulting
13 Distributary channels	Amalgamated channel fills (?)	More common amalgamated channel fills*	Isolated to amalgamated channel fills
14 Along-shore switching of distributary channels	Less common	Minimal	More migration
15 Deformation & slumping	Intense	More intense	Less intense, more stable
16 Tidal sands	??	More common, especially during the early rise	Relatively rare & thin
17 Sand – rich turbidites	Frequent at the clinoform toes	More frequent Delta-front gullies with more sands	Less frequent
18 Type of fluvial source	Likely braided rivers through incised valley	More likely braided rivers through incised valleys	More likely meandering rivers

*With constant sediment influx, channel clustering and amalgamation may be common during early rise and late lowstand just above and below the major sequence boundary (Posamentier & Allen 1993b; Heller & Paola 1996).

thought appropriate to differentiate the lobes of the Lagniappe Delta deposited during the sea-level falls between isotope stages 4 and 2 into a falling stage systems tract.

The most important distinguishing characteristics of the falling stage, the maximum lowstand-early rise, and highstand systems tracts compiled from this and other studies (e.g. Coleman & Roberts 1988*a*, *b*; Posamentier *et al.* 1992; Hunt & Tucker 1992, 1993, 1995; Plint & Nummedal this volume) tracts are listed in Table II and are briefly discussed here. Landward stratal wedging beyond the preceding offlap break with no downward shifts of onlaps and with sediment grainsize decreasing gradually from proximal to distal areas is characteristic of the highstand deltas. They are also likely to be associated with extensive, thick delta-plain strata, whereas the falling stage deltas will have relatively minor fluvial strata, limited to incised valleys. In terms of the palaeobathymetric environment of the encasing shales and position within a sequence, the falling stage deltas have characteristics intermediate between those of the highstand and the maximum lowstand-early rise deltas. Characteristics common to the falling stage and lowstand-early rise deltas include: (i) wedging or pinching of strata landward to or below the offlap break of the preceding, updip units, (ii) proximal sharp-based shoreface sediments, (iii) coarse grained prograding units separated from the updip units by zones of bypassing and (iv) they may be encased in middle- to outer-neritic shales (Table 2).

Characteristic features solely of the falling stage deltas include a series of wedges with predominantly down-shifted landward pinchouts and down-stepped tops and offlap breaks. However, some of the lobes within this stage may show restricted aggradation that mimic the individual lobes deposited during the lowstand–early rise (see Figs 9, 11) as also reported by Hernández-Molina *et al.* (this volume). The falling stage delta complex thus exhibits areally extensive gross regression with minor intervals of aggradation. Falling stage deltas are bounded on the top of the complex by an incised valley floor, that is a major erosive surface and the major sequence boundary, and at the bases by several minor initial sequence boundaries (Table 2).

Deltas deposited during the maximum lowstand–early rise show significant aggradation in addition to progradation in comparison to those deposited during the falling stage. The lowstand–early rise complex is bounded at the base by a major sequence boundary and at the top by a transgressive surface (Table 2). However, as shown by this study, the prograding complex may also contain forced regressive wedges as a result of fifth-order cyclicity. During the sea-level fall(s) within the fourth-order lowstand-early rise, significant erosive surface(s) can occur on the top of prograding complex, as shown in the vicinity of the MP288 site (e.g. Fig. 12). The resulting erosion surface can merge updip with the base of the incised valley formed during the earlier maximum sea-level fall(s). This has implications for both the recognition and distinction of the falling stage and lowstand complexes, and the timing and development of the main incised valley and sequence boundary. Thick transgressive deposits overlie the incised valley that, in turn, overlies both the highstand and falling stage deltas. Transgressive deposits that directly overlie the lowstand-early rise deltas tend to be thin, except in cases where significant incised valley erosion occurred, as at the MP288 site. The highstand deltas are bounded at their base by maximum flooding surface and at the top by an incised valley floor, and may display aggradational and/or progradational stratal geometries (e.g. Fig. 19 and Table 2).

With the full data sets (e.g. integrated cores, logs and high-resolution seismic, preferably a 3D grid), the falling stage deltas can be distinguished from those of the maximum lowstand–early rise and highstand. However, in view of their transitional character, the distinction of the falling stage deltas is not always unambiguous. For example, within even in the high-resolution seismic dip-lines presented herein, precise differentiation between the falling stage and maximum lowstand–early rise delta complexes is often difficult (e.g. Fig. 11a). Some lobes of the falling stage closely resemble those of the lowstand–early rise, due to the composite nature of the eustatic signal, with a higher-frequency fifth-order signal superimposed on the longer term fourth-order sea-level changes. There is also the possibility that the incised valley feature(s) may develop within the lowstand-early rise deltas (as at the MP288 site), and may complicate the distinction. With only multichannel seismic data and a few well logs available, much of the updip falling-stage systems tract is likely to be classified as highstand and the downdip part as lowstand-early rise systems tract. It would appear that the ability to clearly differentiate systems tracts, and correctly interpret their relationship to sea-level changes is highly dependent on data resolution.

Conclusions

The isotope stage 5 to 1 stratigraphic interval of the Lagniappe Delta consists of four systems tracts, highstand, falling stage, maximum lowstand–early rise and transgressive. Of these, the falling stage followed by the lowstand–early rise stage systems tracts form the bulk of the Lagniappe Delta. The falling stage deltaic lobes exhibit a pronounced progradational character and are bounded at the bases of the complex by the several fifth-order initial sequence boundaries and at the top by the final fourth-order (major) sequence boundary. The maximum lowstand–early rise delta complex shows significant aggradation in addition to progradation, and is bounded at the base by the main sequence boundary and at the top by a transgressive surface, or sometimes by another significant erosive surface that may have resulted from sea-level falls within the early rise. Although the results of this study suggest that four systems tracts comprise the Lagniappe Delta, in view of the gradational nature of one systems tract into another (e.g. falling stage to lowstand–early rise), the great variability of natural depositional systems and the varying quality and resolution of different data sets, caution should be exercised in dividing the depositional sequence into four fixed 'compartments'.

Sedimentological, biostratigraphic and oxygen-isotope data utilized in Figs 3a, 5, 10a and 13 and shown in Tables 1 of our paper have been compiled from several unpublished Consortium reports by R.E. Constans, J. Crux, R. Fillon, R.T. Guerra, B. Kohl, G.M. Regan, H. Roberts, and H.W. Spero. We cannot thank enough these individuals for generating an enormous amount of data base which is indispensable for the sequence stratigraphic synthesis of the Lagniappe Delta presented here. R. Winn co-ordinated the Consortium study of the Lagniappe Delta. We thank H.W. Posamentier, D. Nummedal, H. Eichenseer, R. Mitchum, H. Roberts, B. Kohl and P.R. Vail for stimulating discussions. R. Gawthorpe and D. Hunt patiently read the first version of the manuscript and offered many helpful suggestions. D. Hunt also provided editorial assistance in the preparation of the final paper. We thank Elf Exploration/Production for providing funds to meet the costs of drafting the figures for the paper.

References

CHIOCCI, F. L. 2000. Depositional response to Quaternary fourth-order sea-level falls on the Latium margin (Tyrrhenian Sea, Italy). *This volume.*

COLEMAN, J. M. & ROBERTS, H. H. 1988*a*. Sedimentary development of the Louisiana continental shelf related to sealevel cycles, part I: Sedimentary sequences. *Geo-Marine Letters,* **8**, 63–108.

—— & —— 1988*b*. Sedimentary development of the Louisiana shelf related to sealevels: Part II: Seismic response. *Geo-Marine Letters*, **8**, 109–119.

EMILIANI, C. 1966. Paleotemperature analysis of Carribbean cores P6304–8 and P6304–9 and generalized temperature curve for the past 425,000 years. *Journal of Geology*, **74**, 109–126.

—— 1971. The last interglacial paleotemperature and chronology in deep-sea sediments. *Science*, **171**, 571–573.

ERICSON, D. B. & WOLLIN, G. 1968. Pleistocene climates and chronology in deep-sea sediments. *Science*, **162**, 1227–1234.

GARTNER, S. & EMILIANI, C. 1976. Nannofossil biostratigraphy and climatic stages of the Pleistocene. *American Association of Petroleum Geologists Bulletin*, **60**, 1562–1564.

GAWTHORPE, R. L., FRASER, A. J. & COLLIER, R. E. LL. 1994. Sequence stratigraphy in active extensional basins: implications for the interpretation of ancient basin-fills. *Marine and Petroleum Geology*, **11**, 642–58.

GREENLEE, S. M. & MOORE, T. C. 1988. Recognition and interpretation of depositional sequences and calculations of sealevel changes from stratigraphic data, offshore New Jersey and Alabama Tertiary. *In*: WILGUS, C. K., HASTINGS, B. S., KENDALL, C. G. ST. C., POSAMENTIER, H. W., ROSS, C. A. & VAN WAGONER, J. C. (eds) *Sealevel changes: An integrated approach*. Society of Economic Paleontologists and Mineralogists Special Publications, **42**, 329–356.

HELLAND-HANSEN, W. & GJELBERG, J. G. 1994. Conceptual basis and variability in sequence stratigraphy: a different perspective. *Sedimentary Geology*, **92**, 31–52.

HELLER, P. L. & PAOLA, C. 1996. Downstream changes in alluvial architecture: An exploration of controls on channel-stacking patterns. *Journal of Sedimentary Research*, **66**, 297–306.

HERNÁNDEZ-MOLINA, F. J., SOMOZA, I. & LOBO, F. 2000. Seismic stratigraphy of the Gulf of Cádiz continental shelf: a model for late Quaternary very high-resolution sequence stratigraphy and response to sea-level fall. *This volume*.

HUNT, D. & TUCKER, M. E. 1992. Stranded parasequences and the forced regressive wedge systems tract: deposition during baselevel fall. *Sedimentary Geology*, **81**, 1–9.

—— & —— 1993. Sequence stratigraphy of carbonate shelves with an example from the mid-Cretaceous (Urgonian) of southeast. *In:* POSAMENTIER, H. W., SUMMERHAYES, C. P., HAQ, B. U. & ALLEN, G. P. (eds) *Sequence Stratigraphy and Facies Associations*. International Association of Sedimentologists, Special Publications, **18**, 307–341.

—— & —— 1995. Stranded para sequences and forced regressive weddge systems tract: deposition during base-level fall- reply. *Sedimentary Geology*, **95**, 147–160.

IMBRIE, J., HAYS, J. D., MARTINSON, D. H., MCINTYRE, A., MIX, A. C., MORLEY, J. J., PISIAS, N. H., PRELL, W. L. & SHACKLETON, N. J. 1984. The orbital theory of Pleistocene climate: support from a revised chronology of the marine $\delta^{18}O$ record. *In*: BERGER, A., IMBRIE, J., KUKLA, G. & SALTZMAN, B. (eds) *Milankovitch and climate*, Part 1. Riedel, Dordecht, 269–305.

KENNETT, J. P. & HUDDLESTON, D. P. 1972. Late

Pleistocene Paleoclimatology, Western Gulf of Mexico.*Quaternary Research*, **2**, 38–69.

KINDINGER, J. L.1988. Seismic stratigraphy of the Mississippi-Alabama shelf and upper continental slope. *Marine Geology*, **83**, 79–94.

—— 1989*a*. Depositional history of the Lagniappe Delta,northern Gulf of Mexico. *Geo-Marie Letters*, 9, 59–66.

—— 1989*b*. Upper Pleistocene to Recent shelf and upper slope deposits of offshore Mississippi-Alabama. *In*: *Seventh Annual Research Conference Proceedings, Society of Economic Paleontologists and Mineralogists, Gulf Coast Section*, 163–174.

KOHL, B. 1986. Late Quaternary planktonic foraminifers from the Pigmy Basin, Gulf of Mexico Site 619, DSDP LEG 96. *Initial Reports of Deep-Sea Drilling Project*, **96**, 657–670.

KOLLA, V., POSAMENTIER, H.W. & EICHENSEER, H. 1995. Stranded parasequences and the forced regressive wedge systems tract: deposition during base-level fall–discussion. *Sedimentary Geology*, **95**, 139–145.

LOUTIT, T. S., HARDENBOL, J., VAIL, P. R. & BAUM, G. R. 1988. Condensed sections: The key to age-dating and correlation of continental margin sequences. *In*: WILGUS, C. K., HASTINGS, B. S., KENDALL, C. G. ST. C., POSAMENTIER, H. W., ROSS, C. A. & VAN WAGONER, J. C. (eds) *Sea-level changes: An integrated approach*. Society of Economic Paleontologists and Mineralogists Special Publications, **42**, 183–213.

MCMURRAY, L. S. & GAWTHORPE, R. L. G. 2000. Along-strike variability of forced regressive deposits: late Quaternary, northern Peloponnesos, Greece. *This volume*.

MITCHUM, R. M. 1977. Seismic stratigraphy and global changes of sealevel, part 1: Glossary of terms used in Seismic Stratigraphy. *In*: PAYTON, C. E. (ed.) *Seismic Stratigraphy—Applications to hydrocarbon exploration*. American Association of Petroleum Geologists Memoirs, **26**, 117–143.

——, SANGREE, J. B., VAIL, P.R. & WORNARDT, W. W. 1993. Recognising sequences and systems tracts from well-logs, seismic data and biostratigraphy: examples from the Late Cenozoic. *American Association of Petroleum Geologists Memoirs*, **58**, 163–198.

NUMMEDAL, D., RILEY, G. W., COLE, R. D. & TRAVENA, A. S. 1992. The falling stage systems tract in ramp settings (abstract). *In*: *Mesozoic of the Western Interior*. Society of Economic Paleontologists and Mineralogists Theme Meeting, 50.

——, —— & TEPLET, P. L. 1993. High-resolution sequence architecture: a chrono-stratigraphic model based on equilibrium profile studies. *In*: POSAMENTIER, H. W., SUMMERHAYES, C. P., HAQ, B. U & ALLEN, G. P. (eds) *Sequence stratigraphy and facies associations*. International Association of Sedimentologists, Special Publications, **18**, 55–68.

PLINT, A. G. 1988. Global eustasy and the Eocene sequences in the Hampshire Basin, England. *Basin Research*, **1**, 11–22.

—— & NUMMEDAL, D. 2000. The falling stage systems tract: recognition and importance in sequence stratigraphic analysis.*This volume*.

POSAMENTIER, H. W. & ALLEN, G. P. 1993*a*. Variability of the sequence stratigraphic model: effects of local basin factors. *Sedimentary Geology*, **86**, 91–109.

—— & —— 1993*b*. Siliciclastic sequence stratigraphic patterns in foreland basins. *Geology*, **21**, 455–458.

—— & MORRIS, W. S. 2000. Aspects of stratal architecture of forced regressive deposits: late Quaternary, northern Peloponnesos, Greece. *This volume*.

—— & VAIL, P.R. 1988. Eustatic controls on clastic deposition II—Sequence and systems tract models. *In:* WILGUS, C. K., HASTINGS, B. S., KENDALL, C. G. ST. C., POSAMENTIER, H. W., ROSS, C. A. & VAN WAGONER, J. C. (eds) *Sealevel change-An integrated approach*. Society of Economic Paleontologists and Mineralogist Special Publication, **42**, 125–154.

——, ALLEN, G. P., JAMES, D. P. & TESSON, M. 1992. Forced regressions in a sequence stratigraphic framework: concepts, examples and exploration significance.*American Association of Petroleum Geologists Bulletin*, **76**, 1687–1709.

ROBERTS, H., FILLON, R., KOHL, B. & SYDOW, J. 1997. Reinterpretation of a Mobile River shelf edge delta (Lagniappe Delta): the importance of climate and the loop current (abstract). *In*: *American Association of Petroleum Geologists Convention*, 98.

SCHAFER, B. L. 1987. The potential of calcareous nannofossils for recognising Plio-Pleistocene climatic cycles and sequence boundaries on the shelf. *In*: *Gulf Coast Section Society of Economic Paleontologists and Mineralogists Foundation, Eighth Annual Research Conference*, 142–145.

SUTER, J. R. & BERRYHILL JR., H. L. 1985. Late Quaternary shelf margin deltas, Northwest Gulf of Mexico. *American Association of Petroleum Geologists Bulletin*, **69**, 77–91.

SYDOW, J. & ROBERTS, H. H. 1994. Stratigraphic framework of a Late Pleistocene shelf-edge delta, Northeast Gulf of Mexico. *American Association of Petroleum Geologists Bulletin*, **78**, 1276–1312.

TRINCARDI, F. & CORREGGIARI, A. 2000. Quaternary forced-regression deposits in the Adriatic Basin and the record of composite sea-level cycles. *This volume*.

VAIL, P. R. 1987. *Seimic stratigraphy interpretation using sequence stratigraphy: part 1—seismic stratigraphy interpretation procedure*. AAPG Studies in Geology, **27**.

VAN WAGONER, J. C. 1995. Overview of sequence stratigraphy of foreland basin deposits. *In*: VAN WAGONER, J. C. & BERTRAM, G. T. (eds) *Sequence Stratigraphy of Foreland Basin Deposits*. American Association of Petroleum Geologists Memoirs, **64**, ix–xxi.

WINN, R. D., ROBERTS, H. H., KOHL, B., FILLON, R. H., BOUMA, A. & CONSTANS, R. E. 1995. Latest Quaternary deposition on the outershelf, Northern Gulf of Mexico: Facies and sequence stratigraphy from the Mainpass Block 303 shallow core. *Geological Society of America Bulletin*, **107**, 851–866.

WINKER, C. D. 1982, Cenozoic shelf margins, North east Gulf of Mexico: *Gulf Coast Asssociation of Geological Societies Transactions*, **32**, 427–448.

Seismic stratigraphy of the Gulf of Cádiz continental shelf: a model for Late Quaternary very high-resolution sequence stratigraphy and response to sea-level fall

F. J. HERNÁNDEZ-MOLINA[1], L. SOMOZA[2] & F. LOBO[1]

[1] *Facultad de Ciencias del Mar, Universidad de Cádiz (UCA), Polígono del Río San Pedro s/n, 11510 Puerto Real, Cádiz, Spain*

[2] *Geologia Marina, Instituto Tecnologico Geominero de España (ITGE), Rios Rosas 23, 28003 Madrid, Spain*

Abstract: Single-channel, very high-resolution seismic profiles allow detailed study of the Late Quaternary stratigraphic architecture of the Gulf of Cádiz continental margin, Southern Spain. The Late Quaternary stratigraphy of this area comprises fourth-order Type 1 composite depositional sequences, generated by asymmetric relative sea-level changes of 100–110 ka duration. The composite fourth-order sequences consist of forced regressive, lowstand, transgressive and highstand systems tracts. Volumetrically, the forced regressive and lowstand systems tracts are the most important components. The fourth-order composite sequences are themselves comprised of composite fifth-order sequences formed in response to asymmetric relative sea-level changes with a duration of 22–23 ka. Sediments within the forced regressive and lowstand systems tracts dominate the 5th-order sequences; their transgressive and highstand deposits are either (i) perched above present-day sea-level and so not recorded in marine seismic data, (ii) restricted to outer–mid-shelf positions, or (iii) may be absent from the shelf altogether at the resolution of this study (e.g.<0.5 m thick). The fifth-order sea-level falls were themselves modulated by minor cycles, generating very high-frequency (sixth-order) sequences. These very high-order sequences are recognized for the last 80 ka BP, and their development is attributed to asymmetric relative sea-level cycles operating on time scales of: 10–15 ka (Heinrich events), 4–4.5 ka. (P cycles), 2.3–0.97 ka. (Dansgaard–Oeschger oscillations h cycles) and 500–50 a. (c cycles). We have developed a depositional model that accounts for the very high-frequency hierarchy of Late Quaternary depositional sequences observed in the Gulf of Cadiz marine seismic record and incorporates the age of well-constrained highstand coastal deposits that are exposed along the southern Iberian coastline. The model developed serves to illustrate the evolution and importance of depositional systems during falling relative sea-level and forced regression. Development of the forced regressive systems tract appears to be particularly significant within Quaternary strata. This is because the Quaternary was strongly influenced by a high-amplitude, high-frequency glacioeustatic signal characterized by rapid sea-level rises, very short highstands, and gradual relatively long-term sea-level falls suggesting that forced regressive deposits are likely to predominate in continental margin successions subject to low rates of subsidence.

During the Late Pleistocene, changes in climate and relative sea-level were closely linked, with warm periods related to relative sea-level rise and highstand, and cooling-cold periods related to relative sea-level fall and lowstand (Shackleton 1987; Bell & Walker 1992; Dawson 1992). In the Late Quaternary, from 920 ka up to the present, both climatic variation and sea-level change were closely linked to eccentricity cycles of about 100–125 ka, precession cycles of about 20 ka, and less important obliquity cycles of about 40 ka (Hays *et al.* 1976; Broecker 1984; Lowe & Walker 1984; Ruddiman *et al.* 1986; Martison *et al.* 1987, Ruddiman & Raymo 1988; Bell & Walker 1992; Berger & Wefer 1992; Dawson 1992). This hierarchy of cycles acted in concert to form a dominant high-amplitude, high-frequency fourth-order 100–125 ka glacioeustatic signal, with an amplitude of approximately 120 m that was modulated by higher frequency cycles.

The application of sequence stratigraphic concepts and methodology to Quaternary marine deposits has revealed high-frequency changes in the stratigraphic architecture, position, internal stacking patterns and facies of strata deposited in shelf, slope and basin-floor environments. The detailed analysis of Quaternary stratigraphy in marine environments is frequently undertaken by means of high resolution and low-penetration seismic, supported by detailed cores studies that allow facies and stratigraphy to be analyzed in

From: HUNT, D. & GAWTHORPE, R. L. (eds) *Sedimentary Responses to Forced Regressions*. Geological Society, London, Special Publications, **172**, 329–362. 1-86239-063-0/00/$15.00

considerable detail and dated with great accuracy (e.g. Suter *et al.* 1987; Boyd *et al.* 1989; Tesson *et al.* 1990*a*, *b*, 1993; Trincardi & Field 1991; Swift *et al.* 1991; Thomas & Anderson 1991; Gensous & Tesson 1992; Gensous *et al.* 1993; Belloti *et al.* 1994; Chiocci 1994; Hernández-Molina *et al.* 1994,1996; Saito 1994; Sydow & Roberts 1994; Lobo 1995; Tesson & Allen 1995; Torres *et al.* 1995; De Batist & Jacobs 1996; Fernandez-Salas *et al.* 1996; Somoza *et al.* 1994, 1996, 1997, 1998; Tortora 1996; papers in this volume by Chiocci, Kolla *et al.* and Trincardi & Corregiari). The increasing high-resolution knowledge of Late Quaternary stratigraphy, and its relationship to sea-level changes has led to the emergence of data that can have an important bearing on idealized sequence stratigraphic models and concepts (e.g. Sydow & Roberts 1994, papers in this volume by Chiocci, Kolla *et al.* and Trincardi & Corregairi). This paper examines the relationship between the hierarchical Late Pleistocene and Holocene stratigraphy of the Gulf of Cádiz and; (i) high-order sequences deposited in shelf and slope environments, (ii) their relationship to high-frequency cyclicity of sea-level change, and (iii) implications for sequence stratigraphic models and concepts.

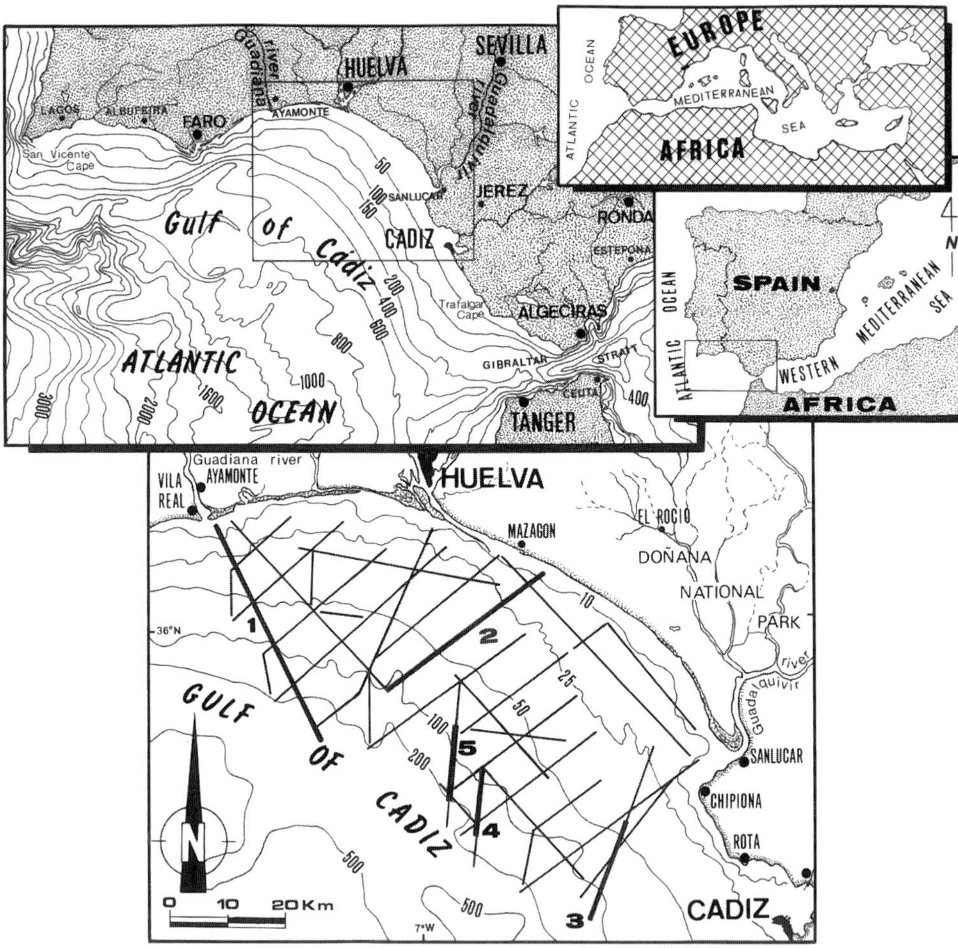

Fig. 1. Sketch maps showing the location of the Gulf of Cadiz continental margin, close to the straits of Gibraltar, separating the Atlantic Ocean and Mediterranean Sea, and the main bathymetric geographical features of the study area. Note the position of the main Guadiana and Guadalquivir rivers that supply sediment to the margin from the adjacent Betic Mountains. The lower map shows locations of high-resolution reflection seismic lines. Line 1 shows the position of Fig. 2; Line 2, the position of Fig. 3; Lines 3, 4 and 5, the position of Fig. 4. The modern shelf break occurs between water depths of 100 and 135 m.

Geological setting

In this paper we present a seismic and sequence stratigraphic analysis based on very high resolution seismic profiles from the continental margin of the Gulf of Cádiz between the Guadiana and Guadalquivir rivers, southern Spain (Fig. 1). The Gulf of Cádiz continental margin has a significant Upper Miocene to Holocene sedimentary cover (up to several kilometres thick). The succession was influenced by evolution of the adjacent Alpine Betic range, so that Cenozoic basin sedimentation occurred under conditions of variable subsidence rates and high sediment supply, the latter related to several important rivers draining the Betic Mountains (e.g. Mélières 1974; Malod & Mougenot 1979; Malod 1982; Martinez del Olmo *et al.* 1996; Maldonado & Nelson 1999; Rodero *et al.* 1999). The Quaternary marine stratigraphy of the Cadiz continental margin is high-energy in nature, and this is attributed to the dynamic interaction of the Atlantic and Mediterranean water masses in proximity to the Strait of Gibraltar (Devaux 1985; Maldonado 1992; Maldonado & Nelson 1988; Nelson *et al.* 1993).

Methods

We have examined Late Quaternary deposits of the Gulf of Cádiz primarily by means of high-resolution seismic reflection profiles, using Geopulse (300 J) and a 3.5 kHz mud penetrator. The main single-channel Geopulse (300 J) seismic source was used to produce a grid of profiles, 800 km in total length, across the continental shelf of the Gulf of Cádiz (Fig. 1) during the cruises Golca 93 (1993) and Golca 94 (1994) on board the *Odon de Buen* of the Spanish Oceanographic Institute. The length and high quality of the lines has allowed us to reconstruct a detailed stratigraphy for the Late Quaternary with great lateral continuity. In addition, the very high vertical resolution of the data (about 0.5 m) allows definition of the internal geometry of sediment bodies in considerable detail.

The interpretation of the seismic data was carried out in two steps. Firstly, seismic units on the shelf and upper slope were identified by studying reflection terminations, internal reflector configuration, reflection character and the external shape and position of the units, in accordance with general seismic stratigraphic methodology (Mitchum *et al.* 1977; Payton 1977; Vail *et al.* 1977; Brown & Fisher 1980; Sheriff 1980; McQuillin *et al.* 1984; Hardage 1987; Vail 1987; Cross & Lessenger 1988). Secondly, sequence stratigraphic analysis was carried out following the main principles established over the past two decades (Vail *et al.* 1977; Vail 1987; Haq *et al.* 1987; Van Wagoner *et al.* 1987, 1988; Posamentier & Vail 1988; Posamentier *et al.* 1988; Mitchum & Van Wagoner 1991; Vail *et al.* 1991; Posamentier *et al.* 1992, 1993; Posamentier & Allen 1993; Christie-Blick & Driscoll 1995; Emery & Myers 1996). We have, however, modified these basic concepts by incorporating the forced regressive systems tract specific to times of falling relative sea-level in the sense of Hunt & Tucker (1992, 1995), equivalent to the falling stage systems tract of Plint & Nummedal (this volume).

We have incorporated a fourth forced regressive systems tract because we believe that deposition during relative sea-level fall is particularly important in Quaternary stratigraphy. This is because the dominant high-amplitude fourth-order glacioeustatic signal during this time was characterized by rapid sea-level rises, very short highstands, gradual sea-level falls and short lowstands. On continental margins undergoing subsidence rates of less than approximately 1.25 m ka^{-1}, and subject to high-frequency, high-amplitude glacioeustasy, forced regressive deposits are likely to be important (Gawthorpe *et al.* 1994). In terms of sequence duration, we have taken sequences of *c.* 100 ka to be fourth-order and those of 20–23 ka to be fifth-order (following Mitchum & Van Wagoner 1991; Bard *et al.* 1992).

An important constraint on the interpretations presented is a lack of direct absolute dating of the offshore marine succession. This requires us to make correlation between the depositional sequences described in this paper and the following.

(a) Highstand coastal deposits exposed along the southern Iberian coastline that are related to the highest position of sea-level at isotopic stages 5e, 5c and 5a; the age of these deposits is well constrained by Th/U dating (Hillaire-Marcel *et al.* 1986; Somoza *et al.* 1987, 1991; Goy *et al.* 1993; Zazo *et al.* 1993, 1996; Zazo 1999). The highstand deposits generated by the last and present interglacial stages in southern Iberia have been studied in detail and are very well known (Lario *et al.* 1993, 1995, 1996; Zazo *et al.* 1993, 1996; Lario 1996; Zazo 1999). The age, vertical structure and correlation of these deposits constrains the updip portion of the stratigraphic model presented here with some precision, especially the correlation and age of seismic units 5–9 (Figs 2, 3).

(b) Similar deposits found on the southern continental shelf of the Gulf of Cádiz and the Alboran Sea shelf (Lobo 1995; Hernández-Molina *et al.* 1994, 1996, 1998; Ercilla & Alonso 1996; Fernandez-Salas *et al.* 1996; Somoza *et al.* 1997; Roque 1998; Rodero *et al.* 1999; Lobo *et al.* in press), that are correlated with seismic units 11–14 recognized here, allowing us to constrain their age.

(c) The high resolution stratigraphy of other Mediterranean shelves (Aloïsi 1986; Maldonado 1990; Tesson *et al.* 1990*b*, 1993; Gensous *et al.* 1993; Chiocci 1994, this volume; Torres *et al.* 1995; Tortora 1996; Somoza *et al.* 1998; Trincardi & Corregiari this volume).

(d) Late Quaternary fourth- and fifth-order sea-level changes taken from published literature.

A key part of this study was to establish criteria to differentiate forced regressive and lowstand systems tracts for each depositional sequence. In all forced regressive and lowstand progradational wedges present on the shelf it is possible to differentiate two minor progradational sedimentary units separated by a minor discontinuity close to the preceding shelf break. The first unit developed on the antecedent inner and middle continental shelf, the second formed

Table 1. *Type of reflection terminations, pattern configuration, geometrical shape and margin position of the seismic units and subunits of the Gulf of Cadiz continental margin*

SEQUENCE STRATIGRAPHY ANALYSIS			SEISMIC STRATIGRAPHY ANALYSIS					
Depositional sequence	Subsequence	Systems tract	SEISMIC UNIT	SUBUNIT	TYPE OF REFLECTION TERMINATIONS	REFLECTION PATTERN CONFIGURATION	GEOMETRICAL SHAPE	MARGIN POSITION
DEPOSITIONAL SEQUENCE -II-	II4	HD	14		T:Concordance Present bottom of the sea floor on the slope B:Onlap	Progradational-divergent	Wedge	Slope
		HD	13	b	T:Concordance Present bottom of the sea floor on the shelf B:Onlap	Sigmoidal to oblique progradational and aggradational	Wedge	Littoral and shelf
				a	T:Concordance B:Onlap	Sigmoidal to parallel progradational and aggradational	Wedge	Littoral and inner shelf
		TD	12	c	T:Erosional truncation landward & Toplap seaward B:Downlap	Oblique to parallel progradational	Lobate to bank	Shelf
				b	T:Concordance landward to toplap latteraly seaward B:Onlap	Weak with aggradational and parallel	Lens	
				a	T:Erosional truncation landward & Toplap seaward B:Downlap	Oblique to parallel progradational	Lobate to bank	
		LD	11	b	T:Erosional truncation landward & Toplap seaward B Downlap	Oblique progradational and aggradational	Wedge	Shelf edge
				a	T:Erosional truncation landward & Toplap seaward B Downlap	Sigmoidal to oblique progradational and aggradational	wedge	
		FRWD	10	a to k: a c e g i k	T:Erosional truncation landward & Toplap seaward B:Downlap	oblique progradational	Wedge	Outer & Shelf edge
				a to k: b d f h j	T:Erosional truncation landward, concordance & Toplap seaward B: Coastal onlap & distal Downlap	Sigmoidal-oblique aggradational and progradational	Wedge	
	II3	SMD	9	b	T:Erosional truncation landward & Toplap seaward B:Downlap	Oblique progradational	Lens	Outer & middle shelf
				a	T:Erosional truncation landward & Toplap seaward B:Downlap	Shingled to parallel progradational	Lobate to lens	
		LD	8	b	T:Erosional truncation landward & Toplap seaward B:Downlap	Oblique progradational	Wedge	Shelf edge
		FRW		a	T:Erosional truncation landward & Toplap seaward B:Downlap	Sigmoidal progradational	Wedge	Outer & shelf edge
	II2	LD	7	b	T:Erosional truncation landward & Toplap seaward B:Downlap	Oblique progradational	Wedge	Shelf edge
		FRW		a	T:Erosional Truncation landward & Toplap seaward B:Downlap	Sigmoidal progradational	Wedge	Outer & shelf edge
	II1	LD	6	b	T:Erosional Truncation landward & Toplap seaward B:Downlap	Weak oblique progradational	Wedge	Shelf edge
		FRW		a	T:Erosional truncation landward & Toplap seaward B:Downlap	Weak sigmoidal-oblique progradational	Wedge	Outer & shelf edge
DEPOSITIONAL SEQUENCE -I-	I2	HD	5		T:Concordance landward to toplap latteraly seaward B:Onlap	Sigmoidal progradational	Wedge	Inner & outer shelf
		TD	4		T:Concordance B:Onlap	Weak aggradational to sigmoidal progradational	Sheet	Shelf
		LD	3	b	T:Erosional truncation landward & Toplap seaward B Downlap	Weak oblique progradational	Wedge	Shelf edge
		FRW		a	T:Erosional truncation landward & Toplap seaward B:Downlap	Weak sigmoidal-oblique progradational	Wedge	Outer & shelf edge
	I1	SMD	2		T:Erosional truncation landward & Toplap seaward B:Downlap	Weak oblique and/or parallel progradational	Lobate to lens	Shelf
		LD	1	b	T:Erosional truncation landward & Toplap seaward B:Downlap	Weak oblique progradational	Wedge	Shelf edge
		FRW		a	T:Erosional truncation landward & Toplap seaward B:Downlap	Weak sigmoidal-oblique progradational	Wedge	Outer & shelf edge

on the previous outer shelf and upper slope. The minor discontinuity most likely represents the boundary between the forced regressive and lowstand systems tracts. Although it is not always possible to determine this minor discontinuity, a change in the internal configuration, from progradational to aggradational, can nevertheless be observed. A well-developed horizon of slumps and slides also appears to mark the boundary present between the two stratal units, and the base of this horizon is considered to mark the boundary between the last forced regressive and lowstand deposits. A similar relationship between forced regressive and lowstand wedges and slump deposits is reported from the Gulf Coast, USA, by Sydow & Roberts (1994) and Kolla *et al.* (this volume).

Finally, in addition to recognizing the existence of the highstand systems tract developed on the shelf, in the sense of Vail *et al.* (1991), we have also recognized the existence of this systems tract as an isolated sedimentary wedge positioned on the middle/outer continental shelf with an aggradational–progradational stacking pattern. Previously, this sedimentary body was interpreted as a 'shelf margin deposit' (Somoza *et al.* 1997). However, we now recognize these packages to represent a 'relative' highstand developed when only part of the shelf was flooded, as a result of the composite, modulated nature of the forcing relative sea-level curve. As such, in this paper we have reinterpreted the shelf margin deposits to represent a 'relative' highstand systems tract.

High-resolution seismic stratigraphy

Seismic stratigraphic analysis has enabled us to define 14 seismic units that can be recognized on *all parts* of the Cadiz continental shelf and upper slope. These seismic units, illustrated in Figs 2 & 3 and summarized in Table 1, are described below in order of their deposition prior to discussing specific aspects of sedimentation during sea-level fall. The seismic lines presented from the Cadiz shelf illustrate the representative stratal architecture from areas characterized by differing subsidence and sedimentation rates.

Seismic unit 1

This progradational wedge-shaped unit is present on the outer shelf and shelf edge and is comprised of two minor subunits. Subunit 1a is a sigmoidal to oblique progradational wedge-shaped unit with an upper erosional surface updip passing seaward to a toplap surface (Fig. 2). It is recognized on the outer shelf and shelf edge, and interpreted to represent a forced regressive systems tract. Subunit 1b is a wedge-shaped seismic subunit that sits basinward and over 1a. It is characterized by oblique progradation and local downlaps onto unit 1a (Fig. 2). The upper boundary of unit 1b is associated with erosional truncation landward and toplap basinward. The downlap onto unit 1a indicates an increase in accommodation space between units 1a and 1b so that seismic subunit 1b is interpreted as a lowstand systems tract that prograded the shelf margin.

Seismic unit 2

As can be seen in Fig. 2, this unit is positioned on the middle–outer shelf updip of unit 1, onto which it downlaps (Fig. 2). Erosional truncation in updip locations at the top of unit 2 passes downdip to a toplap surface. It has a lobate- to lens-shaped external form and is internally characterized by weak parallel and/or oblique progradational reflectors. This shelf margin deposit is separated from the underlying package by a significant flooding surface. It would appear that flooding and transgression was non-accretionary, developing a transgressive ravinement surface (Rs) opposed to a systems tract. Seismic unit 2 is attributed to a 'relative' highstand systems tract when the shelf was only partially flooded at the highest stance of relative sea-level.

Seismic unit 3

This is a shelf edge progradational seismic unit that is comprised of two minor but distinctive subunits. In Fig. 2, subunits 3a and 3b can be observed to sit basinward and below seismic unit 2, and to downlap onto seismic unit 1. The upper surface of both subunits are associated with erosional truncation in a landward direction that passes downward and basinward to a toplap surface. Seismic unit 3 represents a basinward facies shift associated with incision of the shelf. Locally, both subunits are cut through by incised valley fills (Fig. 2). Subunit 3a is situated on the outer shelf and shelf edge where this sigmoidal to oblique progradational wedge-shaped subunit displays low relief clinoforms and is interpreted to represent a forced regressive systems tract. Subunit 3b is a wedge-shaped subunit that sits right at the shelf-edge where it is associated with relatively steep, high-relief clinoforms that display oblique clinoform progradation. Seismic unit 3b is interpreted as a lowstand systems tract.

Seismic unit 4

Seismic unit 4 is a sheet-like deposit that is best developed on the shelf, but does extend to the shelf-edge (Fig. 2). It displays a basal onlap relationship and aggradational to sigmoidal reflectors. This package has a concordant upper surface and is interpreted to represent an accretionary transgressive systems tract (TST) on the basis of the landward shift of the facies it records from seismic unit 3.

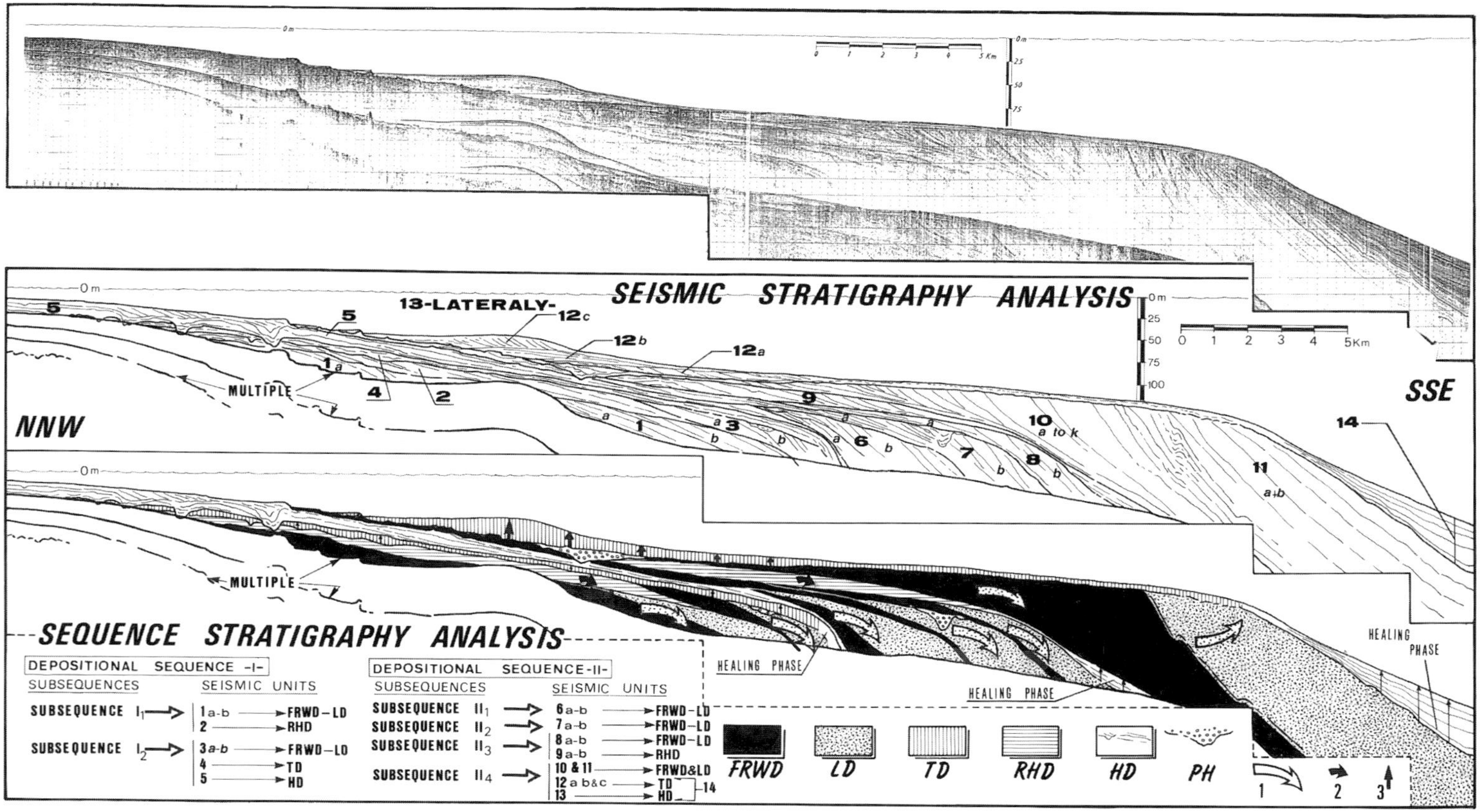

Fig. 2. High-resolution seismic reflection profile 1 (upper) as located on the shelf immediately in front of the Guadiana River (Fig. 1), with accompanying line drawing (middle) showing interpretation of the 14 seismic units identified, and interpretation of the fifth-order sequences and systems tracts that comprise two composite type 1 sequences (lower). The Late Quaternary stratigraphy of the margin is comprised of two composite fourth-order sequences, depositional sequence I and depositional sequence II, that are built from two and four subsequences respectively, as shown on the bottom left of the diagram. Tones in the lower part of the figure show the systems tract components differentiated for each of the six subsequences differentiated. It is apparent that the main phases of outbuilding of the shelf margin are associated with forced regressive and lowstand systems tracts. Also notice the main Healing phase on the upper slope in relation with the main transgressive and highstand sea-level stage. See text for detailed description, interpretation and discussion of the seismic units.

The key to abbreviations is as follows. FRWD: forced regressive systems tract; LD: lowstands systems tract; TD: transgressive systems tract; SMD: 'relative' highstand systems tract; HD: highstand systems tract; PH: palaeo-channels on the shelf, indicators of active fluvial incision. Symbols for arrows as located on the bottom right of the diagram is: 1, major progradation; 2, minor progradation and 3, aggradation.

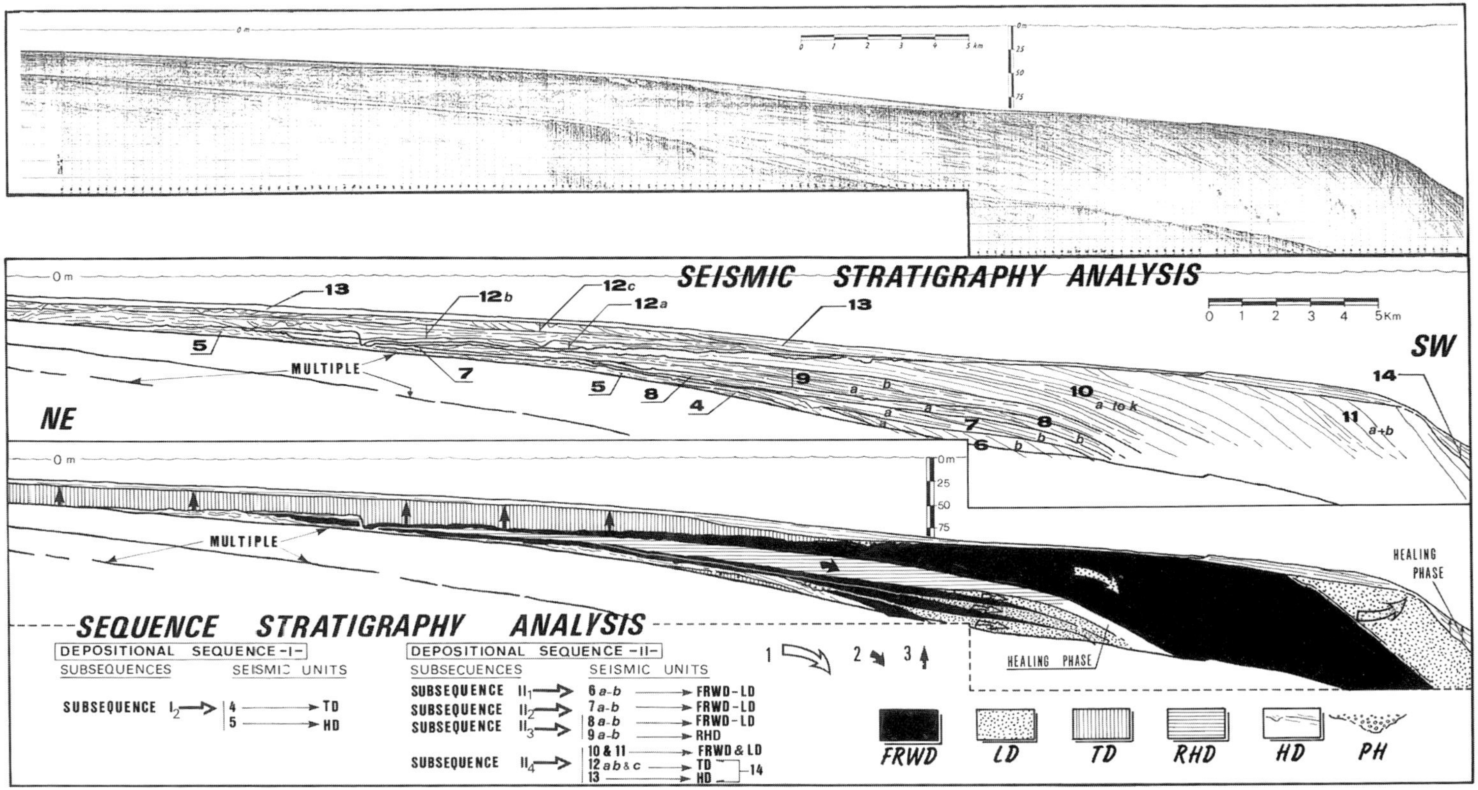

Fig. 3. High-resolution seismic reflection profile 2, with accompanying line drawing (middle) showing interpretation of the seismic units identified, and interpretation of the fifth-order sequences and systems tracts that comprise two composite type 1 sequences (lower). This profile is located some 60–20 km along strike from Fig. 2, at its landward and basinward ends respectively (see Fig. 1 for location). The line is located between the main sediment sources to the Cadiz margin. Despite the considerable distance between the lines and from the main sediment input sites, the stratigraphy is here remarkably similar (Fig. 1). See text for detailed description and interpretation of the seismic and sequence stratigraphy. The key to abbreviations as for Fig. 2.

Seismic unit 5

This unit occurs across the inner and outer shelf, and has a basal onlap surface and an upper surface that is generally concordant up dip, but is associated with toplap in seaward locations (Figs 2, 3). Externally, this unit is wedge-shaped and characterized by sigmoidal progradation on the shelf, passing to oblique progradation basinward. Seismic unit 5 is interpreted to represent a highstand systems tract on the basis of its position on the shelf, external form, basal and upper relationships and internal reflector configuration. We correlate this seismic unit with the Tyrrhenian II highstand deposits exposed on the southern Iberian coastline, deposited 140–120 ka bp (during isotopic stage 5e, e.g. Hillaire-Marcel *et al.* 1986; Somoza *et al.* 1987; Lario *et al.* 1993, 1995, 1996; Zazo *et al.* 1993; Lario 1996; Zazo 1999).

Seismic unit 6

As is evident from Figs 2 and 3, seismic unit 6 sits in a lower and more basinward position than units 4 and 5 that preceded it. It is an outer-shelf to shelf-edge unit that is composed of two subunits. The lower of these subunits 6a, is developed across the outermost shelf (Fig. 2) and is a wedge-shaped progradational unit with sigmoidal-to-oblique reflectors. It downlaps onto seismic unit 5 and has an upper toplap surface that becomes erosional updip. This unit is interpreted as a forced regressive systems tract on the basis of the abrupt basinward shift of the depocentre, its position on the shelf, relationship to unit 5 and its lack of topset preservation. Subunit 6b is an oblique progradational seismic wedge-shaped package that locally downlaps onto the eroded top of unit 6a (Fig. 2). The upper surface of unit 6a is characterized by erosional truncation in its updip reaches, passing to a toplap surface downdip. The position of 6b over and downlapping onto 6a indicates an increase of accommodation so that subunit 6b is interpreted as a lowstand systems tract developed on the shelf edge.

Seismic unit 7

Seismic unit 7 is developed across the outer shelf to the shelf edge (Figs 2, 3) and is divided into two subunits. Subunit 7a occurs on the outer shelf and shelf edge. On the outer shelf it is a relatively thin unit characterized by low-relief clinoforms that downlap onto units 5 and 6 (Figs 2, 3). The extent of this wedge-shaped unit across the shelf suggests a relative base-level rise following deposition of unit 6 for which no deposits are apparent, suggesting that transgression was non-accretionary. No highstand deposits are recognized in the marine seismic record, instead we interpret the narrow strip of coastline deposits associated with the Tyrrhenian III episode 105–90 ka BP to represent the highstand deposits to this sequence (e.g. isotopic stage 5c; Hillaire-Marcel *et al.* 1986; Somoza *et al.* 1987; Lario *et al.* 1993, 1995, 1996; Zazo *et al.* 1993; Lario 1996; Zazo 1999). Accordingly, the highstand present along the coastline is updip of the seismic data considered here. Accordingly, we interpret the sigmoidal to oblique progradational reflectors of seismic unit 7 to represent a forced regressive systems tract associated with a significant basinward facies shift. This forced regressive systems tract is developed on the outer shelf and at the shelf edge. It is physically detached from the antecedent highstand shoreline deposits. By way of contrast, subunit 7b is more local in extent, developed at the shelf-edge and cut by an incised valley fill (Fig. 2). It is typified by an oblique progradational reflector configuration and is wedge-shaped, but displays some along strike variability (Figs 2, 3). In Fig. 3, low-relief reflectors on the outer shelf are seen to have contributed little to shelf-edge progradation, whereas in Fig. 2 high-relief, relatively high-angle reflectors prograded the margin up to 2 km. Subunit 7b is interpreted to represent a lowstand systems tract developed on the shelf edge formed during a low stillstand of relative sea-level, as it typically lacks an aggradational component in its stacking pattern.

Seismic unit 8

Seismic unit 8 shows striking similarities in terms of its position and internal stacking patterns to unit 7 (Figs 2, 3); it is progradational and has limited areal extent, localized to the outer shelf and shelf edge. It is composed of two minor subunits. Subunit 8a has a lower downlap and onlap surface (Figs 2, 3). It is wedge-shaped and characterized by sigmoidal to oblique progradational reflectors. The greater updip extent of unit 8a in comparison to 7b, indicates that there was a base-level rise between these two units. Transgression is represented by a single surface and was non-accretionary in nature (Figs 2, 3). We interpret the basal surface of seismic unit 8 as a maximum flooding surface that underlies a narrow tract of highstand shoreline sediments that are exposed along the Iberian coastline deposited during the Tyrrhenian IV episode 80–70 ka BP during isotopic stage 5a (Lario *et al.* 1993, 1995, 1996; Lario 1996; Zazo 1999). These sediments pass downdip to a condensed section in the marine realm, and are represented by the surface at the base of seismic unit 8 in the data from Cadiz the continental shelf. Accordingly, we interpret seismic subunit 8a to represent a forced regressive systems tract developed on the outer shelf and shelf edge, physically detached from shoreline deposits of the previous highstand. Subunit 8b has at its base a locally erosional surface that distinguishes it from 8a (Fig. 3). Subunit 8b also locally downlaps this surface and seismic unit 7. It is generally a progradational wedge-shaped subunit with oblique clinoforms. Subunit 8b displays similar in along-strike variability in terms of its position on the shelf and shelf-edge progradation to unit 7 (Figs 2, 3). We interpret subunit 8b to represent a lowstand systems tract developed on the shelf edge at a low-stillstand of relative sea-level.

Seismic unit 9

Seismic unit 9 sits in a more landward and higher position in comparison to the shelf-edge location of unit 8. It is located on the middle and outer shelf and is locally split into two subunits (e.g. Fig. 3). Subunit 9a is a lobate to lenticular package with a lower downlap surface and an upper surfaces that is erosional updip and displays toplap downdip. Internally is has shingled to parallel progradational seismic reflectors. When distinguished subunit 9b is a lens-shaped unit characterized by oblique reflectors with a local aggradational component (Fig. 3). It progrades further basinward than its precursor, unit 9a, onto which it downlaps. Where distinguished (e.g. Fig. 3), units 9a and 9b can be interpreted to represent relative highstand systems tracts, overlying a transgressive ravinement surface developed when the shelf was only flooded across its outer part (middle and outer shelves) at the highest stance of relative sea-level. Along strike variability in seismic unit 9 is related to differences of sediment supply.

Seismic unit 10

This seismic unit is a sigmoidal to oblique progradational wedge-shaped unit present on the outer shelf and shelf edge (Figs 2, 3). It is bounded by a lower downlap surface, whereas erosional truncation in updip locations and toplap in downdip locations characterizes the upper surface of this package. Detailed examination allows distinction of 11 minor seismic wedge-shaped subunits within unit 10 (a to k, Fig. 4) that show a complex internal structure. It is possible to recognize subunits which display a purely progradational oblique reflector arrangement (e.g. subunits a, c, e, g, i and k, Fig. 4) or progradational and aggradational stratal pattern (e.g. sigmoidal–oblique clinoform packages in subunits b, d, f, h and j, Fig. 4). Theses subunits (a to k) show significant along-strike variability that can be related to changes in the rates of shelf margin subsidence or uplift (Figs 4, 5). However, at this level of detail no greater internal detail of stacking pattern can be recognized in any of the subunits within seismic unit 10. Seismic unit 10 is interpreted as a forced regressive systems tract associated with downstepping and significant progradation of the shelf margin.

Seismic unit 11

This wedge-shaped shelf edge seismic unit sits basinward of unit 10 and displays an upper toplap surface in its outer reaches that passes updip to an erosional truncation surface (Figs 2, 3). Its lower bounding surface is locally erosive and may show evidence of slumping (Figs 2, 3), although soft sediment deformation is also relatively common within this unit (Fig. 4). In the southern part of the Gulf of Cádiz continental shelf unit 11 is comprised of two minor subunits (Lobo 1995). Subunit 11a is a sigmoidal to oblique progradational and aggradational seismic package whereas subunit 11b is an oblique progradational and aggradational seismic unit. In most cases unit 11 sits basinward of unit 10 and is interpreted to have been deposited during a low stillstand of relative sea-level. However, locally seismic unit 11 downlaps onto the upper toplap surface of unit 10 (e.g. see Figs 4, 6), a relationship that indicates a relative sea-level rise occurred between deposition of seismic units 10 and 11. Along strike variability in unit 11 is attributed to differential subsidence/uplift along the continental margin (Figs 4, 5). Subunits 11a and 11b are correlated with the P1 and P2 sedimentary bodies of Hernández-Molina *et al.* (1994), indicating an age of approximately 20–15 ka BP for seismic unit 11.

Seismic unit 12

Seismic unit 12 can be subdivided into 3 subunits (a–c), all of which occur on the shelf. Subunit 12a is a lobate to bank-shaped oblique to parallel progradational stratal package, subunit 12b displays a weak aggradational and parallel reflectors is lens-shaped and positioned on the mid-shelf, and subunit 12c is characterized by oblique to parallel reflectors and is lobate to bank-shaped. Units 12a–c onlap the relief of

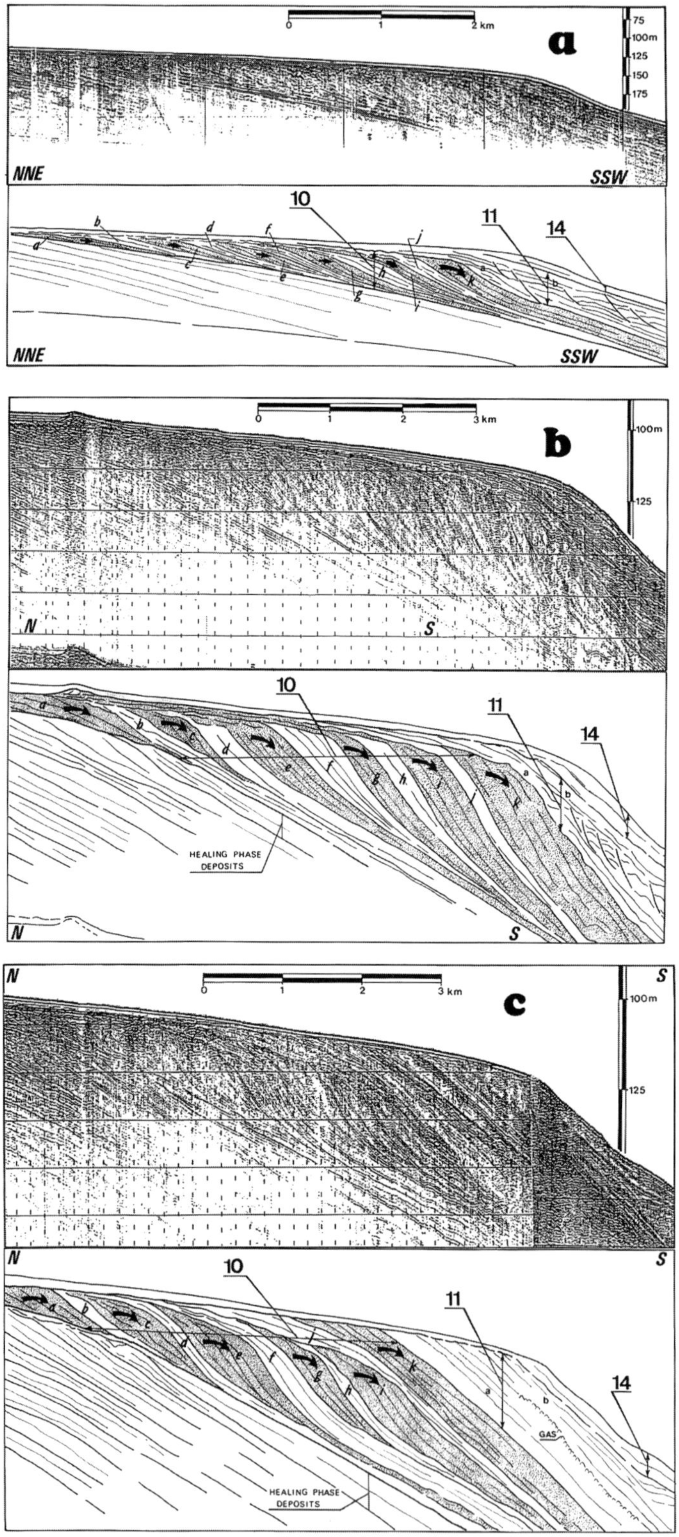
a
0 1 2 km
75
100m
125
150
175
NNE
SSW
10
11
14
b
0 1 2 3 km
100m
125
N
S
HEALING PHASE DEPOSITS
c
GAS

older units landward, and generally record the landward-stepping of the depositional margin from seismic unit 11. Although locally significant progradation is observed within 12c close to major sediment input points, as a whole seismic unit 12 is interpreted to represent an accretionary transgressive systems tract. Subunit 12c is correlated with the P3 sedimentary body of Hernández-Molina *et al.* (1994) deposited during to so called 'Young Dryas' event 13–10 ka BP.

Seismic unit 13

An aggradational and parallel to wedge-shaped unit found the along the modern coastline and shelf. It is characterized by an onlapping baselap reflection termination and a lapout concordance. The upper limit is the present sea floor. In the southern part of the Gulf of Cádiz continental shelf seismic unit 13 is composed of two minor subunits (Lobo 1995). Subunit 13a is a sigmoidal-to-parallel progradational and aggradational seismic wedge-shaped package that occurs on the coastal (deltaic and littoral) and inner shelf sedimentary environments. Subunit 13b is a sigmoidal-to-oblique progradational and aggradational seismic wedge-shaped subunit on the coastal (deltaic and littoral) and inner shelf sedimentary environments. Seismic units 13a and 13b are correlated with the P4 and P5 sedimentary units of Hernández-Molina *et al.* (1994), and are interpreted to represent a highstand systems tract deposited in the coastal and inner/middle-shelf environments during the period from 6.5 ka BP to the present day. We have observed that seismic unit 13 reveals a more complex internal forestepping structure than the highstand of the preceding sequence represented by seismic unit 5. Possible reasons for this include: (1) seismic unit 5 represents a relative long-term and so complete highstand systems tract whereas seismic unit 13 represents an early highstand systems tract in the sense of Vail *et al.* (1991); (2) seismic unit 13 is composed of minor subunits not found in the unit 5, reflecting the influence of minor sea-level cycles.

Seismic unit 14

A progradational–divergent wedge-shaped unit, which represents the last transgressive and highstand deposits on the outer shelf and continental slope. The upper limit is the upper slope, which forms the present sea floor.

High-resolution sequence stratigraphy

The 14 seismic units described in the preceding section are arranged into two major Type 1 depositional sequences (I and II), each comprised of forced regressive, lowstand, transgressive and highstand systems tracts (Table I, Figs 2, 3).

Depositional sequence I

This is a fourth-order composite Type 1 depositional sequence composed of two higher-order sequences referred to as depositional subsequences I1 and I2, respectively (Figs 2, 3).

Depositional subsequence I1. This high-frequency (fifth-order) Type 1 depositional sequence consists of a forced regressive systems tract (seismic unit 1a), a lowstand systems tract (seismic unit 1b), a transgressive surface (discontinuity between seismic units 1 and 2) and a 'relative' highstand systems tract represented by seismic unit 2.

Depositional subsequence I2. A Type 1 high-frequency depositional sequence, comprising a forced regressive systems tract (seismic unit 3a), a lowstand systems tract (seismic unit 3b), a transgressive systems tract (seismic unit 4), and a highstand systems tract (seismic unit 5). The presence of palaeo-channels on the shelf indicates that active fluvial incision continued during the lowstand sea-level stage (Fig. 2). The correlation of seismic unit 5 with the Tyrrhenian II highstand deposits on the southern Iberian coastline dated at 140–120 ka BP (Hillaire-Marcel *et al.* 1986; Somoza *et al.* 1987; Lario *et al.* 1993, 1995, 1996; Zazo *et al.* 1993; Lario 1996; Zazo 1999) constrains the age of the highstand systems tract to this sequence.

Fig. 4. Three detailed views of geopulse very high-resolution reflection profiles and accompanying interpreted line drawings through seismic units 10–14 adjacent to the modern shelf-slope break on the Cadiz margin, from seismic profile seismic profiles 3 (**a**), 4 (**b**) and 5 (**c**), respectively approximately 27 km and 5 km apart (see Fig. 1 for location). Diagrams a–c detail the internal heterogeneity of seismic units 10, an overall forced regressive unit (see Figs 2, 3). At this very high level of detail alternation between progradational oblique (**a**, **c**, **e**, **g**, **i** & **k**) and progradational and aggradational sigmoid clinoform packages (**b**, **d**, **f**, **h** & **j**) is apparent. These changes in stacking pattern are interpreted to reflect hi-amplitude oscillations of sea-level superimposed onto the longer-term fall. See text for further discussion and Fig. 8 for interpretation of cyclicity in seismic unit 10.

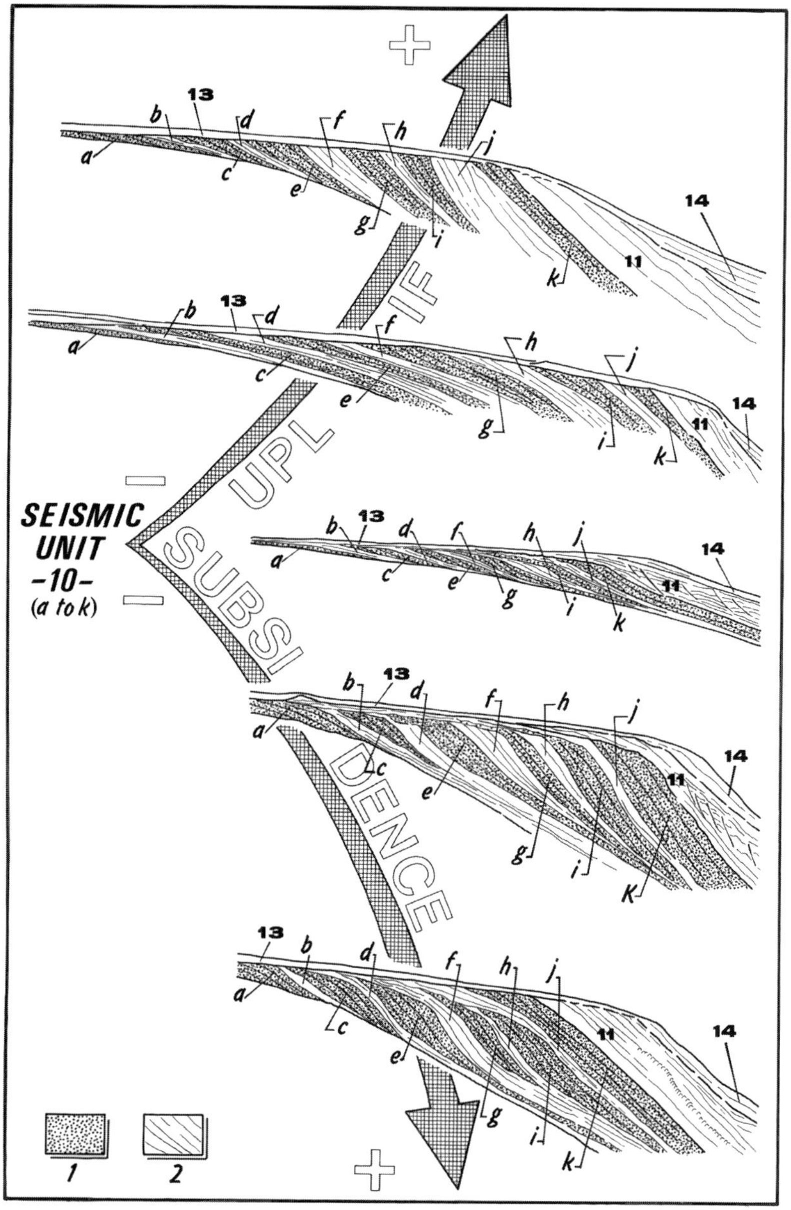

Fig. 5. Interpreted line drawings through seismic unit 10 close to the modern day shelf-slope break. The seismic subunits (**a–k**) in this overall forced regressive wedge displaying along strike variability that is attributed to changes in uplift and subsidence rate. 1, Oblique clinoformed progradational subunits (**a**, **c**, **e**, **g**, **i** & **k**). 2, Sigmoidal clinoformed aggradational subunits (**b**, **d**, **f**, **h** & **j**).

Depositional sequence II

This younger fourth-order composite Type 1 depositional sequence is composed of a forced regressive systems tract (seismic units 6, 7, 8, 9 and 10), a lowstand systems tract (seismic unit 11), a transgressive systems tract (seismic unit 12), and a highstand systems tract (seismic unit 13). Sequence II can itself be divided into four higher-order sequences, referred to as depositional subsequences II1, II2, II3, and II4, respectively (Figs 2, 3). This composite sequence was deposited during the period following the last glacioeustatic highstand, from approximately 120 ka BP, to the present day.

Depositional subsequence II1. A high-frequency fifth-order Type 1 depositional sequence that is characterized by a forced regressive systems tract (seismic unit 6a) and lowstand systems tract at the shelf edge (seismic unit 6b). The transgressive systems tract of this sequence is represented by a single surface; it is non-accretionary (*sensu* Helland-Hansen & Gjelberg 1994). The highstand deposits are not observed in the seismic record of the Gulf of Cadiz, but are interpreted to be represented by a narrow strip of coastal deposits exposed along the Iberian coastline dated at 105–90 ka BP (Lario *et al.* 1993, 1996; Lario 1996; Zazo 1999).

Depositional subsequence II2. On the Cadiz continental shelf, this fifth-order Type 1 sequence is composed of seismic unit 7, and represented by a shoreline-detached forced regressive systems tract (seismic unit 7a) and a lowstand systems tract (seismic unit 7b). The presence of palaeo-channels on the shelf indicates active fluvial incision continued during low-stillstand progradation. The transgressive systems tract of this package is represented by a single surface (Mfs) that along the Iberian coast is interpreted to be overlain by a highstand systems tract composed of coastal sediments deposited during the Tyrrhenian IV episode 80–70 ka BP (Lario *et al.* 1993, 1995, 1996; Lario 1996).

Depositional subsequence II3. This fifth-order Type 1 sequence is composed of a shoreline-detached forced regressive systems tract (seismic unit 8a), a lowstand systems tract (seismic unit 8b). A transgressive systems tract defined by a prominent erosive (ravinement) surface that separates seismic units 8 and 9. The 'relative' highstand systems tract is represented by seismic unit 9.

Depositional subsequence II4. A Type 1 depositional sequence consisting of a forced regressive systems tract (seismic unit 10) comprised of 11 minor aggradational and progradational seismic subunits (a–k). The lowstand systems tract (seismic unit 11) is made up of two minor subunits 11a and 11b that are interpreted to correlate with the P1 and P2 units present on other areas of the Spanish continental shelf (Hernández-Molina, *et al.* 1994). The transgressive systems tract (seismic subunits 12 a, b and c) is accretionary and well developed, and the highstand systems tract is comprised of two minor seismic subunits (13a and 13b) which we have correlated with the P4 and P5 sedimentary bodies found on other Spanish continental shelves (Hernández-Molina *et al.* 1994). The presence of palaeo-channels on the shelf indicates active fluvial incision continued during the lowstand of sea level. Aggradational deposits have also been identified (seismic unit 14) on the outer shelf and upper slope, associated with the drowning of the shelf margin.

Late Quaternary climatic change and asymmetric sea-level cycles

Prior to developing a detailed model of the Late Quaternary stratigraphy of the Cadiz continental shelf, it is pertinent first to review the links between orbital cycles, changes in solar radiation and climatic and sea-level change during the Pleistocene. This section provides an important background for understanding the frequency, amplitude, asymmetry and mechanisms that are interpreted to have controlled the very high-frequency glacioeustatic signal that appears to have played a major role in forming the complex variability of stratigraphy described in this study.

Since the Mid-Pleistocene Revolution, 920 ka BP, climatic and eustatic fluctuations have mainly been forced by Milankovitch eccentricity cycles of *c.* 100 ka and precession cycles of 20–23 ka; obliquity cycles of about 40 ka seem to have been less significant during this time interval, serving to amplify the effects of the 20–23 ka cycles (Ruddiman *et al.* 1986; Martison *et al.* 1987; Ruddiman & Raymo 1988; Berger & Wefer 1992; Berger *et al.* 1994). The *c.* 100–110 ka cycle has caused marked asymmetry in the major trend of eustatic sea-level from the last interglacial peak (isotopic substage 5e) to present, with an amplitude of approximately 120 m (Fig. 6c). Based on detailed studies of deep-sea cores, nine climatic stages have been recognized for the most recent eccentricity cycle that is represented by the Late Pleistocene–Holocene (Fig. 6b; e.g. Shackleton & Opdyke 1973; Shackleton 1987). The 20–23 ka cycle affected sea level during isotopic stage 5, giving rise to the fluctuations between substages 5a to 5d (Figs 6b, c). In addition, very high-frequency climatic and eustatic cycles during the Late Quaternary are inferred from deep-sea cores (Heinrich 1988; Dansgaard *et al.* 1993; Bond *et al.* 1993; Bond & Lotti 1995; Mayewski *et al.* 1996). These cycles are recognized as asymmetric pulses of rapid warming and gradual cooling with frequencies ranging from 15 000 to 1000 years (Fig. 7). The main high-frequency cyclic climatic and eustatic pulses reported for the last 70 ka are described in more detail below (see Figs 6, 7, 8).

High-frequency climatic and eustatic pulses

Asymmetric pulses of sudden warming and gradual cooling with frequencies ranging from 7000 to 14 000 years are attributed to Heinrich

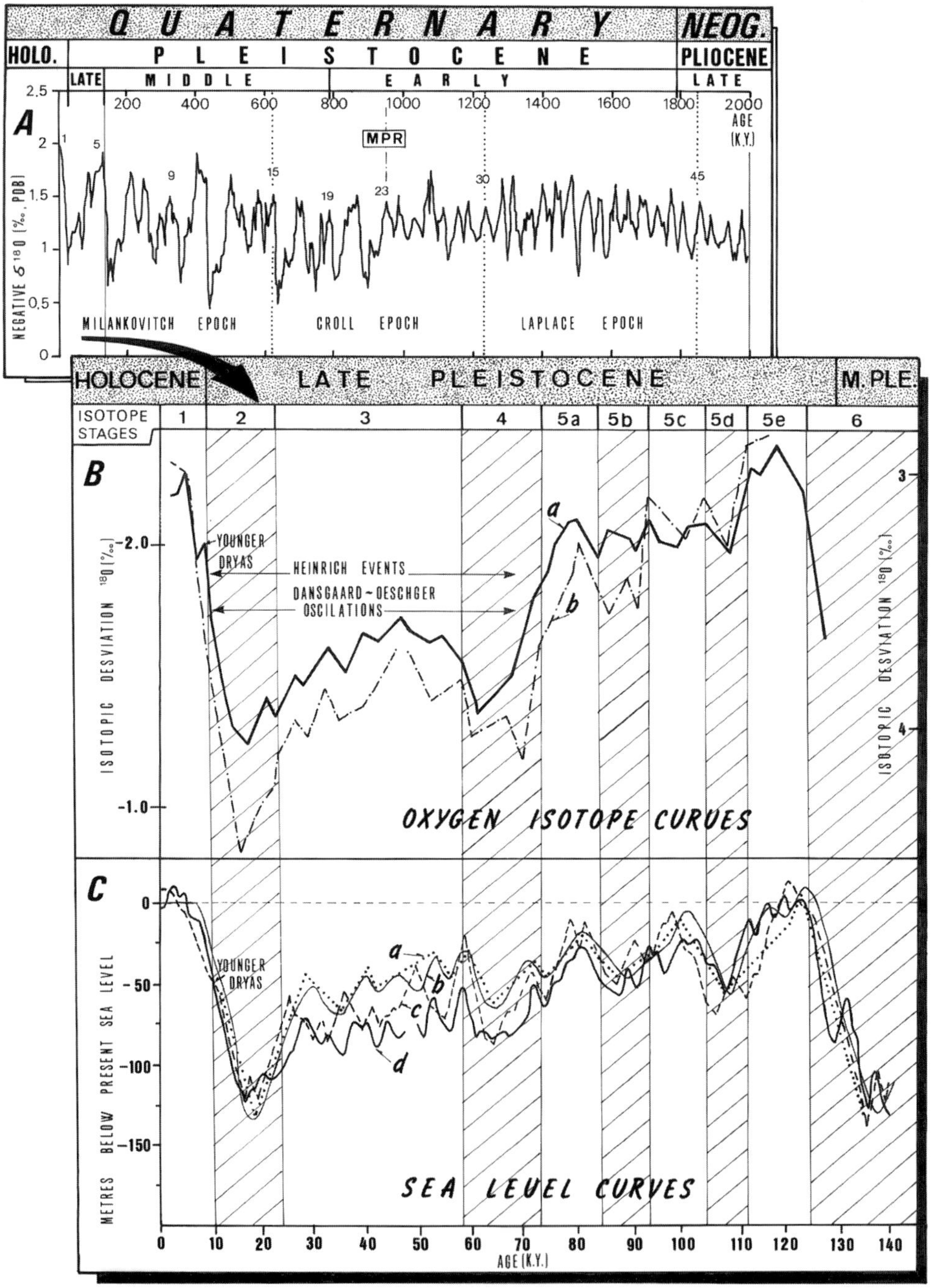

events that occurred at culminations of progressive cooling to almost glacial temperatures, followed by rapid warming (Heinrich 1988; Bond *et al.* 1993; Andrews *et al.* 1994; Verbitsky & Saltzman 1995; Mayewski *et al.* 1996). Warm–cold oscillations of *c.* 1 ka periodicity are referred to as Dansgaard–Oeschger oscillations (Dansgaard *et al.* 1993; Mayewski *et al.* 1996), and may be arranged in asymmetric cycles of *c.* 4.5 ka duration. The Late Pleistocene pulses (10^3–10^4 a scale) are related to slow ocean cooling (latent heat effect) followed by a feedback of accelerated warming (Broecker & Van Donk 1970), giving rise to pulses of gradual fall and rapid rise of sea level.

High-frequency climatic and eustatic cycles during the Late Quaternary have been recognized from the sedimentary record of present deltaic and coastal environments (e.g. Fairbridge 1984, 1996; Swift *et al.* 1991; Zazo *et al.* 1993; Hernández-Molina *et al.* 1994, 1996, 1998; Zazo *et al.* 1994; Lowrie & Hamiter 1995; Lario 1996; Somoza *et al.* 1996, 1997, 1998; Zazo *et al.* 1996). We believe that the Dansgaard–Oeschger cycles may control the P cycles of Hernández-Molina *et al.* (1994) (Figs 7, 8), which are characterized by asymmetric sea-level pulses with a steep rise and gradual fall. Heinrich events and P cycles are both interpreted as sixth-order, and appear to coincide with major warming peaks in the Dansgaard–Oeschger cycles (see Fig. 8).

Hernández-Molina *et al.* (1994) proposed that throughout the Late Pleistocene and Holocene, the lowstand and transgressive segments of the fourth-order sea-level cycle were modified by P cycles (4.5 ka) which, in turn, were modified by both h cycles (2200–900 year) and c cycles (700–500 year). Evidence for the 2200–900 year cyclicity of the h cycles comes from a number of sources. For example, Ruddiman & McIntyre (1981) recognized a 2000 year cyclicity in deep sea waters, and recent high-resolution studies on deep-sea cores have demonstrated that significant increases in iceberg calving occurred at intervals of 2 to 3 ka, with 13 ice-rafting cycles identified for the last 10 to 38 ka (Bond & Lotti 1995). Significantly, this frequency of cyclicity is directly observed in coastal deposits immediately adjacent to the study area. Radiocarbon data show that the main spit-bar system along the Spanish coastline developed during the last 6000 years and is punctuated by a cyclic periodicity of *c.* 2.2–1.2 ka (Somoza *et al.* 1991). This cyclicity gave rise to major episodes of spit-bar progradation related to major deltaic complexes, separated by well-mapped erosive gaps (Zazo *et al.* 1994, 1996; Lario 1996). Within the deltas, the main delta lobe progradation events show prominent periodicity of 2.3 ka (Lowrie & Hamiter 1995). The h cycles (2.2–0.9 ka) show the opposite trend to most, in that they are characterized by steep falls and slow rises which are thought to reflect sudden drops in solar radiation followed by gradual warming recovery, applying the model of the thermo-saline mechanism of ocean currents proposed by Broecker (1984). The c cycles (700–500 years) are related to cyclic gaps within h cycles separating minor progradational sets with *c.* 700–300 year frequencies. Pulses of similar frequencies have been recognized in palaeorecords of El Niño events (Ortlieb *et al.* 1995), giving rise to complete sequences of beach-ridges.

It is evident from the climatic data and sea-level curves reported in the literature that during the Late Quaternary, climatic and sea-level fluctuations were predominantly asymmetric in character (Figs 6, 7, 8). Cyclic changes with a range of periodicities have combined to produce a forced asymmetric eustatic pattern (Fig. 8) with long term and short term relative sea-level variations generally more rapid than rates of regional subsidence. The fourth-order (100–110 ka) gradual base-level fall during the Late Pleistocene shows superimposed cycles of sea-level change related to fifth-order (20–23 ka) precession cycles (Fig. 8). In addition, for the last 70 ka, higher frequency sea-level changes of sixth-order (Heinrich events, 10–15 ka; P cycles, 4.5 ka), seventh-order (Dansgaard-Oeschger oscillations/h pulses) and eighth-order (c cycles, *c.* 700–500 years) appear to have been important as their signatures are recognized from cyclicity in preserved coastal deposits along the southern

Fig. 6. Relationship between variability of oxygen isotopic record, climatic change and high-frequency glacioeustasy in the Late Quaternary. (**a**) Demonstrates the change to asymmetric rapid-rise and gradual stepwise-fall eustatic sea-level cycles of c. 100 ka duration following the are Mid-Pleistocene Revolution (MPR) some 900 ka BP, in comparison to the Early Quaternary and Pliocene climatic cycles (after Berger *et al.* 1994). (**b**) Detail of the oxygen isotope curve for the last 140 ka following the work of Shackleton & Opdyke (1973; line a, bold) and Shackleton (1987; line b, dashed and dotted). The standard divisions of the ^{18}O record into isotopic stages 1–6 are differentiated by means of vertical columns that alternately are plain and have a diagonal hash fill. (**c**) Displays interpreted global glacioeustatic sea-level curves for the same time interval. The four curves are after: a (dotted line) Chappel & Shackleton (1966), Shakleton (1987); b (solid fine line) Bard *et al.* (1990; 1992); c (dashed line) Dansgaard *et al.* (1993); d (solid bold line) Bond *et al.* (1993).

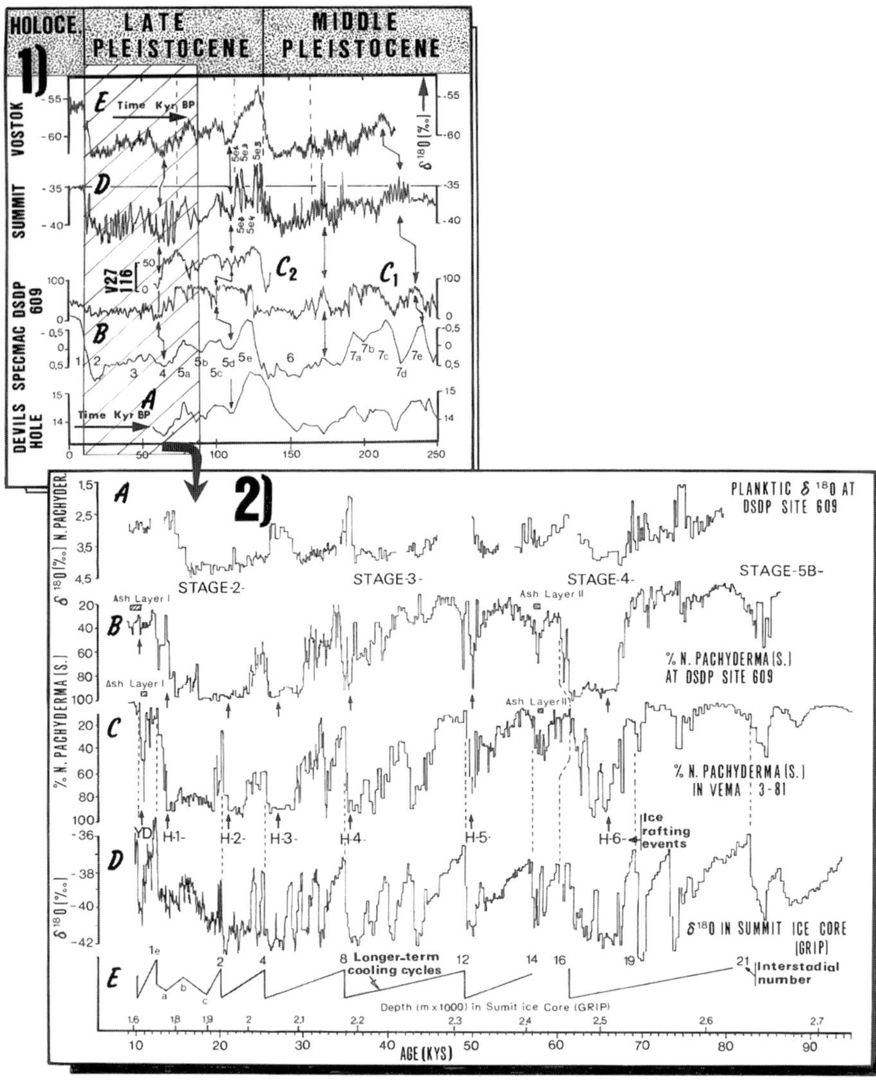

Fig. 7. (1) Detail of very high-frequency Late Quaternary climatic cycles spanning time scales of less than 10 ka during the last 250 ka BP. (2) Detail of the record for the last 70 ka BP. (1) Shows data from a range of sources including ice-cores, and pelagic fauna recovered from deep sea-cores. These short-term cycles modulated the longer-term (fourth-order) eustatic sea-level fall from 60 ka to lowstand conditions at approximately 18 ka (Heinrich events and Dansgaard–Oeschger oscillations), and the rapid rise segment from approximately 18 ka to the present day (e.g. Younger Dryas stade; after Dansgaard *et al.* 1993; Bond *et al.* 1993). (2) Displays the main frequencies that modulated the fourth- and fifth-order signals, namely: Heinrich events, characterized by rapid climatic warming and gradual cooling cycles at frequencies ranging between 7 and 10 ka (A–E; Heinrich 1988; Bond *et al.* 1993); Dansgaard–Oeschger cycles that represent cold-warm events at frequencies of 10 ka, as best seen in B–C (Dansgaard *et al.* 1993); and the 4–5 ka frequency 'P cycles' of Hernández-Molina *et al.* (1994) that may represent (rapid climatic warmings and gradual coolings) related to grouping of the Dansgaard–Oeschger cycles.

coast of the Iberian peninsula. The higher-frequency cycles have served to modify the overall fourth-order trend giving rise to a pattern of stepwise base-level fall, characterized by episodes of rapid deceleration or base-level rise, followed by renewed sea-level fall (Figs 6, 7, 8).

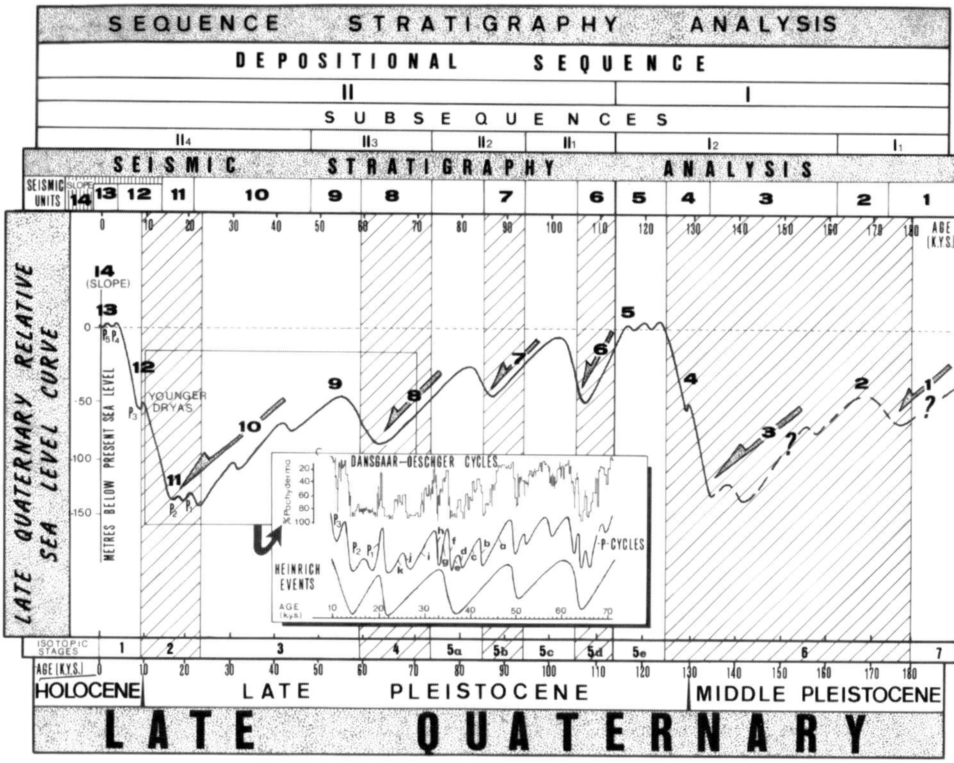

Fig. 8. Proposed correlation of the seismic sequence stratigraphy of the Cadiz margin into a chronological framework for the last 180 ka. The proposed correlation for seismic units (1–14), depositional sequences I and II and subsequences: I1, I2, II1, II2, II3, and II4 is based on the correlation of the seismic stratigraphy with coastal deposits dated by Th/U dating, as discussed in the text. The main relative sea-level curve is modified from global eustatic sea-level curves to take account of the Th/U and radiocarbon dates from Late Pleistocene-Holocene Spanish deposits (Hillaire-Marcel *et al.* 1986; Somoza *et al.* 1987, 1991; Goy *et al.* 1993; Zazo *et al.* 1993, 1996; Lario 1996; Zazo 1999). The inset box displays details of the high-frequency cyclicity of the last 70 ka with: the 1–2 ka Dansgaard–Oeschger cycles (upper); the 4–5 ka 'P cycles' of Hernández-Molina *et al.* (1994) and their proposed correlation with seismic subunits a–k in seismic units 10 (shown in Figs 4, 5); and the 7–10 ka Heinrich events, characterized by rapid warming, gradual cooling. Note that the position of the depositional sequences (I, II), subsequences (I1, I2, II1, II2, II3 and II4) and seismic stratigraphic units (1–14) is shown in the two horizontal rows above the main box containing the interpreted relative sea-level curve for the Cadiz margin. Below the main box absolute ages are given, as well as the boundaries to the standard isotopic stages (differentiated by means of vertical columns, alternately without tone or shaded by diagonal lines).

Interpretation of the Late Quaternary stratigraphy, Gulf of Cadiz

The overall Late Quaternary stratigraphic architecture of the Gulf of Cádiz is interpreted to be comprised of two composite Type 1 depositional sequences that we relate to fourth-order asymmetric relative sea-level changes (Fig. 9a). These composite sequences are comprised of forced regressive, lowstand, transgressive and highstand systems tracts; the forced regressive and lowstand systems tracts being volumetrically the most important (e.g. Figs 2, 3, 9a, c). The fourth-order depositional sequences are composed of higher-order depositional sequences that we relate to fifth-order asymmetric sea-level cycles (e.g. subsequences I1–2, II1–4, Figs 2, 3). These fifth-order sea-level changes were superimposed upon and punctuated the regressive trends of the fourth-order cycles (Figs 8, 9b). The position and development of the fifth-order depositional sequences were controlled by their position within the major fourth-order sea-level cycles. Within the fifth-order depositional sequences the forced regressive and lowstand systems tracts account for the main progradation of the margin (Fig. 9b). Highstand systems tracts either (i) did not develop on the uppermost shelf during deposition of the fifth-order sequences, (ii) the highest level of the last fifth-order cycle are represented by the deposition of 'relative' highstands deposited across the middle-outer shelf (Fig. 9b), or (iii) they are represented by narrow tracts of coastal deposits preserved along the modern Iberian coastline and up dip of the seismic lines examined in this study (e.g. Hillaire-Marcel *et al.* 1986; Somoza *et al.* 1987; Lario *et al.* 1993, 1995, 1996; Zazo *et al.* 1993, 1996; Lario 1996; Zazo 1999). In general, the transgressive systems tract is non-accretionary, and represented by single surface within the fifth-order sequences.

When the forced regressive, lowstand, transgressive and highstand systems tracts of the youngest fifth-order depositional sequence are examined in detail (depositional sequence II4 – seismic units 10–14), it is apparent that each of these systems tracts is itself a composite sequence, composed of higher order asymmetric depositional sequences (Figs 2, 3, 4). The latter are thought to be controlled by Heinrich events, P cycles, h cycles, Dansgaard–Oeschger cycles and c cycles. For example, within the last lowstand, transgressive and highstand systems tracts, the most significant progradation events are attributed to the P cycles of sea-level of approximately 4.5 ka duration, following the arguments forwarded in the previous section and the stratigraphy of Hernández-Molina *et al.* (1994), Lobo (1995) and Somoza *et al.* (1997). Cyclicity detected within these units is attributed to the even higher frequency h cycles, Dansgaard–Oeschger cycles and c cycles.

Sedimentary response to high-frequency sea-level cycles: an evolutionary model

The response of Late Pleistocene to Holocene sedimentation on the Cadiz continental shelf during the last the asymmetric fourth-order sea-level cycle (approximately 120 ka BP to the present day) can be divided into four systems tracts within the composite depositional sequence II (Figs 9c, 10). These are: (1) a forced regressive interval lasting from the last interglacial until the beginning of the Würm glaciation; (2) a lowstand interval during the maximum of the Würm glaciation; (3) a transgressive interval during the Flandrian transgression; (4) a highstand interval from the last eustatic maximum up to the present. Figures 8, 9 and 10, accompany the following section, and respectively show; (i) the exact placements of the isotopic stage boundaries, (ii) seismic units identified on the Cadiz shelf, their stacking patterns and their interpreted relationship to the isotopic stages, and (iii) a series of three-dimensional block diagrams depicting the sedimentary evolution of the Cadiz shelf during deposition of the composite fourth-order sequence II.

Forced regressive interval

The forced regressive interval is volumetrically the most important systems tract within the fourth-order composite sequence II (Fig. 9c, upper). The general trend of falling sea-level was modulated by the fifth- and higher-order cycles so that the fourth-order forced regressive systems tract is comprised of component fifth-order sequences (Fig. 10), as summarized below. During the interval of sea-level fall as a whole,

Fig. 9. (**a**) Proposed stratigraphic model to explain lateral variability of Mid–Late Pleistocene–Holocene stratigraphic architecture on the Cadiz Margin. The stratigraphy of the margin is forced by eccentricity cycles of approximately 100 ka. As shown in the box on the left, these are dominated by gradual fourth-order glacioeustatic falls. After 'Mid-Pleistocene Revolution' (last 920 ka) the geometry of fourth-order sequences forming the shelf margin reflects asymmetric eccentricity cycles with fast rises and stepwise falls in sea-level. Combined with the effects of tectonic uplift or subsidence quite different stratal architectures and development and preservation of systems tracts occurs. However, in either case progradation of the margin is dominated by forced regressive (FRWST) and lowstand systems tracts (LST); transgressive and highstand systems tracts tend to be relatively impoverished. (**b**) Summary interpretative representation of the fifth-order systems tracts that form the fourth-order composite sequence deposited during the last 140 ka on the Cadiz margin. It is clear that outbuilding of the margin is dominated by regressive deposits attributed to the forced regressive (FRW) and lowstand (LST) systems tracts, both of which are often capped by incised valley fills, the reasons for which are discussed in the text. Transgressive and highstand systems tracts (TST, HST) tend to be thin and developed on the shelf only (e.g. seismic units 5, 9, 12, 13). The correlation of the seismic stratigraphic units with the fourth-order relative sea-level curve for the Cadiz margin is shown on the left. (**c**) Summary interpretative representation of the fourth-order sequence deposited during the last 140 ka on the Cadiz margin. At this fourth-order scale, there is profound volumetric partitioning between systems tracts. The proposed correlation with the lower-order relative sea-level curve is shown on the left, and the fifth-order component systems tract are differentiated below. HPD, Healing phase deposit, PCH, incised valley fill. P1–5 correspond to the depositional units differentiated by Hernández-Molina *et al.* (1994).

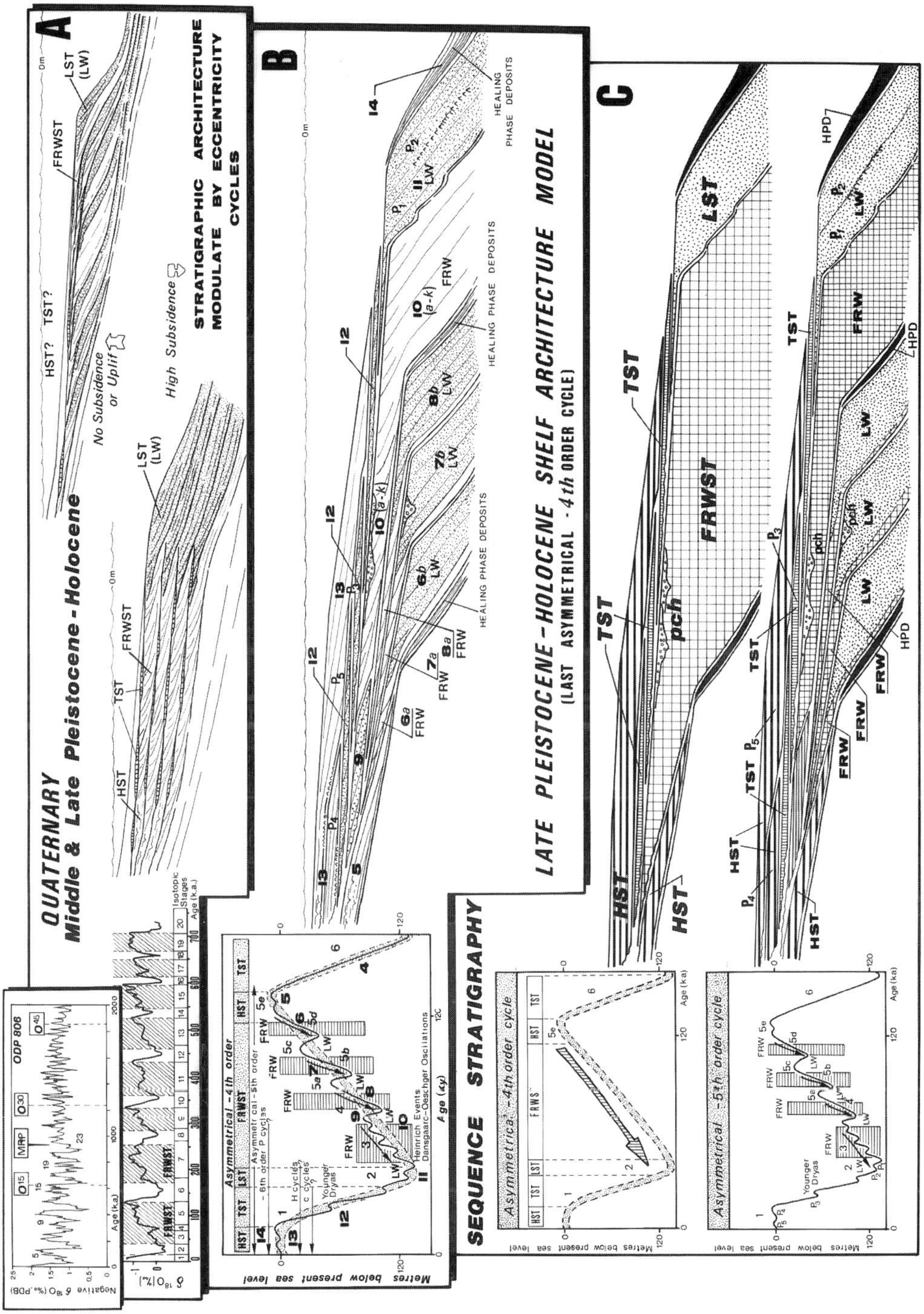
QUATERNARY
Middle & Late Pleistocene - Holocene
STRATIGRAPHIC ARCHITECTURE MODULATE BY ECCENTRICITY CYCLES
No Subsidence or Uplift
High Subsidence
ODP 806
LATE PLEISTOCENE - HOLOCENE SHELF ARCHITECTURE MODEL
(LAST ASYMMETRICAL - 4th ORDER CYCLE)
HEALING PHASE DEPOSITS
Asymmetrical -4th order
Heinrich Events
Dansgaard-Oeschger Oscillations
Metres below present sea level
SEQUENCE STRATIGRAPHY
Asymmetrical -4th order cycle
Asymmetrical -5th order cycle
Younger Dryas
Age (ka)

the higher-order transgressive systems tract tends to be characterized by thin erosional or non-depositional surfaces. The higher-order highstand deposits are generally impoverished and either (i) isolated on the upper shelf, positioned close to modern sea-level, and physically detached from deposits of the preceding and succeeding forced regressive and lowstand systems tracts, or (ii) deposited on the mid-outer shelf and attached to forced regressive and lowstand systems tracts.

(1) During the last interglacial maximum, a

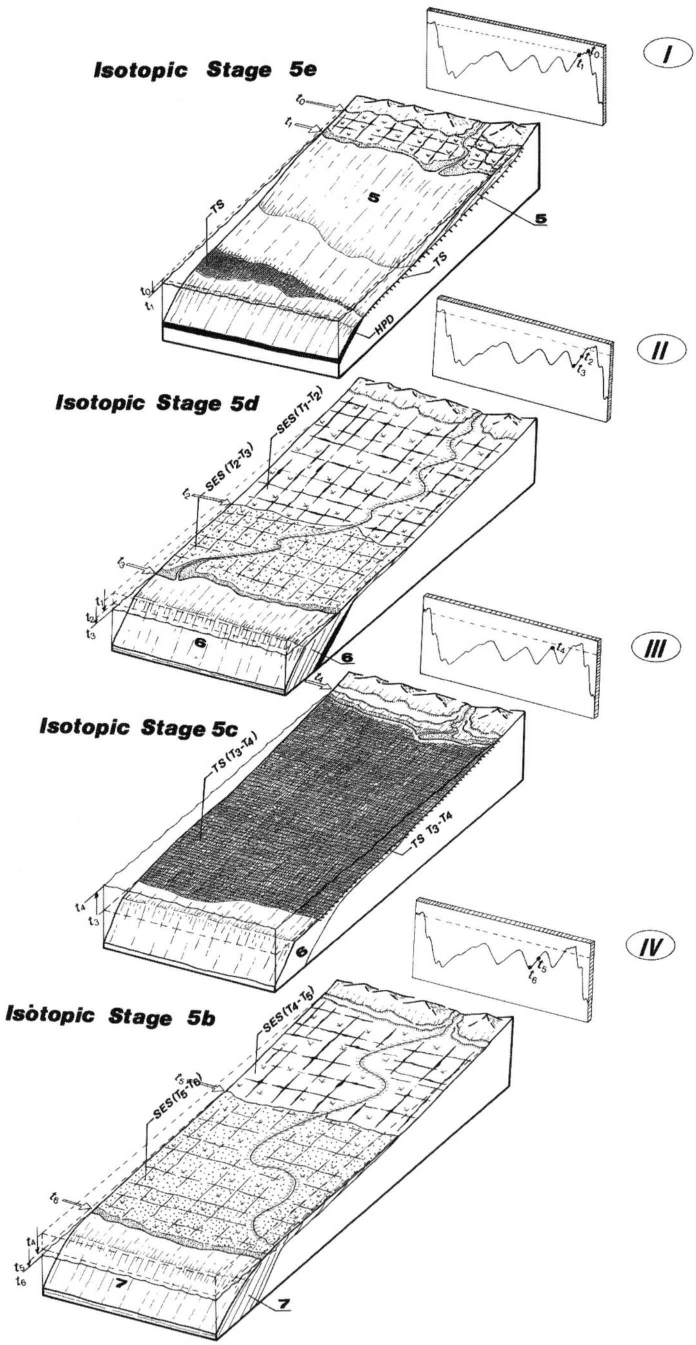

Fig. 10a

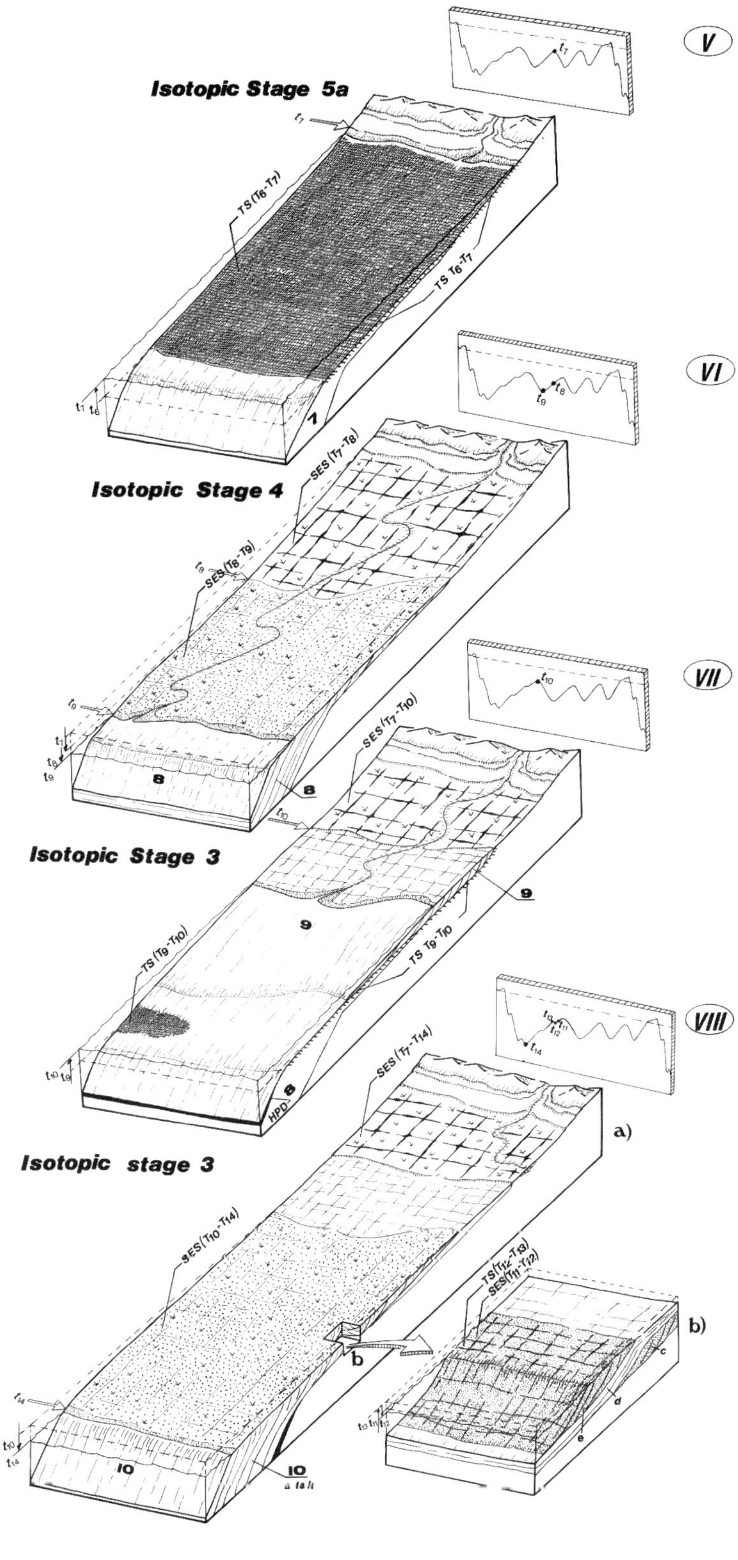
V
Isotopic Stage 5a
TS (T6-T7)
TS T6-T7
7
VI
Isotopic Stage 4
SES (T7-T8)
SES (T8-T9)
8
VII
SES (T7-T10)
Isotopic Stage 3
9
TS (T9-T10)
TS T9-T10
HPD
VIII
SES (T7-T14)
a)
Isotopic stage 3
SES (T10-T14)
TS (T12-T13)
SES (T11-T12)
b)
b
c
d
e
10

Fig. 10b

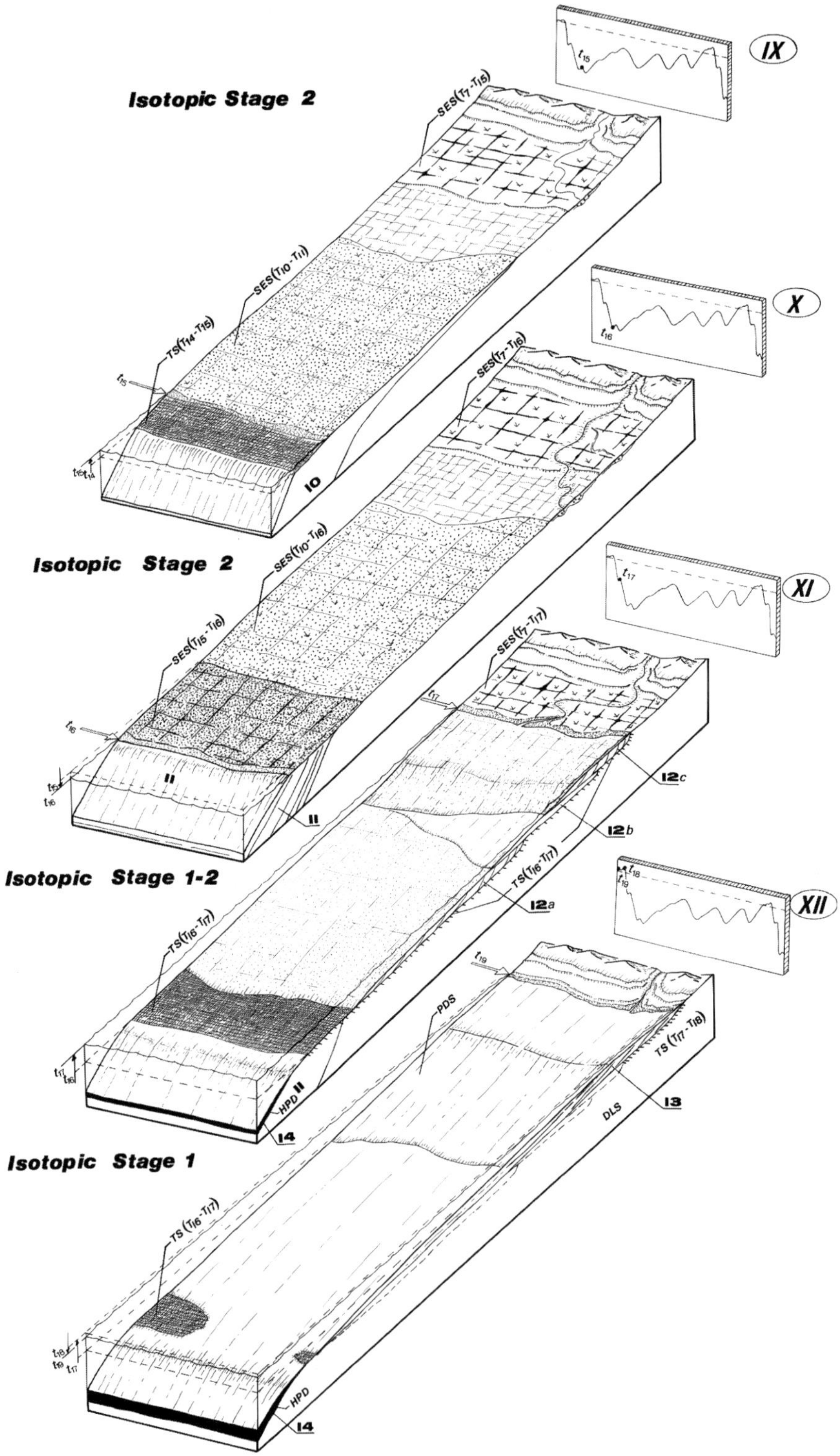
IX
Isotopic Stage 2
SES(T7-T15)
SES(T10-T11)
TS(T14-T15)
10
X
SES(T7-T16)
Isotopic Stage 2
SES(T10-T16)
SES(T15-T16)
11
XI
SES(T7-T17)
12c
12b
TS(T16-T17)
12a
Isotopic Stage 1-2
TS(T16-T17)
HPD
14
XII
PDS
TS(T17-T18)
13
DLS
Isotopic Stage 1
TS(T16-T17)
HPD
14

Fig. 10c

highstand systems tract developed on the upper shelf when sea-level is estimated to have been 4–6 m higher than present (the Tyrrhenian II highstand systems tract between 120 and 140 ka). The highstand deposition of seismic unit 5 is illustrated by Figure 10 (the interval between t_0–t_1 on the relative sea-level curve).

(2) During the ensuing sea-level fall during isotopic stage 5d (Figs 8, 10I–II, t_2–t_3) a dramatic basinward shift of facies occurred resulting in deposition of seismic unit 6. During this fall of sea-level, a subaerial exposure surface expanded across the shelf across which incised valley systems developed (Fig. 10II). At the same time, the forced regressive systems tract (seismic unit 6a) and a lowstand systems tract (seismic unit 6b) built out the outer shelf and shelf edge (Figs 9b,c & 10).

(3) During the latter part of isotopic stage 5d and early part of isotopic stage 5c, a rapid sea-level rise occurred and a transgressive surface developed without any significant deposits (Fig. 10III, time steps t_3–t_4).

(4) Highstand coastal deposits have been determined on land in the south of Iberia (e.g. the Mediterranean Tyrrhenian III episode between 90 and 105 ka; Hillaire-Marcel *et al.* 1986; Somoza *et al.* 1987; Lario *et al.* 1993, 1996; Zazo *et al.* 1993; Lario 1996; Zazo 1999), and these are schematically shown in Fig. 10III. However these strata form a relatively narrow strip in proximity to the coast, and are not believed to be volumetrically significant offshore where they are represented by a single seismic surface (e.g. Figs 2, 3).

(5) During the latter part of isotopic stage 5c and the first part of isotopic stage 5b, a relative sea-level fall took place that resulted in a major basinward facies shift, subaerial exposure of the shelf and its incision by fluvial systems (time steps t_3–t_4, Fig. 10IV). The sea-level fall led to deposition of a forced-regressive systems tract on the middle-outer shelf (seismic unit 7a) that is physically detached from the highstand strata up dip. The lowstand systems tract (seismic unit 7b) is interpreted to have been deposited during the subsequent lowstand and built the outer shelf margin into deep water.

(6) During the early part of isotopic stage 5a, a rapid sea-level rise developed a transgressive surface that lacks any significant deposits across the shelf (Fig. 10V). Coastal deposits have been determined on land that equate to the Tyrrhenian IV episode between 70 and 80 ka (Lario *et al.* 1993; Lario 1996; Zazo 1999). These strata are interpreted to represent a rather narrow tract (in dip sense) of highstand deposits perched up-dip on the upper shelf (Fig. 10V, time step t_7).

(7) During the latter part of isotopic stage 5a and stage 4, further relative sea-level fall is interpreted to have resulted in the subaerial exposure of the shelf (Fig. 10VI, t_{7-9}). Forced regressive and lowstand systems tracts represented by seismic units 8a and 8b respectively, were deposited on the outer shelf and at the shelf edge (Figs 2, 3, 9b–c, 10). The highstand deposits of isotopic stage 5a and the forced regressive systems tract of isotopic stage 4 are physically detached from one another by a zone of non-deposition and sediment bypass (Fig. 10VI). As sea-level fell the shelf is interpreted to have become incised by palaeochannels, through which sediments were fed across the shelf to the shelf-edge.

(8) During the latter part of isotopic stage 4, and the beginning of isotopic stage 3, a rapid sea-level rise occurred. This is interpreted to have developed a transgressive surface that lacked significant deposition on the outer and middle shelf (Figs 2 & 3, 10VII). In the subsequent highest position of sea-level, highstand deposits prograded across the middle to the outer shelf, depositing seismic unit 9 (t_{10}, Fig. 10VII).

(9) During the remaining part of isotopic stage 3, relative sea-level fall is related to the progradation of a very substantial forced regressive systems tract (seismic unit 10a–k) across the outer shelf and of shelf edge (Fig. 10VIII, t_{11-14}). This was a time of the most widespread subaerial exposure and palaeochannel development across the shelf (Fig. 9c). Minor aggradational and progradational sedimentary bodies within the forced regressive systems tract time are thought to record higher-frequency (fifth-order) relative sea-level fluctuations superimposed on the longer term fall (Fig. 10VIIIb).

Fig. 10. Twelve sequential three dimensional block diagrams showing key stages of the sedimentary evolution of the Cadiz shelf in response to relative sea-level changes taking into account that the asymmetric fourth-order sea-level curve was modulated by higher-order cyclicity (fifth- to seventh-order). For each of the twelve diagrams the various seismic stratigraphic units are labelled in bold, and the time interval with reference to the standard isotopic stages is given. The interpreted position of sea-level at and between the 19 time steps that are differentiated (t_{1-19}) is shown both (i) on the block diagram and (ii) with reference to the interpreted relative sea-level curve developed from the Cadiz margin (in the box behind each block diagram). t_{1-19}: time interval for each specific point of sea-level curve; SES, subaerial erosion surface; TS, transgressive surface. See text for detailed explanation.

Lowstand interval

Between deposition of seismic units 10 and 11, there is interpreted to have been a relative sea-level rise (time steps t_{14-15}; Fig. 10IX). The lowstand systems tract is represented by seismic unit 11 (Fig. 9b, c), itself composed of two progradational seismic units 11a and 11b that are correlated with cycles P1 and P2 of Hernández-Molina *et al.* (1994; Figs 8, 9c). The lowstand systems tract is thought to have been deposited at the beginning of isotopic stage 2 when sea-level was approximately 120 m below its present level (Figs 8, 10X, time steps t_{15-16}).

Transgressive interval

The transgressive interval was punctuated by stillstands and minor sea-level cycles. Several aggradational and progradational sedimentary bodies can be recognized (Figs 9b, 10XI). The Younger Dryas stage at 10.5 ka BP (Fig. 8) is one of the most significant progradational events during the transgression. During this stage a brief sea-level fall of about 3 m took place, and an important progradational sediment body was deposited on the shelf (seismic unit 12C, Fig. 9, t_{17}, Fig. 10XI) that is equivalent to unit P3 of Hernández-Molina *et al.* (1994).

Highstand interval

This interval lies between the eustatic maximum of 6.5 ka BP and the present, with the formation of a highstand deposit on the coastal and inner shelf (Fig. 10XII). This deposit is composed of two minor progradational sediment bodies, P4 and P5 of Hernández-Molina *et al.* (1994).

Discussion

Several important differences are apparent when comparing the Late Quaternary stratigraphy of the Gulf of Cadiz with the idealized model of the Exxon group (with lowstand, transgressive and highstand systems tracts) and the modification of this model to incorporate the fourth forced regressive or falling stage systems tract of Hunt & Tucker (1992, 1995) and Plint & Nummedal (this volume), respectively. The Quaternary fourth-order composite depositional sequences are comprised of sets of fifth-order composite depositional sequences that are themselves built from the stacking of sixth-, seventh- and even higher-order sequences. In this 'Russian dolls' or fractal model, there are practical problems concerning the expression and stratigraphic significance of the boundaries to the sequences of the different orders. As a consequence, there is overlap between the systems tracts, (stranded) parasequences and high-order sequences, as discussed below.

Systems tracts

One of the most significant differences with respect to the idealized Exxon models is the profound asymmetry of the systems tracts developed on the Cadiz continental shelf. Idealized stratigraphic models assume a sinusoidal sea-level curve, constant sediment supply and portray the highstand and lowstand systems tracts as most volumetrically significant, and therefore associated with the most significant progradation (e.g. Mitchum *et al.* 1977; Vail 1977; Van Wagoner *et al.* 1987, 1988, Posamentier & Vail 1988).

In this study, we have recognized two types of highstand systems tracts; those developed as narrow tracts along the Iberian coastline with no deposits recognized downdip in the seismic lines presented here, and 'relative' highstands developed on the middle-outer shelf. The latter are developed when relative sea-level rises and highstands were characterized by only partial flooding the shelf and in preliminary interpretations, were initially mistaken for shelf margin deposits (Somoza *et al.* 1997). The position of the fifth-order highstand systems tracts on the shelf is believed to reflect the composite nature of the forcing glacioeustatic signal during the Late Quaternary, with fifth-order cycles superimposed on the dominant 100–110 ka signal (e.g. Figs 8, 9, 10). This means that the highstand systems tracts (fourth- and fifth-order) deposited during the fourth-order interglacials, when sea level was at or close to its highest position in the fourth-order cycle (Figs 8, 9) are positioned furthest up-dip. In contrast, those fifth-order highstands deposited during the latter part of the long-term fourth-order falls and lowstands tend to be developed on the middle and outer shelf and tend to be more extensive in dip section (e.g. seismic unit 9, Figs 9, 10).

Our data also allow us to compare the volumetric significance of the highstand systems tract and amount of highstand progradation to that of other systems tracts (e.g. Fig. 9c). Our data illustrate that the fourth- and fifth-order highstand systems tracts on the Cadiz shelf are (i) less volumetrically significant and (ii) tend *not* to be associated with major basinward progradation across the shelf and/or of the shelf edge when compared against the forced regressive and lowstand systems tracts (e.g. Figs, 2, 3, 9b, c). It seems that the fourth- and fifth-order highstand systems tracts recognized in this study are quite different to those portrayed in idealized sequence stratigraphic models (e.g. Mitchum *et al.* 1977; Vail 1977; Van Wagoner *et al.* 1987, 1988, Posamentier & Vail 1988) in terms of their relative position on the shelf, volume and basinward progradation. The paucity in the development of the highstand systems tracts on the Cadiz shelf is believed to be a reflection of the very short time interval spent within the highstand condition (e.g. Figs 8, 9, 10).

The transgressive systems tract is also rather poorly represented within the Late Quaternary stratigraphy of the Cadiz shelf (Figs 2, 3, 9c). In most parts of the succession, relative sea-level rise is represented by a single surface (e.g. non-accretionary transgression following Helland-Hansen & Gjelberg 1994), rather than by deposition of a retrogradational package(s) of strata (e.g. accretionary following Helland-Hansen & Gjelberg 1994). A notable exception is the accretionary fourth-order transgressive systems tract to depositional sequence II, represented by seismic units 12a–c (Figs 2, 3, 9). In general terms, the low volume of strata attributed to this systems tract is thought to reflect to a combination of relatively rapid rises and reduced sediment supply during intervals of glacial warming.

In terms of volume and basinward progradation (across the shelf and of the shelf margin) forced regressive systems tracts, deposited during times of falling sea-level, dominate the Late Quaternary fourth-order stratigraphy of the Cadiz margin (e.g. Figs 2, 3, 9c, upper). Within fifth-order sequences most progradation is attributed to the forced regressive *and* lowstand systems tracts (Fig. 9c, lower). A similar relationship has been observed on other continental margins (e.g. Sydow & Roberts 1994; Chiocci this volume; Kolla *et al.* this volume; Trincardi & Corregiari this volume). This is one of the most significant difference between our data and the idealized Exxon models. In the earliest models of sequence stratigraphy, periods of sea-level fall were assumed instantaneous, with all sediment bypassed across the shelf to the basin-floor during periods of rapid relative sea-level fall (e.g. Mitchum *et al.* 1977; Vail *et al.* 1977). In later models, sea-level fall was not thought instantaneous, although deposition on the shelf during sea-level fall remained largely overlooked (e.g. Posamentier & Vail 1988). It is only since the late 1980s that shelf deposition during sea-level fall and the importance of differentiating a forced regressive or falling stage systems tract has been appreciated (e.g. Hunt & Tucker 1992, 1995; Plint & Nummedal this volume – see their historical review).

We believe that the incorporation and distinction of a fourth systems tract to be particularly important in the Late Quaternary sediments deposited on continental margins undergoing low rates of subsidence (<1.25 m ka^{-1}; Hunt & Gawthorpe this volume) or uplift (e.g. McMurray & Gawthorpe this volume). The greater importance of the forced regressive and lowstand systems tracts in comparison to the relatively impoverished transgressive and highstand systems tracts on the Cadiz shelf during the Late Quaternary develops a profound asymmetry of fourth-order sequences and systems tracts (e.g. Fig. 9). The asymmetric nature of systems tracts development is interpreted to reflect the importance of an asymmetric glacioeustatic sea-level signal, that is comprised of rapid rises with long gradual falls and short periods of time spent in the high and low stillstand conditions. Figure 9a attempts to show the effects of varying subsidence and uplift rates on the asymmetry of systems tract development and the stacking in during the late Quaternary based on our observations along the Gulf of Cadiz continental shelf and also in the Alboran Sea Shelf (Hernández-Molina *et al.* in prep.).

Attached and detached forced regressions

Within the fourth-order depositional sequences (I and II, Figs 2, 3), we have been able to distinguish between forced regressive systems tracts that are physically attached from the previous highstand deposits and those that are detached (following the concepts introduced by Ainsworth & Pattison 1994). Those forced regressive systems tracts detached from the previous highstand deposits generally follow short intervals when the highstand systems tract is represented by a narrow wedge developed on the upper shelf and today exposed along the Iberian coastline (e.g. prior to development of the forced regressive systems tracts represented by seismic units 7a and 8a, Figs 9, 10). On the other hand, the highstand-attached forced regressive deposits occur when either (i) highstand deposits had prograded to mid-outer shelf positions (e.g. seismic units 5), or (ii) when fifth-order 'relative' highstands are developed on the middle-outer shelf as a result of the modulation of the fourth-order signal by fifth-order cycles (e.g. seismic unit 9).

Sedimentation and sequence boundary formation in response to stepwise sea-level fall

In depositional sequences I and II, we have observed that the main erosion surfaces associated with deposition *during* a modulated fourth-order relative sea-level fall are (i) sharp basal surfaces that underlie the forced regressive systems tract, (ii) sharp surfaces at the base of component downstepping fifth-order sequences, and (iii) the transgressive to maximum flooding surfaces of the component fifth-order sequences. Within the fifth-order sequences that comprise the composite fourth-order forced regressive systems tracts we have observed that erosion surfaces developed from the combined effects (i) marine erosion during base-level fall, (ii) subaerial exposure, (iii) fluvial incision, and (iv) transgressive ravinement (Figs 2, 3, 4, 9, 10). It appears that the development of erosional surfaces during a stepwise relative sea-level fall are not solely associated with the fall itself (cf. Posamentier *et al.* 1992), but are in detail much more complex in origin (e.g. Sonnenfeld & Cross 1993; Mellere & Steel 1995, this volume).

Analysis of the high resolution architecture of composite depositional sequence II shows that this composite fourth-order forced regressive systems tract is underlain by a sharp basal surface and overlain by an erosional discontinuity on the shelf (e.g. Figs 2, 3, 9b). Internally, the fourth-order forced regressive systems tract is arranged into sharp-based 5th-order sequences, that are interpreted to be mainly comprised of forced regressive and lowstand deposits; the highstand and transgressive systems tracts of these fifth-order sequences are poorly or not developed (Fig. 4). At higher resolution still, within the forced regressive and lowstand systems tracts of the youngest fifth-order sequence several units can be differentiated; these we attribute to very-high-order sequences likely controlled by sixth- and seventh-order sea-level cycles. Such a conclusion is backed by detailed study of the Late Quaternary sediments exposed along the Iberian coastline where erosive events between stratigraphic units of 2–3 ka duration are bounded by erosive surfaces interpreted to be related to cycles of sea-level rise and fall (Hernández-Molina *et al.* 1994). In this setting the concept of parasequences (units bounded by flooding surfaces) is generally not appropriate or applicable as sea-level falls bound cycles across the spectrum of depositional units (from the fourth-order to eighth-order, as further discussed below).

The classical interpretation of sequence boundaries is that they are due to the initiation of a rapid relative sea-level fall (Posamentier & Vail 1988; Vail *et al.* 1991), the sequence boundary being placed below the forced regressive deposits at the onset of sea-level fall (Posamentier *et al.* 1992; Kolla *et al.* 1995). Following this definition, a sequence boundary would be placed at the base of the fourth-order forced regressive systems tract (e.g. below seismic unit 10, Fig. 9a) when sea-level is interpreted to have begun to fall, following the major fourth-order highstand, and deposition began to migrate basinward in sequence II. Yet this initial erosional surface does not seem to have any special characteristics, such as a greater extent or truncation, than any of the sequence boundaries that bound each of the component fifth-order sequences that comprise the forced regressive wedge (Fig. 4), and has a relatively low preservation potential being formed in an up dip location. In addition, the idea of the instantaneous development of the sequence boundary is somewhat at odds with the progressive expansion of the subaerial exposure surface and deposition of forced regressive wedges during prolonged fall that we have documented.

In contrast to the idea of instantaneous sequence boundary formation, Hunt & Tucker (1992, 1995) and Plint & Nummedal (this volume) argued that the 'master' sequence boundary (e.g. fourth-order in depositional sequences I and II here) should be placed *above* sediments deposited during forced regression because; (i) of the progressive nature of surface development and deposition during relative sea-level fall, (ii) the potential cannibalisation of sediments and surfaces and (iii) the greatest basinward development and areally extensive subaerial exposure occurs at the lowest point of relative sea-level. In this case, the fourth-order sequence boundary would be placed between seismic units 10 and 11 (Figs 2, 3, 9a). This choice, and separation of the fourth systems tract makes sense when considering the progressive development of the sequence boundary and deposition during relative sea-level fall. On the shelf, the upper surface of the forced regressive systems tract clearly shows incised fluvial channels on the inner shelf and mass gravity-flow deposits on the slope. However, in the upper slope and continental shelf deposits on the Gulf of Cadiz we have found it difficult to recognize the boundary between the forced regressive systems tract and lowstand systems tract in all sequences. Commonly two unconformities are defined (Fig. 9a); (i) the sharp basal surface of the fourth-order forced regressive systems tract, and (ii) the erosional unconformity above the forced regressive systems tract *and* lowstand systems tract. The development of the erosional surface *over* the lowstand wedge may, in some cases, reflect the deposition of this systems tract and expansion of subaerial exposure during a low-stillstand of sea level following forced regression, prior to onset of any relative rise in sea level.

These observations mean that there are practical problems concerning the position of the fourth-order sequence boundary, depending on whether it is taken (i) at the onset of sea-level fall (*sensu* Posamentier *et al.* 1992; Kolla *et al.* 1995; Posamentier & Morris this volume) or (ii) at the lowest point of sea-level fall (*sensu* Hunt & Tucker 1992, 1995; Plint & Nummedal this volume). During sea-level fall, two main surfaces can be differentiated: (1) the basal surface of the forced regressive systems tract, marking the onset of the sea-level fall, and (2) the erosional unconformity (the sequence boundary in fast-falling sea-level cycles) above the forced regressive and lowstand systems tract and related to the lowest point of sea-level. The maximum flooding surface should be taken into account by dividing arrays of the main systems tracts (forced regressive systems tract and lowstand systems tract) into fast-rise gradual-fall sea-level cycles. In the Late Pleistocene–Holocene deposits, the clearest discontinuity is the transgressive surface because of the rapidity of relative sea-level rise. This boundary may represent a useful stratigraphic surface for the fourth- and fifth-order depositional sequences and also for the higher-order sequences. Finally, it is worth reiterating that although traditionally, the sequence boundary has been considered to be an isochronal chronostratigraphic surface, but this interpretation is inconsistent with the evolutionary development of Late Quaternary depositional sequences. The sequence boundary does not form at a specific time, but develops over a time span corresponding to the relative sea-level curve.

High-order sequences v. stranded parasequences

The term 'stranded parasequence' (Van Wagoner *et al.* 1990; Hunt & Tucker 1992) defines a sediment body deposited on the upper slope during decelerations in sea-level fall modulated by high-order sea-level cycles (10 ka to 0.5 Ma). Mitchum & Van Wagoner (1991) proposed a hierarchical scheme under which fourth-order sequences (0.1–0.2 Ma cyclicity) have all the stratal attributes of conventional sequences, whereas fifth-order cyclicity (0.01–02 Ma) results in the formation of parasequences within fifth-order sequences. In the high resolution stratigraphic architecture presented in this paper, the fourth-order forced regressive systems tract is composed of several higher order sequences (Fig. 9). Every higher order sequence is marked by a basal surface of the forced regressive deposits, there being the same number of basal surfaces as higher order sequences. The existence of relative highstands within some higher order sequences seems to indicate that the high-frequency cyclicity punctuating the general trend should be considered in terms of incomplete or complete sequences rather than stranded parasequences. The term 'stranded parasequence' reflects a relative sea-level curve that shows small stillstands (Hunt & Tucker 1995) but not complete oscillations of sea-level (Figs 7, 8). Tectonic or isostatic uplift seems likely to be responsible for modifying the eustatic curve (complete cycles) into minor stillstands punctuated by relative curves (e.g. stranded beaches). Thus, stranded parasequences reflect high-order cyclicity modified by tectonic or isostatic processes, giving rise to incomplete sequences. We consider the term 'stranded parasequence' to be of practical use in some cases but feel that it is not an all-embracing paradigm that can account for the correlation between high-frequency cyclicity and high-order sequences.

Deposition triggered by the hierarchical higher frequency cyclicity is often assumed to consist of depositional sequences made up of parasequences and parasequence sets. Since higher frequency cyclic fluctuations (e.g. P cycles) affect deposition of the fourth-order forced regressive, lowstand, transgressive and highstand systems tracts, deposition triggered by the higher-order signals must in turn be considered in terms of depositional sequences rather than parasequences. Based on our detailed analysis of high resolution seismic data, we suggest that the term 'parasequence', defined as upward-shoaling successions bounded by flooding surfaces, should be applied to low-order sequences defined by fast-fall stepwise-rise sea-level cycles. High-order cyclicity will form depositional (sub)sequences defined principally by flooding surfaces. The preservation of these high-frequency sequences depends on the rates of relative sea-level change and sediment supply (Swift *et al.* 1991), being observed mostly in the highest sedimentary-rate environments (e.g. influencing delta-lobe switching, Lowrie & Hamiter 1995).

Factors controlling Late Quaternary depositional sequence development

Quaternary environmental changes were controlled by complex feedback processes involving three main factors: (1) high- and very high-frequency climatic changes, (2) high- and very high-frequency eustatic or relative sea-level changes and (3) variability in oceanographic conditions (Fig. 11). All these factors are related and a change in one produces a change in the others. During the Late Quaternary, eustatic changes resulted chiefly from climatic and glacioeustatic mechanisms, generating mainly fourth- and fifth-order cyclicity.

Seismic and sequence stratigraphic analyses carried out in the Gulf of Cádiz allow us to conclude that Late Quaternary depositional sequences were mainly controlled by relative, rather than purely eustatic, sea-level changes. Depositional sequence boundaries were mainly generated by the fourth- and fifth-order cyclicity, but other factors were also important in the spatial evolution and vertical stacking of depositional sequences. In this respect, neotectonic features and sediment supply controlled the spatial and vertical stacking of depositional sequences on a regional/local scale. Vertical subsidence or uplift controls the backstepping or forestepping relationship of the depositional sequences and systems tracts (Fig. 9a). Neotectonic and geodynamic evolution controls on the largest scale the physiographic characteristics of the continental margin, lateral distribution and preservation potential of depositional sequences (Fig. 11). Finally, variability in oceanographic conditions (waves, tides and currents) across the shelf controls the lateral distribution of sedimentary input.

Conclusions

The application of seismic and sequence stratigraphic concepts to high resolution seismic profiles from the Late Quaternary of the Gulf of Cádiz has allowed us to draw the following conclusions.

(1) Fourteen seismic units have been defined and are evident in all parts of the continental shelf and upper slope. These seismic units are arranged into two major asymmetric fourth-order composite depositional sequences, themselves composed of higher order depositional sequences.

(2) The Late Quaternary sedimentary evolution of the continental margin took place under the influence of asymmetric relative sea-level cycles of various frequencies that were superimposed upon one another. These cycles were largely forced by linked climatic and glacioeustatic sea-level changes, characterized by rapid sea-level rises, very short highstand and lowstand stages, and gradual sea-level falls.

(3) The stratigraphic architecture of the Late Pleistocene–Holocene continental margin comprises depositional sequences that are related to fourth-order asymmetric sea-level changes. These sequences are dominated by forced regressive and lowstand systems tracts. The transgressive and highstand systems tracts are volumetrically much less significant. This makes

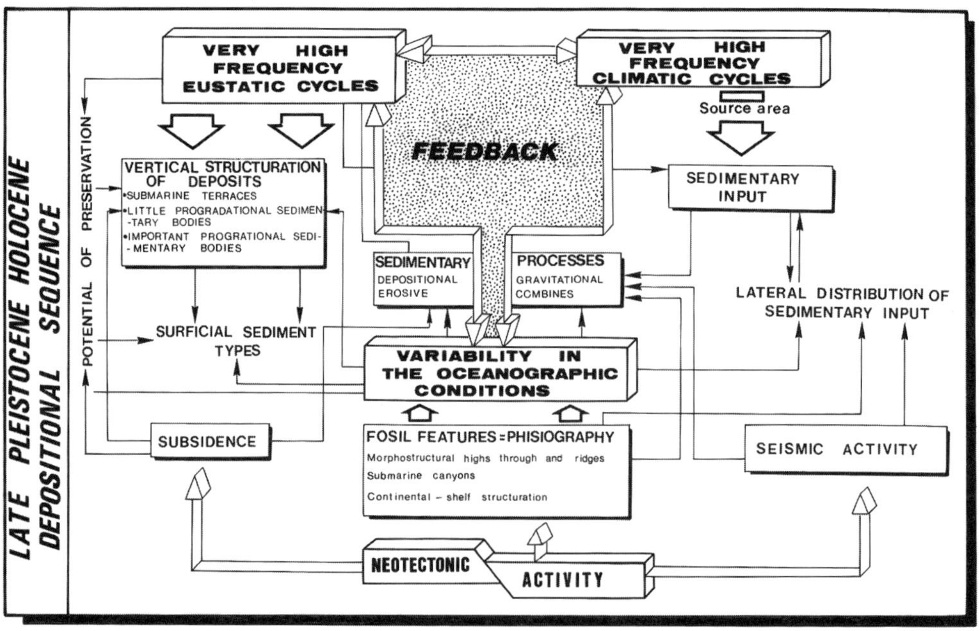

Fig. 11. Interplay between the main processes that controlled the Late Pleistocene–Holocene Depositional sequence and subsequences on the Cadiz margin.

for a very different volumetric partitioning of systems tracts compared to idealized Exxon group models. We suggest that the forced regressive systems tract is particularly relevant to Quaternary strata because deposition took place under conditions of asymmetrical relative sea-level change, with sudden sea-level rises, very short highstands, and gradual sea-level falls so that, in continental margin evolution, forced regressive deposits are likely to predominate.

(4) The fourth-order composite depositional sequences are comprised of component depositional sequences related to fifth-order asymmetric sea-level changes that were superimposed upon the dominant fourth-order cycles. The fifth-order depositional sequences are also mainly composed of forced regressive and lowstand systems tracts.

(5) Within the fourth-order forced regressive systems tracts, the formation of erosional surfaces is complex because of the modulation of the fourth-order fall by the fifth-order signal and can be attributed to the combined effects (i) marine erosion during base-level fall, (ii) subaerial exposure, (iii) fluvial incision, and (iv) transgressive ravinement within component sequences.

(6) The youngest high-order depositional subsequences (last 48 ka) are composed of a forced regressive systems tract, a lowstand systems tract, a transgressive systems tract, and a highstand systems tract. Each systems tract shows complex stacking patterns, and it is possible to differentiate sedimentary units which themselves constitute minor depositional sequences produced by higher frequency relative changes in sea-level. These sea-level changes may have been controlled by Heinrich events (10–15 ka), P cycles (4–4.5 ka), Dansgaard–Oeschger oscillations-h cycles (2300–970 years) and c cycles (500–50 years).

(7) The Holocene and Late Pleistocene sea-level cycles seem to be related to cyclic frequencies higher than the Milankovitch orbital parameter band, giving rise to composite sequences. These cycles are assumed to be externally controlled; similar cyclicity might therefore be expected in other low-order sequences and may, in some situations, be encompassed by the generic term 'parasequences' in low resolution conventional seismic/stratigraphic data. Where asymmetric cycles show rapid rises and gradual falls, flooding surfaces are likely to be the major bounding surface in high-order sequences. In terms of the Late Quaternary, the parasequence seems to be an abstract concept, used when there is insufficient resolution to see higher order depositional sequences produced by minor sea-level cycles.

(8) In the Late Pleistocene–Holocene strata, the clearest discontinuity is the transgressive surface, reflecting the very rapid relative sea-level rise. This surface could be a useful stratigraphic level as a depositional sequence boundary for the Late Quaternary fourth-, fifth- and higher-order depositional sequence.

This work has been supported by projects DGICYT PB91–0622-C03, DGICYT PB94–1090-C03 and CICYT MAR-98-0209 (TASVO) of the Spanish Research Programme. The authors wish to acknowledge the helpful review of the manuscript by H. Posamentier (ARCO Exploration and Production Technology) and J. A. Howe (British Antarctic Survey). We also thank D. J. Swift (Old Dominion University), for his constructive criticism of an earlier idea and draft of this paper. We are also grateful for the comments, corrections and ideas of both referees, F. L. Chiocci and I. Sharp, and both editors, D. Hunt and R. Gawthorpe, which have helped considerably in improving this paper. We also thank the support of the Oceanographic Spanish Institute Ship '*Odon de Buen*', especially its crew. This work is part of the 396 IGCP Project 'Continental Shelves in the Quaternary'.

References

AINSWORTH, B. R. & PATTISON, S. 1994. Where have all the lowstands gone? Evcidence for attached lowstand systems tracts in the western Interior of North America. *Geology*, **22**, 215–218.

ALOSI, J. C. 1986. *Sur un modele de sedimentation deltaique contriution a la connaissance des marges passives*. Thèse, Unversité Perpignan.

ANDREWS, J. T., ERLENKEUSER, H., TEDESCO, K., AKSU, A. E. & JULL, A. J. T: 1994. Late Quaternary stage 2 and 3: meltwater and Heinrich events, northwest Labrador Sea. *Quaternary Research*, **41**, 26.

BARD, E., FAIRBANKS, R. G. & HAMELIN. B. 1992. How accurate are the U-th ages obtained by mass spectometry on coral terraces? In: KUKLA, G. J. & WENT, E. (eds) *Start of a glacial*. Nato ASI Series I: Global Environmental Changes, **3**, 15–21.

——, HAMELIN, B., FAIRBANKS, R. G. 1990. U/Th ages obtained by mass spectrometry in corals from Barbados sea level during the past 130,000 years. *Nature*, **346**, 456–8.

BELL, M. Y. & WALKER, M. J. C. 1992. *Late Quaternary Environmental Change. Physical and Human Perspectives*. Longman Scientific & Technical.

BELLOTI, P., CHIOCCI, F. L., MILLI, S., TORTORA, P. & VALERI, P. 1994. Sequence Stratigraphy and depositional setting of the Tiber Delta. Integration of high-resolution seismic, well logs, and archeological data. *Journal of Sedimentary Research*, **364**, 416–432.

BERGER, W. H. & WEFER, G. 1992. Klimageschichte aus tiefseesedimenten: Neues vom Ontong Java Pateau (Westpazifik). *Naturwissenscheften*, **79**, 1–541.

——, YASUDA, M. K., BICKERT, T., WEFER, G. & TAKAYAMA, T. 1994. Quaternary time scale for

Ontong Java Plateau: Milankovitch template for Ocean Drilling Program Site 806. *Geology*, **22**, 463–7.

BOND, G. & LOTTI, R. 1995. Iceberg Discharges into the North Atlantic on Millenial Time Scales During the Last Glaciation. *Science*, **267**, 1005–1010.

——, BROECKER, W., JOHNSEN, J., MCMANUS, L, LABEYRIE, J., JOUZEL, J. & BONANI, G. 1993. Correlations between climate records from North Atlantic sediments and Greenland ice. *Nature*, **265**, 143–167.

—— & FISHER, W. C. 1980. *Seismic stratigraphy interpretation and petroleum exploration*. AAPG Continuing Education Course Notes Series **16**.

BOYD, R., SUTER, J. & PENLAND, S. 1989. Relation of sequence stratigraphy to modern sedimentary environments. *Geology*, **17**, 926–929.

BROECKER, W. S. 1984. Therminations. *In*: BERGER, A., IMBRIE, J., HAYS, G., KUKLA, G. & SALTZMAN, B. (eds) *Milankovitch and climate – understanding the response to Astronomical Forcing*. NATO ASI Series C. Reidel, Dortrecht, 687–698.

—— & VAN DONK, J. 1970. Insolation changes, ice volumes, and the ^{18}O record in deep-sea cores. *Reviews of Geophysics and Space Physics*, **8**, 169–198.

BROWN, L. F. JR. 1969. Geometry and distribution of fluvial and deltaic sandstones (Pennsylvanian and Permian), north-central Texas. *Gulf Coast Association Geol. Soc. Transactions*, **19**, 23–47.

CHAPPEL, J. & SHACKLETON, N. J. 1986. Oxygen isotopes and sea level. *Nature*, **234** (13): 137–140.

CHIOCCI, F. L. 1994. Very high-resolution seismics as a tool for sequence stratigraphy applied to oucrop scale. Examples from Eastern Tyrrenian Margin Holocene/Pleistocene deposits. *American Association of Petroleum Geologists Bulletin*, **78**, 378–395.

—— 2000. Depositional response to Quaternary fourth-order sea level falls on the northern Latium margin (Tyrrhenian Sea, Italy). *This volume.*

CHRISTIE-BLICK, N. & Driscoll, N. W. 1995. Sequence stratigraphy. *Annual Reviews in Earth and Planetary Science*, **23**, 451–78.

CORREGGIARI, A., ROVERI, M. & TRINCARDI, F. 1992. Regressioni 'fozate', regressioni 'deposizionali' e fenomeni di instabilità in unità progradazionali tardo-quaternarie (Adriatico centrale). *Giornale di Geologia*, **3** (54/1),19–36.

CROSS, T. A. & LESSENGER, M. A. 1988. Seismic Stratigraphy. *Reviews of Earth and Planetary Science*, **16**, 319–354.

DANSGAARD, W., JOHNSEN, S. J., CLAUSEN, H. B., DAHL-JENSEN, D., GUNDESTRUP, N. S., HAMMER, C. U., HVIDBERG, C. S., STEFFENSEN, J. P., SVEINBJÖRNSDOTTIR, J., JOUZEL, J. & BOND, G. 1993. Evidence for general instability of past climate from a 250-Kyr ice record. *Nature*, **364**, 218–220.

DAWSON, A. G. 1992. *Ice Age Earth. Late Quaternary Geology and Climate*. Routledge.

DE BATIST, M. & JACOBS, P. (eds) 1996. *Geology of Siliciclastic Shelf Seas*. Geological Society, London, Special Publications, **117**.

DEVAUX, M. 1985. *Foraminifères et isotopes lègers, indicateurs stratigraphiques et environmentaux de la dernière Quaternaire. Golfe de Cadix-Mer D'Alboran*. Thèse, Université Bordeaux.

ERCILLA, G. & ALONSO, B. 1996. Quaternary siliciclastic sequence stratigraphy of western Mediterranean and tectonically active margins: the role of global versus local controlling factors. *In*: DE BATIST, M. & JACOBS, P. (eds) *Geology of Siliciclastic Shelf Seas*. Geological Society, London, Special Publications, **117**, 125–137.

EMERY, D. & MYERS, K. 1996. *Sequence Stratigraphy*. Blackwell Science.

FAIRBRIDGE, R. W. 1987. The Spectra of Sea Level in a Holocene Time Frame. *In*: RAMPINO, M., SANDERS, J. E., NEWMAN, W. S. & KINIGSSON, L. K. (eds) *Climate: History, Peridocity and Predictability*. Van Nostrand Reinhold Company.

—— 1996. *Subboreal Climatic chronology*. NATO Report, Turkey.

FERNANDEZ-SALAS, L. M., HERNÁNDEZ-MOLINA, F. J., SOMOZA, L. VILLALAIN, J. J., GUERRERO, B., ALONSO, J. J. & DÍAZ DEL RÍO, V. 1996. Estructuración sedimentaria de los depósitos de alto nivel Holoceno del delta del río Guadalhorce (Malaga, Spain). *Geogaceta*, **20**, 412–415.

GAWTHORPE, R. L., FRASER, A. J. & COLLIER, R. E. LL. 1994. Sequence stratigraphy in active extensional basins: implication for the interpretation of ancient basin-fills. *Marine and Petroleum Geology*, **11**, 642–658.

GENSOUS, B. & TESSON, M. 1992. High-resolution sequence stratigraphy of the Late Quaternary Rhone Shelf deposits: relations with tectonic and eustacy; reservoirs implications. *In*: *Symposium on Sequence Stratigraphy of European Basins, Abstract Volume*, 446–447.

——, WILLIAMSON, D. & TESSON, M. 1993. Late Quaternary Transgressive and Highstand Deposits of a Deltaic shelf (Rhone Delta, France). *In*: POSAMENTIER, H. W., SUMMERHAYES, C. P., HAQ B. U. and ALLEN, G. P. (eds) *Sequence Stratigraphy and Facies Associations*. International Association of Sedimentologists Special Publications, **18**, 197–211.

GOY, J. L., ZAZO, C., BARDAJI, T., SOMOZA, L., CAUSSE, C. & HILLAIRE-MARCEL, C. 1993. Elements d'une chronostratigraphie du Tyrrhenien des regions de Alicante–Murcia, Sud-Est de l'Espagne. *Geodinamica Acta,* **6**, 103–119.

HAQ, B. U., HARDENBOL, J. & VAIL, P. R. 1987. Chronology of fluctuating sealevels since the Triassic. *Science*, **235**, 1156–1167.

HARDAGE, B. A. 1987. Seismic Stratigraphy. *In*: HELBIG, K. & TREITEL, S. (eds) *Handbook of Geophysical Exploration, section I; seismic exploration.* Geophysical Press.

HAYS, J. D., IMBRIE, J. & SHACKLETON, N. J. 1976. Variations in the Earth's Orbit – Pacemaker of the Ice Ages. *Science*, **194**, 1121–1132.

HEINRICH, R. 1988. Origin and Consequences of Cyclic Ice Rafting in the Northeast Atlantic Ocean during the Past 130,000 Years. *Quaternary Research*, **29**, 142–152.

HELLAND-HANSEN, W. & GJELBERG, J. G. 1994. Conceptual basis and variability in sequence stratigraphy: a different perspective. *Sedimentary Geology*, **92**, 31–52.

HERNÁNDEZ-MOLINA, F. J. & SOMOZA, L. 1995. Late Quaternary continental shelf high resolution sequence stratigraphy model responses to punctuated sea-level asymmetric cycles. *In*: HUNT, D., GAWTHORPE, R. & DOCHERTY, M. (eds) *Sedimentary Responses to Forced Regression: Recognition, Interpretation and Reservoir potential*. Conference abstracts, Geological Society, London.

——, —— & REY, J. 1996. Late Pleistocene–Holocene high-resolution sequence analysis on the Alboran Sea continental shelf. *In*: DE BATIST, M. & JACOBS, P. (eds) *Geology of Siliciclastic Shelf Seas*. Geological Society, London, Special Publications, **117**, 139–154.

——, ——, —— & POMAR, L. 1994. Late Pleistocene-Holocene SEDIMENTS ON on the Spanish continental shelves: Models for very high-resolution sequence stratigraphy. *Marine Geology*, **120**, 129–174.

——, ——, LOBO, F. J., DÍAZ DEL RÍO, V. & ROQUE, C. 1998. Seismic Stratigraphy of the South Iberian Continental Shelf and Slope: an Hypothesis for Quaternary High Resolution Sequence Stratigraphy. Strata and Sequences on Shelves and Slopes. *In*: SEPM-IAS Sicily, Italy, September 15–19, Abstract volume.

HILLAIRE-MARCEL, CL., CARRO, O., CAUSSE, CH, GOY, J. L. & ZAZO, C. 1986. Th/U dating of Strombus bubonius-bearing marine terraces in southeastern Spain. *Geology*, **14**, 613–616.

HUNT, D. & TUCKER, M. E. 1992. Stranded parasequences and the forced regressive wedge systems tract: deposition during base-level fall. *Sedimentary Geology*, **81**, 1–9.

—— & —— 1995. Stranded parasequences and the forced regressive wedge systems tract: deposition during base-level fall.-Reply. *Sedimentary Geology*, **95**, 147–160.

JERVEY, M. T. 1988. Quantitative Geological Modeling of Siliciclastic rock sequences and their seismic expression. *In*: WILGUS, C. K., HASTINGS, B. S., KENDALL, C. G. S. C., POSAMENTIER, H. W., ROSS, C. A. & VAN WAGONER, J. C. (eds) *Sea level changes- an integrated approach*. SEPM, Special Publication, **42**, 47–69.

KOLLA, V., P. BIONDI, P., LONG, B. & FILLON, R. 2000. Sequence stratigraphy and architecture of the late Pleistocene Lagniappe delta complex, northeast Gulf of Mexico. *This volume*.

——, POSAMENTIER, H. W. & EICHENSEER, H. 1995. Stranded parasequences and the forced regressive wedge systems tract: deposition during base-level fall.-Discussion. *Sedimentary Geology*, **95**, 147–160.

LARIO, J. 1996. *Ultimo y presente interglaciar en el area de conexión Atlantico-Mediterraneo (Sur de España) Variaciones del nivel del mar, paleoclima y paleoambientes*. Univ. Complutense Madrid, PhD Dissertation.

——, ZAZO, C., DABRIO, C. J., SOMOZA, L., GOY, J. L., BARDAJI, T. & SILVA, P. G. 1995. Record of Recent Holocene Sediment input on Spit Bards and Deltas of South Spain. *In*: CORE, B. (ed.) *Holocene Cycles: climate, Sea Levels, and Sedimentation. Journal of Coastal Research Special Issue*, **17**, 241–245.

——, ——, PLATER, A. J., GOY, J. L., DABRIO, C. & BORJA, F. 1996. Late Holocene evolution of the Guadalquivir marshes (SW Spain). *In*: *28th International Geographical Congress*, The Hague, August 1996. Abstract volume 249.

——, ——, SOMOZA, L., GOY, J. L., HOYO, M., SILVA, P. G. & HERNÁNDEZ-MOLINA, F. J. 1993. Los episodios marinos cuaternarios de la costa de Malaga (España). Revista Sociedad Geología España, **6**, 41–46.

LOBO, F. 1995. *Estructuración y evolución morfosedimentaria de un sector del margen continental septentrional del Golfo de Cádiz durante el Cuaternario terminal*. Dissertation. Universidad de Cádiz.

——, HERNÁNDEZ-MOLINA, F. J., SOMOZA, L. & DÍAZ DEL RÍO, V., in press. Paleoenvironments, relative sea level changes and tectonic influence on the Quaternary seismic units of the Huelva continental shelf (Gulf of Cadiz, Southwestern Spain). *In*: *Special Publications of the Spanish Oceanographic Institute*.

LOWE, J. J. & WALKER, M. J. C. 1984. *Reconstructing Quaternary Environments*. Longman, Essex, U. K.

LOWRIE, A. & HAMITER, R. 1995. Fifth and sixth order eustatic events during Holocene (fourth order) highstand influencing Mississippi delta-lobe switching. *Journal of Coastal Research, Special issue*, **17**, 225–230.

MALDONADO, A. 1992. El Mar de Alboran y el Golfo de Cadiz conexiones Atlántico-Mediterráneo. Una introducción. *In*: *III Congreso Geológico de España*, Simposios, Salamanca, **2**, 239–246.

—— & NELSON, C. H. 1988. Dos ejemplos de margenes continentales de la Peninsula Ibérica: el margen del Ebro y el Golfo de Cadiz. *Revista Sociedad Geología España* **1**, 317–325.

—— & —— (eds) 1999. *Marine Geology of the Gulf of Cadiz. Marine Geology*, **155**.

MALOD, J. A. 1982. *Comparaison de l'évolution des marges continentales au nord et au sud de la Peninsule Ibèrique*. Thèse de doctorat d'etat. Memoires Sciences de la Terre Université, P & M Curie, Paris.

—— & MOUGENOT, D. 1979. L'Histoire géologique néogène du Golfe de Cadiz. *Bulletin de la Société Géologique France*, **7**, 603–611.

MARTINEZ DEL OLMO, W., GARCIA-MALLO, L., LERET-VERDU, G., SERRANO-O—ATE, A. & SUAREZ-ALBA, J. 1984. Modelo tectosedimentario del bajo Guadalquivir. I Congreso Español de Geología, **1**, 199–213.

MARTISON, D. G., PISIAS, N. G., HAYS, J. D., IMBRIE, J., MOORE, T. C. & SHACKLETON, N. J. 1987. Age, dating and orbital theory of Ice Ages: Development of a high resolution 0 to 300.000-year chronostratigraphy. *Quaternary Research*, **27**, 1–29.

MAYEWSKI, P. A., MEEKER, L. D., TWICKLER, M. S., WHITLOW, S., YANG, Q. & PRENTICE, M. 1996. Major Features and Forcing of High Latitude Northern Hemisphere Atmospheric Circulation Over the Last 110,000 Years. *Journal of Geophysical Research*, **102**, **C12**, 26345–26366.

MCMURRAY, L. S. & GAWTHORPE, R. L. G. 2000. Along-strike variability of forced regressive deposits: late Quaternary, northern Peloponnesos, Greece. *This volume.*

MCQUILLIN, R., BACON, M. & BARCLAY, W. 1984. *An Introduction to Seismic Interpretation: Reflection Seismics in Petroleum Exploration*. Graham & Trotman Limited.

MÉLÌRES, F. 1974. *Recherches sur le dynamique sèdimentaire du Golfe de Cadix (Espagne)*. Thesis, Univ. Paris.

MELLERE D. & STEEL, R. 1995. Variability of lowstand wedges and their distinction from the forced-regressive wedges in the Mesaverde Group, southeast Wyoming. *Geology*, **23**, 803–806.

—— & —— 2000. Style contrast between forced regressive and lowstand/transgressive wedges in the Campanian of south-central Wyoming. *This volume.*

MITCHUM, R. M. Jr. 1977. Seismic stratigraphy and global changes of sea level, part 1: glossary of terms used in seismic stratigraphy. *In*: PAYTON, C. E. (ed.) *Seismic Stratigraphy – Aplications to Hydrocarbon Exploration*. AAPG Memoirs, **26**, 53–62.

—— & VAN WAGONER, J. C. 1991. High-frequency sequences and their stacking patterns: sequence-stratigraphy evidence of high-frequency eustatic cycles. *Sedimentary Geology*, **70**, 131–160.

——, VAIL, P. R. & THOMPSOM, S. III, 1977. Seismic stratigraphy and global changes of sea level,part 2: the depositional sequence as a basic unit for stratigraphic analysis. *In*: PAYTON, C. E. (ed.) *Seismic Straigraphy – Applications to Hydrocarbon Exploration*. AAPG Memoirs, **26**, 53–62.

NELSON, C. H. & MALDONADO, A. 1990. *The Ebro Delta continental shelf. Special Issue Marine Geology,* **95**.

——, BARAZA, J. & MALDONADO, A. 1993. Mediterranean undercurrent sandy contours, Gulf of Cádiz, Spain. *In*: STOW, D. A. V. & FAUGERES, J. C. (eds) *Contourites and Hemipelagites in the deep sea. Sedimentary Geology,* **82**, 103–131.

——, ——, ——, RODERO, J., ESCUTIA, C. & BARBER, J. H., JR. 1999. Influence of the Atlantic inflow and Mediterranean outflow currents on Late Quaternary sedimentary facies of the Gulf of Cadiz continental margin. *Marine Geology*, **155**, 99–129.

ORTLIEB, L., FOURNIER, M. & MACHAR, J. 1995. Beach ridges and major late Holocene El Niño events in northern Peru. *Journal of Coastal Research, Special issue*, **17**, 109–118.

PAYTON, C. 1977. *Seismic Stratigraphy – Applications to Hydrocarbon Exploration*. American Association of Petroleum Geologists, Memoirs, **26**.

PLINT, A. G. & NUMMEDAL, D. 2000. The falling stage systems tract: recognition and importance in sequence stratigraphic analysis. *This volume.*

POSAMENTIER, H. W. & ALLEN, G. P. 1993. Variability of the sequence stratigraphic model: effects of local basin factors. *Sedimentary Geology*, **86**, 91–109.

——, ALLEN, G. P., JAMES, D. P. & TESSON, M. 1992. Forced regressions in a sequence stratigraphic framework: concepts, examples, and exploration significance. *AAPG Bulletin*, **76**, 1687–1709.

—— & JAMES, D. P. 1993. An overview of sequence-stratigraphy concepts: uses and abuses. *In*: POSAMENTIER, H. W., SUMMERHAYES, C. P., HAQ, B. & ALLEN, G. P. (eds) *Sea-level changes – an integrated approach*. Special Publications, International Association of Sedimentologists, **18**, 3–18.

—— & VAIL, P. R. 1988. Eustatic controls on clastic deposition II.- Sequence and systems tract models. *In*: WILGUS, C. K., HASTINGS, B. S., KENDALL, C. G. S. C., POSAMENTIER, H. W., ROSS, C. A. & VAN WAGONER, J. C. (eds) *Sea level changes – an integrated approach*. SEPM, Special Publications, **42**, 125–154.

——, JERVEY, M. T. & VAIL, P. R. 1988. Eustatic controls on clastic deposition I.- Conceptual Framework. *In*: WILGUS, C. K., HASTINGS, B. S., KENDALL, C. G. S. C., POSAMENTIER, H. W., ROSS, C. A. & VAN WAGONER, J. C. (eds) *Sea level changes- an integrated approach*. SEPM, Special Publications, **42**, 109–124.

—— & MORRIS, W. S. 2000. Aspects of the stratal architecture of forced regressive deposits. *This volume.*

——, SUMMERHAYES, C. P., HAQ, B. U. & ALLEN, G. P. (eds) 1993. *Sequence Stratigraphy and Facies Associations*. International Association of Sedimentologists, Special Publication, **18**, 197–211.

RIAZA, C. & MARTINEZ DEL OLMO, W. 1996. Depositional Model of the Guadalquivir-Gulf of Cádiz Terttiary Basin. *In*: FRIEND, P. F. & DABRIO, C. J. (eds) *Tertiary Basins of Spain. The Stratigraphic Record of Crustal kinematics*.

RUDDIMAN, W. & MCINTYRE, A. 1981. The north Atlantic ocean during the last deglaciation. *Palaeogeography Palaeoclimatology Palaeoecology*, **35**, 145–214.

—— & RAYMO, 1988. Northern Hemisphere climate regimes during the past 3 ma: possible tectoni connections. *Philosophical Transactions of the Royal Society, London*, **B318**, 411–430.

——, MCINTYRE, A. & SHACKLETON, N. J. 1986. North Atlantic sea-surface temperatures for the last 1.1 million years. *In*: SUMMERHAYES, C. P. & SHACKLETON, N. J. (eds) *North Atlantic Paleoceanography*. Geological Society, London, Special Publications, **21**, 155–173.

RODERO, J., PALLARÉS, L. & MALDONADO, A. 1999. Late Quaternary seismic facies of the Gulf of Cadiz Spanish margin: depositional processes influenced by sea-level change and tectonic controls. *Marine Geology*, **155**, 131–156.

ROQUE, A. C. E. P. 1998. *Análise morfosedimentar da sequência deposicional do Quaternário Superior da plataforma continental Algarvia entre Faro e a foz do Rio Guadiana*. Dissertation, University of Lisbon.

SAITO, Y. 1994. Shelf sequences and characteristic boundary surfaces in a wave dominated setting; Latest Pleistocene-Holocene samples from the North East Japan. *Marine Geology*, **120**, 105–127.

SHACKLETON, N. J. 1987. Oxygen isotopes, ice volume and sea level. *Quaternary Science Review*, **6**, 183–190.

—— & OPDYKE, N. K. 1973. Oxygen isotope and paleomagnetic stratigraphy of equatorial Pacific core V28-238: Oxygen istope temperatures and ice-volumes on a 105 and 106 year scale. *Quaternary Research*, **3**, 39–55.

SHERIFF, R. E. 1980. *Seismic Stratigraphy*. International Human Resources Development Corporation. Boston.

SOMOZA, L., ANDRÉS, J. R., REY, J., HERNÁNDEZ-MOLINA, F. J., RODRÍGUEZ-VIDAL, J., CLEMENTE, L., RODRÍGUEZ-RAMÍREZ, A. & DÍAZ DEL RÍO, V. 1994. Arquitectura morfo-deposicional de la plataforma continental del Golfo de Cádiz: Proyecto Golca. *Gaia*, **9**, 79–82.

——, BARNOLAS, A., ARASA, A., MAESTRO, A. & HERNÁNDEZ-MOLINA, F. J. 1998. Architectural staking patterns of the Ebro Delta controlled by Holocene high-frequency eustatic fluctuations, delta-lobe switching and subsidence procesees. *Sedimentary Geology*, **117**, 11–32.

——, GOY, J. L., ZAZO, C., BARDAJI, T., DABRIO, C. J., SILVA, P. & LARIO, J. 1991. Holocene Sea-Level Highstands in the Western Mediterranean (SE Spain). A computer-statistical analysis of radiocarbon dating. Mediterranean and Black Sea Shorelines *INQUA Subcommision Newsletter*, **13**, 61–64.

——, HERNÁNDEZ-MOLINA, F. J., DE ANDRÉS, J. R. & REY, J. 1997. Continental shelf architecture and sea level cycles: Late Quaternary high-resolution stratigraphy of the Gulf of Cádiz, Spain. *Geo-Marine Letters*, **17**, 133–139.

——, ——, LOBO, F. J. & DE ANDRÉS, J. R. 1996. Modelo de estratigrafía secuencial de alta resolución de la plataforma continental del Golfo de León. *Geogaceta*, **20**, 142–145.

——, ZAZO, C., BARDAJI, T., GOY, J. L. & DABRIO, C. J. 1987. Late Pleistocene sea-level changes in a tectonically active coast (SE Spain). *In*: ZAZO, C. (ed.) *Late Quaternary Sea Level Changes in Spain*. Trabajos sobre Neogeno-Cuaternario, Museo de Ciencias Naturales-CSIC, **10**, 49–77.

SONNENFELD, M. D. & CROSS, T. A. 1993. Volumetric partionning and facies differentiation within the Permian Upper San Andres Formation of Last Chance Canyon, Guadalupe Mountains. New Mexico. *In*: LOUCKS, R. G. & SARG, J. F. (eds) *Carbonate Sequence Stratigraphy*. AAPG. Memoirs, **57**, 435–74.

SUTER, J. R., BERRYHILL, H. L. & PENLAND, S. 1987. Late Quaternary sea-level fluctuations and depositional sequences, Southwest Louisiana continental shelf. *In*: NUMMEDAL *ET AL.*, (eds) *Sea-Level Fluctuation and Coastal Evolution*. SEPM, Special Publications, **41**, 199–222.

SWIFT, D. J. P., OERTEL, G. F., TILLMAN, R. W. & THORNE, J. A. (eds) 1991. *Shelf sand and sandstone bodies: Geometry, facies and sequence stratigraphy*. Special Publications of the International Association of Sedimentologists, **14**.

SYDOW, J. & ROBERTS, H. H. 1994. Stratigraphic Framework of a late Pleistocene shelf-Edge Delta northeast Gulf of Mexico. *AAPG Bulletin*, **78**, 1276–1312.

TESSON, M. & ALLEN, G. P. 1995. Contrôle tectonique et eustatique haute-fréquence de l'architecture et de la stratigraphie de dépôts de plate-forme péricratonique. Exemple du Golfe du Lion (Méditerranée, France) et des dépôts quaternaires. *Comptes Rendus Academie del'Sciences, Paris*, **320**, 39–46.

——, ALLEN, G. P. & RAVENNE, C. H. 1993. Late Pleistocene shelf-perched lowstand wedges on the Rhone continental shelf. *In*: Special Publication of International Association of Sedimentologists, **18**, 183–196.

——, GENSOUS, B., ALLEN, G. P. & RAVENNE, CH 1990 b. Late Quaternary Deltaic Lowstand Wedges on the Rhône Continental shelf, France. *Marine Geology*, **91**, 325–332.

——, RAVENNE, CH. & ALLEN, G. P. 1990*a*. Application des concepts de l'analyse en stratigraphie séquentielle et un profil sismique haute résolution transverse á la plataforme rhôdanienne. *Comptes Rendus de Academie des Sciences, Paris*, **310** (II), 565–570.

THOMAS, M. A. & ANDERSON, J. B. 1991. Late Pleistocene Sequences Stratigraphy of the Texas Continental Shelf relationship to $\delta^{18}O$ Curves. *In*: GCS SEPM Foundation Annual Research Conference, 265–270.

—— & FIELD, M. E. 1991. Geometry, lateral variability, and preservation of downlapped regressive shelf deposits: eastern Tyrrhenian margin, Italy. *Journal of Sedimentary Petrology*, **61**, 775–790.

TORRES, J., SAVOYE, B. & COCHONAT, P. 1995. The effects of Late Quaternary sea-level changes on the Rhône slope sedimentation (north-western Mediterranean), as indicated by seismic stratigraphy. *Journal of Sedimentary Research*, **B65**, 368–397.

TORTORA, P. 1996. Depositional and erosional coastal processes during the last postglacial sea-level rise: an example from the central Tyrrhenian continental shelf (Italy). *Journal of Sedimentary Research*, **66**, 391–405.

TRINCARDI, F. & CORREGGIARI, A. 1997. Quaternary forced-regression deposits in the Adriatic basin and the record of composite sea-level cycles. *This volume*.

VAIL, P. R. 1987. Seismic stratigraphy interpretation procedure. *In*: BALLY, A. W. (ed.) *Atlas of Seismic Stratigraphy*. AAPG Studies in Geology, **27 (1)**, 1–11.

——, ANDERMARD, F., BOWMAN, S. A., EISNER, P. N. & PEREZ-CRUZ, G. 1991: The Stratigraphic Signatures of Tectonics, Eustasy and Sedimentology-an Overview. *In*: EINSELE, G., RICKEN, W. & SEILACHER, A. (eds) *Cycles and Events in Stratigraphy*. Springer-Verlag, 617–659.

——, MITCHUM, R. M. JR. & THOMPSON, M. S. 1977. Seismic stratigraphy and global changes of sea level, part 4: global cycles of relative changes of sea level. *In*: PAYTON, C. E. (ed.) *Seismic Stratigraphy-Applications to Hydrocarbon Exploration*. AAPG Memoirs, **26**, 63–81.

VAN WAGONER, J. C., MITCHUM, R. M. JR, CAMPION, K. M. & RAHMANIAN, V. D. 1990. *Siliciclastic sequence stratigraphy in well logs, cores and outcrops*. AAPG, Methods, Exploration Services, **7**.

——, ——, POSAMENTIER, H. W. & VAIL, P. R. 1987. Seismic stratigraphy interpretation using sequence stratigraphy. Part II: Key definitions of sequence stratigraphy. *In*: BALLY, A. W. (ed.) *Atlas of seismic stratigraphy 1*. AAPG Studies in Geology, **27**, 11–14.

——, POSAMENTIER, H. W., MITCHUM, R. M., VAIL, P. R., SARG, J. F., LOUTIT, T. S. & HARDENBOL, J. 1988. An overview of the fundamentals of sequence stratigraphy and key definitions. *In*: WILGUS, C. K., ROSS, CH. A., POSAMENTIER, H. W., VAN WAGONER, J. & KENDALL (eds) *Sea Level Changes: An Integrated Approach*. SEPM Special Publication, **42**, 39–45.

VERBITSKY, M. & SALTZMAN, B. 1995. A diagnostic analysis of Heirich glacial surge events. *Paleoceanography*, **10**, 59.

ZAZO, C. 1999. Interglacial sea levels. *Quaternary International*, **55**, 101–113.

——, GOY, J. L., DABRIO, C. J., BARDAJI, T., SOMOZA, L. & SILVA, P. G. 1993. The last interglacial in the Mediterranean as a model for the present interglacial. *Global and Planetary Change*, **7**, 109–117.

——, ——, LARIO, J. & SILVA, P. G. 1996. Littoral zone and rapid climatic changes during the last 20,000 years. The Iberian study case. Zeitschrift für Geomorphologie, N. F. Suppl.-Bd. **102**, 119–134.

——, ——, SOMOZA, L., DABRIO, C. J., BELLUOMINI, G., IMPROTA, S., LARIO, J., BARDAJI, T. & SILVA, P. G. 1994. Holocene Sequence of Sea-Level Fluctuations in Relation to Climatic Trends in the Atlantic-Mediterranean Linkage Coast. *Journal of Coastal Research*, **10**, 933–945.

Along-strike variability of forced regressive deposits: late Quaternary, northern Peloponnesos, Greece

LESLEY S. McMURRAY[1] & ROBERT L. GAWTHORPE

Basin & Stratigraphic Studies Group, Department of Earth Sciences, University of Manchester, Manchester M13 9PL, UK

[1]Present address: BP Amoco Company, Farburn Industrial Estate, Dyce, Aberdeen AB21 7PB, UK

Abstract: Late Quaternary depositional sequences in northern Peloponnesos, central Greece, occur as a number of offlapping and downstepping forced regressive wedges. Major along-strike variability is evident, with three environmental responses to forced regression: (i) attached shoreface deposits; (ii) detached shoreface deposits; (iii) fan-deltas. All three depositional systems are cut by incised valleys. Relative sea-level change was responsible for similarities in key stratal surfaces and stacking patterns; regional uplift drove the overall forced regression, whereas individual sequences relate to fourth- and fifth order glacio-eustatic cycles. Variations in basin physiography and the amount and type of sediment led to along-strike variability of depositional sequences. Fan-deltas developed at the mouths of incised valleys in the west of the area, where supply of coarse-grained sediment was high and slope gradients were steep. In contrast, limited supply of coarse-grained sediment and low slope gradients over most of the study area promoted the development of shoreface systems. Forced regressive wedges in the shoreface systems attach and detach along strike. Detached wedges developed where both slope gradients and coarse-grained sediment supply were low, away from the axes of major incised valleys.

Sequence stratigraphic models tend to emphasise variations in accommodation space as the control on stratal geometry and facies stacking patterns (e.g. Posamentier & Vail 1988; Van Wagoner *et al.* 1990). Recently, however, additional factors such as sediment supply variability and basin physiography have been increasingly recognized as important controls on the temporal and spatial evolution of depositional systems (e.g. Posamentier & Allen 1993; Schlager 1991; Gawthorpe *et al.* 1994; Church & Gawthorpe 1997). In particular, basin margin physiography has been cited as a major control on the nature of depostion during relative sea-level fall (e.g. Posamentier & Allen 1993). Basin margins with pronounced shelf-slope breaks commonly exhibit major incised valleys and canyons, with sediment deposited as deep water turbidites on the basin floor (e.g. Posamentier & Vail 1988; Van Wagoner *et al.* 1990). In contrast, ramp-type margins typically lack turbidites; instead, shallow-marine forced regressive wedges characterize times of relative sea-level falls (Plint 1988; Hunt & Tucker 1992; Posamentier *et al.* 1992; Ainsworth & Pattison 1994).

Despite such recent advances, sequence stratigraphic models still tend to be two-dimensional and dominated by dip-oriented sections; in particular, relatively few studies document the strike variability of depositional sequences. Notable exceptions include Gawthorpe *et al.* (1994) who document strike variation in sequence development resulting from changes in sediment supply and accommodation space around segmented normal fault zones, and Martinsen & Helland-Hansen (1995) and Church & Gawthorpe (1997) who illustrate the influence of strike-variable sediment supply on facies stacking patterns during relative sea-level rise.

The aim of this paper is to illustrate the strike-variability of a forced regressive sequence set composed of shallow marine and deltaic depositional systems and to discuss the processes responsible for the observed variations in facies and stratal geometry. Our study is based on detailed outcrop studies of late Quaternary sediments exposed along an approximately 50 km strike section of the northern Peloponnesos coast, central Greece. The area has been subject to regional uplift during the late Quaternary and this has led to punctuated relative sea-level fall and pronounced forced regression. Although the processes and products of forced regression have received attention in recent years (e.g. Hunt & Tucker 1992, 1995; Posamentier *et al.* 1992; Ainsworth & Pattison 1994), the strike variability of forced regressive wedges and the factors controlling their variability have not been fully addressed.

From: HUNT, D. & GAWTHORPE, R. L. (eds) *Sedimentary Responses to Forced Regressions.* Geological Society, London, Special Publications, **172**, 363–377. 1-86239-063-0/00/$15.00

Geological setting

Our data are taken from the late Quaternary deposits of the northern Peloponnesos coast, central Greece (Fig. 1). Although central Greece is undergoing active N–S extension, resulting in the development of a series of half grabens and asymmetric grabens (e.g. the Gulf of Corinth; Fig. 1), the northern Peloponnesos is undergoing active uplift at a rate of 0.3–1.3 m ka^{-1} (Collier 1990; Collier *et al.* 1992; Gawthorpe *et al.* 1994; Armijo *et al.* 1996). There is still debate regarding the mechanism generating uplift. Collier *et al.* (1992) suggest that the uplift is regional in nature, whereas Armijo *et al.* (1996) conclude that the observed uplift between Xylokastro and Corinth is due to local uplift in the footwall of the Xylokastro fault (Fig. 1). However, the presence of uplifted marine terraces between Xylokastro and Akrata, in the hanging wall of the Xylokastro fault (as mapped by Armijo *et al.* 1996), suggests that the Xylokastro fault cannot be the only cause of uplift in the northern Peloponnesos.

When combined with late Quaternary glacio-eustatic sea-level change, the uplift results in long-term destruction of accommodation space, with short intervals of accommodation creation associated with rapid post-glacial eustatic sea-level rise (Fig. 2). The magnitude of sea-level fall in the Gulf of Corinth is also moderated by a structural high, the Rion Sill, at the western end of the Gulf of Corinth which has a present-day elevation of –60 m below sea level (e.g. Perissoratis *et al.* 1993).

A further aspect of this study is that spatial (along-strike) variations in sediment supply and physiography can be assessed through analysis of drainage basin catchment areas and geomorphology. Thus it is possible to link the stratigraphy of the forced regressive deposits directly to both regional relative sea-level change and local variations in physiography and sediment supply.

The late Pleistocene deposits that are the subject of this study form a coast-parallel outcrop belt extending up to 5 km inland from the present day southern shore of the Gulf of Corinth. Stratigraphic relationships (Gawthorpe *et al.* 1994; and documented here) and U-series dating of corals (e.g. Collier 1990; Collier *et al.* 1992), indicate that the progressively younger late Quaternary deposits are situated at lower elevations and closer to the present coastline. This basinward shifting and stratigraphically

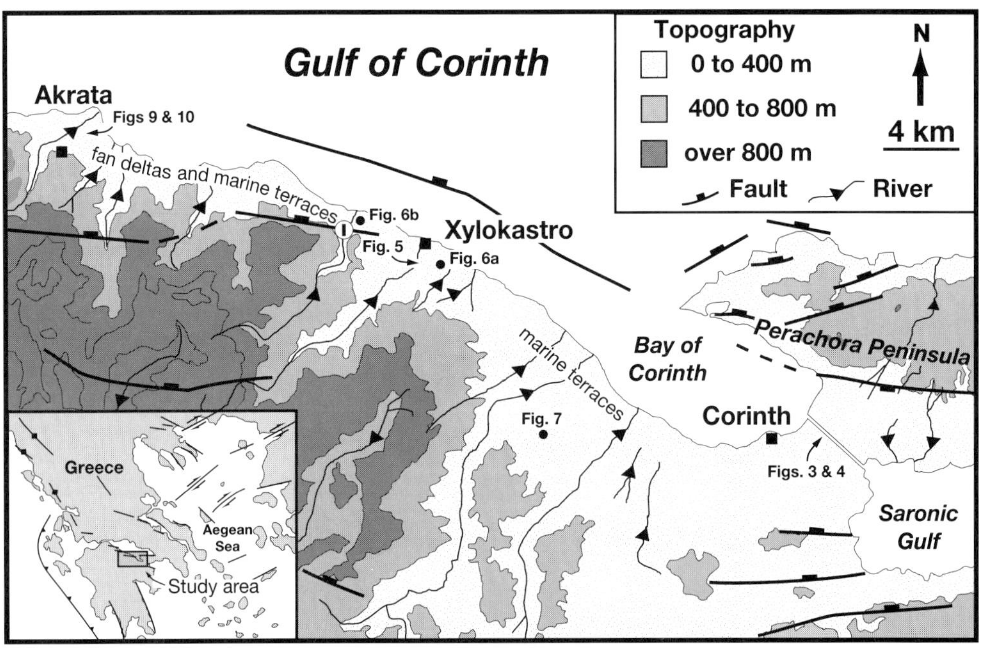

Fig. 1. Simplified topographic map of the study area. The late Quaternary deposits discussed in this paper form most of the low ground (below 400 m elevation) along the coastal belt between the towns of Corinth and Akrata. The location of fluvial gorge incision through Mesozoic limestone basement is highlighted (I). The inset map shows the location of the study area in central Greece.

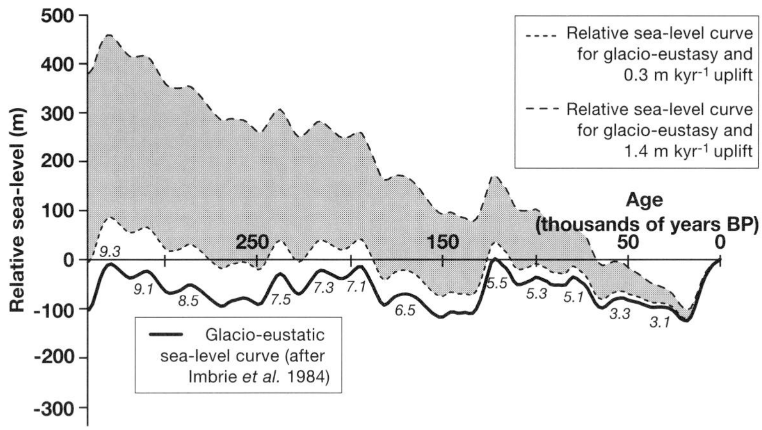

Fig. 2. Glacio-eustatic sea-level curve (after Imbrie *et al.* 1984), and relative sea-level curves for the study area. The range of relative sea-level in the study area is shaded; numbers in italics are oxygen isotope stages of the main interglacials and interstadials for the last 350 ka. Relative sea-level curves are based on estimates of range of uplift rates of between 0.3 and 1.3 m ka^{-1} (Collier 1990; Collier *et al.* 1992; Armijo *et al.* 1996).

falling character of the deposits reflects forced regression during long-term relative sea-level fall. Although the deposits are cut by a series of incised valleys, which are presently the sites of the main rivers entering the Gulf of Corinth, it is possible to trace coeval depositional systems along strike for approximately 50 km between the towns of Corinth in the east and Akrata in the west (Fig. 1). Over this strike distance, three main styles of forced regressive deposit can be identified, based on the dominant depositional environment and their connectivity. The three main styles of forced regressive deposit are: (i) attached shoreface deposits near the town of Corinth in the eastern part of the study area; (ii) detached shoreface deposits in the central part of the study area; (iii) fan-deltas in the west (Fig. 1). Descriptions and interpretation of the sedimentology and sequence stratigraphy of the three types of depositional system are outlined in the following sections, together with a discussion of the regional and local controls on the evolution of the forced regressive deposits.

Attached shoreface deposits

The attached shoreface deposits are superbly exposed in the Corinth Canal (Collier 1990; Gawthorpe *et al.* 1994). The sediments exposed in the canal consist of marls, sandstones and conglomerates that reflect deposition in offshore, shoreface, foreshore and alluvial environments (see Collier (1990) for a full description of these facies). In addition, bioclastic and oolitic grainstones represent shallow marine deposition in shoreface and offshore bar environments under clastic-starved conditions. Clast lithologies include serpentinite and red chert, which have a distinctive provenance, sourced from serpentinite-rich drainage catchments in the footwall of the Loutraki fault to the east, on the Perachora Peninsula; (Fig. 1). Sediment was transported westwards by longshore currents from an alluvial fan/fan-delta, the Smarpsi fan, to the Corinth Canal area.

At least six depositional sequences can be recognized in the Corinth Canal (the 'sub-sequences' of Collier 1990) each of which attains a maximum preserved thickness of up to 30 m (Fig. 3). U-series dating of *Acropora* corals from these sequences (Collier 1990) gives ages coincident with 100 ka interglacial highstands (Fig. 3), suggesting they are fourth-order sequences, related to variations in the eccentricity of the Earth's orbit. Within these 4th order sequences higher frequency sequences, up to 10 m thick, can be recognized. Based on the existing U-series dates, and correlation with the detached shoreface deposits forming terraces further west, these fifth- and/or sixth-order sequences are inferred to be related to obliquity or precessional changes in the Earth's orbit.

The internal facies geometries and bounding surfaces of the fourth- and higher-order sequences display a number of similarities. Internally the sequences coarsen and shoal upward (Fig. 4a) and display a proximal to distal fining from foreshore conglomerates, closest to the Central Horst in the SE, to bioturbated lower shoreface sandstones towards the NW

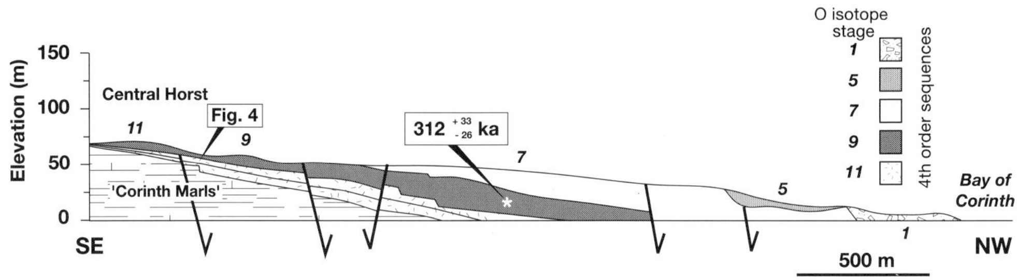

Fig. 3. Cross-section showing the fourth-order sequences and their stacking pattern exposed in the northern section of the Corinth Canal (after Collier 1990; Gawthorpe *et al.* 1994) (see Fig. 1 for location). The cross-section illustrates the basinward offlapping and downstepping stacking pattern of successive sequences. Note how successively younger sequences are connected to one another and thus form an attached forced regressive sequence set. The location of the U-series age from the stage 9 sequence is also indicated (Collier *et al.* 1992).

(Fig. 3). The surfaces between individual sequences are characterised by evidence of subaerial exposure (calcrete and minor karst development) and local fluvial incision in up-dip locations (Fig. 4b). In more basinward settings, the surfaces lack evidence of subaerial exposure and are generally planar and seaward dipping with spectacular steps (palaeocliffs) up to several metres in height (Fig. 4b). The evidence for subaerial exposure and incision suggests that the surfaces formed initially during a phase of relative sea-level fall and, as such, are interpreted as sequence boundaries. The down-dip portions of these surfaces were subsequently modified by transgressive wave ravinement erosion, which largely removed the evidence for subaerial exposure and created the seaward dip and stepped profile of the composite sequence boundary/transgressive surface (Gawthorpe *et al.* 1994).

The stacking pattern of the sequences in the canal is distinctive, with younger sequences located further basinward and at a lower elevation (Fig. 3). They exhibit an offlapping and downstepping stacking pattern suggesting that they formed during long-term relative sea-level fall (Fig. 3). Up-dip, the individual sequence boundaries converge and become coincident as

Fig. 4. Facies stacking patterns and key stratal surface characteristics of the attached shoreface sequences from the Corinth Canal (see Figs 1 and 3 for location). (**a**) Shoaling and coarsening upward facies stacking pattern indicated by transition from lower shoreface sandstones to upper shoreface and foreshore conglomerates. Note sharp contact between the lower shoreface facies and the underlying foreshore conglomerates. This surface is interpreted as a coincident transgressive surface/sequence boundary (Ts/Sb) due to remnant calcrete indicating subaerial exposure prior to flooding. (**b**) Detail of basal contact of the first sequence overlying Corinth Marls showing a prominent palaeocliff approximately 1.75 m high. Note seaward dipping wave ravinement surface on left (NW) side of palaeocliff. On the right (SE) of the palaeocliff calcrete and minor karst provide evidence of subaerial exposure prior to transgression.

a composite surface (Fig. 3). The exposures in the sides of the canal show that the shoreface deposits of individual sequences are connected to each other, and that they are not isolated by zones of by-pass. Thus, overall, the succession is interpreted as an attached forced regressive sequence set.

Detached shoreface deposits

The central part of the study area is characterised by a series of thin, generally <5 m thick, marine shoreface deposits that form a prominent staircase of terraces stepping down to the present coastline (Fig. 5). These shoreface deposits can be correlated along strike from Corinth in the east, past the town of Xylokastro, towards Akrata in the west (Fig. 1). The morphology of the terraces has been well documented (e.g. Keraudren & Sorel 1987; Doutsos & Piper 1990; Armijo *et al.* 1996), although limited work has been undertaken on their sedimentology and sequence stratigraphy. Up to 12 terrace levels can be mapped, but several of these merge or become indistinct along strike. Around the town of Xylokastro, prominent terrace levels occur at 20–30, 60, 90, 120, 150, 240, 300 and 350 m above present sea-level (Fig. 5). Individual terrace levels are separated by steep, seaward-dipping scarps, 10–30 m high. The terraces themselves are generally 0.4-1.5 km wide, with terrace width decreasing systematically from east to west.

The marine terraces are cut by modern drainage networks that form a series of incised valleys and provide spectacular exposures of the facies, stratal geometry and stacking patterns of the shoreface deposits (e.g. Fig. 5). The shoreface deposits that form the terraces unconformably overlie older, tilted basin-fill sediments, which have structural dips of up to 50°. The tops of the terraces are often extensively calcretised and incised. Locally, pedogenic calcrete is preserved on the basal unconformity surfaces in up-dip locations, particularly around palaeocliffs that mark the landward edge of the terraces (Fig. 6a). However, *in situ* calcrete is generally absent over the sub-horizontal basal contact of the terraces, yet it is present as clasts within *Lithophaga*-bored boulder lags immediately overlying the basal surface (Fig. 6b). These relationships are very similar to those exposed in the Corinth Canal and suggest a two-stage development for the basal surfaces of the terraces; an initial stage of subaerial exposure, followed by a second stage of marine erosion. The first stage of subaerial exposure is interpreted to have occurred during relative sea-level fall and would have been associated with fluvial incision localised along what are now the present-day valleys. The second stage of marine erosion is interpreted to correspond with transgressive ravinement erosion during the subsequent sea-level rise. It is this transgressive marine erosion that created the planar, seaward-dipping basal contact to the terrace deposits.

Above each basal conglomerate lag, the shoreface deposits display an overall aggradational to progradational stacking pattern, dominated by conglomerates, sandstones and marls

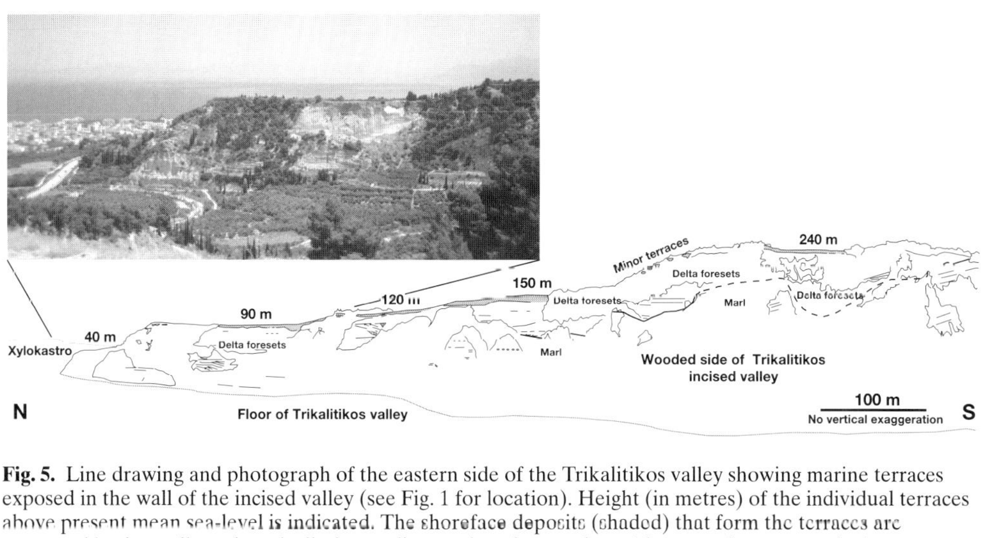

Fig. 5. Line drawing and photograph of the eastern side of the Trikalitikos valley showing marine terraces exposed in the wall of the incised valley (see Fig. 1 for location). Height (in metres) of the individual terraces above present mean sea-level is indicated. The shoreface deposits (shaded) that form the terraces are separated horizontally and vertically form adjacent shoreface wedges. Also note that progressively younger wedges occur more basinward (towards the N) and at lower elevations.

Fig. 6. Characteristics of the bounding surfaces and stacking pattern of the detached shoreface deposits that form the terraces (see Fig. 1 for location). (**a**) View to NW showing the relationships between the 110 m and 60 m terraces S of Xylokastro. The individual terraces are separated vertically and horizontally by a palaeocliff formed by a combination of subaerial slope failure/erosion and transgressive ravinement erosion (Ts/Sb = composite transgressive surface/sequence boundary). (**b**) Detail of composite transgressive surface/sequence boundary showing bored cobble lag overlying soil profile with calcrete nodules. Note bioturbated shoreface sandstones above basal lag. Pen, 12 cm long, for scale is circled.

deposited in shallow-marine, shoreface to offshore-transition environments. Internally, these deposits are composed of 0.5–2 m thick shoaling upward units (e.g. Fig. 7). The tops of these shoaling-upward units mark an abrupt return to deeper water facies and are often associated with intense bioturbation by *Skolithos* and/or *Thalassinoides*. In proximal parts of the terraces, these shoaling-upward units are composed of trough cross-bedded, upper shoreface sandstones that pass upwards into open framework, foreshore conglomerates (Fig. 7). In more distal settings, the shoaling-upward units are composed of bioturbated marls and sandstones with remnants of hummocky cross stratification (lower shoreface) that grade into cross-bedded,

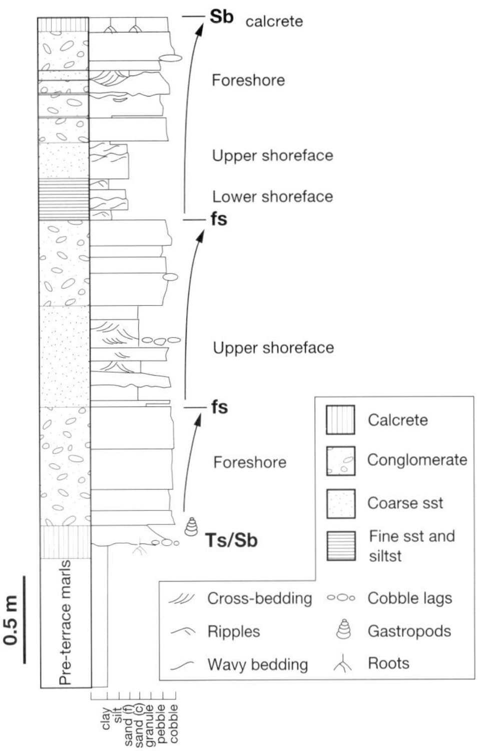

Fig. 7. Simplified measured section of the 150 m marine terrace level (see Fig. 1 for location). The succession is composed of three shoreface-foreshore parasequences, bounded by marine flooding surfaces (fs), forming an aggradational to progradational parasequence set. The base of the terrace is marked by an erosive surface and calcrete development, overlain by a cobble lag, which is interpreted as a combined sequence boundary and transgressive surface (Ts/Sb). The calcrete top of the terrace is the overlying sequence boundary (Sb) and marks the stranding of this marine unit.

upper shoreface sandstones. This facies stacking pattern and the surfaces indicating abrupt deepening are interpreted as parasequences and marine flooding surfaces respectively.

Mapping the terrace levels along strike and U-series dating of corals from the terraces (e.g. Collier *et al.* 1992) together suggest that the lower terrace levels are contemporaneous with the attached shoreface deposits exposed in the Corinth Canal. U-series dating of corals from the 80 m terrace level south of the town of Corinth indicates that shallow marine deposition on the terraces corresponds with oxygen isotope stage 7 to early isotope stage 6 (Collier *et al.* 1992), suggesting that progradation occurred during highstand and early eustatic sea-level fall. This interpretation is consistent with that of other workers who have suggested that deposition of the shallow marine facies forming the terraces occurred during interglacials and major interstadials (e.g. Keraudren & Sorel 1987; Gawthorpe *et al.* 1994; Armijo *et al.* 1996).

The overall downstepping and offlapping stacking pattern of the terraces is similar to that of the sequences exposed in the Corinth Canal (see Figs 3, 5). However, there are differences between the architecture of the sequences that form the terraces and those exposed in the Corinth Canal. In particular, the terraces form discrete, generally isolated tabular bodies that are commonly separated by a few metres to tens of metres vertically and several tens of metres horizontally (e.g. Figs 5, 6). Thus the shoreface deposits that comprise the terraces are interpreted as detached forced regressive wedges that stack to create a forced regressive sequence set.

Fan-delta deposits

In the western part of the study area, between the coastal towns of Xylokastro and Akrata, a series of five uplifted coarse-grained fan-deltas are exposed (Fig. 8). The deposits occur as discrete fan-deltas, each of which has an areal extent of <8 km^2, with an average of 2 km^2. Prominent delta topsets occur at elevations of 140–160 m above sea level, with older topsets exposed up to 600 m above sea level (e.g. landward of the Akrata delta, Figs 2, 8, 9). Incision of modern drainage networks through these uplifted fan-deltas provides excellent exposure of topset and foreset facies associations along the walls of incised valleys (Figs 8, 10).

The topset facies are composed of crudely horizontally stratified, poorly to moderately sorted cobble-grade conglomerates, which have a high proportion of coarse sand matrix. Like delta topsets described from older deltas in the area (e.g. Ori 1989; Doutsos & Piper 1990; Ori *et al.* 1992; Poulimenos *et al.* 1993; Dart *et al.* 1994), these are interpreted to have been deposited in a delta plain environment dominated by braided streams. The foresets of the deltas are composed of alternations of clast- and matrix-supported, pebble to cobble grade conglomerates with a sandy matrix. These foreset conglomerates form sub-planar beds that dip basinward at angles of up to 40° and may be cut by erosional scours up to several metres deep. Sandier units within the foresets exhibit cross stratification and, in addition, dune-like forms are observed, interpreted as 'backsets' infilling scours (e.g. Massari

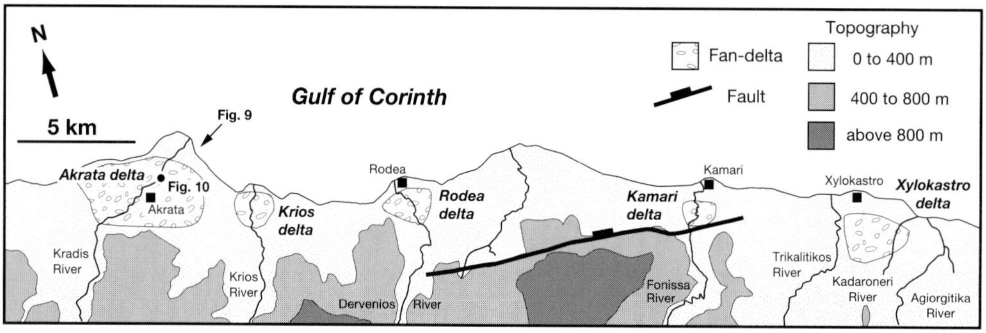

Fig. 8. Map of the western part of the study area showing the location of the stranded late Quaternary fan-deltas and the position of modern drainage networks. The location of the Trikalitikos valley, and the Kamari and Akrata fan-deltas are also shown.

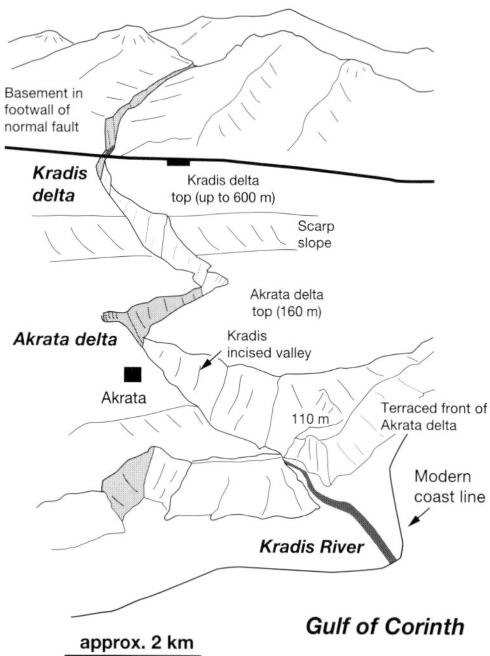

Fig. 9. Sketch oblique aerial view looking toward the SW showing the relationships between the modern Kradis river and the Kradis and Akrata deltas. Heights of the topsets of the Akrata and the older Kradis delta. Note that the deltas exhibit an offlapping and downstepping stacking pattern, and are incised by the Kradis River that supplies sediment to the modern delta at the coast.

& Parea 1990). Occasionally very fine sandstone or marl units, up to a metre thick, drape the conglomerate foresets.

The foreset facies record the different processes supplying sediment to the delta front. The better sorted, clast supported units are interpreted to represent avalanching of grains down from the delta top, possibly as a result of slope failure. The more poorly sorted matrix-supported units represent deposition from underflows during periods of high fluvial discharge, as recorded from modern deltas (e.g. Heezen *et al.* 1966; Ferentinos *et al.* 1988). Periods of delta front abandonment are recorded by the fine sandstone and marl drapes. The erosional scours that cut the foresets appear to be similar to chutes seen on side scan sonar images of the modern deltas in the Gulf of Corinth (Ferentinos *et al.* 1988).

The fan-delta deposits are generally over 100 m thick but, as no bottomsets are exposed, this represents their minimum thickness. The internal stacking of the fan-deltas records an initial phase of topset aggradation followed by predominantly progradation and toplap. The surfaces that bound the fan-delta deposits are in general poorly exposed; however, a steep erosional contact is observed at the lateral margins of the Kamari fan-delta (Fig. 8). In addition, the terminal foresets of the fan-deltas may also show a stepped, or terraced, profile suggesting that subsequent sea-level changes reworked the front of the deltas. The present day incised valleys potentially represent a good modern analogue for the steep erosional surfaces that bound the Kamari fan-delta. The modern drainage system cuts valleys through older delta deposits and supplies sediment to the present day deltas at the coastline (e.g. Figs 2, 8, 9, 10).

The stacking pattern of successive fan-delta bodies is consistent with that observed in the attached and detached shallow marine deposits to the east. Younger fan-deltas offlap and downstep, such that the topsets of successively younger fan-deltas are located in progressively

Fig. 10. Photograph looking to the NW along the incised Kradis valley which has cut through the Akrata fan-delta (see Figs 1 & 8). The Kradis River supplies sediment to a modern fan-delta at the coast (see Fig. 8). Excellent exposures of the stranded fan delta foresets are seen along the walls of the valley, which are 90 m high.

more basinward positions and at lower elevations (Fig. 9). However, their position along strike remains more or less fixed over time because they are fed by incised valleys. The topsets of successively younger deltas onlap onto the foresets of the previous delta, so that the deltas are connected and form an attached forced regressive sequence set.

Controls on sequence development

The coeval depositional systems exposed along the northern Peloponnesos coast all exhibit very similar stacking patterns and stratal surface development, yet the facies, stratal geometry and degree of facies shifts vary markedly along strike. The influence of accommodation development, sediment supply and physiography in controlling the similarities in stratigraphic evolution and the along-strike variability are discussed in this section.

Regional controls: accommodation development

Throughout the study area the combination of tectonic uplift and glacio-eustatic sea-level change led to long-term relative sea-level fall, with each successive eustatic highstand generally lower than the preceding one (Figs 2, 11). This led to incision and 'stranding' of the older marine deposits and the creation of an overall forced regressive stacking pattern over the entire study area (Fig. 11).

The bounding surfaces of each of the shoreface packages and fan deltas also display marked similarities along strike. Up-dip, the surfaces are characterized by subaerial exposure and/or incision, whereas down-dip they have been subsequently modified by marine ravinement erosion. Thus, the bounding surfaces record evidence of erosion during both relative sea-level fall and rise. Internally the stacking pattern of the individual sequences is also similar, with all three styles of deposition displaying initial aggradation followed by predominantly progradation. These observations suggest that development of the key stratal surfaces and the internal stacking patterns of individual sequences was driven by regional changes in accommodation space, which we interpret to have been dominated by the glacio-eustatic signal.

During intervals of glacio-eustatic fall, previously formed shorelines were progressively exposed and incised, and sediment was ultimately shed directly into the Gulf of Corinth (Fig. 11). Unfortunately, relatively little is known in detail about the basinal deposits, although discrete packages of stratified seismic

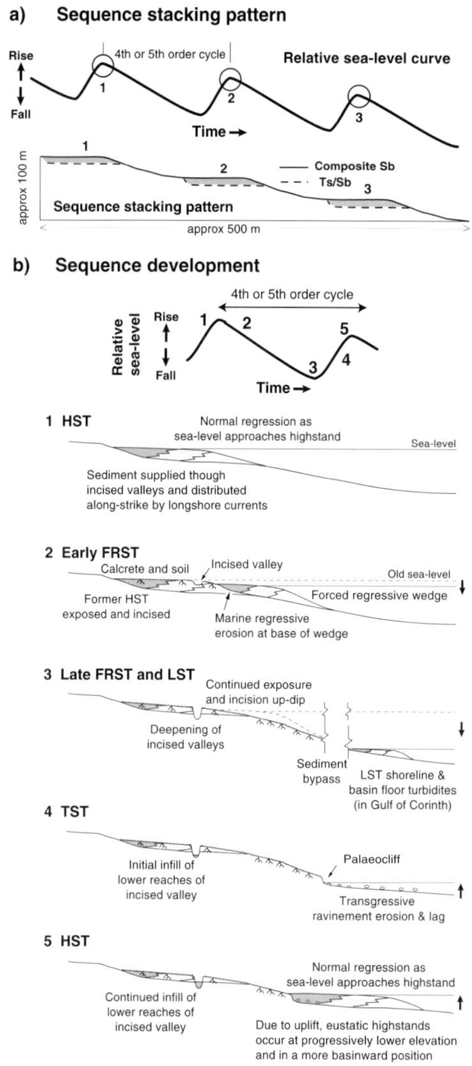

Fig. 11. Regional relative sea-level control on sequence development. (**a**) Semi-quantitative cartoon illustrating offlapping and downstepping sequence stacking pattern of the successive highstand deposits which develops in response to fourth- and higher-order glacio-eustatic highstands of sea-level and tectonic uplift. (**b**) Sequence and systems tract development showing the stratigraphic response to different portions of the relative sea level curve. Note that individual sequences form as a result of fourth- and fifth-order (*c*. 100 ka and 20 ka, respectively) glacio-eustatic cycles.

facies imaged from the Gulf of Corinth have been interpreted to reflect deposition of 'lowstand' turbidites (e.g. Perissoratis *et al.* 1993). It is likely that regression of the shoreline continued during the early stages of relative sea-level fall, so that each fourth- and higher-order sequence itself contains a forced regressive systems tract. The U-series dating from the terraces supports deposition during early eustatic fall (Collier *et al.* 1992), but transgressive ravinement and/or subaerial exposure and incision severely modify the upper parts of individual sequences. As a result, evidence of forced regression within individual sequences is generally equivocal.

Following lowstand, the subsequent glacio-eustatic rise outpaced the rate of uplift and transgressive erosion associated with shoreface retreat formed the planar to stepped erosion surface, which removed any evidence for subaerial exposure, except in the up-dip portions of the succession (Fig. 11). Subsequently, the depositional systems aggraded and prograded at times of low rates of accommodation development as sea-level approached highstand.

Thus we interpret the similarities in sequence stratigraphic development of the late Quaternary marine deposits exposed on the northern Peloponnesos coastal belt as reflecting regional-scale changes in accommodation space. Available U-series dating (e.g. Collier 1990; Collier *et al.* 1992) supports our interpretation that the development of individual sequences and their bounding surfaces was driven by 10–100 ka (fourth- and higher-order) glacio-eustatic sea-level changes. In contrast, tectonic uplift dominated the longer-term sequence stacking pattern, leading to development of the forced regressive sequence set (Fig. 11).

Local controls

Whereas regional accommodation development acted as the dominant control on the late Quaternary sequences of the northern Peloponnesos, the marked variations in facies and stratal geometry both within and between sequences indicates that local controls also exerted a significant influence on sequence development. Here we concentrate on how local changes in uplift rates, sediment supply (amount and composition) and basin physiography played an important role in generating the along strike-variability of the sequences.

Uplift rates. Although accommodation development led to similarities in stratigraphic style along strike, there are some local variations in the number of detached shoreface deposits and associated terraces. In particular, fewer terraces are recorded in the east of the area, where uplift rates are lowest (Armijo *et al.* 1996). In this area, prominent terraces record 4th order interglacial highstands of sea-level (particularly oxygen isotope stage 5.5, 7.5 and 9.3), and some intervening higher order interstadials are missing (e.g. stage 5.3 and 5.1) (Collier *et al.* 1992; Gawthorpe *et al.* 1994; Armijo *et al.* 1996). The absence of terraces associated with the relatively low interstadial highstands may be because the low uplift rates were insufficient to cause vertical separation of the interstadial and subsequent interglacial highstands. As a result, sediments deposited during relatively low interstadial highstands were reworked by the next major transgression associated with post-glacial sea-level rise.

Sediment supply. Analysis of the modern drainage catchments (e.g. Seger & Alexander 1993) provides some indication of the along-strike variability of the volume and the type of sediment supplied to the study area. The size and type of drainage basin varies across the study area, as do the bedrock lithologies drained by the river systems (Fig. 12). Regional uplift has resulted in widespread stream incision throughout the area, but major incised valleys are only associated with large, generally antecedent drainage systems (Fig. 12). Once incised, the drainage network is effectively fixed in position, especially where drainage networks have incised through resistant bedrock in the footwall of normal faults. The larger river systems occupy incised valleys which are in excess of 100 m deep and up to 1–2 km wide.

In the western part of the area, where the fan-delta deposits crop out, large antecedent drainage catchments (up to 200 km^2) dominate (Fig. 12). These drainage systems have maintained their courses despite regional uplift and the propagation of major normal faults (Seger & Alexander 1993). They have cut deep gorges through Mesozoic basement in the footwall of normal faults (e.g. I, Fig. 1), and now cannibalize conglomeratic and sand-dominated Neogene basin fill deposits. As a result, these large antecedent catchments supply relatively large volumes of coarse-grained sediment to the present coastline where modern fan-deltas are being deposited. In a similar way, the outlets from these antecedent catchments also supplied sediment to the late Quaternary fan deltas described in this paper.

Drainage catchments in the central and eastern part of the study area where the marine terrace deposits are developed are also large,

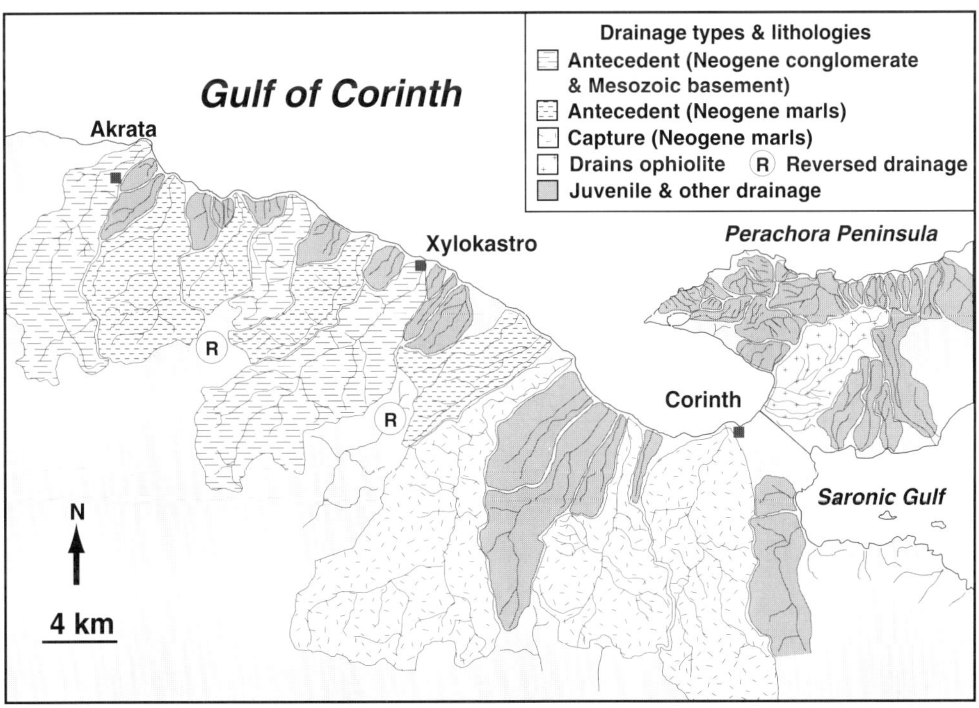

Fig. 12. Map showing drainage catchments in the study area, and the predominant lithology drained. Drainage reversal is denoted by 'R'. Modified from Seger & Alexander (1993).

often >100–200 km^2. However, these catchments erode bedrock which is dominated by Neogene marls (Fig. 12). Although easy to erode, compared to Mesozoic basement carbonates, the grade of sediment is such that it is transported in suspension out into the Gulf of Corinth and thus the supply of coarse-grained sediment to the coastal zone is low. The limited volume of coarse-grained sediment in this part of the area is one of the factors responsible for the development of the thin (<10 m) shorefaces that form the marine terrace deposits between Xylokastro and Corinth. In this area most of the coarse-grained sediment constituting the shoreface deposits is derived from adjacent fan-deltas by longshore drift.

In the extreme east of the study area, a moderate sized drainage basin (68 km^2) supplies sediment from a serpentinite body on the Perachora Peninsula into the Bay of Corinth (Fig. 12). The serpentinite is easily eroded, compared to Mesozoic basement carbonates, and provides sand and gravel grade sediment (Leeder *et al.* 1991). The relatively large volume of coarse-grained sediment sourced from this catchment has constructed an alluvial fan/fan-delta (the Smarpsi fan) to the east of Corinth. The development of the relatively thick attached shoreface deposits exposed in the Corinth Canal is partially attributed to the proximity of this sediment source, with the sediment transported from the Smarpsi fan by longshore currents (Collier & Dart 1991; Gawthorpe *et al.* 1994).

Physiography. The location of normal faults and variations in bedrock lithology have an appreciable influence on the geomorphology of the study area. In particular, the location of resistant Mesozoic carbonate basement at the Earth's surface strongly influences slope gradients. In the western part of the study area, where Mesozoic carbonates are uplifted in the footwall of normal faults, slopes are steep, with elevations of up to 800 m being reached in less than 2 km from the modern shoreline. These steep gradients produce a rapid offshore increase in water depth which results in the development of relatively narrow coastal plains and, where coarse-grained sediment supply is high enough, fan-deltas. In contrast to the other depositional systems, the fan-deltas have a pronounced offlap break, with well developed foresets in excess of 100 m high.

In contrast, the central and eastern parts of the study area have gentle slopes, generally <4°, and elevations remain below 400 m up to 6 km inland of the present shoreline. Over this part of the study area the bedrock is dominated by Neogene marls and sandstones and any normal faults breaking the surface have not exposed resistant basement lithologies. The depositional systems developed in this area are mainly shallow marine shoreface deposits which have a ramp-like morphology compared to the fan-deltas further to the west.

The strike variation in physiography also has a major impact on the amount of shoreline migration associated with relative sea-level change. The steep slope gradients associated with the fan-deltas result in a relatively small basinward shift in the position of the shoreline for a given sea-level fall. For a foreset slope of 30°, a fall of 100 m would only produce a horizontal shift in shoreline position of 175 m. This limited shift, combined with the height of the foresets (which may be several hundred metres if the modern fan-deltas are used as an analogy; Ferentinos *et al.* 1988), produces successive offlapping fan-deltas that are attached. This situation contrasts markedly with that of the shorefaces in the central part of the study area. Here the low slope gradients result in major translations of the shoreline during relative sea-level fall. Using our example of a 100 m sea-level fall, the low gradients of the central part of the study area would result in an approximately 1500 m basinward shift in the position of the shoreline. This large basinward shift, together with low sediment supply, is thought to be one of the main factors responsible for the development of the isolated (detached) architecture of the offlapping shorefaces that form the marine terraces.

Summary and conclusions

This paper documents the similarities and variations in contemporaneous shallow marine and deltaic depositional systems along a *c.* 50 km segment of the northern Peloponnesos coastline, central Greece, which has been subject to punctuated relative sea-level fall during the late Quaternary. The sedimentary response to punctuated relative sea-level fall resulted in the development of a forced regressive sequence set, the exposed portion of which is expressed as fan-deltas in the west of the study area, detached shallow marine shorefaces in the central area and attached shallow marine shorefaces in the east (Fig. 13). All these depositional systems are cut by major incised valleys which focused sediment supply to the coastline.

The coeval shallow marine and deltaic deposits all exhibit very similar stacking patterns and stratal surface development, yet the facies associations, facies shifts and stratal geometry

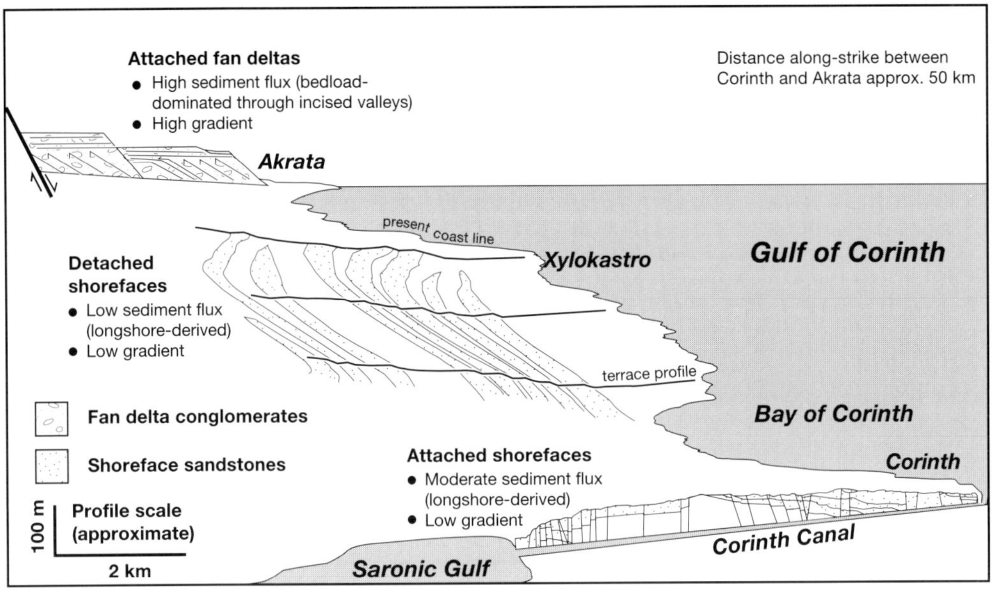

Fig. 13. Diagram summarising the along-strike variability in the late Quaternary sequences developed along the northern Peloponnesos coastline.

vary significantly along strike. The similarities in stacking patterns and stratal surface development are interpreted to have resulted from regional-scale relative sea-level fluctuations, with regional uplift leading to the overall forced regression. Individual sequences and their bounding surfaces are inferred to reflect 10–100 ka (fourth- and higher-order) cycles of glacio-eustatic sea-level change (Fig. 11).

Local factors, in particular sediment supply and physiography, played an important role in creating major strike variability. A combination of coarse-grained sediment, high sediment flux, and steep slopes resulted in the deposition of fan-deltas in the west of the study area. The location of the deltas was fixed by major incised valley axes, with steep gradients and the height of the delta front leading to the development of an attached forced regressive sequence set (Fig. 13). In contrast, the marine terraces in the centre of the study area represent the opposite end member, consisting of detached shallow marine shorefaces (Fig. 13). Low gradients resulted in major shifts in facies associated with both relative sea-level fall and rise. Although drainage catchments are as large as the ones feeding the fan-deltas in the west, they predominantly drain Neogene marls. Thus, the volume of coarse-grained sediment supplied from these catchments was low, with the result that only thin (<10 m) highstand shorefaces developed. The attached shoreface deposits exposed in the Corinth Canal represent an intermediate scenario (Fig. 13). This locality, situated at the eastern end of the northern Peloponnesos coastline, is close to a major fan-delta sourced from a serpentinite body yielding large volumes of coarse-grained detritus. The higher sediment supply appears to be one of the main factors controlling the development of thicker shorefaces and an attached forced regressive sequence set compared to the thin, detached shorefaces immediately to the west.

The results of this study clearly show the along-strike variability of forced regressive deposits and the range of depositional and erosional responses to forced regression. Forced regressive deposits may alternate from attached to detached geometries along strike. Local factors such as physiography, sediment supply and the calibre of sediment have a major impact on the stratigraphic response to forced regressions. These local controls may create major strike variations in facies architecture, stratal geometry and the magnitude of facies shifts over distances of less than 10 km. This has significant implications for sequence stratigraphic interpretations in general.

The authors would like to thank IGME for permission to undertake field studies in Greece and for logistical help. L.S.McM would like to thank DENI for funding her PhD studies in Greece. The manuscript greatly benefited from comments by D. Pirrie, R. Collier and an anonymous referee.

References

AINSWORTH, R. B. & PATTISON, S. A. J. 1994. Where have all the lowstands gone? Evidence for attached lowstand systems tracts in the Western Interior of North America. *Geology*, **22**, 415–418.

ARMIJO, R., MEYER, B., KING, G. C. P., RIGO, A. & PAPANASTASSIOU, D. 1996. Quaternary evolution of the Corinth Rift and its implications for the late Cenozoic evolution of the Aegean. *Geophysical Journal International,* **126**, 11–53.

CHURCH, K. D. & GAWTHORPE, R. L. 1997. Sediment supply as a control on the variability of sequences: an example from the late Namurian of northern England. *Journal of the Geological Society, London*, **154**, 55–60.

COLLIER, R. E. LL. 1990. Eustatic and tectonic controls upon Quaternary coastal sedimentation in the Corinth Basin, Greece. *Journal of the Geological Society, London*, **147**, 301–314.

—— & DART, C. J. 1991. Neogene to Quaternary rifting, sedimentation and uplift in the Corinth Basin, Greece. *Journal of the Geological Society, London*, **148**, 1049–1065.

——, LEEDER, M. R., ROWE, P. J. & ATKINSON, T. C. 1992. Rates of tectonic uplift in the Corinth and Megara Basins, Central Greece. *Tectonics*, **11**, 1159–1167.

DART, C. J., COLLIER, R. E. LL., GAWTHORPE, R. L., KELLER, J. V. A. & NICHOLS, G. 1994. Sequence stratigraphy of (?)Quaternary syn-rift fan deltas, northern Peloponnesos, Greece. *Marine and Petroleum Geology*, **11**, 545–560.

DOUTSOS, T. & PIPER, D. J. W. 1990. Listric faulting, sedimentation, and morphological evolution of the Quaternary eastern Corinth rift, Greece: First stages of continental rifting. *Geological Society of America Bulletin*, **102**, 812–829.

FERENTINOS, G., PAPATHEODOROU, G. & COLLINS, M. B. 1988. Sediment transport processes on an active fault escarpment: Gulf of Corinth, Greece. *Marine Geology*, **83**, 43–61.

GAWTHORPE, R. L., FRASER, A. J. & COLLIER, R. E. LL. 1994. Sequence stratigraphy in active extensional basins: implications for the interpretation of ancient basin fills. *Marine and Petroleum Geology*, **11**, 642–658.

HEEZEN, B. C., EWING, M. & JOHNSON, J. L. 1966. The Gulf of Corinth floor. *Deep Sea Research*, **13**, 381–411.

HUNT, D. & TUCKER, M. E. 1992. Stranded parasequences and the forced regressive wedge systems tract: deposition during base-level fall. *Sedimentary Geology*, **81**, 1–9.

—— & —— 1995. Stranded parasequences and the forced regressive wedge systems tract: deposition during base-level fall-reply. *Sedimentary Geology*, **95**, 147–160.

IMBRIE, J., HAYS, J. D., MARTINSON, D. G., MCINTYRE, A., MIX, A. C., MORLEY, J. J., PISIAS, N. G., PRELL, W. L. & SHACKLETON, N. J., 1984. The orbital theory of Pleistocene climate: support from a revised chronology of the marine $\partial^{18}O$ record. *In:* BERGER, A. L., IMBRIE, J., HAYS, J., KUKLA, G. & SALTZMAN, B. (eds) *Milankovitch and Climate: Understanding the Response to Astronomical Forcing*. Reidel, Dordrecht, 269–305.

KERAUDREN, B. & SOREL, D. 1987. The terraces of Corinth (Greece) – a detailed record of eustatic sea-level variations during the last 500,000 years. *Marine Geology*, **77**, 99–107.

LEEDER, M. R., SEGER, M. J. & STARK, C. P. 1991. Sedimentation and tectonic geomorphology adjacent to major active and inactive normal faults, southern Greece. *Journal of the Geological Society, London*, **148**, 331–343.

MARTINSEN, O. J. & HELLAND-HANSEN, W. 1995. Strike variability of clastic depositional systems; does it matter for sequence-stratigraphic analysis. *Geology*, **23**, 439–442.

MASSARI, F. & PAREA, G. C. 1990. Wave-dominated Gilbert-type gravel deltas in the hinterland of the Gulf of Taranto (Pleistocene, southern Italy). *In:* COLELLA, A. & PRIOR, D. B. (eds) *Coarse-Grained Deltas*. International Association of Sedimentologists, Special Publications, **10**, 311–331.

ORI, G. G. 1989. Geologic history of the extensional basin of the Gulf of Corinth (?Miocene-Pleistocene), Greece. *Geology*, **17**, 918–921.

——, ROVERI, M. & NICHOLS, G. 1992. Architectural patterns in large-scale Gilbert-type delta complexes, Pleistocene, Gulf of Corinth, Greece. *In:* MIALL, A. D & TYLER, N (eds) *The Three-dimensional Facies Architecture of Terrigenous Clastic Sediments and its Implications for Hydrocarbon Discovery and Recovery*. Society of Economic Paleontologists and Mineralogists, Concepts in Sedimentology and Paleontology, **3**, 207–216.

PERISSORATIS, C., PIPER, D. J. W. & LYKOUSIS, V. 1993. Late Quaternary sedimentation in the Gulf of Corinth: the effects of marine-lake fluctuations driven by eustatic sea level changes. *In*: *Special Publications of the Technical University of Athens,* 693–744.

PLINT, A. G. 1988. Sharp-based shoreface sequences and 'offshore bars' in the Cardium formation of Alberta; their relationship to relative changes in sea level. *In*: WILGUS, C. K., HASTINGS B. S., KENDALL, C. G. ST.C., POSAMENTIER, H. W., ROSS, C. A. & VAN WAGONER, J. C. (eds) *Sea-level Change – An Integrated Approach*. Society of Economic Mineralogists and Paleontologists, Special Publications, **42**, 357–370.

POSAMENTIER, H. W. & ALLEN G. P. 1993. Variability of the sequence stratigraphic model: effects of local basin factors. *Sedimentary Geology*, **86**, 91–109.

—— & VAIL, P. R. 1988. Eustatic controls on clastic deposition II: sequence and systems tracts models. In: WILGUS, C., HASTINGS, B. S., KENDALL, C. G. ST.C., POSAMENTIER, H. W., ROSS, C. A. & VAN WAGONER, J. C. (eds.) *Sea-level Change – An Integrated Approach*. Society of Economic

Paleontologists and Mineralogists, Special Publications, **42**, 126–154.

——, ALLEN, G. P., JAMES, D. P. & TESSON, M. 1992. Forced regressions in a sequence stratigraphic framework: concepts, examples and exploration significance. *American Association of Petroleum Geologists Bulletin*, **76**, 1687–1709.

POULIMENOS, G., ZELILIDIS, A., KONTOPOULOS, N. & DOUTSOS, T. 1993. Geometry of trapezoidal fan deltas and their relationship to extensional faulting along the south-western active margin of the Corinth rift, Greece. *Basin Research,* **5**, 179–192.

SCHLAGER, W. 1991. Depositional bias and environmental change – important factors in sequence stratigraphy. *Sedimentary Geology*, **70**, 109–130.

SEGER, M. J. & ALEXANDER, J. 1993. Distribution of Plio-Pleistocene and Modern coarse-grained deltas south of the Gulf of Corinth, Greece. *In:* FROSTICK, L. & STEEL, R. (eds) *Tectonic controls and signatures in sedimentary successions.* International Association of Sedimentologists, Special Publications, **20**, 37–48.

VAN WAGONER, J. C., MITCHUM, R. M., CAMPION, K. M. & RAHMANIAN, V. D. 1990. *Siliciclastic Sequence Stratigraphy in Well Logs, cores and Outcrop.* American Association of Petroleum Geologists Methods in Exploration Series, **7**.

Index

Page numbers in italic e.g. *247* signify references to figures. Page numbers in bold e.g. **32** denote references to tables.